Larentiinae II

Larentiinae II

(*Perizomini and Eupitheciini*)

By

Vladimir Mironov

BRILL

LEIDEN | BOSTON

Reprint of *The Geometrid Moths of Europe*. Volume 4: Larentiinae II (Perizomini and Eupitheciini) (ISBN: 87 88757-40-4). Published by Apollo Books, Stenstrup, Denmark, 2004.

This publication should be cited as:
Mironov, V.: Larentiinae II (Perizomini and Eupitheciini). – In A. Hausmann (ed.): The Geometrid Moths of Europe 4: 1–463.

The studies on the tribes *Perizomini* and *Eupitheciini*, subject of the present volume, have been supported by Peder Skou (Apollo Books, Stenstrup, Denmark) and the Russian Foundation for Basic Research (grant no. 01-04-49637).

Cover illustration: *Eupithecia absinthiata* (Cl.). © Michael Leipnitz.

Editor: Axel Hausmann, Zoologische Staatssammlung, München, Germany.

Colour plates and text figures 5–6, 12–14, 17–85: David Wilson, Wicken, UK.

Text figures 69–70. © Sir Anthony Galsworthy.

All other drawings (venation and genitalia) are by the author.

The Library of Congress Cataloging-in-Publication Data is available online at https://catalog.loc.gov

ISSN 2588-9990
ISBN 978-90-04-49958-4 (hardback)
ISBN 978-90-04-69161-2 (e-book)
DOI 10.1163/9789004691612

This book is printed on acid-free paper and produced in a sustainable manner.

Contents

Preface

Thanks to the diligence and the competence of the author, only 8 years after initiation of the series 'The Geometrid Moths of Europe', the fourth volume on Perizomini and Eupitheciini, which is the first of the 'thicker' volumes containing a large number of species, can proudly be presented.

To date 151 European geometrid moths of the tribes Eupitheciini and Perizomini are known. This is about 8% of the world's fauna. Whilst the Perizomini are a comparatively small and heterogeneous group that usually do not offer any problems of identification and taxonomy, the tribe Eupitheciini is rich in species, very homogenous, the species are often very small and inconspicuous in wing pattern and colours, hence not aesthetically attractive. They are said to be difficult to identify, and there are indeed many species that cannot be properly identified without turning to refined techniques of anatomical preparation. Apart from some old publications (DIETZE 1910-1913; PETERSEN 1910), only some regional treatments were available (WEIGT 1976-1993; GOATER et al. 1981) together with very scattered information widely distributed in many single publications. Whilst the *Eupithecia*-parts in standard literature on central and northern European Geometridae (e.g. FORSTER & WOHLFAHRT 1981, SKOU 1986) provide a certain data basis, there has been a dramatic lack of information on southern and eastern European species.

Therefore until today, only a very small number of specialists has been doing serious research on these organisms. The consequences are frequent misidentifications and unnamed moths in collections that show how urgently this volume is needed. I often get large boxes of European determinanda-material of *Eupithecia* from colleagues and amateurs 'in despair'. The publication of this volume will surely have an immense impact for both amateur and professional entomologist, and will help all of us to be able to study this group in more detail.

Therefore, of the four aims of the book series (see preface of Vol. 1) aim 1, the identification aid, is by far the most important in this volume: colour plates, exact genitalia drawings, enlarged photos with indicated differential features and valuable differential diagnoses in the text will, for the first time, allow and facilitate the identification of all European Perizomini and Eupitheciini. The fourth aim, the updating (see also vol. 1: p. 250) will be electronically realised by the Fauna Europaea project (see www.faunaeur.org).

Taxonomically this group needs worldwide revision. As editor I would be pleased if this volume will be the starting point for joint revisional efforts, in order to realize this goal before tropical biodiversity is destroyed. The research initiative 'Forum Herbulot' (ZSM; 1991) has discussed perspectives to start with this task (see www.herbulot.de).

This volume brings an almost new order of species (but see earlier publications MIRONOV 1990; 1991). Usually the GME book series prefers traditional, conservative systems. Here, however, considering the lack of revisionary work until today, the author has done well to establish and improve a system of species groups that is purely based on morphological differential analysis. I am sure that this system will be widely accepted by European lepidopterists.

Last but not least I have to acknowledge the merits of Peder Skou, who initiated this book series and who enthusiastically promotes this publication on a 'cinderella group' of moths in order to improve the attractiveness of this group. Thanks also to the Systematics of Insects Laboratory (Zoological Institute RAN; St. Petersburg) and the Russian Foundation for Basic Research (grant No. 01-04-49637) that enabled the author to do such extensive studies.

Munich, 1.2.2003 — Axel Hausmann *(Editor-in-Chief)*

Abstract

This fourth volume of the book series 'The Geometrid Moths of Europe' presents a modern revision of the two closely related Larentiinae tribes Perizomini and Eupitheciini.

The tribe Perizomini is represented in Europe by 18 species belonging to five genera *Pseudobaptria* INOUE, 1982, *Mesotype* HÜBNER, 1825, *Perizoma* HÜBNER, 1825, *Martania* MIRONOV, 2000 and *Gagitodes* WARREN, 1893. The tribe Eupitheciini is represented by 133 species belonging to four genera: *Gymnoscelis* MABILLE, 1868, *Chloroclystis* HÜBNER, 1825, *Pasiphila* MEYRICK, 1883 and *Eupithecia* CURTIS, 1825. In Europe, the extremely species-rich genus *Eupithecia* may be divided into 35 species-groups, according to characters of male and female genitalia. In the diagnoses of *Eupithecia* species-groups closely related species from the rest of the Palaearctic and some species from the Nearctic region are included.

One new species is described: *Eupithecia senorita* sp. n.. 27 taxonomic changes are formally proposed, including: One taxon is raised to species rank: *Eupithecia weissi* PROUT, 1938, stat. n. (formerly subsp. of *E. euphrasiata*). One taxon is downgraded from species to subspecies rank: *Eupithecia graphata olympica* TULESCHKOV, 1951, stat. n. 25 species are downgraded to synonymy: *Eupithecia golearia* LUCAS, 1940 (syn. n. to *E. ultimaria*); *E. kaszabi* VOJNITS, 1974 (syn. n. to *E. dissertata*); *E. certa* VOJNITS, 1981 (syn. n. to *E. dissertata*); *E. atlanticata* PINKER, 1969 (syn. n. to *E. massiliata*); *E. atlanticata pinkeriata* MOBERG, 1983 (syn. n. to *E. massiliata*); *E. peterseni* WAGNER, 1914 (syn. n. to *E. sardoa*); *E. euxinata* BOHATSCH, 1893 (syn. n. to *E. ericeata*); *E. laboriosa* VOJNITS, 1977 (syn. n. to *E. lariciata*); *E. tsushimensis* INOUE, 1980 (syn. n. to *E. groenblomi*); *E. fujisana* INOUE, 1980 (syn. n. to *E. groenblomi*); *E. kunashiriensis* VIIDALEPP & MIRONOV, 1988 (syn. n. to *E. groenblomi*); *E. divina* VOJNITS, 1979 (syn. n. to *E. selinata*); *E. maspalomae* PINKER, 1961 (syn. n. to *E. orana*); *E. olympica* TULESCHKOV, 1951 (syn. n. to *E. graphata*); *E. wehrlii* WAGNER, 1931 (syn. n. to *E. mystica*); *E. pliniata* STAUDER, 1929 (? syn. n. to *E. indigata*); *E. mandarinca* MIRONOV, 2001 (syn. n. to *E. nachadira*); *E. antaggregata* INOUE, 1977 (syn. n. to *E. pernotata*); *E. catharinae* VOJNITS, 1969 (syn. n. to *E. absinthiata*); *E. barbaria* SCHWINGENSCHUSS, 1955 (syn. n. to *E. weissi*); *E. wettsteini* VOJNITS, 1974 (syn. n. to *E. millefoliata*); *E. salami* BRANDT, 1938 (syn. n. to *E. fuscicostata*); *E. domogledana* VOJNITS & SZABÓ, 1988 (syn. n. to *E. fuscicostata*); *Catarina dolosa* VOJNITS, 1977 and *C. necessaria* VOJNITS, 1977 (syn. n. to *Eupithecia subbrunneata* DIETZE, 1913).

Lectotypes have been designated for *Larentia albulata* var. *dissoluta* STRAND, 1902; *Eupithecia cocciferata* var. *semitinctaria* MABILLE, 1868, *Larentia sinuosaria* EVERSMANN, 1848; *Eupithecia olympica* TULESCHKOV, 1951; *Eupithecia nachadira* BRANDT, 1941; *Eupithecia mekrana* BRANDT, 1941; *Eupithecia mekrana khorassana* BRANDT, 1941; *Eupithecia scopariata* var.? *orientata* STAUDINGER, 1901 (identity: *Eupithecia sutiliata*); *Eupithecia millefoliata* var. *maeoticaria* BOHATSCH, 1893 (identity: *Eupithecia millefoliata*).

The species are presented with their full name and reference to the original description, fully referenced junior synonyms, unavailable names (including misidentifications), diagnosis, distribution, phenology, biology including host-plants, habitats (with vertical distribution), parasitoids (in certain cases) and differences from similar species. Special emphasis is given to the descriptions of the shape of uncus, valva, everted vesica and shape of sternum A8 of the male, as well as to the bursa copulatrix, shape of tergum A8 and apophyses of the female. For each species line drawings of male and female genitalia, colour photographs of adults, and distribution maps for Europe and adjacent countries are provided.

Three species are new for the fauna of Europe: *Eupithecia usbeca* VIIDALEPP, 1992; *E. repentina* VOJNITS & DE LAEVER, 1978 and *E. sutiliata* CHRISTOPH, 1876.

The following *Perizoma, Pasiphila, Chloroclystis* and *Eupithecia* species are recorded as new for the European and adjacent countries listed below:

Portugal: *Perizoma hydrata* (TREITSCHKE, 1829); *Eupithecia extraversaria* HERRICH-SCHÄFFER, 1852
Spain: *Eupithecia extensaria occidua* PROUT, 1914
France: *Eupithecia weissi* PROUT, 1938
Corsica: *Eupithecia schiefereri* BOHATSCH, 1893
Italy: *Eupithecia addictata* DIETZE, 1908; *E. spadiceata* ZERNY, 1933
Austria: *Eupithecia scopariata* (RAMBUR, 1833)
Hungary: *Eupithecia pernotata* GUENÉE, 1858; *E. addictata* DIETZE, 1908; *E. spadiceata* ZERNY, 1933
Romania: *Eupithecia mystica* DIETZE, 1910
Bulgaria: *Eupithecia spadiceata* ZERNY, 1933
Macedonia: *Perizoma lugdunaria* (HERRICH-SCHÄFFER, 1855); *Eupithecia mystica* DIETZE, 1910; *E. fuscicostata* CHRISTOPH, 1887
Greece: *Perizoma hydrata* (TREITSCHKE, 1829); *P. lugdunaria* (HERRICH-SCHÄFFER, 1855); *Pasiphila chloerata* (MABILLE, 1870); *Eupithecia tenuiata* (HÜBNER, 1813); *E. mystica* DIETZE, 1910; *E. indigata* (HÜBNER, 1813); *E. vulgata* (HAWORTH, 1809); *E. sutiliata* CHRISTOPH, 1876; *E. fuscicostata* CHRISTOPH, 1887
Ukraine: *Eupithecia pauxillaria* BOISDUVAL, 1840; *E. nachadira* BRANDT, 1941; *E. irriguata staudingeri* BOHATSCH, 1893
Russia (European part): *Eupithecia usbeca* VIIDALEPP, 1992; *E mekrana* BRANDT, 1941; *E. scalptata* CHRISTOPH, 1885; *E. sutiliata* CHRISTOPH, 1876; *Eupithecia spadiceata* ZERNY, 1933
Canary Islands: *Eupithecia pusillata* (DENIS & SCHIFFERMÜLLER, 1775)
Morocco: *Eupithecia abbreviata* STEPHENS, 1831; *E. chalikophila* WEHRLI, 1926
Algeria: *Perizoma flavosparsata* (WAGNER, 1926); *Chloroclystis v-ata* (HAWORTH, 1809); *Eupithecia minusculata* ALPHÉRAKY, 1882
Tunisia: *Eupithecia minusculata* ALPHÉRAKY, 1882; *Eupithecia abbreviata* STEPHENS, 1831
Turkey (Asian part): *Eupithecia karadaghensis* MIRONOV, 1988; *E. satyrata* (HÜBNER, 1813); *E. fuscicostata* CHRISTOPH, 1887.

Acknowledgements

I express my sincere gratitude to all who assisted in this study over many years. I thank especially my mentor, colleague and friend Dr. Jaan VIIDALEPP (Institute Zoology and Botany, Tartu, Estonia). I also thank many colleagues from the Systematics of Insects Laboratory in Zoological Institute of St. Petersburg as follows: Dr. S.V. BARYSHNIKOVA, Dr. S. A. BELOKOBYLSKIJ, Dr. A.F. EMELJANOV, Dr. M.I. FALKOVITCH, Dr. D.R. KASPARYAN, Dr. I.M. KERZHNER, Dr. V.I. KUZNETZOV, Dr. A.L. LVOVSKY, A.Yu. MATOV, Dr. S.Yu. SINEV, Dr. I.L. SUKHAREVA and Dr. A.K. ZAGULJAEV.

I thank all those who have generously loaned material and types from their personal collections or from institutional and museum collections of which they are in charge:

Dr. Leif AARVIK, Ås (NO): Provision of faunistic data (revision of maps for Norway).

Dr. Matti AHOLA, Koski (FIN): Loan of material.

Dr. Vasilij ANIKIN, Saratov (RUS): Loan of material.

Elena ANTONOVA, Moscow (RUS): Loan of material. Kind support of the author when visiting the Moscow museum. Provision of literature. Discussion of taxonomic problems.

Günter BAISCH, Mettenberg (D): Loan of material.

Dr. Edward BARANIAK, Poznan (PL): Provision of literature.

Dr. Bengt BENGTSSON, Stokholm (SW): Loan of material (types of BRANDT).

Roland BERARD, Saint-Etienne (F): Provision of many valuable comments and revision of maps for France.

Dr. Stoyan BESHKOV, Sofia (BU): Loan of material (type of *E. olympica* TULESCHKOV, 1951). Provision of literature and faunistic data. Discussion of taxonomic problems.

Dr. Alexey BIDZILYA, Kiev (UKR): Loan of material.

Lavr BOLSHAKOV, Tula (RUS): Loan of material. Provision of literature and faunistic data. Discussion of taxonomic problems.

Dr. Klaus BOLTE, Ottawa (CN): Provision of literature. Discussion of taxonomic problems.

Dr. Yurij BUDASHKIN, Karadagh Nature Reserve (UKR): Loan of material. Discussion of taxonomic problems.

Rainer BUSSE, Zerpenschleuse (D): Provision of faunistic data on some *Eupithecia* for France.

Dr. Jaroslav BUSZKO, Torun (PL): Provision of literature and faunistic data (revision of maps for Poland). Discussion of taxonomic problems.

Martin CORLEY, Oxfordshire (GB): Provision of faunistic data. Linguistic editor.

Dr. Jorge DANTART, Barcelona (E): Provision of literature.

Michael FIBIGER, Sorø (DK): Loan of material. Kind support on private visits. Provision of literature and faunistic data. Discussion of taxonomic problems.

Egbert FRIEDRICH, Jena (D): Loan of material. Provision of faunistic data (revision of maps for Germany and some other countries).

Dr. Jörg GELBRECHT, Berlin (D): Loan of material. Provision of literature and faunistic data (revision of maps for Germany). Discussion of taxonomic problems.

Barry GOATER, Hampshire (GB): Provision of faunistic data (revision of maps for United Kingdom and Ireland).

Dr. Stanislav GOMBOC, Ljubljana (SLO): Provision of literature, faunistic and geographic data. Discussion of taxonomic problems.

Dr. Carlos GÓMEZ de AIZPÚRUA, Madrid (E): Provision of literature.

David GREEN, Hampshire (GB): Provision of faunistic data (revision of maps for United Kingdom).

Prof. Dr. Hans-Joachim HANNEMANN, Berlin (D): Loan of material (types of STAUDINGER and DIETZE).

Dr. Axel HAUSMANN, Munich (D): Provision of literature and faunistic data (revision of maps for Germany). Revision of the whole manuscript with many precious comments and advices. Kind support of the author when visiting the Munich museum. Loan of material (also types of SCHÜTZE). Discussion of taxonomic problems and various valuable help.

Claude HERBULOT, Paris (F): Loan of material (also types of MABILLE). Provision of literature. Discussion of taxonomic problems.

Nils HYDEN, Enköping (SW): Discussion of taxonomic and biological problems. Provision of literature, phenological and faunistic data (revision of maps for Sweden).

Prof. Dr. Hiroshi INOUE, Iruma (JP): Loan of material (also types). Provision of materials and literature. Discussion of taxonomic problems.

Dr. Povilas IVINSKIS, Vilnjus (LT): Loan of material. Provision of literature and faunistic data (revision of maps for Lithuania).

Gert JEPPESEN, Elkenøre, Falster (DK): Loan of material.

Dr. Lauri KAILA, Helsinki (FIN): Loan of material (also types). Provision of literature.

Ole KARSHOLT, Copenhagen (DK): Loan of material (also types). Kind support of the author when visiting the Copenhagen museum and other very valuable and important help. Provision of materials and literature. Discussion of taxonomic and faunistic problems.

Maxim KLEPIKOV, Jaroslavl (RUS): Loan of material.

Andrzej KOKOT, Wroclaw (PL): Revision of maps for Poland.

Dmitrij KOMAROV, Volgograd (RUS): Loan of material.

Igor KOSTJUK, Kiev (UKR): Kind support of the author when visiting the Kiev museum. Loan of material. Provision of literature and faunistic data (revision of maps for Ukraine). Discussion of taxonomic problems and other very valuable and important help. Revision of the whole manuscript with precious comments and advices.

Dr. Mikhail KOZLOV, Turku (FIN): Loan of material. Provision of literature and other kind and very important help.

Dr. František KRAMPL, Prague (CZ): Loan of material. Provision of literature and faunistic data (revision of maps on some species for Czech Republic and Slovakia). Discussion of taxonomic problems.

Dr. Anatolij KULAK, Minsk (BY): Loan of material. Provision of literature and faunistic data (revision of maps for Belorussia). Discussion of taxonomic problems.

Albert LASTUKHIN, Cheboksary (RUS): Loan of material. Provision of literature and faunistic data (revision of maps for Chuvashia Republic, Russia). Discussion of taxonomic problems.

Dr. Hans LÖBEL, Sondershausen (D): Loan of material. Provision of faunistic data (revision of the maps for Germany).

Dr. Martin LÖDL, Vienna (A): Loan of material (types of PINKER).

David LONG, Edinburgh (GB): Assistance with plant nomenclature.

Dr. Karl-Erik LUNDSTEN, Helsinki (FIN): Provision of literature and faunistic data.

Dr. Adam MALKIEWICZ, Wroclaw (PL): Loan of material. Provision of literature and faunistic data (revision of maps for Poland). Discussion of taxonomic problems.

Geoff MARTIN, London (GB): Valuable information about type-specimens deposited in BMNH. Loan of material (types of PROUT and WARREN).

Dr. Wolfram MEY, Berlin (D): Loan of material (types of STAUDINGER and DIETZE).

Dr. Kauri MIKKOLA, Helsinki (FIN): Loan of material (also types). Provision of literature. Discussion of taxonomic problems.

Dr. Bernd MÜLLER, Berlin (D): Loan of material. Discussion of taxonomic problems.

Kari NUPPONEN, Espoo (FIN): Loan of material. Provision of literature.

Michael OCHSE, Bad Dürkheim (D): Loan of material.

Dr. Zuhal OKYAR, Edirne (TR): Provision of literature.

Prof. Dr. Paolo PARENZAN, Bari (I): Provision of literature.

Dr. Mark PARSONS, London (GB): Provision of faunistic data (revision of maps for United Kingdom).

† Rudolf PINKER, Vienna (A): Loan of material (also types). Provision of literature. Discussion of taxonomic problems.

Azif PIRIEV, Baku (AZ): Loan of material.

Dr. Linda PITKIN, London (GB): Provision of important bibliographical and other information.

Dr. Valter RAINERI, Genova (I): Provision of literature.

Alex RIEMIS, Turnhout (B): Provision of faunistic data for Turkey.

Dr. Luis ROMERA, Madrid (E): Provision of faunistic data.

Dr. László RONKAY, Budapest (H): Kind support of the author when visiting the Budapest museum. Loan of material (also types of TREITSCHKE and VOJNITS). Provision of literature and faunistic data, and other very valuable help.

Jan ROTH, Rösselsheim (D): Loan of material. Provision of faunistic data (revision of maps for Germany). Discussion of taxonomic problems.

Dr. Eugeny RUTJAN, Kiev (UKR): Loan of material.

Nikolaj SAVENKOVS, Riga (LV): Provision of faunistic data (revision of maps for Latvia).

Dr. Stefano SCALERCIO, Cosenza (I): Loan of material.

Christoph SCHÖNBORN, Benzingerode (D): Provision of faunistic data (revision of maps for Germany).

Dr. Malcolm SCOBLE, London (GB): Loan of material (also types of PROUT). Provision of literature.

Nazar SHAPOVAL, St. Petersburg (RUS): Loan of material.

Dr. Irina SHMYTOVA, Kaluga (RUS): Loan of material.
Pasi SIHVONEN, Helsinki (FIN): Provision of literature.
Peder SKOU, Stenstrup (DK): Idea and first moves to initiate the GME-series. Kind support during private visits. Provision of literature, materials and faunistic data. Revision of the whole manuscript with many precious comments and advices. Various valuable and important help. Discussion of taxonomic problems.
Manfred SOMMERER, Munich (D): Loan of material. Provision of literature.
Dr. Olavi SOTAVALTA, Vantaa (FIN): Provision of literature and faunistic data (revision of maps for Finland and NW Russia). Discussion of taxonomic problems.
Dr. Wolfgang SPEIDEL, Bonn (D): Loan of material. Provision of literature.
Dirk STADIE, Eisleben (D): Provision of faunistic data (revision of maps for Germany).
Dr. Mihai STANESCU, Bucharest (RO): Provision of literature and faunistic data (revision of maps for Romania). Discussion of taxonomic problems.
Dr. Anatolij STEKOL'NIKOV, St. Petersburg (RUS): Loan of material.
Dr. Dieter STÜNING, Bonn (D): Kind support of the author when visiting the Bonn museum. Loan of material (also types of WEHRLI and VOJNITS). Provision of literature and faunistic data, and other very valuable help. Discussion of taxonomic problems.
Dr. Valerij SHCHUROV, Krasnodar (RUS): Loan of material.
Giedrius ŠVITRA, Ukmerge (LT): Loan of material.
Levente SZÉKELY, Brašov (RO): Provision of literature, geographical and faunistic data for maps of Romania.
Claude TAUTEL, Saint-Etienne (F): Loan of material. Provision of literature, phenological, biological and faunistic data (revision of maps for France). Discussion of taxonomic problems; many important and valuable comments.
Dr. Valentin TIKHONOV, Pjatigorsk (RUS): Loan of material.
Dr. Robert TRUSCH, Karlsruhe (D): Kind support of the author when visiting the Karlsruhe museum. Loan of material (also types of HAUSMANN). Valuable help in computer questions.
Dr. Jaan VIIDALEPP, Tartu (EST): Kind support of the author during private and institutional visits. Loan of material (also types). Provision of innumerable literature and faunistic data (revision of maps for Estonia). Discussion of many taxonomic problems.
Dr. Andras VOJNITS, Budapest (H): Loan of material (types). Discussion of taxonomic problems.
† Dr. Seda WARDIKJAN, Erivan (AM): Loan of material (also types).
Hartmut WEGNER, Adendorf (D): Provision of faunistic, phenological and biological data (revision of maps for northern Germany). Discussion of taxonomic problems.
Hans-Joachim WEIGT, Schwerte (D): Provision of literature.
Dr. Werner WOLF, Bindlach (D): Provision of literature.
Dr. Norbert ZAHM, Schmelz (D): Provision of faunistic and phenological data (revision of maps for some European countries). Discussion of taxonomic problems.
Dr. Vadim ZOLOTUHIN, Uljanovsk (RUS): Loan of material. Provision of faunistic data. Discussion of taxonomic problems.
Dr. Kees ZWAKHALS (NL): Kind help in nomenclatural problems concerning Ichneumonidae and elimination of non-reliable data.

Finally, but not least, I thank my wife Olga, who endured with much understanding the very long period over which this book was compiled.

I shall be very grateful to everyone for information about depositions of other type-specimens of *Eupithecia*.

Introduction

This volume treats the European species of Geometrid moths belonging to two closely related tribes: Perizomini (5 genera and 18 species) and Eupitheciini (4 genera and 133 species).

In the current interpretation, the species-rich genus *Perizoma* is highly diverse in genital mophology. Exhibiting a number of common features in the structure of genitalia, many of its representatives nevertheless differ considerably from the type-species *P. albulata*. The former genus *Perizoma* (s.l.) has been divided into four genera on the basis of the sexual dimorphism, expressed in the structure of antennae, the shape and pattern of wings, and differential features in the structure of genitalia (MIRONOV 2000). In addition the genus *Pseudobaptria* has been transferred to the Perizomini, as these species show all the characteristic morphological features of the tribe.

The genus *Eupithecia* is one of the most species-rich genera of Lepidoptera. The majority of the so-called 'pugs' are small, unattractive grey or brown coloured moths, that can hardly be distinguished from each other by means of external features. Sometimes, they do not show clear morphological differences either. Therefore, the identification of many Eupitheciini species including European species is extremely difficult.

In earlier publications (HÜBNER 1825; MABILLE 1869; PACKARD 1876; HAMPSON 1895; MEYRICK 1928), various classifications of this genus based principally on the external characters such as wing venation, pattern and colour were proposed. The sequence of *Eupithecia* species given in the catalogue of STAUDINGER & WOCKE (1871) which has often been used until now in faunistic publications, does not correspond well with the morphological data sets found during the last century. The publications of PETERSEN (1909) and PIERCE (1914) were of great importance for comprehensive studies of the tribe Eupitheciini. These authors used both the external features and structures of the male and female genitalia for taxonomic purposes. PETERSEN, based on studies of the genitalia, considered Eupitheciini to include three genera: *Eupithecia*, *Gymnoscelis* and *Chloroclystis*. This tribe was also treated in the extensive monograph of DIETZE (1910, 1913), where the new genus *Calliclystis* DIETZE was established based on three European species that earlier had been attributed to *Chloroclystis*. The most detailed survey of Palaearctic pugs was published by PROUT (1914; 1938), where the traditional species order was maintained and no sub-grouping of the extremely species-rich genus *Eupithecia* was proposed.

For the first time, MCDUNNOUGH (1949) distinguished complexes of allied species (species-groups) in his revision of the North American *Eupithecia* on the basis of the structure of male and female genitalia. Attempts to divide the Palaearctic species of the huge genus *Eupithecia* into subgenera, such as *Dietzea*, *Petersenia*, *Bohatschia* and *Delaeveria* (SCHÜTZE 1956, 1958, 1960, 1961) based on genitalic structures were important for the systematic exploration of the genus, but they have not been accepted by the scientific community.

The classification proposed here for the European representatives of the genus *Eupithecia* (35 species-groups) is based primarily on the structure of the male and female genitalia; from my point of view the other features are not of value for characterizing various species-groups, especially those containing a large number of species.

For the studies on this volume, many publications dealing with the systematics and faunistics of geometrid moths in most of the European countries have been traced and studied, as well as atlases, catalogues and original descriptions of all European species. The synonymy has been partly checked, some misidentifications and incorrect (casual) subsequent spellings of specific names have been mentioned. In the nomenclatural lists type deposition is unknown unless stated otherwise. Subsequent names are junior synonyms unless stated otherwise.

The author has examined material in the collections of Munich, Copenhagen, Budapest, Bonn, Saint Petersburg, Tartu, Kiev, Moscow, Vladivostok, the types and other specimens of some species from the Museums of London, Berlin, Vienna, Paris, Barcelona, Prague, Stockholm, Helsinki, Erevan, and some private collections as well.

The diagnoses of species are generally based on serial material, only exceptionally on one or just a few specimens.

The phenology presented mainly refers to literature and collection data (using primary data, i.e. original observations). The flight period of many species, especially the more widespread, are diverse in the different European countries. Therefore, the earliest and latest dates for imago and larvae are added in brackets. The larval host-plants and periods of larval development were taken mainly from literature. The upper limits of vertical distribution are given for the larger mountain massifs, such as the Alps, Pyrenees and sometimes for the Carpathians, the Apennines and the mountains of the Balkan peninsula.

Abbreviations

AMNH	American Museum of Natural History, New York, U.S.A
AZM	Amsterdam Zoological Museum, Amsterdam, The Netherlands
BMNH	The British Museum (Natural History), London, United Kingdom
DEUU	Department of Entomology, University of Uppsala, Sweden
EIHU	Entomological Institute, Hokkaido University, Sapporo, Japan
FBUB	Facultat de Biologia, Universitat de Barcelona, Spain
HNHM	Termeszettudomanyi Muzeum Allattara (Hungarian Natural History Museum), Budapest, Hungary.
ITZA	Instituut voor Taxonomisch Zoölogie, Universiteit van Amsterdam, The Netherlands
IZA	Institute of Zoology, Academy of Sciences, Erivan, Armenia
IZB	Institute of Zoology National Academy of Sciences of Belarus, Minsk, Belorussia
IZBE	Institute of Zoology and Botany of the Academy of Sciences, Tartu, Estonia
MCZ	Museum of Comparative Zoology, Harvard University, Cambridge, Massachusetts, U.S.A
MGAB	Muzeul de Istorie Naturala 'Grigore Antipa', Bucharest, Romania
MNCN	Museo Nacional de Ciencias Naturales, Madrid, Spain
MNHN	Museum National d'Histoire Naturelle, Paris, France
MNHU	Museum für Naturkunde, Zentralinstitut der Humboldt-Universitat zu Berlin, Germany
MRSN	Museo Regionale di Scienze Naturali, Torino, Italy
MSND	Museo civico di Storia naturale 'Giacomo Doria', Genova, Italy
MZLU	University Museum of Zoology, Lund, Sweden
MZSF	Museo Zoologico Universitario 'La Specola', Firenze, Italy
NHMW	Naturhistorisches Museum Wien, Austria
NHRS	Naturhistoriska Riksmuseet, Stockholm, Sweden
NLHW	Niederösterreichisches Landesmuseum, Vienna, Austria
NMBS	Naturhistorisches Museum, Bern, Switzerland
NMNH	National Museum (Natural History), Praha, Czech Republic
NSMW	Naturwissenschaftliche Sammlung, Museum Weisbaden, Germany
USNM	United States National Museum, Washington, U.S.A.
ZFMK	Zoologisches Forschungsinstitut und Museum A. Koenig, Bonn, Germany
ZISP	Zoological Institute, Russian Academy of Sciences, Saint-Petersburg, Russia
ZMH	Zoological Museum of the University, Helsinki, Finland
ZMKU	Zoological Museum, National Shevchenko University, Kiev, Ukraine
ZMMU	Zoological Museum, Moscow State University, Russia
ZMTU	Zoological Museum of Tartu University, Estonia
ZMUC	Zoological Museum of the University, Copenhagen, Denmark
ZMUH	Zoologische Museum Universität, Hamburg, Germany
ZSM	Zoologische Staatssammlung München, Germany

Systematic Account

Tribus **Perizomini** HERBULOT, 1962

At least five genera and more than 170 species in the Palaearctic, Nearctic, Oriental and Afrotropical regions.

DIAGNOSIS: Wingspan 11-36 mm. Antennae bipectinate or multiply indented to filiform in male and filiform in female. Forewing venation diagnostic for genera. Hindwing: cell with inwardly projecting angle between M1 and M2.

MALE GENITALIA: Uncus broad, membranous, usually its apical part heavily sclerotized, narrower, acidotus uni- or biapical. Valva broad, usually without ventral projection, but in some mainly Asian species with digitus, harpa or ventral process. Transtilla or hemitranstilla present. Posterior arms of labides with united or bifurcate, setose papilla; anterior arms of labides sometimes with minute papillae on their apices, connected with juxta by a membrane. Shape of juxta diverse. Vinculum from short to long and broad, bladder-like. Size and shape of aedeagus diverse, with caulis in some species. Vesica often armed with cornuti or without them. Sternum A8 not modified.

FEMALE GENITALIA: Bursa copulatrix saccate, spherical or ovate, membranous, usually with one or two compact scobinate or rasp-like signa, or with numerous minute granules, or without signa in some species. Ductus seminalis usually attached to ductus bursae. Colliculum frequently well developed (collar-like, funnel-shaped or lamelliform). Lamella antevaginalis present in some species. Anterior apophyses usually with spurs connected with each other by a narrow, heavily sclerotized band, which encircles the antrum.

IMMATURE STAGES: Pupal morphology discussed by PATOČKA (1995).

REMARKS: The representatives of the tribe Perizomini can usually be distinguished from all other Geometrid moths by the presence of labides (twin sclerotized arms, springing from the basal points of the costae of valvae) in the male genitalia and by the presence of a narrow, heavily sclerotized band, which encircles the antrum bursae in the female genitalia (absent in some species).

Pseudobaptria INOUE, 1982

TYPE-SPECIES: *Cidaria corydalaria* GRAESER, 1889. Four species in the Palaearctic region (SATO 2002).

DIAGNOSIS: Labial palpi short, pointed, in both male and female extending beyond front of eyes, about 0.6-0.9 times diameter of eye. Eyes with thick white collars behind. Antennae in male bristle-like, bipectinate; female antennae filiform. Forewing venation: areole with oblique transverse vein. Hindwing: Sc+R1 arising at short distance from upper angle of cell. Sexual dimorphism well defined only in structure of antennae.

MALE GENITALIA: Uncus long, somewhat clavate, sclerotized, with a few setae near apex and with short and broad membranous base. Tegumen without papillae. Valvae small, short and broad, with sclerotized median part, and with minute, blunt digitus. Arms of hemitranstilla long and narrow, curved caudad, connected with bifurcate setaceous papil-

Text-fig. 1: Venation of *Pseudobaptria bogumilaria* (REBEL, 1904).

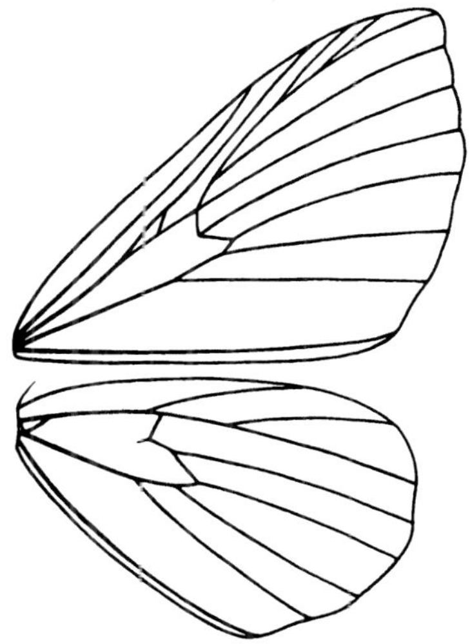

la. Anterior arms of labides narrow, smoothly tapered, papillae lacking, connected with each other by a membrane via juxta; posterior arms connected with papilla. Juxta elongate, peg-like, broadened at base and tapered toward apex. Vinculum short, with heavily sclerotized, digitiform saccus. Aedeagus slim, relatively short and thin. Vesica armed with some rows of small inward-facing spines forming one elongate, dense patch of spines.

FEMALE GENITALIA: Bursa copulatrix membranous with slightly sclerotized ductus bursae, armed with a large, heavily sclerotized trilobed signum. Antrum funnel-shaped, sclerotized. Spurs of anterior apophyses not connected with each other by heavily sclerotized band.

IMMATURE STAGES: Larval and pupal morphology unknown.

REMARKS: Structural details of the male genitalia suggest that this unique depauperate and probably relatively ancient genus is closely related to the genera *Mesotype* and *Perizoma*. The representatives of these genera show similar shape of valva; anterior arms of labides are connected with each other by a membrane via the juxta; posterior arms of labides and arms of hemitranstilla are connected with setose papilla. Nevertheless, the presence of a digitus on the valva, the shape of uncus and juxta, the structure of the vinculum with dorsal digitiform appendage, as well as the female genitalia easily distinguish *Pseudobaptria* from these genera.

1. *Pseudobaptria bogumilaria* (REBEL, 1904)

Larentia corydalaria var. *bogumilaria* REBEL, 1904: Annln naturh. Hofmus. Wien 19: 267, pl. 5: 16, 17 (Bosnia: Bocac, between Jaice-Banjaluka). Syntypes ♂ ♀ (NHMW).

Larentia corydalaria f. *eurytaenia* REBEL, 1908: Verh. zool.-bot. Ges. Wien 58 (1): (30). Valid at subspecific rank.

Unavailable names (misidentification): *corydalaria*: sensu REBEL (1901) nec GRAESER, 1889 (identity: *Pseudobaptria corydalaria*, eastern Asia). - (incorrect subsequent spelling): *aurytaenia*: TOMIČ & al. (1990).

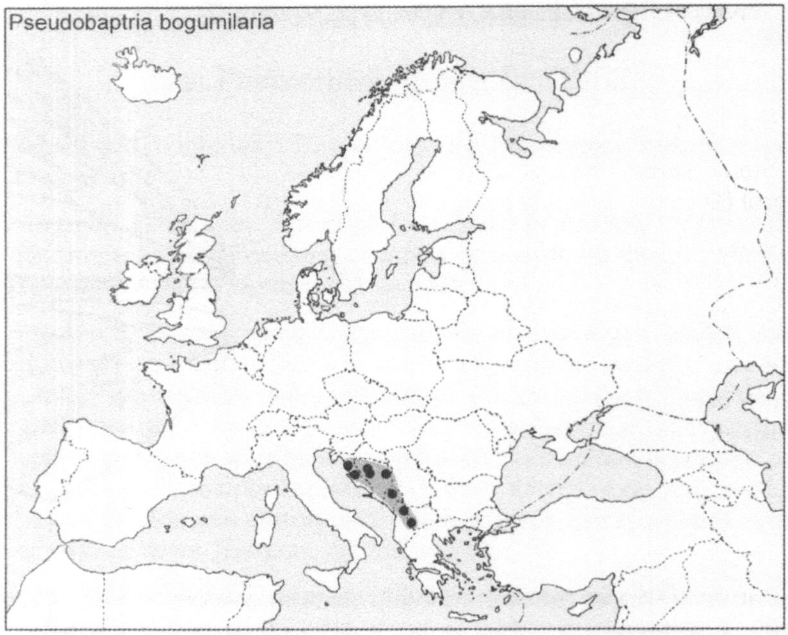

DIAGNOSIS: Wingspan 14-26 mm. Labial palpi with first segment covered by long white or pale orange scales; second segment white; third segment covered by close, black scales. Frons, vertex and notum black. Forewing black with basal stripe broken into three white spots; antemedial line slightly sinuate, unbroken, white; medial area has three white spots (one large, elongate costal and two smaller, rounded near dorsum); terminal area with five small separate white dots. Hindwing black, usually with three small white spots: one near dorsum and two others in terminal area (one medial and the other tornal). Fringe black with four small white spots on each wing. Abdomen without tufts, black with narrow white rings at posterior margin of segments.

The f. *eurytaenia* REBEL, 1908 is usually smaller; frons and vertex covered with mixture of black and white scales, one or other predominating; forewing with broader antemedial line and with larger medial spots; hindwing with broad, medial, transverse white stripes broken into two large, elongate spots, and also with paler, whitish basal area.

The f. *montenegrica* nova is slightly larger (wingspan 22-25 mm) than *eurytaenia* REBEL, 1908. It is distinguished from *eurytaenia* by presence of a continuous and wide white medial fascia on both wings, by the frequently well expressed black, white-edged discal spot on the forewing, and by the fringe evenly chequered black and white. In contrast, form *eurytaenia* has five small white spots on the fringe of both fore- and hindwing. The male and female genitalia of this new form are identical with those of the typical form.

MALE AND FEMALE GENITALIA: See generic diagnosis.

DISTRIBUTION: Endemic to the western mountains of Balkan peninsula (Croatia, Bosnia, Montenegro, Albania and Macedonia).

PHENOLOGY: Probably bivoltine: mid-May to late June; mid-July to August.

BIOLOGY: Larva monophagous (2). Recorded probably on *Pseudofumaria alba* (*Corydalis ochroleuca*) (Albania: REBEL & ZERNY 1934) and „*C. eurytaenia*" (Fumariaceae), the identity of the latter name remains unclear (Montenegro: TOMIČ & al. 1990).

HABITAT: Stony slopes. Up to about 2,500 m above sea-level in Durmitor Mts (Montenegro) (TOMIČ & al. 1990).

SIMILAR SPECIES: *P. bogumilaria* is unlikely to be confused with any other European Geometrid species. Very similar to East Asian *P. corydalaria* but distinguished by broader wings, less angulated white antemedial line, narrower white medial fascia, broken into three spots on the forewing, as well as by absence of unbroken, broad, white medial fascia on hindwing (except in *P. corydalaria japonica* HORI, 1926). The male and female genitalia of *P. bogumilaria* can be difficult to distinguish from those of *P. corydalaria*. However, the Balkan species has the valva narrower, less sclerotized, digitus shorter and broader; papilla smaller and narrower; small appendage of vinculum absent, in comparison with equivalents in the male genitalia of *P. corydalaria*. According to SATO (2002) the shape of sternite A8 in males of both allied species is also different. The female genitalia are distinguished by ductus bursae more elongate; signum larger, trilobous, tergum A8 larger and broader, without deep and narrow, membranous medial hollow in posterior margin.

REMARKS: This species was reported from Turkey (OKYAR-GOBEKCIOGLU & AKTAC 1997) probably in error. Possibly the genitalia of *Trichodezia haberhaueri* (LEDERER 1864) were figured in this article.

Mesotype HÜBNER, 1825

TYPE-SPECIES: *Geometra parallelaria* DE VILLERS, 1789 (junior subjective synonym of *Phalaena parallelolineata* RETZIUS, 1783). Three species in the Palaearctic region.

DIAGNOSIS: Labial palpi short and blunt; extending beyond front of eyes about 0.4-0.8 times diameter of eye in ♂, about 0.6-1.0 times diameter of eye in ♀. Eyes without thick collars behind. Forewing venation: areole with or without (in *M. didymata*) oblique transverse vein. Hindwing: Sc+R1 arising at short distance from upper angle of cell or from upper angle of cell (as in *M. parallelolineata*). Sexually dimorphic: Male usually larger, with antennae bristle-like, bipectinate or multiply indented. Female smaller, with antennae according to species from multiply indented to filiform. Forewing of female narrower, with apex more pointed than in male, ground colour of head, body and wings lighter.

MALE GENITALIA: Uncus long, narrow, spine-like, pointed at apex, sclerotized, with short and broad, weakly sclerotized base. Tegumen without papillae. Valva small or medium size. Arms of hemitranstilla usually long and narrow, curved caudad and connected with papilla. Papilla united, elongated, setaceous. Labides: posterior arms broad; anterior arms short, broad, sharply tapered to apex; connected with each other by membrane via juxta. Juxta broad, usually with two short, broad and blunt lateral processes (and with two short apical processes in *M. didymata*); covered with sparse short setae on the ventral surface. Vinculum short and narrow, slightly prominent dorsad. Saccus not present. Aedeagus thin, from short to relatively long, with spinose surface at apex. Vesica armed with a patch of spinules and one small dentate cornutus (in *M. parallelolineata*).

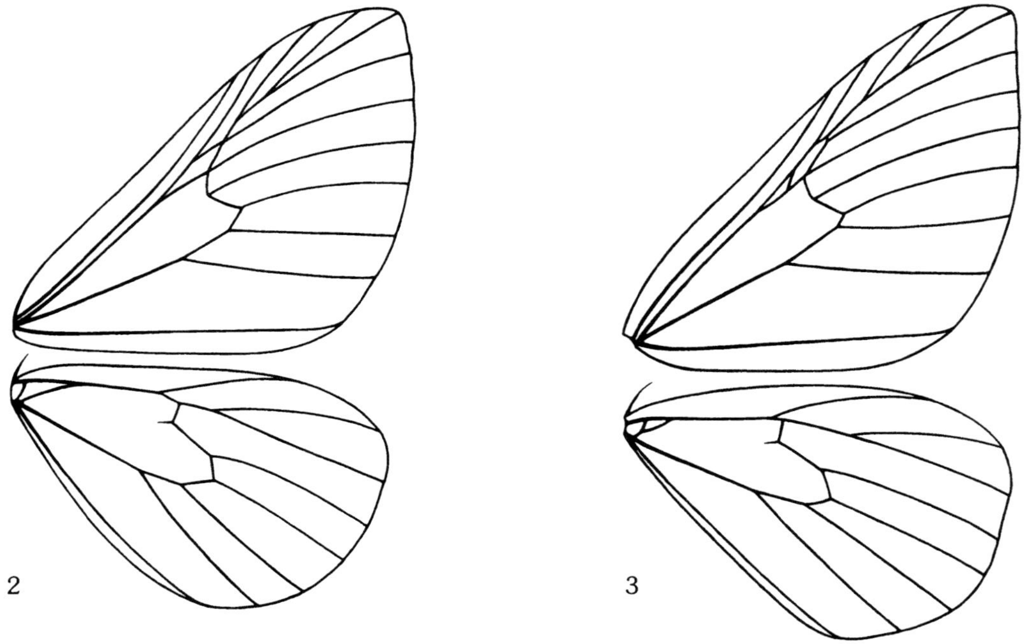

Text-figs 2-3: Venation of European *Mesotype* species. Text-fig. 2: *M. didymata* (LINNAEUS, 1758). Text-fig. 3: *M. parallelolineata* (RETZIUS, 1783).

FEMALE GENITALIA: Bursa copulatrix membranous; armed with various sclerites: a discoid scobinate signum, or a patch of scattered sclerotized pips, or without signum. Colliculum collar-like, short and narrow or large, long and broadened at base (as in *M. parallelolineata*). Antrum membranous. Spurs of anterior apophyses connected with each other by narrow heavily sclerotized band, which encircles antrum (except in *M. didymata*).

IMMATURE STAGES: Pupal morphology of *Perizoma didymata* and *P. parallelolineata* discussed by PATOČKA (1995).

REMARKS: This genus is closely related to *Pseudobaptria* on the one hand and to *Perizoma* on the other hand. *Mesotype didymata* appears to be most closely related to *Pseudobaptria*, suggested by bristle-like, bipectinate antennae with long branches in male, by digitiform saccus and elongate, dense patch of small spinules in the vesica, as well as by the short apophyses and short spurs of anterior apophyses and the very broad, obtuse and angulated papillae anales in the female genitalia. *Mesotype verberata* appears to be closer to the genus *Perizoma*, a relationship suggested by multiply indented antennae in male and filiform antennae in female.

I have included three species in the genus *Mesotype* based on the presence of distinct sexual dimorphism in the structure of antennae and the shape and ground colour of body and wings, and also on the presence of setae on the ventral surface of juxta in the male genitalia (MIRONOV 2000). Moreover, all three species of *Mesotype* overwinter as egg instead of pupa; their larvae are polyphagous on various plants.

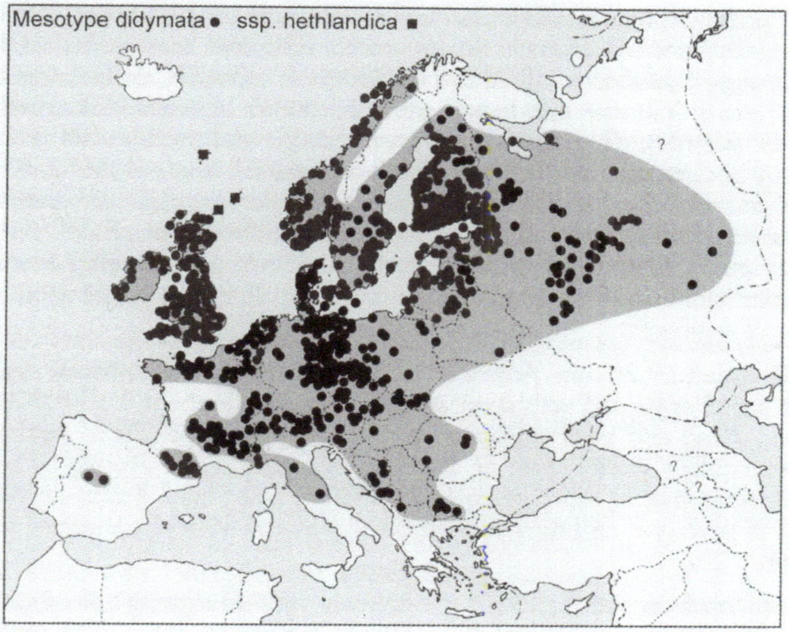

2. *Mesotype didymata* (LINNAEUS, 1758)

Phalaena (Geometra) didymata LINNAEUS, 1758: Syst. Nat. (Ed. 10) 1: 528 (Europe: Sweden). Syntype(s) (LSL).

? *Geometra angustata* DENIS & SCHIFFERMÜLLER, 1775: Syst. Verz. Schmett. Wien.: 315 (Austria: Vienna district). Syntype(s) lost.

Geometra inundaria DENIS & SCHIFFERMÜLLER, 1775: Syst. Verz. Schmett. Wien.: 316 (Austria: Vienna district). Syntype(s) lost.

Geometra scabrata HÜBNER, 1799: Samml. Eur. Schmett. 5 Geometrae (2), pl. 44: 229 (Europe). Syntype(s) lost.

Geometra alpestrata HÜBNER, 1809: Samml. Eur. Schmett. 5 Geometrae (1 & 2), pl. 62: 320 (Europe). Syntype(s) lost.

Cidaria didymata var. *ochroleucata*: AURIVILLUS, 1891: Nordens Fjärilar. Macrolepidoptera: 244 (Sweden: Uppland; southern Finland). Syntype(s).

Larentia didymata var. *hethlandica* REBEL, 1910: Fr. Berge's Schmetterlingsbuch (Edn. 9): 344 (United Kingdom, Scotland: Shetland Islands). Syntype(s) (NHMW). Valid at subspecific rank.

Larentia didymata attenuata CULOT, 1919: Noctuelles et Géomètres d'Europe 3: 189, pl. 27: 558 (United Kingdom, Scotland: Shetland Islands). Holotype ♀. Valid at subspecific rank.

Larentia didymata var. *pyrenaeata* BUBACEK, 1926: Verh. zool.-bot. Ges. Wien., 74/75: (9) (Europe: Pyrenees, type locality not stated). Syntypes ♂ ♀. Valid at subspecific rank.

Cidaria didymata gedrensis PROUT, 1937: in SEITZ, Macrolep. 4 (Suppl.): 134 (France: Gèdre district). Syntype(s) (BMNH). Valid at subspecific rank.

Unavailable names (infrasubspecific): *koehni*: WARNECKE (1936); *trifasciata*: WARNECKE (1936); *planicolor*: LEMPKE (1950); *contrasta*: LEMPKE (1950); *bilineata*: LEMPKE (1969). - (incorrect subsequent spellings): *scabraria*: HÜBNER 1825; *didymaria*: DOUBLEDAY (1849); *dydimata*: BUBACEK (1926).

DIAGNOSIS: Labial palpi brown tipped with white. Antennae in both sexes white ringed light brown. Forewing without oblique transverse vein in areole; R1 arising from middle of areole; common stalk of R2+R4 short.

Male. Wingspan 21-32 mm. Frons, vertex and notum covered with mixture of ash-grey and brown scales. Antennae bristle-like, bipectinate, with long branches. Forewing pale ash-grey, mottled and irrorate with brown scales; transverse lines brown, wavy, slightly curved, not angled near costa; basal and medial areas brownish, darker near costa; terminal area broad, its inner half brown with two darker blotches (one costal and one between M1 and M3), its outer half ash-grey, mottled and irrorate with brown scales; whitish wavy line distinct; terminal line broad, broken, forming a series of minute black dots near vein ends; discal spot absent. Hindwing unicolorous, pale grey; transverse lines obsolete; terminal area broad, slightly darker, greyish; discal spot absent. Fringe short, unicolorous, pale ash-grey with brownish tinge. Abdomen pale ash-grey or whitish grey with scattered light brown scales and with a row of small, brown dorsal spots at posterior margins of segments.

Female. Wingspan 19-26 mm. Antennae multiply indented, each antennal segment with some long apical setae and with short sparse setae, length about 0.15-0.2 times width of flagellum. Head and body smoky-white. Forewing with apex more pointed than in male; ground colour smoky-white or nearly white, mottled and irrorate by light brown scales; transverse lines as those in male but distinctly paler, light brown. Hindwing nearly white, with narrower, slightly darker terminal area and without transverse lines; discal spot absent.

Subsp. ***hethlandica*** (REBEL, 1910) is extremely variable in wing colour; some specimens little different from habitus of the nominate subspecies, but in most specimens the forewings have the central band dark or suffused with blackish or all wings have brownish basal half and lighter terminal part. The females often pale ochreous, with a more or less distinct brownish central band, and on the whole with a definite and darker pattern on the wings than in females of nominate subspecies.

MALE GENITALIA: Vinculum with small, digitiform appendage dorsad. Aedeagus shorter than length of valva, about 4-5 times longer than its medial width. Vesica armed with some rows of small, delicate, inward-pointing spines forming an elongate dense patch of small spinules.

FEMALE GENITALIA: Bursa copulatrix armed with one small discoid scobinate signum. Antrum not encircled by narrow sclerotized band. Anterior and posterior apophyses very short. Spurs of anterior apophyses short, thin, pointed. Papillae anales short, broad, obtuse and angulated.

DISTRIBUTION: Euro-Caucasian (boreal). Almost throughout Europe from Iberian peninsula to Urals (except Iceland, Mediterranean islands, Greece and European Turkey). On the Shetland and Faroe Islands replaced by subsp. *hethlandica*. – Outside Europe reported for Transcaucasus (Georgia and Armenia) (WARDIKJAN 1985; VIIDALEPP 1996), but I have not seen the Caucasian specimens.

PHENOLOGY: Univoltine: late June (late May) to early September (late September). The subsp. *hethlandica* from early August to mid-September. Larva: early May (mid-April) to mid-June. Overwintering as egg.

BIOLOGY: Larva polyphagous (1). Found on *Vaccinium myrtillus, Calluna vulgaris* (Ericaceae), *Quercus robur* (Fagaceae), *Geum rivale, Filipendula ulmaria* (Rosaceae), *Lathyrus pratensis* (Fabaceae), *Geranium sylvaticum* (Geraniaceae), *Anthriscus sylvestris* (Apiaceae), *Fraxinus excelsior* (Oleaceae), *Scrophularia nodosa* (Scrophulariaceae) (Finland: SEPPÄNEN 1970), also on *Primula* (Primulaceae), *Oxalis ace-*

tosella (Oxalidaceae), *Stellaria*, *Silene dioica* (Caryophyllaceae) (United Kingdom: BARRETT 1902), *Gentiana lutea* (Gentianaceae), *Cardamine* (Brassicaceae), *Anemone* (Ranunculaceae) (Switzerland: VORBRODT 1928). Recorded also on *Cerastium* (Caryophyllaceae) (Germany: SPULER 1910), *Chaerophyllum temulum* (Apiaceae), *Anemone nemorosa* (Ranunculaceae), *Minuartia (=Alsine)* (Caryophyllaceae), *Prunus* (Rosaceae) (France: LHOMME 1935), *Senecio nemorensis* subsp. *fuchsii* (Asteraceae), *Rumex acetosella* (Polygonaceae), *Saxifraga granulata* (Saxifragaceae) (Germany: BERGMANN 1955), *Anemone ranunculoides* (Ranunculaceae) (Germany: KOCH 1984), *Salix herbacea* (Salicaceae) (United Kingdom: SKINNER 1984), especially feeding on flowers.

PARASITOIDS: Trichogrammidae, Hymenoptera: *Trichogramma evanescens* (HERTING 1976).

HABITAT: Deciduous, spruce- and mixed forests, scrub, usually on slightly wet soils, moors, mountain hillsides, rocks and stone walls, roadside verges, waste ground, cliffs, sandhills and also on drier terrain. The subsp. *hethlandica* in grassland (WOLFF 1929a). From 0 m up to about 1,700 m above sea-level in Apennines (Mt. Majella: ZAHM pers. comm.), to about 2,000-2,200 m in Alps (FAVRE 1899; VORBRODT & MÜLLER-RUTZ 1914; FORSTER & WOHLFAHRT 1981; LUQUET 1982), Balkan peninsula (TOMIČ & al. 1990) and Pyrenees (MAZEL & PESLIER 1997).

SIMILAR SPECIES: It is readily recognized and can be easily distinguished from any other species by the distinctive maculation and bipectinate antennae in male. The genitalia in both sexes can be easily distinguished from those of relatives

3. *Mesotype parallelolineata* (RETZIUS, 1783)

Phalaena parallelolineata RETZIUS, 1783: Genera et Species Insectorum: 42 (Europe: type locality not stated). Syntype(s).
Phal(aena) Geometra parallelaria DE VILLERS, 1789: Linn. ent. 2: 322 (Italy: Brescia). Syntype(s).
Eubolia sudduplicaria COSTA, 1849: Fauna Regno Napoli, Lepidotteri: 88 (402), pl. 12: 2 (Italy: Napoli region, Majella, Orfenda valley). Syntype(s).
Unavailable names (misidentification): *vespertaria*: sensu DENIS & SCHIFFERMÜLLER (1775) nec LINNAEUS, 1767 (identity: *Epione vespertaria*); *brunnearia*: sensu LEDERER (1853) nec DE VILLERS, 1789 (identity: *Selidosema brunnearia*). - (incorrect subsequent spellings): *subduplicaria*: COSTA 1849; *vespertarea*: FIXEN (1849).

DIAGNOSIS: Labial palpi light brown. Antennae in both sexes unicolorous. Forewing with oblique transverse vein in areole; R1 arising from near upper angle of areole; common stalk of R2+R4 long.

Male. Wingspan 21-30 mm. Frons, vertex and notum pale ash-grey or whitish grey, sometimes with sparse light brown scales. Antennae bristle-like, bipectinate, with short branches. Forewing pale ash-grey; transverse lines ochreous brown; three basal lines inconspicuous, right-angled near costa; antemedial and subterminal lines straight, slightly sinuate, perpendicular to dorsum, dark brown with broad, light ochreous brown shade; outer half of terminal area ash-grey, mottled and irrorate with dark scales, with one to three darkish spots or lacking these, always with a short darkish apical streak; terminal line broken, forming a series of minute black dots near vein ends; discal spot absent. Hindwing pale smoky white or pale grey; transverse lines obsolete; terminal area broad, slightly darker, greyish; discal spot absent. Fringe short, unicolorous, pale ash-grey with

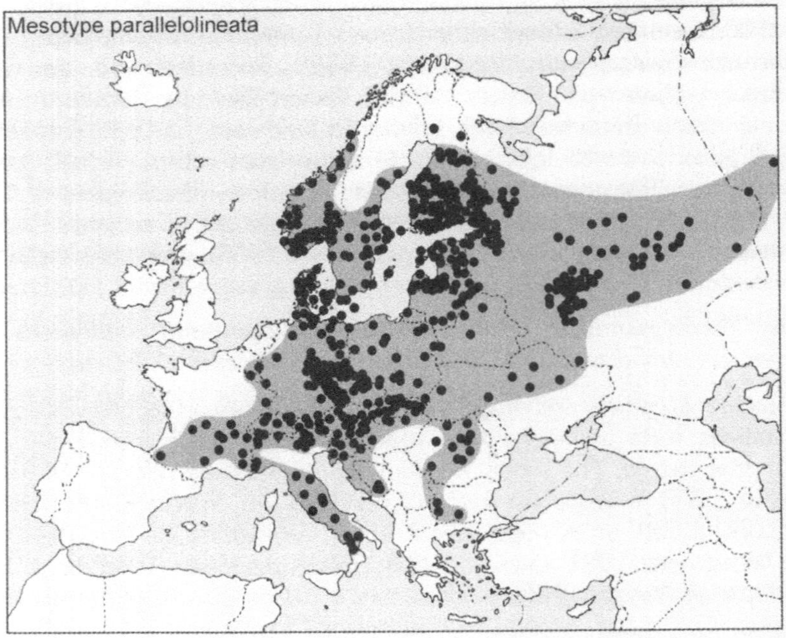

brownish tinge. Abdomen pale ash-grey or whitish grey with scattered light brown scales and with a row of small, brown dorsal spots at posterior margins of segments.

Female. Wingspan 23-30 mm. Frons and vertex whitish grey. Antennae multiply indented; setae length about 0.3-0.4 times width of flagellum. Wings and body lighter than in male, smoky-white or almost white; all transverse lines paler, light ochreous brown; pattern of wings similar to those in male.

MALE GENITALIA: Aedeagus slender. Vesica armed with a patch of spinules and one small, weakly sclerotized dentate cornutus.

FEMALE GENITALIA: Bursa copulatrix armed with a large patch of scattered sclerotized pips in the right side. Colliculum large and elongated, flat and broad at base and collar-like and narrow at apex, over half as long as of corpus bursae. Papillae anales large, elongated, rounded at apex.

DISTRIBUTION: European ? (boreal). Almost throughout Europe from western France to Urals (except Iceland, Ireland, United Kingdom, Netherlands, Portugal, Spain, Mediterranean islands, Albania, Macedonia, Greece and European Turkey). – Outside Europe reported for Altai Mts (southern Siberia) (VIIDALEPP 1996). I have Asian specimens from western Siberia (Tjumen' region) only.

PHENOLOGY: Univoltine: late July (early July) to early September (early October), in the Alps until late October. Larva: early April (mid-March) to late May (early July). Overwintering as egg.

BIOLOGY: Larva polyphagous (1). Recorded on *Galium* (Rubiaceae), *Saxifraga* (Saxifragaceae) (France: LHOMME 1935), *Pteridium aquilinum* (Polypodiaceae), *Prunus spinosa* (Rosaceae) (Germany: SÄLZL 1935), *Rumex* (Polygonaceae), *Taraxacum,*

Prenanthes (Asteraceae), *Campanula* (Campanulaceae) (Germany: BERGMANN 1955), *Plantago* (Plantaginaceae) (Germany: KOCH 1984), *Rhamnus* (Rhamnaceae), *Populus* (Salicaceae), *Corylus* (Betulaceae) (Austria: HABELER 1969).

HABITAT: Deciduous forests and scrub, uncultivated open habitats, field hedges, ditches and other similar areas. From 0 m up to about 1,600-1,700 m above sea-level in Apennines (Mt. Pollino: coll. ZSM), and to about 2,000-2,100 m in Alps (FAVRE 1899; VORBRODT & MÜLLER-RUTZ 1914; VORBRODT 1928).

SIMILAR SPECIES: Unlikely to be confused with any other European Geometrid species. It is distinguished by the two broad, ochreous brown transverse stripes on the forewing, the bipectinate antennae with short branches in male and multiply indented antennae in female. Also, distinguished from allied species by the structure of male and female genitalia, especially by the presence of a patch of scattered sclerotized pips in corpus bursae in the female genitalia (as in *Scopula* species, Sterrhinae). The species is most frequently caught at light.

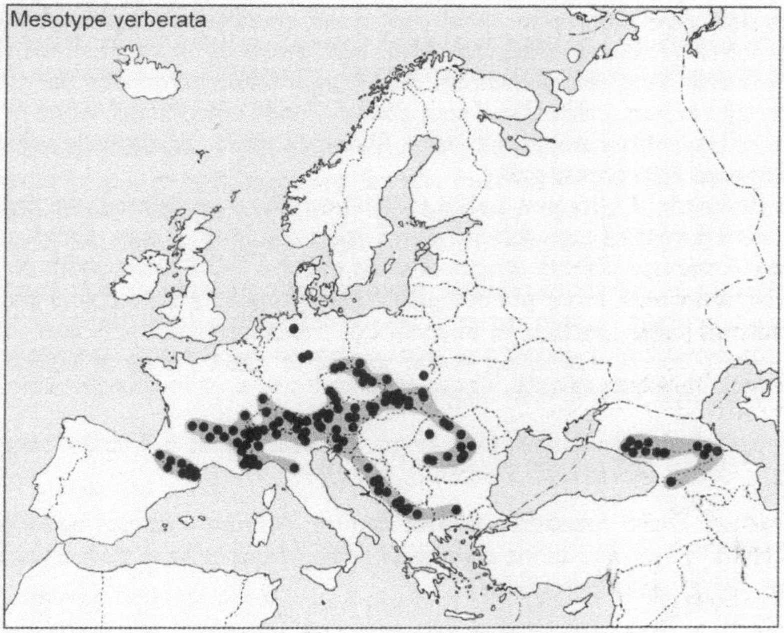

4. *Mesotype verberata* (SCOPOLI, 1763)

Phalaena verberata SCOPOLI, 1763: Ent. Carniolica, Ordo 3: 223, fig. 554 (Slovenja: Carniolia). Syntype(s) lost.

Geometra rupestrata DENIS & SCHIFFERMÜLLER, 1775: Syst. Verz. Schmett. Wien. (2): 109 (Austria: Vienna district). Syntype(s) lost.

Acidalia bassiaria DE FEISTHAMEL, 1835: Annls Soc. ent. France 4 (4): 133, pl. 1A: 2 (Italy, Piémont, Domo d'Ossola, on the edges of la Toccia, near Mazera). Syntype(s) (MNHN?).

Cidaria verberata var. *vogesiaria* PEYERIMHOFF, 1880: Catalogue des Lépid. d'Alsace (2me édition) 1 (Macrolépidoptères): 142 (France), Alsace: Rhinkopf; Hoheneck; Lachapelle-sous-Rougemt near Belfort; Ballon de Guebwiller). Syntype(s) (MNHN?). Valid at subspecific rank.

Cidaria verberata var. *pawlitscheki* HORMUZAKI, 1899: Verh. zool.-bot. Ges. Wien. 49 (1): 71 (Romania: (Bukowina), Lutschina (Mt), Rareu Mts). Syntypes. Valid at subspecific rank.

Cidaria verberata judicariae FIORI, 1931: Studi trent. Sci. nat. 12 (3): 193, figs. 13-30 (Italy: Pinzolo, Rendena). Syntypes ♂ ♀. Valid at subspecific rank.

Cidaria verberata juraphila PROUT, 1939: in SEITZ, Macrolep. 4 (Suppl.): 250 (Switzerland: Jura). Syntype(s) (BMNH). Valid at subspecific rank.

Unavailable names (infrasubspecific): *jurassica* WEHRLI (1918). - (incorrect subsequent spellings): *rupestraria*: BOISDUVAL (1840); *verbereta*: WEHRLI (1924).

DIAGNOSIS: Antennae unicolorous in both sexes. Forewing with oblique transverse vein in areole; R1 arising from near upper angle of areole; common stalk of R2+R4 short.

Male. Wingspan 24-32 mm. Labial palpi light brown; extending beyond front of eyes about 0.4-0.5 times diameter of eyes. Frons, vertex and notum white. Antennae multiply indented; setae length about 0.3-0.5 times width of flagellum. Forewing white, usually mottled and irrorate with brownish scales, especially along costa; transverse lines narrow, slightly sinuate, light brown or light ochreous brown; basal and postbasal lines angled or evenly curved near costa; ante- and postmedial lines nearly straight, wavy; subterminal line double, parallel with termen, slightly wavy, angled near apex between R5 and M1; white wavy line broad, clouded; terminal line broken, forming a series of small dark brown dots near vein ends; discal spot small, rounded or slightly elongate, light brown or obsolete. Hindwing white with weak, pale ochreous brown postmedial and subterminal transverse lines, also sometimes with slightly darker terminal area, mottled and irrorate with light brown scales; discal spot absent. Fringe unicolorous, white or with narrow patches of brownish scales at vein ends. Abdomen white, usually with yellowish tinge and with scattered light brown scales.

Female. Wingspan 23-26 mm. Labial palpi white with a few light brown apical scales; extending beyond front of eyes about 0.6-0.7 times diameter of eyes. Frons, vertex and notum white. Antennae filiform, length of setae about 0.2-0.3 times width of flagellum. Forewing narrower, with more pointed apex than in male; ground colour paler, white, without brownish tinge; pattern as in male but transverse lines narrower, discal spot minute or usually absent. Hindwing similar to that of male. Fringe unicolorous, white. Abdomen white, without yellowish tinge.

MALE GENITALIA: Aedeagus shorter than length of valva, about 6-7 times longer than its medial width. Vesica armed with a patch of spinules.

FEMALE GENITALIA: Bursa copulatrix without signum. Antrum encircled by long, narrow, sclerotized band. Spurs of anterior apophyses arising from close to their apices. Papillae anales large, broad, apically rounded.

DISTRIBUTION: Euro-Caucasian (alpine). Mainly in high mountains of Central and southern Europe (Pyrenees, Alps, Carpathians and mountains of Balkan peninsula). – Outside Europe in Caucasus, Transcaucasus (Georgia and Azerbaijan) and north-east of Asian Turkey.

PHENOLOGY: Univoltine: late June (early June) to late August (late September). Larva: early April (late March) to mid-June. Overwintering as egg. According to VORBRODT (1928) larvae from late August until mid-June, overwintering (!).

BIOLOGY: Larva polyphagous (1). Found on *Taraxacum, Leontodon, Hieracium* (Asteraceae) (France: CHRÉTIEN 1906). Recorded also on *Pinus, Abies* (Pinaceae) (France: LHOMME 1935) and *Vaccinium* (Ericaceae) (Austria: HABELER 1969).

HABITAT: In a variety of moist montane habitats on clay soils or wet soils with siliceous substrata, including alpine meadows, mires, and damp grassland on sunny and semi-shaded slopes. From 1,500 m up to about 1,900 m above sea-level in Pyrenees (MAZEL & PESLIER 1997), from 800 m up to about 2,000 m in Carpathians (POPESCU-GORJ, 1964; coll. ZMUC), from 600 m up to about 2,700 m in Alps (REBEL 1904; VORBRODT 1928; FEICHTENBERGER 1962; HABELER 1969; FORSTER & WOHLFAHRT 1981).

SIMILAR SPECIES: It is only remotely similar to *M. parallelolineata*, being easily distinguished by the multiply indented antennae in ♂ and filiform antennae in ♀; by the lighter, white head, body and wings; the curved and wavy transverse lines, without broad ochreous shade beyond ante- and postmedial lines, and without dark apical streak on the forewing. The male and female genitalia can be easily distinguished from those of other species.

Perizoma HÜBNER, 1825

=*Emmelesia* STEPHENS, 1829; *Zerynthia* CURTIS, 1830

TYPE-SPECIES: *Geometra albulata* DENIS & SCHIFFERMÜLLER, 1775. According to SCOBLE (1999) 166 species in the Palaearctic, Nearctic and Afrotropical regions. In Europe 12 species.

DIAGNOSIS: Labial palpi short and blunt; extending beyond front of eyes about 0.3-0.8 times diameter of eye in both male and female. Eyes without collars behind. Forewing venation: areole with oblique transverse vein (except *P. incultaria*). Hindwing: Sc+R1 arising at short distance from upper angle of cell. No sexual dimorphism.

MALE GENITALIA: Uncus relatively long, narrow, spine-like, pointed at apex, sclerotized, with short and broad, weakly sclerotized base. Tegumen without papillae. Valva relatively short and narrow. Arms of hemitranstilla short and usually wide, placed apex to apex; connected with each other by a membrane via papilla. Papilla united or in some species

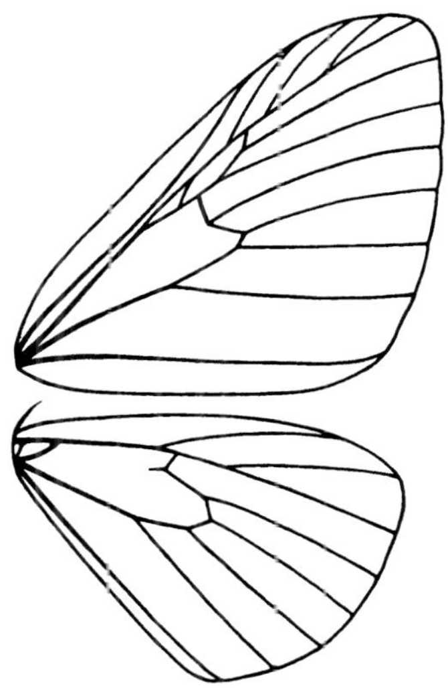

Text-fig. 4: Venation of *Perizoma albulata* (DENIS & SCHIFFERMÜLLER, 1775).

bifurcate, setaceous. Labides: posterior arms narrow; anterior arms short, narrow and tapered to apices; connected with each other by a membrane via juxta. Juxta broad, diverse in shape, however not distinctly bifurcate; without setae on the ventral surface (except *P. affinitata*). Vinculum short, narrow, slightly prominent dorsad. Saccus absent. Aedeagus thin, short and straight, usually with spiniferous surface at apex. Vesica armed with one or two diverse cornuti, or with numerous spinules, or without cornuti.

FEMALE GENITALIA: Bursa copulatrix membranous; armed with one or two discoid or elongate ovate-acuminate scobinate signa. Colliculum collar-like, small, short and narrow (except *P. flavofasciata*). Antrum membranous. Spurs of anterior apophyses connected with each other by a narrow sclerotized band which encircles antrum or without this band.

IMMATURE STAGES: Pupal morphology discussed by PATOČKA (1981, 1995).

REMARKS: All species of *Perizoma* hibernate in the pupal stage; their larvae are monophagous or oligophagous on the flowers and seeds of various herbaceous plants, especially on Caryophyllaceae and Scrophulariaceae.

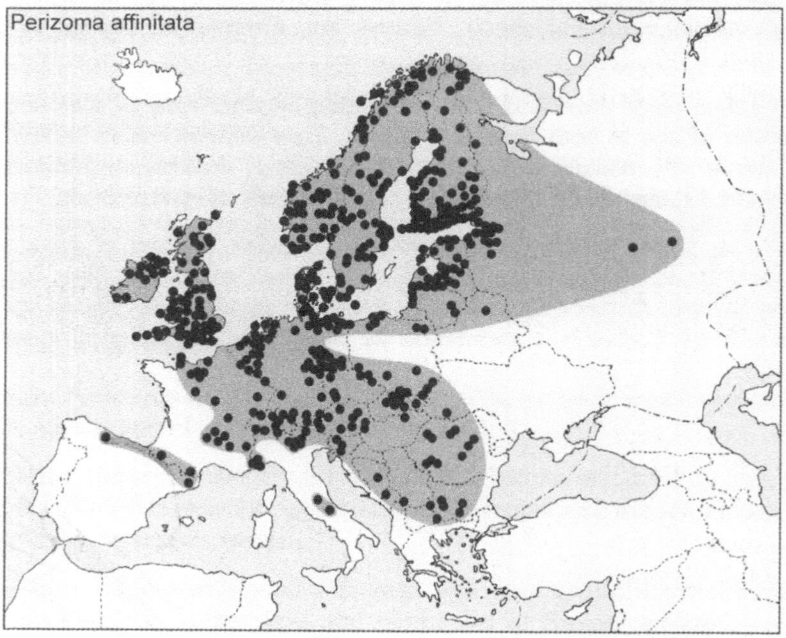

5. *Perizoma affinitata* (STEPHENS, 1831)

Emmelesia affinitata STEPHENS, 1831: Illustr. Br. Ent. (Haustellata) 3: 297 (United Kingdom: England, Hertfordshire, Copenhagen fields; vicnity of Hertford; Epping; Davidson's bank; Meldon Park). Syntype(s) (BMNH).

Emmelesia turbaria STEPHENS, 1831: Illustr. Br. Ent. (Haustellata) 3: 298 (United Kingdom: near Kentish town; Raehills; Epping; Flisk, N.B.; Cheshire, Alderley; Hill-ends). Syntypes (BMNH).

Cidaria rivinata FISCHER VON RÖSLERSTAMM, 1838: Abbildungen zur Berichtigung und Ergänzung der Schmetterlingskunde (8): 100 (Europe: type locality not stated). Holotype ♀.

Acidalia inciliata ZETTERSTEDT, 1839: Insecta Lapponica 4 (6): 961 (Sweden: Lappland, Dalecarlia). Syntype(s) ♂ (MZLU).

Cidaria turbulata STANDFUSS, 1851: Z. Ent. 5 (Lepidoptera): 81 (Border between Poland and Czech Republic: Riesen Mountains [=Western Sudety]). Syntypes 2♂4♀.

Larentia affinitata magistraria TURATI & VERITY, 1912: Boll. Soc. ent. ital. 43 (1-4): 198 (Italy: Valle del Gesso, Luglio, Terme). Syntypes. Valid at subspecific rank.

Unavailable names (infrasubspecific): *effusa*: SCHNEIDER (1933); *lineata*: LEMPKE (1950); *albofasciata*: LEMPKE (1969). - (incorrect subsequent spellings): *affinitaria*: HERRICH-SCHÄFFER (1848); *rivularia*: DOUBLEDAY (1849) (nec BOISDUVAL, 1840); *affinaria*: SIEVERS (1856); *affinata*: FEICHTENBERGER (1965).

DIAGNOSIS: Wingspan 21-30 mm; the male usually smaller than female. Labial palpi umber, sometimes with white basal and apical scales. Frons and vertex covered with mixture of pale grey and brown scales. Notum greyish brown. Forewing dark umber; basal and medial areas dark umber with inconspicuous, narrow, much indented, whitish stripe between them; postmedial line much waved with one blunt outward medial prominence between CuA1 and CuA2; the broad, white rivulet stripe between medial and terminal areas bisected along its length by a fine indented brown line or sometimes without it; terminal area broad, dark umber, usually becoming darker near apex; wavy line white, sometimes broken; terminal line relatively broad, dark brown, interrupted by white dots at vein ends or unbroken; discal spot absent. Hindwing white; basal area pale smoky-brown; terminal area more or less broad, distinct, brown or pale brown; wavy line usually absent or ill-defined, white; terminal line dark brown, interrupted by veins; discal spot minute, rounded, brownish, inconspicuous or obsolete. Some specimens have hindwing smooth smoky brown with broad clouded whitish medial stripe, bisected by a clouded smoky brown line. Fringe unicolorous brown or slightly spotted with dark brown at vein ends. Abdomen slender, brownish grey or pale brown.

MALE GENITALIA: Uncus relatively short. Anal tube very short. Valva narrow, elongate, with narrowly rounded apex; costa short and broad, not extending beyond apex of valva. Papilla small, globular; covered with medium length setae. Anterior arms of labides short and broad. Manica membranous. Juxta almost hexagonal, with three pairs of short and blunt processes (basal, lateral and apical); with one or two pairs of setae on the ventral surface. Vesica armed with one sclerotized pointed hooked cornutus.

FEMALE GENITALIA: Bursa copulatrix armed with two scobinate signa (one dorsal, elongate, ovate or ovate-acuminate, the other ventral, smaller, discoid or ovate-acuminate). Tergum A8 rectangular, elongated.

DISTRIBUTION: European. Almost throughout Europe, except Iceland, southern Iberian peninsula, southern Apennines and Balkan peninsula and Mediterranean islands; eastward to central European Russia.

PHENOLOGY: Univoltine: late May (late April) to mid-July (late August); sometimes there is a partial second generation in August to early September. Larva: mid-June (late May) to late August (early September). Overwintering as pupa, sometimes two years.

BIOLOGY: Larva oligophagous (1). Found in the seed capsules of *Silene dioica (Melandrium rubrum)* (Austria: HABELER & REBEL 1883) and *S. latifolia (Melandrium album)* (Caryophyllaceae) (United Kingdom: BUCKLER 1899). Recorded on *Silene nutans* (Caryophyllaceae), *Melampyrum* (Scrophulariaceae) (France: LHOMME 1935), *Silene vulgaris* (Caryophyllaceae) (Netherlands: LEMPKE 1969).

HABITAT: Margins and clearings of forests and plantations, roadside verges, wet woodland

and scrub, mountain moors and rocks among scrub. From 0 m up to about 1,600 m above sea-level in Apennines (RACHELI et al. 1978; PROLA & RACHELI 1980), to about 1,700 m in mountains of Balkan peninsula (REBEL 1917; coll. FIBIGER), to about 1,800 m in Pyrenees (MAZEL & PESLIER 1997), and to about 2,000-2,100 m in Alps (VORBRODT 1928, 1931; WOLFSBERGER 1965; FORSTER & WOHLFAHRT 1981).

SIMILAR SPECIES: *P. affinitata* can usually be easily distinguished from other European Geometrid moths by coloration and maculation. However, it can be confused with the three following species. The relatively short uncus, the very small anal tube, the small globular papilla, the short and broad anterior arms of labides, the narrow, elongated valva with narrowly rounded apex and with short and broad costa, the shape of juxta, as well as the elongated female genitalia with two scobinate signa in corpus bursae, should distinguish this species.

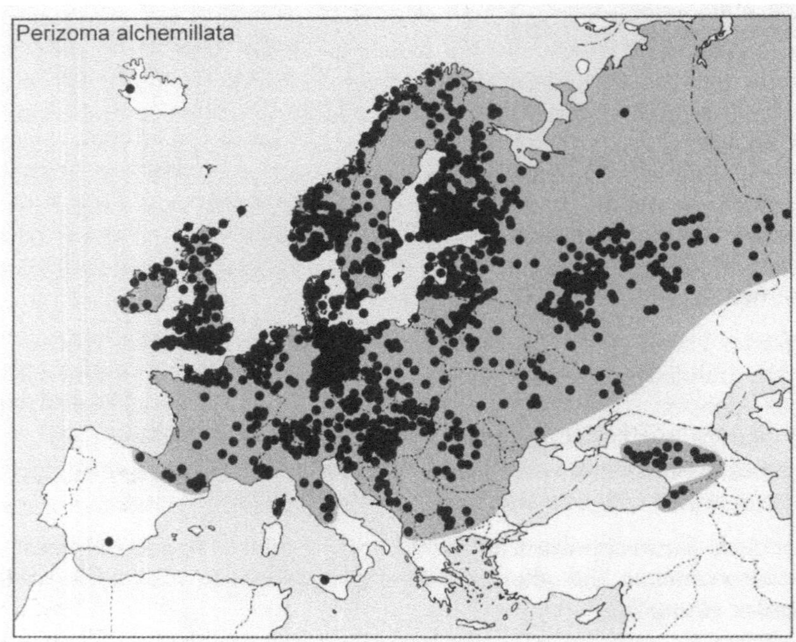

6. *Perizoma alchemillata* (LINNAEUS, 1758)

Phalaena (Geometra) alchemillata LINNAEUS, 1758: Syst. Nat. (Ed. 10) 1: 526 (probably Sweden). Syntype(s) (LSL).

Geometra rivulata DENIS & SCHIFFERMÜLLER, 1775: Syst. Verz. Schmett. Wien. 2: 109 (Austria: Vienna district). Syntype(s) lost.

Phalaena nassata FABRICIUS, 1787: Mantissa Insect. 2: 212 (Austria: Vienna). Syntype(s).

Epirrhoe aparinata HÜBNER, 1825: Verz. bek. Schmett. (21): 328 (Europe). Syntype(s) lost.

Emmelesia silenata STEPHENS, 1850: List Spec. Br. Anim. Colln Br. Mus. 5: 216 (United Kingdom: type locality not stated). Syntype(s) not traced in BMNH (L. PITKIN pers. comm.).

Cidaria fennica REUTER, 1893: Acta Soc. Fauna Flora fenn. 9 (6): 75 (Finland: Pargas, Lilltervo; Nyslott [=Savonlinna]). Syntype 1♂1♀ (ZMH).

Cidaria alchemillata peterseni PROUT, 1914: in SEITZ, Macrolep. 4: 260 (Estonia; Scandinavia; Scotland). Syntypes (BMNH?). Valid at subspecific rank.

Unavailable names (misidentification): *hydraria*: sensu DOUBLEDAY (1849) nec BOISDUVAL, 1840 (identity: *Perizoma hydrata*). - (infrasubspecific): *peterseni*: PROUT (1915); *interrupta*: BOLDT (1927); *albofasciata*: LEMPKE (1969); *obscura*: LEMPKE (1969). - (incorrect subsequent spellings): *alchemillae*: RETZIUS (1783); *alchemillaria*: BORKHAUSEN (1794); *rivularia*: BOISDUVAL (1840); *achlemillata*: KUDYN & CERVINKA (1995); *alchelillata*: POPESCU-GORJ (1995).

DIAGNOSIS: Wingspan 15-24 mm. Labial palpi brown tipped by white and usually with white basal scales. Frons and vertex covered with mixture of grey and dark brown scales. Notum brown with scattered grey scales. Forewing umber; basal and medial areas umber with more or less distinct, broad, much indented, whitish double stripe between them, and sometimes with similar but narrower stripe beyond basal line or only with small white spot near dorsum in place of it; postmedial line very wavy with two blunt outward medial prominences (one between M3 and CuA1, the other between CuA1 and CuA2); the broad, white rivulet stripe beyond postmedial line always bisected longitudinally by a fine, indented, brown, sometimes broken line; terminal area broad, umber, usually gradually darker near apex; wavy line white, unbroken but sometimes broken or even disintegrated into separate minute white dots; terminal line relatively broad, dark brown, interrupted by veins; discal spot minute, rounded, dark brown, ill-defined or absent. Hindwing smooth shining smoky brown with lighter, broad transverse medial stripe and usually with broad and slightly darker terminal area; pale wavy line usually absent; terminal line broad, dark brown, conspicuous or clouded, and unbroken or interrupted by veins; discal spot minute, rounded, dark brown, indistinct or sometimes obsolete. Fringe pale brown or pale brownish grey, spotted with brown at vein ends. Abdomen brownish grey.

MALE GENITALIA: Uncus relatively short, covered with sparse long setae near apex. Anal tube short. Valva short, wide, with broadly rounded apex; costa narrow and long, extending a short distance beyond apex of valva. Papilla trilobate; these lobes elongate and narrow, covered with long setae on apices. Anterior arms of labides rather long and thin. Manica membranous. Juxta tapered caudad, with three pairs of short processes (blunt basal, pointed lateral and apical). Vesica without cornuti.

FEMALE GENITALIA: Bursa copulatrix armed with one discoid scobinate signum. Spurs of anterior apophyses short but slightly broadened to apices. Tergum A8 almost square.

DISTRIBUTION: Holarctic. Widespread in Europe except Portugal, Balearics, Sardinia, Malta, Crete, southern Greece and European Turkey. – Outside Europe in north-eastern Turkey, Caucasus and Transcaucasus (Georgia, Armenia) (WARDIKJAN 1985; VIIDALEPP 1996), eastward throughout northern Kazakhstan and Siberia to Transbaikalia and Mongolia. Recorded for North America (HODGES & al. 1983).

PHENOLOGY: Uni- or bivoltine: early June (early May) to mid-August (early September), sometimes a partial second generation from late September to October (LEMPKE 1969). In the south bivoltine: early May (late April) to late June; mid-July (early July) to late August (VORBRODT 1931; WOLFSBERGER 1965). Larva: July to September; in the south: mid-June to mid-July; late July to late September. Overwintering as pupa.

BIOLOGY: Larva oligophagous (1). Found on the flowers and in the seed capsules of *Galeopsis tetrahit* (United Kingdom: BUCKLER 1899), more rarely on *Galeopsis ladanum, Stachys sylvatica, S. alpina* (United Kingdom: BARRETT 1902), *Lamium purpureum*, *Galeopsis speciosa* (Lamiaceae) (Finland: SEPPÄNEN 1970); recorded also on *Ballota nigra* (Lamiaceae), *Dianthus superbus* (Caryophyllaceae), *Linaria repens* (Scrophulariaceae), *Alchemilla xanthochlora* (Rosaceae) (France: LHOMME 1935),

Lamium maculatum and *Lamiastrum galeobdolon* (Lamiaceae) (Germany: BERGMANN 1955).

PARASITOIDS: Hymenoptera, Ichneumonidae: *Campoplex difformis*, *Agripon clandestinum*, *Campoletis femoralis*, *Triclistus congener*, *T. lativentris* - Hymenoptera, Braconidae: *Apanteles juniperatae*, *A. praepotens (=sericeus)*, *A. spurius* (BUCKLER 1899; THOMPSON 1946; SHENEFELT 1972).

HABITAT: Forest edges and clearings, commons, marshes, waste ground, downland, along field hedges and ditches and other similar uncultivated areas, preferably on slightly wet soil, rarely around plantations and on mountain moors. From 0 m up to about 1,500 m above sea-level in Apennines and Balkan peninsula (TEOBALDELLI 1994; TOMIČ & al. 1990), to about 2,000 m in Pyrenees (MAZEL & PESLIER 1997), to about 2,100 m in Alps (VORBRODT 1928; HELLMANN & al. 1999), and to about 2,500-2,700 m in Asia Minor (RIEMIS pers. comm.).

SIMILAR SPECIES: A rather invariable species, similar to *P. affinitata* but can be easily distinguished from it by the usually smaller size, much more indented postmedial line with two blunt outward medial prominences and usually darker, smooth smoky brown hindwing. The unique trilobed papilla, the short and wide valva, with costa slightly extending beyond apex of valva, the long and thin anterior arms of labides in the male genitalia, as well as discoid scobinate signa in bursa copulatrix in the female genitalia should distinguish this species from the closely related and similar *P. affinitata*.

7. ***Perizoma hydrata*** (TREITSCHKE, 1829)

Acid(alia) hydrata TREITSCHKE, 1829: Die Schmett. Eur. 7: 217 (border between Bohemia (Czech Republic) and Saxonia (Germany). Syntype ♂ ♀ (HNHM?).

Cidaria flexuosaria BOHEMAN, 1852: K. svenska Vetensk. Akad. Handl. 1851: 135 (Sweden: Ronneby). Syntype(s).

Cidaria hydrata var. *clarior* SCHAWERDA, 1928: Z. Öst. Ent. Ver. 13 (5): 43 (Corsica: Mt. Incudine). Holotype ♀ (NHMW). Valid at subspecific rank.

Unavailable names (misidentification): *fluctuata malleola*: sensu BELJAEV & OH (2001, fig. 19) nec INOUE, 1955 (identity: *Xanthorhoe fluctuata malleola*). – (infrasubspecific): *fasciata* TURATI & VERITY (1912). - (incorrect subsequent spelling): *hydraria*: BOISDUVAL (1840).

DIAGNOSIS: Wingspan 15-25 mm. Labial palpi dark grey with tips white. Frons and vertex covered with mixture of pale and dark grey scales. Notum covered with mixture of light and dark grey scales, with dark grey anterior transverse band. Forewing grey, sometimes with brownish tinge; basal and medial areas grey with inconspicuous, much indented, lighter stripe between them; postmedial line very wavy with one or two blunt, outward medial prominences; the broad smoky white rivulet stripe beyond postmedial line longitudinally bisected by a fine, indented grey line, the outer half of the rivulet stripe usually pale grey; terminal area broad, grey, sometimes mottled and irrorate with whitish scales near costa; wavy line, unbroken, white, usually well defined near costa; terminal line relatively broad, dark grey or black, interrupted by veins; discal spot usually absent but in some specimens present, minute, rounded, darkish. Hindwing smooth, pale grey, unicolorous or with slightly darker terminal area; terminal line broad, dark grey, conspicuous or clouded, and unbroken or interrupted by veins; discal spot obsolete. Fringe pale grey, spotted with grey at vein ends, especially on the forewing. Abdomen pale grey.

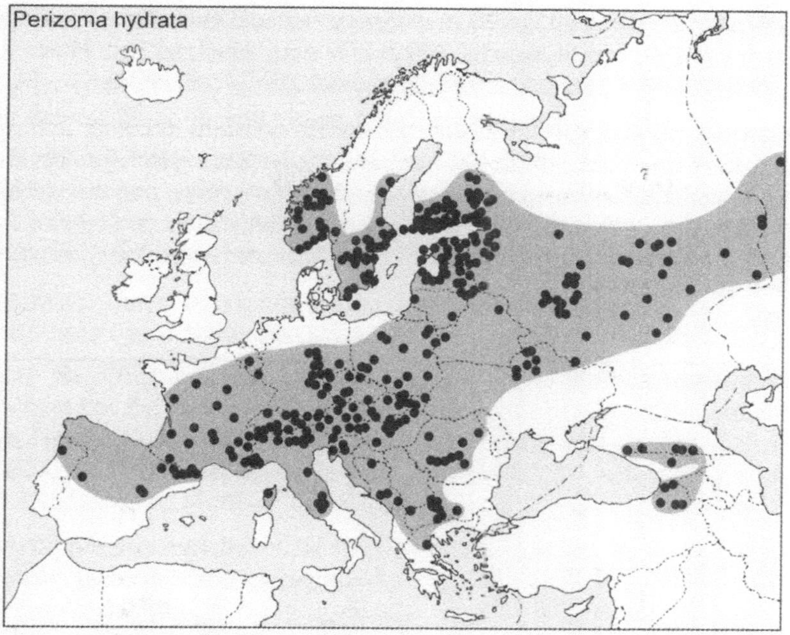

MALE GENITALIA: Anal tube small. Anterior arms of labides short. Valva small, narrow, with narrowly rounded apex and shallow ventral hollow near base; costa narrow, tapered to apex. Papilla small, globular; covered with short setae. Anterior arms of labides short. Manica finely spined. Juxta with two small basal processes, medially broadened. Vesica armed with one hooked cornutus.

FEMALE GENITALIA: Bursa copulatrix armed with two scobinate signa (one small, discoid on the ventral side, the other larger, elongate on the dorsal side). Spurs of anterior apophyses arising from close to their base.

DISTRIBUTION: Euro-Siberian. Almost all Europe from Portugal to Urals except Iceland, Ireland, United Kingdom, Netherlands, Denmark, Balearics, Sardinia, Sicily, Malta, Crete and European Turkey. – Outside Europe in Caucasus, Transcaucasus and north-eastern Turkey, eastward through Urals, northern Kazakhstan and Siberia to Magadan oblast' and Korea. – New for the faunas of Portugal and Greece: 4 exx., Portugal, Santo Tirso, 14.IV.1991, leg. J. LENZ (ZSM); 1♂1♀, Gr(eece), Olympos S, 30 km NW. Karia, 1850 m, 29.-30.VII.1990, leg. M. FIBIGER (coll. FIBIGER).

PHENOLOGY: Uni- or bivoltine: late May (mid-April) to late June (mid-July), sometimes a second generation from early July to mid-August (late August). Larva: late July (late June) to late August (late September). Overwintering as pupa.

BIOLOGY: Larva oligophagous (1). Found in the seed capsules of *Silene nutans* (Austria: HABICH & REBEL 1883), *S. vulgaris* (Switzerland: PÜNGELER 1889), *Lychnis viscaria* and *Silene latifolia (Melandrium album)* (Caryophyllaceae) (Finland: VALLE 1946; SEPPÄNEN 1970).

HABITAT: Predominantly along coastlines on warm and dry rocky terrain, and on rocks among scrub. From 0 m up to about 1,800 m above sea-level in Pyrenees (MAZEL &

PESLIER 1997), to about 1,900-2,000 m in Apennines and Balkan peninsula (RACHELI & al. 1978; coll. FIBIGER); and to about 2,200-2,300 m in Alps and Asia Minor (VORBRODT & MÜLLER-RUTZ 1914; VORBRODT 1928; BURMANN 1957; coll. ZSM).

SIMILAR SPECIES: A variable species in size; yet usually constant in colour and maculation. *P. hydrata* can be confused with the two preceding species, especially with *P. alchemillata*, but distinguished from them by the greyish colour of wings, and the weakly defined, light rivulet stripe beyond postmedial line. The male and female genitalia of this species are readily distinguished from those of *P. affinitata* and especially from those of *P. alchemillata*, as illustrated.

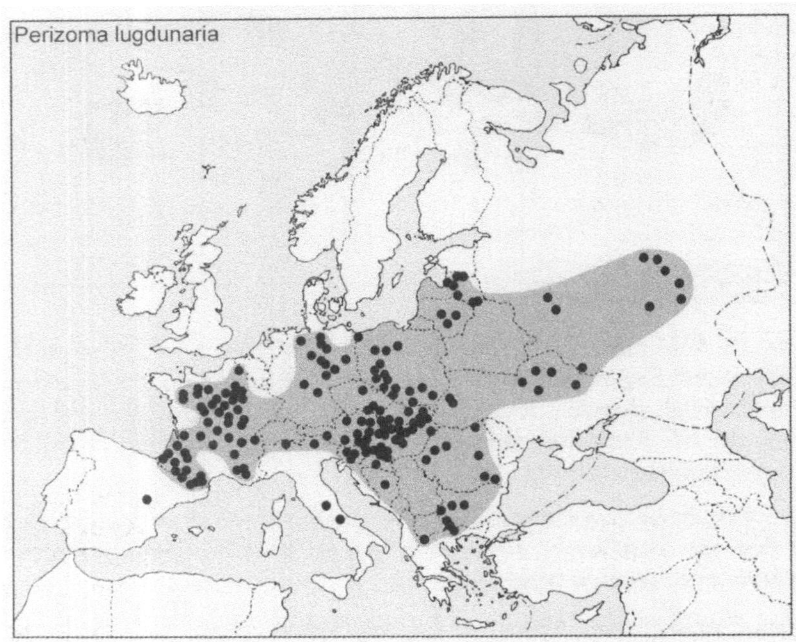

8. Perizoma lugdunaria (HERRICH-SCHÄFFER, 1855)

Larentia lugdunaria HERRICH-SCHÄFFER, 1855: Syst. Bearb. Schmett. Eur. 6 (68): 139; ibidem (1856), 3 (suppl.), pl. 91: 565 (non binominal), (France: Lyon). Syntype(s) lost.

DIAGNOSIS: Wingspan 16-22 mm. Labial palpi black or dark brown, with white apical scales. Frons and vertex dark brown, sometimes with scattered greyish brown scales. Notum greyish brown with a black transverse band at front and another behind. Forewing brown or greyish brown; transverse lines dark brown; basal, postbasal and antemedial lines rather indistinct, dentated; basal and especially antemedial lines well indicated at dorsum and marked by dark brown spots; postmedial line sinuate, much waved, with narrow, parallel outer white stripe from M3 to costa and with small wedge-shaped white spots on CuA1 and CuA2 and A; terminal area broad, brown, darker towards apex; wavy line, indistinct, pale brown, near apex, white and sharply dentated; outer part of terminal area near apex whitish brown with black or dark brown apical streak; terminal line broad, black, interrupted by veins; discal spot absent. Hindwing pale brownish grey or smoky grey with paler, broad, evenly curved medial stripe and with slightly darker terminal area; terminal line distinctly paler than on forewing; discal spot absent. Fringe light brown, slightly spotted with brown or blackish brown at vein ends. Abdomen greyish brown with narrow black or dark brown bands on the segment ends.

MALE GENITALIA: Anal tube small. Valva rather elongated, narrow, with narrowly rounded apex; costa narrow, tapered to apex. Papilla small, narrow and elongate; covered with short setae. Anterior arms of labides short. Manica membranous. Juxta with two broad and blunt lateral processes. Vesica armed with one minute hooked cornutus.

FEMALE GENITALIA: Bursa copulatrix armed with two signa, one narrow, elongate, the other discoid. Spurs of anterior apophyses arising from the medial part.

DISTRIBUTION: European (nemoral). Central and southern Europe; westward to Spain, eastward to south of European Russia (Middle Volga), northward to Latvia and southward to northern Greece. – New for the faunas of Macedonia and Greece: 1♂, Andartiko, Mak(edonia), 1.VIII.(19)87, leg. E.v. MENTZER (coll. SKOU); 1♀, Greece, Livadakion, Kato N., 10.VIII.1982, leg. A. MOBERG (coll. SKOU).

PHENOLOGY: Univoltine: mid-June (late May) to mid-August (late August). Larva in August and September. Overwintering as pupa.

BIOLOGY: Larva monophagous (1). Found in the fruits of *Cucubalus baccifer*, devouring the unripe seeds (France: LHOMME 1935; LUQUET 1991, 1992; Lithuania: SULCS & SULCS 1981; Germany: WEGNER pers. comm.).

PARASITOIDS: Hymenoptera, Ichneumonidae: *Pimpla calobata* - Hymenoptera, Braconidae: *Bracon (=Microbracon) guttator*, *B. variator* (THOMPSON 1945, SHENEFELT 1978).

HABITAT: Warm slopes, forests, forest edges and flood plain forests, river banks and scrub near water. From 0 m up to about 1,000 m above sea-level in Apennines (PROLA & RACHELI 1980), and to about 1,500 m in Pyrenees (MAZEL & PESLIER 1997). Found in Asia Minor (Prov. Kars) at altitude 2,500 m above sea-level (RIEMIS pers. comm.).

SIMILAR SPECIES: Very constant in colour and markings. It is remotely similar to *P. affinitata* but can be easily distinguished from it by the darker, chocolate-brown ground colour of the forewing; the narrower and darker rivulet stripe and darker wavy line are both white only near costa; the distinct ash-grey apex and presence of dark apical streak on the forewing, as well as the darker, smoky brown hindwing. The male genitalia with smaller, narrower papilla, with narrower arms of hemitranstilla and labides, also with much smaller hooked cornutus on the vesica than in the male genitalia of *P. affinitata*. The female genitalia usually smaller on the whole, yet both the scobinate signa in bursa copulatrix are larger and longer than in the female genitalia of related species.

9. *Perizoma bifaciata* (HAWORTH, 1809)

Phalaena bifaciata HAWORTH, 1809: Lepid. Br. (2): 334 (as *bifasciata* in index), (United Kingdom). Syntype(s) (BMNH).

Phalaena unifasciata HAWORTH, 1809: Lepid. Br. (2): 335 (United Kingdom, England: Cantio [=Kent], Westerham). Syntype(s) (BMNH). Junior homonym of *(Phalaena) unifasciata* GMELIN, 1790.

Eubolia scitularia RAMBUR, 1833: Annls Soc. ent. France 2 (1): 42, pl. 2: 8 (Corsica: Ajaccio; Bastia). Syntype(s) (MNHN).

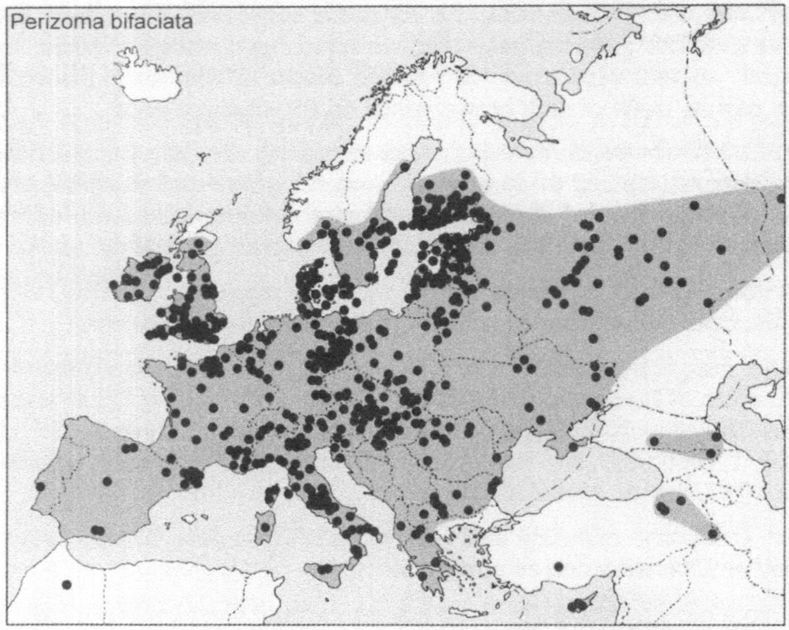

Larentia aquilaria HERRICH-SCHÄFFER, 1848: Syst. Bearb. Schmett. Eur. 3 (32): 163; ibidem (1847), pl. 55: 336 (non binominal), (Germany: Föhren near Regensburg). Syntypes 3 (at least 2♀) lost.

Emmelesia linulata GUENÉE, 1858: in BOISDUVAL & GUENÉE, Hist. nat. Insectes (Spec. gén. Lépid.) 10 (2): 294 (Europe: Pyrenees). Holotype ♀.

Emmelesia unifasciata var. *euphrasiata* MILLIÈRE, 1870: Iconogr. Descr. Chenilles Lepid. inedits 3 (26): 148 (Europe: type locality not stated). Syntype(s) (MNHN?). Valid at subspecific rank.

Emmelesia unifasciata var. *odontata* MILLIÈRE, 1870: Iconogr. Descr. Chenilles Lepid. inedits 3 (26): pl. 114: 13 (3: 148 as var. *euphrasiata*), (Europe: type locality not stated). Syntype(s) (MNHN?). Valid at subspecific rank.

Perizoma bisfasciata var. *pujoli* AGENJO, 1950: EOS 26 (3-4): 332, pl. 9: 25-27 (Spain: Madrid). Holotype ♂ (MNCN). Valid at subspecific rank.

Unavailable names (incorrect subsequent spellings): *bifasciata*: HAWORTH (1828) nec CRAMER, 1779 (identity: *Eupseudosoma bifasciata*, Arctiidae); *bifasciaria*: DOUBLEDAY (1849); *unifasciaria*: DOUBLEDAY (1849); *unifsciata*: DANNEHL (1927).

DIAGNOSIS: Wingspan 16-26 mm. Labial palpi brown or dark brown. Frons and vertex covered with mixture of grey and brown scales. Notum brown with ochreous tinge. Forewing ground colour ochreous brown; basal area ochreous or ochreous brown, usually darker near dorsum, with slender, angulated, pale grey stripe beyond basal line; medial area usually relatively narrow, broadest on the costa, brown or dark brown, margined by narrow white stripes; antemedial line straight or slightly sinuate, dark brown; postmedial line also dark brown, almost straight or slightly sinuate, sometimes with two short and blunt outward prominences; outer white transverse line followed by parallel fine brown line and broader greyish yellow stripe, these lines absent in some specimens; terminal area wide, ochreous or ochreous brown, usually darker towards costa, mottled and irrorate with white scales near apex and sometimes between wavy and terminal lines, and also usually with black or dark brown short oblique apical streak and similar spot near it; wavy line white, well defined, unbroken or sometimes interrupted, or even absent in

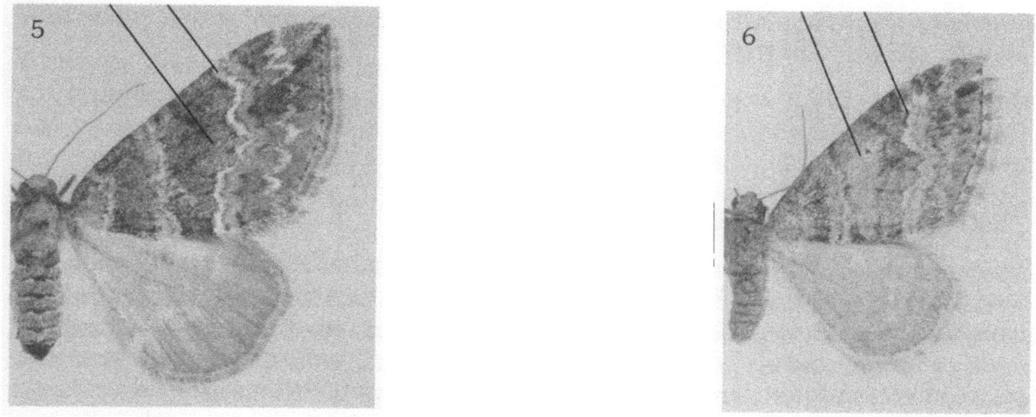

Text-figs 5-6: Diagnostic characters (indicated) of European *Perizoma* species. Text-fig. 5: *P. bifaciata* (HAWORTH, 1809). Text-fig. 6: *P. flavosparsata* (WAGNER, 1926).

some specimens; terminal line narrow, black or dark brown, interrupted by veins, sometimes forming a series of minute black dots; discal spot absent. Hindwing pale smoky grey or smoky brown with slightly darker, evenly curved postmedial line; terminal area usually slightly darker, clouded; terminal line narrow, brown, interrupted by veins or absent; discal spot minute, rounded, brown, absent in some specimens. Fringe pale brownish grey, slightly spotted with brown at vein ends. Abdomen brownish grey with brown and white dorsal bands on posterior margin of each segment.

MALE GENITALIA: Anal tube small. Valva rather elongated, narrow, with narrowly rounded apex; costa narrow, tapered to apex, not extending beyond apex of valva. Papilla small, ovate, broadened caudad; covered with sparse long setae only on the apical surface. Anterior arms of labides long, relatively broad. Manica membranous. Juxta broad, with three pairs of short and blunt processes (basal, lateral and apical). Vesica armed with one narrow, elongated, weakly sclerotized cornutus.

FEMALE GENITALIA: Bursa copulatrix ovate; signum scobinate, discoid, posteriorly produced or ovate. Anterior and posterior apophyses slightly expanded near apices; posterior apophyses with short spurs. Papillae anales narrow, tapered to apices.

DISTRIBUTION: European-Westasiatic. Widespread in Europe from Portugal to Urals. – Outside Europe occurs in North Africa (Morocco), Cyprus, Asia Minor, Caucasus, Transcaucasus (Georgia and Azerbaijan) and northern Kazakhstan.

PHENOLOGY: Uni- or bivoltine: mid-July (early July) to mid-August (late August). In the south bivoltine: early May to mid-June: early August to late October. Larva: mid-August to early October. Overwintering as pupa, sometimes two or even three years.

BIOLOGY: Larva oligophagous (1). Found on the unripe seeds of *Odontites lutea* (France: MILLIÈRE 1870) and *Odontites verna* (Scrophulariaceae) (United Kingdom: BARRETT 1902). Recorded also on *Bartsia alpina*, *Euphrasia* (France: LHOMME 1935), *Euphrasia stricta* (Germany: SÄLZL 1935), *E. nemorosa* (Germany: BERGMANN 1955).

PARASITOIDS: Hymenoptera, Ichneumonidae: *Ectopius (=Platylabus) rubellus;*

Platylabus thedenii - Hymenoptera, Braconidae: *Apanteles spurius* (BUCKLER 1899; THOMPSON 1945; SHENEFELT 1972; HERTING 1976).

HABITAT: Downland, commons, rough pasture, waste ground, along the coast at the upper margins of salt marshes and beaches. From 0 m up to about 1,300 m above sea-level in Apennines (Mt. Majella: ZAHM pers. comm.), to about 1,500 m in Pyrenees and mountains of Balkan peninsula (MAZEL & PESLIER 1997; coll. ZMUC), and to about 2,300-2,400 m in Alps (EMBACHER 1998).

SIMILAR SPECIES: Variable in the depth of colour and the breadth of transverse bands, especially in medial and terminal areas. In spite of this variability *P. bifaciata* is readily recognized and can be easily distinguished from allied species by external features. The ovate papilla, broadened caudad, covered with dense long setae only at the apex in the male genitalia and short spurs on the posterior apophyses in the female genitalia should distinguish this species.

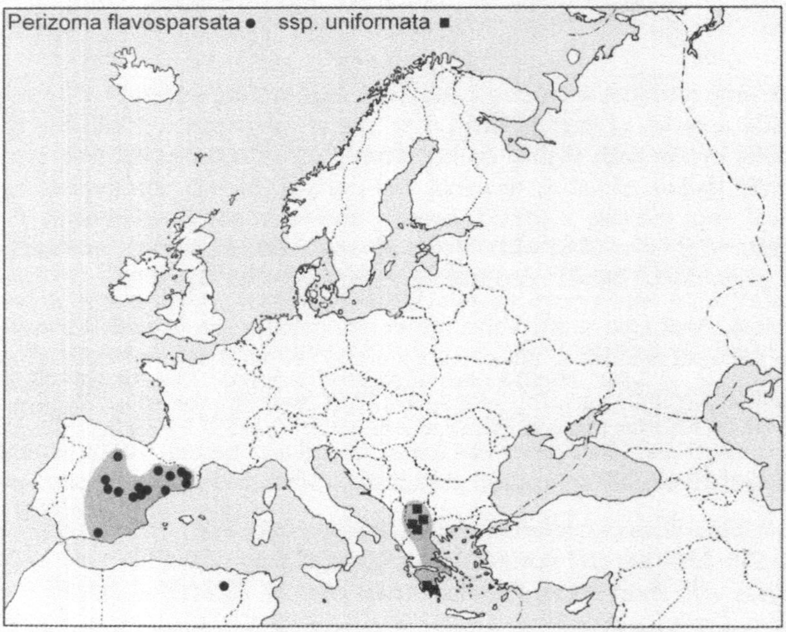

10. *Perizoma flavosparsata* WAGNER, 1926

Larentia flavosparsata WAGNER, 1926: Z. Öst. Ent.-Ver. 11 (12): 114 (Spain: Teruel, Albarracin). Syntypes ♂ ♀ (NHMW).

Cidaria perpusillata FERNÁNDEZ, 1931: EOS 7 (2): 217, pl. 2: 11 (Spain: Cuenca, Uclés). Syntype(s) (MNCN?).

Perizoma flavosparsata var. *subdesertica* AGENJO, 1950: EOS 26 (3-4): 328, pl. 9: 10-12 (Spain: Madrid, de Montarco). Holotype ♂ (MNCN). Valid at subspecific rank.

Perizoma flavosparsata var. *pseudobisfasciata* AGENJO, 1950: EOS 26 (3-4): 329, pl. 9: 14-15 (Spain: Madrid, de Montarco). Holotype ♂ (MNCN). Valid at subspecific rank.

Perizoma flavosparsata var. *pseudoscitularia* AGENJO, 1950: EOS 26 (3-4): 329, pl. 9: 16-18 (Spain, Cuenca: Sierra de Altomira, Vellisca). Holotype ♂ (MNCN). Valid at subspecific rank.

Perizoma flavosparsata uniformata MENTZER, 1974: Ent. Tidskr. 95 (3/4): 196, figs 1c-f, 2c-d (Macedonia: Mt. Vodno near Skopje). Holotype ♂ (EMS). Valid at subspecific rank.

Unavailable names (misidentification): *minorata*: sensu WEISS (1920) nec TREITSCHKE 1828 (identity: *Perizoma minorata*); *euphrasiata*: sensu ZERNY (1927) nec MILLIÈRE, 1870 (identity: *Perizoma bifaciata*). - (incorrect subsequent spelling): *pseudobifasciata*: AGENJO (1950).

DIAGNOSIS: Wingspan 15-21 mm. Labial palpi dark grey. Frons and vertex covered with mixture of whitish grey and blackish grey scales. Notum whitish grey with wide black transverse band at first. Forewing pale grey; transverse lines slightly sinuate, blackish grey; basal line fine, angled near costa; postbasal and antemedial lines gently curved, well marked at dorsum by small dark spots; medial area between antemedial and medial lines pale grey, almost straight and gently angled near costa beyond discal spot; postmedial line twice angled outwards (at middle and near costa); terminal area relatively narrow, darker near apex, with short apical streak, with longer and broader blackish streak and dark blotch near apex of wing; terminal line narrow, broadened to apex, black, interrupted by veins; discal spot minute, rounded, black. Hindwing smoky white with inconspicuous paler transverse stripe before terminal area; terminal line rather obsolete; discal spot absent. Fringe whitish, narrowly spotted by dark grey at vein ends on the forewing, unicolorous on the hindwing. Abdomen whitish grey, usually with narrow rows of dark grey or brownish scales on the posterior margins of segments.

The Balkan subsp. ***uniformata*** MENTZER, 1974 has more distinct pattern of wings than the nominate subspecies. Forewing with more conspicuous white wavy line in terminal area, with broader and more distinct dark apical streak, and with clearer white wavy line and two bands separating basal, medial and terminal areas; hindwing with more definite postmedial line than in nominate Spanish subspecies.

MALE GENITALIA: Anal tube short and blunt. Valva with apically slightly broadened costa. Papilla entire, globular or reniform, completely covered with short setae. Juxta with paired lateral and posterior projections. Vesica armed with two small elongated minutely dentated cornuti.

FEMALE GENITALIA: Bursa copulatrix armed with one small ovate or rounded scobinate signum. Anterior and posterior apophyses relatively short and thick. Tergum A8 short and broad. Papillae anales rather elongated and tapered to the apices.

DISTRIBUTION: Mediterranean (disjunct). Iberian peninsula (Spain, Andorra) and Balkan peninsula (Macedonia and Greece: subsp. *uniformata*) (MENTZER 1974, 1981; HACKER & DERRA 1985). – Outside Europe in Algeria. – New for the fauna of North Africa: 3♂ 4♀, Algérie,Lambèse, September 1913, H. Powell (ZFMK).

PHENOLOGY: Univoltine: early August (mid-July) to late September (early October).

BIOLOGY: Unknown.

HABITAT: Open uncultivated land with scattered bushes and trees (SKOU pers. comm.), rocks and mountain slopes. Up to about 2,500 m above sea-level in Sierra Nevada (coll. ZMUC); subsp. *uniformata* from 800 m up to about 1,500 m above sea-level in the mountains of Balkan peninsula (MENTZER 1974; coll. FIBIGER).

SIMILAR SPECIES: *E. flavosparsata* can be confused with *P. bifaciata* but distinguished from it by the slightly smaller size, the paler greyish ground colour and absence of white edges to the medial area on the forewing. This last character allows accurate distinction from paler Spanish, Balkan and Turkish specimens of *P. bifaciata*. *P. flavosparsata* can occasionally be confused with *P. minorata*. Distinguished from it by the narrower medial area, the less waved ante- and postmedial lines and the rather inconspicuous and less sinuate whitish wavy line in terminal area of forewing. The globular or reniform papilla covered with short setae, the juxta with prominent pairs of lateral and posterior projections, and the vesica which is armed with two small minutely dentated cornuti in the male genitalia, as described and illustrated, should distinguish this species. The female genitalia are very similar to those of *P. bifaciata* and *P. minorata*; however, in *P. bifaciata* there are spurs on the posterior apophyses.

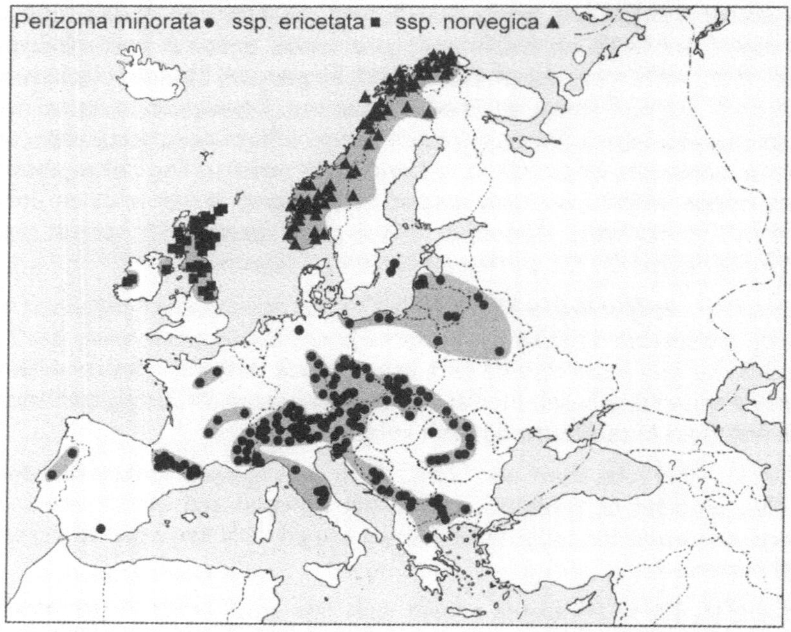

11. Perizoma minorata (TREITSCHKE, 1828)

Cidaria minorata TREITSCHKE, 1828: Die Schmett. Eur. 6 (2): 143 (Austria: Styria, Alps). Syntype(s) (HNHM).
Emmelesia ericetata STEPHENS, 1831: Illust. Br. Ent. (Haustellata) 3: 298, pl. 32: 3 (nec 2) (United Kingdom: N of England; Cumberland; Scotland; Flisk N.B.). Syntypes (BMNH?).
Emmelesia monticola STEPHENS, 1850: List Spec. Br. Anim. Colln Br. Mus. 5: 217 (United Kingdom: type locality not stated). Syntype(s) not traced in BMNH (L. PITKIN pers. comm.).
Perizoma minorata var. *norvegica* PROUT, 1906: Trans. Lond. ent. nat. Hist. Soc. 15: 40 (Norway: Hammerfest). Syntypes (BMNH). Valid at subspecific rank.
Cidaria minorata var. *albida* SOHN-RETHEL, 1929: Dt. ent. Z. Iris 43 (1): 17 (Italy: Abruzzo Mountains). Syntype(s). Valid at subspecific rank.
Unavailable names (infrasubspecific): *albidella* PROUT (1938). - (incorrect subsequent spellings): *minoraria*: BOISDUVAL (1840); *norvegicola*: STRAND (1920); *mimorata*: EXPOSITO HERMOSA (1978).

DIAGNOSIS: Wingspan 15-20 mm. Labial palpi light brown, with a few white apical scales. Frons and vertex white or whitish grey, with scattered light brown scales. Notum whitish grey, with scattered light brown scales. Forewing greyish white or white, with mottled slate-grey or brownish grey transverse lines; basal area pale brownish grey with narrow, whitish medial stripe; basal and postbasal lines slender, angulated; antemedial line slightly curved; postmedial line sinuate or slightly wavy, with two obtuse angles; medial area usually pale brownish grey, broadened to costa, partially interrupted by small white clouds in the middle, in which is visible a small, rounded or ovate discal spot; terminal area grey or brown-grey, usually darker at apex; wavy line unbroken, white; terminal line black or dark brown, unbroken or interrupted by veins. Hindwing unicolorous, smoky white, with two irregular whiter transverse stripes beyond the middle and with slightly darker terminal area; discal spot small, rounded, rather indistinct. Fringe white or pale grey, spotted with grey or brownish grey at vein ends. Abdomen whitish grey, with scattered light brown scales and with light brown bands on posterior margin of each segment.

The British subsp. ***ericetata*** (STEPHENS, 1831) larger (wingspan 18-23 mm), usually has more distinct and darker basal and terminal areas, narrower dark central band, more conspicuous white multiply indented wavy line, and white stripe separating central band and terminal area on the forewing than in nominate subspecies.

The Scandinavian subsp. ***norvegica*** PROUT, 1906 with smoky white or pale grey forewings, sometimes with yellowish tinge, the pattern less distinct, the dark medial bands usually paler than in both preceding subspecies.

MALE GENITALIA: Anal tube small. Valva small, narrow, with blunt apex and with shallow ventral hollow near base; costa narrow, tapered to apex. Papilla small, narrow and elongate; covered with short setae. Anterior arms of labides short. Manica finely spined. Juxta with two broad and blunt lateral processes. Vesica armed with numerous spinules.

FEMALE GENITALIA: Bursa copulatrix rather elongate, pear-shaped; armed with one small, discoid, ovate or ovate-acuminate scobinate signum. Anterior and posterior apophyses short, with pointed apexes. Papillae anales small.

DISTRIBUTION: European (alpine). Nominate subspecies mainly in mountains of Central and southern Europe; westward to Portugal, eastward to Byelorussia, northward to Gotland Isl. and Latvia, southward to Sierra Nevada Mts, central Apennines and Mount Olympus in Greece. In British Isles (Scotland, England and western Ireland) replaced by subsp. *ericetata*, in Scandinavia (except Gotland Isl.) by subsp. *norvegica*. Outside Europe reliably recorded for Caucasus and Transcaucasus (WARDIKJAN 1985; VIIDALEPP 1996), but I have not seen Caucasian specimens.

PHENOLOGY: Univoltine: late June (mid-May) to late August (early September). In the south probably bivoltine. Larva: early July (mid-June) to late August (late September). Overwintering as pupa, sometimes two years.

BIOLOGY: Larva disjunct oligophagous. Recorded in the seed capsules of *Euphrasia* sp. (Scrophulariaceae) (Germany: SPULER 1910; Switzerland: VORBRODT 1928; France: LHOMME 1935) and also on *Calluna vulgaris* (Ericaceae) (Austria: HABELER 1969). Larva of subsp. *ericetata* found in the seed capsules of *Euphrasia* sp., devouring the seeds (United Kingdom: BARRETT 1902; HAGGETT 1968) and *Euphrasia salisburgensis* (Scrophulariaceae) (Ireland: HAGGETT 1968).

PARASITOIDS: Hymenoptera, Ichneumonidae: *Tycherus nigridens* (HERTING 1976).

HABITAT: Birch zones, dry, grassy areas, mountain meadows, limestone hills, grassy moorland and upland pasture. From 1,200 m up to about 2,500 m above sea-level in Apennines (PROLA & RACHELI 1980; coll. ZSM; ZAHM pers. comm.), from 1,000 m up to about 2,100 m in Balkan peninsula (TOMIČ & al. 1990; coll. ZSM), from 2,000 m up to about 2,300 m in Pyrenees (MAZEL & PESLIER 1997), and from 800 m up to about

2,500-2,600 m in Alps (FEICHTENBERGER 1962; WOLFSBERGER 1965; HABELER 1969; FORSTER & WOHLFAHRT 1981; HELLMANN & al. 1999).

SIMILAR SPECIES: Usually constant in colour and markings; only a little variable in the breadth of the medial band on forewing. *P. minorata* can usually be distinguished from other allied species by the wing colour and maculation. The small genitalia with ovate-oblong papilla and with obtusely apical valva in male, and the female with one minute scobinate signum in the corpus bursae should distinguish this species.

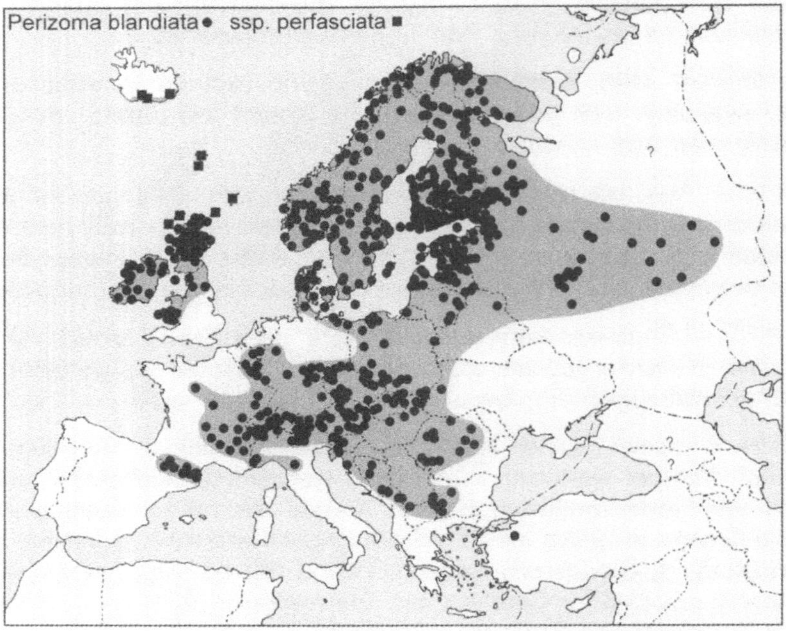

12. *Perizoma blandiata* (DENIS & SCHIFFERMÜLLER, 1775)

Geometra blandiata DENIS & SCHIFFERMÜLLER, 1775: Syst. Verz. Schmett. Wien.: 316 (Austria: Vienna district). Syntype(s) lost.
Phalaena (Geometra) adaequata BORKHAUSEN, 1794: Natur. Eur. Schmett. 5: 444 (Europe). Syntype(s).
Phalaena derasata SCHRANK, 1802: Fauna boica 2 (2): 49 (Germany: Bavaria, Burghausen). Syntype(s).
Phalaena trigonata HAWORTH, 1809: Lepid. Br. (2): 338 (United Kingdom: Westerham). Syntype(s) (BMNH).
Zerene dilacerata ZETTERSTEDT, 1839: Insecta Lapponica 4 (6): 967 (Sweden: 'Lapponia'). Holotype ♀ (MZLU).
Melanthia jucundaria BOISDUVAL, 1840: Gen. Index meth. Europ. Lepid.: 217 (Switzerland: 'Helvetia'). Syntype(s).
Zerene albidata EVERSMANN, 1842: Bull. Natur. Moscou 3 (12): 557, pl. 6: 10 (Russia: Kasan). Syntypes 1♂ 4♀ (ZISP, examined).
Cidaria blandiata f. *perfasciata* PROUT, 1914: in SEITZ, Macrolep. 4: 262 (United Kingdom, Scotland: Hebrides Islands). Syntype(s) (BMNH). Valid at subspecific rank.
Unavailable names (incorrect subsequent spelling): *blandiaria*: BOISDUVAL (1840).

DIAGNOSIS: Wingspan 16-24 mm. Labial palpi grey or black-grey, with white tips. Frons and vertex white. Notum white with blackish grey frontal spots. Forewing white; basal line black or dark brown, evenly curved or multiply indented; basal area with greyish black or blackish brown costal blotch, white with yellowish tinge beyond the basal line; ante- and postmedial lines sinuate; medial area broadened to costa, with large, greyish black or black cloudy costal blotch and smaller similar blotch near dorsum, with small black dots on M3, CuA1 and CuA2; beyond the postmedial line is a broad white stripe; terminal area grey or yellowish grey, usually darker near apex with one or two small black apical spots; wavy line broad, white; terminal line interrupted, forming a series of small black dots; discal spot large, ovate, intensely black. Hindwing white or smoky white with broad white transverse stripe between medial and terminal areas; terminal area broad, unicolorous, pale grey; wavy line inconspicuous, clouded, white; dark terminal line absent; discal spot minute, rounded or ovate, greyish. Fringe white or pale grey, spotted with dark grey at vein ends, especially on the forewing. Abdomen white with blackish grey dorsal spots on the segment ends.

Subsp. ***perfasciata*** (PROUT, 1914) is distinguished from the nominate subspecies by the more or less complete dark medial band on the forewing and frequently by the lighter, white terminal areas on both fore- and hindwings.

MALE GENITALIA: Anal tube small. Valva small, relatively narrow, with blunt apex; costa narrow, with slightly expanded and blunt apex, not extended beyond apex of valva. Papilla large, broad, bilobed; covered with long setae. Anterior arms of labides medium length. Manica finely spined. Juxta almost rectangular. Vesica without cornuti.

FEMALE GENITALIA: Bursa copulatrix elongate, ovate-orbicular; armed with one discoid scobinate signum. Anterior apophyses tapered to tips; posterior apophyses slightly expanded to tips.

DISTRIBUTION: European-Westasiatic. Almost throughout Europe from Pyrenees to Urals (except Portugal, Mediterranean islands, Albania and Greece). The subsp. *perfasciata* in the Outer Hebrides, Faroe Islands and Iceland. – Outside Europe found in Asia Minor. Recorded for Transcaucasus (Georgia, Akhaltisikhe: ROMANOFF 1887), Caucasus and western Siberia (VIIDALEPP 1996).

PHENOLOGY: Univoltine: mid-June (mid-May) to late August (late September). The subsp. *perfasciata* from early June to mid-July (late July). Larva: mid-August (late July) to mid-September (early October). Overwintering as pupa, often two years.

BIOLOGY: Larva monophagous (2). Found on the flowers and in the seed capsules of *Euphrasia* species (Switzerland: BUCKLER 1899) including *E. micrantha* (Scrophulariaceae) (Germany: BERGMANN 1955).

HABITAT: Moors and well-drained grassland in lowland and mountain meadows, on limestone hills and in rocky places. From 0 m up to about 1,900 m above sea-level in Balkan peninsula (TOMIČ & al. 1990), to about 2,200 m in Alps (VORBRODT 1928), and to about 2,400 m in Pyrenees (coll. FIBIGER).

SIMILAR SPECIES: The coloration and maculation of this species can usually be used to distinguish it successfully from other European Geometrid moths. *P. blandiata* is most similar and closely related to *P. minorata*, but distinguished by lighter, white ground colour; by absence of dark band between basal and postbasal lines; by presence of distinct blackish grey costal blotch in medial area with larger discal spot and broader pale grey or pale yellowish grey terminal area with small black spot near apex of the forewing. In the male genitalia of *P. blandiata* the uncus is broader, the papilla larger, broader, bilobate, covered with long setae, the anterior arms of labides usually longer, the valva without shallow ventral hollow near base, with shorter sclerotized costa and with more pointed apex than in *P. minorata*. In the female genitalia the signum is larger.

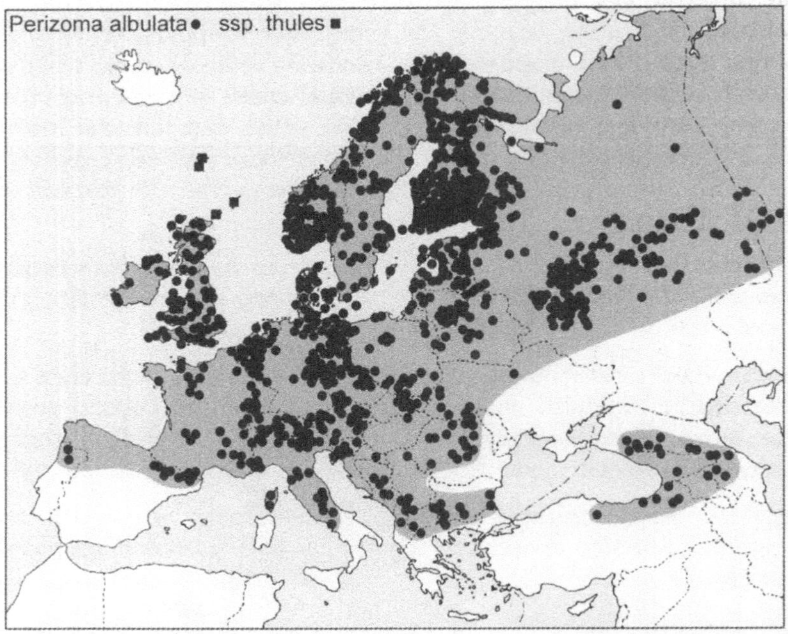

13. *Perizoma albulata* (DENIS & SCHIFFERMÜLLER, 1775)

Geometra albulata DENIS & SCHIFFERMÜLLER, 1775: Syst. Verz. Schmett. Wien. (2): 109 (Austria: Vienna district). Syntype(s) lost.

Cleogene niveata STEPHENS, 1831: Illust. Br. Ent. (Haustellata) 3: 291 (United Kingdom: Scotland). Holotype (BMNH).

Acidalia ablutata EVERSMANN, 1844: Fauna Lepid. Volgo-Uralensis: 398 (Russia: Kasan, Orenburg and Saratov). Syntypes 11♂ 11♀ (ZISP, examined).

Acidalia subfasciaria BOHEMAN, 1853: K. svenska VetenskAkad. Handl. (1851): 133 (Sweden: Ålhem, Bröms). Syntypes.

Cidaria albulata var. *griseata* STAUDINGER, 1871: Cat. Lepid. Eur. Faunengeb. 1: 190 (United Kingdom: England). Syntype(s) (MNHU). Valid at subspecific rank.

Emmelesia albulata var. *thules* WEIR, 1880: The Entomologist 13: 290, pl. 4: 4, 5 (United Kingdom, Scotland: Shetland Islands). Syntype(s). Valid at subspecific rank.

Emmelesia albulata var. *hebrideum* ROBSON & GARDNER, 1886: Young Naturalist 7 (Suppl.): 40 (United Kingdom: type locality not stated). Syntype(s) not traced in BMHN (L. PITKIN pers. comm.). Valid at subspecific rank.

Larentia albulata var. *dissoluta* STRAND, 1902: Nyt Mag. Naturvid. 40 (2): 167 (Norway: Bosekop). Lectotype ♂, herewith designated to stabilize nomenclature (ZFMK, examined). Valid at subspecific rank.

Unavailable names (infrasubspecific): *hebudium*: WEIR (1881); *mundata*: KLEMENSIEWICZ (1913); *reducta*: LEMPKE (1950); *planicolor*: LEMPKE (1969); *flavescens*: LEMPKE (1969). - (incorrect subsequent spelling): *albularia*: BOISDUVAL (1840).

DIAGNOSIS: Wingspan 16-26 mm. Labial palpi white, sometimes with light brown upper scales. Frons, vertex and notum white. Forewing white with pale yellow or pale yellow-brown transverse lines; basal area from white to pale yellow, always with yellowish outer half; postbasal and antemedial lines slightly curved; antemedial and postmedial lines sometimes indicated on the veins by minute brown spots; postmedial line slightly sinuate, waved; medial area from narrow to broad, usually paler on inner side; terminal area broad, pale yellow with distinct, wide, unbroken white wavy line; terminal line light brown, interrupted by veins or obsolete; discal spot absent. Hindwing white with clouded, yellowish terminal area and with broad, indistinct, white, wavy line; discal spot absent. Fringe white, spotted with pale yellow-brown at vein ends. Abdomen white, sometimes with yellowish tinge.

Subsp. ***thules*** (WEIR, 1880) is usually darker than the nominate subspecies, but variable in amount of dark suffusion. Forewing from pale ochreous brown to shining pale olive-brown, sometimes almost leaden brown; basal, medial and terminal areas broad, unicolorous, more or less distinct, especially medial area; these areas or bands separated by fine pale stripes; but in some specimens the dark bands almost or totally obsolete. Hindwing unicolorous, smoky grey or smoky brown.

MALE GENITALIA: Anal tube large, broad. Valva small, narrow and elongated, with broadly rounded apex and with shallow ventral hollow near apex; costa narrow, tapered to apex, slightly produced. Papilla large, broad, lunate or bilobed; covered with short setae. Anterior arms of labides elongate. Juxta with two narrow blunt basal processes and two short blunt apical processes. Vesica armed with numerous spinules.

FEMALE GENITALIA: Bursa copulatrix large, pouch-like, elongate; armed with one small, ovate scobinate signum. Anterior and posterior apophyses slightly expanded at apices.

DISTRIBUTION: Euro-Siberian. Widespread in Europe, except Iceland, south of Appenines, Iberian and Balkan peninsulas, Balearics, Sardinia, Sicily, Malta, Crete and European Turkey. The subsp. *thules* in the Shetland and Faroe Islands. – Outside Europe in north-eastern Asian Turkey, Caucasus and Transcaucasus; eastward through northern Kazakhstan and western Siberia to Lake Baikal (near Irkutsk).

PHENOLOGY: Uni- or bivoltine: late May (early May) to mid-July (late July), occasionally a partial second generation in August and September. In southern Europe bivoltine: mid-April to late June: early July to early October. Larva: late June to late September (in the south from early June to early July, and late August to late October). Overwintering as pupa in a loose web on the ground.

BIOLOGY: Larva oligophagous (1). Found on the flowers and ripening seeds of *Rhinanthus minor,* concealing itself entirely within the seed capsules (United Kingdom: BUCKLER 1899; Finland: SEPPÄNEN 1970; Russia: original data), and recorded also on *R. alectorolophus* (Scrophulariaceae) (France: LHOMME 1935).

HABITAT: Grassland where the foodplant grows including chalk and limestone hills, sand-hills, cliffs, shingle beaches, grassy embankments, wet and dry meadows, mountain meadows, the edge of marshes and oak scrub. The subsp. *thules* in damp places and shingle beaches, where the food-plant *Rhinanthus minor* occurs (GOATER pers. comm.). From 0 m up to about 1,800-1,900 m above sea-level in Pyrenees (MAZEL & PESLIER 1997), to about 2,000 m in Carpathians (POPESCU-GORJ 1964), and to about 2,200-2,300 m in Alps and Apennines (DANNEHL 1927; FEICHTENBERGER 1962; HELLMANN & al. 1999; coll. ZMUC; ZAHM pers. comm.).

SIMILAR SPECIES: Usually showing slight variation in the intensity of colour of yellowish transverse lines and in the size of medial area on the forewing. However, a few specimens occur ranging from clear white, with pattern obsolete to olive-brown or grey-brown ground colour, with the stripes distinct and sharply marked, as in f. *hebudium* WEIR, 1881 from the Orkneys or the Hebrides (BARRETT 1902; LEMPKE 1969). Typical race usually impossible to confuse with other European Geometrid moths.

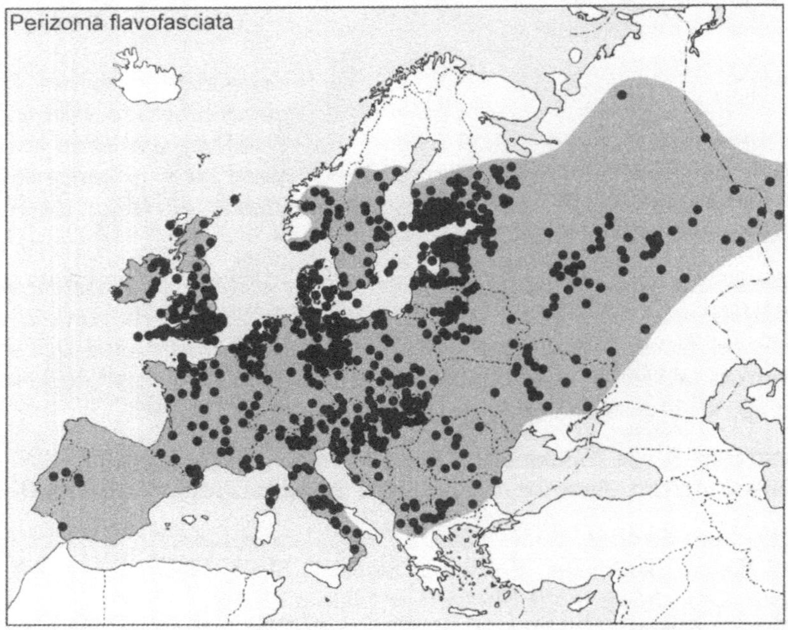

14. *Perizoma flavofasciata* (THUNBERG, 1792)

Phalaena flavofasciata THUNBERG, 1792: Diss. ent. sistens Insecta Suecica 4: 62, pl. 4 (Sweden: type locality not stated). Lectotype, without head, thorax and abdomen (DEUU, examined: KARSHOLT & NIELSEN 1986).

Phalaena chryso-pallida FOURCROY, 1785: Ent. parisiensis 2: 310 (Europe: type locality not stated). Syntype(s) in Mus. Autun? (F).

Geometra decolorata HÜBNER, 1799: Samml. Eur. Schmett. 5 Geometrae (2), pl. 47: 243 (Europe). Syntype(s) lost. Junior homonym of *Phal(aena) Geom(etra) decolorata* DE VILLERS, 1789.

Unavailable names (infrasubspecific): *interrupta*: LEMPKE (1950); *divisa*: LEMPKE (1950); *dissoluta*: LEMPKE (1950); *continua*: LEMPKE (1950); *aequilimbata*: LEMPKE (1950); (incorrect subsequent spelling): *decoloraria*: BOISDUVAL (1840).

DIAGNOSIS: Wingspan 19-32 mm. Labial palpi brown with a few white apical scales. Frons, vertex and notum white, sometimes with a soft ochreous tinge. Forewing white with yellowish brown or ochreous, much waved and slightly curved transverse lines; basal area yellowish brown or ochreous, with fine, white medial fascia; medial area yellowish brown or ochreous; postmedial line prominent outwards between M3 and CuA2, connecting medial area with broad yellowish brown or ochreous terminal area; wavy line distinct, white, sometimes broken; terminal line broken, forming a series of minute dark ochreous dots near vein ends; discal spot absent. Hindwing silky white, with inconspicuous ochreous medial transverse line; terminal area darker, pale ochreous; discal dot absent. Fringe white, distinctly spotted with yellowish brown at vein ends. Abdomen white with weak ochreous tinge and sometimes with ochreous or yellowish brown lateral stripes on segments four to seven.

MALE GENITALIA: Anal tube small. Valva medium length, rather narrow and elongated, with broadly rounded apex; costa relatively broad, tapered to apex. Papilla large, broad, globular or slightly bilobed; covered with short setae. Anterior arms of labides short. Manica finely spined. Juxta with four wide and blunt processes (two basal and two lateral). Vesica armed with two small, pointed cornuti and with numerous spinules.

FEMALE GENITALIA: Bursa copulatrix ovate; armed with one narrow, elongate scobinate signum. Colliculum almost flat, large, elongate and broad.

DISTRIBUTION: Euro-Siberian. Widespread in Europe except Iceland, northern Scandinavia, Balearics, Sardinia, Sicily, Malta, southern Greece, Crete and European Turkey. – Outside Europe eastward through Urals and northern Kazakhstan to southern Siberia (Sajan Mts).

PHENOLOGY: Uni- or bivoltine: early May (early April) to early July (mid-July); early August to mid-September. In the north univoltine: July and August. Larva: late June to July; August to mid-September. In northern Europe from July to early September. Overwintering as pupa in the ground.

BIOLOGY: Larva oligophagous (1). Found within the calyx feeding on the stamens of the male flowers, and also in the seed capsules on the unripe seeds of the female plants of *Silene dioica* (United Kingdom: BUCKLER 1899), also *S. latifolia (Melandrium album)* (United Kingdom: BARRETT 1902) (Caryophyllaceae). Recorded also in seed capsules of *Silene nutans* (Caryophyllaceae) and *Melampyrum* (Scrophulariaceae) (France: LHOMME 1935).

PARASITOIDS: Hymenoptera, Braconidae: *Lissogaster (=Microgaster) tibialis* (BUCKLER 1899; THOMPSON 1946; SHENEFELT 1973).

HABITAT: Light and open forests, scrub on wet soil, downland, lanes, hedges, commons, roadside verges, corners and margins of fields and forests, embankments, and sandhills. From 0 m up to about 1,200 m above sea-level in the Carpathians (POPESCU-GORJ 1964), to about 1,300 m in Pyrenees (MAZEL & PESLIER 1997), and to about 1,400-1,500 m in Alps (VORBRODT 1931; WOLFSBERGER 1965; FORSTER & WOHLFAHRT 1981).

SIMILAR SPECIES: It is a rather delicate species, varying little in the depth of colour of the yellowish brown stripes of the forewing. This species is impossible to confuse with any other European Geometrid moth. *P. flavofasciata* can be easily recognized by the bicoloured white with brownish yellow or ochreous wings with much waved transverse lines; white head and body. Ovate papilla and short anterior arms of labides in the male genitalia, narrow elongate scobinate signum in corpus bursae and large, broad colliculum in the female genitalia should distinguish this species.

15. *Perizoma obsoletata* (HERRICH-SCHÄFFER, 1838)

Larentia obsoletata HERRICH-SCHÄFFER, 1838: in PANZER, Faunae Insect. Germ. initia (Dtsch. Insect.) (Hf. 163): pl. 13 (Italy: Sicily). Syntype(s).

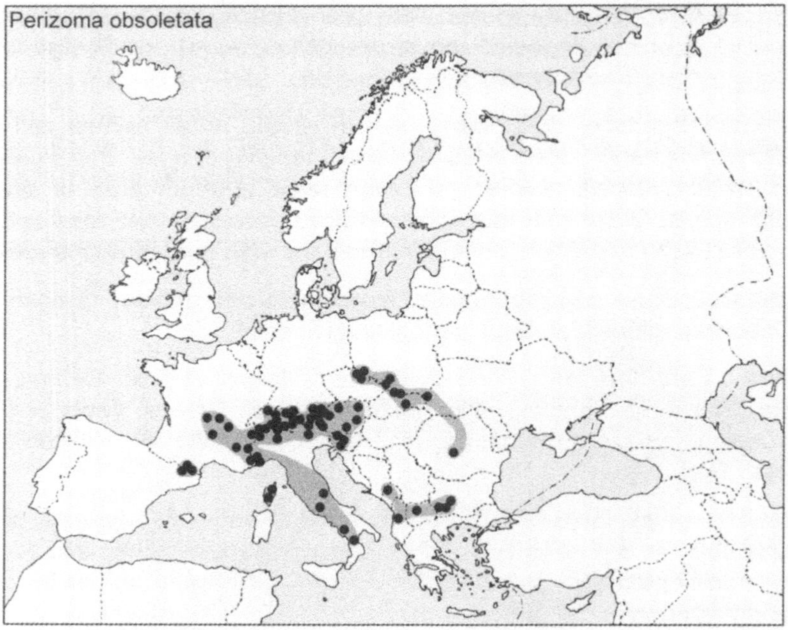

Larentia alpicolaria HERRICH-SCHÄFFER, 1848: Syst. Bearb. Schmett. Eur. 3 (32): 153; ibidem (1845), pl. 20: 119, 120 (as *abstersaria*) (non binominal), (Austria: Schneeberge). Syntype(s).
Melanippe gentianata MILLIÈRE, 1873: Revue Mag. d'Zool. (3) 1: 1 (Switzerland: Engadine valley). Syntype(s) ♂ ♀ (MNHN?).
Larentia reisseri SCHAWERDA, 1932: Int. ent. Z. 26 (26): 283, figured (Corsica: Monte Rotondo). Syntype(s) (NHMW).
Cidaria avilaria REISSER, 1936: Ent. Rund. 53 (7): 108, pl. 3: A7, A8 (Spain: Garganta de las Pozas). Holotype ♀ (NHMW).
Unavailable names (infrasubspecific): *juracolaria* WEHRLI (1919).

DIAGNOSIS: Wingspan 23-29 mm. Labial palpi brown, usually with a few white apical scales. Frons, vertex and notum covered with brown or with mixture of grey and brown scales. Forewing pale grey, mottled and irrorate with brown scales; inner part of basal area brown near costa; outer part of basal area brown or dark brown; postbasal and antemedial lines oblique, nearly straight, dentated; medial area broadened to costa, brown near costa and dorsum; postmedial line wavy, twice angled near costa and on M3, with inward-facing obtuse angle between; subterminal line wavy; terminal area brown, darker toward apex; wavy line white, distinct, sometimes broken; terminal line broad, dark brown, often broken, forming a series of elongate dots near vein ends; discal spot small, rounded, or ovate, or narrow and slightly elongate, black. Hindwing pale grey or smoky white, with slightly darker, brownish, clouded terminal area and with pale, wavy, rather indistinct subterminal line; terminal line brown, clouded, broadened to tornus, interrupted by veins; discal spot conspicuous, small, rounded, dark brown. Fringe pale grey or brownish grey, spotted with brown at vein endings. Abdomen covered with mixture of grey and brown scales, with brown rings at segment ends.

MALE GENITALIA: Uncus relatively broad, with some short setae near apex. Anal tube large, broad. Valva medium length, slightly broadened medially, with broadly rounded apex; costa broad, with apex broader and blunt, not extended beyond apex of valva. Papilla large, elongated and pointed; covered with short setae. Anterior arms of labides elongated and narrow. Manica finely spined. Juxta with two broad and blunt lateral processes. Vesica without cornuti.

FEMALE GENITALIA: Bursa copulatrix ovate; armed with one ovate scobinate signum. Anterior and posterior apophyses tapered at apices. Papillae anales medium size, slightly elongate.

DISTRIBUTION: European (alpine). High mountains of Central and southern Europe.

PHENOLOGY: Univoltine: early July (mid-June) to late July (mid-August). Larva: mid-August to late September. Overwintering as pupa.

BIOLOGY: Larva monophagous (3). Found in the seed capsules of *Gentiana lutea* (Germany (Alps): DIETZE 1874a, as *Eupithecia* sp., 1877), *G. punctata*, *G. purpurea* (Switzerland: VORBRODT 1921, 1928), also recorded on *Gentiana asclepiadea* (Gentianaceae) (SPULER 1910).

HABITAT: Alpine meadows, by mountain streams, among rocks and on stony slopes. From 1,000 m up to about 1,600 m above sea-level in Apennines (RACHELI & al. 1978; PROLA & RACHELI 1980; ZAHM pers. comm.), to about 2,200 m in Balkan peninsula (TOMIČ & al. 1990), from 800 m up to about 2,400 m in Alps (VORBRODT 1928; WOLFSBERGER 1954-1955; FEICHTENBERGER 1962; URBAHN 1966; HELLMANN & al. 1999), and from 2,000 m up to about 2,400 m in Pyrenees (MAZEL & PESLIER 1997).

SIMILAR SPECIES: Very constant in colour and markings. This species is very well distinguished from other European *Perizoma* species by external features. The male genitalia of this species can be easily distinguished from those of all other *Perizoma* species by the combination of long and broad uncus, the large anal tube, the simple, ovate-acuminate papilla, the thin anterior arms of labides and the wide sclerotized costal margin of valva. The female genitalia are typical for *Perizoma*.

16. *Perizoma incultaria* (HERRICH-SCHÄFFER, 1848)

Larentia incultaria HERRICH-SCHÄFFER, 1848: Syst. Bearb. Schmett. Eur. 3 (32): 162; ibidem, pl. 60: 372 (non binominal), (Austria: Gastein). Syntype(s) lost.
Larentia incultaria var. *latifoliata* MILLIÈRE, 1874: Iconogr. Descr. Chenilles Lépid. inédits 3 (35): 432, pl. 153: 16-19 (Switzerland). Syntype(s) (MNHN?). Valid at subspecific rank.
Unavailable names (incorrect subsequent spelling): *incultraria*: WEHRLI (1924).

DIAGNOSIS: Wingspan 19-25 mm. Labial palpi dark grey. Frons and vertex white. Notum pale grey. Forewing pale grey; basal area broad, pale grey with indistinct basal line; antemedial line straight or slightly curved, finely waved, broader and darker near costa; medial area usually narrow, broader to costa, pale grey; postmedial line strongly waved, prominent outwards between M3 and CuA2, usually broader and darker near costa; terminal area pale grey; whitish wavy line distinct, sometimes broken; terminal line absent; discal spot minute, inconspicuous and clouded, rounded or slightly elongated. Hindwing pale silky smoky white, without transverse lines; terminal area usually slightly darker; discal spot minute, indistinct and clouded, rounded. Fringe greyish white, spotted with clouded pale grey at vein ends. Abdomen pale grey.

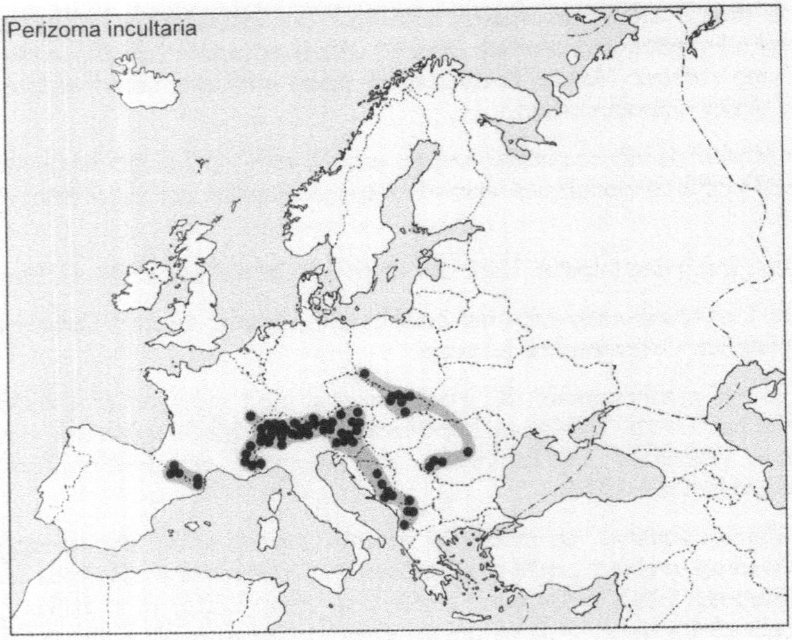

MALE GENITALIA: Anal tube small. Valva short and wide, with broadly rounded apex and with shallow ventral hollow near base; costa broad, with apex broader and blunt, slightly produced. Papilla small, slightly elongated; covered with long setae. Anterior arms of labides short, broad and curved. Juxta with two wide and blunt basal processes. Vesica armed with a row of slightly curved, slim, horn-like cornuti and one small, pointed and finely spined apical cornutus.

FEMALE GENITALIA: Bursa copulatrix small, globular; armed with one discoid scobinate signum. Colliculum elongate and broad. Anterior and posterior apophyses tapered to apices.

DISTRIBUTION: European (alpine). High mountains of Central and southern Europe.

PHENOLOGY: Univoltine: mid-June (mid-May) to early August (mid-September). Larva: late July to late August (early September). Overwintering as pupa.

BIOLOGY: Larva polyphagous (1). First instar mines leaves of *Primula auricula* (Austria: HOFMANN 1871; HERING 1935-1937, 1957), *P. latifolia* (Primulaceae) (Switzerland: PÜNGELER 1889); however, larvae of the later instars feed in the seed capsules of *Primula* and also on *Saxifraga bryoides* (Saxifragaceae) (Switzerland: PÜNGELER 1889), *Bartsia alpina* (Scrophulariaceae) (Alps: SPULER 1910), *Saxifraga paniculata (=aizoon), S. rotundifolia* (Saxifragaceae) (Switzerland: VORBRODT 1928, 1931) and *Fagus* (!) (Fagaceae) (France: LHOMME 1935).

This appears to be the only case of leaf mining among European *Geometridae* (HOFMANN 1871; HERING 1935-1937, 1957).

HABITAT: Rocks, stony slopes, screes, alpine meadows, sometimes in mountain pine forests. From 1,500 m up to about 1,800 m above sea-level in mountains of Balkan peninsula (REBEL 1917), from 1,600 m up to about 2 100-2,200 m in Pyrenees (AGENJO 1971; MAZEL & PESLIER 1997), and from 1,000 m up to about 3,200 m in Alps (VORBRODT 1928; FEICHTENBERGER 1962; URBAHN 1966).

SIMILAR SPECIES: It is a pale coloured species rather similar to *Nebula salicata* (HÜBNER, 1799) and *N. achromaria* (DE LA HARPE, 1853); but *P. incultaria* is not mottled and irrorated with whitish scales; it has broader basal area and more distinct and darker terminal area on the forewing, but lacks the darkish, angled medial line on the hindwing. The male and female genitalia, especially the shape of valva and juxta, the small papilla with long setae, the row of spine-like cornuti on the vesica, the combination of small globular corpus bursae armed with one discoid scobinate signum, the elongate and broad collar-like colliculum and short tergum A8 in female, can be used to determine this species successfully.

Martania MIRONOV, 2000

TYPE-SPECIES: *Emmelesia taeniata* STEPHENS, 1831. More than 30 species in the Palaearctic and Oriental regions. Only one species in Europe. The vast majority of South- and East Asian *Perizoma* species described from Nepal, India (Sikkim and West Bengal), and south-western China can also be included in *Martania* (INOUE 2000; MIRONOV 2000).

DIAGNOSIS: Labial palpi long, pointed; extending beyond front of eyes about 0.7-1.2 times diameter of eyes in ♂, and about 1.1-1.5 times diameter of eyes in ♀. Eyes without thick collars behind. Antennae in both sexes filiform, setaceous. Forewing venation: areole with oblique transverse vein. Hindwing: Sc+R1 arising at short distance from upper angle of cell. Sexual dimorphism not present.

MALE GENITALIA: Uncus weakly sclerotized. Tegumen without papillae. Valva usually large, narrower at base or with shallow ventral hollow near base. Transtilla present; it is not

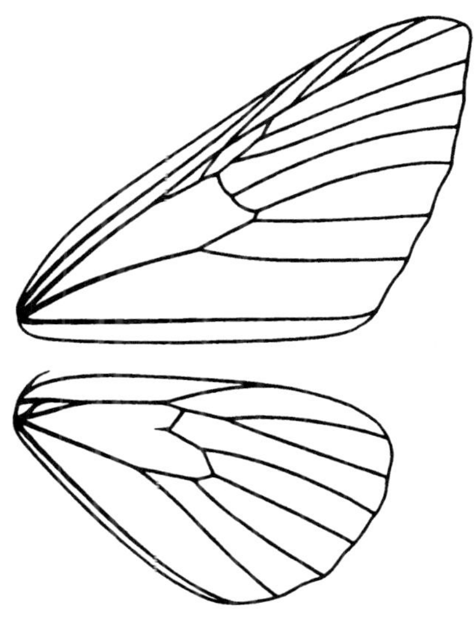

Text-fig. 7:
Venation of *Martania taeniata* (STEPHENS, 1831).

connected with papilla. Papilla united, elongate, setaceous or glabrous. Labides: posterior arms broad, sometimes characteristically curved; anterior arms long, narrow, tapered to apex, connected with juxta, at which point there are minute setaceous papillae (or with one papilla at juxta base, as in East Asian *M. minimata* STAUDINGER, 1897). Juxta broad, bifurcate, with two basal processes and with two long and wide apical processes; without setae on the ventral surface. Vinculum very long and broad, bladder-like, distinctly prominent dorsad. Saccus absent. Aedeagus with caulis in some species (*M. minimata*, *M. argentipuncta* INOUE, 1982), usually with spinules near apex. Vesica unarmed or with one or two horn-like cornuti.

FEMALE GENITALIA: Bursa copulatrix membranous, with scobinate signum or without it. Colliculum short, narrow, collar-like to long, broad, funnel-shaped. Spurs of anterior apophyses connected with each other by narrow, long, heavily sclerotized band, which encircles antrum.

IMMATURE STAGES: Larval and pupal morphology unknown.

REMARKS: The genus *Martania* is distinguished from *Perizoma* by several features such as: longer labial palpi; male genitalia with short, weakly defined, almost membranous uncus; anal tube long and sclerotized, probably assuming the function of uncus; papilla glabrous or covered with short setae; transtilla present; anterior arms of labides connected with juxta and usually with minute papillae with long, sparse setae at this junction; juxta bifurcate; vinculum bladder-like, large, long and broad; aedeagus slim, thin and relatively long, 8-14 times longer than it medial width.
The two genera show close relationship in the absence of sexual dimorphism in the structure of antennae, shape and coloration of wings, and the absence of setae on the ventral side of juxta in the male genitalia. The female genitalia are very similar. However, the female genitalia in some representatives of the genus *Martania* have something in common with those in *Pseudobaptria*. For example, the bursa copulatrix of *Martania seriata* (MOORE, 1888) has a heavily sclerotized, circular signum with a large, inward-pointing central spine, similar to the large trilobous signum in the female genitalia of *Pseudobaptria* species. A remotely similar signum in the form of two large semicircular, weakly sclerotized lateral discs is found in the bursa copulatrix of East Asian *Gagitodes parvaria* (LEECH, 1891). Tergum A8 in the female of *Martania taeniata* has a convex and sclerotized anterior margin, much as in the representatives of the genus *Pseudobaptria*.

17. *Martania taeniata* (STEPHENS, 1831)

Emmelesia taeniata STEPHENS, 1831: Illustr. Br. Ent., (Haustellata) 3: 299, pl. 32: 2 (not 3) (United Kingdom, England: Cumberland). Syntypes 4 (BMNH).
Cidaria arctata LIENIG, 1846: in LIENIG & ZELLER, Isis 1846 (3/4): 199 (Germany: Riesengebirge; Schmiedeberg near Dresden). Syntypes 2♀ (BMNH).
Acidalia albimacularia FREYER, 1849: Neuere Beitr. Schmettkde 6 (89): 87, pl. 534: 5 (Italy: Alps). Syntype(s).
Larentia intrusata HERRICH-SCHÄFFER, 1841: in PANZER, Faunae Insect. Germ. initia (Dtl. Insect.) (Hf. 176): 140 (Germany: Alps). Syntype(s) lost.
Cidaria (Perizoma) obsoleta DJAKONOV, 1929: Ark. Zool. 21A (1): 13 (Russia: Kamchatka: Petropavlovsk, Tarja, Avatsha Bay). Syntypes ♂♀ (NHRS).
Unavailable names (incorrect subsequent spellings): *arctaria*: HERRICH-SCHÄFFER (1848); *taeniaria*: DOUBLEDAY (1849); *teniata*: SIEVERS (1866).

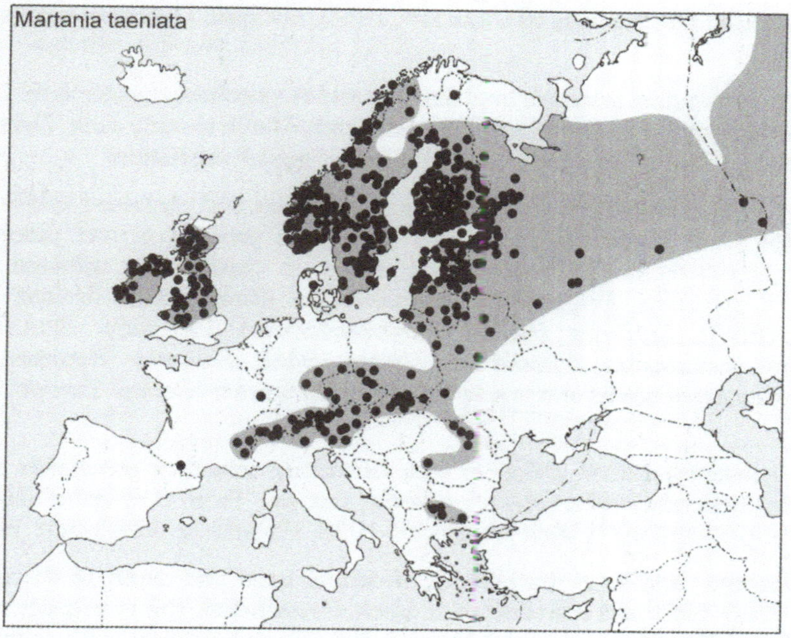

DIAGNOSIS: Wingspan 18-30 mm. Labial palpi dark brown, sometimes with a few whitish apical scales. Frons, vertex and notum brown or light brown. Forewing ochreous brown; basal and medial areas brown, bordered by narrow white stripes; basal line rather curved; postbasal line obsolete; antemedial line slightly sinuate and evenly curved near costa; postmedial line slightly sinuate, usually wavy, marked on the veins with dark brown wedge-shaped dashes; terminal area with large, clouded, brown blotch at apex and with large paler, whitish medial blotch; wavy line broken, dividing into one, two, or three small white spots near apex and one or two white tornal spots; discal spot small, ovate or narrow and elongate, black or dark brown. Hindwing pale smoky grey or pale smoky brown with clouded, slightly darker terminal area; discal spot usually well defined, minute, rounded, brownish. Fringe pale ochreous grey, spotted with brown at vein ends. Abdomen pale grey with black dorsal spots or bands on posterior margin of each segment end; three initial segments usually with ochreous tinge.

MALE GENITALIA: Papilla elongate, glabrous. Valva with broad, rather angulated apex and minute, sclerotized apical process. Vesica armed with one slim and narrow horn-like cornutus.

FEMALE GENITALIA: Bursa copulatrix subglobular or ovate, sharply curved at first to right and then to left; armed with small, narrow, elongate, minutely dentated signum, placed on the right side under first bend. Ductus bursae large, long and broad, funnel-shaped, heavily sclerotized. Ductus seminalis arising from middle of ductus bursae. Antrum membranous.

DISTRIBUTION: Eurasiatic (boreal). Widespread mainly in northern Europe and mountains of Central and southern Europe (Pyrenees, Alps, Carpathians and Rodopi). – Outside Europe eastwards through Siberia to Kamchatka peninsula, Kuril Islands, Sakhalin and

Japan. Recorded for Caucasus by VIIDALEPP (1996), however, I have not seen specimens from there.

PHENOLOGY: Univoltine: late June (mid-June) to mid-August (early September). Larva: late July to mid-October, and after hibernation from early March to early June. Overwintering as small larva. Pupation in a loose web in moss or similar vegetation.

BIOLOGY: Larva oligophagous (?). Feeds on the capsules and contained spores of mosses, perhaps most frequently on species of *Bryum* growing in wet places (United Kingdom: HODGKINSON 1895). Recorded feeding in captivity on withered leaves of *Polygonum aviculare* (Polygonaceae), *Nasturtium* (Brassicaceae) (United Kingdom: BUCKLER 1899), *Stellaria media* (Caryophyllaceae) (Germany: SPULER 1910), *Taraxacum* (Asteraceae) (Finland (?): VALLE 1946), *Fragaria* (Rosaceae), *Alsine* (Caryophyllaceae), *Cymbalaria muralis* (Scrophulariaceae) (Central Europe: SEPPÄNEN 1970).

HABITAT: In heavily shaded, damp or wet, deciduous, mixed or sometimes coniferous forests, also open deciduous forests on fertile, wet soil. From 0 m up to about 1,900-2,000 m above sea-level in Alps (VORBRODT 1928; FORSTER & WOHLFAHRT 1981).

SIMILAR SPECIES: Variable in the breadth of central band and depth of its colour from brown to wholly black and even deep blue-black. Nevertheless, this is an easily recognized species, distinguished by the long labial palpi, distinctive maculation and colour. The male and female genitalia of *M. taeniata* are unlikely to be confused with those of other European *Geometridae* species.

Gagitodes WARREN, 1893

TYPE-SPECIES: *Anticlea schistacea* MOORE, 1888. At least 7 species in the Palaearctic and Oriental regions (DAYONG & HONGFU 1999; MIRONOV 2000).

DIAGNOSIS: Labial palpi short and blunt; extending beyond front of eyes about 0.5-0.6 times diameter of eye. Eyes without thick collars behind. Antennae in both male and female filiform or multiply indented. Forewing venation: areole with oblique transverse vein. Hindwing: Sc+R1 arising at short distance from upper angle of cell. Sexual dimorphism not present.

MALE GENITALIA: Uncus membranous. Tegumen with two lateral setaceous papillae. Valva relatively small with or without harpe near base. Arms of hemitranstilla short and very broad, curved caudad and connected with papilla base. Papilla bifurcate, with two elongated, tapered, setaceous appendages, joined at base. Labides: posterior arms broad; anterior arms elongated, tapered to apices, connected with juxta, without papillae. Juxta broad, without setae on the ventral surface, with two small lateral processes and usually with two pairs of large and broad processes (basal and apical). Vinculum broad, usually elongate, slightly prominent dorsad. Saccus absent. Aedeagus has spiniferous surface at apex. Vesica armed with a patch of spinules or with a few elongated and slim horn-like cornuti.

FEMALE GENITALIA: Bursa copulatrix membranous, with scobinate signum. Colliculum collar-like, narrow, from short to elongate. Spurs of anterior apophyses connected with each other by sclerotized band, which encircles antrum.

Text-fig. 8:
Venation of *Gagitodes sagittata* (FABRICIUS, 1787).

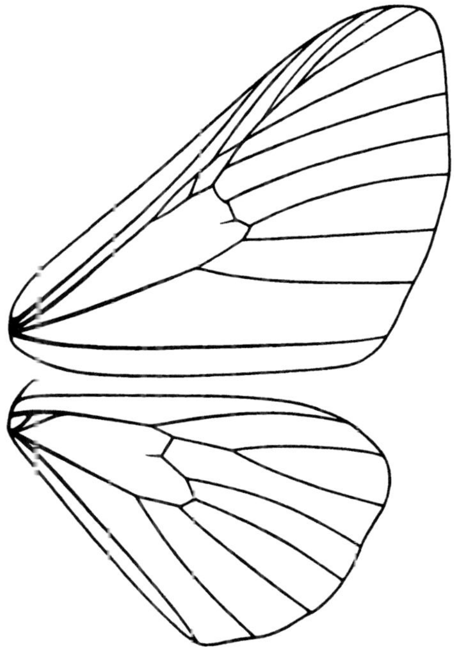

IMMATURE STAGES: Pupal morphology of *Perizoma sagittata* discussed by PATOČKA (1995).

REMARKS: After carefully examining the genitalia of *Gagitodes schistacea* (MOORE, 1888) and *G. olivacea* (WARREN, 1893), as well as studying the literature on species and genus descriptions, I believe that the palaearctic *Perizoma sagittata* must be included in the genus *Gagitodes* together with six other palaearctic species (DAYONG & HONGFU 1999; MIRONOV 2000). The genus *Gagitodes* shows close relationship to *Martania* in the broad, membranous uncus and the long anterior arms of the labides, which are connected with the bifurcate juxta. These features and also the papillae on the tegumen and bifurcate papilla in the male genitalia easily distinguish representatives of the genus *Gagitodes* from *Perizoma*.

18. *Gagitodes sagittata* (FABRICIUS, 1787)

Phalaena sagittata FABRICIUS, 1787: Mantissa Insect. 2: 210 (Austria). Syntype(s) lost.
Phalaena bidentata HUFNAGEL, 1767: Berlin Mag. 4 (6): 600 (Germany: Berlin region). Syntype(s) lost. Junior homonym of *(Phalaena) bidentata* CLERCK, 1759.
Geometra comitata HÜBNER, 1800: Samml. Eur. Schmett. 5 Geometrae (2), pl. 60: 310 (Europe) nec LINNAEUS, 1758 (identity: *Pelurga comitata*, Geometridae). Junior primary homonym.
Unavailable names (infrasubspecific): *approximata*: LEMPKE (1950). - (incorrect subsequent spellings): *sagittaria*: BOISDUVAL (1840); *sagittania*: FIXEN (1849); *sigittata*: SAVENKOVS (1984); *sagittatus*: MIRONOV (2000); *sagittatum*: BELJAEV & KURANISHI (2000).

DIAGNOSIS: Wingspan 22-36 mm. Labial palpi black. Frons black or blackish brown, usually with white frontal scales. Vertex white, sometimes with ochreous tinge. Notum ochreous brown. Forewing delicate smooth satiny fawn-colour or ochreous brown; basal line oblique, slightly angled near dorsum; basal area black or blackish brown, edged outwardly

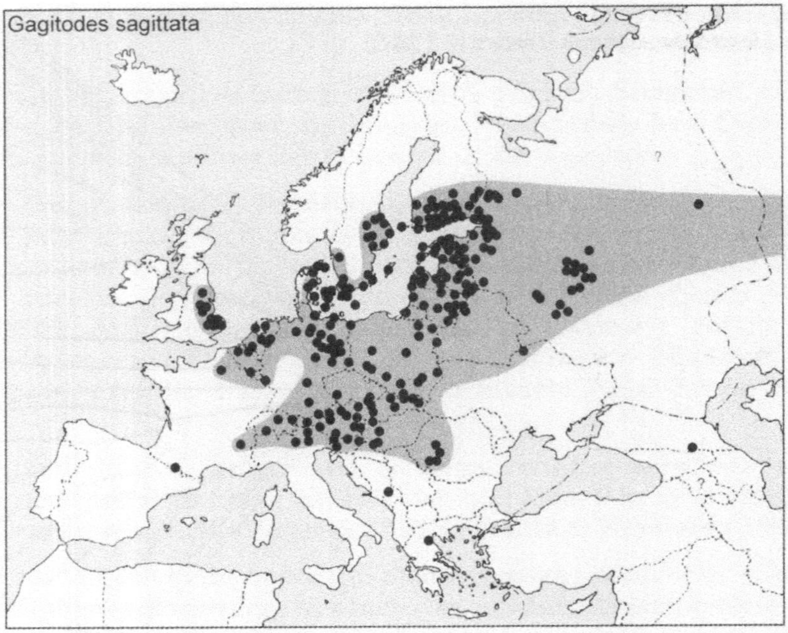

with white; antemedial line much curved outwards, minutely indented or slightly wavy; postmedial line perpendicular to costa, then curved out into a very long pointed projection in the middle of the wing and indented curved to the dorsum; medial area black or blackish brown, usually narrower between CuA2 and A, edged on the inner and outer sides by white lines, which are usually faintly black-margined; terminal area broad; wavy line inconspicuous, forming a narrow, oblique, cloudy white streak near apex and a large, ovate or wedge-shaped, white medial blotch; discal spot absent. Hindwing paler, brownish white, faintly shaded with smoky brown to termen, and with slender, medially angled smoky brown medial transverse line, followed by a whitish stripe; discal spot small, rounded, blackish. Fringe brownish white, distinctly chequered with dark brown at vein ends. Abdomen ochreous brown, sometimes with dorsal row of small black or dark brown spots; first segment always black or blackish brown.

MALE GENITALIA: Valva with heavily sclerotized nipple-shaped harpe near base; sacculus heavily sclerotized. Vinculum large, long and broad. Vesica armed with a patch of spinules.

FEMALE GENITALIA: Colliculum short. Spurs of anterior apophyses connected with each other by broad sclerotized band.

DISTRIBUTION: Eurasiatic. Widespread in Palaearctic region but local and rare, westward to Pyrenees and England, eastward to Kamchatka peninsula, southern Kuril Islands and Japan, northward to southern Finland, and southward to a line from Greece to Japan, passing through Caucasus and Transcaucasus (Armenia), mountains of Central Asia, southern Siberia and Korea.

PHENOLOGY: Univoltine: late June (mid-June) to early August (mid-August). Larva: July and August (to September). Overwintering as pupa in the ground, sometimes twice.

BIOLOGY: Larva monophagous (3) on *Thalictrum* species (Ranunculaceae). Found on the flowers and upon the unripe seeds of *Thalictrum aquilegiifolium*, *T. flavum* (United Kingdom: KNAGGS 1864; BUCKLER 1899; Finland: SEPPÄNEN 1970; Germany: WEGNER pers. comm.; Russia: original data), *T. minus* (United Kingdom: BARRETT 1902), *T. simplex* (Germany: WOLFSBERGER 1954-1955) and *T. lucidum* (GELBRECHT pers. comm.). In captivity eat the old dry-looking leaves of *Aquilegia vulgaris* (Ranunculaceae) (United Kingdom: KNAGGS 1864; BARRETT 1902).

PARASITOIDS: Hymenoptera, Ichneumonidae: *Mesochorus confusus*, *Platylabus pumilio*, *P. transversus*, *Exetastes notatus* (BUCKLER 1899; THOMPSON 1946; HERTING 1976).

HABITAT: Restricted to places where its food-plants grow, such as wet meadows, open areas in forests with wet soil and other marshy places, gardens, fenlands, river banks, near the coasts. From 0 m up to about 1,300-1,400 m above sea-level in Pyrenees (DANTART 1988), to about 1,400-1,500 m in Alps and mountains of Balkan peninsula (VORBRODT & MÜLLER-RUTZ 1914; FORSTER & WOHLFAHRT 1981; TOMIČ & al. 1990).

SIMILAR SPECIES: This handsome, elegant and rather invariable species is unlikely to be confused with any other European Geometrid moths. The male and female genitalia can be reliably used to distinguish *G. sagittata* from other species.

Tribus **Eupitheciini** PIERCE, 1914

At least 39 genera and more than 1,700 species in the Whole World. Four genera and 133 species in Europe.

DIAGNOSIS: Small moths. Wingspan 7-35 mm. Wings usually more or less broad. Ground colour of wings from white to black, often greyish or brownish. Discal spot frequently present on both fore- and hindwing. Pattern of wings usually inconspicuous, with numerous transverse lines; medial area of the forewing not usually darker than the rest of wing. Forewing venation diverse in each genus. Hindwing: cell without inward angle between M1 and M2.

MALE GENITALIA: Uncus entirely membranous or plate-like at base, tapering towards apex (hook-like). Valva from narrow to broad, with sclerotized costal margin, and with ventral process in some species. There are labides with bifurcate setaceous papillae on their posterior arms and usually with a pair of smaller membranous and setaceous papillae on apices of their anterior arms. The shape of juxta and vinculum varies. The vesica of aedeagus usually richly armed with variable cornuti (plate-like, horn-like, dentate plates or numerous teeth or spines), but sometimes lacking all these. The sternum A8 modified; it has a complicated and characteristic shape in each species.

FEMALE GENITALIA: Bursa copulatrix diverse in shape, from membranous to heavily sclerotized; it is usually covered with numerous spine-like signa, or rasp-like signa or without signa in some species. Colliculum usually present. Spurs of anterior apophyses not connected with each other, and they do not form a continuous sclerotized ring.

IMMATURE STAGES: Morphology and surface sculpturation of eggs described and figured in DRAUDT (1905), DIETZE (1910, 1913), WEIGT (1988). Larval morphology in SINGH (1951) based on one specimen of *Axinoptera* (*Chloroclystis*) *subcostalis* HAMPSON, 1893 and MCGUFFIN (1958). Pupal morphology discussed by PATOČKA (1996).

Text-fig. 9:
Venation of *Gymnoscelis rufifasciata* (HAWORTH, 1809).

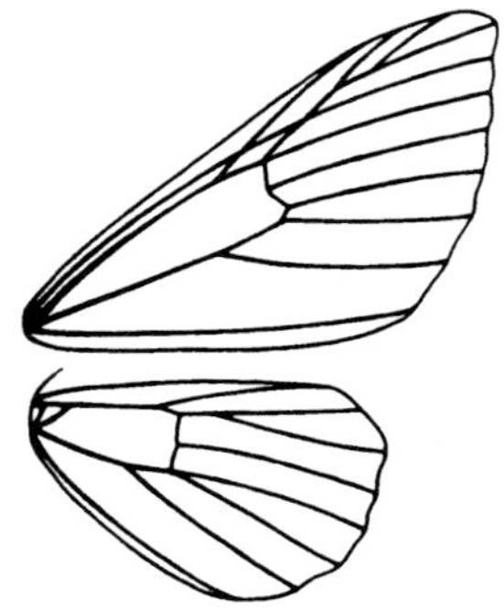

Gymnoscelis MABILLE, 1868

TYPE-SPECIES: *Geometra pumilata* HÜBNER, 1813 (=*Phalaena rufifasciata* HAWORTH, 1809). A total of over 90 species in the Palaearctic, Afrotropical, and mainly in the Oriental and Australian regions (SCOBLE 1999). In Europe one species only.

DIAGNOSIS: The pattern on the forewing usually with clear basal, ante- and postmedial transverse lines; postmedial line often twice angled near costa; other transverse lines and discal spot rather inconspicuous. Forewing venation: common stalk of R2+R4 anastomoses with Sc; R1 absent (it is completely united with Sc); areole without transverse vein. Hindwing: M3 and CuA1 arising from one point at lower angle of cell. Hindtibia with proximal pair of spurs, but in some tropical species there is a distal pair of short spurs.

MALE GENITALIA: Uncus usually short, weakly sclerotized, spine-like, uniapical; it is not sharply separated from tegumen. Valva usually narrow, tapered to apex, without ventral processes. Vinculum often elongated, with a pair of sclerotized lobes and coremata at each apex. Anterior arms of labides with small membranous papillae or without them. Juxta triangular with two short and narrow apical lobes. Vesica armed with spine-like cornuti (from one to numerous). Sternum A8 broad, weakly sclerotized, with narrow linear base and a pair of narrow apical rods.

FEMALE GENITALIA: Bursa copulatrix diverse in size and shape, membranous to heavily sclerotized; it is often chaotically covered with numerous spine-like signa. The other characters in the female genitalia varying.

IMMATURE STAGES: Morphology and surface sculpturation of eggs described and figured in DRAUDT (1905), DIETZE (1910, 1913), WEIGT (1988). Pupal morphology see in PATOČKA (1996).

19. Gymnoscelis rufifasciata (HAWORTH, 1809)

Phalaena rufifasciata HAWORTH, 1809: Lepid. Br. (2): 361 (United Kingdom). Syntype(s) (BMNH).
Geometra pumilata HÜBNER, 1813: Samml. Eur. Schmett. 5 Geometrae (1), pl. 75: 388 (nec 389) (Europe). Syntype(s) lost.
Eupithecia recictaria BOISDUVAL, 1840: Gen. Index meth. Eur. Lepid.: 209 (France: Paris). Syntype(s).
Larentia improbata LIENIG & ZELLER, 1846: Isis 3: 196 (Latvia: ?Koknese [=Kokenhusen]). Syntypes, including 1♂ (BMNH).
Larentia tempestivata ZELLER, 1847: Isis 7: 502 (Italy: Messina; Sicily; Greece: Rhodos Isl.). Syntypes 1♂1♀ (BMNH).

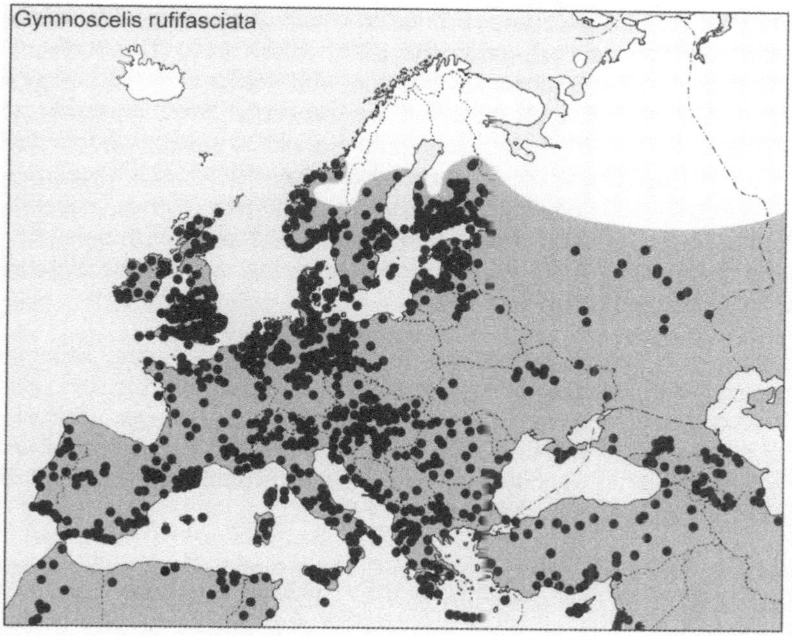

Eupithecia parvularia HERRICH-SCHÄFFER, 1848: Syst. Bearb. Schmett. Eur. 3 (32): 124, 140; ibidem (1846), pl. 30, fig. 187 (non binominal), (Europe?). Syntype(s).
Eupithecia globulariata MILLIÈRE, 1861: Iconogr. Descr. Chenilles Lepid. inedits 1 (4): 206, pl. 3: 1-7 (France: d'Hyères region, Toulon and Marseille). Syntypes ♂ ♀ (MNHN?).
Eupithecia incertata MILLIÈRE, 1876: Cat. raisonne Lepid. Alpes-Maritimes 3: 410 (France: Cannes). Syntype(s) (MNHN?).
Eupithecia bucovinata HORMUZAKI, 1893: Soc. ent. 8 (6): 41 (Ukraine: Bukowina, Czernowitz). Syntype(s).
Gymnoscelis pumilata f. *postgenitata* DIETZE, 1910: Biol. Eupithecien 1, pl. 69: 13, 14; ibidem (1913), 2: 17 (Kirghizstan: Osch; Uzbekistan: Syr Darja). Syntypes (MNHU). Valid at subspecific rank.
Gymnoscelis pumilata f. *nigrostriata* DIETZE, 1910: Biol. Eupithecien 1, pl. 69: 15, 16, 19; ibidem (1913), 2: 17 (Austria: Graz; France: Digne; Spain: Catalonia). Syntypes (MNHU). Valid at subspecific rank.
Gymnoscelis pumilata f. *nigrofasciata* DIETZE, 1910: Biol. Eupithecien 1, pl. 69: 18; ibidem (1913), 2: 17 (France: Digne). Syntype(s) (MNHU). Valid at subspecific rank.
Gymnoscelis pumilata f. *tenebrata* DIETZE, 1910: Biol. Eupithecien 1, pl. 69: 20; ibidem (1913), 2: 17 (Germany: Mittelrhein). Syntype(s) (MNHU). Valid at subspecific rank.
Unavailable names (misidentification): *bistrigata*: sensu HAWORTH (1809) nec BORKHAUSEN, 1790 (identity: *Cymatophorima diluta* DENIS & SCHIFFERMÜLLER, 1775; Thyatiridae); *strobilata*: sensu STEPHENS (1829) nec BORKHAUSEN, 1794 (identity: *Eupithecia abietaria*); *pauxillaria*: sensu HERRICH-SCHÄFFER (1848) nec BOISDUVAL, 1840 (identity: *Eupithecia pauxillaria*). - (infrasubspecific): *basifasciata*: DIETZE (1913); *contrastata*: LEMPKE (1951); *albescens*: LEMPKE (1951); *puncta*: LEMPKE (1951); *nigra*: LEMPKE (1969). - (incorrect subsequent spellings): *pumilaria*: BOISDUVAL (1840); *rufifasciaria*: DOUBLEDAY (1849); *pumilata* f. *pauxillata*: MABILLE (1869); *pumillata*: THURNER (1938); *rufofasciata*: CIFUENTES (1989).

DIAGNOSIS: Wingspan 11-19 mm. Labial palpi grey with light apices. Frons, vertex and notum pale grey. Forewing narrow, costa straight, apex pointed, termen oblique; ground colour greyish white with brownish tinge; transverse lines usually slightly sinuate, dark brown; basal area from pale brownish grey to reddish brown; postbasal and antemedial lines sharply angled near costa; medial area between antemedial and medial lines brownish grey to brick-red, often darker than rest of wing; postmedial line thickened, usually twice slightly angled and marked with blackish or reddish black dashes on all veins, bordered with double pale stripe; terminal area brownish grey, often with reddish or brick-red inner part, and with pale ash-grey outer part; whitish wavy line usually well defined, multiply indented; discal spot absent. Hindwing with shallow medial hollow in termen; ground colour greyish white with rather strong transverse lines, especially with distinct postmedial line, which is medially angulated and marked by a series of dark dashes on the veins; terminal area usually darker, in some specimens with reddish tinge; pale wavy line conspicuous, multiply indented; discal spot minute or obsolete. Fringe brownish grey, narrowly spotted with pale grey or whitish at vein ends. Abdomen from pale grey to brownish grey, sometimes with reddish tinge; second abdominal segment usually darker, sometimes brick-red, bordered by black.

MALE GENITALIA: Uncus short and narrow, uniapical, with broad base. Tegumen not projected before uncus. Papillae very small, covered by short setae. Vinculum relatively short and broad; processes of coremata pointed toward apices. Aedeagus relatively small and slim, shorter than length of valva, about 5.0-6.0 times longer than its medial width. Vesica armed with one elongated, slightly curved, horn-like cornutus. Sternum A8 weakly sclerotized, its apical rods without setae.

FEMALE GENITALIA: Bursa copulatrix ovate, small; more than half covered with small spines. Ductus bursae absent. Ductus seminalis narrow, attached to corpus bursae near colliculum. Colliculum weakly defined, short and narrow. Anterior apophyses without spurs.

DISTRIBUTION: Palaearctic. Widespread from Canary Islands and Madeira to north-western China.

PHENOLOGY: Bi- or probably polyvoltine: mid-April (late March) to early June (early July); late June to late August (mid-September). In the south of all the year round. Larva: late April to mid-June; early July to early October (usually from early August to late September in Central Europe). Overwintering as pupa.

BIOLOGY: Larva polyphagous (2). Found on the flowers and seeds of a wide variety of herbaceous plants and shrubs: *Zea mays* (Poaceae), *Citrus* (Rutaceae), *Olea europaea* (Oleaceae), *Berberis vulgaris* (Berberidaceae), *Myrica faya* (Myricaceae), *Calluna vulgaris*, *Erica arborea*, *E. tetralix*, *Arbutus unedo* (Ericaceae), *Salix* (Salicaceae), *Buxus* (Buxaceae), *Clematis vitalba* (Ranunculaceae), *Odontites lutea*, *O. verna*, *Digitalis purpurea*, *D. grandiflora*, *Euphrasia*, *Melampyrum*, *Verbascum* (Scrophulariaceae), *Dianthus caryophyllus* (Caryophyllaceae), *Thymelaea hirsuta* (Thymelaeaceae), *Diospyros kaki* (Ebenaceae), *Vitex agnus-castus* (Verbenaceae), *Mercurialis annua* (Euphorbiaceae), *Satureja*, *Hyssopus*, *Thymus*, *Origanum vulgare*, *Rosmarinus officinalis*, *Mentha sylvestris* (Lamiaceae), *Anthriscus sylvestris*, *Angelica sylvestris*, *Daucus carota* (Apiaceae), *Suaeda vera* (Chenopodiaceae), *Rubus idaeus*, *Crataegus monogyna*, *C. laevigata*, *Sorbus aucuparia*, *Rosa* (Rosaceae), *Tamarix* (Tamaricaceae), *Ipomoea batatas* (Convolvulaceae), *Lycopersicon* (Solanaceae), *Epilobium (=Chamaenerion) angustifolium* (Onagraceae), *Ilex aquifolium* (Aquifoliaceae), *Gentiana pneumonanthe* (Gentianaceae), *Valeriana officinalis* (Valerianaceae), *Ulex europaeus*, *Genista corsica*, *Vicia faba*, *Cytisus scoparius*, *Calicotome* (Fabaceae), *Globularia alypum* (Globulariaceae), *Cynara cardunculus*, *C. scolymus*, *Eupatorium cannabinum*, *Solidago virgaurea*, *Santolina*, *Achillea millefolium*, *A. ageratum*, *Artemisia* (Asteraceae), *Urginea maritima* (Liliaceae) and flowers of *Laurus azorica*

(Lauraceae) (United Kingdom: CREWE 1861c; BARRETT 1904; Corsica: MABILLE 1868; Germany: DIETZE 1910, 1913; WEIGT 1988; WEGNER pers. comm.; Italy: CARRARA 1928; France: MILLIÈRE 1861; LHOMME 1935; Sardinia: HARTIG 1951; Finland: SEPPÄNEN 1970; Azores Islands: VIEIRA 1997; CORLEY pers. comm.). Recorded also on *Achras sapota (Sapotaceae), Anthemis, Chrysanthemum, Dahlia, Gerbera jamesonii, Leucanthemum, Sonchus oleraceus* (Asteraceae), *Asphodelus* (Liliaceae), *Borago officinalis* (Boraginaceae), *Cistus* (Cistaceae), *Eriobotrya japonica* (Rosaceae), *Gladiolus* (Iridaceae), *Gossypium barbadense*, *Lavatera* (Malvaceae), *Pancratium maritimum* (Amaryllidaceae), *Papaver somniferum* (fruits) (Papaveraceae), *Pelargonium* (Geraniaceae), *Ricinus communis* (inflorescence and fruits) (Euphorbiaceae), *Antirrhinum majus, Scrophularia sambucifolia* (Scrophulariaceae), *Sempervivum* (Crassulaceae), *Phalaris, Sorghum* (Poaceae) (Morocco: RUNGS 1981), on *Malus domestica, Pyrus communis, P. bucharica* (Rosaceae) and *Atraphaxis pyrifolia* (Polygonaceae) (Tajikistan: DEGTJAREVA 1973).

PARASITOIDS: Hymenoptera, Ichneumonidae: *Agrypon clandestinum, Trichionotus (=Labrorychus) clandestinus* - Hymenoptera, Braconidae: *Aleiodes circumscriptus, Rogas* sp. (THOMPSON 1946; SHENEFELT 1975).

HABITAT: Open deciduous, mixed and coniferous forests, edges of forest, hedges, scrub, meadows, grasslands, moors, heaths, dunes, waste ground, rough verges, gardens, parks, cultivated areas and edges of fallow fields, warm and dry slopes, rocks and screes. From 0 m up to about 1,800 m above sea-level in Apennines (Mt. Pollino: coll. ZSM), to about 2,300-2,400 m in Alps (EMBACHER 1998), and to about 2,700-2,800 m in Great Atlas (coll. ZSM).

SIMILAR SPECIES: One of the more easily recognized species of the European *Eupitheciini* in spite of its variability. Distinguished by presence of one pair of long spurs on the hind tibia, by the venation and shape of wings, by usually dull reddish tone and blackened postmedial line. The male and female genitalia can be used to determine this species with accuracy.

REMARKS: Very varied in size, in depth of colour and intensity of markings. There are some forms and aberrations in this species (DIETZE 1910, 1913; LEMPKE 1951, 1969); in some specimens the forewing is rather pale brownish grey without distinct postmedial line; in others brick-red colour predominates and the postmedial line is well marked with black dashes on the veins. The large specimens from southern Russia usually have alternating blackish grey and whitish grey transverse stripes on all wings, without reddish tinge.

Chloroclystis HÜBNER, 1825

=*Dyserga* PETERSEN, 1909

TYPE-SPECIES: *Geometra coronata* HÜBNER, 1813 (=*Phalaena v-ata* HAWORTH, 1809). A total of over 150 species, mainly from Oriental and Australian regions. However, tropical representatives of the genus '*Chloroclystis*' are in need of expert revision (SCOBLE 1999). In Europe only one species occurs.

DIAGNOSIS: Pattern on the forewing usually with clear basal, ante- and postmedial transverse lines; postmedial line often twice angled near costa; discal spots frequently incon-

Text-fig. 10:
Venation of *Chloroclystis v-ata* (HAWORTH, 1809).

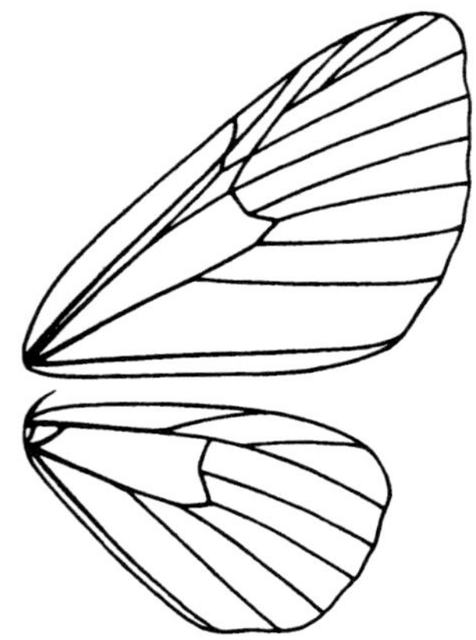

spicuous. Forewing venation: Sc anastomosed with R1; areole without transverse vein. Hindwing: M3 and CuA1 arising separately from cell. Hindtibia with two pairs of spurs, but in some tropical representatives one pair of spurs only.

MALE GENITALIA: Uncus separated from tegumen, weakly sclerotized, short, hook-like and uniapical in its apical part. Valva usually narrow or sharply tapered to apex; with medial process in some tropical species. Vinculum without pair of elongated lobes and coremata on each apex. Membranous papillae on the anterior arms of labides absent. Juxta elongated, triangular, with short and narrow apical lobes. Vesica unarmed or with numerous horn-like cornuti (for example in tropical *C. macroaedeagus* HOLLOWAY, 1979). Sternum A8 broad, usually with narrow band-like base and pair of narrow, often heavily sclerotized and pointed apical rods.

FEMALE GENITALIA: Bursa copulatrix diverse in shape and size, usually armed with numerous spine-like signa. Ductus bursae usually elongated. Ductus seminalis attached to ductus bursae. Colliculum absent.

IMMATURE STAGES: Morphology and surface sculpturation of eggs described and figured by DRAUDT (1905), DIETZE (1910, 1913), WEIGT (1988). Pupal morphology discussed by PATOČKA (1996).

20. Chloroclystis v-ata (HAWORTH, 1809)

Phalaena v-ata HAWORTH, 1809: Lepid. Br. (2): 364 (United Kingdom). Syntype(s) (BMNH).

Geometra coronata HÜBNER, 1813: Samml. Eur. Schmett. 5 Geometrae (1), pl. 72: 372, 373 (Europe). Syntype(s) lost.

C(hloroclystis) coronata lanceolata VORBRODT & MÜLLER-RUTZ, 1914: Schmett. Schweiz 2 (6-8): 643 (Switzerland: Frauenfeld). Syntype(s) (NMBS). Valid at subspecific rank.

Eupithecia stabiensis STAUDER, 1929: Ent. Anz. 9 (6): 105 (Italy: Sorrento, Monte Faito region, near Castellammare di Stabia). Holotype ♀ (BMNH?).

Chloroclystis v-ata relicta KROGERUS, 1996: Ent. Fennica 7 (2): 64, figs 1, 2 (Finland: Karislojo). Holotype ♂ (ZMUH; paratype ♀ in coll. FIBIGER examined). Valid at subspecific rank.

Unavailable names (infrasubspecific): *bistrigata*: LEMPKE (1951); *obsoleta*: LEMPKE (1969); *obscura*: LEMPKE (1969); *mediofasciata*: LEMPKE (1969). - (incorrect subsequent spelling): *coronaria*: HERRICH-SCHÄFFER (1848).

DIAGNOSIS: Wingspan 12-19 mm. Labial palpi yellowish green, usually with scattered black or dark brown scales. Frons, vertex and notum yellowish green. Antennae yellow-

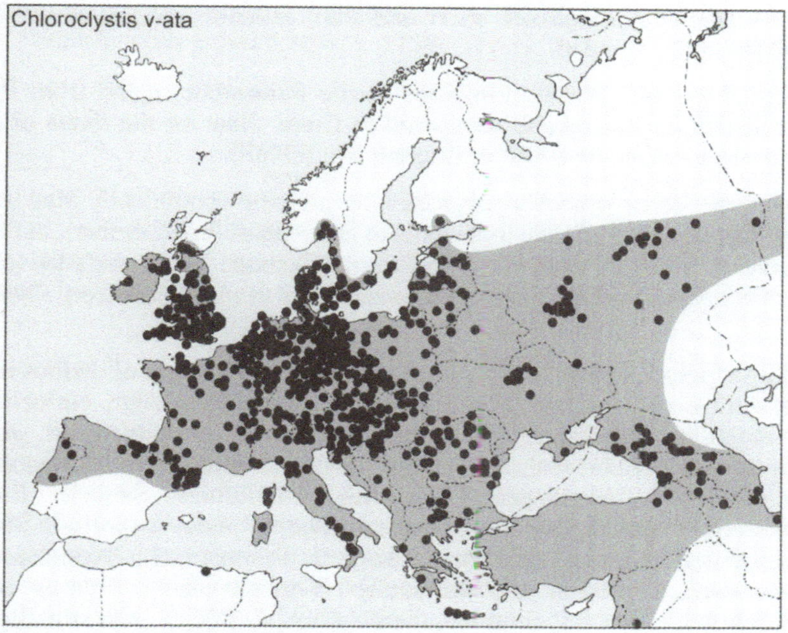

ish green, distinctly ringed brown. Forewing wide with rounded apex and convex termen; ground colour light yellowish green; transverse lines dark brown; basal line well marked at costa; postbasal line well marked only at dorsum; antemedial line usually twice sharply angled with wide brownish outer shade; postmedial line slightly sinuate, indicated on CuA1, CuA2 and A with dark brown points or wedge-shaped dashes, it is twice angled at costa forming black or dark brown V-shaped or zigzag mark with dark inner shade; terminal area usually with minute dark points on the veins and with indiscernible pale wavy line; discal dot absent. Hindwing more or less pale grey; basal, antemedial and postmedial transverse lines weakly discernible, with smooth greyish inner shades; terminal area sometimes slightly darker, greyish, with inconspicuous pale wavy line; discal dot absent. Fringe pale grey, spotted with brownish between vein ends, sometimes unicolorous pale grey only on the hindwing. Abdomen yellowish green with grey tinge; three initial segments black or dark brown, especially third abdominal segment, or with concolorous large lateral spots; sometimes with minute dark lateral dots on each segment and always with such dots on the seventh segment.

MALE GENITALIA: Uncus short and narrow, uniapical, with broadened apical part. Papillae very small, covered with short setae. Transtilla with two slender bifurcate processes. Valva narrow, abruptly broadened near base. Vinculum short and narrow. Aedeagus relatively small and slim, shorter than length of valva. Vesica without cornuti. Sternum A8 weakly sclerotized, broadened basally, basal hollow broad and shallow, apical hollow narrow and deep.

FEMALE GENITALIA: Bursa copulatrix membranous, almost spherical, covered by longitudinal striae and small spines. Ductus bursae elongated. Ductus seminalis narrow, broadly attached to ductus bursae from left side. Colliculum absent. Antrum broad and sclerotized.

Anterior and posterior apophyses short and thin; anterior apophyses without spurs. Papillae anales small, rounded.

DISTRIBUTION: Eurasiatic. Widely distributed in the Palaearctic region from Portugal to Japan, northward to Norway and southward to Crete. New for the fauna of Algeria: 4 ♀♀, Algérie, Cap Aokas, IV.1909, J. Dayrem leg. (ZFMK).

PHENOLOGY: Uni-, bi- or probably polyvoltine (in extreme south): early May (early April) to mid-June (late June); early July (mid-June) to late August (mid-October). In the north in June and July; in the south from early April to mid-October. Larva: mid-May to early July (mid-July); early August to late September (sometimes to mid-December). Overwintering as pupa.

BIOLOGY: Larva polyphagous (1). Found on the flowers and seeds of various herbaceous plants and shrubs, such as *Lathyrus* (Fabaceae), *Laserpitium siler*, *Heracleum sphondylium*, *Angelica sylvestris*, *Peucedanum palustre*, *Aegopodium podagraria*, *Pimpinella saxifraga* (Apiaceae), *Thalictrum flavum*, *Clematis vitalba* (Ranunculaceae), *Valeriana officinalis* (Valerianaceae), *Achillea millefolium*, *Senecio*, *Eupatorium cannabinum*, *Solidago virgaurea*, *Artemisia vulgaris*, *Aster tripolium* (Asteraceae), *Calluna vulgaris* (Ericaceae), *Lysimachia vulgaris* (Primulaceae), *Hypericum perforatum* (Clusiaceae), *Gentiana pneumonanthe* (Gentianaceae), *Thymus serpyllum*, *Origanum vulgare*, *Mentha aquatica* (Lamiaceae), *Lythrum salicaria* (Lythraceae), *Mercurialis* (Euphorbiaceae), *Ligustrum* (Oleaceae), *Rubus*, *Crataegus*, *Sorbus aucuparia* (Rosaceae), *Sambucus nigra* (Caprifoliaceae), *Polygonum* (Polygonaceae), *Callicarpa* (Verbenaceae), in capivity feeds on leaves of *Tilia* (Tiliaceae) (United Kingdom: CREWE 1861c, 1862, 1863; BUCKLER 1899; BARRETT 1904; Corsica: MABILLE 1869; Germany: DIETZE 1910, 1913; SÄLZL 1935; WEIGT 1988; WEGNER pers. comm.; France: LHOMME 1935; Finland: SEPPÄNEN 1970).

PARASITOIDS: Hymenoptera, Ichneumonidae: *Endromopoda detrita* (?), *Agrypon clandestinum*, *Triclistus spiracularis* - Hymenoptera, Braconidae: *Aleiodes (=Rogas) testaceus* (THOMPSON 1944; SHENEFELT 1975; HERTING 1976).

HABITAT: Open deciduous, mixed and coniferous forests, forest fringes, hedges, scrub and scrub-covered meadows, fields, coastal grassland, gardens, parks, cultivated and uncultivated areas, rocks and screes. From 0 m up to about 2,000-2,100 m above sea-level in southern Alps and Pyrenees (HELLMANN & al. 1999; coll. FIBIGER).

SIMILAR SPECIES: Resembles *Pasiphila rectangulata* in the postmedial line, which is strongly toothed (or twice angled) near costa on the forewing, but the smaller size, venation and light yellowish green colour of the forewing with concolorous body, pale grey hindwing and absence of discal dots on the wings in *C. v-ata* separate it from all European representatives of the genus *Pasiphila*. The structure of the male and female genitalia should distinguish this species.

Pasiphila MEYRICK, 1883

=*Rhinoprora* WARREN, 1895; *Cithecia* STAUDINGER, 1897; *Calliclystis* DIETZE, 1910

TYPE-SPECIES: *Eupithecia bilineolata* WALKER, 1862. A total about 40 species in the Palaearctic, Oriental and Australian regions, mainly in New Zealand (SCOBLE 1999). In Europe only three species occur.

Text-fig. 11:
Venation of *Pasiphila rectangulata* (LINNAEUS, 1758).

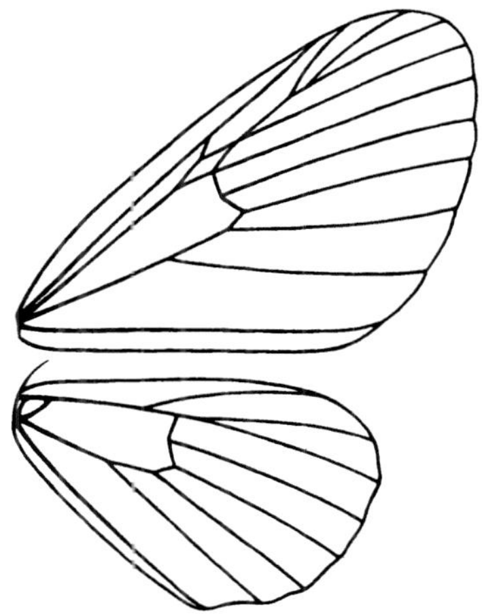

DIAGNOSIS: Pattern on the forewing usually with clear basal, ante- and postmedial transverse lines; postmedial line usually twice angled near costa; discal spots often clearly visible. Forewing venation: R1 short, almost completely united with Sc; areole without transverse vein. Hindwing: M3 and CuA1 arising separately from cell. Hindtibia with two pairs of spurs.

MALE GENITALIA: Uncus usually broad, membranous. Valva usually broad and elongated; sacculus with numerous dense setae. Vinculum wide. Anterior arms of labides with small membranous papillae or without them, but these arms covered with setae on each apex. Juxta broad, heart-shaped or trapeziform, with broad apical lobes. Vesica armed with two apical horn-like cornuti and numerous minute spines. Sternum A8 broad, usually with narrow, heavily sclerotized basal margin and two more or less broad apical rods, which are tapered and sclerotized to their apices.

FEMALE GENITALIA: Bursa copulatrix membranous, usually large, diverse in shape (spherical, ovoid or heart-shaped), with apical membranous diverticulum in some species. Bursa copulatrix armed with one or two rasp-like signa (lunate, triangular, ovoid or other shape) or without signa. Ductus bursae usually elongated, membranous or sclerotized, sometimes with rasp-like surface or granulated. Ductus seminalis attached to ductus bursae. Colliculum absent, or short and weakly sclerotized. Lamella antevaginalis present in many species. Surface of tergum A8 covered with short setae (in some species).

IMMATURE STAGES: Morphology and surface sculpturation of eggs described and figured in DRAUDT (1905), DIETZE (1910, 1913), WEIGT (1988). Pupal morphology see PATOČKA (1996).

21. *Pasiphila chloerata* (MABILLE, 1870)

Eupithecia chloerata MABILLE, 1870: Petites Nouv. Ent. 1 (24): 96 (France: near Paris). Syntype(s) (coll. HERBULOT in ZSM, examined).
Eupithecia consueta BUTLER, 1879: Ann. Mag. nat. Hist. (5) 4: 442 (Japan: Yokohama). Syntype(s) (BMNH).
Eupithecia horticolaria FUCHS, 1892: Jahrb. Nassau Ver. Naturkde 45 (2): 102 (Germany: Loreley district). Syntypes 2♂2♀ (NSMW?).

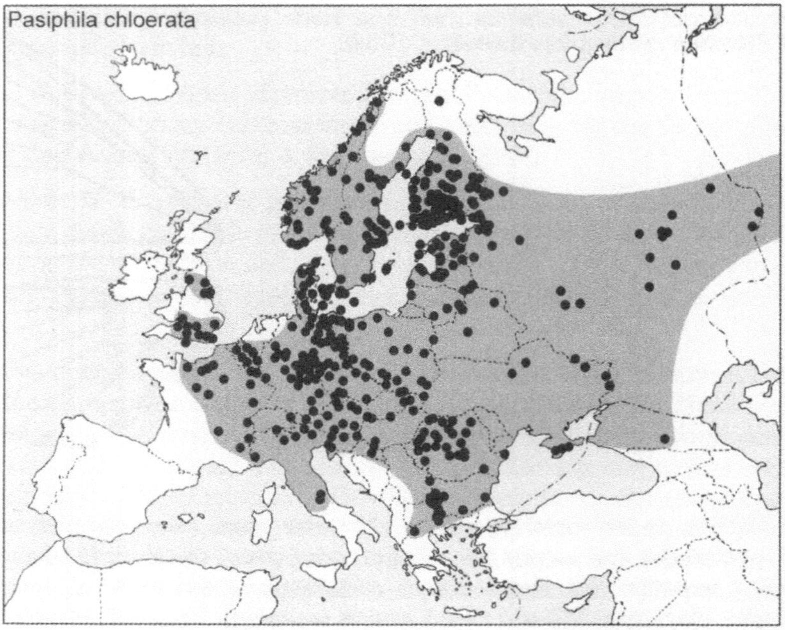

(Calliclystis) chloerata f. *nigrofasciata*: DIETZE, 1913: Biol. Eupithecien 2: 19 (Austria: Styria, Mecklenburg). Syntypes 2 (MNHU). Valid at subspecific rank.
Unavailable names (infrasubspecific): *hadenata*: FUCHS (1900). - (incorrect subsequent spellings): *chlorata*: PROLA & RACHELI (1980); *cloerata*: UHERKOVICH (1981); *chloreata*: PIZZETTI & RAINERI (1998).

DIAGNOSIS: Wingspan 16-21 mm. Labial palpi grey or brownish grey. Frons, vertex and notum from dark grey to pale grey with brownish tinge. Forewing grey, often with brownish tinge; costal margin usually darker; transverse lines fine, slightly sinuate; basal line evenly curved or angled in the middle; postbasal and antemedial lines angled near costa; postmedial line evenly curved or obtusely angled once or twice, with dark dashes on the veins on the innerside and with paler outer border; pale wavy line usually distinct; terminal margin often mottled and irrorate with white scales, especially near apex of wing; discal spot small, round-ovoid, blackish. Hindwing grey; transverse lines inconspicuous; postmedial line medially slightly angled, with darker inner shade and/or with short dark dashes on the veins; pale wavy line distinct; terminal area often mottled with white scales, especially near apex of wing. Abdomen grey or pale grey, usually with reddish tinge; a series of dark grey or black dorsal spots present, as well as sometimes short black lateral stripes on the anterior segments.

MALE GENITALIA: Uncus decorated with short setae. Papillae absent but anterior tips of labides covered by short setae. Valva with smoothly curved ventral margin; medially broadened. Sternum A8 with completely sclerotized basal margin; with two narrow, parallel and straight sclerotized bands.

FEMALE GENITALIA: Bursa copulatrix heart-shaped with large, broad, globular diverticulum at base; signa: one lunate, very long, transverse band of rasp-like teeth, one smaller, triangular, compact patch of similar teeth near ductus bursae and one small, weak patch of teeth between corpus bursae and diverticulum. Ductus bursae partly sclerotized, with numerous granules. Antrum large, sclerotized. Lamella antevaginalis large, triangular, sclerotized. Tergum A8 almost square, not decorated by short setae.

DISTRIBUTION: Eurasiatic. Widely distributed in Palaearctic region from United Kingdom to Japan. In Europe not found in Iceland, Ireland, Portugal, Spain, Albania, Turkey and Mediterranean islands. – New for the fauna of Greece: 1♀, Greece, Kastoria, Eptaho´ri, 12.VI.1996, leg. Z. Laštuvka (coll. FIBIGER); 1 specimen, Greece, Drama, Sidirokastron near Kapnophron, 600 m, B VII (coll. FIBIGER).

PHENOLOGY: Univoltine: mid-May (late April) to mid-July (late July); usually mid-June to mid-July. Larva: mid-April (mid-March) to late May (early June). Overwintering as egg.

BIOLOGY: Larva oligophagous (1). Found in the buds, flowers or rarely on the leaf-buds and leaves of *Prunus spinosa* (France: GOOSSENS 1871; Germany: DIETZE 1910, 1913; WEIGT 1988; WEGNER pers. comm.; Denmark: JUUL 1948), also on *P. padus*, *P. virginiana* (Rosaceae) (Finland: SEPPÄNEN 1970).

PARASITOIDS: Hymenoptera, Encyrtidae: *Litomastix chalconotus* (HERTING 1976).

HABITAT: Open deciduous, and mixed forests, edges of forest, hedges, scrub, gardens, parks, warm and dry slopes, rarely on the rocks. From 0 m up to about 1,000 m above sea-level in Alps (coll. ZSM), and to about 1,300 m in Apennines (coll. ZSM).

SIMILAR SPECIES: *P. chloerata* can be easily confused with *P. rectangulata* but distinguished from it by the more evenly curved, not twice sharply angled postmedial line on the forewing, and more evenly curved postmedial line on the hindwing also (Text-fig. 12). The male genitalia, the shape of sternum A8 of the male and especially the female genitalia readily distinguish this species from *P. rectangulata*.

REMARKS: This species shows little variation in the depth of colour and intensity of the markings. However, specimens with brassy olive-green ground colour are often found as well as grey-coloured specimens. Extreme melanism is unknown in this species; but much darker specimens occur moderately rarely (ab. *hademata* FUCHS, 1900). The f. *nigrofasciata* DIETZE, 1913 has darker medial area only on the forewing.

22. *Pasiphila rectangulata* (LINNAEUS, 1758)

Phalaena (Geometra) rectangulata LINNAEUS, 1758: Syst. Nat. (Ed. 10) 1: 528 (Europe). Syntype(s) (DEUU?).
Phalaena viridulata HUFNAGEL, 1767: Berlin Mag. 4 (5): 524 (Germany: Berlin region). Syntype(s) lost.
Phalaena (Geometra) cydoniata BORKHAUSEN, 1794: Naturgesch. Eur. Schmett. 5: 354 (Europe). Syntype(s).
Pyralis mediana PANZER, 1804: D. J. C. Schaefferi Icones Insectorum: 162, pl. 182: 5, 6 (Germany: Regensburg). Syntypes.
Phalaena nigrosericeata HAWORTH, 1809: Lepid. Br. (2): 362 (United Kingdom). Syntype(s) (BMNH).
Phalaena sericeata HAWORTH, 1809: Lepid. Br. (2): 363 (United Kingdom: England). Syntype(s) (BMNH).
Geometra subaerata HÜBNER, 1817: Samml. Eur. Schmett. 5 Geometrae (2), pl. 90: 463 (Europe). Syntype(s) lost?
Geometra bischoffaria GEYER, 1838: Samml. Eur. Schmett. 5 Geometrae (1): pl. 112: 586-588 (Europe). Syntype(s).
Eupithecia debiliata var. (ab.) *griseata* STAUDINGER, 1897: Dt. ent. Z. Iris 10 (1): 110 (Russia: Amur, Sutschan). Syntypes 3 (MNHU). Valid at subspecific rank.
Unavailable names (infrasubspecific): *antrax*: DIETZE (1910); *bistrigata*: DIETZE (1913); *joannisata*: CULOT (1919); *grisescens*: LEMPKE (1951); *effusa*: COCKAYNE (1953). - (incorrect subsequent spelling): *rectangularia*: BOISDUVAL (1840).

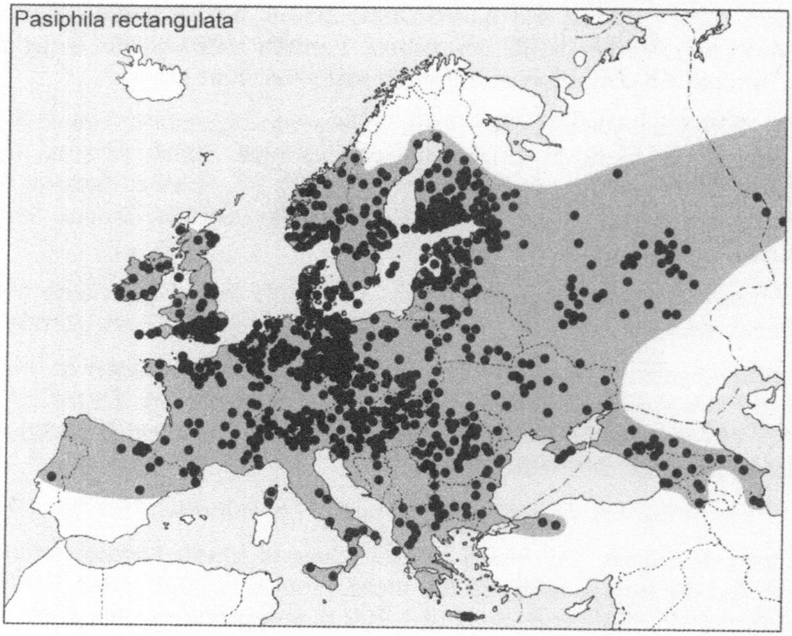

DIAGNOSIS: Wingspan 16-21 mm. Labial palpi dark brownish grey or brown, sometimes with brassy green tinge. Frons, vertex and notum from brownish grey to brassy olive-green. Forewing grey, in some specimens with brassy olive-green tinge; transverse lines fine, slightly sinuate; ante- and postmedial lines usually thickened to costa with dark costal shade; basal line multiply indented, evenly curved; postbasal line ill-defined; antemedial line oblique, usually evenly curved but sometimes sharply angled near costa; postmedial line twice toothed near costa, often with dark dashes on the veins and with lighter double stripe beyond it; pale wavy line usually conspicuous, multiply indented; outer half of terminal area usually lighter, mottled and irrorate with whitish or greenish scales, especially near apex; discal spot very small, round, ovate or elongated, dark, sometimes obsolete. Hindwing grey with more or less distinct, fine dark lines; postmedial line medially right-angled, sometimes with short dark dashes on the veins; pale wavy line multiply indented, from distinct to obsolete; discal spot minute, round, dark. Fringe unicolorous or slightly chequered pale and dark grey. Abdomen pale grey or pale brownish grey, in some specimens with brassy green tinge; a row of small black dorsal dots present; second and part of third segment from brown to dark brown, usually darker than other segments.

MALE GENITALIA: Papillae short and narrow. Valva with almost parallel ventral and dorsal margins. Sternum A8 with sclerotized basal margin, and with two narrow, slightly incurved, sclerotized rods.

FEMALE GENITALIA: Bursa copulatrix large, subglobular, without membranous diverticulum at base; signa two lunate bands of small rasp-like teeth, one larger than other. Ductus bursae membranous. Antrum short, sclerotized. Lamella antevaginalis absent. Tergum A8 elongated, longer than wide; its posterior margin with deep medial hollow, laterally covered by short setae. Anterior apophyses longer than tergum A8; posterior apophyses slightly expanded at apices.

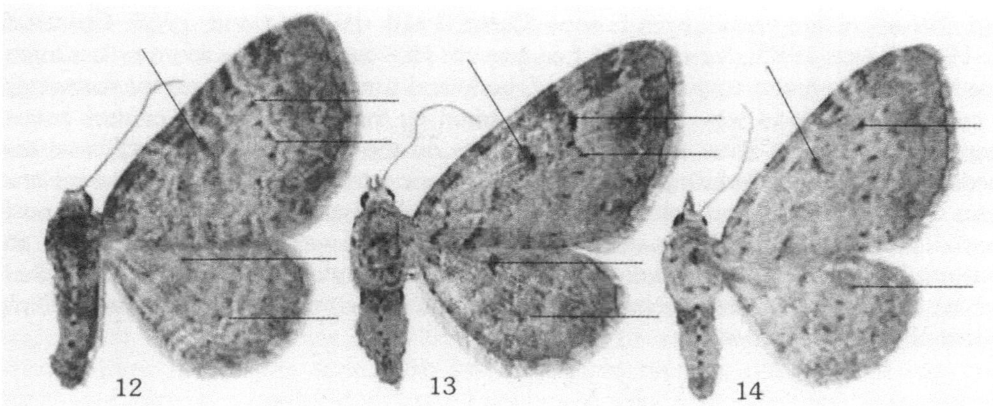

Text-figs 12-14: Diagnostic characters (indicated) of European *Pasiphila* species. Text-fig. 12: *P. chloerata* (MABILLE, 1870). Text-fig. 13: *P. rectangulata* (LINNAEUS, 1758). Text-fig. 14: *P. debiliata* (HÜBNER, 1817).

DISTRIBUTION: Eurasiatic. Widely distributed in Palaearctic region from Portugal to Kuril Islands and Japan. All Europe except Iceland, northern Scandinavia, Balearics, Sardinia and Malta.

PHENOLOGY: Univoltine: mid-June (late April) to mid-July (mid-August). Larva: late April to mid-May (early June). Overwintering as egg.

BIOLOGY: Larva oligophagous (1). Found on the leaf-buds and on the flowers of *Malus domestica* (United Kingdom: CREWE 1861c; BUCKLER 1899; Russia: original data), also *Pyrus communis, Prunus padus* (Germany: DIETZE 1910, 1913; Finland: SEPPÄNEN 1970), *Crataegus* (Denmark: JUUL 1948), *Prunus spinosa* (Corsica: MABILLE 1868), *Malus sylvestris, Amelanchier ovalis* (Rosaceae) (Germany: WEIGT 1988).

PARASITOIDS: Hymenoptera, Ichneumonidae: *Tranosemella coxalis* (?), *Homotherus locutor* (=*albicinctus*), *Phobocampe unicincta* - Hymenoptera, Braconidae: *Apanteles popularis* (BUCKLER 1899; THOMPSON 1944; CLAUSEN & LA TOUR-DE-PEILZ 1945; SHENEFELT 1972).

HABITAT: Deciduous and mixed forests, forest edges, hedges, scrub, gardens, parks, rocks. From 0 m up to about 1,400 m above sea-level in Apennines (Mt. Sirino: coll. ZSM), and to about 1,500 m in Alps (VORBRODT 1921; WEIGT 1988).

SIMILAR SPECIES: The typical form of *P. rectangulata* is distinguished from the very similar *P. chloerata* by the toothed postmedial line on the forewing, the more sharply angulated postmedial line on the hindwing and also on the underside of all wings (Text-fig. 13). Valva longer and narrower than in *P. chloerata*, and vesica not armed with a row of small spines. Female genitalia distinguished by the different shape of corpus bursae and shape of scobinate signum, by absence of scobinate signum in ductus bursae and triangular sclerotized lamella antevaginalis; the longer apophyses, the longer and narrower tergum A8 and papillae anales compared to *P. chloerata*.

REMARKS: This is a very variable species in coloration and markings. Many different forms and aberrations are known from Europe (DIETZE 1910, 1913; LEMPKE 1969; GOATER & al. 1981; WEIGT 1988), for example: f. *subaerata* HÜBNER, 1817 has more or less intensive brassy olive-green colour of the head, body and wings, especially on the forewings; f. *cydoniata* BORKHAUSEN, 1794 is distinguished by the darker, brown or dark brown medial area between ante- and postmedial lines on the forewing and darker basal and medial areas on the hindwing as well; ab. *nigrosericeata* HAWORTH, 1809 is a melanic form with blackish ground colour, with intensely black discal spots, fine ante- and postmedial lines and with conspicuous, fairly complete, pale wavy lines on all wings; ab. *anthrax* DIETZE, 1910 is the darkest melanic form with uniform black wings, unmarked, except only for an intensely black discal spot and sometimes weakly defined multiply indented, pale wavy lines.

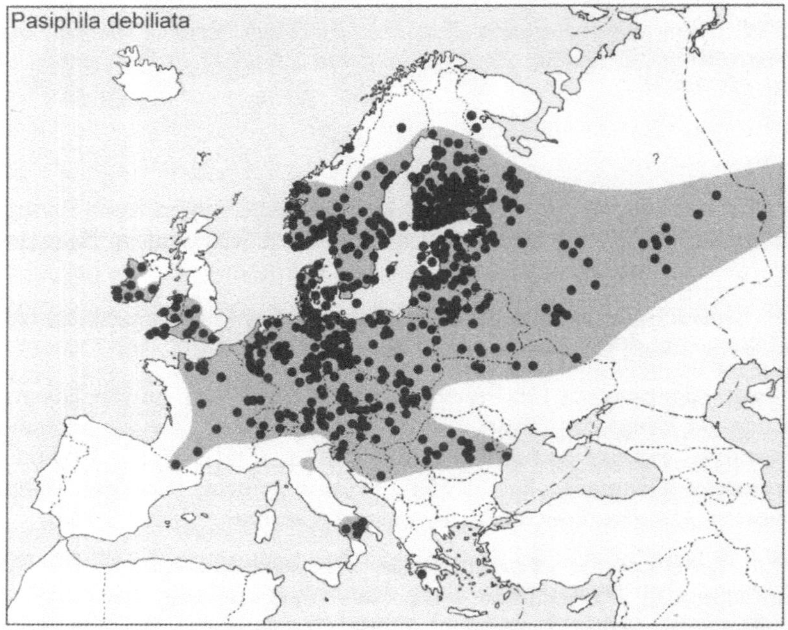

23. *Pasiphila debiliata* (HÜBNER, 1817)

Geometra debiliata HÜBNER, 1817: Samml. Eur. Schmett. 5 Geometrae (2), pl. 90: 462, pl. 91: 466 (Europe). Syntype(s) lost.

Eupithecia nigropunctata CHANT, 1833: in CHANT & BENTLEY, Ent. Mag. 1 (2): 184, pl. 24 (United Kingdom: England, Devon, Spitchwick). Syntype(s) (BMNH?).

Eupithecia clerci KRULIKOVSKY, 1891: Zap. ural'. Obshch. Lyub. Estest. 12 (2): 80 (Russia: Wiatka province, Sarapul). Syntype(s) (ZMKU?).

(Calliclystis) debiliata f. *grisescens*: DIETZE, 1913: Biol. Eupithecien 2: 22 (Europe ?: type locality not stated). Syntype(s) (MNHU). Valid at subspecific rank.

(Calliclystis) debiliata f. *mediofasciata*: DIETZE, 1913: Biol. Eupithecien 2: 22; ibidem (1910), 1, pl. 85: 902 (Austria: Styria, Mürzthal). Holotype ♂ (MNHU). Valid at subspecific rank.

Chloroclystis suspiciosa INOUE, 1982: in INOUE & al., Moths of Japan 1: 515, pl. 79: 2 (Japan: Honshu, Hyogo Pref., Kaibara-machi). Holotype ♀ (BMNH).

Unavailable names (infrasubspecific): *grisescens*: DIETZE (1913). - (incorrect subsequent spellings): *debiliaria*: BOISDUVAL (1840); *debilata* LASTUKHIN (1998).

DIAGNOSIS: Wingspan 16-20 mm. Labial palpi grey or dark grey. Frons, vertex and notum greyish white. Forewing whitish grey or grey; basal line evenly curved or sometimes angled near costa; postbasal line inconspicuous; antemedial line straight, oblique, right-angled and broadened near costa with outer dark shade, forming three dark dots on the stalks of R, Cu and A; postmedial line evenly curved, usually twice slightly angulated, broadened to costa with dark inner costal shade, very often formed from close set series of small dark grey dots; terminal area darker, sometimes with dark dashes on the veins on inner part and with outer part beyond subterminal line mottled and irrorate with white scales; wavy line usually distinct, much indented, whitish; discal spot small, round or ovoid. Hindwing whitish grey, with more or less distinct transverse lines; postmedial line angled, with broad dark inner shade and often formed from a series of dark dots on the veins; terminal area slightly darker with fine, continuous, multiply indented pale wavy line; discal spot minute, rounded, dark grey. Fringe unicolorous, whitish grey or sometimes slightly spotted with dark grey at vein ends. Abdomen whitish grey with a row of small, dark brown or black dorsal spots; first abdominal segment whitish, lighter than posterior segments, second and third segments dark brown or black, darker than others.

MALE GENITALIA: Uncus almost completely covered by short setae. Papillae appressed, short and broad. Valva with almost parallel ventral and dorsal margins. Sternum A8 medially sclerotized at base; with two narrow, slightly incurved, sclerotized bands.

FEMALE GENITALIA: Bursa copulatrix relatively small, subglobular, without membranous diverticulum at base; signum a weak, triangular patch of small spines. Ductus bursae membranous with small sclerotized part at apex. Antrum lightly sclerotized. Lamella antevaginalis absent. Tergum A8 rectangular, slightly longer than wide, without setae. Anterior apophyses slightly longer than tergum A8.

DISTRIBUTION: Eurasiatic. Widely distributed from Pyrenees to Japan. In Europe not found in Iceland, Portugal, Montenegro, Albania, Macedonia, Bulgaria and Mediterranean islands. Recorded for European Turkey by OKYAR & AKTAÇ (1999) probably erroneously (*P. chloerata* ?).

PHENOLOGY: Univoltine: early June (mid-May) to early July (early August). Larva: late March to early June (usually from late April to late May). Overwintering as egg.

BIOLOGY: Larva rather monophagous (2). Found between spun up leaves and on the flowers of *Vaccinium myrtillus* (Ericaceae) (United Kingdom: CREWE 1862, 1863, 1865; BUCKLER 1899; France: GOOSSENS 1869; MILLIÈRE 1870; Germany: DIETZE 1910, 1913; WEGNER pers. comm.) and *V. uliginosum* (Denmark: JUUL 1948; Finland: SEPPÄNEN 1970; Germany: WEIGT 1988).

PARASITOIDS: Hymenoptera, Ichneumonidae: *Diadegma majale* (THOMPSON 1944; HERTING 1976).

HABITAT: Spruce-, pine-, mixed-, deciduous and paludal forests where food-plants carpet the forest floor, forest paths, edges of forest, moors, heaths. From 0 m up to about 1,500 m above sea-level in Alps (VORBRODT & MÜLLER-RUTZ 1914; VORBRODT 1931; FORSTER & WOHLFAHRT 1981).

SIMILAR SPECIES: Similar to the two preceding species but can usually be distinguished from them by the lighter, whitish grey colouration, by presence of series of small dark dots representing the antemedial and postmedial lines and also by the thicker and darker postmedial lines on the underside of all wings (Text-fig. 14). In the male genitalia the vesica without rows of small spines as in *P. chloerata*, the valva shorter than in *P. rectangulata*, the base of sternum A8 heavily sclerotized only in the middle. The female genitalia of *P. debiliata* is easily distinguished from those of the other European *Pasiphila* species by the smaller bursa copulatrix and by presence of only one patch of microscopic spinules in corpus bursae.

REMARKS: Usually not very variable, yet there is a little variation in colour. Some specimens have head, notum and forewing with slight lime-green tinge and also abdomen with more or less well defined greenish tinge. Other forms include f. *grisescens* DIETZE, 1913 with silver-grey ground colour without greenish tinge; f. *nigropunctata* CHANT, 1833 with distinct rows of dark dots instead of ante- and postmedial lines on the forewing and also similar dots instead of postmedial line on the hindwing; f. *mediofasciata* DIETZE, 1913 with darker medial areas on the forewings.

Eupithecia CURTIS, 1825

=*Tephroclystia* HÜBNER, 1825

TYPE-SPECIES: *Phalaena absinthiata* CLERCK, 1759. More than 1,300 species worldwide (SCOBLE 1999). 128 species occur in Europe. It is the most species-rich genus of *Geometridae* and one of the most species-rich genera among all *Lepidoptera*.

DIAGNOSIS: Pattern on the forewing usually with somewhat indistinct basal, ante- and postmedial transverse lines; postmedial line often not twice sharply angled near costa; discal spots usually definite. Forewing venation: R1 long, not uniting or anastomosing with Sc; areole with/without oblique transverse vein. Hindwing: M3 and CuA1 arising separately from cell. Hindtibia with two pairs of spurs.

MALE GENITALIA: Uncus usually membranous, broadest in basal part and sclerotized, rigid, hook-like and pointed apically, with one or two tips (uniapical or biapical); in some species

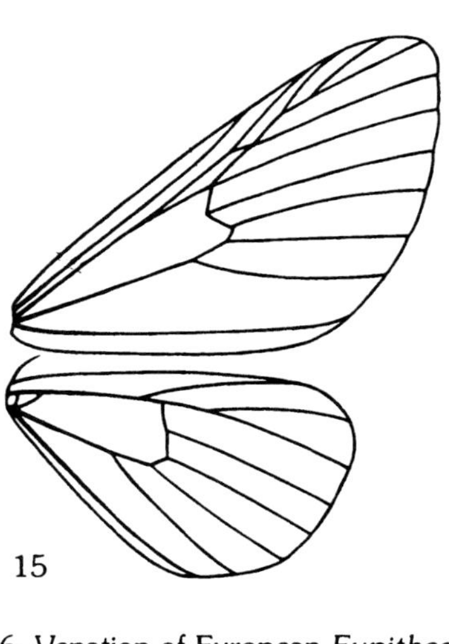

15

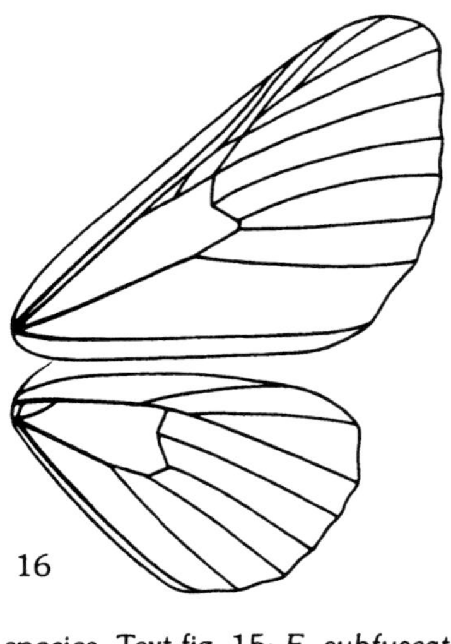

16

Text-figs 15-16: Venation of European *Eupithecia* species. Text-fig. 15: *E. subfuscata* (HAWORTH, 1809). Text-fig. 16: *E. simpliciata* (HAWORTH, 1809).

uncus completely membranous, spatulate. Valva variable in size and shape (usually shaped like a segment of an orange), with ventral process in some species. Anterior arms of labides with apical small digitiform, membranous papillae. Juxta broadest transversely, oval or reniform, with two narrow apical lobes. Vesica of aedeagus armed with one or more variously shaped cornuti (horn-like, tubular, irregular, dentate or flat and twisted) and sometimes with a patch of small spines. Sternum A8 usually broadest at base, with shallow basal hollow; it is usually of a consistent shape within a species but can vary slightly in some species.

FEMALE GENITALIA: Bursa copulatrix membranous to strongly sclerotized; variable in size and shape (globular, subglobular, pouch-like, pear-shaped, etc.), in some species with membranous diverticulum at base. Bursa copulatrix covered with numerous signa, which can be variable in position. Ductus seminalis attached to corpus bursae. Colliculum usually well defined, sclerotized, collar-like or plate-like. Lamella antevaginalis lacking. Spurs of anterior apophyses can be prominent with a very enlarged apex, or be reduced in the size.

IMMATURE STAGES: Morphology and surface sculpturation of eggs described and figured in DRAUDT (1905), DIETZE (1910, 1913), WEIGT (1988, 1990, 1991, 1993). Discussion of larval morphology in DIETZE (1913) and MCGUFFIN (1958). Pupal morphology see PATOČKA (1981, 1996).

The ***haworthiata*** species-group

Group introduced by VOJNITS (1972). This is one of the most species-rich groups, but represented by only one species in Europe: *E. haworthiata*. Nearly 30 species are known from Asia (VOJNITS 1976, 1981, 1983, 1984; INOUE 1988, 2000; MIRONOV 1990). Some African species belong to this group as follows: *E. jacksoni* FLETCHER, 1956 (Kenya, Uganda), *E. alticomora* HERBULOT, 1980 (Comores Islands), *E. sporobola* HERBULOT, 1988, *E. tranquilla* HERBULOT, 1988 (Madagascar). One North American species *E. coloradensis* (HULST, 1896) belongs to the *haworthiata* group also.

MALE GENITALIA: Uncus membranous, without stick-like process. Papillae short and often wide, length about 0.3-0.9 (usually 0.4-0.7) times uncus length. Valva without ventral process, broad, with curved ventral margin and broadly rounded apex; sacculus weakly sclerotized. Vinculum short and wide, usually with medial hollow. Aedeagus with more or less developed lateral hirsute lobes, relatively thin and short, slightly shorter than length of valva. Vesica armed with numerous minute spinules and with one elongated, thin cornutus, pointing almost caudad and one small cornutus at ductus ejaculatorius base. Sternum A8 with two long and narrow rods.

FEMALE GENITALIA: Bursa copulatrix membranous, globular or egg-shaped, with two large patches of small spines (one dorsal and one ventral) and one separate patch of spines near colliculum. Ductus bursae usually ill-defined. Ductus seminalis medium size or long, narrow. Colliculum collar-like, short and narrow, but it is elongated in some species. Antrum membranous. Anterior and posterior apophyses short to very long, often apically expanded and rounded. Papillae anales wide to narrow, setaceous or spinuliferous (in some Asian species).

REMARKS: The species of this group can usually be distinguished from all other *Eupithecia* species by membranous uncus and apical hirsute lobes of aedeagus in male genitalia. On the base of this difference, the *haworthiata* species-group is considered to be closely allied to genus *Pasiphila* MEYRICK, 1883.

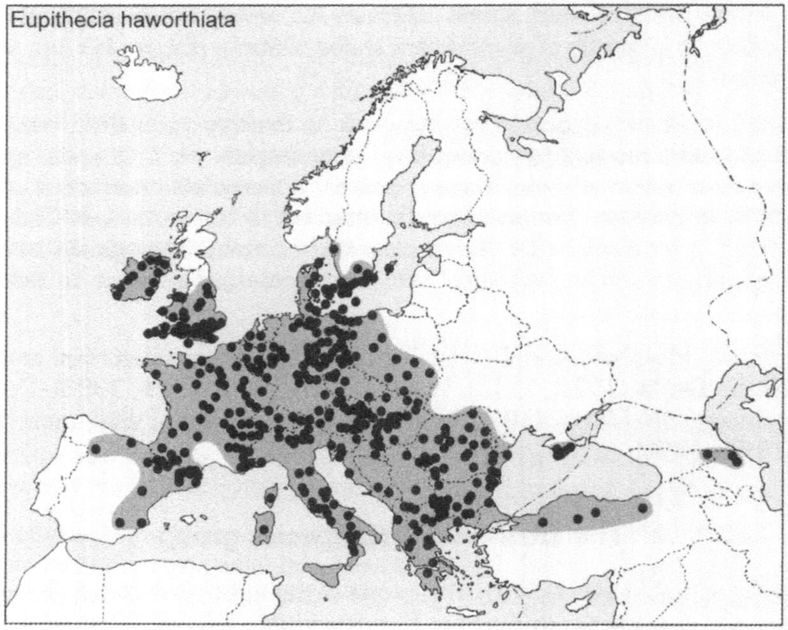

24. *Eupithecia haworthiata* DOUBLEDAY, 1856

Eupithecia haworthiata DOUBLEDAY, 1856: The Zoologist 14: 5139, 5141 (United Kingdom). Syntype(s) (BMNH).

Eupithecia gozmanyi VOJNITS, 1972: Ann. Hist.-Nat. Mus. Nat. Hung. 64: 299 (Bulgaria: Stanimaka). Holotype ♀ (ZFMK, examined).

Unavailable names (misidentification): *valerianata*: sensu TREITSCHKE (1828) nec HÜBNER, 1813 (identity: *Eupithecia valerianata*); *isogrammaria*: sensu HERRICH-SCHÄFFER (1848) nec BOISDUVAL, 1840 (identity: *Eupithecia plumbeolata*); *isogrammata*: sensu BREYER (1864) nec TREITSCHKE, 1828 (identity: *Eupithecia plumbeolata*). - (infrasubspecific): *obscura*: LEMPKE (1969). - (incorrect subsequent spellings): *haworthiaria*: MORRIS (1861); *haworthiana*: EXPOSITO HERMOSA (1978).

DIAGNOSIS: Wingspan 12-18 mm. Labial palpi dark grey. Frons covered with projecting dark grey scales. Vertex and notum dark grey. Forewing dark grey or brownish grey; transverse lines fine, indistinct, slightly sinuate, obtuse angled near costa, pale grey; the double medial and also double pale stripe between medial and terminal areas more distinct than other transverse markings; postmedial line evenly curved; pale wavy line unbroken; discal spot usually absent but in some specimens it is minute, elongated or rounded. Hindwing dark grey, usually slightly paler than forewing; transverse lines inconspicuous, but double pale stripe between postmedial and subterminal line more distinct; discal spot absent. Fringe grey, slightly spotted with dark grey at vein ends. Abdomen dark grey; first segment always reddish, usually some other anterior segments reddish also, especially laterally.

Text-figs 17-20: Diagnostic characters (indicated) of European *Eupithecia* species. Text-fig. 17: *E. haworthiata* DOUBLEDAY, 1856. Text-fig. 18: *E. plumbeolata* (HAWORTH, 1809). Text-fig. 19: *E. pygmaeata* (HÜBNER, 1799). Text-fig. 20: *E. valerianata* (HÜBNER, 1813).

MALE GENITALIA: Vinculum short and wide, with shallow medial hollow. Sternum A8 small, with two narrow rods, apically slightly sclerotized, rounded and expanded.

FEMALE GENITALIA: Colliculum short and narrow. Tergum A8 almost square. Anterior and posterior apophyses relatively short and thin, but apically slightly expanded and rounded. Papillae anales narrow and tapered.

DISTRIBUTION: Euro-Caucasian. Western, Central and southern Europe (except Portugal, Malta and Crete). Disjunct population in southern Crimea. – Outside Europe in Asia Minor and eastern Caucasus.

PHENOLOGY: Univoltine: early June (early May) to late July (early August); however one specimen was found in the middle of September (coll. ZSM). Larva: late June (mid-June) to mid-August (early September). Overwintering as pupa.

BIOLOGY: Larva monophagous (1). Found in the flower buds and flowers of *Clematis vitalba* (Ranunculaceae) (United Kingdom: CREWE 1861c, 1862; BUCKLER 1899; Germany: HERRICH-SCHÄFFER 1861a; DIETZE 1910, 1913; WEIGT 1988; WEGNER pers.

comm.; Austria: HABICH & REBEL 1883; Switzerland: PÜNGELER 1889; Denmark: JUUL 1948).

PARASITOIDS: Hymenoptera, Braconidae: *Lissogaster (=Microgaster) stictica* (THOMPSON 1945; SHENEFELT 1973).

HABITAT: Open deciduous forests, edges of forest, hedges, bushy thickets, scrub and scrub-covered meadows, gardens, parks, warm and dry localities, chalk districts, rocks and screes. From 0 m to about 1,400 m above sea-level in Apennines (coll. ZSM), to about 1,500 m in Balkan peninsula (coll. ZMUC; coll. FIBIGER), and to about 1,800 m in Alps (WEIGT 1988).

SIMILAR SPECIES: *E. haworthiata* can be confused with *E. plumbeolata* but distinguished from it by the usually narrower wings of brownish colour, by the sharply angled medial line on the forewing, and by the pale brick-reddish abdominal segments, this coloration sometimes restricted to first abdominal segment (Text-fig. 17). The combination of membranous uncus, vinculum with medial hollow, cornuti on the vesica in the male genitalia, as well as egg-shaped bursa copulatrix with characteristic areas covered by spines in the female genitalia should easily distinguish this species from all other European *Eupithecia*.

The *tenuiata* species-group

Group introduced by MIRONOV (1990). This species-poor group is represented in Europe by two species: *E. tenuiata* and *E. inturbata*. The East Asian species *E. praepupillata* WEHRLI, 1927 and North American species *E. bryanti* TAYLOR, 1906 also belong to the *tenuiata* group.

MALE GENITALIA: Uncus biapical, short, normal or broad; widened near apex, when viewed laterally. Papillae short, length about 0.3-0.7 times uncus length. Valva without ventral process, broad or normal width, with curved ventral margin and rounded apex; sacculus weakly sclerotized. Vinculum short and medium width. Aedeagus thin and short, slightly shorter than length of valva. Vesica armed with one small, elongated U-shaped cornutus at ductus ejaculatorius base. Sternum A8 with two symmetrical, long and narrow rods, which are rounded and slightly expanded apically (except *E. inturbata*) and connected by narrow band at base.

FEMALE GENITALIA: Bursa copulatrix membranous, ovate or elliptical; almost completely covered with small spines (except small membranous area in some species). Ductus seminalis short and thin, attached to bursa copulatrix near colliculum base. Colliculum collar-like, narrow, and elongated in some species. Antrum membranous. Anterior and posterior apophyses long, apically rounded and expanded. Anterior apophyses longer than or nearly equal to length of sternum A8. Tergum A8 long, about 1.4-1.7 times longer than medial width. Papillae anales relatively narrow, tapered.

REMARKS: The species of this group can usually be distinguished from other European species by the vesica armed with only one small, elongated U-shaped cornutus at ductus ejaculatorius base and the shape of sternum A8 in male; as well as the long, apically rounded and expanded anterior and posterior apophyses; the elongated tergum A8 and the narrow, tapered papillae anales in the female genitalia.

The *tenuiata* species-group appears to be closely related to the *haworthiata* species-group suggested by the external features of adults and some minor characters in the genitalia.

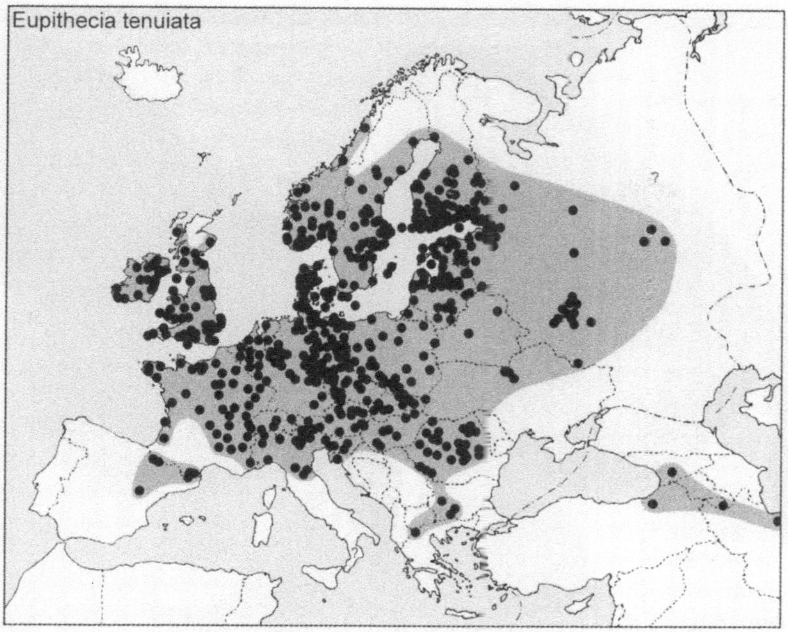

25. Eupithecia tenuiata (HÜBNER, 1813)

Geometra tenuiata HÜBNER, 1813: Samml. Eur. Schmett. 5 (Geometrae 1), pl. 76: 394 (Europe). Syntype(s) lost.

Eupithecia tenuiata var. *cinerae* GREGSON, 1888: Young Naturalist 9: 104 (United Kingdom: Morayshire). Syntype(s) not traced in BMNH (L. PITKIN pers. comm.). Valid at subspecific rank.

Eupithecia macellata FISCHER VON RÖSLERSTAMM, 1829: in THON, Ent.Archiv. 1 (2): 27 (Russia). Syntype(s).

Unavailable names (misidentification): *inturbata*: sensu TREITSCHKE (1828) nec HÜBNER, 1817 (identity: *Eupithecia inturbata*); *inturbaria*: sensu FREYER (1858) nec BOISDUVAL, 1840 (identity: *Eupithecia inturbata*). – (infrasubspecific): *niveipicta*: BASTELBERGER (1908). - (incorrect subsequent spellings): *tenularia*: HERRICH-SCHÄFFER (1848); *tenuiaria*: HERRICH-SCHÄFFER (1848); *tenujata*: NOWICKI (1860); *tenuita*: PETERSEN (1902); *tenuata*: HUGGINS, 1953 (nec HULST, 1880).

DIAGNOSIS: Wingspan 14-19 mm. Labial palpi dark grey. Frons and vertex pale grey with scattered dark grey scales. Notum pale grey. Forewing pale grey, in some specimens with weak ochreous tinge; costal margin usually with slightly darker spots; transverse lines fine, rather indistinct, slightly sinuate, grey; three medial lines usually more conspicuous, especially the first line, which is sharply angled, touching discal spot or crossing beyond it; postmedial line sometimes indicated by dark dashes on the veins; terminal area slightly darker but often mottled and irrorate with white scales, especially near apex of wing; pale wavy line inconspicuous; discal spot small, rounded or ovate, black. Hindwing only slightly paler than forewing; usually with ill-defined transverse lines and with clouded, slightly darker terminal area; pale wavy line indistinct or obsolete; discal spot minute, dark, rounded. Fringe slightly chequered pale grey and grey. Abdomen pale grey, often with ochreous tinge, with small dorsal tufts and narrow rows of white scales on posterior margins of segments, and usually with lateral black stripes also.

MALE GENITALIA: Uncus medium stoutness. Papillae length about 0.7 times uncus length.

Text-figs 21-22: Diagnostic characters (indicated) of European *Eupithecia* species. Text-fig. 21: *E. tenuiata* (HÜBNER, 1813). Text-fig. 22: *E. inturbata* (HÜBNER, 1817).

FEMALE GENITALIA: Bursa copulatrix ovate, without spines in membranous anterior part. Colliculum elongated, length about 0.7 times that of corpus bursae. Antrum very elongate. Tergum A8 with slightly curved posterior margin and deeply notched anterior margin.

DISTRIBUTION: Euro-Caucasian. Widely distributed in Europe except Iceland in the north; Corsica, Sardinia, Sicily and Albania in the south. – Outside Europe in north-eastern Turkey, Georgia, Armenia, Azerbaijan (WARDIKJAN 1985; MIRONOV 1990) and northern Iran (Mts Elburs). – New for the fauna of Greece: 1♀, Graecia, Florina, Agios Ilias, 900 m, 15.IX.1991, leg. F. SCHEPLER (ZMUC).

PHENOLOGY: Univoltine: early July (mid-June) to mid-August (mid-September). Larva: mid-April (early March) to late May (early June). Overwintering as egg.

BIOLOGY: Larva monophagous (2). Found in the catkins, preferring the male catkins of *Salix caprea* (United Kingdom: CREWE 1861c, 1862; BUCKLER 1899; Austria: HABICH & REBEL 1883; Germany: DIETZE 1910, 1913; Finland: SEPPÄNEN 1970), and also on *S. cinerea* (France: LHOMME 1935), *S. aurita* (Salicaceae) (Germany: WEIGT 1988).

PARASITOIDS: Hymenoptera, Ichneumonidae: *Itoplectis maculator* (THOMPSON 1945).

HABITAT: Hedges, bushy thickets, river banks, edges of forests and fields, forest paths, ditches, marshes, gardens and parks, rocks. From 0 m up to about 1,500-1,600 m above sea-level in Alps and Apennines (DANNEHL 1927; coll. ZSM; BUSSE pers. comm.), and to about 2,000 m in Pyrenees (MAZEL & PESLIER 1997).

SIMILAR SPECIES: This rather invariable species can be distinguished from other small

Eupithecia by the full and rounded forewing with arched costa, by presence of small, conspicuous discal spot, and by absence of three dark blotches on the terminal area of forewing (Text-fig. 21), as in *E. inturbata*. One small U-shaped cornutus on the vesica and shape of sternum A8 in male; ovate corpus bursae completely covered with spines and elongated apophyses, much expanded at apices in the female genitalia clearly distinguish this species from all others.

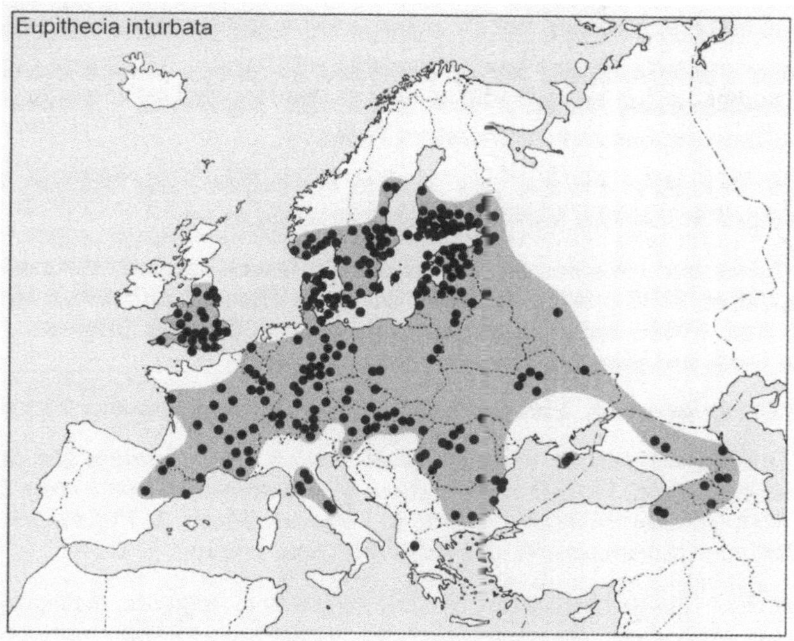

26. *Eupithecia inturbata* (HÜBNER, 1817)

Geometra inturbata HÜBNER, 1817: Samml. Eur. Schmett. 5 Geometrae (2), pl. 90: 461 (Europe). Syntype(s) lost.

Eupithecia neglectata HERRICH-SCHÄFFER, 1848: Syst. Bearb. Schmett. Eur. 3 (32): 123 (Europe?). Syntype(s) lost.

Eupithecia subciliata DOUBLEDAY, 1856: The Zoologist 14: 5143 (United Kingdom, England: Cambridgeshire). Syntypes (at least 7) (BMNH).

Unavailable names (incorrect subsequent spellings): *inturbaria*: BOISDUVAL (1840); *subciliaria*: MORRIS (1861).

DIAGNOSIS: Wingspan 13-18 mm. Labial palpi grey. Frons and vertex pale grey. Notum grey with dark grey anterior transverse band. Forewing pale grey, sometimes with yellowish tinge; transverse lines more or less distinct, slightly sinuate, evenly curved or obtuse angled near costa, also broadened at costa, dark grey or brownish grey; antemedial line usually touching discal spot; postmedial line fine-wavy, well defined; terminal area with three darker clouded blotches (inner apical, outer medial and inner tornal); pale wavy line much indented, usually ill-defined and broken, sometimes forming a small pale tornal spot; discal spot small, dark, rounded or slightly elongated. Hindwing paler, usually with well defined, darker postmedial and fine-wavy terminal lines, and with pale stripe between them, which is bisected longitudinally by fine, dark line; terminal area narrow, darker; pale wavy line usually indistinct; discal spot minute, dark, rounded. Fringe pale grey, distinctly spotted with dark grey or brownish grey at vein ends. Abdomen grey with black lateral spots on each segment; the last segment pale grey in male.

MALE GENITALIA: Papillae length about 0.5 times uncus length. Aedeagus thin. Sternum A8 with two rods, apically sclerotized and narrow, connected by band at base.

FEMALE GENITALIA: Bursa copulatrix rather egg-shaped, covered with spines exept small membranous area nearly at base of corpus bursae. Colliculum short. Tergum A8 with deep medial notch in posterior margin, about 0.4-0.5 times length of tergum.

DISTRIBUTION: Euro-Caucasian. Widely distributed in Europe (except Iceland, Ireland, Portugal, Mediterranean islands, Albania and former Yugoslavia). – Outside Europe in Caucasus, Transcaucasus and north-eastern Turkey.

PHENOLOGY: Univoltine: mid-July (late June) to mid-August (early October). Larva: late April (mid-April) to mid-May (early June). Overwintering as egg.

BIOLOGY: Larva monophagous (2). Found on the flowers of *Acer campestre* (United Kingdom: CREWE 1872a; BUCKLER 1899; Germany: DIETZE 1910, 1913; WEIGT 1988; Denmark: JUUL 1948), and *A. platanoides* (Aceraceae) (Finland: SEPPÄNEN 1970), preferring the pistils and stamens; mainly on the lower trees.

PARASITOIDS: Hymenoptera, Ichneumonidae: *Virgichneumon tergenus* (HERTING 1976).

HABITAT: Broad-leaved forests, edges of forest, hedges, scrub, gardens and parks, warm and dry localities, rocks. From 0 m up to about 600 m above sea-level in Alps (REZBANYAI-RESER & MAIER 2000), to about 1,000 m in Pyrenees (MAZEL & PESLIER 1997), and to about 1,100 m in Apennines (Mt. Majella: ZAHM pers. comm.).

SIMILAR SPECIES: This species is only remotely similar to *E. tenuiata,* distinguished by the lighter colour with usually weak yellowish tinge, by the forewing with more pointed apex and straighter costal margin, by the presence of three dark clouds in terminal area of forewing, by the transverse lines well marked, thicker and postmedial line more wavy, and by the fringe more chequered than in *E. tenuiata* (Text-fig. 22). The narrower aedeagus with smaller cornutus on the vesica in the male genitalia, and smaller spherical bursa copulatrix with shorter colliculum in the female genitalia should distinguish *E. inturbata* from *E. tenuiata*.

The ***abietaria*** species-group

This species-group was introduced by INOUE (1979) as group 'A' with first species *E. abietaria*. Later BOLTE (1990) described the similar *mutata* group. Represented in Europe by two species: *E. abietaria* and *E. analoga*. Some Asian species belong to this group also, such as: *E. rufescens* BUTLER, 1878; *E. gigantea* STAUDINGER, 1897; *E. interrubrescens* (HAMPSON, 1902) and *E. larentimima* VOJNITS, 1974. Five North American species belong here: *E. albicapitata* PACKARD, 1876; *E. helena* TAYLOR, 1906; *E. cocoata* PEARSALL, 1908; *E. mutata* PEARSALL, 1908 and *E. spermaphaga* DYAR, 1917 (BOLTE 1990).

MALE GENITALIA: Uncus long, thin, uniapical. Papillae small, short and wide, length about 0.2-0.4 times uncus length. Valva without ventral process; long, narrow, with ventral and dorsal margins almost parallel; with widely rounded apex; sacculus weakly sclerotized.

Vinculum short and relatively broad. Aedeagus large, relatively long, length about 0.7 times valva length. Vesica armed with one prominent tapered, elongated semi-cylindrical cornutus at ductus ejaculatorius base. Sternum A8 long and broad with relatively short apical horns terminating in thin prongs; basal hollow shallow; apical hollow about 0.3 times sternum length; sternum length about 0.6-0.8 times valva length.

FEMALE GENITALIA: Bursa copulatrix more or less ovate, membranous, with patches of spines at base and in posterior part. Ductus seminalis relatively short and narrow, widest at base. Colliculum collar-like, short and relatively broad. Antrum membranous or slightly sclerotized. Tergum A8 trapeziform, with well defined rows of setae on posterior margin. Anterior and posterior apophyses medium length. Papillae anales large, rounded or slightly tapered.

REMARKS: The species from this group can usually be distinguished from most other *Eupithecia* species by the long and uniapical uncus, the elongated and narrow valva, the armature of vesica, the several distinct patches of spines in the bursa copulatrix and well defined rows of setae on the posterior margin of tergum A8 in female. The *abietaria* species-group is closed to the *palpata* species-group from North America (BOLTE 1990).

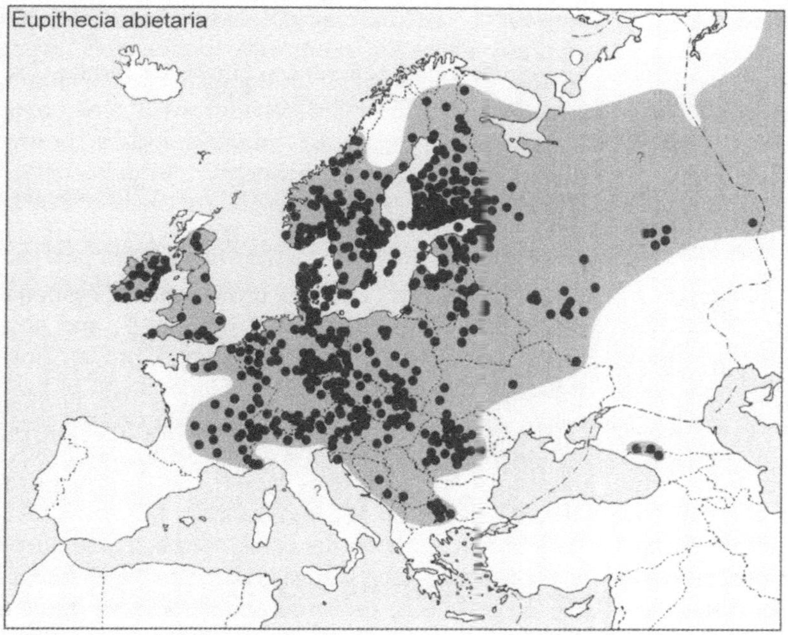

27. *Eupithecia abietaria* (GOEZE, 1781)

Phalaena (Geometra) abietaria GOEZE, 1781: Ent. Beytr. 3 (3): 439 (Europe). Syntype(s).
Phalaena (Tinea) pini RETZIUS, 1783: Genera et Species Insectorum: 50 (Europe) nec LINNAEUS, 1758 (identity: *Dendrolimus pini*, Lasiocampidae). Junior primary homonym.
Phalaena (Geometra) strobilata BORKHAUSEN, 1794: Natur. Eur. Schmett. 5: 352 (Europe). Syntype(s).
Geometra togata HÜBNER, 1817: Samml. Eur. Schmett. 5 (Geometrae 2), pl. 90: 464 (Europe). Syntype(s) lost.
Acidalia bilunulata ZETTERSTEDT, 1839: Insecta Lapponica 4 (6): 960 (Sweden?: Lappland). Syntype(s) ♀ (MZLU).

Text-figs 23-24: Diagnostic characters (indicated) of European *Eupithecia* species. Text-fig. 23: *E. abietaria* (GOEZE, 1781). Text-fig. 24: *E. analoga* DJAKONOV, 1926.

Eup(ithecia) togata var. *debrunneata* STAUDINGER, 1897: Dt. ent. Z., Iris 10 (1): 109 (Russia: Ussuri). Syntypes 2 (MNHU). Valid at subspecific rank.
Unavailable names (incorrect subsequent spellings): *togaria*: BOISDUVAL (1840); *abjetaria*: KITT (1912).

DIAGNOSIS: Wingspan 17-26 mm. Labial palpi pointed, brown; extending beyond front of eyes about 1.3-1.8 (the average 1.6) times diameter of eyes in ♂, and about 1.4-1.5 times diameter of eyes in ♀. Frons, vertex and notum light greyish brown. Antennal setae about 0.3-0.4 times width of flagellum in ♂, and about 0.2 times width of flagellum in ♀. Forewing with oblique transverse vein in areole; light liver-brown or light greyish brown; transverse lines slightly sinuate, dark brown; basal line angled near costa, with broad purplish brown outer stripe; antemedial line with dark brown costal spot; postmedial line broadened at costa, usually weakly angled on M1 and between M3 and CuA1, and with a series of short, triangular dark brown spots on the veins; terminal area purplish brown, usually ash-grey near apex or on outer side and with slightly paler blotch between M3 and CuA1; pale wavy line usually conspicuous, dentated and unbroken; discal spot large, ovate or ovate-acuminate, dark brown or black. Hindwing slightly paler; transverse lines darkish, clouded, well marked only near dorsum; postmedial line well defined along its entire length, twice curved, broad, clouded, usually with a series of short, triangular spots; terminal area darker, usually light purplish brown; pale wavy line more or less distinct; discal spot large, ovate-acuminate or ovate. Fringe pale brownish grey, spotted with dark brownish grey at vein ends. Abdomen light greyish brown, usually with ochreous tinge, and with small, dark brown dorsal and lateral spots on each segment.

MALE GENITALIA: Valva longer than 2 mm. Sternum A8 with wide basal lobes and apical horns with rounded apices.

FEMALE GENITALIA: Bursa copulatrix with patches of large spines at base and in its apical part. Anterior and posterior apophyses with expanded and rounded apices.

DISTRIBUTION: Eurasiatic. Widespread in Europe except Iceland, Iberian peninsula, Albania and Greece. Absent from all Mediterranean islands. Doubtful record from southern Italy (WOLFSBERGER 1971; PARENZAN 1994). – Outside Europe in Caucasus, western Transcaucasus (Georgia) and Urals, southern Siberia, Kamchatka, Sakhalin, Kuril Islands and Japan (LEMPKE 1969; VIIDALEPP 1978, 1996; INOUE 1979; MIRONOV 1990).

PHENOLOGY: Univoltine: early June (mid-April) to late July (mid-August). Larva: mid-July (early July) to late August (late September). Overwintering as pupa.

BIOLOGY: Larva oligophagous (1). Found in the ripening cones of *Picea abies (=excelsa)* (United Kingdom: CREWE 1872c; BUCKLER 1899), *Pinus cembra, Abies alba* (Germany: DIETZE 1910, 1913; Denmark: JUUL 1948) and *Pinus sylvestris* (Pinaceae) (Germany: WEIGT 1988; Finland: SEPPÄNEN 1970), often on the diseased or damaged trees, eating unripe seeds. It has been found feeding in the cones of *Picea jezoensis* (Pinaceae) (Japan: INOUE 1979) and *Pinus pumila* (Pinaceae) (Kamchatka peninsula: KURENTZOV & IVLIEV 1960).

PARASITOIDS: Hymenoptera, Ichneumonidae: *Diadegma* sp., *Scambus sagax* - Hymenoptera, Braconidae: *Apanteles petrovae (=dioryctriae), Baeacis abietis* (?), *Bracon pineti* (?) - Hymenoptera, Encyrtidae: *Tyndarichus scaurus* (hyperparasite), *Copidosoma* sp., *Lithomastix truncatulus* - Hymenoptera, Trichogrammidae: *Trichogramma embryophagum* (in eggs) - Diptera, Tachinidae: *Lydina aenea, Craspedhotrix aristata, Craspedhotrix* sp. (=*Elfia*) (BUCKLER 1899; BAKKE 1955; YAKOVLEV1961; STADNITZKY 1971; HERTING 1976; STADNITZKY & al. 1978).

HABITAT: Spruce, fir, pine and mixed forests and plantations, old parks. From 0 m up to about 2,800 m above sea-level in Alps (coll. ZSM).

SIMILAR SPECIES: *E. abietaria* can usually be easily distinguished from other European *Eupithecia* species by the distinctive maculation and colour. Nevertheless, this species can be confused with *E. analoga* from which it differs in the larger size and shorter labial palpi (Text-figs 23, 25; see also LEMPKE 1969; WEIGT 1988; MIKKOLA & al. 1989). The male genitalia and especially the female genitalia reliably separate *E. abietaria* from *E. analoga*, as well as from any other European *Eupithecia* species.

28. *Eupithecia analoga* DJAKONOV, 1926

Eupithecia bilunulata analoga DJAKONOV, 1926: Ezheg. gosud. Muz. N.M. Mart'anova, Minussinsk 4 (1): 48 (Russia: southern Siberia, Lake Tiberkul; Abakan range). Syntypes not traceable in ZISP, probably lost; yet the genital preparate of one syntype (♀) in ZISP. Raised to species rank by LEMPKE (1969).

Thephroclystia [sic!] *nageli* SKALA, 1929: Ent. Z., Frankfurt a/M. 42 (24): 318 (Czech Republic: Moravia, Fulnek). Syntype(s) (NLHW?).

Eupithecia analoga europaea LEMPKE, 1969: Tijdschr. Ent. 112 (2): 30, pl. 3: 7-11 (Netherlands: Naarden, Oud-Valkeveen). Holotype ♂ (ITZA). Valid at subspecific rank.

Unavailable names (misidentifications): *strobilata*: sensu HÜBNER (1813) nec BORKHAUSEN, 1794 (identity: *Eupithecia abietaria*); *bilunulata*: PROUT (1914) nec ZETTERSTEDT, 1839 (identity: *Eupithecia abietaria*). - (incorrect subsequent spellings): *bilunata*: FAVRE (1899); *analogia*: INOUE (1979); *analogo*: FIBIGER & SVENDSEN (1981).

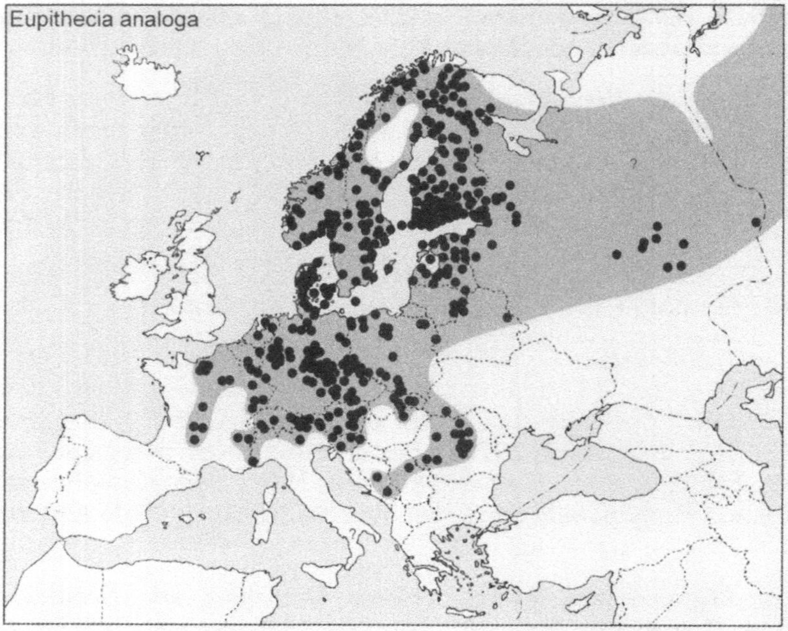

DIAGNOSIS: Wingspan 15-24 mm. Labial palpi pointed, brown; extending beyond front of eyes about 1.7-2.1 (the average 1.9) times diameter of eyes in ♂, and about 1.9-2.5 (the average 2.2) times diameter of eyes in ♀. Frons and vertex ash-grey, sometimes with brownish tinge. Antennal setae about 0.3-0.4 times width of flagellum in ♂, and about 0.2-0.3 times width of flagellum in ♀. Notum ash-grey, sometimes with brownish tinge. Forewing light brownish grey; transverse lines slightly sinuate, dark brown; basal line angled near costa, outer side with broad, clouded, purplish grey or purplish brown stripe; antemedial line with dark brown costal spot; postmedial line with fuscous costal spot, usually weakly angled on M1 and between M3 and CuA1, and with a series of short, triangular dark brown spots on the veins; terminal area purplish brown or purplish grey, outer part ash-grey with slightly paler blotch between M3 and CuA1; pale wavy line usually conspicuous, dentated and unbroken or obsolete; discal spot dark brown or black, large, rounded but sometimes ovate. Hindwing slightly paler; transverse lines darkish, clouded, well indicated only near dorsum; postmedial line well defined along its entire length, clouded, without triangular spots on the veins; terminal area from light to dark brownish grey, sometimes with purplish tinge; pale wavy line usually inconspicuous; discal spot darkish, small, ovate-acuminate or ovate. Fringe grey, spotted with dark brownish grey at vein ends. Abdomen ash-grey, sometimes with brownish tinge, and usually with dorsal row and lateral rows of small, dark brown spots; second segment darker, ochreous brown or dark brown.

MALE GENITALIA: Valva shorter than 2 mm. Sternum A8 with narrower basal lobes than in *E. abietaria* and with tapered apical horns.

FEMALE GENITALIA: Bursa copulatrix with patches of small spines: one large belt-shaped patch at the middle and one rounded patch near colliculum. Anterior and posterior apophyses with tapered apices.

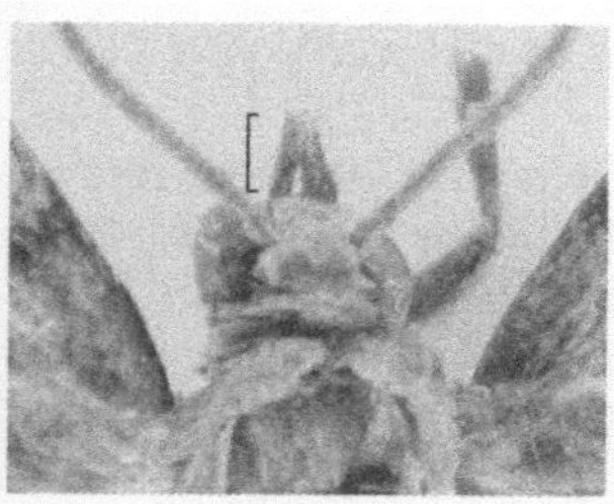

25

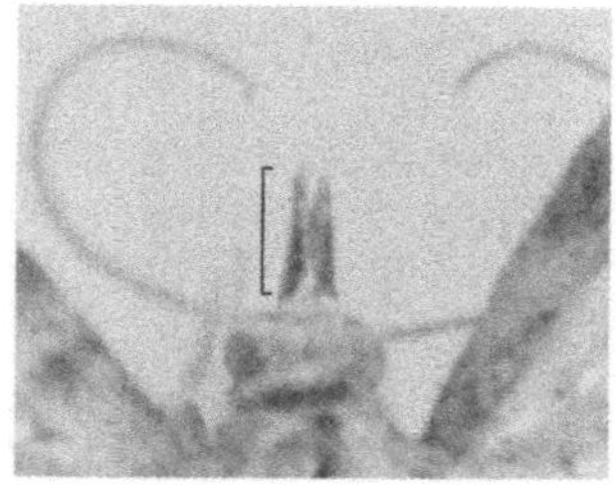

26

Text-figs 25-26: Head parts (indicated) of European *Eupithecia* species. Text-fig. 25: Head parts of *E. abietaria* (GOEZE, 1781). Text-fig. 26: Head parts of *E. analoga* DJAKONOV, 1926 (Photo Sir ANTHONY GALSWORTHY).

DISTRIBUTION: Eurasiatic. Widely distributed in Europe from France to Urals except southern Europe and Mediterranean islands, but recorded on mountains of the Balkan peninsula. Doubtful record from southern Italy (WOLFSBERGER 1971; PARENZAN 1994). – Outside Europe eastward to Altai Mts in southern Siberia.

PHENOLOGY: Univoltine: mid-May (early May) to late June (late July). Larva: mid-July (late June) to late August (early September). Overwintering as pupa.

BIOLOGY: Larva monophagous (2). Found in the galls of chermes *Sacchiphantes (Adelges) abietis* LINNAEUS, *S. (Chermes) viridis* RATZ. and *Aphrastasia pectinatae* CHOL. (=*Chermes coccineus* CHOL.), commonly known as 'false cones' of *Picea abies (=excelsa)* (Pinaceae) (Germany: DIETZE 1910, 1913; WEIGT 1988; Denmark: JUUL 1948; Finland: SEPPÄNEN 1970). Nevertheless, forest protection specialists have reported that larvae of this species feed solely within spruce cones (Sweden: SPESSIVTSEFF 1924; Slovakia: PATOČKA 1960; Russia: YAKOVLEV1961; STADNITZKY 1971; STADNITZKY & al. 1978).

PARASITOIDS: Hymenoptera, Ichneumonidae: *Pimpla instigator* - Hymenoptera, Encyrtidae: *Lithomastix truncatulus* (BAKKE 1955; GYÖRFI 1956; HERTING 1976).

HABITAT: Spruce forests, forest fringes, parks. From 0 m up to about 1,400-1,500 m above sea-level in Balkan peninsula (TOMIČ & al. 1990), and to about 1,600-1,700 m in Alps (WOLFSBERGER 1965; FORSTER & WOHLFAHRT 1981; WEIGT 1988).

SIMILAR SPECIES: *E. analoga* can be easily confused with *E. abietaria* but usually distinguished from it by the smaller size, the longer labial palpi, the more rounded discal spot on the forewing and by the absence of dark triangular spots on the veins along the postmedial line on the hindwing, as well as the dark brown tergum A2 (Text-figs 24, 26; see also LEMPKE 1969; WEIGT 1988; MIKKOLA & al. 1989).

REMARKS: I have not discovered any specimen of *E. analoga* in the collection of A.M. DJAKONOV (ZISP). However, there are genitalia of one of the syntypes of *E. analoga* in a glass test-tube, with label 'Microprep. No. 304♀, *Eupithecia abietaria* Goeze, Altai, Minussinsk'. This preparation number is correctly indicated by DJAKONOV for *E. analoga* in the original description.

The *linariata* species-group

Introduced by WEIGT (1988). Represented in Europe by four species: *E. linariata, E. pulchellata, E. pyreneata* and *E. laquaearia*. In my opinion the following species belong to this group as well: *E. albidulata* STAUDINGER, 1892 from Caucasus; *E. mirificata* BRANDT, 1938 from Iran; *E. linariatoides* MIRONOV, 1989 from Central Asia, and three species from North America: *E. johnstoni* McDUNNOUGH, 1945; *E. dichroma* McDUNNOUGH, 1945 and *E. rindgei* McDUNNOUGH, 1949.

MALE GENITALIA: Uncus short, biapical. Papillae small, length about 0.5-0.7 times uncus length. Valva without ventral process, with rounded ventral margin, slightly curved dorsal margin and broadly rounded apex; sacculus slightly sclerotized. Vinculum short, narrow and rounded. Aedeagus relatively short and thin, length about 0.6-0.8 times valva length. Vesica armed with one elongated, thin and slightly sclerotized tube. Sternum A8 small, with two long, narrow, tapered rods and with shallow basal hollow.

FEMALE GENITALIA: Bursa copulatrix large, elliptic or ovate, membranous, with multitudinous microscopic slightly sclerotized spinules. Ductus seminalis short and narrow, attached to bursa copulatrix posteriorly. Colliculum absent. Antrum membranous, long and narrow. Tergum A8 almost square. Anterior and posterior apophyses short and thin. Papillae anales short and wide.

REMARKS: The male genitalia of species of the *linariata* group can be distinguished from those of other European species by the relatively short and thin aedeagus, with one long, thin and slightly sclerotized tube in the vesica and the shape of sternum A8, with two long, narrow and apically tapered rods; and the female genitalia can be distinguished by the large, elliptic, membranous bursa copulatrix, with a multitude of minute spinules; absence of colliculum; long, narrow and membranous ductus bursae.

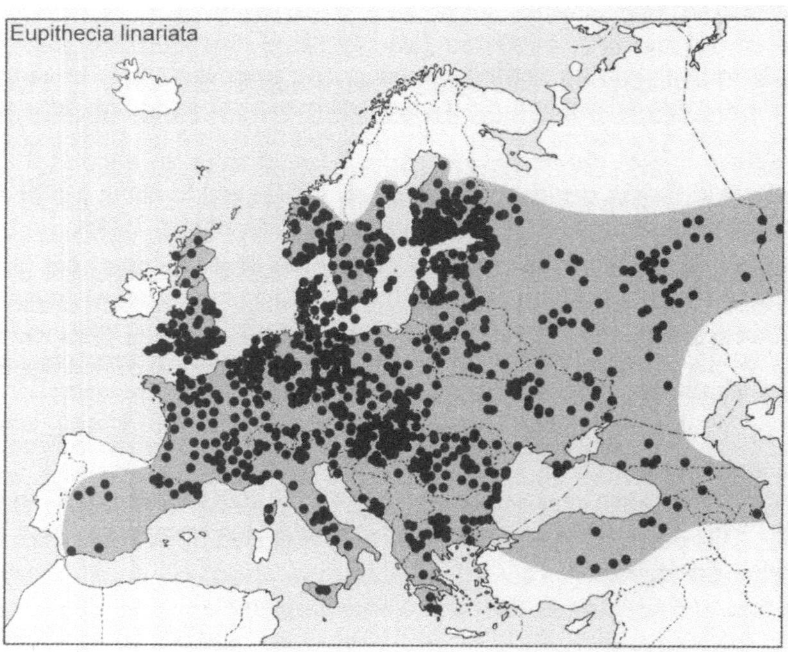

29. *Eupithecia linariata* (DENIS & SCHIFFERMÜLLER, 1775)

Geometra linariata DENIS & SCHIFFERMÜLLER, 1775: Syst. Verz. Schmett. Wien. 2: 113 (Austria: Vienna district). Syntype(s) lost.

Unavailable names (infrasubspecific): *aestiva*: DIETZE (1910); *nigrofasciata*: DIETZE (1913); *approximata*: LEMPKE (1951); *reducta*: LEMPKE (1951). - (incorrect subsequent spellings): *linariaria*: BORKHAUSEN (1794); *linaria*: BOISDUVAL (1840).

DIAGNOSIS: Wingspan 14-23 mm. Labial palpi black or dark grey. Frons black to whitish grey with black transverse band between antennal bases. Vertex whitish grey. Notum pale yellowish grey. Forewing yellowish ochreous to basal line, with dark grey costal margin; warm brownish ochreous between basal and antemedial lines; antemedial line slightly sinuate, oblique, right-angled near costa, bordered by narrow white stripe; medial area ash-grey, mottled and irrorate with white scales, often with indistinct greyish white medial lines; postmedial line evenly curved, on inner side with short black dashes on the veins, bordered on outer side by narrow white stripe and wider pale greyish yellow stripe beyond it; terminal area with two blackish clouds (one between M1 and M3, the other between CuA2 and A); inner part of terminal area warm brownish ochreous; outer part of terminal area ash-grey with yellowish tinge; white wavy line continuous, sinuate or very multiply indented; discal spot small, ovate-oblong, intensely black. Hindwing light grey; basal area dark ash-grey along dorsum; medial area with transverse lines conspicuous only near termen; pale yellowish grey double stripe between medial and terminal areas usually distinct; terminal area with small yellowish ochreous blotch; pale wavy line rather indistinct, much indented; discal spot minute, round or ovate, darkish. Fringe light ochreous grey, spotted with dark grey at vein ends. Abdomen whitish grey with yellowish tinge, usually with dorsal row of black dots and with blackish scales on the ending of last segment; first segment whitish grey or pale ochreous; second segment blackish brown or dark ochreous brown.

MALE AND FEMALE GENITALIA: See species-group diagnosis.

DISTRIBUTION: Euro-Siberian. Widely distributed in Europe except northern Scandinavia, Iceland, Ireland, Portugal, Balearics, Sardinia, Malta and Crete. – Outside Europe in Asia Minor, Caucasus, Transcaucasus (WARDIKJAN 1985), Urals, northern Kazakhstan, western and southern Siberia, Altai Mts (MIRONOV 1989) and northern Mongolia (VOJNITS 1977).

PHENOLOGY: Uni- or bivoltine: late May (late April) to early July (mid-July); late July (mid-July) to early September (late September). In the north univoltine. Larva: mid-June (early June) to late July (early August); mid-August (mid-July) to early October; in the north from July to September. Overwintering as pupa.

BIOLOGY: Larva almost monophagous (1). Found in the flowers and seed capsules of *Linaria vulgaris* (Scrophulariaceae) (United Kingdom: CREWE 1861c; BUCKLER 1899; Corsica: MABILLE 1869; France: MILLIÈRE 1870; Germany: DIETZE 1872, 1910, 1913; WEIGT 1988; WEGNER pers. comm.; Denmark: JUUL 1948; Finland: SEPPÄNEN 1970; Russia: original data), often obtainable in quantity by gathering capsule-bearing stems. Recorded also on *Digitalis purpurea* (Corsica: MABILLE 1869) and cultivated *Antirrhinum* (Scrophulariaceae) (SCOBLE 1999).

PARASITOIDS: Hymenoptera, Ichneumonidae: *Diadegma armillatum*, *Campoletis* sp., *Heteropelma amictum*, *Mesochorus tachypus*, *Scambus brevicornis*, *Gregopimpla inquisitor*, *Agrypon clandestinum*, *T. flexorius* - Hymenoptera, Braconidae: *Aleiodes*

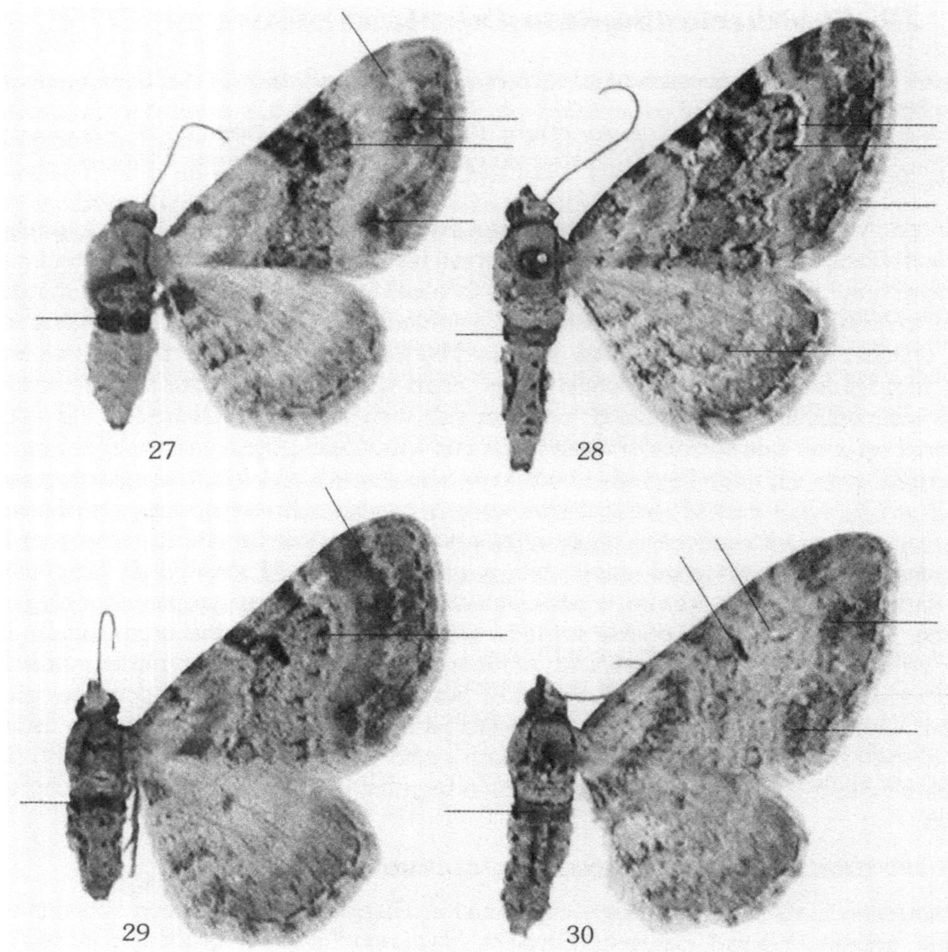

Text-figs 27-30: Diagnostic characters (indicated) of European *Eupithecia* species. Text-fig. 27: *E. linariata* (DENIS & SCHIFFERMÜLLER, 1775). Text-fig. 28: *E. pulchellata* STEPHENS, 1831. Text-fig. 29: *E. pyreneata* MABILLE, 1871. Text-fig. 30: *E. laquaearia* HERRICH-SCHÄFFER, 1848.

(=Rogas) circumscriptus, *Apanteles* sp., *Lissogaster (=Microgaster) globatus*, *M. tibialis*, *Microplitis tuberculifer* (BUCKLER 1899; THOMPSON 1945; SHENEFELT 1973, 1975; HERTING 1976; original data).

HABITAT: All places where its food-plant abounds: edges of forest preferably on sandy soil, sandy slopes, sandhills, dunes, sunny borders of fields, sunny roadsides and railway embankments, ditch-banks and gravel pits, coasts, warm and dry uncultivated areas, gardens, rocks, stony slopes and screes. From 0 m up to about 1,600-1,700 m above sea-level in Apennines (Mt. Majella: ZAHM pers. comm.), to about 1,700-1,800 m in Pyrenees (DANTART & ROCHE 1992), to about 1,800-1,900 m in Balkan peninsula (coll. FIBIGER), to about 2,000-2,100 m in Alps (VORBRODT 1928, 1931; HELLMANN & al. 1999), and to about 2,200 m in central Spain (coll. FIBIGER).

SIMILAR SPECIES: Usually constant in colour and breadth of medial area on the forewing. Similar to *E. pyreneata* and *E. pulchellata* with which it could be confused. *E. pyreneata* is usually paler, especially medial area of forewing, antemedial line often only evenly curved near costa, two dark clouds in terminal area on the forewing smaller and shorter, dark band at abdominal base much paler than in *E. linariata*. The second allied species *E. pulchellata* has sinuate antemedial line and more curved and wavy postmedial line on the forewing, fringe more distinctly chequered and is on the whole brighter coloured than *E. linariata* (Text-fig. 27). No clear differences can be seen in the male and female genitalia from those of the two closely related compared species.

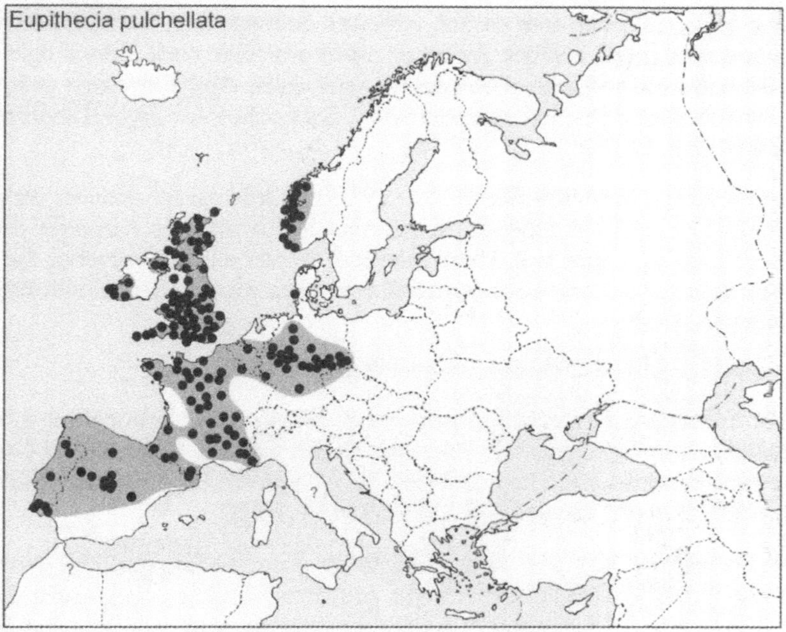

30. *Eupithecia pulchellata* STEPHENS, 1831

Eupithecia pulchellata STEPHENS, 1831: Illust. Br. Ent. (Haustellata) 3: 280 (United Kingdom: England, Coombe and Birch Woods). Syntypes (BMNH).

Eupithecia pulchellata var. *hebudium* SHELDON, 1899, The Entomologist's Rec. J. Var. 11 (12): 344 (United Kingdom: Scotland, Inner Hebrides). Syntypes.

Eupithecia pulchellata f. *intermedia* DIETZE, 1913 (nec DIETZE, 1872): Biol. Eupithecien 2: 36; ibidem (1910) 1, pl. 70: 97 (Germany: Taunus). Syntype(s) (MNHU). Valid at subspecific rank.

(Eupithecia) pulchellata f. *iberica* DIETZE, 1913: Biol. Eupithecien 2: 36; ibidem (1910) 1, pl. 70: 100 (Spain: Castilia). Syntype(s) (MNHU). Valid at subspecific rank.

Unavailable names (infrasubspecific): *reducta*: BASTELBERGER (1908); *approximata*: LEMPKE (1951). - (incorrect subsequent spellings): *pulchellaria*: DOUBLEDAY (1849); *digitaliata*: DIETZE, 1910 (nec DIETZE, 1875); *digitaliaria*: DIETZE, 1913 (nec DIETZE, 1872); *pulchellalla*: GANEV (1985).

DIAGNOSIS: Wingspan 18-24 mm. Labial palpi black to dark grey. Frons and vertex greyish white with black transverse band between antennal bases. Notum pale brownish ochreous. Forewing pale ochreous brown; basal area yellowish ochreous with dark grey costal margin and distinct, oblique, multiply indented basal line; outer part of basal area warm brownish ochreous; antemedial line angulated or much indented, curved or right-angled near costa, bordered by narrow white stripe along inner edge; medial area ash-grey, mottled and irrorate with white scales, usually darker towards costa, with fine, rather inconspicuous whitish medial line and with lighter, pale ochreous streak under discal spot; postmedial line multiply wavy, evenly curved, forming a series of inward-pointing dark dashes on the veins, and bordered by narrow white stripe and pale greyish yellow stripe beyond it; terminal area with rather warm brownish ochreous tinge, darkened near costa, with two blackish clouds (medial and tornal); white wavy line continuous, sinuate or gently indented, usually with three white expansions (costal, medial and tornal); discal spot small, ovate or narrow and elongated, intensely black. Hindwing light grey with conspicuous dark transverse lines near dorsum; postmedial line usually forming a row of dark dashes on the veins; terminal area darker, with well defined, fine, much indented, whitish wavy line; discal spot small, distinct, round or ovate-orbicular, dark. Fringe light ochreous grey, very distinctly spotted with dark grey at vein ends. Abdomen pale yellowish grey with dorsal row of black dots; first segment paler, light ochreous; second segment brownish grey, darker than others.

The continental subsp. ***intermedia*** DIETZE, 1913 is a little darker and less variegated on the whole than the nominate subspecies; distinguished from it by narrower dark medial area with more indistinct light (not white) marginal stripes on the forewing, the hindwing slightly darker with brown ochreous terminal area near tornus, and fringe distinctly less chequered as well.

MALE AND FEMALE GENITALIA: See species-group diagnosis.

DISTRIBUTION: European. Nominate subspecies in British Isles. Continental subsp. *intermedia* in western Norway, Denmark, Portugal, Spain, France, Belgium and Germany. In other European countries probably replaced by *E. pyreneata*. – Outside Europe questionably recorded from Rif range in Morocco (RUNGS 1981).

PHENOLOGY: Univoltine: May and June. The subsp. *intermedia*: late May (early May) to late July (early August). In southern Europe probably bivoltine: from early July to late September. Larva: late June to early August; elsewhere on continent: late July (early July) to mid-August (early September). Overwintering as pupa, occasionally twice.

BIOLOGY: Larva monophagous (2). Found in the flowers of *Digitalis purpurea* (Scrophulariaceae), feeding on the stamens and closing the mouth of the flower with silk (United Kingdom: CREWE 1864a, 1865; BUCKLER 1899). Larva of subsp. *intermedia* found in the flowers and seed capsules of *Digitalis purpurea, D. grandiflora* (France: MILLIÈRE 1870; GOOSSENS 1871; Germany: DIETZE 1871b, 1872, 1910, 1913; GELBRECHT pers. comm.) and *D. thapsi* (Portugal: CANDIDO MENDES d'AZEVEDO 1903).

PARASITOIDS: Hymenoptera, Ichneumonidae: *Enytus apostata*, *Deuterolabops eupitheciae*, *Venturia deficiens*, *Platylabus tricingulatus* - Hymenoptera, Braconidae: *Apanteles praepotens (=sericeus)* (BUCKLER 1899; THOMPSON 1945; SHENEFELT 1972; HERTING 1976).

HABITAT: Gardens, plantations, deciduous and mixed forests, forest clearings, forest edges, warm slopes and among rocks. From 0 m up to about 200 m above sea-level in United Kingdom (BARRETT 1904). Recorded to about 800 m in Alps (WEIGT 1988), and to about 2,400 m in Pyrenees (MAZEL & PESLIER 1997).

SIMILAR SPECIES: The nominate British subspecies can be confused only with *E. linariata*. Distinguished by the usually larger wingspan, the more indented ante- and postmedial lines, the more obvious discal spot and by the presence of three whitish spots along wavy line on the forewing, as well as by the paler hindwing with less distinct pale medial band, the more obviously chequered fringe, and by the less dark tergum A2 compared with *E. linariata* (Text-fig. 28). The male and female genitalia are typical for species of the *linariata* group and it is practically impossible to differentiate them from those in other close relatives.

REMARKS: A species with rather little variation. The lighter, greyer, less ochreous brown f. *hebudium* SHELDON, 1899 occurs in western Wales, Outer Hebrides and western Ireland (GOATER & al. 1981).

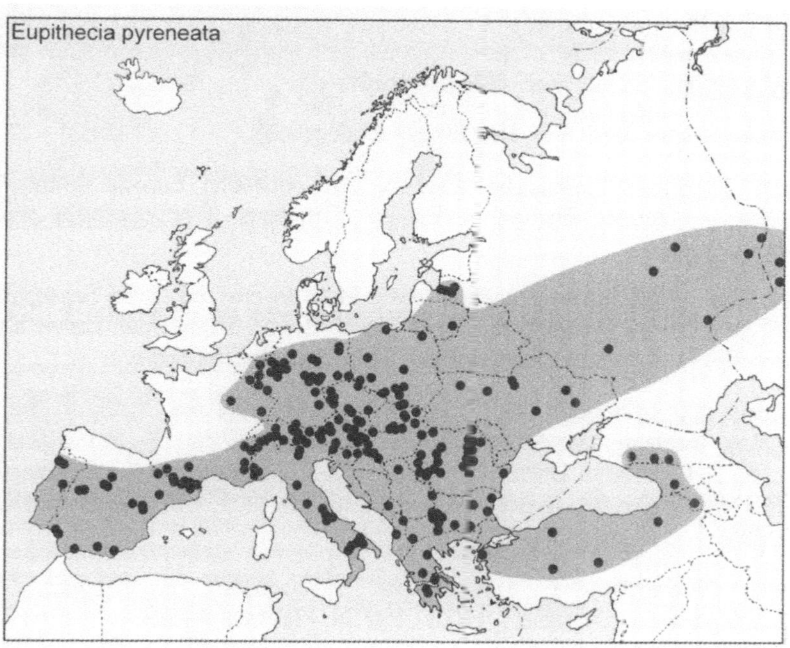

31. *Eupithecia pyreneata* MABILLE, 1871

Eupithecia pyreneata MABILLE, 1871: Petites Nouv. Ent. 1 (42): 168 (France: Pyrenees). Lectotype ♀ (coll. HERBULOT in ZSM, examined).
Eupithecia linariata digitaliaria DIETZE, 1872: Stett. Ent. Z. 33 (4-6): 194 (Germany). Syntype(s) (MNHU).
Eupithecia pulchellata digitaliata DIETZE, 1875b: Stett. Ent. Z. 36 (4-6): 244 (Germany: Höllensteig; St.Blasien). Syntypes (MNHU).
Eupithecia pulchellata var. *granadensis* BUBAČEK, 1926: Verh.. Zool.-Bot. Ges. Wien 74/75: (9) (Spain: Granada). Syntypes (ZFMK?). Valid at subspecific rank.
Unavailable names (incorrect subsequent spellings): *grenadensis*: WEHRLI (1927); *pyrenaeata* OSTHELDER (1929).

DIAGNOSIS: Wingspan 15-19 mm. Labial palpi dark grey. Frons and vertex whitish grey, usually with dark grey transverse band between antennal bases. Notum pale ochreous grey. Forewing: basal area yellowish ochreous with ash-grey costal margin; warm brownish ochreous ground colour beyond basal line; antemedial line slightly sinuate, evenly curved or sometimes right-angled near costa, bordered by narrow whitish stripe; medial area ash-grey, mottled and irrorate with white scales, also with two indistinct whitish medial lines; postmedial line evenly curved near costa, with inward-pointing short dark dashes on the veins, bordered by narrow white stripe and broader pale greyish yellow stripe beyond it; terminal area with two blackish clouds (one between M1 and M3, the other between CuA2 and A); inner part of terminal area warm brownish ochreous; outer part of terminal area ash-grey, mottled and irrorate with white scales; white wavy line unbroken, sinuate or multiply indented; discal spot small, ovate-oblong, black or sometimes obsolete. Hindwing light grey; basal and other transverse lines grey, well defined only along dorsum; terminal area usually darker, with yellowish ochreous tinge near tornus; pale wavy line much indented, inconspicuous; discal spot minute, round or ovate, darkish. Fringe pale ochreous grey, spotted with dark grey at vein ends. Abdomen whitish grey with yellowish tinge, usually with dorsal row of dark brown dots and with dark grey scales on the posterior margin of last segment; first segment pale ochreous grey; second segment light brown, darker than other segments.

MALE AND FEMALE GENITALIA: See species-group diagnosis.

DISTRIBUTION: European-Westasiatic. Central and southern Europe from Portugal to southern Urals. – Outside Europe in Turkey, Caucasus, Transcaucasus and northern Kazakhstan.

PHENOLOGY: Uni- or bivoltine: late June (early May) to late July (mid-September). In the south in two generations probably from late March to mid-September. Larva: early August (mid-July) to mid-August (early September). Overwintering as pupa.

BIOLOGY: Larva monophagous (2). Found in the seed capsules, rarely in the flowers of *Digitalis lutea* (France: GOOSSENS 1871; MABILLE 1872), *D. grandiflora* (France: GOOSSENS 1871; Germany: DIETZE 1910, 1913), *D. lanceolata* (Germany: GELBRECHT pers. comm.), *D. ferruginea* (eastern Mediterrain: WEIGT 1988) and *D. obscura* (Spain: BUBAČEK 1926).

PARASITOIDS: Hymenoptera, Braconidae: *Ascogaster canifrons*, *Lissogaster (=Microgaster) globata* (SHENEFELT 1973, 1973a; HERTING 1976).

HABITAT: Deciduous and mixed forests, edges of forests, hedges, scrub and scrub-covered meadows, warm and dry slopes, steppes, rocks, stony slopes and screes. From 0 m up to about 2,200 m above sea-level in Alps (VORBRODT & MÜLLER-RUTZ 1914), and from 600 m up to about 2,800-2,900 m in Pyrenees (WEHRLI 1927).

SIMILAR SPECIES: *E. pyreneata* can be confused with *E. linariata* and *E. pulchellata intermedia*, but is usually paler than both these species (Text-fig. 29). The male and female genitalia are not clearly different from those of other *linariata* group species.

REMARKS: The moths from Sierra Nevada, described as var. *granadensis* BUBACEK, 1926, are distinguished from typical *E. pyreneata* by more greyish colour, darker rusty brown bands (one between basal and antemedial lines, and other along inner part of terminal area) on the forewing, less wavy lines, especially postmedial line, and broader medial area on the forewing. I consider them to be the high-mountain form of *E. pyreneata* in spite of well expressed external differences.

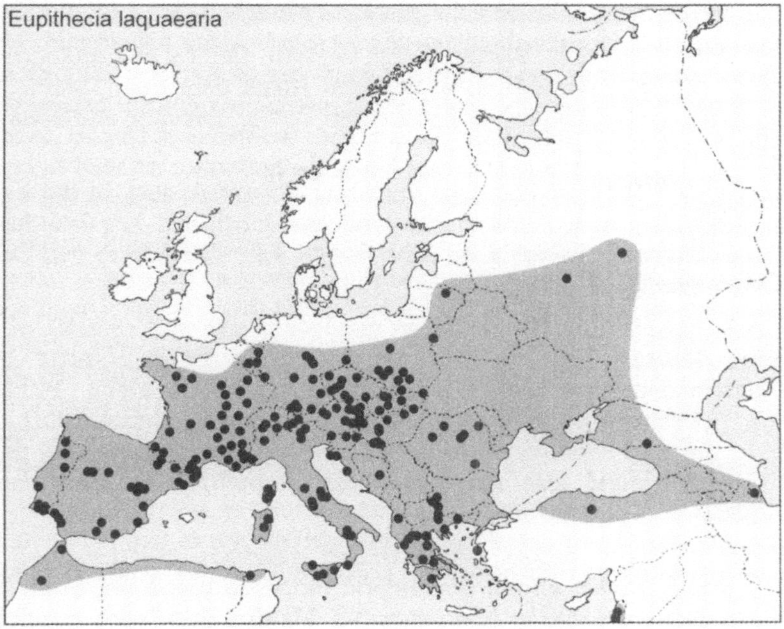

32. *Eupithecia laquaearia* HERRICH-SCHÄFFER, 1848

Eupithecia laquaearia HERRICH-SCHÄFFER, 1848: Syst. Bearb. Schmett. Eur. 3: 124, 139, pl. 29: 181, 182 (Germany: Regensburg). Syntype(s) lost.

Eupithecia perfidata MANN, 1855: Verh. Zool.-bot. Ver. Wien 5: 547 (Corsica). Syntype(s) (NHMW?).

Eupithecia merinata GUENÉE, 1858: in BOISDUVAL & GUENÉE, Hist. nat. Insectes (Spec. gén. Lépid.) 10 (2): 326 (Corsica and France). Syntypes 2♂ 1♀.

Eup(ithecia) laquaearia f. *istriaca* DIETZE, 1910: Biol. Eupithecien 1, pl. 69: 83, 87; ibidem (1913), 2: 37 (Croatia: Istria, Pola). Syntypes (MNHU). Valid at subspecific rank.

Unavailable names (misidentification): *subumbrata*: sensu TREITSCHKE (1828) nec DENIS & SCHIFFERMÜLLER, 1775 (identity: *Eupithecia subumbrata*). - (incorrect subsequent spellings): *laquearia*: GUENÉE (1858); *merinuta*: MABILLE (1868); *lacquaearia*: TURATI (1913); *laquaaria*: KOSTROWICKI (1953).

DIAGNOSIS: Wingspan 17-19 mm. Labial palpi grey with light apices. Frons and vertex greyish white. Notum light grey with brown ochreous tinge. Forewing whitish brown-grey; transverse lines fine, rather indistinct, brown; basal area proximally brownish grey and distally reddish brown; antemedial line almost straight or slightly sinuate, sharply or right-angled, broadened and darker near costa; medial area not darker than rest of wing, except slightly darker area between antemedial and first medial lines; postmedial line sinuate, more or less evenly curved, well marked on the veins as a series of dark brown dashes; terminal area reddish brown in inner half and brown in outer half, without dark clouds; whitish wavy line usually clouded, lacking the tornal dot and in some specimens obsolete; discal spot relatively large, ovate or ovate-oblong, black or dark brown. Hindwing paler, whitish brownish grey; transverse lines evident only along dorsum and postmedial line sometimes forming a row of elongate dark dashes on the veins; terminal area usually slightly darker, pale wavy line inconspicuous; discal spot minute, distinct, elongate or comma-like. Fringe light brownish grey, spotted with brownish grey at vein ends. Abdomen light grey, usually with brown ochreous tinge, and with dark lateral stripes.

MALE AND FEMALE GENITALIA: See species-group diagnosis.

DISTRIBUTION: West-Palaearctic. Widespread in western, Central and southern Europe. – Outside Europe separate populations in northern Africa (Morocco, Tunisia), Caucasus, eastern Transcaucasus (Talysh Mts), north of Asian Turkey, Lebanon and Syria (WILTSHIRE 1936, 1939).

PHENOLOGY: Uni- or bivoltine: late June (mid-June) to mid-August. In the south in two generations: early May (early April) to mid-June (late June); mid-July (late June) to mid-September (mid-October). Larva: early June to late July; late July to mid-October (late October). Overwintering as pupa.

BIOLOGY: Larva oligophagous (1). Found on the flowers and seed capsules of *Odontites lutea* (Corsica: MABILLE 1868, 1869; France: MILLIÈRE 1870; MABILLE 1872), also *Euphrasia stricta* (Germany and Switzerland: DIETZE 1871b, 1872, 1974a, 1975a; FAVRE 1905), *Hypericum perforatum* (Clusiaceae) (Germany: DIETZE 1910, 1913) and occasionally in the seed capsules of *Rhinanthus minor* (Scrophulariaceae) (Switzerland: VORBRODT 1928). Found also on *Hypericum serpyllifolium* (Clusiaceae) (Syria: WILTSHIRE 1936).

HABITAT: Edges of forest, hedges, scrub-covered meadows, limestone and sandstone slopes, stony slopes, rocks, screes, gorges and mountain meadows. From 0 m up to about 1,100 m above sea-level in Apennines (Mt. Majella: ZAHM pers. comm.), to about 1,800-1,900 m in Alps, Pyrenees and central Spain (VORBRODT 1928, 1931; WOLFSBERGER 1965; FORSTER & WOHLFAHRT 1981; coll. ZMUC; coll. FIBIGER; coll. GELBRECHT).

SIMILAR SPECIES: This species is only remotely similar to *E. linariata*, and can easily be distinguished by the considerably lighter medial area lacking white marginal stripes, the less warm brownish ochreous basal and terminal areas and by the absence of two large blackish terminal clouds on the forewing and the paler hindwing as well (Text-fig. 30). No distinct differences can be seen in the male and female genitalia between *E. laquaearia* and its relatives in the *linariata* species-group. Adults of *E. laquaearia* also resemble *E. analoga*, but are readily distinguished by the shorter labial palpi in both male and female, the smaller discal spot and evenly curved postmedial line on the forewing.

The ***ultimaria*** species-group

Group introduced by MIRONOV (1990). This distinctive species-group is represented in Europe by three species only: *E. ultimaria*, *E. usbeca* and *E. minusculata*. There are several Asian species in this species-group also: *E. inconspicuata* BOHATSCH, 1893; *E. opistographata* DIETZE, 1906; *E. djakonovi* STSHETKIN, 1956; *E. dominaria* STSHETKIN, 1956; *E. inquinata* SCHÜTZE, 1961; *E. mima* MIRONOV, 1989 and *E. vicina* MIRONOV, 1989. One African species: *E. depasta* VOJNITS, 1994 (Gambia) and two North American species, *E. miserulata* GROTE, 1863 and *E. niveifascia* (HULST, 1898) can also be included in the *ultimaria* species-group.

MALE GENITALIA: Uncus short, narrow to broad, biapical. Papillae various in size and shape, length about 0.5-1.5 times uncus length. Valva short, medium width to narrow, usually widest at base, without ventral process (in some species valva with slight medial bulge or slight medial hollow in ventral margin), with curved ventral margin, and usually narrowly rounded apex; sacculus weakly sclerotized. Vinculum short, broad or very broad.

Aedeagus thin to thick, relatively long, about 0.7-1.2 times length of valva, about 3.0-5.0 times longer than medial width. Vesica usually wrinkled, sclerotized; armed with a numerous minute spinules, with one to four horn-like cornuti or without these, usually with one elongated V-shaped cornutus at ductus ejaculatorius base. Sternum A8 with two long, relatively narrow, usually symmetrical rods, heavily sclerotized to apex and connected by narrow band at base or connected by broad membranous area along their entire length (in some species sternum A8 with broad, asymmetrical and apically incurved rods).

FEMALE GENITALIA: Bursa copulatrix membranous or posteriorly sclerotized, pear-shaped, nearly globular or irregular; a greater or lesser area of surface covered with spines, the remainder without spines or with two or three large patches of small spines and in some species the apical part also armed with larger spines and/or with a narrow band with marginal spines. Ductus seminalis usually short and narrow, attached to medial part of corpus bursae on right side. Colliculum collar-like, short and narrow. Antrum membranous or sclerotized. Anterior and posterior apophyses short and thin, not expanded at apices. Spurs of anterior apophyses usually expanded at the apex. Tergum A8 various in shape.

REMARKS: This group is close to the nearctic *palpata* species-group (MCDUNNOUGH 1944, 1949; BOLTE 1990) and to the *undata* and *venosata* species-groups, due to similarities with sternum A8 which consists of two long rods connected by narrow band at base.

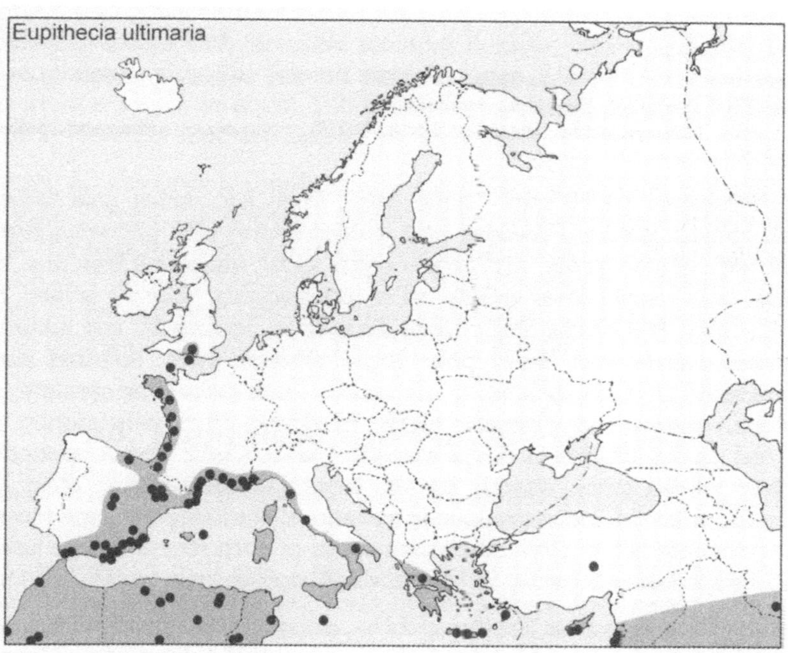

33. *Eupithecia ultimaria* BOISDUVAL, 1840

Eupithecia ultimaria BOISDUVAL, 1840: Genera Index meth. eur. Lepid.: 210 (France: Massilia, Galloprov [=Marseille]). Syntype(s).
Eupithecia tamariciaria HEYDENREICH, 1851: Syst. Verz. Eur. Schmett. 57 (Europe). Syntype(s).
Tephroclystia tornifascia ROTHSCHILD, 1913: Novit. zool. 20: 132 (Algeria: Sahara, between Onargla and El-Golen). Holotype ♀ (BMNH).

Text-figs 31-32: Diagnostic characters (indicated) of European *Eupithecia* species. Text-fig. 31: *E. ultimaria* BOISDUVAL, 1840. Text-fig. 32: *E. minusculata* ALPHÉRAKY, 1882.

Eupithecia golearia LUCAS, 1940: Lambillionea (1939) 12: 206 (Algeria: El Goléa). Lectotype ♂ (coll. HERBULOT in ZSM, examined). New junior synonym.

Unavailable names (infrasubspecific): *dilucida*: DIETZE (1910). - (incorrect subsequent spelling): *ultimata* MILLIÈRE (1870).

DIAGNOSIS: Wingspan 14-19 mm. Labial palpi, frons and vertex pale brownish grey. Notum pale greyish brown. Forewing pale greyish brown with ochreous tinge, mottled and irrorate with whitish scales, with brown costal spots; transverse lines fine, slightly sinuate, brown; two medial lines usually more conspicuous, sharply angled near costa beyond discal spot; first medial line crosses discal spot; postmedial and subterminal lines sharply curved outside with broad paler stripe between them; terminal area narrow, sharply broadened to costa; pale wavy line usually weak defined or obsolete; discal spot large, oblique, narrow and elongate, black. Hindwing often only slightly paler than forewing, and always whitish at costal margin; CuA1, CuA2 and A spotted with dark brown; lighter medial stripe, sharply curved, more well defined than other lines; paler wavy line much indented, inconspicuous or sometimes obsolete; discal spot large, distinct, narrow and elongate but in some specimens ovate or even round. Fringe pale brownish grey, distinctly spotted with brown at vein ends. Abdomen pale greyish brown.

MALE GENITALIA: Uncus narrow. Papillae clublike, elongated, length about 1.5 times longer than length of uncus, with sclerotized apical area, covered by short setae. Valva with shallow hollow in ventral margin nearly at base. Vinculum with shallow medial hollow. Vesica armed with numerous minute spinules; with one elongated V-shaped cornutus at ductus ejaculatorius base; one wide, blunt horn-like cornutus with basal wrinkles and one small, blunt cornutus near apex. Sternum A8 two long, narrow, symmetrical rods, with medial expansion, sclerotized at apex and connected by very narrow band at base.

FEMALE GENITALIA: Bursa copulatrix membranous, pear-shaped, relatively large; basal 0.7

covered with uniform sized spines; with row of large spines between ductus seminalis base and colliculum; with long band of smaller spines opposite ductus seminalis on left side. Ductus seminalis widest and sclerotized at base. Colliculum strongly inclined to right. Antrum membranous. Tergum A8 trapeziform, with rounded apical corners and deep medial hollow in posterior margin, about 0.4-0.5 length of tergite.

DISTRIBUTION: Mediterranean-Turanian. In Europe occurs in Portugal, Spain, western and southern France (near coasts), Italy, Mediterranian islands and Greece. Also, discovered recently in southern England and on Channel Islands (RILEY 1985). – Outside Europe in Morocco, Algeria, Tunisia, Libya, Egypt, Cyprus, Lebanon, Israel, Iraq and Iran (ANDRES 1913; ROTHSCHILD 1913; ANDRES & SEITZ 1924; WILTSHIRE 1939, 1949, 1957, 1985b; BRANDT 1941; SCHÜTZE 1959, 1961; HAUSMANN 1997; coll. ZMUC).

PHENOLOGY: Bivoltine: late April (mid-March) to late June (early August); mid-September (early September) to late October (early November). In extreme south possibly polyvoltine. Larva found in August. Overwintering as pupa.

BIOLOGY: Larva oligophagous (2). Found on the flowers and leaves of *Tamarix gallica* (France: MILLIÈRE 1871; MABILLE 1872; DIETZE 1913), *T. africana* (Tamaricaceae) (Italy: ZOCCHI 1971) and *Myrica gale* (Myricaceae) (France: LHOMME 1935). Reported also on *Tamarix nilotica, T. tetragyna* (Egypt: ANDRES 1913; ANDRES & SEITZ 1924) and *T. smyrnensis* (Tamaricaceae) (SCOBLE 1999).

HABITAT: Bushy riversides with tamarisk, tamarisk thickets, oases, and river beds in the mountains at moderate altitudes. From 0 m up to about 1,000-1,100 m above sea-level in Morocco (RUNGS 1981), to about 1,200-1,300 m in Saudi Arabia (WILTSHIRE 1986), and to about 1,400 m in southern Spain (coll. ZMUC).

SIMILAR SPECIES: *E. ultimaria* is usually readily recognized and can be easily distinguished from other small European *Eupithecia* species by the large narrow elongate discal spots on the wings, the sharply curved postmedial and subterminal lines with broad paler stripe between them and distinctly chequered fringe. Nevertheless, this species can be easily confused with *E. minusculata* (Text-fig. 31). The structure of genitalia in both male and female should distinguish *E. ultimaria* from all other species.

34. *Eupithecia usbeca* VIIDALEPP, 1992

Eupithecia usbeca VIIDALEPP, 1992: in MYARTZEVA, Ecology and distribution of insects in Turkmenia, Askhabad: 116, fig. 3: A-E (Turkmenia: Repetek). Holotype ♀ (ZISP, examined).

DIAGNOSIS: Wingspan 15.5-21 mm. Labial palpi, frons and vertex pale grey. Notum ash-grey. Forewing narrow and elongated, costa slightly arched, apex rather pointed, termen obliquely rounded; ground colour ash-grey or whitish grey; transverse lines numerous, fine, multiply indented or wavy, dark grey; basal, postbasal and antemedial lines not angulated; medial line fine, oblique, very wavy, sharply angled onto costa beyond discal spot; postmedial line inconspicuous; medial and terminal areas separated by fine, pale, multiply wavy line; pale wavy line narrow, distinct, indented, but sometimes inconspicuous; discal spot obliquely elongate, narrow, black. Hindwing ovate-oblong, pale grey with darker anal and terminal margins; alternating dark grey and whitish transverse lines developed only along posterior margin to discal cell and CuA1; discal spot narrow and elongate or

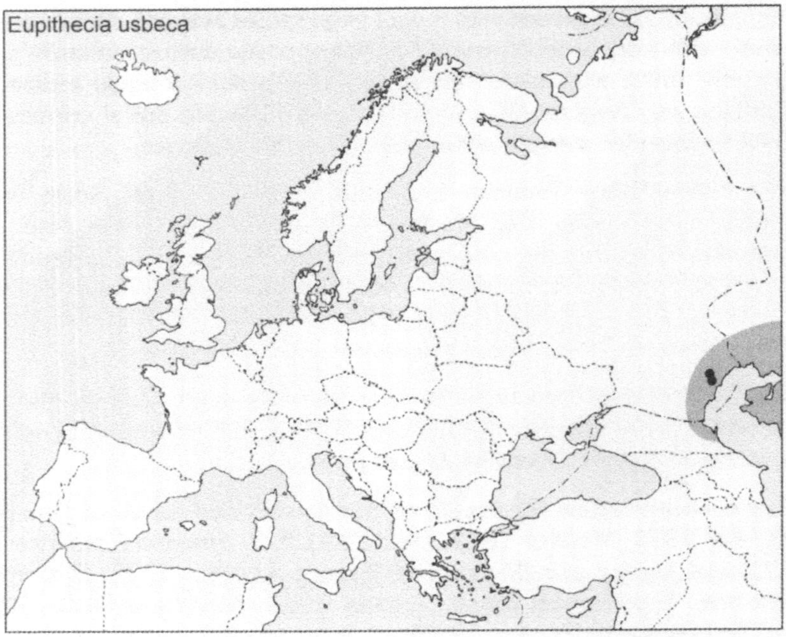

wedge-shaped, dark. Fringe dirty white or whitish grey, distinctly spotted with dark grey at vein ends. Abdomen grey with dorsal row of triangular grey spots, bordered by white.

MALE GENITALIA: Uncus relatively narrow. Papillae clavate, elongated, length about 1.5-1.7 times uncus length. Valva narrow. Aedeagus thick and long, length nearly equal or slightly longer than length of valva. Vesica armed with two horn-like cornuti (one apical cornutus shorter than the other). Sternum A8 with two separate long and narrow symmetrical rods, connected to each other by thin membrane only.

FEMALE GENITALIA: Bursa copulatrix posteriorly sclerotized, pear-shaped; covered with various spines along left side, in the middle part and between ductus seminalis base and colliculum along right side. Antrum membranous. Tergum A8 with medial hollow on the posterior margin.

DISTRIBUTION: Turanian. Southern European Russia (Lower Volga: Astrakhan province). New for the fauna of Europe: 1♂, (Russia), Astrakhan Province, Aksaray district, Molodezhny near Dosang, 8.V.1998, leg. ANIKIN & ZOLOTUHIN; 2♂1♀, Russia, Astrakhan Province, near Stepnoe, at light, sandy steppes, 6.-10. and 6.-14.05.1998, leg. V. ANIKIN (ZISP). – Outside Europe in Transcaspian area in Central Asia (western Kazakhstan?, Uzbekistan and Turkmenistan).

PHENOLOGY: Univoltine: late March (early March) to mid-May in Central Asia.

BIOLOGY: Unknown; the food-plant of larvae may be *Tamarix* (Tamaricaceae).

HABITAT: Sandy steppes, deserts and bushy riversides where tamarisk grows.

SIMILAR SPECIES: Similar to *E. minusculata* but larger and paler. In comparison, *E. usbeca* has more elongated forewings with numerous transverse lines and narrower discal spots; lighter hindwings with smaller discal dots, and underside of wings with inconspicuous pattern and small, narrow, elongate discal spots. The genitalia should be examined for accurate determination. The male is easily recognized by the presence of long clavate papillae, the two large horn-like cornuti on the vesica, absence of medial expansions of sternum A8 rods and the female can be distinguished by disposition of spiniferous areas in bursa copulatrix and shape of tergum A8.

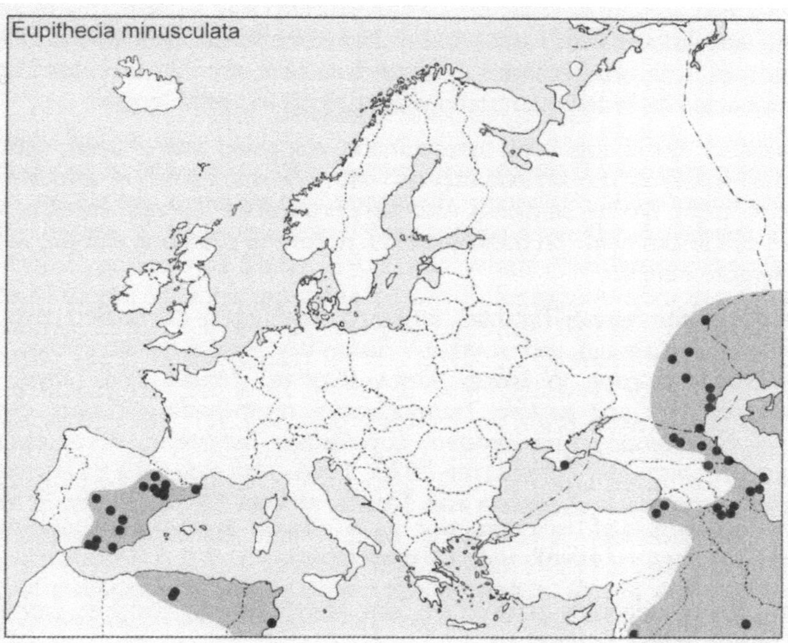

35. *Eupithecia minusculata* Alphéraky, 1882

Eupithecia minusculata Alphéraky, 1882: Horae Soc. Ent. Ross. 17 (3/4): 225, pl. 9: 91 (China: Kuldja [=Yining (Gulja)]). Holotype ♀ (ZISP, examined).

Eupithecia inversaria Turati, 1930: Atti Soc. ital. Sci. nat. 69 (1): 63, pl. 2: 1-2 (Libya: Cyrenaica, Zuara). Syntype(s) (MRSN?).

Eupithecia penultimaria Wiltshire, 1985: The Entomologist's Gaz. 36 (4): 265, pl. 9: 2, 3, 5a-c, 6 (Iraq: Baghdad, on a Tigris Isl.). Holotype ♂ (BMNH).

Eupithecia herrenschmidti Mentzer & Moberg, 1992: Nota lepid. 14 (4): 313, fig. 2, 4, 6, (Spain: Andalusia (Sierra Nevada), (between Motril and Granada), Orjiva). Holotype ♀ (EMS).

Diagnosis: Wingspan 12-15 mm. Labial palpi, frons, vertex and notum pale brownish grey. Forewing brownish grey, mottled and irrorate with pale grey scales, especially near apex and along termen; transverse lines fine, slightly sinuate, dark grey, broadened at costa; two medial lines sharply curved near costa beyond discal spot; first medial line usually touching discal spot; postmedial and basiterminal lines sharply curved outside with more or less broad paler stripe between them; postmedial line fine-wavy, well indicated by dark dashes on the veins; terminal area narrow, sharply broadened to costa; pale wavy line fine, much indented, usually weakly defined or obsolete; discal spot large, narrow, elongate or ovate-oblong, black. Hindwing the same colour as forewing; transverse lines weakly defined except paler medial stripe, sharply angulated in the middle; terminal area slightly darker, narrow, broadened to apex; pale wavy line multiply indented, often ill-defined or obsolete; discal spot large, ovate-oblong or linear, blackish grey. Fringe pale brownish grey, distinctly spotted with dark brownish grey at vein ends. Abdomen brownish grey.

MALE GENITALIA: Uncus broadened near apex, when viewed laterally. Papillae relatively long and wide, length about 0.8-0.9 times uncus length. Valva with curved ventral margin. Vesica armed with three sclerotized strips and with one elongated, V-shaped cornutus at ductus ejaculatorius base; without horn-like cornuti. Sternum A8 two long, narrow, symmetrical rods, with wavy margins in its medium part, strongly sclerotized and pointed to apex, connected by broad membranous area along its entire length.

FEMALE GENITALIA: Bursa copulatrix membranous, elongated pear-shaped, with one large patch of small spines on the left side and also with one small patch of spinules on the left side near colliculum. Ductus seminalis with base broadened. Antrum strongly sclerotized. Tergum A8 spade-like, with prominent medial hollow in posterior margin, length about 0.7 times tergum heighth.

DISTRIBUTION: Mediterranean-Turanian (disjunct). In Europe discovered only in Spain, southern Ukraine (Crimea) and southern European Russia (Lower Volga, Stavropol Prov.). – Outside Europe in North Africa (Algeria, Tunisia and Libya), Caucasus (Daghestan), southern and eastern Transcaucasus, north-eastern Turkey, Syria, Israel, Jordan, Iraq, Kazakhstan, Turkmenistan, Uzbekistan, Tadjikistan, north-western China and Mongolia (VOJNITS 1977; VIIDALEPP 1978, 1996; MIRONOV 1990; HAUSMANN 1991, 1997). New for the fauna of Algeria and Tunisia; at least 35♂♀, Algier, El Mesrane, 4-14.5.1931, PREDOTA leg., 1♀, Algeria centr., Hassi Bahbah, Plantation, V.1930, SCHWINGENSCHUSS leg.; 2♂, 1♀, Tunisia, Sfax (ZFMK).

PHENOLOGY: Bi- or probably polyvoltine: late March to late October. Larva: April to October. Overwintering as pupa.

BIOLOGY: Larva monophagous (2). Found on the inflorescences and leaves of *Tamarix gallica* (Tamaricaceae) and its hybrids (Spain: GOMEZ DE AIZPURUA 1994). According to my data the larvae feed at night on the flowers and rarely on the leaves of *Tamarix leptostachys* (Tamaricaceae) in Central Asia (Turkmenistan).

HABITAT: Places where its food-plant abounds, such as: bushy riverside places with tamarisk and tamarisk thickets. From 0 m up to about 400 m above sea-level in Spain (coll. ZSM), and to about 1,200-1,300 m in Turkey and Uzbekistan (coll. GELBRECHT; WEISERT 1998).

SIMILAR SPECIES: Similar to *E. ultimaria* and can easily be confused with it (Text-fig. 32). Distinguished by the smaller size, usually more pointed apex of forewing, mostly thicker discal spots, darker and more conspicuous underside of wings and more sharply angled pale medial stripe. Valva narrower with more pointed apex, papillae shorter and broader, aedeagus smaller and thinner and also sternum A8 shorter with rods narrower than in *E. ultimaria*. Papillae on the labides show very good differences from both compared species especially: they are short, broad, ovate-oblong, about equal in length to uncus in *E. minusculata*, and they are long, narrow, clavate, covered by short setae only at apex, about 1.5 times uncus length in *E. ultimaria* (see also MENTZER & MOBERG 1992). The female genitalia can be easily distinguished by the smaller bursa copulatrix with smaller spiniferous area on the left side and without row of large spines near ductus seminalis and without row of smaller spines on the left side, as well as the smaller and narrower colliculum, the wider and heavily sclerotized antrum, and also the broader tergum A8 compared with *E. ultimaria*.

REMARKS: The holotype of *E. minusculata* is deposited in ZISP (see also DIETZE 1906; pl. 1, fig. 14, and classic folio by DIETZE 1910; pl. 80, fig. 899). After carefully examining the external features and genitalia of the holotype, as well as studying the types and descriptions in the literature of *E. minusculata, E. penultimaria* and *E. herrenschmidti*, it is clear that they are conspecific (GOMEZ DE AIZPURUA 1994; SCOBLE 1999).

The *undata* species-group

Group introduced by Mironov (1990). Represented mainly by northern and high mountain species. In Europe there are five species: *E. pentellata, E. plumbeolata, E. pygmaeata, E. undata, E. fennoscandica*. Two species from Canary Islands: *E. boryata* (REBEL, 1906), *E. gomerensis* (REBEL, 1917) and several Asian species have been examined and could be placed in this group, as follows: *E. leptogrammata* STAUDINGER, 1882; *E. consortaria* LEECH, 1897; *E. homogrammata* DIETZE, 1908; *E. assulata* BASTELBERGER, 1911; *E. puengeleri* DIETZE, 1913; *E. alishana* INOUE, 1970; *E. holti* VIIDALEPP, 1973; *E. pacifica* INOUE, 1980; *E. schachdarensis* VIIDALEPP, 1988 and *E. kostjuki* MIRONOV, 1989. The following species from North America can also be included to the *undata* group: *E. subcolorata* (HULST, 1898); *E. georgii* MCDUNNOUGH, 1929; *E. appendiculata* MCDUNNOUGH, 1945 and *E. lafontaineata* BOLTE, 1990.

MALE GENITALIA: Uncus short, usually stout and uniapical (in some species biapical). Papillae length about 0.5-0.9 times uncus length. Valva without ventral process; short and broad, widest medially, with evenly curved ventral margin, almost straight dorsal margin and with apex rounded; sacculus weakly sclerotized. Vinculum short and broad. Aedeagus stout and relatively long, length about 0.8-1.3 times valva length. Vesica armed with one to four long and slim horn-like cornuti, usually with one thin, elongated cornutus and one smaller cornutus at ductus ejaculatorius base. Sternum A8 usually sclerotized, elongated and broad, with two parallel, long apical rods, often with each one tipped with a short incurving sclerotized spine and connected by narrow band at base; basal hollow shallow; apical hollow deep; sternum length about 0.6-1.0 times valva length.

FEMALE GENITALIA: Bursa copulatrix membranous or heavily sclerotized, usually elongated, medially broadened; irregularly covered with cap-like spines (in some species the spines lacking) and with large, globular, membranous diverticulum at base. Ductus seminalis narrow. Colliculum absent, except in some species. Antrum membranous or lightly sclerotized. Anterior and posterior apophyses short, thin and tapered or very long, thick and apically expanded. Spurs of anterior apophyses usually short and thin, not expanded apically, except in *E. pygmaeata* and *E. kostjuki*. Papillae anales usually short and wide or elongated, sclerotized, apically broadly rounded or tapered.

REMARKS: The *undata* species-group is closely related to the *venosata* species-group mainly on genitalic characteristics, as follows: uniapical uncus, vesica usually armed with some horn-like cornuti and sternum A8 with two long rods in male, also often with defined membranous diverticulum at base of corpus bursae in the female genitalia. However, the male can be recognized by the short valva; the rods of sternum A8, usually with each one tipped by a short incurved sclerotized spine and the female can be recognized by the elongated bursa copulatrix, usually covered with cap-like spines, without colliculum.

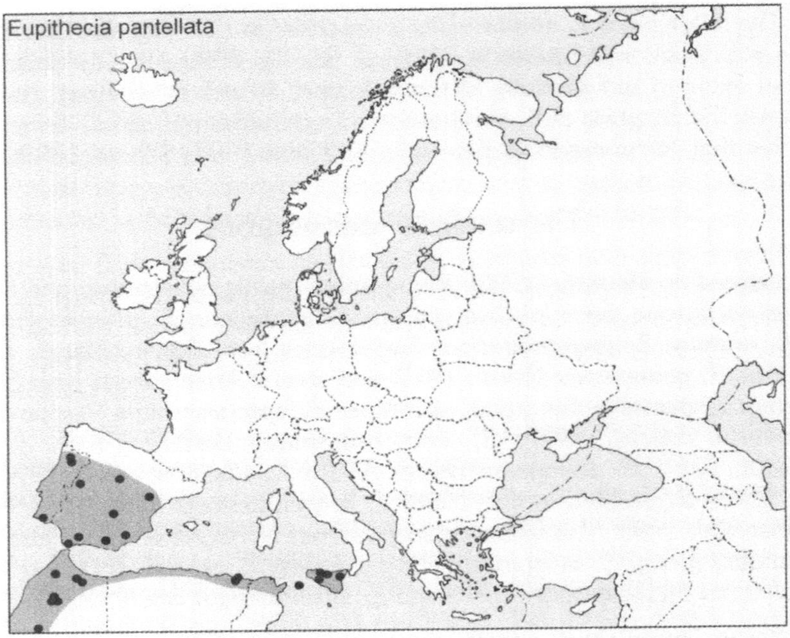

36. *Eupithecia pantellata* MILLIÈRE, 1875

Eupithecia pantellata MILLIÈRE, 1875: in RAGUSA: Bull. Soc. Ent. Ital. 7 (4): 253 (Italy: Pantelleria Isl.). Syntype(s) ♂ (MNHN?).

Eupithecia luteostrigata STAUDINGER, 1876: Stett. Ent. Z. 37 (4-6): 142 (Sicily: Palermo, Malaspina). Holotype ♂ (MNHU).

Eupithecia illuminata JOANNIS, 1891: Bull. Soc. Ent. France 6: 80 (Algeria: Philippeville). Syntypes 2 (MNHN).

Eupithecia tedaldiata FUCHS, 1901: Stett. Ent. Z. 62 (1-2): 125 (Sicily). Syntypes (NSMW?).

Eup(ithecia) pantellaria f. *lusitanica* DIETZE, 1910: Biol. Eupithecien 1, pl. 69: 71, 72; ibidem (1913), 2: 39 (Portugal: San Fiel). Syntypes, at least 2 (MNHU). Valid at subspecific rank.

Eupithecia pantellaria f. *andalusica* WEHRLI, 1926: Intern. Ent. Z. 20 (2): 10 (Spain: Sierra Nevada). Syntypes 2♂2♀ (ZFMK, examined). Valid at subspecific rank.

Unavailable names (misidentification): *liguriata*: sensu REDONDO & GASTÓN (1999, pl. 4: 153) nec MILLÈRE, 1884 (identity: *Eupithecia liguriata*). - (incorrect subsequent spellings): *pantellaria*: MILLIÈRE (1877); *pantelaria*: MOBERG (1983).

DIAGNOSIS: Wingspan 14-18 mm. Labial palpi brown. Frons and vertex ochreous white. Notum pale ochreous. Forewing with broadly arched costa and obliquely rounded termen; pale yellowish ochreous ground colour; transverse lines minutely indented, broadened and darker, brown or dark brown at costa; basal line separated from broad pale yellowish ochreous postbasal by wide whitish stripe; antemedial line angled or curved near costa; postbasal line very much angulated, especially onto costa, usually marked with small dark brown dashes on the veins; medial area pale yellowish ochreous, darker, brown or dark brown at costa and dorsum, bordered with broad whitish stripes; terminal area broadened to costa with pale apical streak; whitish wavy line minutely indented, often broken; pale medial subterminal dot and a distinct, larger, whitish tornal spot, are shaded on inner side with scattered brownish scales; discal spot large, oblique, ovate-oblong, intensely blackish brown. Hindwing costa whitish; transverse lines shaded with dark brown; postmedial line very much angulated; broad, double whitish stripe separates medial and terminal areas; terminal area relatively wide, pale yellowish ochreous; white wavy line fine, much indented, frequently broken and forming small, whitish medial and tornal spots; discal spot large, narrow and elongate, often comma-like. Fringe ochreous white, spotted with ochreous brown at vein ends. Abdomen pale ochreous, usually with a series of small, dark brown dorsal tufts and lateral dots; first and second segments darker, ochreous brown.

MALE GENITALIA: Uncus biapical. Aedeagus slim. Vesica armed with one long and slim horn-like cornutus, about 0.8 times aedeagus length. Sternum A8 with two thin symmetrical pointed rods joined narrowly at base.

FEMALE GENITALIA: Bursa copulatrix small, irregular, rather pouch-like, with numerous longitudinal striations; covered with small spinules in posterior part; its anterior part membranous. Colliculum collar-like, short and broad. Anterior and posterior apophyses relatively short and thin, not expanded to apices. Tergum A8 almost square, with slightly sclerotized, deep medial hollow in the posterior margin. Papillae anales short, broad and rounded.

DISTRIBUTION: West-Mediterranean. Portugal, Spain, Sicily and Pantelleria Island. – Outside Europe in Canary Islands (subsp. *canariata* PINKER, 1965) and northern Africa (Morocco, Algeria).

PHENOLOGY: Univoltine: early May (late April) to early June (mid-June), on mountains somewhat later. Larva in June. Overwintering as pupa, often twice.

BIOLOGY: Larva monophagous (1). Found on the seeds (?) of *Umbilicus* (Crassulaceae) (Algeria: DIETZE 1913).

HABITAT: Rocks, shaded walls, dry and warm mountain slopes (?). In Andalusia from 0 m up to about 1,000 m above sea-level (coll. ZSM), at 2,500 m (coll SKOU), in Great Atlas from 2,600 to 2,700 m (RUNGS 1981).

SIMILAR SPECIES: Brindled pale yellowish ochreous colour, the medial area bordered with wide whitish stripes and presence of the large, oblique, ovate-oblong discal spot on the forewing serves to distinguish this species from other small European congeners.

REMARKS: Variable in the depth of colour. The f. *andalusica* WEHRLI, 1926 from Spain only slightly darker than Sicilian f. *luteostrigata* STAUDINGER, 1876. North African moths, which were described as *E. illuminata* JOANNIS, 1891, usually darker, with warm ochreous brown ground colour. However, ab. *lusitanica* DIETZE, 1910 from Portugal is dark also. Subsp. *canariata* PINKER, 1965, which was described from Canary Islands, is dark, warm ochreous brown colour with larger discal spots on the forewings, and without the ochreous colour of terminal area of the hindwings, which is found in the other subspecies.

37. Eupithecia plumbeolata (HAWORTH, 1809)

Phalaena plumbeolata HAWORTH, 1809: Lepid. Br. (2): 360 (United Kingdom). Syntype(s) (BMNH?).
Larentia isogrammata TREITSCHKE, 1828: Die Schmett. Eur. 6 (2): 100 (Hungary: Ofen). Holotype ♀ (HNHM).
Eupithecia begrandaria BOISDUVAL, 1840: Genera Index meth. eur. Lepid.: 211 (France: Paris). Syntype(s).

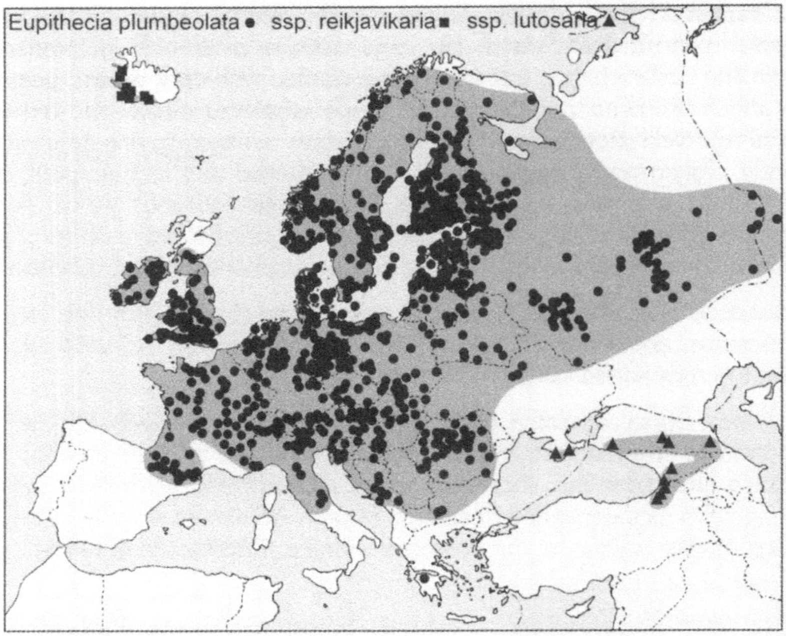

Eupithecia singularia HERRICH-SCHÄFFER, 1848: Syst. Bearb. Schmett. Eur. 3 (32): 121, 132, ibidem (1846), pl. 23: 141, 142 (non binominal), (Germany). Syntypes lost.

Eupithecia immundata var. *reikjavikaria* STAUDINGER, 1871: in STAUDINGER & WOCKE, Cat. Lepid. Eur. Faunengeb. 1: 197 (Iceland: Reikjavik). Syntypes 2♂2♀ (MNHU).

Eupithecia plumbeolata var. *lutosaria* BOHATSCH, 1893: Dt. ent. Z. Iris 6: 29 (Georgia: Achalzich). Syntypes 2♂ (MNHU). Valid at subspecific rank.

Eup(ithecia) plumbeolata f. *uralensis* DIETZE, 1910: Biol. Eupithecien 1, pl. 72: 246; ibidem (1913), 2: 31 (Russia: Ural centr., Sojmonowsk). Syntypes 3, including ♀ (MNHU). Valid at subspecific rank.

Eup(ithecia) plumbeolata f. *enucleata* DIETZE, 1910: Biol. Eupithecien 1, pl. 72: 248-250; ibidem (1913), 2: 30 (Germany: Bavaria, Allgäu, Oberdorf). Syntypes 5, ♂♀ (MNHU). Valid at subspecific rank.

Eupithecia dshungarica VIIDALEPP, 1988: Fauna pjadenits gor Srednej Azii: 91, pl. 2: 5; text-pl. 14: 5, 6 (Kazakhstan: Dzhungarskij Alatau Mts, Tekeli). Holotype ♂ (IZBE, examined).

Unavailable names (misidentifications): *valerianata*: sensu STAUDINGER (1857) nec HÜBNER, 1813 (identity: *Eupithecia valerianata*); ? *immundata*: sensu WOLFF (1929) nec LIENIG & ZELLER, 1846 (identity: *Eupithecia immundata*). - (infrasubspecific): *obscura*: LEMPKE (1969). - (incorrect subsequent spellings): *isogrammaria*: BOISDUVAL (1840); *plumbeolaria*: DOUBLEDAY (1849); *begrandata*: BRUAND (1850).

DIAGNOSIS: Wingspan 12-20 mm. Labial palpi covered with brownish grey scales with light brown apices. Frons, vertex and notum unicolorous, brownish grey. Forewing pale brownish grey or grey; transverse lines brownish grey, fine, slightly sinuous, usually evenly curved and slightly more emphasised at costa; pale wavy line indistinct; tornal spot absent; discal dot minute, inconspicuous, brownish grey or absent. Hindwing unicolorous brownish grey with inconspicuous transverse lines; discal dot as on forewing. Fringe unicolorous brownish grey or slightly spotted with brownish. Abdomen unicolorous grey.

The subsp. ***reikjavikaria*** (STAUDINGER, 1871) on the whole larger (wingspan 17-20 mm) than adults of nominate subspecies. The unicolorous, shiny brownish grey wings, lacking clear transverse lines distinguish this Icelandic subspecies from the nominate subspecies.

The subsp. ***lutosaria*** BOHATSCH, 1893 has the habitus and wing maculation in nominate subspecies, but differs in labial palpi covered with pale yellowish ochreous scales with light brown apices; frons, vertex and notum unicolorous pale yellowish ochreous; wings pale yellowish ochreous with brownish yellow, fine, slightly sinuous transverse lines, discal dots lacking; abdomen unicolorous yellowish grey

MALE GENITALIA: Uncus stout, biapical. Aedeagus stout, relatively large. Vesica armed with three cornuti, one small, biapical, one elongated, twisted, broadened at base and a small one at ductus ejaculatorius base. Sternum A8 elongated, with two parallel apical rods; the tips not heavily sclerotized.

FEMALE GENITALIA: Bursa copulatrix heavily sclerotized, elongated pouch-like, covered almost completely with small spines except the membranous anterior part. Ductus bursae short with longitudinal striations. Ductus seminalis with evenly curved and sclerotized base. Colliculum short, collar-like. Anterior and posterior apophyses thin, rather short, not expanded apically. Papillae anales short and wide.

DISTRIBUTION: Euro-Siberian. Widely distributed in Europe except Portugal, Albania, Macedonia and Mediterranean islands. In Iceland replaced by subsp. *reikjavikaria*, in southern Crimea (Ukraine) by subsp. *lutosaria*. – Outside Europe nominate subspecies in Urals, Central Asia (Dzhungarsky Alatau Mts, northern Tien-Shan), north-western Siberia, Altai and Sajan Mts, replaced by subsp. *lutosaria* in Asia Minor (north-eastern Turkey), Caucasus and western Transcaucasus (Georgia).

PHENOLOGY: Univoltine: late May (late April) to mid-July (late August). The subsp. *reikjavikaria* from mid-June to early July. Larva: mid-July (early July) to early September (late September). Overwintering as pupa.

BIOLOGY: Larva oligophagous (1). Found on the flowers and in the seed capsules of *Melampyrum pratense*, *M. sylvaticum*, *M. arvense*, *Rhinanthus alectorolophus*, *R. minor* (United Kingdom: CREWE 1865; BUCKLER 1899; BARRETT 1904; Germany: DIETZE 1910, 1913; WEIGT 1980, 1988; GELBRECHT and WEGNER pers. comm.; France: LHOMME 1935; Denmark: JUUL 1948; Finland: SEPPÄNEN 1970) and *Melampyrum nemorosum* (Scrophulariaceae) (southern Europe: DIETZE 1913; France: LHOMME 1935), probably occasionally on *Clematis vitalba* (Ranunculaceae), *Calluna vulgaris* (Ericaceae) (France: LHOMME 1935) and *Galeopsis* (Lamiaceae) (Finland SEPPÄNEN 1970). According to my data the larva feeds in the seed capsules of *Rhinanthus songaricus* (Scrophulariaceae) in Tien-Shan Mts (Central Asia).

HABITAT: Bushy thickets and scrub-covered meadows, open and dry deciduous and mixed forests, glades, forest edges, country roads and roadsides, hedges, heaths, dunes, sandhills, sandy fields and slopes, as well as wet meadows and marshes. From 0 m up to about 1,900-2,000 m above sea-level in Alps (VORBRODT 1931; REZBANYAI-RESER & MAIER 2000), and to about 2,400 m in Pyrenees (coll. FIBIGER). The subsp. *lutosaria* inhabits open mountain birch forests, from 1,300 m up to about 2,100 m above sea-level in Great Caucasus Mts and north-eastern Turkey (coll. ZISP; coll. GELBRECHT).

SIMILAR SPECIES: The nominate subspecies of *E. plumbeolata* can usually be easily separated from the similar *E. valerianata* by the more greyish colour and clearer maculation, as well as the indistinct wavy line and absence of pale tornal dot on the forewing, and broader pale stripe between medial and terminal areas on the fore- and hindwing. This species is also very similar to *E. haworthiata* but can be separated from it by the lighter yellowish grey colour and greyish segment A2 (Text-fig. 18). Valva with broadly rounded apex, vesica without dentate cornuti, the shape of sternum A8 in male and elongated pouch-like bursa copulatrix with membranous base, and position of the spiniferous areas in corpus bursae should distinguish *E. plumbeolata*.

The adults of subsp. *lutosaria* are very similar to *E. spissilineata*. However, on the whole it is smaller; less bright ground colour with less dentated pale transverse lines; pale wavy lines on the fore- and hindwings inconspicuous; whitish transverse stripe on the hindwing broader than in *E. spissilineata* (Text-fig. 58). The male and female genitalia easily distinguish these two species.

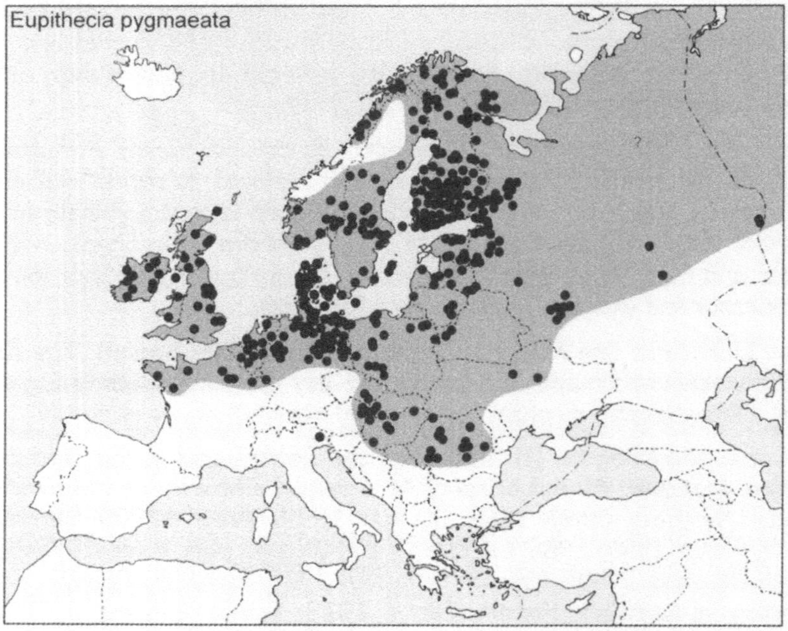

38. Eupithecia pygmaeata (HÜBNER, 1799)

Geometra pygmaeata HÜBNER, 1799: Samml. Eur. Schmett. 5 Geometrae (2), pl. 45: 234 (Europe). Syntype(s) lost.

Eupithecia palustraria DOUBLEDAY, 1850: The Zoologist 8 (App., Art. 8): 55 (United Kingdom, England: Huntingdonshire). Syntypes (BMNH).

Eup(ithecia) pygmaeata f. *pseudozibellianata* DIETZE, 1910: Biol. Eupithecien 1, pl. 72: 267; ibidem (1913) 2: 46 (as *pseudozibellinata*) (Finland: Kuusamo). Holotype ♀ (MNHU). Valid at subspecific rank.

Eupithecia palustraria grabei CORNELSEN, 1920: Intern. Ent. Z. (Guben) 14 (11): 87 (Germany: Herne, Rhein-Herne Kanal district). Syntypes 1♂1♀. Valid at subspecific rank.

Eupithecia sidemii VOJNITS, 1973: Annales hist.-nat. Mus. natn. hung. 65: 253, figs. 4a, b, only holotype ♀; nec paratypes ♂ (Russia: Askold Isl.). Holotype ♀ (ZFMK; examined).

Unavailable names (incorrect subsequent spellings): *pygmaearia*: BOISDUVAL (1840); *pygmacata*: DIETZE (1910); *pseudozibellinata*: DIETZE (1913); *pygmeata* BAYNES (1964); *pygmaeta*: LAINE (1977).

DIAGNOSIS: Wingspan 12-18 mm. Labial palpi brown or with mixture of brown and grey scales. Frons, vertex and notum from dark brown to ochreous brown with scattered grey scales. Forewing dark brownish fuscous or chocolate-brown; all transverse lines slightly sinuate, usually indistinct, greyish edged; tornal dot greyish, small, rather indistinct; discal spot usually absent, when present minute, elongated, indistinct, dark brown. Hindwing chocolate-brown with brownish grey, inconspicuous transverse lines; discal spot minute, elongated, rather indistinct, dark brown, often absent. Fringe brown, slightly spotted with grey. Abdomen mainly brown with scattered grey scales.

MALE GENITALIA: Uncus elongated, thin, uniapical. Papillae small, length about 0.4-0.5 times uncus length. Valva with narrow apex. Aedeagus nearly equal to or slightly longer than length of valva. Vesica armed with five cornuti, three elongated, delicate, horn-like, one elongated and broad and one small, irregular, curved cornutus at ductus ejaculatorius base. Sternum A8 with two narrow, tapered rods, without heavily sclerotized spiniform apices.

FEMALE GENITALIA: Bursa copulatrix sclerotized, elongated, rather pouch-like, medially convex; with membranous, large, broad, pouch-like diverticulum near base on the left side. Ductus seminalis long, narrow, broadly attached to corpus bursae near base on the right side. Colliculum weakly defined, collar-like, short and broad. Antrum membranous. Anterior and posterior apophyses short, thin, pointed at apices. Tergum A8 short and broad, with deep membranous medial hollow on the posterior margin. Papillae anales small, short and narrow.

DISTRIBUTION: Holarctic (boreal). Widespread mainly in northern and Central Europe. Disjunct population in north-eastern Italy. Reputedly also in southern Italy (PARENZAN & HAUSMANN 1992), specimens and genitalia figured in PARENZAN (1994), identification uncertain (see Remarks below). In the Palaearctic region from Ireland to Far East Russia (Kamtchatka) and Japan (Hokkaido). The subsp. *obumbrata* TAYLOR, 1906 occurs in North America (BOLTE, 1990).

PHENOLOGY: Uni- or bivoltine: late May (mid-May) to mid-June (late June); mid-July to early September (mid-September). In the north univoltine: late May to late June. Larva: early July to late July; late July to late September (in the north in July to mid-August). Overwintering as pupa, sometimes up to three times.

BIOLOGY: Larva oligophagous (1). Found on the flowers and usually in the seed capsules of *Stellaria holostea* (United Kingdom: CREWE 1865, 1872b; HELLINS 1872; BUCKLER 1899; Germany: WEIGT 1985; WEGNER pers. comm.), *Myosoton aquaticum, Cerastium fontanum* (Germany: DIETZE 1874a, 1910, 1913; WEIGT 1985, 1990), *Stellaria nemorum, Cerastium arvense* (Finland: SEPPÄNEN 1970; Germany: WEIGT 1985; WEGNER pers. comm.) and *C. holosteoides* (Caryophyllaceae) (Germany: GELBRECHT pers. comm.). Recorded also on *Stellaria media, Minuartia verna* (Caryophyllaceae) (France: LHOMME 1935). In captivity also feed on *Cerastium tomentosum* (Caryophyllaceae) (United Kingdom: CREWE 1872b; BARRETT 1904).

PARASITOIDS: Hymenoptera, Ichneumonidae: *Deuterolabops eupitheciae* (THOMPSON 1945; HERTING 1976; ZWAKHALS pers. comm.).

HABITAT: Fringes of deciduous and mixed forests, marshy and flood plain forests, ditches and shady roadsides, wet meadows, grassland near water, fields, cultivated areas, moors, very rarely on warm and dry slopes. From 0 m up to about 1,800 m above sea-level in southern Tirol (WEIGT 1990).

SIMILAR SPECIES: This species appears to be most similar to *E. plumbeolata* and *E. vale-*

rianata. However, *E. pygmaeata* can usually be distinguished easily from *E. plumbeolata* by the darker brown ground colour, by the presence of a small pale tornal spot on the forewing and usually distinctly chequered fringe. It can be distinguished from *E. valerianata* by the darker brown ground colour and less distinct pale wavy line and tornal spot on the forewing (Text-fig. 19). Uniapical uncus, broad valva and vesica with three slim horn-like cornuti connected at base, and irregular shaped bursa copulatrix with membranous diverticulum and narrow ductus seminalis near its base should distinguish this species.

REMARKS: Usually a little variable species. However, some specimens have unicolorous chocolate-brown unmarked wings except a few greyish costal spots (which are sometimes absent), pale small and round tornal dot on the forewing, smaller tornal spot on the hindwing and fringe unicolorous fuscous brown.
Recorded for Puglia and Basilicata in southern Italy (PARENZAN & HAUSMANN 1992; PARENZAN 1994) in error probably. It appears not to be *E. pygmaeata* (but might be *E. valerianata* according to the wings and female genitalia) illustrated in figs. 15a and 16d in PARENZAN (1994).

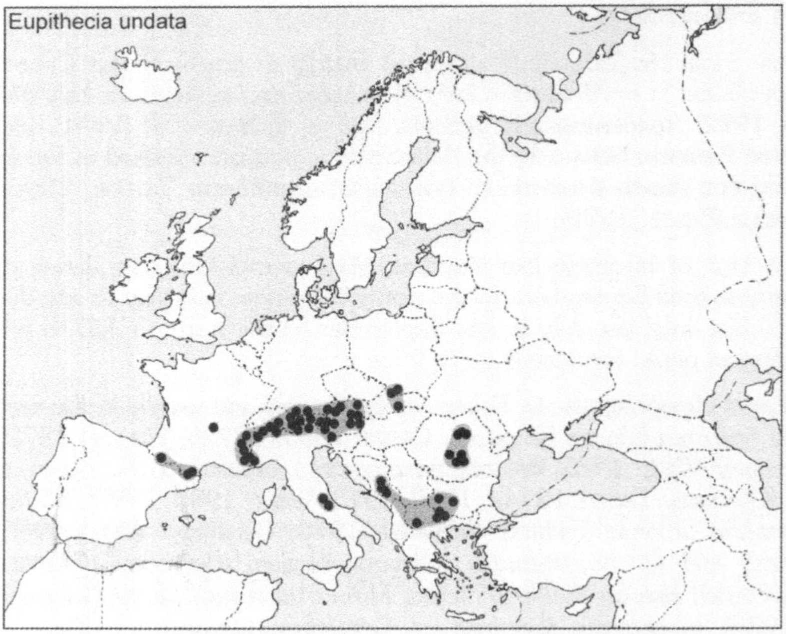

39. Eupithecia undata (FREYER, 1842)

Larentia undata FREYER, 1842: Neuere Beitr. Schmett. 4: 54, pl. 323: 2 (Slovenia: Laybach [=Ljubljana]). Syntype(s).

Eupithecia scriptaria HERRICH-SCHÄFFER, 1848: Syst. Bearb. Schmett. Eur. 3: 121, 132, ibidem (1845), pl. 20: 123 (non binominal), (Austria: Schneeberg Mt.). Syntype(s) lost.

Eup(ithecia) undata f. *nigerrima* DIETZE, 1910: Biol. Eupithecien 1, pl. 72: 262; ibidem (1913), 2: 47 (Austria: Stilfser Joch). Holotype ♂ (MNHU). Valid at subspecific rank.

Eup(ithecia) undata f. *abruzzensis* DIETZE, 1910: Biol. Eupithecien 1, pl. 77: 637; ibidem (1913), 2: 47 (Italy: Majella Mts). Holotype ♂ (MNHU). Valid at subspecific rank.

DIAGNOSIS: Wingspan 14-22 mm. Labial palpi covered with mixture of brown and grey scales, concolorous with frons, vertex and notum. Forewing brownish grey; transverse lines slightly sinuous, rather inconspicuous, dark grey or greyish brown; postmedial line slightly angled near costa; wavy line usually fine, light grey; discal dot brown, faint, slightly elongated and oblique or absent. Hindwing slighty paler, weakly marked, brownish grey with darker, greyish brown terminal area and fine-dentated, light grey wavy line; discal dot absent or minute, rounded, inconspicuous. Fringe chequered grey and brown or dark brown. Abdomen covered with grey and light brown scales.

MALE GENITALIA: Uncus uniapical. Papillae nearly equal to length or slightly smaller than uncus length. Vinculum tapered. Aedeagus stout, long, about 1.3 times length of valva. Vesica armed with four large horn-like cornuti and one small, irregular and bowed cornutus at base. Sternum A8 almost equal to valva length.

FEMALE GENITALIA: Bursa copulatrix covered with spines except at base and left side. Anterior and posterior apophyses long and thin. Tergum A8 rectangular; slightly longer than wide. Papillae anales elongated and narrow.

DISTRIBUTION: European (alpine). The mountains of Central and southern Europe: Poland, Slovakia, Germany, France, Italy, Switzerland, Austria, Romania and Bulgaria.

PHENOLOGY: Univoltine: late May (early May) to late July (early August). Larva: late July (early July) to late August (mid-September). Overwintering as pupa, occasionally twice.

BIOLOGY: Larva monophagous (2). Found in the flowers and seed capsules of *Silene rupestris* (Switzerland: PÜNGELER 1889; DIETZE 1910, 1913), *S. alpestris* and *Minuartia austriaca* (Caryophyllaceae) (Austria and Italy: WEIGT 1990).

PARASITOIDS: Diptera, Tachinidae: *Ceranthia tristella* (ANDERSEN 1996).

HABITAT: Alpine meadows, warm and dry slopes, especially on limestone, rocks, screes and stony south-facing slopes. From 1,000 m up to about 2,700 m above sea-level in Alps (VORBRODT 1928, 1931; WEIGT 1990).

SIMILAR SPECIES: Adults of *E. undata* are similar to those of *E. graphata*. However, they can be usually be separated by the absence of conspicuous light and dark rings on the antennae, by the slightly darker, browner and less bright ground colour of all wings and abdomen, and the forewing with postmedial stripe widened and less dentated than those in *E. graphata*. The male and female genitalia are considerably distinct in these two species.

40. *Eupithecia fennoscandica* KNABEN, 1949

Eupithecia fennoscandica KNABEN, 1949: Ent. Tidskr. 70 (1-2): 78, pl. 1: 1-4, 8 (Finland: Lapland, Saana [=Kilpisjarvi, Saanatunturi]). Holotype ♀.

Unavailable names (misidentification): *undata*: sensu GRÖNBLOM (1936) nec FREYER, 1842 (identity: *Eupithecia undata*). - (incorrect subsequent spelling): *fennoscandia*: WEIGT (1990).

DIAGNOSIS: Wingspan 14-18 mm. Labial palpi greyish brown. Frons, vertex and notum covered with mixture of pale grey and greyish brown scales. Forewing brownish grey; transverse lines greyish brown, slightly sinuous, rather inconspicuous and obscure, widened and slightly darker at costa; postmedial line obtuse angled near costa; pale wavy line indistinct, sometimes with small pale tornal spot; discal dot minute, dark brown,

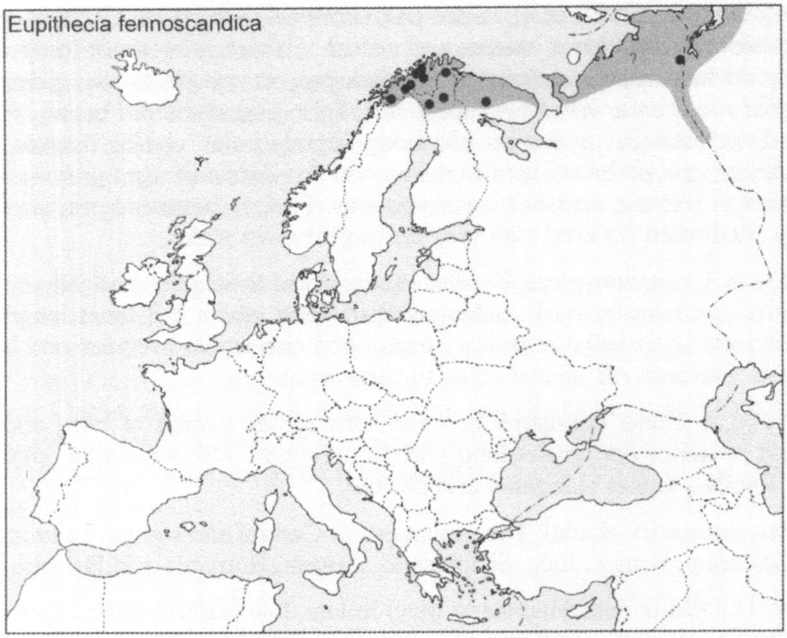

rounded or weakly elongated. Hindwing slightly paler, brownish grey with inconspicuous transverse lines, which are marked along posterior margin, and with terminal shade; discal dot minute, rounded, indistinct or absent. Fringe pale grey, spotted with brown at vein ends. Abdomen greyish brown.

MALE GENITALIA: Uncus uniapical. Papillae about 0.7-0.8 times uncus length. Vinculum blunt. Aedeagus length about 1.2-1.3 times length of valva. Vesica armed with four cornuti, two long, delicate, horn-like, one elongated, narrow, flat, bowed, and one small, irregular and curved, at ductus ejaculatorius base. Sternum A8 with two long and broad rods, with heavily sclerotized and pointed tips; these rods have wavy inner margins; sternum length about 0.7-0.9 times valva length.

FEMALE GENITALIA: Bursa copulatrix strongly sclerotized, completely covered with large, cap-like spines except small area at base; with one globular, membranous diverticulum and one pouch-like, membranous diverticulum at base of corpus bursae. Anterior and posterior apophyses long and thin. Tergum A8 rectangular, heighth about 1.5 times longer than width. Papillae anales elongated and narrow.

DISTRIBUTION: Euro-Siberian (arctic-alpine). Northern Scandinavia (Norway, Sweden, Finland), northern European Russia and Polar Urals. – Outside Europe in southern Siberia (Altai and Sajan Mts) and northern Mongolia.

PHENOLOGY: Univoltine: early July (late June) to mid-July (late July). Larva: mid-July to mid-August. Overwintering as pupa.

BIOLOGY: Larva monophagous (1). Recorded in the flowers and seed capsules of *Lychnis alpina* (Caryophyllaceae) (Northern Scandinavia: SEPPÄNEN 1970; SKOU 1984, 1986; MIKKOLA & al. 1989).

HABITAT: Rocky cliffs, stony slopes and screes. From 400 m up to about 800 m above sea-level in Lapland (KNABEN 1977).

SIMILAR SPECIES: This species is similar to and closely related to *E. undata*. However, all wings usually slightly narrower and paler, with less conspicuous pattern than in *E. undata*. In the male genitalia the second large horn-like cornutus is shorter, and in the female genitalia the corpus bursae is heart-shaped, shorter and broader than in *E. undata*. Also, the position of the membranous diverticulum is different in the two species: in *E. undata* it is attached to corpus bursae simply at base; in *E. fennoscandica* it is attached to corpus bursae at a short distance from base.

The ***silenata*** species-group

This small species-group is introduced here for the first time. Represented in Europe by one species only: *E. silenata*. One closely related species from North America, *E. misturata* (HULST, 1896) also belongs to the *silenata* group and it is included in the diagnosis of this species-group.

MALE GENITALIA: Uncus biapical, narrow, with slightly curved apex. Papillae small, length about 0.3-0.5 times uncus length. Valva evenly broad from base to middle then narrowing to rounded apex; with small spine-like medial ventral process; sacculus weakly sclerotized. Vinculum short, medium width. Aedeagus slim, about 0.6-0.7 times valva length; about 4.5-5.0 times longer than medial width. Vesica armed with one elongated, tapered, semi-cylindrical cornutus and one short twisted cornutus at ductus ejaculatorius base. Sternum A8 with shallow basal hollow; shaft narrow, broadened slightly near apex; apex blunt with narrow, short or medium length apical notch; sternum length about 0.5 times valva length.

FEMALE GENITALIA: Bursa copulatrix membranous, pouch-like with few scattered spinules or small patch of spinules mid-dorsally. Ductus seminalis membranous, short and thin, attached mid-ventrally (or at right side) to corpus bursae. Colliculum collar-like, sclerotized, relatively narrow and long, length about 0.2-0.4 times bursa copulatrix length. Antrum membranous. Anterior and posterior apophyses short and thin. Spurs of anterior apophyses prominent and thin. Tergum A8 almost square. Papillae anales relatively small, short and rounded.

REMARKS: Representatives of the *silenata* species-group can usually be separated from other species of *Eupithecia* by the combined characters of unusual, narrow sternum A8, spine-like medial ventral process on the valva and armature of vesica, and membranous corpus bursae with small patch of spinules and with narrow and long colliculum.

41. *Eupithecia silenata* ASSMANN, 1848

Eupithecia silenata ASSMANN, 1848: Z. Ent. Breslau 2 (5): 11, pl. 1: 3, 3a-o (Poland: Silesia, Breslau [=Wroclaw]). Syntype(s).

Eupithecia silenata var. *pseudolariciata* STAUDINGER, 1871: Catalogue Enum. Méth. Lépid. (Ed. 2) 1: 199, No: 2834 (Austria). Syntype(s) (MNHU). Valid at subspecific rank.

Eupithecia silenata var. *geroldiata* FUCHS, 1900: in PAGENSTECHER, Jahrb. nassau. Ver. Naturkde 53: 63 (Germany: Hessen, Valley of Wisper, Geroldstein). Syntypes (NSMW?). Valid at subspecific rank.

Unavailable names (incorrect subsequent spellings): *silenaria*: HERRICH-SCHÄFFER (1852); *sileneata*: GARTHE (1976).

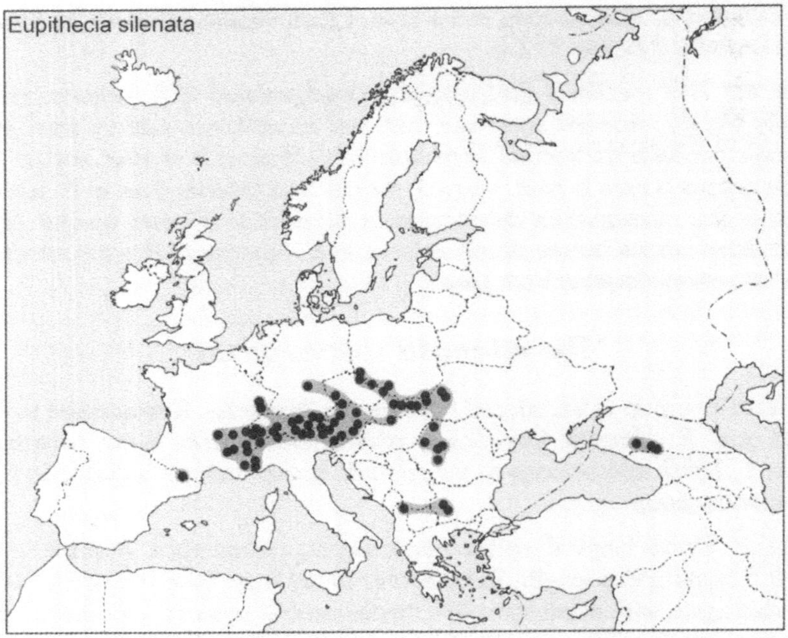

DIAGNOSIS: Wingspan 17-21 mm. Labial palpi greyish brown. Frons, vertex and notum covered with brownish grey scales with grey apices. Antennae in both male and female grey, ringed greyish brown. Forewing brownish grey; transverse lines brown, relatively broad and obscure, slightly sinuate, usually broader and right-angled near costa; terminal area slightly darker, greyish brown; pale tornal spot minute or absent; discal dot dark brown, large and rounded. Hindwing paler brownish grey; postmedial line brownish, indistinct, usually broken; terminal area darker brownish-grey with inconspicuous pale wavy line; discal dot always conspicuous, minute, usually elongated, light brown. Fringe grey or brownish grey, distinctly spotted with brown at vein endings. Abdomen greyish brown.

MALE AND FEMALE GENITALIA: See species-group diagnosis.

DISTRIBUTION: Euro-Caucasian (alpine). Mountains of Central and southern Europe. – Outside Europe in western Caucasus.

PHENOLOGY: Univoltine: early May (late April) to early July (mid-August). Larva: late July (early July) to early September (mid-September). Overwintering as pupa.

BIOLOGY: Larva monophagous (1). Found in the flowers, seed capsules and also on the leaves of *Silene vulgaris (=inflata)* (Caryophyllaceae) (France: MILLIÈRE 1870; Switzerland: FAVRE 1905; Germany: DIETZE 1910, 1913; WEIGT 1990).

PARASITOIDS: Diptera, Tachinidae: *Ceranthia anomala*, *C. tristella* (HERTING 1960; ANDERSEN 1996).

HABITAT: Forest edges, warm and dry slopes, rocks and screes. From 400 m up to about 2,200-2,300 m above sea-level in Alps and Pyrenees (DUFAY 1972; DANTART 1988; WEIGT 1990).

SIMILAR SPECIES: Similar to *E. lariciata*. However, it is brighter coloured on the whole, basal half of the forewing usually lighter, terminal line broader, hindwing more elongated than in *E. lariciata*, and abdomen without black lateral stripes as in *E. lariciata*. Shapes of valva and sternum A8, cornuti on the vesica, as well as membranous corpus bursae with small patch of spines can be used to confirm determinations.

The *venosata* species-group

Introduced by DIETZE (1913). Subsequently described by SCHÜTZE (1960) as subgenus *Bohatschia* including 14 Palaearctic species. This is one of the most species-rich mainly Palaearctic *Eupithecia* species-groups, represented in Europe by six species, as follows: *E. carpophagata*, *E. venosata*, *E. schiefereri*, *E. silenicolata*, *E. alliaria* and *E. dissertata*. In Canary Islands occurs *E. rosai* PINKER, 1962 which is also from this species-group. Many species of the *venosata* group occur in Asia, such as *E. demetata* CHRISTOPH, 1885; *E. stigmaticata* CHRISTOPH, 1885; *E. amasina* BOHATSCH, 1893; *E. carpophilata* STAUDINGER, 1897; *E. vacuata* DIETZE, 1903; *E. ponderata* DIETZE, 1906; *E. hilariata* DIETZE, 1908; *E. achyrdaghica* WEHRLI, 1929; *E. cheituna* BRANDT, 1938; *E. sectila* BRANDT, 1938; *E. montanata* BRANDT, 1938; *E. pfeifferata* SCHÜTZE, 1960; *E. problematicata* SCHÜTZE, 1960; *E. pengata* SCHÜTZE, 1961; *E. comparanda* VOJNITS, 1981; *E. kondarana* VIIDALEPP, 1988; *E. christophi* MIRONOV, 1988 and *E. pinkeri* MIRONOV, 1991 (SCHÜTZE 1960; VIIDALEPP 1988; MIRONOV 1990). The North American species *E. borealis* (HULST, 1898) belongs to this group also.

MALE GENITALIA: Uncus short to long, uniapical or biapical. Papillae usually short, broad or very short, length about 0.2-0.6 times uncus length. Valva without ventral process, of medium length and broadened medially or relatively elongated, narrow, with almost parallel ventral and dorsal margins, usually widened at base; sacculus weakly sclerotized. Vinculum short and broad (except in *E. silenicolata*). Aedeagus usually long and narrow (except in *E. pinkeri*), length about 0.8-0.9 times valva length. Vesica armed with one to three delicate horn-like cornuti. Sternum A8 consists of two long, narrow, symmetrical rods, connected by narrow band at base or broad with two more or less long rods, apically pointed; basal hollow from shallow to deep; apical hollow about 0.3-0.6 or 0.8-0.9 times sternum length; sternum length about 0.5-0.8 times valva length.

FEMALE GENITALIA: Bursa copulatrix diverse: pear-shaped, pouch-like, irregular or elongated, fusiform, almost completely covered with small spines; with membranous diverticulum near base (except in some species). Ductus seminalis narrow; attached to corpus bursae near colliculum. Colliculum collar-like or semicircular (sometimes plate-like), usually short and relatively narrow. Antrum membranous. Anterior and posterior apophyses often short and thin, not expanded at apices; spurs of anterior apophyses broadened to ends. Tergum A8 square or rectangular. Papillae anales usually large, broad and rounded.

42. *Eupithecia carpophagata* STAUDINGER, 1871

Eupithecia carpophagata STAUDINGER, 1871: in STAUDINGER & WOCKE, Cat. Lepid. eur. Faunengeb. 1: 193 (Spain: Andalusia ?). Holotype ♀ (coll. HERBULOT in ZSM, examined).

Acidalia cassandrata MILLIÈRE, 1874: Iconogr. Descr. Chenilles Lépid. inédits 3 (34): 382, pl. 148: 3 (France: Saint-Martin-Lantosque). Holotype ♀ (MNHN?).

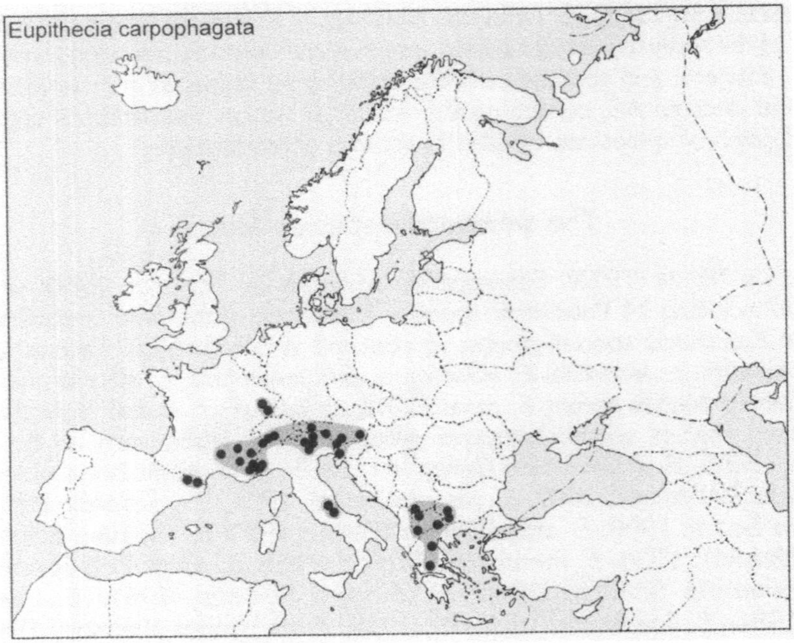

Eup(ithecia) cassandrata f. *teriolensis* DIETZE, 1910: Biol. Eupithecien 1, pl. 70: 153, 154; 85: 929, 930; ibidem (1913), 2: 54 (Italy: Tyrol, Mendel-Pass). Syntypes, at least 2, ♂ ♀ (MNHU). Valid at subspecific rank.

T(ephroclystia) carpophagata benacaria DANNEHL, 1933: Ent. Z., Frankfurt a/M. 47 (14): 123 (Italy: Mt. Baldo). Syntype(s) (ZSM). Valid at subspecific rank.

DIAGNOSIS: Wingspan 18-22 mm. Labial palpi light grey, with scattered brown scales; concolorous with frons, vertex and notum. Forewing light ash-grey, with scattered brown scales; costal margin slightly darker, brownish grey; all transverse lines faint, indicated on costa by five divided greyish brown spots; terminal area slightly darker, brownish grey; discal dot absent. Hindwing paler, light ash-grey, with scattered brown scales; transverse lines inconspicuous; terminal area slightly darker, brownish grey; discal dot absent. Fringe shiny light grey, spotted with greyish brown opposite vein ends. Abdomen light ash-grey, with scattered brown scales; third segment darker, greyish brown.

MALE GENITALIA: Uncus short, uniapical. Papillae small, length about 0.2-0.3 times uncus length. Valva medium length and broadest medially. Vesica armed with one short, horn-like cornutus with wavy basal margin at ductus ejaculatorius base. Sternum A8 consists of two long, narrow, slightly curved, sclerotized and apically tapered rods, connected by short and narrow band at base; basal hollow deep; apical sinus about 0.8 times sternum length; sternum length about 0.8-0.9 times valva length.

FEMALE GENITALIA: Bursa copulatrix sclerotized, subglobular in basal part, completely covered with small spines, with short, very broad, transverse diverticulum in medial part; apical part of corpus bursae short, narrower, with short longitudinal striations and without spines. Ductus seminalis attached to corpus bursae between spiniferous basal and spineless posterior part from ventral side. Colliculum collar-like, relatively narrow but apically broadest. Anterior and posterior apophyses very short. Tergum A8 large, broad, rather trapeziform.

DISTRIBUTION: European (alpine). Mountains of Central and southern Europe: Spain, France, Italy, Switzerland, Austria, Slovenia, Macedonia, Bulgaria and Greece.

PHENOLOGY: Univoltine: early June (late May) to mid-July (late August). The f. *teriolensis* DIETZE, 1913 from late April to early July (WOLFSBERGER 1965). Larva: mid-July to mid-August, but recorded also from early May to early June (VORBRODT 1928). Overwintering as pupa in insecure cocoon in the ground.

BIOLOGY: Larva monophagous (1). Found in the flowers and seed capsules of *Silene saxifraga* (Caryophyllaceae) (Italy (Dolomites Alps): DIETZE 1900, 1910, 1913; France: WEIGT 1990), recorded also on *Silene rupestris* (Switzerland: VORBRODT 1928).

HABITAT: Forest edges, warm and dry slopes, rocks and screes. From 900 m up to about 2,400 m above sea-level in Alps (VORBRODT 1928; coll. ZSM), and from 1,400 m up to about 2,000 m in Apennines (PROLA & RACHELI 1980).

SIMILAR SPECIES: *E. carpophagata* can be easily separated from other *Eupithecia* species by the light ash-grey colour, with scattered brown scales, without or with faint transverse lines, absence of discal dots and with slightly darker, brownish grey terminal areas on all wings. The male and female genitalia of this species are very similar to those of *E. venosata* and *E. schiefereri*. However, the apical rods of sternum A8 are without hook-like apices, and the spineless, narrower apical part of corpus bursae is shorter, and membranous diverticulum of corpus bursae broader than in allied species.

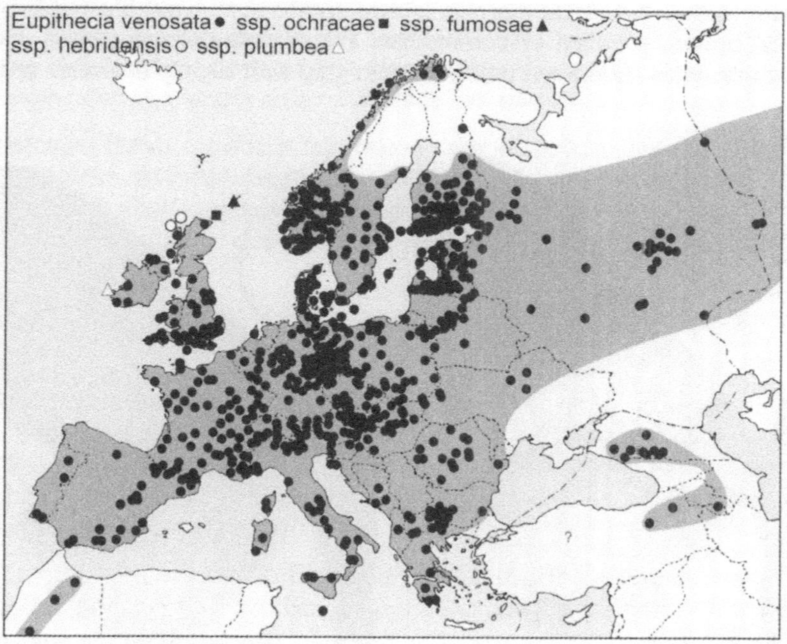

43. *Eupithecia venosata* (FABRICIUS, 1787)

Phalaena venosata FABRICIUS, 1787: Mantissa Insect. 2: 209 (Austria). Syntype(s) lost.
Ph(alaena) Geom(etra) insignata HÜBNER, 1789: Beitr. Ges. Schmett. 1(4): 9, pl. (4) 2: G (Austria: Vienna district). Syntype(s) lost.

Phalaena decussata DONOVAN, 1799: The Natural Hist. Brit. Insects 8: 38, pl. 266: 3 (United Kingdom, England: Kent, Faversham) nec DENIS & SCHIFFERMÜLLER, 1775 (identity: *Schistostege decussata*, Geometridae). Junior primary homonym.

Eupithecia venosata var. *ochracae* GREGSON, 1886: Young Naturalist 7: 128 (United Kingdom: Orkney Islands, Hoy; Wales). Syntypes (BMNH). Valid at subspecific rank.

Eupithecia venosata var. *fumosae* GREGSON, 1887: Young Naturalist 8: 111 (United Kingdom: Scotland: Shetland Islands). Syntypes (BMNH). Valid at subspecific rank.

(Eupithecia) venosata f. *grisea* DIETZE, 1913: Biol. Eupithecien 2: 56; ibidem (1910), 1, pl. 71: 164 (United Kingdom: Shetland Islands, Cunningsburg). Syntype(s) (MNHU). Valid at subspecific rank.

Eupithecia venosata f. *hebridensis* CURTIS, 1944: J. Soc. Br. Ent. 2 (5): 169 (United Kingdom: Scotland, Isle of Lewis). Holotype not traced in BMNH (L. PITKIN pers. comm.). Valid at subspecific rank.

Eupithecia venosata plumbea HUGGINS, 1962: The Entomologist's Rec. and J. Var. 74 (8): 171 (Ireland: Co(unty) Kerry, The Blaskets, Inishvickilaun [=Inishvickillane]). Holotype ♀ (BMNH?). Valid at subspecific rank.

Unavailable names (infrasubspecific): *confluens*: DIETZE (1913). - (incorrect subsequent spellings): *venosaria*: BOISDUVAL (1840); *ochracea*: DIETZE (1910).

DIAGNOSIS: Wingspan 18-27 mm. Labial palpi light brown. Frons, vertex and notum pale yellowish grey with slightly brownish tinge. Forewing unicolorous, delicate pale yellowish grey with slightly brownish tinge; costal margin slightly lighter; transverse stripes between basal, medial and terminal areas white; transverse lines very distinct, dark brown or black, sinuate, slightly widened and evenly curved near costa; postbasal and antemedial lines outwards sharply angled between stalk of Cu and A; medial line touching or passing through discal dot; postmedial line wavy; terminal area with black dashes on all M and CuA veins, without pale wavy line; discal dot black, dash-like, oblique. Hindwing paler, yellowish-grey; transverse lines brownish, rather conspicuous; terminal area darker, concolorous with forewing, with dark brown dashes on M3 to A, wavy line lacking; discal dot conspicuous, brown, rounded or comma-like. Fringe pale yellowish grey, spotted with brownish at vein ends. Abdomen pale yellowish grey with slightly brownish tinge; second

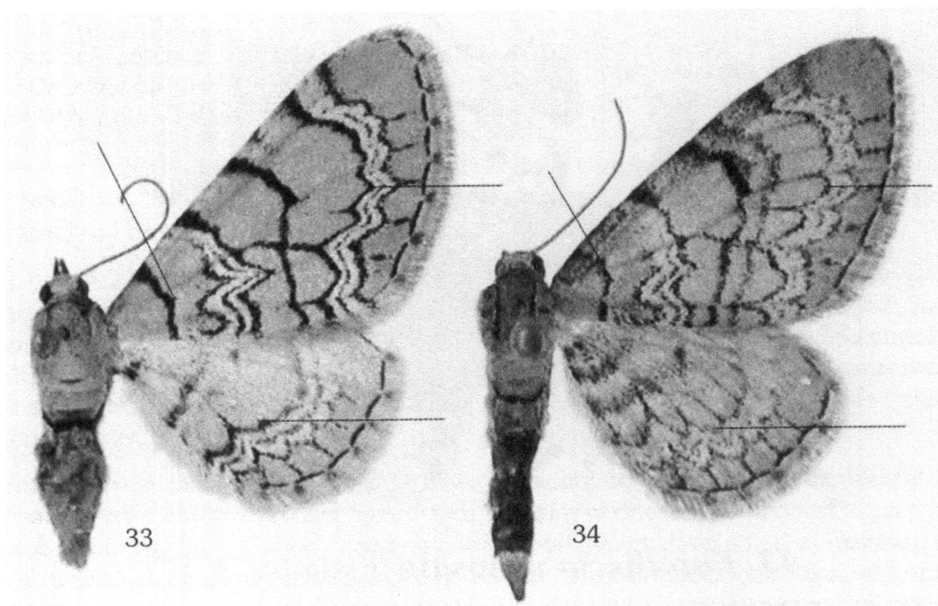

Text-figs 33-34: Diagnostic characters (indicated) of European *Eupithecia* species. Text-fig. 33: *E. venosata* (FABRICIUS, 1787). Text-fig. 34: *E. schiefereri* BOHATSCH, 1893.

segment with dark brown or black ring and with similarly coloured lateral stripes, which are also present on the next one to three segments.

In British Isles the following subspecies (or colour forms ?) occur:

Subsp. ***ochracae*** GREGSON, 1886 from the Orkney Islands, has ground colour rich yellowish ochreous. It is only slightly lighter than the following subspecies.

Subsp. ***fumosae*** GREGSON, 1887 from the Shetland Islands, with head, body and forewing darker, ochreous brown; medial line usually more sharply angled and closer to antemedial line on costa; postmedial line more wavy; two whitish double rivulet stripes between basal and medial areas, and between medial and terminal ones are more visible against background of the forewing than in nominate subspecies.

Subsp. ***hebridensis*** CURTIS, 1944 from the Hebrides Islands, is most similar to subsp. *fumosae*, but distinguished from it by the unicolorous warm ochreous brown forewings with bolder black transverse lines.

Subsp. ***plumbea*** HUGGINS, 1962 from the Blasket Islands, Ireland, has ground colour deep leaden grey on head, body and wings; postmedial line usually more sharply angled and closer to antemedial line on costa; transverse lines wider on the forewing costa than in nominate and other subspecies.

MALE GENITALIA: Uncus short, uniapical. Papillae length about 0.3-0.4 times uncus length. Valva medium length and broadest medially. Vinculum slightly tapered. Vesica armed with one small, delicate horn-like cornutus at ductus ejaculatorius base. Sternum A8 consists of two long, narrow, slightly curved and sclerotized rods, connected by narrow band at base, with incurved and very pointed tips; basal hollow deep; apical sinus about 0.8 times sternum length; sternum length about 0.8 times valva length.

FEMALE GENITALIA: Bursa copulatrix sclerotized, almost globular basally, completely covered by small spines, rounded diverticulum attached to medial part of corpus bursae from right side. Ductus bursae membranous, elongated and narrow, without spines. Colliculum collar-like, narrow. Anterior and posterior apophyses very short. Tergum A8 almost square.

DISTRIBUTION: West-Palaearctic. Widespread in Europe, except Iceland, Crete and European Turkey. – Outside Europe in North Africa (RUNGS 1981), Caucasus, Transcaucasus, north-eastern Turkey, to east through southern Urals, western and southern Siberia eastward to Baikal Lake.

PHENOLOGY: Uni- or probably bivoltine: late May (late March) to mid-July (late August). Larva: mid-July (late May) to mid-August (early September). Overwintering as pupa in the ground, sometimes two years.

BIOLOGY: Larva oligophagous (1). Found in the flowers and especially seed capsules of *Silene vulgaris, S. dioica* (United Kingdom: CREWE 1861c), *S. nutans* (France: MABILLE 1872), *S. latifolia (Melandrium album)* (Germany: DIETZE 1874a), *Silene uniflora (=maritima)* (United Kingdom: BARRETT 1904), *S. italica* (Italy: CARRARA 1928), *S. muscipula, Cucubalus*, (France: LHOMME 1935), *Lychnis viscaria* (Finland: SEPPÄNEN 1970) and *Silene sericea* (Malta: VALLETTA 1983). Larvae of all British subspecies recorded in the flowers and seed capsules of *Lychnis viscaria* and *Silene uniflora* (Caryophyllaceae) (United Kingdom: SCOBLE 1999).

PARASITOIDS: Hymenoptera, Ichneumonidae: *Dusona nidulator*, *D. angustifrons* - Hymenoptera, Braconidae: *Meteorus luridus* (THOMPSON 1945; SHENEFELT 1969).

HABITAT: Preferring dunes, dry hills, chalk or limestone districts, old gravel pits, hedges, sunny roadsides, sea cliffs, warm and dry forest fringes, sandy fields and slopes, rocks and stony slopes. From 0 m up to about 1,600 m above sea-level in Balkan peninsula (PINKER 1968), to about 1,700 m in Pyrenees and Apennines (coll. FIBIGER; PARENZAN 1994), to about 2,200 m in Alps (VORBRODT & MÜLLER-RUTZ 1914; FORSTER & WOHLFAHRT 1981; REZBANYAI-RESER & MAIER 2000) and to about 2,600-2,700 m in Middle Atlas (RUNGS 1981).

SIMILAR SPECIES: A handsome species, which is impossible to confuse with any other. *E. venosata* can usually be easily separated from other closely related *Eupithecia* species by the distinctive maculation and delicate colour of wings (Text-fig. 33). Sternum A8 in male, and shape of corpus bursae, area covered by spines and position of membranous diverticulum in the female genitalia should distinguish this species from other representatives of the *venosata* species-group.

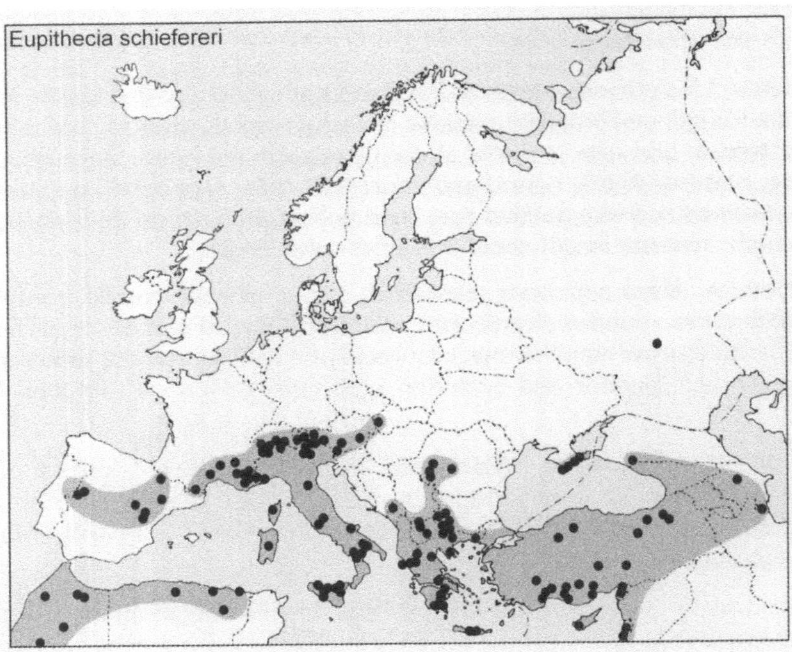

44. *Eupithecia schiefereri* BOHATSCH, 1893

Eupithecia venosata var. *schiefereri* BOHATSCH, 1893: Dt. ent. Z. Iris 6: 31 (Austria: Graz). Syntype(s) (NHMW). Raised to species rank by DIETZE (1904).

Eupithecia venosata var. *caeruleata* FAVRE, 1899: Faune des Macro-Lépid. du Valais: 305 (Switzerland: Mt. Chemin, Mt. Ravoire, Sion). Syntypes. Valid at subspecific rank.

Unavailable names (infrasubspecific): *confluens*: DIETZE (1910). - (incorrect subsequent spellings): *schiefererie*: PINKER (1958); *schieferi*: ARNSCHEID (2000).

DIAGNOSIS: Wingspan 18-23 mm. Labial palpi brown. Frons, vertex and notum pale yellowish grey with slightly brownish tinge. Forewing unicolorous, pale yellowish grey with slightly brownish tinge; costal margin slightly lighter; transverse stripes between basal, medial and terminal areas white; transverse lines very distinct, dark brown or black, sinuate, slightly widened and evenly curved near costa; postbasal and antemedial lines twice sharply angled outwards between stalks of R and Cu, and between stalk of Cu and A; medial line touches or passes through discal dot; postmedial line wavy; terminal area with black dashes on all M and CuA veins, without pale wavy line; discal dot black, dash-like, oblique. Hindwing paler, yellowish grey; transverse lines brownish, rather conspicuous; terminal area darker, concolorous with forewing, with dark brown dashes on M3 to A, without pale wavy line; discal dot distinct, brown, usually minute, rounded or comma-like. Fringe pale yellowish grey, spotted with brownish at vein ends. Abdomen pale yellowish grey with slightly brownish tinge; second segment with dark brown or black ring and concolorous lateral stripes, which are also present on the next one to three segments.

MALE GENITALIA: Uncus short, uniapical. Papillae broad and very short, less than 0.2 times uncus length. Valva medium length, broadest medially. Aedeagus relatively small and delicate. Vesica armed with two cornuti, not connected at base, one very small, the other larger, horn-like. Sternum A8 consist of two long, relatively broad, curved, apically sclerotized and blunt rods, bearing subapical incurved, tapered process; rods connected by short and narrow band at base; basal hollow broad and deep; apical sinus very deep, about 0.8 times sternum length; sternum length about 0.8 times valva length.

FEMALE GENITALIA: Bursa copulatrix elongated, irregular, broadest and rounded at base, covered with small spines, except 0.3-0.4 broadest basal part; diverticulum bursae membranous, short and very broad, attached to medial part of corpus bursae; ductus bursae elongated and sclerotized. Ductus seminalis attached to medial part of corpus bursae between spiniferous corpus bursae and ductus bursae. Colliculum collar-like, relatively elongated and narrow. Tergum A8 large, rectangular, with slightly curved posterior margin.

DISTRIBUTION: Mediterranean. Central and southern Europe. – Outside Europe in North Africa (RUNGS 1981), Asia Minor, Cyprus, Lebanon, Israel, Caucasus and Transcaucasus. – New for the fauna of Corsica: 1♀, Corsica, Haut Asco, 1,200-1,400 m, 23.VI.1994, leg. B. SKULE & P. SKOU (coll. SKOU); 1♀, Corse, 0-30 m, Cargèse, 15.-16.V.1999, leg. O. KARSHOLT (ZMUC).

PHENOLOGY: Univoltine: early May (early April) to late June (early August). Larva: early June (mid-May) to mid-July (early August). Overwintering as pupa, occasionally twice.

BIOLOGY: Larva oligophagous (1). Found in the flowers and seed capsules of *Silene italica* (Switzerland: FAVRE 1905), also *Saponaria ocymoides* (France: DIETZE 1910, 1913), *Silene otites* (Switzerland: VORBRODT 1921), *S. nutans, Lychnis alpina* (Caryophyllaceae) (Italy: WEIGT 1986, 1990).

HABITAT: Warm and dry forest fringes, meadows, steppes, rocks and stony slopes. From 0 m up to about 1,800 m above sea-level in Apennines (coll. ZSM), to about 2,000 m in Alps (WEIGT 1990), and to about 2,300 m in Great Atlas (RUNGS 1981).

SIMILAR SPECIES: This species is most similar and closely related to *E. venosata*. However, the wings are usually slightly paler and more greyish tinged, all transverse lines narrower, basal line medially always right-angled, postbasal and antemedial lines twice sharply angled outwards, postmedial line more wavy than in *E. venosata* (Text-fig. 34). The genitalia of *E. schiefereri* are similar to those of *E. venosata* but very clearly distinguished by the different shape of sternum A8 in male, and the shape and sclerotization of bursa copulatrix, the area covered by spines and the position of membranous diverticulum in the female genitalia.

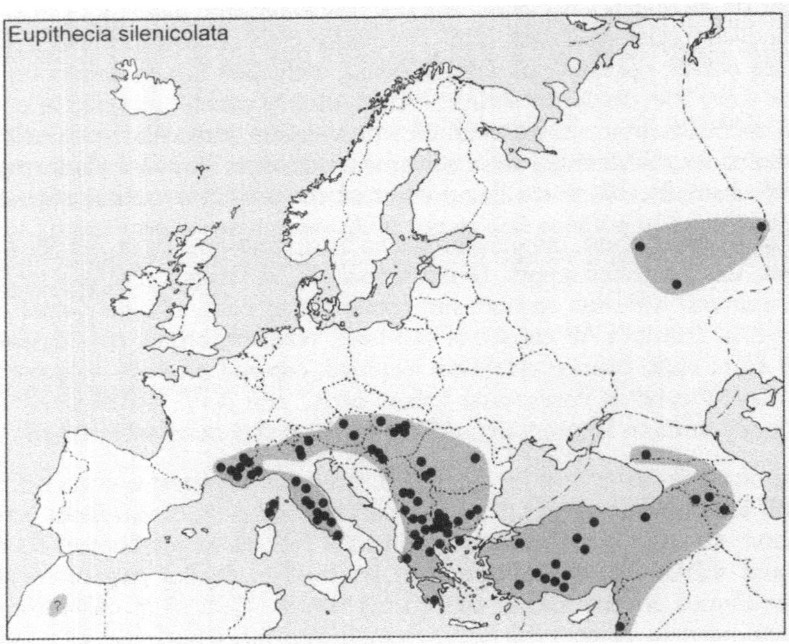

45. *Eupithecia silenicolata* MABILLE, 1867

Eupithecia silenicolata MABILLE, 1867: Annls Soc. ent. France (4) 6 (4): 562, pl. 8: 9a, b (Corsica: Bastia). Lectotype ♀ (coll. HERBULOT in ZSM, examined).

Eupithecia silenicolata perdistincta WEHRLI, 1933, in ZERNY, Dt. ent. Z. Iris 47 (2): 95, pl. 1: 21 (Lebanon: Beyruth). Syntypes 6 (ZFMK and NHMW). Valid at subspecific rank.

Eupithecia silenicolata soultanabadi BRANDT, 1941: Mitt. münch. ent. Ges. 31 (3): 875, pl. 29: 27 (Iran: Soultanabad). Holotype ♀ (NHRS). Valid at subspecific rank.

Eupithecia silenicolata zengoensis FAZEKAS, 1979: Linn. belg. 7 (11): 406, figs 1-3 (Hungary: Mecsek Mt., Püspökszent-lázló). Holotype ♂ (HNHM?). Valid at subspecific rank.

DIAGNOSIS: Wingspan 17-22 mm. Labial palpi brown. Frons pale greyish brown with scattered brown scales. Vertex and notum unicolorous pale greyish brown. Forewing unicolorous pale greyish brown; transverse lines distinct, dark brown, slightly sinuous, evenly curved and slightly widened near costa; postbasal and antemedial lines sharply angled outwards between stalk of Cu and A; first medial line touches or passes through crosses discal dot; second medial line usually parallel with it and inconspicuous; postmedial line slightly wavy; terminal area without pale wavy line; discal spot dark brown, dash-like or elongated ovate, oblique. Hindwing paler, pale greyish brown; brown transverse lines distinctly marked near inner margin; terminal area darker, narrow, without pale wavy line; discal spot conspicuous, brownish, rounded or elongated. Fringe pale greyish brown, spotted with darker brown at vein ends. Abdomen pale greyish brown usually with small dark brown dorsal spots; second segment with dark brown band and dark brown lateral stripes on the second to fourth segments.

MALE GENITALIA: Uncus elongated, uniapical. Papillae length about 0.2 times uncus length. Valva elongated, narrow, with almost parallel ventral and dorsal margins. Vinculum relatively elongated and narrow. Aedeagus delicate. Vesica armed with two long, narrow, delicate horn-like cornuti, connected at base. Sternum A8 consist of two long, narrow, slightly sclerotized, apically rounded rods, connected by short, narrow band at base; basal hollow shallow; apical sinus very deep, more than 0.8 times valva length; sternum length about 0.5-0.6 times valva length.

FEMALE GENITALIA: Bursa copulatrix pear-shaped, almost completely covered with small spines, with small membranous, globular diverticulum near base and with small, sclerotized diverticulum in medial part of corpus bursae on right side. Ductus seminalis attached to bursa copulatrix near posterior margin from right side. Colliculum collar-like, short and narrow. Antrum lightly sclerotized. Posterior apophyses slightly expanded medially. Tergum A8 large, rectangular.

DISTRIBUTION: Sub-Mediterranean. Southern France, Corsica, Italy, Switzerland, Austria, Slovakia, Hungary, Serbia, Macedonia, Romania, Bulgaria, Greece and south-eastern European Russia. – Outside southern Europe in Morocco (RUNGS 1981), Asia Minor, Lebanon (subsp. *perdistincta* WEHRLI, 1933), western Caucasus and Transcaucasus.

PHENOLOGY: Bivoltine: early May (late April) to mid-June (late June); mid-July (early July) to mid-August (late August). Larva: late June to late July. Overwintering as pupa.

BIOLOGY: Larva monophagous (3). Found in the flowers and seed capsules of *Silene paradoxa* (Corsica: MABILLE 1867, 1868), also *S. otites, S. nutans* (Switzerland: FAVRE 1905), *S. italica (=nemoralis)* (Austria: DIETZE 1910, 1913; France: WEIGT 1990), *S. viridiflora* (Caryophyllaceae) (Hungary: FAZEKAS 1979), and probably on *Silene dichotoma, S. nicaeensis, S. colorata* and *S. sericea* (Caryophyllaceae) (WEIGT 1990).

HABITAT: Warm and dry forest edges and meadows, dry slopes, rocks and screes. From 0 m up to about 1,300 m above sea-level in Balkan peninsula and Asia Minor (NESTOROVA 1998; coll. ZSM), to about 1,400 m in Apennines (Gran Sasso Mts: coll. ZSM), to about 1,600 m in Middle Atlas (RUNGS 1981), and to about 1,800 m in Alpes Maritimes (HAUSMANN pers. comm.).

SIMILAR SPECIES: Similar to *E. venosata* but labial palpi longer; male antennae with longer setae; wing colour browner without yellow tinge; transverse lines narrower and less distinct; transverse stripes between basal, medial and terminal areas not white, concolorous with forewing. *E. silenicolata* can usually be easily distinguished from other species in the *venosata* group by the longer uncus, narrower valva and vinculum, two long and slim cornuti in the vesica, shape of sternum A8 in male, as well as by the shape of corpus bursae and position of membranous diverticulum in the female genitalia.

46. *Eupithecia alliaria* STAUDINGER, 1870

Eupithecia alliaria STAUDINGER, 1870: Berl. Ent. Z. 14: 129 (Hungary: Ofen (near Budapest)). Syntype(s) (MNHU).

Eup(ithecia) alliaria f. *notata* DIETZE, 1910: Biol. Eupithecien 1, pl. 70: 135, 136; ibidem (1913), 2: 61 (France: Digne). Syntypes (MNHU). Valid at subspecific rank

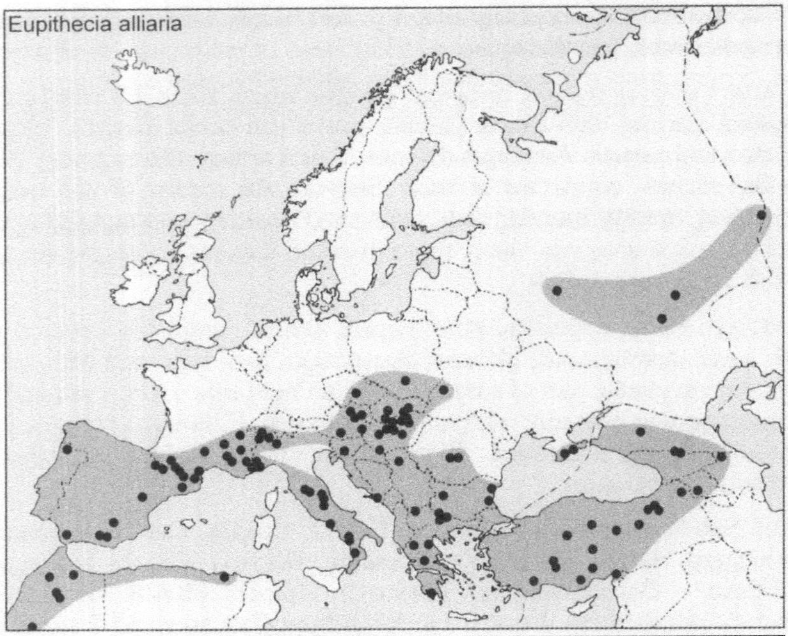

DIAGNOSIS: Wingspan 17-21 mm. Labial palpi brown. Frons, vertex and notum greyish brown. Forewing unicolorous greyish brown; transverse lines dark brown, very narrow, slightly widened near costa, rather inconspicuous except postmedial line; postbasal and antemedial lines obtuse angled outwards between stalks of R and Cu, and between stalk of Cu and A; double medial line touches or passes through discal spot; postmedial line usually evenly curved near costa; pale wavy line evenly curved with small pale tornal spot; discal spot minute, dark brown, ovate or elongated, oblique, sometimes inconspicuous. Hindwing slightly paler, greyish brown; transverse lines indistinct, marked only near dorsum; terminal area darker and obscure; discal spot brownish, minute, rounded, indistinct or absent. Fringe greyish brown, rather unicolorous. Abdomen greyish brown, without dark lateral stripes or dark ring on the second segment.

MALE GENITALIA: Uncus short, biapical. Papillae length about 0.5-0.6 times uncus length. Valva medium length, medially broadened. Vinculum with shallow medial hollow. Vesica armed with one small, delicate horn-like cornutus at ductus ejaculatorius base. Sternum A8 broad with two elongated, slightly curved, sclerotized, apically pointed horns; basal hollow deep; apical sinus deeper, about 0.5 times sternum length; sternum length about 0.7-0.8 times valva length.

FEMALE GENITALIA: Bursa copulatrix pear-shaped, almost completely covered with small spines, with small, globular, membranous diverticulum near base from right side. Colliculum collar-like, short and narrow. Tergum A8 square. Papillae anales small, slightly tapered.

DISTRIBUTION: Sub-Mediterranean. Central and southern Europe, except Mediterranean islands. – Outside Europe in North Africa (Morocco and Algeria), Asia Minor, Caucasus and Transcaucasus.

PHENOLOGY: Univoltine: late June (late May) to early August (early September). Larva: late July to late September (early October). Overwintering as pupa, occasionally twice.

BIOLOGY: Larva rather monophagous (3). Found on the inflorescences and seeds of *Allium flavum* (Hungary: STAUDINGER 1870), also *Allium paniculatum* (France: DIETZE 1910, 1913), *A. oleraceum, A. lineare, A. vineale, A. senescens (=montanum)* (Alliaceae) (Austria and Italy: WEIGT 1990) and probably occasionally on flowers and seeds of *Camphorosma monspeliaca* (Chenopodiaceae) (France: LHOMME 1935).

PARASITOIDS: Hymenoptera, Braconidae: *Aleiodes circumscriptus* (SHENEFELT 1975).

HABITAT: Forest edges, warm and dry slopes, rocks and screes. From 0 m up to about 1,500 m above sea-level in Pyrenees (coll. ZMUC), to about 1,600 m in Apennines (Gran Sasso Mts: coll. ZSM), to about 1,700 m in Balkan peninsula (coll. FIBIGER), to about 1,800 m in southern Alps (HELLMANN & al. 1999), to about 1,700-2,000 m in Sierra Nevada (coll. ZSM), to about 2,300 m in Asia Minor (coll. ZSM). and to about 2,600-2,700 m in Great Atlas (RUNGS 1981).

SIMILAR SPECIES: *E. alliaria* can usually be easily distinguished from the similar and closely related *E. silenicolata* by the shorter labial palpi and shorter setae on the male antennae, browner wing colour, less distinct transverse lines and less wavy postmedial line, by the presence of pale wavy line and tornal spot on the forewing, smaller, narrower and paler discal spots and darker, brown, almost unicolorous fringe. The male and female genitalia can be used to separate *E. alliaria* from *E. silenicolata* with accuracy.

47. *Eupithecia dissertata* (PÜNGELER, 1905)

Tephroclystia dissertata PÜNGELER, 1905: Soc. Ent. 20 (18): 138 (Switzerland: Wallis, Zermatt). Syntype(s), including ♀ (MNHU).
Eupithecia kaszabi VOJNITS, 1974: Acta Zool. Acad. Sci. Hung. 20 (3-4): 445, figs. 1, 2 (Mongolia: SO of Somon Bajanzogt, Central aimak). Holotype ♂ (HNHM, examined). New junior synonym.
Eupithecia certa VOJNITS, 1981: Acta Zool. Acad. Sci. Hung. 27 (3-4) 420, fig. 31 (China: Shansi Province, Mien-Shan). Holotype ♀, nec paratype ♂ (ZFMK, examined). New junior synonym.

DIAGNOSIS: Wingspan 18-24 mm. Labial palpi, frons and vertex pale brownish grey. Notum reddish brown. Forewing rather narrow, with straight costa and obliquely rounded termen; reddish brown, suffused at termen and ash-grey along costa; transverse lines fine, dark brown, broadened on to costa; basal line once angled at stalk of Cu; postbasal and antemedial lines twice angled; two parallel lines sinuate, crosses on the short distance with each other, and they are sharply angled near discal spot on to costa; postmedial line sinuate, curved near costa; pale wavy line usually inconspicuous; discal spot small, obliquely ovoid or narrow, black. Hindwing brownish grey with terminal shade; three fuscous transverse lines well marked along dorsum only; discal spot minute, pale, indistinct. Fringe slightly chequered whitish brown and brown. Abdomen pale brownish grey with a series of small brown dorsal dots.

MALE GENITALIA: Uncus short, biapical. Papillae narrow, relatively long. Vesica armed with two horn-like cornuti (one large, elongated, slightly curved, the other smaller, hook-like near apex) and one small, elongated, lightly sclerotized cornutus in medial part of aedeagus. Sternum A8 broad with two long, pointed, pincer-like apical horns; basal and apical hollows deep, especially apical hollow.

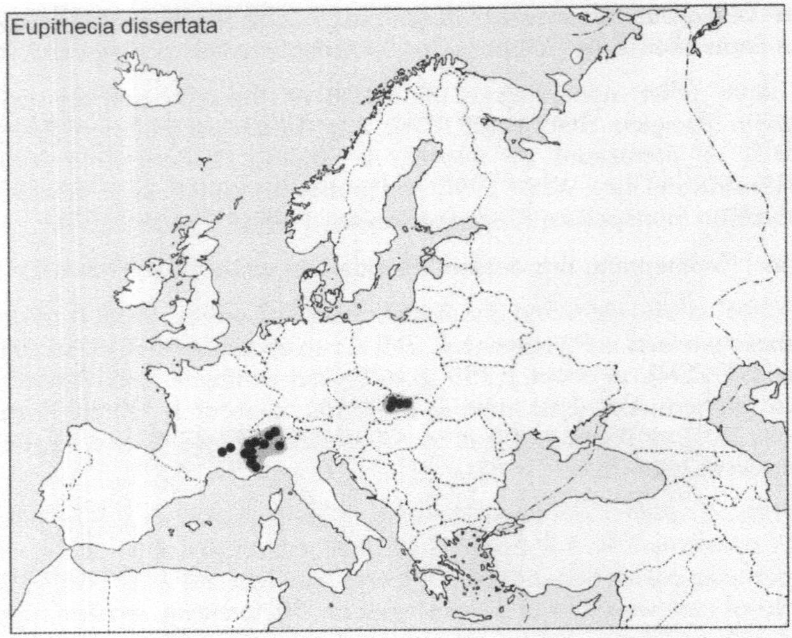

FEMALE GENITALIA: Bursa copulatrix elongated pouch-like or rather fusiform, almost half covered with small spines; diverticulum large, membranous. Ductus seminalis on the dorsal side of corpus bursae near colliculum. Colliculum rather plate-like, short and broad.

DISTRIBUTION: Euro-Siberian (disjunct alpine). This species occurs in the Alps (south-eastern France, Switzerland and north-western Italy), the French Central Massif (two specimens from dept. Gard and dept. Ardèche) and the Carpathians (Slovakia). – Outside Europe in Ili district? (Central Asia), southern Siberia (Altai, Sajan, Transbaikalia), northern Mongolia and north-eastern Siberia (Magadan region) and China (Shansi Province) (VOJNITS 1977; VIIDALEPP 1978, 1996; MIRONOV 1989, 1990).

PHENOLOGY: Univoltine: late June to early August (late August).

BIOLOGY: Unknown.

HABITAT: Rocky limestone slopes. From 800 m up to about 2,000 m above sea-level in Alps and Carpathians (DUFAY 1976; MIKAT 1998; HELLMANN & al. 1999; coll. ZSM).

SIMILAR SPECIES: Variable in the depth of colour from light grey to brown. This species is usually larger than *E. alliaria*, the only species with which it could be confused, with more elongated and pointed forewings. In *E. dissertata*, the transverse lines are more distinctly marked with dark costal spots, postmedial line strongly wavy on the forewing, fringe usually more chequered. The male and female genitalia can be used to separate *E. dissertata* from similar species with accuracy, especially cornuti in the vesica and bursa copulatrix with large anterior diverticulum, as illustrated.

The ***interruptofasciata*** species-group

This group was introduced and described in detail by BOLTE (1990) for the first time after carefully examining two North American species, *E. interruptofasciata* PACKARD, 1873 and *E. niphadophilata* (DYAR, 1904) and one European species *E. pusillata*. It is a large Holarctic species-group, most numerous in the West Palaearctic region, especially in the Mediterranean area. The *interruptofasciata* group is represented in Europe by fourteen species: *E. cocciferata*, *E. abbreviata*, *E. lentiscata*, *E. dodoneata*, *E. reisserata*, *E. massiliata*, *E. extremata*, *E. scopariata*, *E. pusillata*, *E. sardoa*, *E. ericeata*, *E. phoeniceata*, *E. oxycedrata* and *E. rosmarinata*. The North African *E. rusicadaria* DIETZE, 1910, *E. rhoisata* CHRÉTIEN, 1917 and at least four Asian species, *E. dubiosa* DIETZE, 1910, *E. maerkerata* SCHÜTZE, 1938, *E. kozlovi* VIIDALEPP, 1973 and *E. masuii* INOUE, 1980 can be included in this group also.

MALE GENITALIA: Uncus short, biapical. Papillae usually narrow and short, length about 0.6-0.8 times uncus length. Valva broadest medially, diverse in size and shape, with more or less small, blunt medial ventral process or without it; sacculus heavily sclerotized only in some species. Vinculum short and broad. Aedeagus usually large, long and thick, broader at base; length about 0.9-1.3 times valva length, about 3.0-5.5 times longer than medial width. Vesica armed with one or two usually long and slim horn-like cornuti at apex (often one being longer and more sclerotized than other), with one thin, flat, elongated cornutus and usually with one smaller, irregular, curved cornutus at ductus ejaculatorius base. Sternum A8 peg-like, short or elongate, with base broadened and with two small or minute narrow apical lobes; basal and apical hollows shallow; sternum length about 0.5-0.9 times valva length.

FEMALE GENITALIA: Bursa copulatrix usually more or less elongated, twisted, sclerotized, and also usually broader and rounded at base; covered with spines at base, with or without a row of more or less large spines along left side and with a row of spinules on right side in some species. Ductus bursae with longitudinal striations. Ductus seminalis narrow, from short to long; attached to medial part of corpus bursae on right side. Colliculum usually weakly defined. Antrum membranous or slightly sclerotized. Anterior and posterior apophyses thin and short, tapered to apexes. Spurs of anterior apophyses usually slightly broadened at apices or thin along entire length. Tergum A8 diverse in shape: square, trapeziform or rectangular. Papillae anales usually small, rounded or slightly tapered.

REMARKS: The combination of features including shape of valva and sternum A8; large aedeagus; long, slim horn-like cornuti in the vesica; sclerotized and elongated bursa copulatrix with characteristic spiniferous areas, can be used to separate this species-group from others initially. Some features of the *egenaria*, *lanceata*, *lariciata* and *tripunctaria* species-groups actually suggest these groups are closely related to representatives of the *interruptofasciata* species-group. See also corresponding species-group diagnosis and remarks.

48. *Eupithecia cocciferata* MILLIÈRE, 1864

Eupithecia cocciferata MILLIÈRE, 1864: Iconogr. Descr. Chenilles Lépid. inédits 2 (12): 45, pl. 56: 1-4 (France: Provence; Celles-les-Bains, Ardèche). Syntypes ♂ ♀ (MNHN?).

Eupithecia cocciferata var. *semitinctaria* MABILLE, 1868: Annls Soc. ent. France 7 (4): 651, pl. 14: 2 (Corsica). Lectotype ♂ herewith designated to stabilize nomenclature (coll. HERBULOT in ZSM, examined). Valid at subspecific rank.

Unavailable names (incorrect subsequent spelling): *coccifera*: MABILLE (1868).

DIAGNOSIS: Wingspan 16-22 mm. Labial palpi dark brownish grey. Frons and vertex light

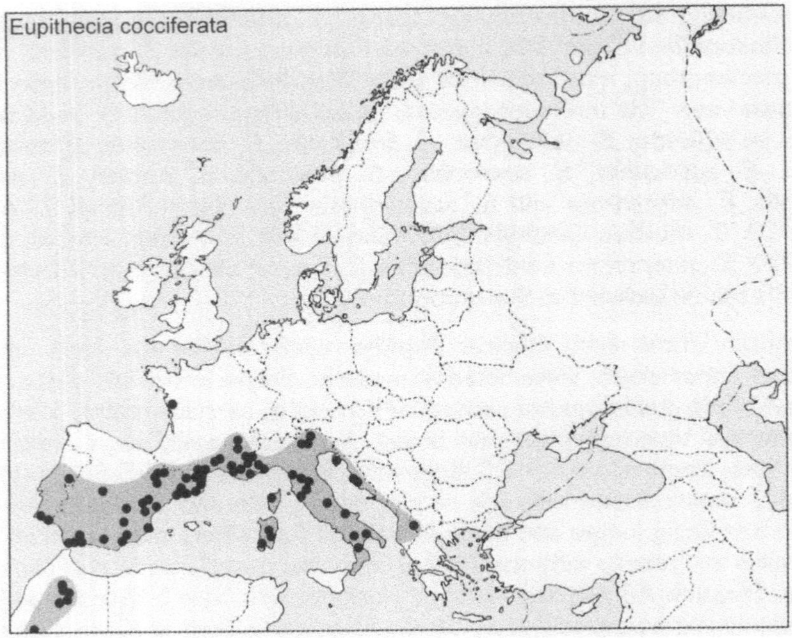

brownish grey. Notum dark brownish grey. Forewing elongated; costa arched; apex pointed; termen rounded, oblique; ground colour brownish grey; stalk of Cu usually chequered black and pale grey; transverse lines dark brownish grey to black-brown; basal oblique, twice indented; postbasal and antemedial lines sinuous, evenly curved near costa; first medial line strongly sinuous, touching discal spot and sharply angled from it towards costa; second medial line fine, more or less distinct; postmedial line multiply wavy, twice right-angled near costa, darker on the veins; medial area between antemedial and medial lines usually darker, but lighter under discal spot; terminal area narrow; pale wavy line clouded, broken, often invisible, forming a small, double, whitish tornal spot; discal spot small, circular, intensely black. Hindwing angulated with shallowly concave termen; ground colour paler than on the forewing, especially in anterior part; transverse lines distinct along dorsum to discoidal cell; terminal area narrow, slightly darker; discal spot minute, usually rounded, ill-defined. Fringe light brownish grey, spotted with dark brownish grey at vein ends. Abdomen brownish grey; narrow black lateral stripes present.

MALE GENITALIA: Papillae broad, nearly equal to uncus length. Valva without medial ventral process, broader at base and evenly tapered toward apex; apex narrowly rounded. Vesica armed with two apical horn-like cornuti (one large, strongly hooked, the other minute, straight), with one small nipple-shaped cornutus and also with one elongated, irregular and curved cornutus at ductus ejaculatorius base. Sternum A8 elongate, with two minute apical lobes.

FEMALE GENITALIA: Bursa copulatrix rather pear-shaped, not twisted, more than half covered by spines, with a patch of larger spines on the left side. Ductus seminalis very long, evenly tapered along its entire length. Colliculum well defined, short and narrow. Spurs of anterior apophyses broader at apexes. Tergum A8 almost square. Papillae anales tapered, covered with long setae.

Text-figs 35-37: Diagnostic characters (indicated) of European *Eupithecia* species. Text-fig. 35: *E. cocciferata* MILLIÈRE, 1864. Text-fig. 36: *E. abbreviata* STEPHENS, 1831. Text-fig. 37: *E. dodoneata* GUENÉE, 1858.

DISTRIBUTION: West-Mediterranean. Portugal, southern Spain, southern France, Italy, Corsica, Sardinia and Balkan peninsula along Adriatic coast. – Outside Europe there is a separate population in North Africa (Morocco).

PHENOLOGY: Univoltine: early March to mid-May (late June). Larva: late April to early June (late June). Overwintering as pupa in fairly dense cocoon in the ground.

BIOLOGY: Larva monophagous (2). Found on the male flowers and young leaves of *Quercus coccifera, Q. suber* and *Q. ilex* (Fagaceae) (France: MILLIÈRE 1864; WEIGT 1993; Portugal: CANDIDO MENDES d' AZEVEDO 1903; 'Mauretania': DIETZE 1910, 1913).

HABITAT: Oak forests and scrub, cultivated areas, gardens and orchards with scattered oaks. From 0 m up to about 1,300 m above sea-level in southern Alps (coll. ZSM), to about 1,500-1,600 m in Pyrenees and Sierra Nevada (MAZEL & PESLIER 1997; coll. ZSM) and to about 1,800-2,000 m in Middle Atlas and Great Atlas (coll. ZSM; coll. ZMUC).

SIMILAR SPECIES: *E. cocciferata* is similar to *E. abbreviata*, the only species with which it could be confused, but on the whole darker, with less curved, broader and darker post-medial line, with small double whitish tornal spot and with larger, circular, black discal spot on the forewing (Text-fig. 35). The male genitalia of *E. cocciferata* clearly differ from those of *E. abbreviata* in the narrower valva without ventral process, rounded vinculum, shorter aedeagus with smaller and sharply curved horn-like cornutus on the vesica. The female genitalia differ in the different shape of the bursa copulatrix, different disposition of spines in corpus bursae and narrower colliculum.

49. *Eupithecia abbreviata* STEPHENS, 1831

Eupithecia abbreviata STEPHENS, 1831: Illustr. Br. Ent. (Haustellata) 3: 283 (United Kingdom: England, Kent, Darenth; York). Syntypes (BMNH).

Phalaena nebulata HAWORTH, 1809: Lepid. Br. (2): 360 (United Kingdom). Syntype(s) (BMNH). Junior primary homonym of *Phalaena nebulata* SCOPOLI, 1763.

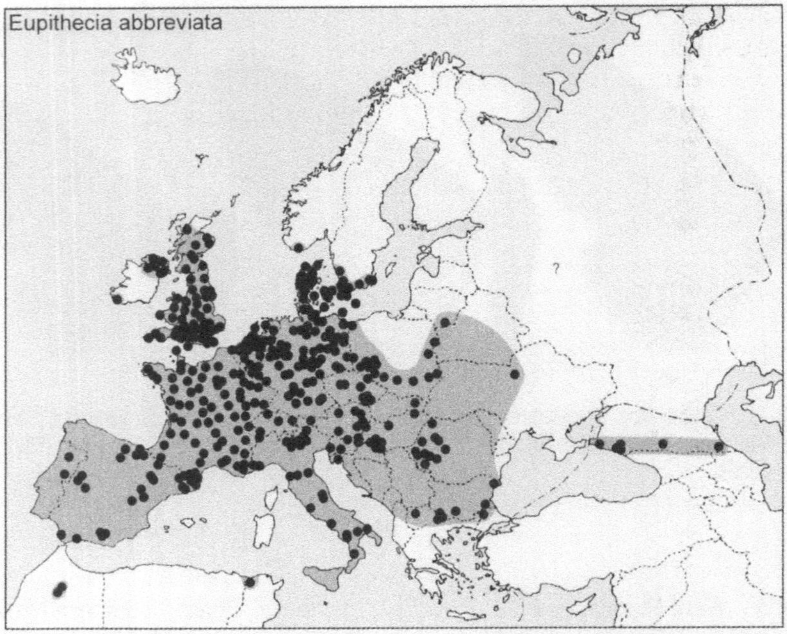

Eupithecia subfasciata STEPHENS, 1831: Illustr. Br. Ent. (Haustellata) 3: 284 (United Kingdom: England, Kent, Darenth Wood). Holotype ♀ (BMNH).

Unavailable names (misidentification): *guinardaria*: sensu DUPONCHEL (1842) nec BOISDUVAL, 1840 (identity: *Eupithecia scopariata*). – (infrasubspecific): *hirschkei*: BASTELBERGER (1908); *striata*: LEMPKE (1951); *nigra*: COCKAYNE (1953); *pallifasciata*: LEMPKE (1969); *impuncta*: LEMPKE (1969); *obsoleta*: LEMPKE (1969). - (incorrect subsequent spellings): *abbreviaria*: DOUBLEDAY (1849); *abreviata* MILLIÈRE (1870).

DIAGNOSIS: Wingspan 18-25 mm. Labial palpi brownish grey. Frons and vertex pale brownish grey. Antennae ringed. Notum light brownish grey with wide brown transverse band on the pronotum. Forewing rather elongate; costa arched; apex pointed; termen gently rounded and oblique; ground colour light brownish grey; transverse lines more or less conspicuous, dark brownish grey; basal, postbasal and antemedial lines oblique, indented, sharply angled near costa; strongly sinuous, touching discal spot and obtusely curved from it toward costa; medial area between antemedial and medial lines darker and between medial and postmedial lines usually lighter than rest of wing; postmedial line multiply wavy, rather irregularly curved and twice right-angled near costa, on its inner side with series of elongated, black dashes on the veins; terminal area narrow, broadest near apex; pale wavy line broad, clouded, slightly wavy, sometimes forming a small pale tornal spot; discal spot small, often indistinct, very narrow, obliquely elongate or sometimes ovoid. Hindwing rather angulated with weakly concave termen; ground colour distinctly paler than forewing; transverse lines represented by series of dark dashes or wedge-shaped marks, well marked along dorsum; terminal area narrow, usually darker; pale wavy line invisible; discal spot minute, round. Fringe chequered light and dark brownish grey. Abdomen brownish grey, with narrow black lateral stripes.

MALE GENITALIA: Valva with large, apically blunt medial ventral process; sacculus heavily sclerotized. Aedeagus equal to valva length. Vesica armed with two horn-like cornuti (one very long, slim, usually slightly sinuate and other short, broad, rather nipple-like), and with two flat, curved cornuti (one elongated and one smaller, irregular) at ductus ejaculatorius base. Sternum A8 with two small, tapered apical lobes.

FEMALE GENITALIA: Bursa copulatrix strongly twisted; covered with spinules at base, with larger marginal spines, and with elongated patch of spinules in ductus bursae. Spurs of anterior apophyses broadened throughout almost their entire length, attached to anterior apophyses base. Tergum A8 almost square, with shallow medial hollow and medial tooth on the posterior margin.

DISTRIBUTION: Euro-Caucasian. Western, Central and southern Europe, except Balearics, Corsica, Sardinia, Malta, Crete and south of Balkan peninsula. Recorded for central Russia by ANTONOVA (1981) based on one very old specimen captured near Moscow. – Outside Europe in North Africa (Morocco, Tunisia), and northern Caucasus. - New for the fauna of North Africa: 1♂, Tunisia, 7 km S of Ain Draham, Les Chênes, 22.III.1986; 2♂3♀, 13 km N. of Ain Draham, 22.-24.III.1986; 1♂3♀, Ain Draham area, 5.-18.V.1988, Zool. Mus. Copenh. Exp. (ZMUC); 2♀, Nord-Africa, Marokko, Moyen Atlas, Azrou, ca. 1300 m, 14.IV.1997, leg. J.GELBRECHT & E.SCHWABE; Nord-Africa, Marokko, Moyen Atlas, Ifrane, 1600-1700 m, 15.-17.IV.1997, leg. J. GELBRECHT, T. DRECHSEL & E. SCHWABE (coll. GELBRECHT).

PHENOLOGY: Univoltine: early April (early March) to late May (mid-June). Larva: mid-May (early May) to mid-June (in British Islands found to July). Overwintering as pupa.

BIOLOGY: Larva monophagous (3). Found on the flowers and leaves of *Quercus robur (=pedunculata), Q. ilex, Q. suber, Q. coccifera* (France: MILLIÈRE 1870; LHOMME 1935) and *Quercus petraea* (Fagaceae) (Germany: WEIGT 1980, 1993; WEGNER pers. comm.), mainly on the young shoots.

PARASITOIDS: Hymenoptera, Ichneumonidae: *Dusona pugillator, Netelia tarsata* - Hymenoptera, Braconidae: *Apanteles cajae, A. tibialis (=congestus), Meteorus pulchricornis* (BUCKLER 1899; THOMPSON 1945; SHENEFELT 1969, 1972; ZWAKHALS pers. comm.).

HABITAT: Deciduous forests, scrub, hedges, gardens and parks where there are mature oaks. From 0 m up to about 1,000 m in Apennines (PROLA & RACHELI 1980), to about 1,300 m in Pyrenees (MAZEL & PESLIER 1997), and to about 1,800-1,900 m in southern Alps (coll. FIBIGER).

SIMILAR SPECIES: On the whole *E. abbreviata* is an easily recognised species, but it could occasionally be confused with some of its relatives. Usually larger than *E. dodoneata*, with more pointed apex and less rounded termen of the forewing, discal spot narrower, paler and less visible, dark dashes on the veins of forewing shorter, discal spot on the hindwing more evident (Text-fig. 36). Valva with large ventral process, broad blunt vinculum and considerably larger cornutus on the vesica in the male genitalia; and bursa copulatrix longer with different disposition of spines and wider colliculum in the female genitalia can be used successfully to distinguish *E. abbreviata* from *E. dodoneata*.

REMARKS: Variable in the depth of ground colour from browner to whiter and in distinctness of markings (LEMPKE 1969; GOATER & al. 1980). The melanic ab. *hirschkei* BASTELBERGER, 1908 has almost unicolorous, glossy coppery black ground colour of the forewing with intensely black veins and discal spot.

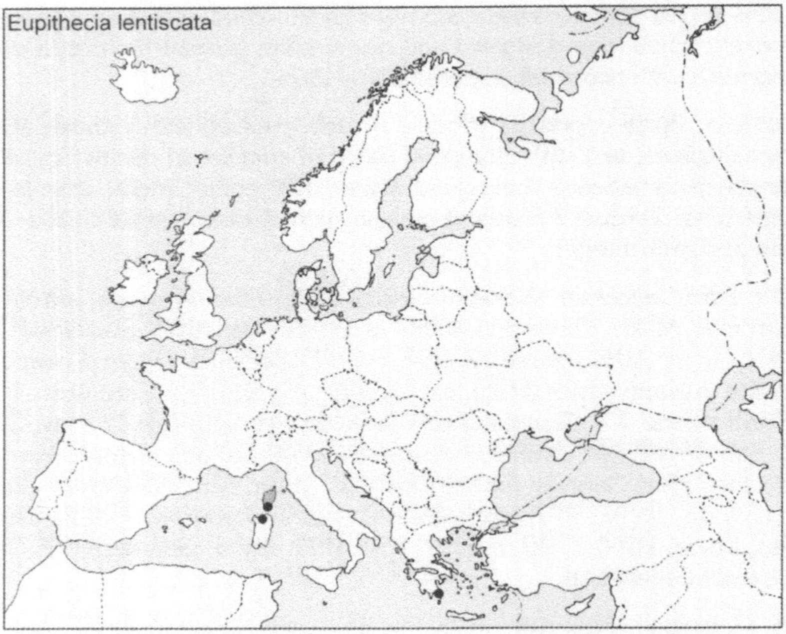

50. *Eupithecia lentiscata* MABILLE, 1869

Eupithecia lentiscata MABILLE, 1869: Annls Soc. ent. France (4) 9 (1): 75 (Corsica: Bonifacio). Holotype ♂ (coll. HERBULOT in ZSM, examined).

DIAGNOSIS: Wingspan 17.5-21 mm. Labial palpi light brown with whitish apices. Frons and vertex covered with whitish grey scales. Notum pale brownish grey with brown transverse band in front. Forewing rather short and broad; costa arched near base and apex; termen evenly curved; apex bluntly rounded; ground colour brownish grey, costal margin slightly darker; stalk of Cu covered with black scales; transverse lines (basal, postbasal, antemedial and medial) sinuate, not sharply angled near costa, distinctly marked with dark narrow costal spots; medial line crosses discal spot; postmedial line inconspicuous, with right angle near costa, well marked with two pairs of short black dashes (on M1 and M2 and as well as on CuA2 and below it); terminal area with distinct pale minutely waved line; pale tornal spot absent; discal spot obliquely elongate, very narrow, blackish. Hindwing paler, whitish grey with delicate brownish tinge; transverse lines indistinct, marked with minute dark dots on stalk of Cu; terminal area darker, with waved inner border and with a series of small dark wedge-shaped dots; pale waved line fine, indented; discal spot large, rounded. Fringe light brownish grey, with brown at vein ends.

MALE GENITALIA: Valva without ventral process, with apex rather pointed; sacculus weakly sclerotized. Vinculum with shallow medial hollow. Aedeagus large, long and thick. Vesica armed with three long and stout horn-like cornuti (one smaller almost straight, one slightly curved along its length and sharply curved at base, third cornutus about twice as long as others, with broadened irregular base) and with one short twisted cornutus at ductus ejaculatorius base. Sternum A8 elongate, broadened at base, with two small apical rods; basal and apical hollows are shallow.

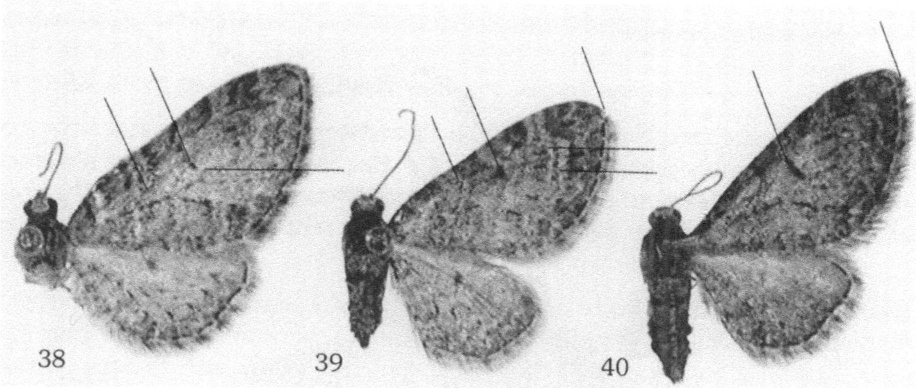

Text-figs 38-40: Diagnostic characters (indicated) of European *Eupithecia* species. Text-fig. 38: *E. lentiscata* MABILLE, 1869. Text-fig. 39: *E. reisserata* PINKER, 1976. Text-fig. 40: *E. massiliata* MILLIÈRE, 1865.

FEMALE GENITALIA: Bursa copulatrix large, ovoid, with numerous weakly sclerotized spines and heavily sclerotized marginal spines, and with membranous diverticulum at medial part of bursa copulatrix on left side. Ductus bursae broad with a row of large and heavily sclerotized spines along left side, with a diametrical row of smaller spines in medial part, and with two small patches of tiny spines near colliculum. Colliculum broad.

DISTRIBUTION: Mediterranean. Corsica, Sardinia and southern Greece (Peloponnese peninsula).

PHENOLOGY: Univoltine: January to early March. Larva in April. Overwintering as pupa in dense cocoon in the ground.

BIOLOGY: Larva monophagous (1). Found on flowers and buds of *Pistacia lentiscus* (Anacardiaceae) (Corsica: MABILLE 1869, 1872; DIETZE 1910, 1913; Sardinia: SKOU pers. comm.).

HABITAT: Rocks, cliffs, warm and dry stony slopes where *Pistacia lentiscus* grows, in Sardinia on sand close to the coast (SKOU pers. comm.).

SIMILAR SPECIES: Closely related to *E. abbreviata* and *E. dodoneata*, the only two species with which it could be confused (Text-fig. 38). *E. abbreviata* has the forewing more pointed and elongated in proportion, the ground colour more brown with distinctly paler medial area between discal spot and postmedial line, the small pale tornal dot is present; the hindwing browner with termen more deeply concave and with more conspicuous transverse lines than those in *E. lentiscata*. In the male genitalia of *E. abbreviata*: the valva with large ventral process, the sacculus heavily sclerotized, the vesica armed with two horn-like cornuti (one very long and other short, basally broadened) and with a single elongated curved cornutus. *E. lentiscata* can be distinguished from *E. dodoneata* by the more obtuse forewing, the less angulated basal, postbasal and antemedial lines near the costa, the smaller and narrower discal spots on the forewings, and by the less deeply concave termen of the hindwing. The very short sternum A8, the smaller heavily sclerotized horn-like cornutus, pointed weakly sclerotized cornutus and elongated twisted cornutus at ductus ejaculatorius base in the vesica render it easy to distinguish the male genitalia of *E. dodoneata* from those of *E. lentiscata*.

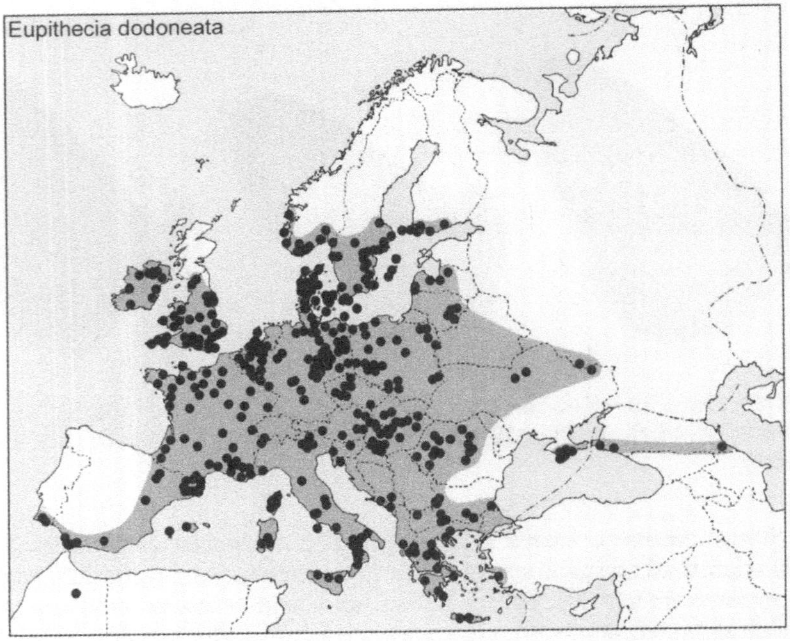

51. *Eupithecia dodoneata* GUENÉE, 1858

Eupithecia dodoneata GUENÉE, 1858: in BOISDUVAL & GUENÉE, Hist. nat. Insectes (Spec. gén. Lépid.) 10 (2): 344, pl. 11: 7 (England, France and Belgie). Lectotype ♀ (BMNH; France: Paris, designated by Lempke, 1969).
Eupithecia quercifoliata BANG-HAAS, 1875: Naturhist. Tidsskr. (3) 9: 514 (Denmark: Sjaelland, Jaegersborg, Dyrehave; Fyen, Odense; Falster, Omegnen; Jylland, Hald Egeskov; Riis Skov; Aarhuus). Syntypes (MNHU?).
Eupithecia dodoneata austrina HERBULOT, 1961: Alexanor 2 (4): 123; 2 (5): 152 (France: Var, St.-Raphaël). Holotype ♀ (coll. HERBULOT in ZSM, examined). Valid at subspecific rank.
Unavailable names (infrasubspecific): *meridionalis*: DIETZE (1910); *mediopallens*: LEMPKE (1969); *approximata*: LEMPKE (1969). - (incorrect subsequent spellings): *dodonearia*: MORRIS (1861); *dodonnaeata*: MABILLE (1872); *dodonaeata*: CREWE (1862).

DIAGNOSIS: Wingspan 15-21 mm. Labial palpi grey. Frons and vertex whitish grey. Notum whitish grey with brownish grey transverse band in front. Forewing rather narrow; costa slightly arched; apex rather pointed; ground colour whitish grey, often with brownish tinge; transverse lines slightly waved, dark grey; medial line crosses discal spot, curving from it to costa; medial area between antemedial and medial lines usually darker; lighter between medial and postmedial lines; postmedial line well marked by a series of elongated black dashes on the veins; terminal area relatively narrow, grey or brownish grey, with conspicuous pale wavy line; discal spot black, ovate, ovate-oblong or rarely circular. Hindwing with narrow apex and concave termen, grey or whitish grey; transverse lines conspicuous along dorsum only; pale wavy line bordered with shade on outer side or with a series of dark dashes on the veins; discal spot small, distinct, circular or ovate. Fringe distinctly chequered whitish grey and blackish grey. Abdomen grey with a dorsal row of blackish tufts posteriorly bordered by white scales and with blackish lateral stripes.

MALE GENITALIA: Valva without medial ventral process, with evenly curved ventral margin.

Aedeagus large. Vesica armed with one small slim horn-like cornutus, one elongated twisted cornutus and one small, irregular, curved cornutus at ductus ejaculatorius base. Sternum A8 short, broadened at base, with two minute apical lobes.

FEMALE GENITALIA: Bursa copulatrix small, elongated, twisted and sclerotized, uniformly covered with small spines in basal part, with short row of spines extending to medial part of corpus bursae. Ductus bursae without spines. Colliculum relatively short and wide. Spurs of anterior apophyses thin along their entire length. Papillae anales small, rounded.

DISTRIBUTION: Euro-Caucasian. Widespread in Europe, except northern Scandinavia and a large part of eastern Europe. Outside Europe separate populations in North Africa (Morocco), west of Asia Minor and Caucasus. Doubtful record for southern Turkey by GELBRECHT (pers. comm.).

PHENOLOGY: Univoltine: mid-April (late March) to late May (mid-June); in the south from February to April. Larva: early May (early April) to late June (late July), in United Kingdom found to early August. Overwintering as pupa.

BIOLOGY: Larva monophagous (3). Found on the young and tender leaves, buds and flowers of *Quercus robur* (France: MILLIÈRE 1870; GOOSSENS 1871; DIETZE 1910, 1913; Finland: SEPPÄNEN 1970; Germany: WEGNER pers. comm.), also *Q. ilex* (United Kingdom: BARRETT 1904), *Q. petraea* (Central Europe: WEIGT 1993), *Q. coccifera, Q. suber* (Corsica: GOOSSENS 1871; Italy: DIETZE 1872; Mediterranean: WEIGT 1993), *Q. pubescens* (Fagaceae) (Italy: WOLFSBERGER 1966), *Q. pyrenaica* (Spain: GOOSSENS 1871) and also on *Crataegus* (Rosaceae) (France: LHOMME 1935; France and Ireland: JUUL 1948). Often heavily parasitised.

PARASITOIDS: Hymenoptera, Ichneumonidae: *Casinaria morionella* - Hymenoptera, Braconidae: *Apanteles vitripennis* (THOMPSON 1945; SHENEFELT 1972; HERTING 1976).

HABITAT: Oak- and broad-leaved forests, copses, edges of mature Hawthorn thickets and hedges, scrub and parks. From 0 m up to about 1,000 m above sea-level in Alps and Pyrenees (MAZEL & PESLIER 1997; HELLMANN & al. 1999), to about 1,200 m in Balkan peninsula (TOMIČ & al. 1990), and to about 1,800 m in Apennines (Mt. Pollino: coll. ZSM).

SIMILAR SPECIES: Usually smaller than *E. abbreviata*. Distinguished from it by the rather fuller and rounder wings, the more sinuate postmedial line and larger, blacker, less narrow, usually ovate discal spot on the forewing, and stronger discal spot on hindwing (Text-fig. 37). The smaller valva with narrower apex, without medial ventral process, the smaller slim horn-like cornutus on the vesica, the shorter sternum A8 in the male, and the smaller and shorter bursa copulatrix with different disposition of spines in the female genitalia distinguish *E. dodoneata* from closely related *E. abbreviata* with accuracy.

52. *Eupithecia reisserata* PINKER, 1976

Eupithecia reisserata PINKER, 1976: Z. Arb.Gem. Österr. Ent. 28 (1-3): 2, figs 5, 6, 19 (Turkey, Anatolia: Kizilcahamam). Holotype ♂ (NHMW, 4 paratypes ♂♀ examined).

‡ *Eupithecia danilevskyi*: sensu WARDIKJAN, 1961: Dokl. Acad. Nauk Arm. SSR 32 (1): 61, fig. 1, only paratype ♀; misidentification nec WARDIKJAN, 1961, holotype ♂ (Azerbaijan, Nakhichevan: Chananab) (identity: *Eupithecia millefoliata*).

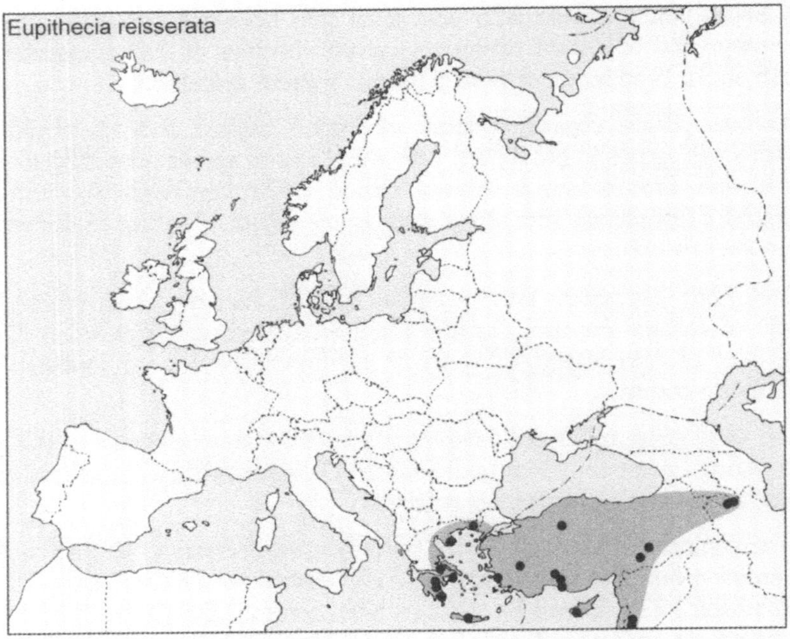

DIAGNOSIS: Wingspan 14-18 mm. Labial palpi, frons, vertex and notum brownish grey. Forewing pale brownish grey with small wedge-shaped brown marks on all veins along postmedial line midway between discal spot and terminal area; transverse lines brownish, slightly sinuous, obscure, inconspicuous, usually slightly widened and darker at costa; stalk of Cu darker, covered with brown scales; pale line between medial and terminal areas usually distinct, rather angled between M3 and CuA1, twice angled near costa; terminal area narrow with inconspicuous pale wavy line; discal spot distinct, dark brown, small and slightly elongated. Hindwing paler, brownish grey with darker CuA1 and CuA2; transverse lines indistinct; terminal area slightly darker with pale finely dentated wavy line; discal spot brownish, small, rounded or slightly elongated. Fringe brownish grey, spotted with brown at vein ends.

MALE GENITALIA: Valva relatively narrow, without ventral process, with evenly curved ventral margin. Aedeagus small. Vesica armed with two small apical patches of spines, one short medial row of small spines and a minute irregular, curved cornutus at ductus ejaculatorius base. Sternum A8 small, slightly elongated, broadened basally and apically, with two small and short apical lobes and shallow but broad hollow between them.

FEMALE GENITALIA: Bursa copulatrix pear-shaped, covered with large spines; the marginal spines larger and broader than other spines in the spiniferous area. Ductus bursae with longitudinal striations and patch of small spines. Ductus seminalis attached to corpus bursae near colliculum. Colliculum collar-like, short and broad. Papillae anales relatively large (short and wide), rounded.

DISTRIBUTION: East-Mediterranean. Eastern and southern Greece and Crete. – Outside Europe in Cyprus, Asia Minor (Anatolia), Jordan (subsp. *levarii* HAUSMANN, 1991) and southern Transcaucasus (Nakhichevan region) (PINKER 1976; MIRONOV 1990).

PHENOLOGY: Univoltine: late March (mid-March) to mid-May (late May).

BIOLOGY: Unknown.

HABITAT: Light oak woods on the slopes along riversides. From 0 m up to about 700 m above sea-level in Greece (coll. ZSM), and to about 1,200-1,300 m in southern Turkey (Taurus Mts) (coll. GELBRECHT).

SIMILAR SPECIES: This is one of the smaller species and is most similar to *E. dodoneata*. However, the wings are usually broader with less pointed apices, less conspicuous pattern, pale line between medial and terminal areas angled, rather ribbon-like, narrow but more distinct, the discal spot on the forewing less conspicuous and less elongated than in *E. dodoneata* (Text-fig. 39). The shape of sternum A8, the cornuti on the vesica, and the corpus bursae with larger and broader marginal spines should distinguish this species.

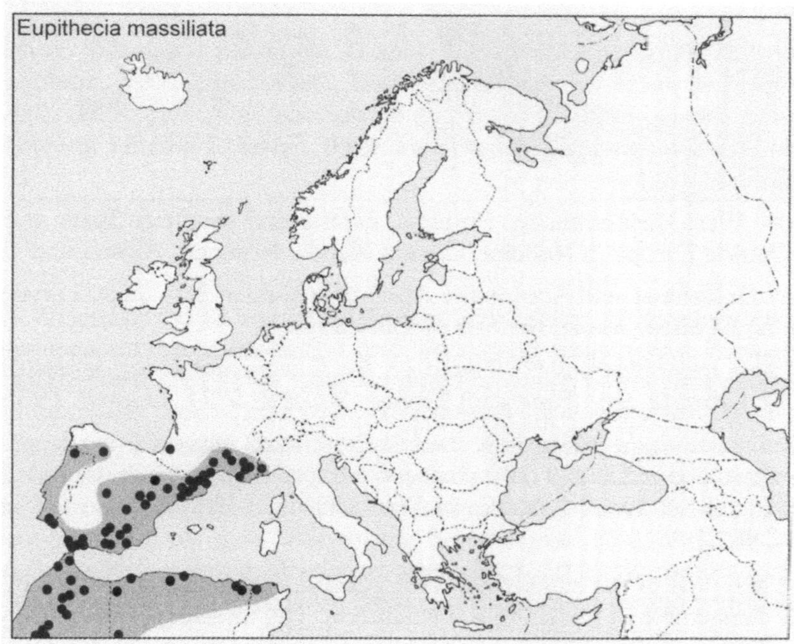

53. *Eupithecia massiliata* DARDOIN & MILLIÈRE, 1865

Eupithecia massiliata DARDOIN & MILLIÈRE, 1865: in MILLIÈRE, Iconogr. Descr. Chenilles Lépid. inédits 2 (15): 145, pl. 67: 1-2 (France: Marselle). Syntypes many, including 1♂ (MNHN?).
Eupithecia peyerimhoffata MILLIÈRE, 1870: Petites Nouv. Ent. 1 (22): 88 (Spain). Syntypes ♂ ♀ (MNHN?).
Eupithecia atlanticata PINKER, 1969: Z. Wien. Ent. Ges. 54 (80; 9-12): 109, pl. 12: E7; text-fig. 5 (Madeira: Funchal). Holotype ♀ (NHMW). New junior synonym.
Eupithecia atlanticata pinkeriata MOBERG, 1983: SHILAP Revta. lepid. 11 (42): 141, figured (Spain: Huelva Province, Valvarde del Camino). Holotype ♀ (EMS). Valid at subspecific rank. New junior synonym.

DIAGNOSIS: Wingspan 14-17 mm. Labial palpi grey or brownish grey. Frons and vertex pale brownish grey. Notum grey to brownish grey. Forewing rather elongate and narrow: costa slightly arched, apex pointed, termen obliquely rounded; ground colour dark, brownish grey fuscous; transverse lines inconspicuous, slightly sinuate, evenly curved;

medial line fine, usually indistinct, crossing discal spot, twice right-angled onto costa; post-medial line sometimes marked by a series of small dark dashes on the veins, it is twice right-angled near costa; pale minutely wavy line weakly developed; discal spot small, obliquely elongate, blackish. Hindwing rather angulated: termen slightly concave medially; ground colour paler near base and along costa; transverse lines inconspicuous; pale wavy line ill-defined; discal spot relatively large, rounded, dark. Fringe chequered whitish grey and greyish fuscous or dark brownish grey. Abdomen greyish fuscous (sometimes with brownish tinge) with fine blackish lateral stripes.

MALE GENITALIA: Valva with small blunt ventral process; sacculus heavily sclerotized. Vinculum usually with shallow medial hollow. Aedeagus small. Vesica armed with two slim horn-like cornuti (one larger than the other), one thin elongated cornutus and one smaller, irregular, curved cornutus at ductus ejaculatorius base. Sternum A8 short, broadened at base, with two small, blunt, pincer-like sclerotized apical lobes; apical hollow shallow and narrow.

FEMALE GENITALIA: Bursa copulatrix small, twisted, elongated pouch-like, completely covered with spinules, exept for membranous base. Ductus bursae not developed. Ductus seminalis very narrow, attached to corpus bursae near colliculum. Colliculum relatively broad, plate-like with two small apical lateral teeth. Spurs of anterior apophyses broadened to apexes.

DISTRIBUTION: West-Mediterranean. Portugal, central and southern Spain and southern France. – Outside Europe in Madeira, Canary Islands, Morocco, Algeria and Tunisia.

PHENOLOGY: Univoltine: mid-April (early April) to mid-June (late July). Larva in March. Overwintering as pupa, sometimes two or three times.

BIOLOGY: Larva monophagous (2). Found on the flowers and young leaves of *Quercus coccifera, Q. suber, Q. ilex* (Fagaceae) (France: MILLIÈRE 1871; LHOMME 1935).

HABITAT: Open and warm oak forests, scrub and cultivated areas with scattered oak trees. From 0 m up to about 1,000-1,100 m above sea-level in central Spain (coll. ZSM; coll. ZMUC), and to about 1,900-2,000 m in Great Atlas and Middle Atlas in Morocco (coll. ZSM; coll. ZMUC).

SIMILAR SPECIES: One of the smaller darker European *Eupithecia* species, with little variation in the distinctness of markings and coloration. The forewing is narrower, with more pointed apex and less arched costal and terminal margins, ground colour usually darker, than in *E. tenuiata*, the only species with which it could be confused (Text-fig. 40). No significant differences in wing maculation between these two species could be found. However, the genital features such as presence of ventral process on the valva, two horn-like cornuti in the vesica and different shape of sternum A8 in male, very small bursa copulatrix of different shape, with short and wide colliculum, the shorter and tapered apophyses and the shorter tergum A8 in the female can be successfully used to distinguish *E. massiliata* from *E. tenuiata*.

54. *Eupithecia extremata* (FABRICIUS, 1787)

Phalaena extremata FABRICIUS, 1787: Mantissa Insect. 2: 211 (Austria). Syntype(s) lost.

Eupithecia glaucomictata MANN, 1855: Verh. zool.-bot. Ver. Wien 5: 547 (Corsica: near Cavro). Syntypes (NHMW).

Unavailable names (incorrect subsequent spellings): *extremaria*: BOISDUVAL (1840); *extrema*: RIEMIS (1994).

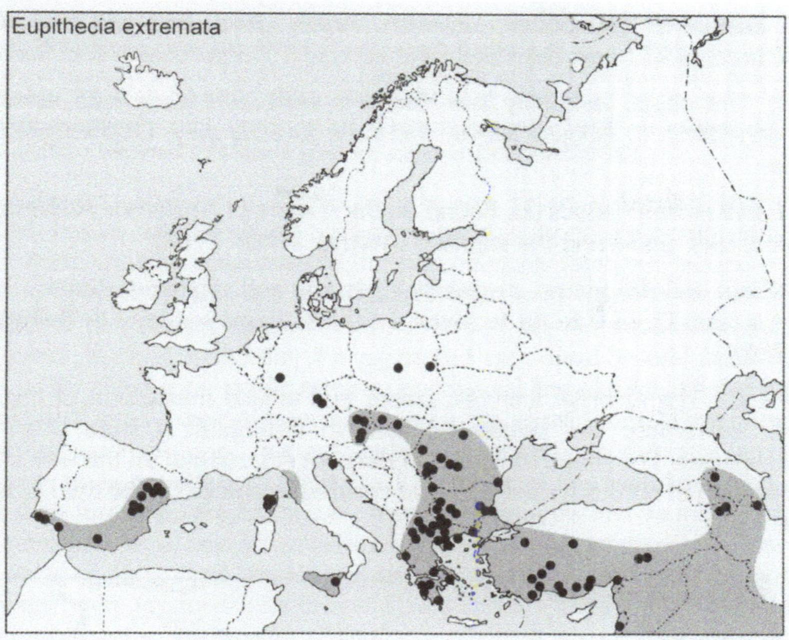

Diagnosis: Wingspan 16-18 mm. Labial palpi light brown or ochreous brown. Frons white, sometimes with brownish medial spot. Vertex and notum white. Forewing white or yellowish white; basal area white with elongated brownish costal spot and with slightly brownish outer shade; medial area between antemedial line and discal dot relatively narrow, slightly sinuate, broader near costa, brown or brownish ochreous; outer part of medial area yellowish white with dark brown dashes on the veins; postmedial line indistinct, slightly sinuate and evenly curved near costa; terminal area narrow, ochreous with large darker brown apical spot, with smaller brown tornal spot, and with conspicuous white wavy line, which is only toothed near apex; discal dot black, narrow, elongated and oblique. Hindwing white with brownish basal and terminal areas and with darker, brown tornal spot; discal dot brown, elongated or rounded, always present. Fringe white, distinctly spotted with brown at vein ends. Abdomen reddish brown with a few dark brown or black lateral spots; the last segment whitish, especially in male.

Male genitalia: Valva with large, broad and blunt medial ventral process; sacculus heavily sclerotized. Vesica armed with numerous spinules, one slim horn-like cornutus, two small spinescent cornuti, one elongated flat cornutus and one minute, irregular, curved cornutus. Sternum A8 short and broad, with waved margins; apical rods broad, elongated and tapered to apices; apical hollow deep and narrow.

Female genitalia: Bursa copulatrix large, spherical, completely covered with spines. Ductus bursae sclerotized, with longitudinal striations, three sclerotized rows of spines and with patch of small spines near colliculum. Colliculum short and broad. Spurs of anterior apophyses slightly broadened at apexes.

Distribution: Mediterranean-Turanian. Scattered through Central and southern Europe.

– Outside Europe in Asia Minor, Lebanon, Israel, Libya, Iraq and Iran. The subsp. *hangayi* VOJNITS, 1977 was described from southern Transcaucasus and Syria.

PHENOLOGY: Univoltine: early May (late March) to early June (early July), sometimes with a second generation in August. Larva: mid-June to early July. Overwintering as pupa, often twice.

BIOLOGY: Larva monophagous (2). Found on the flowers of *Anthemis ruthenica, A. tinctoria* (Asteraceae), preferring the stamens (Austria: WEIGT 1988).

HABITAT: Warm and dry places, especially limestone and sandstone slopes, rocks, stony slopes and screes. From 0 m up to about 1,600 m above sea-level in Balkan peninsula (PINKER 1968).

SIMILAR SPECIES: The external features, colour and unique maculation of the wings and body of *E. extremata* can usually be used to distinguish it from other European *Eupithecia* species. The shape of valva and sternum A8, cornuti on the vesica, and shape and sclerotization of corpus bursae and area covered by spines can be used to make determinations.

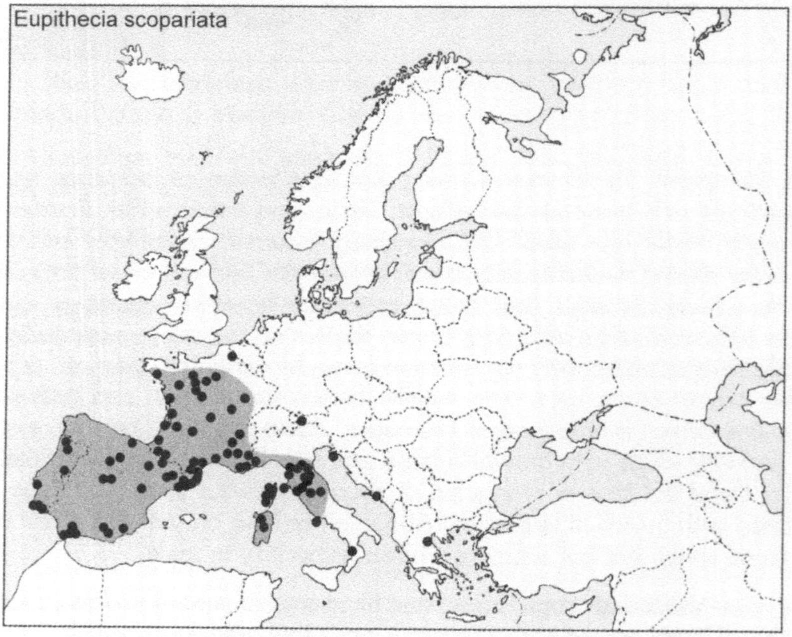

55. *Eupithecia scopariata* (RAMBUR, 1833)

Larentia scopariata RAMBUR, 1833: Annls Soc. ent. France 2 (1): 49, pl. 2: 13 (Corsica: Bogognano [=Bocognano]). Lectotype ♂ (coll. HERBULOT in ZSM, examined).

Eupithecia guinardiaria BOISDUVAL, 1840: Gen. Index meth. Eur. Lepid.: 211 (France: Monspelii; Corsica). Syntype(s).

Eupithecia tenebrosaria HERRICH-SCHÄFFER, 1848: Syst. Bearb. Schmett. Eur. 3 (31): 120, 130; ibidem (1846), pl. 25: 157 (non binominal), (Italy: Montenero). Syntype(s).

Eupithecia multiflorata MILLIÈRE, 1866: Iconogr. Descr. Chenilles Lépid. inédits 2 (16): 194, pl. 71: 8-13 (France: near Cannes; Sainte-Marguerite). Syntypes (MNHN?).

Eupithecia scopariata var. *graslinaria* STAUDINGER, 1871: in STAUDINGER & WOCKE, Cat. Lepid. Eur. Faunengeb. 1: 195 (France). Syntypes (MNHU). Valid at subspecific rank.

Eup(ithecia) scopariata f. *matertera* DIETZE, 1910: Biol. Eupithecien 1, pl. 74: 437, 438; ibidem (1913), 2: 78 (Spain: Castilia). Syntypes (MNHU). Valid at subspecific rank.

Unavailable names (misidentification): *nanata*: sensu GÓMEZ DE AIZPURÚA (1989) nec HÜBNER, 1813 (identity: *Eupithecia nanata*). - (incorrect subsequent spelling): *scoparia*: BOISDUVAL (1840).

DIAGNOSIS: Wingspan 15-23 mm. Labial palpi brown. Frons and vertex whitish brown. Notum whitish brown with brownish transverse band in front. Forewing rather narrow and pointed: costa straight, apex pointed, termen obliquely rounded; ground colour light rusty brown to greyish brown; the veins usually chequered blackish brown and white, especially stalk of Cu; transverse lines fine, numerous, slightly sinuate, oblique, sharply angled near costa; the first medial line sharply angled onto costa beyond discal spot; small white blotch under discal spot usually present; postmedial line obtusely angled onto costa, forming a series of small, blackish brown dashes on the veins; terminal area narrow; pale wavy line rather inconspicuous, minutely indented, forming an elongate whitish tornal spot; discal spot small, rounded, intensely black, sometimes obsolete. Hindwing elongate, ovoid; paler than forewing, greyish brown; transverse lines distinct mainly along dorsum; pale stripe between medial and terminal areas usually visible; terminal area narrow with very weak trace of wavy line; discal spot minute, circular, dark or ill-defined. Fringe whitish brownish grey, spotted with brown at vein ends. Abdomen whitish brown with dorsal row of white tufts or with white dorsal stripe and also with small black lateral dots or fine lines; second segment darker, brown.

MALE GENITALIA: Valva with sclerotized, short and pointed ventral process. Vinculum short and narrow. Aedeagus short and thick. Vesica armed with two long, stout horn-like cornuti, one small flat, elongated cornutus and one small irregular, curved cornutus at ductus ejaculatorius base. Sternum A8 elongate, tapered to apex, with two short pointed apical lobes and shallow, broad hollow between them.

FEMALE GENITALIA: Bursa copulatrix pear-shaped, sclerotized, covered with small spines at base. Ductus bursae with longitudinal striations and with small medial patch of spines. Ductus seminalis narrow. Colliculum short and broad. Papillae anales large, short but very wide.

DISTRIBUTION: West-Mediterranean. Portugal, Spain, France, Corsica, Sardinia, Italy, Austria, Croatia and Greece. - New for the fauna of Austria: 1♀, (Austria), Ötztaler Alpen, Kaunertal/Tirol, Umg. Fliess, 1500 m, 14.VII.1991, leg. W. PAVLAS (ZSM). An allied darker species *E. rusicadaria* DIETZE, 1910 occurs in North Africa (Morocco, Algeria and Tunisia).

PHENOLOGY: Bi- or polyvoltine: mid-May (mid-April) to early June (late July); late September to November. Larva: February and March; September and October. Overwintering as pupa.

BIOLOGY: Larva oligophagous (1). Found on *Erica arborea*, *E. scoparia*, *E. multiflora*, *Calluna vulgaris* (Ericaceae) (France: MILLIÈRE 1866, 1870; DIETZE 1910, 1913; Corsica: MABILLE 1868) and *Cytisus scoparius* (Fabaceae) (France: TAUTEL pers. comm.).

HABITAT: Open steppes, Atlantic coasts, mountain slopes, heaths and woodland where *Erica* grows. From 0 m up to about 1,500 m above sea-level in Alps (coll ZSM), to about 2,000 m in Pyrenees (coll. FIBIGER), and to about 2,200 m in central Spain (coll. FIBIGER).

SIMILAR SPECIES: *E. scopariata* is somewhat similar to *E. intricata helveticaria*, but easily distinguished from it by the more elongate, narrower and more pointed wings; forewing with straight costa; lighter ground colour; more oblique transverse lines; smaller discal spots on all wings; abdomen usually with whitish dorsal stripe. The stout uncus, presence of small spine-like ventral process on valva, two large, elongate and broad horn-like cornuti on the vesica and the shape of sternum A8 in male; the elongate, sclerotized bursa copulatrix with small patch of spinules at base and additional smaller separate patches of spines in the middle part of bursae; the narrow ductus seminalis and broad colliculum should distinguish this species.

REMARKS: Variable in size and in the depth of colour and distinctness of markings, as figured. There are several forms of this species (DIETZE 1913).
This West-Mediterranean species was questionably recorded in the Catalogue of STAUDINGER & REBEL (1901: 314) for Armenia and Tura (f. *orientata* STAUDINGER, 1901) in error. After carefully examining the type-specimen of *E. scopariata orientata*, I have come to the conclusion that this moth is conspecific with *E. sutiliata* CHRISTOPH, 1876 (MIRONOV 1991).

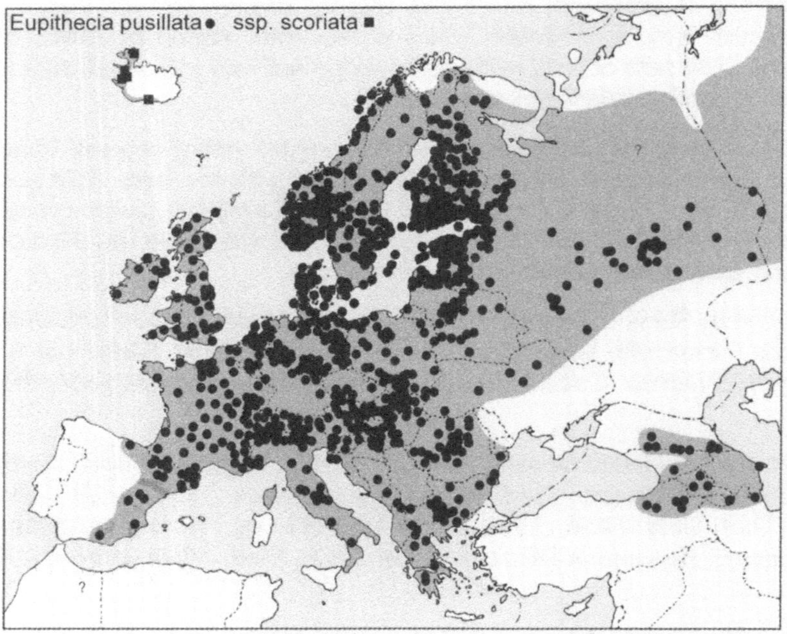

56. *Eupithecia pusillata* (DENIS & SCHIFFERMÜLLER, 1775)

Geometra pusillata DENIS & SCHIFFERMÜLLER, 1775: Syst. Verz. Schmett. Wien.: 110 (Austria: Vienna district). Syntype(s) lost.

Phalaena laevigata HAWORTH, 1809: Lepid. Br. (2): 362 (United Kingdom). Syntype(s) (BMNH). Junior primary homonym of *Phalaena laevigata* SCOPOLI, 1763.

Geometra sobrinata HÜBNER, 1817: Samml. Eur. Schmett. 5 Geometrae (2), pl. 90: 465 (Europe). Syntype(s) lost.

Eupithecia expressaria HERRICH-SCHÄFFER, 1848: Syst. Bearb. Schmett. Eur. 3 (32): 121; pl. 46: 284, 285 (Europe). Syntypes.

Eupithecia scoriata STAUDINGER, 1857: Stett. Ent. Z. 18 (7-9): 265 (Iceland). Syntypes ♂ ♀ (MNHU).
Eupithecia anglicata HERRICH-SCHÄFFER, 1863: KorrespBl. zool.-min. Ver. Regensburg 17: 23 (United Kingdom: Kent, Dover). Syntype(s).
Eupithecia latoniata MILLIÈRE, 1882: Lépidoptérologie 8: 13, pl. 2: 10 (France: near Saint-Martin-Lantosque). Syntypes ♂ ♀ (MNHN?).
Eupithecia sobrinata var. *graeseriata* RÄTZER, 1882: in FREY, Mitt. Schweiz. ent. Ges. 6 (7): 361 (Switzerland: Gadmen). Syntype(s). Valid at subspecific rank.
Eupithecia stevensata WEBB, 1896: The Entomologist's Rec. J. Var. 8: 10 (United Kingdom: Kent, Dover). Syntypes.
Eupithecia lavicaria FUCHS, 1902: Stett. Ent. Z. 63 (1): 318 (Norway: Lavik). Syntypes 2. Junior synonym (AARVIK pers. comm.).
Eup(ithecia) sobrinata f. *luneburgensis* DIETZE, 1910: Biol. Eupithecien 1, pl. 77: 679, 680; ibidem (1913), 2: 147 (Germany: Lüneburg). Syntypes, at least 2 (MNHU). Valid at subspecific rank.
Eup(ithecia) sobrinata f. *scotica* DIETZE, 1910: Biol. Eupithecien 1, pl. 77: 681-683; ibidem (1913), 2: 147 (United Kingdom: Scotland, Archimore; England). Syntype(s) (MNHU). Valid at subspecific rank.
Eup(ithecia) sobrinata f. *castiliana* DIETZE, 1910: Biol. Eupithecien 1, pl. 77: 684, 685; ibidem (1913), 2: 147 (Spain: Castilia; Andalusia, Sierra Nevada). Syntypes, at least 2 (MNHU). Valid at subspecific rank.
(Eupithecia) sobrinata f. *confluens* DIETZE (1910): Biol. Eupithecien 1, pl. 77: 689, 690; ibidem (1913), 2: 146 (Austria: Stainz). Syntypes, at least 2 (MNHU). Valid at subspecific rank.
Unavailable names (misidentification): *satyrata*: sensu FRISTRUP (1943) nec HÜBNER, 1813 (identity: *Eupithecia satyrata*). – (infrasubspecific): *conjuncta*: DIETZE (1913); *nigrofasciata* DIETZE (1910); *impuncta*: LEMPKE (1951). - (incorrect subsequent spellings): *pusillaria*: HÜBNER, 1825; *sobrinaria*: BOISDUVAL (1840); *variata* PAGENSTECHER (1902; misspelling of *scoriata* STAUDINGER, 1857); *graessariata*: DIETZE (1910): *pussillata*: DANTART (1989).

DIAGNOSIS: Wingspan 16-22 mm. Labial palpi light brown. Frons, vertex and notum pale greyish brown. Forewing rather narrow, costa slightly arched, apex narrow and rounded, termen full and rounded; ground colour whitish brown to blackish fuscous; stalk of Cu and sometimes stalk of R spotted with short white and long black dashes; basal and postbasal lines inconspicuous, oblique, much indented, sharply angled onto costa; antemedial line more distinct, straight, oblique, sharply angled near costa also; first medial line less oblique, straight or slightly sinuate, crossing discal spot, and right-angled to faintly curved at this point towards costa; narrow medial area between antemedial and medial lines usually darker, brown; postmedial line much sinuate, twice obtuse angled near costa, forming a series of elongate, black or blackish brown dashes on the veins; whitish wavy line fine, usually much indented, complete to broken or not marked in some specimens; small whitish tornal spot with dark shade on inner side usually present; discal spot small, black, rounded, or ovoid, or even faintly crescent-shaped. Hindwing paler, greyish brown; transverse lines weakly marked; terminal area narrow, slightly darker with trace of whitish indented wavy line; discal spot minute, rounded, dark or absent. Fringe light brownish grey, spotted with dark, greyish brown at vein ends. Abdomen pale greyish brown, usually with narrow black lateral stripes, second segment often darker, brown or blackish brown.

The Icelandic subsp. ***scoriata*** STAUDINGER, 1857 has rather less pointed forewing with unicolorous, brown ground colour compared with nominate subspecies, especially specimens from southern Europe. Postmedial line of forewing usually evenly curved, without outward tooth on M1.

MALE GENITALIA: Valva broad with a small, sclerotized ventral process; sacculus heavily sclerotized. Vesica armed with two elongated, slim horn-like cornuti, one elongated and broad flat cornutus and one smaller, irregular, curved cornutus at ductus ejaculatorius base. Sternum A8 elongate, broadened at base, with two elongated, narrow and blunt apical horns.

FEMALE GENITALIA: Bursa copulatrix elongated, twisted and sclerotized, covered with small spines at base; marginal spines larger than others. Ductus bursae with short row of large spines on the left side. Colliculum short and broad. Spurs of anterior apophyses slightly broadened to apexes. Papillae anales large.

DISTRIBUTION: Euro-Siberian. Widely distributed in the Palaearctic region westward to Canary Islands, eastward to Kamchatka and Sakhalin. In Iceland and south-western Greenland replaced by subsp. *scoriata*. – New for the fauna of Canary Islands: 1♂, Canary Islands, Gran Canaria, Santa Lucia, 1000 m, 10.VIII.1979, P. STADEL NIELSEN leg. (HNHM).

PHENOLOGY: Univoltine: mid-July (early July) to late September (mid-October). Recorded for Pyrenees as bivoltine: early June to late August; early October to late November (MAZEL & PESLIER 1997). Icelandic population univoltine: late July to early August. Larva: late April (late March) to mid-June (mid-July); in Iceland in June and July. Overwintering as egg.

BIOLOGY: Larva monophagous (2). Found on the needles and flowers of *Juniperus communis* (United Kingdom: CREWE 1861c; BARRETT 1904; France: MILLIÈRE 1870; Germany: DIETZE 1910, 1913; WEIGT 1980, 1993; WEGNER pers. comm.; Denmark: JUUL 1948; Finland: SEPPÄNEN 1970), *J. communis* ssp. *alpina* (Iceland: STAUDINGER 1857; LINDROTH 1931; Switzerland: DIETZE 1910, 1913; VORBRODT 1928) and *J. oxycedrus* (Cupressaceae) (France: MILLIÈRE 1870).

PARASITOIDS: Hymenoptera, Ichneumonidae: *Campoletis zonata*, *Chorinaeus cristator*, *Ch. subcarinatus*, *Mesochorus brevipetiolatus*, *Netelia thomsonii*, *Platylabus iridipennis*, *P. pactor*, *P. pedatorius*, *Campoletis raptor* - Hymenoptera, Braconidae: *Aleiodes (=Rogas) testaceus*, *Apanteles cajae (=difficilis)*, *A. impurus*, *A. juniperatae*, *A. liparidis (=nemoriae)*, *A. spurius*, *Meteorus abdominator (=brunnipes)*, *Rogas* sp. (THOMPSON 1945; SHENEFELT 1969, 1972, 1975; HERTING 1976; ZWAKHALS pers. comm.).

HABITAT: Preferring heaths, pine and open mixed forests, forest edges but it is also found in gardens, chalk hills, and on mountains on cliffs and among rocks wherever the food plants grow. The subsp. *scoriata* inhabits landscapes on the black lava-ground. From 0 m up to about 1,800-1,900 m above sea-level in Apennines and Asia Minor (coll. ZSM, coll. ZISP, coll. SOMMERER; PROLA & RACHELI 1980), to about 2,000-2,200 m in Alps and mountains of Balkan peninsula (DANNEHL 1927; VORBRODT 1928, 1931; FORSTER & WOHLFAHRT 1981; HELLMANN & al. 1999; REZBANYAI-RESER & MAIER, 2000; coll. FIBIGER), to about 2,300 m in Pyrenees (coll. ZMUC), and to about 2,500 m in Sierra Nevada (coll. ZMUC). Also recorded to about 2,500 m above sea-level in Alps (WEIGT 1993).

SIMILAR SPECIES: Variable in the depth of coloration and distinctness of markings, with transverse lines variably expressed (DIETZE 1910, 1913). In spite of its variability, *E. pusillata* is an easily recognised species on the whole, but it could occasionally be confused with some other closely related species, such as *E. ericeata* and *E. oxycedrata*. See Text-fig. 41 and differential diagnosis under these species.

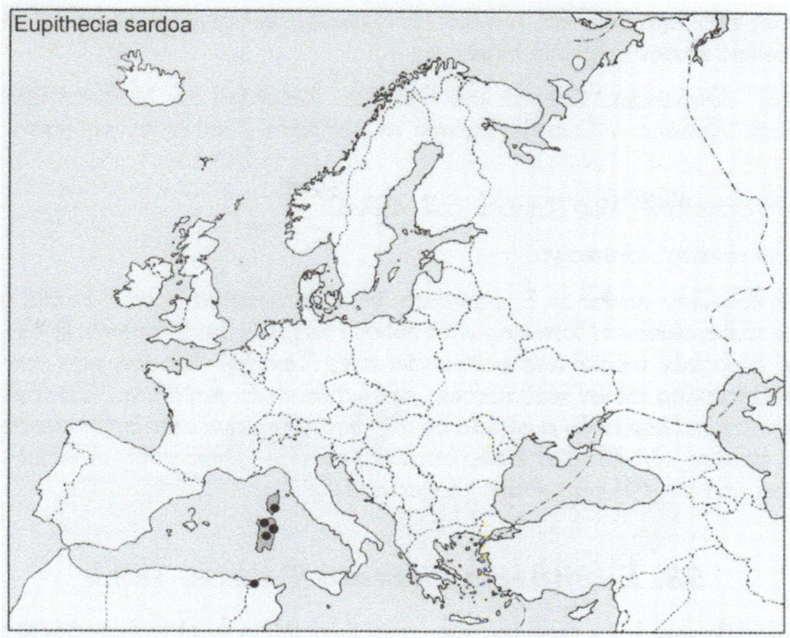

57. *Eupithecia sardoa* DIETZE, 1910

Eup(ithecia) sardoa DIETZE, 1910: Biol. Eupithecien 1, pl. 77: 703, 704; ibidem (1913), 2: 151 (Sardinia: Sassari). Syntypes 3♂ (MNHU).

Eupithecia peterseni WAGNER, 1914: Jber. Wien. ent. Ver. 24: 176, figs 1-5 (Tunisia). Syntypes ♂ ♀ (ZSM). New junior synonym.

DIAGNOSIS: Wingspan 17-20 mm. Labial palpi greyish brown. Frons and vertex brownish white. Male antennae bipectinate, with dense setae. Notum pale greyish brown. Forewing rather narrow and elongated, costa slightly arched, apex pointed, termen obliquely rounded; ground colour greyish brown; basal and antemedial lines sharply angled near costa; medial line crossing discal spot and slightly curved from it onto costa; area between antemedial and medial lines darker; postmedial line sinuate, twice sharply angled near costa, with elongated blackish dashes on M1, M2, M3, CuA2 and between CuA2 and A; terminal area with two elongated blackish dashes near apex; pale wavy line faintly sinuate, usually with inner shade, especially near tornus; discal spot small, ovate-oblong, black. Hindwing paler; transverse lines conspicuous near dorsum; pale wavy line invisible; discal speck small, rounded, weakly marked. Fringe chequered pale and dark brown. Abdomen pale greyish brown, with blackish lateral stripes; tergum A2 weakly darker, brownish.

MALE GENITALIA: Valva with relatively large and elongated ventral process; sacculus rather heavily sclerotized. Vesica armed with two slim horn-like cornuti, considerably differing in size (one three times longer than the other). Sternum A8 peg-like, elongated, with two rather pointed basal lobes and very shallow apical hollow.

FEMALE GENITALIA: Bursa copulatrix with lightly sclerotized base, covered with small spines, medial part of corpus bursae with small but heavily sclerotized spines. Ductus bursae long, narrow, sclerotized, with longitudinal striations. Colliculum absent. Anterior and posterior apophyses thick and pointed. Tergum A8 trapeziform with heavily sclerotized anterior margin. Papillae anales relatively large.

DISTRIBUTION: Tyrrhenian. Corsica and Sardinia. Recorded for Mallorca (Balearics) by GOATER (pers. comm.). – Outside Europe in northern Tunisia (as *peterseni* WAGNER, 1914).

PHENOLOGY: Univoltine: May (Tunisia: coll. ZMUC).

BIOLOGY AND HABITAT: Unknown.

SIMILAR SPECIES: Very similar to *E. pusillata*, but distinguished from it by the bipectinate antennae in male, narrower forewing with oblique termen; the transverse lines less angled near costa, especially medial and postmedial lines (Text-fig. 42); the pale wavy line less sinuate; the hindwing rather unicolorous, paler than in *E. pusillata*. The male genitalia distinguished from those of *E. pusillata* by the narrower valva with larger ventral process, the shorter papillae of labides, the two horn-like cornuti considerably differing from each other in size, and the different shape of sternum A8.

58. Eupithecia ericeata (RAMBUR, 1833)

Larentia ericeata RAMBUR, 1833: Annls Soc. ent. France 2 (1): 50, pl. 2: 14 (Corsica: Bastia). Lectotype ♀ (coll. HERBULOT in ZSM, examined).
Eupithecia millierata STAUDINGER, 1871: in STAUDINGER & WOCKE, Cat. Lepid. Eur. Faunengeb. 1: 200 (France Gallia: Ardèche). Syntype(s) (MNHU).
Eupithecia oxycedrata var. *euxinata* BOHATSCH, 1893: Dt. Ent. Z. Iris 6: 4 (Turkey: Amasia). Syntypes 54, ♂ ♀ (MNHU; 2 syntypes examined). Valid at subspecific rank. New junior synonym.
Unavailable names (incorrect subsequent spelling): *ericearia*: BOISDUVAL (1840).

DIAGNOSIS: Wingspan 14-22 mm. Labial palpi brownish grey or brown; frons and vertex whitish brown. Notum pale greyish brown, with broad, brown transverse band in front. Forewing rather narrow and pointed with costa slightly arched and termen obliquely rounded; ground colour from yellowish brown to greyish brown; stalk of Cu blackish brown; basal, postbasal and antemedial lines brown, almost straight, oblique, sharply

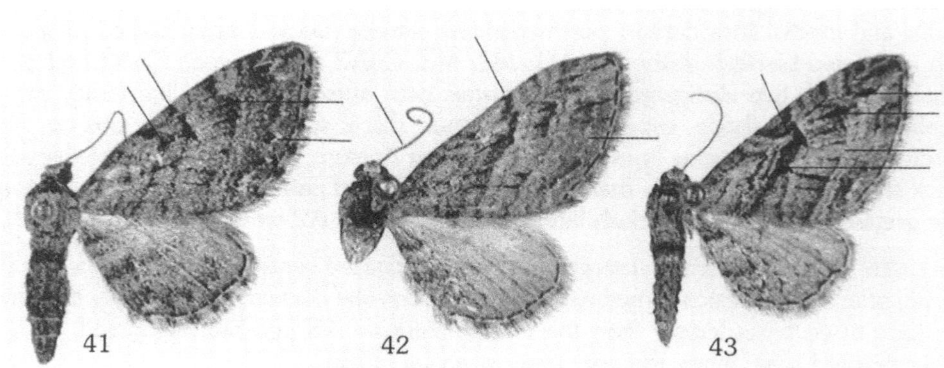

Text-figs 41-43: Diagnostic characters (indicated) of European *Eupithecia* species. Text-fig. 41: *E. pusillata* (DENIS & SCHIFFERMÜLLER, 1775). Text-fig. 42: *E. sardoa* DIETZE, 1910. Text-fig. 43: *E. ericeata* (RAMBUR, 1833).

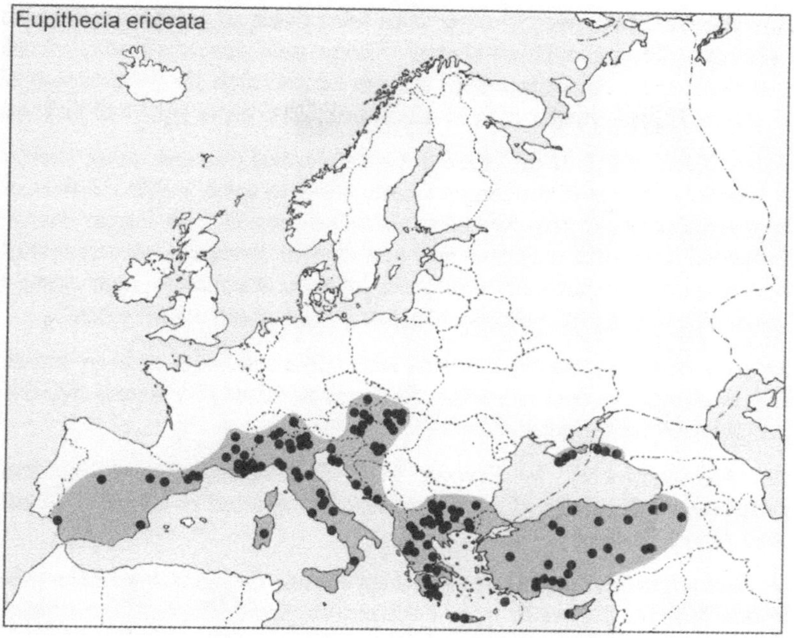

angled on to costa; the first medial line less oblique, slightly sinuate, right-angled from discal spot to costa; postmedial line evenly curved and faintly angled on to costa; medial area with two pairs of blackish streaks on M1, M2 and on CuA2 and between CuA2 and A; terminal area dark brown near costa, usually with fine blackish apical streak; pale wavy line weakly defined, minutely wavy, much toothed near apex or obsolete; often small whitish tornal spot is present; discal spot obliquely ovoid, or rounded or T-shaped, intensely black. Hindwing with medially slightly concave termen, paler, brownish grey; transverse lines marked only along dorsum, especially basal and postmedial; terminal area

Text-figs 44-46: Diagnostic characters (indicated) of European *Eupithecia* species. Text-fig. 44: *E. phoeniceata* (RAMBUR, 1834). Text-fig. 45: *E. oxycedrata* (RAMBUR, 1833). Text-fig. 46: *E. rosmarinata* MILLIÈRE, 1865.

narrow, slightly darker than rest of wing, with faint trace of pale wavy line near tornus; discal spot minute, circular, dark or absent. Fringe pale greyish brown, clearly spotted with brown at vein ends. Abdomen pale greyish brown with blackish brown or blackish grey second segment and usually with black lateral lines, especially well defined in male.

MALE GENITALIA: Valva with a large tapered medial ventral process; apex narrowly rounded; sacculus heavily sclerotized. Aedeagus nearly equal to valva length. Vesica with numerous transverse striae; armed with two slim horn-like cornuti (one longer than the other), one thin, elongated, whip-like cornutus and one smaller, irregular, curved cornutus at ductus ejaculatorius base. Sternum A8 elongate, laterally sclerotized near apex; with base broadened and with two short, narrow, blunt and sclerotized apical lobes.

FEMALE GENITALIA: Bursa copulatrix covered with spines at base, with or without rows of minute spines in elongated ductus bursae. Spurs of anterior apophyses slightly expanded at apices. Tergum A8 rectangular.

DISTRIBUTION: Mediterranean. In Europe from Spain eastwards to southern Russia (Krasnodar region); northward to Czech Republic; southward to Crete. – Outside Europe in Cyprus and Asian Turkey.

PHENOLOGY: Univoltine: early August (mid-July) to late October (mid-November). Larva: late April to late May (in the south from February to April). Pupa in flimsy cocoon in the ground. Overwintering as egg.

BIOLOGY: Larva disjunct oligophagous. Found on the flowers of *Erica arborea* (Ericaceae) (France: RAMBUR 1833; MILLIÈRE 1870; DIETZE 1910, 1913), on the needles of *Juniperus oxycedrus* (Italy: DIETZE 1910, 1913; WEIGT 1993), *J. communis* (Italy: CARRARA 1928), *J. foetidissima* (Cupressaceae) (Russia (ZISP): STSHUROV pers. comm.). Recorded also on *Cytisus scoparius* (Fabaceae) (France: LHOMME 1935).

HABITAT: Dry and light scrub, where junipers grow, light coniferous forests, edges of forests, rocks and stony slopes. From 0 m up to about 1,400 m above sea-level in Apennines (PROLA & RACHELI 1980), to about 1,600 m in Pyrenees (MAZEL & PESLIER 1997), to about 1,700 m in mountains of Balkan peninsula (coll. ZMUC; coll. FIBIGER), to about 1,800-1,900 m in southern Alps (coll. FIBIGER), and to about 2,000 m in southern Turkey (Taurus Mts) (coll. GELBRECHT).

SIMILAR SPECIES: This species is most similar to *E. pusillata*, yet usually slightly smaller, lighter, usually with delicate yellowish suffusion, the postmedial line often more curved and without black dashes on M3 and CuA1, the dark apical streak always present, the hindwing lighter, the distinct pattern lacking (Text-fig. 43). The narrower valva with larger ventral process, the shorter aedeagus, the shorter and narrower horn-like cornuti and presence of a whip-like cornutus on the vesica and the different shape of sternum A8 in the male should distinguish this species from *E. pusillata*. The female genitalia of *E. ericeata* are distinguished by the slim corpus bursae and the presence of medial rows of minute spines, but there are no large spines, as in the female genitalia of *E. pusillata*.

REMARKS: Very variable in depth of colour and distinctness of markings. The forewings are almost unicolorous, brownish grey, without pattern, except for the black streaks on the veins, discal spots and pale wavy lines in some specimens.

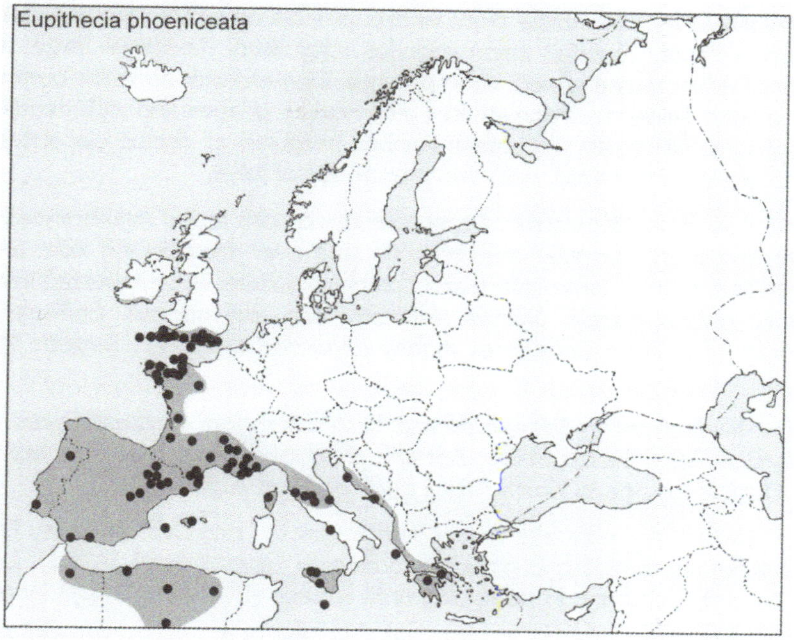

59. *Eupithecia phoeniceata* (RAMBUR, 1834)

Larentia phoeniceata RAMBUR, 1834: Annls Soc. ent. France 3 (2): 392, pl. 8: 6 (France: Marseille). Lectotype ♂ (coll. HERBULOT in ZSM, examined).

Eupithecia mnemosynata MILLIÈRE, 1876: Cat. Lépid. Alpes-Maritimes (3/4): 409 (France: Cannes). Syntype(s) (MNHN?).

Eup(ithecia) phoeniceata f. *albescens* DIETZE, 1910: Biol. Eupithecien 1, pl. 77: 716, 717; ibidem (1913), 2: 152 (France: Hyères). Syntypes 2 (MNHU). Valid at subspecific rank.

Unavailable names (incorrect subsequent spellings): *phoenicearia*: BOISDUVAL (1840); *phoenicata*: BUSTILLO & VARELA (1981).

DIAGNOSIS: Wingspan 18-24 mm. Labial palpi, frons and vertex light brownish grey. Notum light brownish grey with broad brown transverse stripe on the pronotum. Forewing elongate, costa slightly arched, termen oblique, apex rather pointed; ground colour pale brownish grey; all transverse lines brown, straight and oblique; basal and post-basal lines sharply angled at costa, rather inconspicuous; two distinct antemedial and medial lines medially connected; antemedial line sharply angled near costa; medial line crossing discal dot and slightly angled near costa or not angled at all; postmedial line oblique, slightly curved, sharply angled at costa, inner side with two dark brown dashes on M1 and M2; terminal area with similar dark brown apical dash; pale wavy line inconspicuous; discal dot black or intensely dark brown, usually rounded. Hindwing a little paler, light brownish grey; transverse lines indistinct except two darkish lines (basal and postmedial), well indicated only near dorsum reaching one third of way across wing; terminal area narrow, pale, inconspicuous, wavy line lacking; discal dot minute, dark, rounded and rather indistinct. Fringe pale grey, weakly spotted pale greyish brown at vein ends. Abdomen light grey or light brownish grey with narrow, dark brown second segment and with more or less distinct black or dark brown lateral stripes.

MALE GENITALIA: Valva with minute blunt ventral process or rather with angulated ventral margin, with narrowly rounded apex; sacculus sclerotized. Aedeagus large, about 1.3 times length of valva. Vesica armed with two large, long and slim horn-like cornuti, curved at apex; one lightly sclerotized and striated diverticulum at apex and with two flat, curved cornuti (one large, elongate, the other smaller, irregular) at ductus ejaculatorius base. Sternum A8 slightly sclerotized, with two minute apical lobes.

FEMALE GENITALIA: Bursa copulatrix heavily sclerotized, with broad membranous diverticulum at base and small triangular membranous area near apex on left side; armed with some longitudinal rows of large spines and numerous smaller spines situated irregularly in basal part of corpus bursae. Ductus seminalis long and narrow. Colliculum weakly defined. Spurs of anterior apophyses slightly expanded at apices. Tergum A8 trapeziform.

DISTRIBUTION: Mediterranean-Atlantic. Ireland, United Kingdom, Portugal, Spain, Balearics, France, Corsica, Italy, Sicily, Malta, Adriatic coast of Balkan peninsula and southern Greece. – Outside Europe in North Africa (Morocco and Algeria).

PHENOLOGY: Univoltine: early September (late August) to mid-October (early December). 'May to October' in southern England according to GOATER & al. (1981). Larva from December to early February. Pupa resting until autumn. Overwintering as larva.

BIOLOGY: Larva oligophagous (1). Found on the needles of *Juniperus phoenicea* (France: MILLIÈRE 1870; DIETZE 1910, 1913), *Cupressus macrocarpa* (United Kingdom: HAGGETT 1968), recorded also on *Cupressus sempervirens* (Cupressaceae) (France: LHOMME 1935) Reared in captivity on *Chamaecyparis lawsoniana* (United Kingdom: HAGGETT 1968) and recorded as probable on hybrid *Cupressocyparis* x *leylandii* (Cupressaceae) (United Kingdom: GOATER & al. 1981). Usually on the older, denser foliage of mature trees.

HABITAT: In habitats with juniper or cypress trees including Mediterranean thorn forests, dry and light scrub, sea coasts, rocks and mountain slopes, gardens and hedges. From 0 m up to about 1,400-1,500 m above sea-level in Spain (coll. ZMUC), and to about 1,600 m in Rif Mts (RUNGS 1981).

SIMILAR SPECIES: Adults of *E. phoeniceata* can usually be difficult to distinguish from two other closely related narrow-winged species, *E. oxycedrata* and *E. rosmarinata*. However, *E. oxycedrata* has less rich and less distinct antemedial and medial lines, usually not connected medially; postmedial line lighter and less distinct at costa and near dorsum, with two distinct parallel dark dashes, one on CuA2, the other between CuA2 and A. *E. rosmarinata* can be easily distinguished by medial line sharply angled beyond discal spot on the costa, by obliquely straight postmedial line twice sharply angled near costa, by absence of dark dashes on M1 and M2 and by clavate discal spot on the forewing (Text-fig. 44). The male genitalia of *E. phoeniceata* can usually be easily distinguished by the large aedeagus with two long horn-like cornuti slightly curved near their apices. The female genitalia can be easily distinguished by the large and heavily sclerotized bursa copulatrix with broad membranous appendage near its base and by the dense rows of large and smaller spines placed in the corpus bursae.

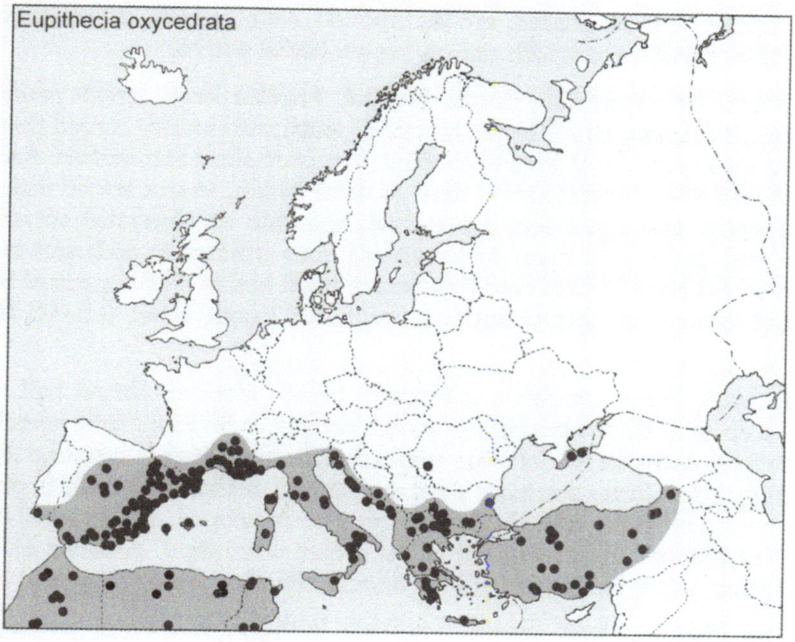

60. *Eupithecia oxycedrata* (Rambur, 1833)

Larentia oxycedrata Rambur, 1833: Annls Soc. ent. France 2 (1): 47, pl. 2: 12 (Corsica: Corté). Syntype(s) (MNHN?).

Eupithecia oxycedrata var. *adscriptaria* Staudinger, 1871: in Staudinger & Wocke, Cat. Lepid. Eur. Faunengeb. 1: 200 (Turkey: Bithynia). Syntype(s) (NHMW?). Valid at subspecific rank.

Eupithecia adscriptaria macedoniata Pinker, 1976: Z. Arb.Gem. Österr. Ent. 28 (1-3): 1 (Macedonia: Ochrid). Holotype ♂ (NHMW). Valid at subspecific rank.

Eupithecia provinciata Millière, 1872: Revue Mag. Zool. (2) 23: 61 (France: near Cannes). Syntypes (MNHN?).

Unavailable names (incorrect subsequent spellings): *oxycedraria*: Boisduval (1840); *exycedrata*: Dietze (1910).

Diagnosis: Wingspan 14-20 mm. Labial palpi greyish brown; frons and vertex whitish brown. Notum pale greyish brown with brown transverse band in front. Forewing narrow and elongate, costa slightly arched, apex pointed, termen faintly arched, oblique; light greyish brown ground colour; basal, postbasal and antemedial lines slightly sinuate, very oblique, sharply angled onto costa; the first medial line less oblique, crossing discal spot, not angled onto costa; medial area between antemedial and first medial lines slightly darker, with two parallel, fine, elongate blackish brown dashes on the stalk of Cu and between stalk of Cu and A; the outer half of medial area with four similar dashes on M1, M2, CuA2 and between CuA2 and A; postmedial line fine, rather inconspicuous, almost straight, oblique, twice right-angled near costa; terminal area narrow, usually with blackish brown dashes between veins; pale wavy line minutely undulate along its length, much dentated near costa; discal spot small, sickle-shaped to ovate-orbicular, black. Hindwing elongate, pointed; termen slightly concave medially; distinctly paler than the forewing, smoky-brownish grey; transverse lines well marked along dorsum, especially basal and postmedial; terminal area narrow with trace of pale wavy line; discal spot small, rounded, dark. Fringe whitish greyish brown, spotted with greyish brown at vein ends. Abdomen pale greyish brown with narrow brown lateral stripes.

MALE GENITALIA: Uncus medium length, biapical. Papillae large, length nearly equal to uncus length. Valva with small, blunt, ventral process; with weakly curved dorsal margin and tapered apex; sacculus slightly sclerotized. Vinculum short and narrow. Aedeagus relatively large and stout, length nearly equal to valva length. Vesica armed with two short horn-like cornuti, one apical and one medial, and with an elongated cornutus with a smaller irregular piece at its base; ductus ejaculatorius attached to aedeagus in its medial part. Sternum A8 peg-like with base broadened; basal hollow shallow; apical hollow very small; apical lobes short and pointed; sternum A8 length about 0.5-0.6 times valva length.

FEMALE GENITALIA: Bursa copulatrix sclerotized, basally globular, almost half covered by small spines at base; ductus bursae elongated, narrow, cylindrical, with numerous longitudinal striations, and with row of large spines on left side and also, with one row of small spines on right side. Ductus seminalis short and narrow, attached medially to globular part of corpus bursae on right side. Colliculum not well developed, nearly equal in width to apical part of corpus bursae. Antrum short, broad, sclerotized. Anterior and posterior apophyses short and thin. Tergum A8 rectangular. Papillae anales small and rounded.

DISTRIBUTION: Mediterranean. Spain, southern Portugal, Balearics, southern France, Corsica, Sardinia, Italy, Sicily, Slovenia, Croatia, Herzegovina, Montenegro, Macedonia, Bulgaria, Greece, Crete. Local populations in western Romania and southern Crimea. - Outside Europe in North Africa (Morocco to Tunisia) and Turkey.

PHENOLOGY: Bivoltine: early April (mid-March) to late June (mid-July); early September (early August) to mid-October (late November). Larva: late May to early July; November to early January. Overwintering as pupa.

BIOLOGY: Larva monophagous (2). Found on the needles and flowers of *Juniperus oxycedrus* (Corsica: RAMBUR 1833; MABILLE 1868; France: MILLIÈRE 1870).

PARASITOIDS: Hymenoptera, Ichneumonidae: *Diadegma chrysostictos*, *Deuterolabops eupitheciae*, *Platylabus iridipennis*, *P. pactor* (HERTING 1976).

HABITAT: Found where junipers are present, in dry and light scrub, (macchia, aquis), among rocks and in forest steppe slopes. From 0 m up to about 1,100 m above sea-level in Apennines (Mt. Majella: ZAHM pers. comm.), to about 1,700 m in Greece (coll. FIBIGER), to about 2,300 m in Sierra Nevada (coll. ZMUC), and to about 2,600-2,700 m in Great Atlas (RUNGS 1981; coll. SOMMERER).

SIMILAR SPECIES: A species showing little variation, similar to *E. phoeniceata*, however, distinguished from it by the usually darker ground colour, by two dark dashes on CuA2 and between CuA2 and A, the faintly visible postmedial line, and the sickle-shaped discal spot on the forewing (Text-fig. 45). The shorter aedeagus with two smaller horn-like cornuti on the vesica in the male genitalia; and the smaller and lighter sclerotized bursa copulatrix decorated by a series of large spines on the prominent left side in the female genitalia easily distinguish *E. oxycedrata* from *E. phoeniceata*.

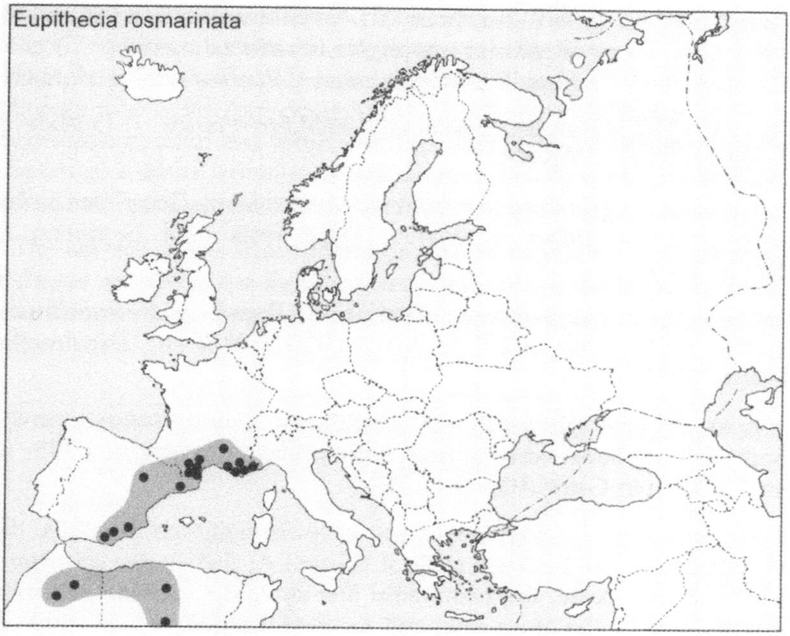

61. *Eupithecia rosmarinata* DARDOIN & MILLIÈRE, 1865

Eupithecia rosmarinata DARDOIN & MILLIÈRE, 1865: in MILLIÈRE, Iconogr. Descr. Chenilles Lépid. inédits 2 (14): 104, pl. 63: 4-8 (France: near Marseille). Syntypes including ♀, (MNHN?).

DIAGNOSIS: Wingspan 19-22 mm. Labial palpi greyish brown; frons and vertex pale greyish brown. Notum pale greyish brown with dark transverse band in front. Forewing narrow, elongate; costa straight; apex pointed; termen oblique, slightly arched; ground colour light greyish brown; transverse lines oblique, sharply angled near costa; postbasal and antemedial lines twice sharply angled along their length; elongated, narrow blackish brown dash between second angle of antemedial line and discal spot present; medial line straight and oblique, touching discal spot on outer side and sharply angled beyond it onto costa; postmedial line straight, oblique, with acute and right angles near costa; terminal area narrow with darker inner half and with elongate blackish brown apical streak; pale wavy line fine, very slightly sinuate, broadened near tornus; discal spot small, ovate-orbicular, black. Hindwing ovate, noticeably paler, smoky-greyish brown; transverse lines well marked only along dorsum; terminal area narrow with trace of pale wavy line; discal spot minute, rounded, dark, placed nearer to costa or obsolete. Fringe light, whitish grey-brown, faintly spotted with dark at vein ends. Abdomen pale greyish brown with slightly darker second segment, and with rows of small, dark brown lateral dots.

MALE GENITALIA: Valva without ventral process, with evenly curved ventral margin, and with apex narrowly rounded and curved dorsad; sacculus lightly sclerotized. Aedeagus shorter than valva. Vesica armed with two rather short and stout horn-like cornuti, and two flat, curved cornuti (one elongated and one smaller, irregular) at ductus ejaculatorius base. Sternum A8 strongly broadened at base, without apical hollow.

FEMALE GENITALIA: Bursa copulatrix small, pouch-like, membranous, without spines.

Colliculum weak defined, broad, membranous. Anterior and posterior apophyses relatively long and thick. Spurs of anterior apophyses broadened medially. Tergum A8 with deep membranous medial hollow in posterior margin. Papillae anales relatively large.

DISTRIBUTION: West-Mediterranean. Spain and southern France. – Outside Europe in North Africa (Morocco and Algeria).

PHENOLOGY: Univoltine: mid-November (mid-October) to early December (in Morocco to March). Larva from December to March. Pupa rests until beginning of winter. Overwintering as larva.

BIOLOGY: Larva disjunct oligophagous. Found on the flowers of *Rosmarinus officinalis* (France: MILLIÈRE 1865; MABILLE 1872; DIETZE 1910, 1913) and *Thymus* (Lamiaceae) (France: LHOMME 1935).

HABITAT: Steppes, dry and light scrub and coastal sites, where rosemary grows. From 0 m up to about 400 m above sea-level in Pyrenees (MAZEL & PESLIER 1997; coll. ZSM), and to about 1,900 m in Great Atlas (coll. ZSM).

SIMILAR SPECIES: Similar to *E. phoeniceata* but easily distinguished from it by the outward angle in the antemedial line between stalk of Cu and A; the medial line sharply angled beyond discal spot onto costa, the postmedial line straighter, oblique and more sharply angled near costa, and by the more even and rounded margins of hindwings. *E. rosmarinata* is also closely related and similar to *E. oxycedrata*, but can be easily distinguished from it by the less pointed apex of forewing, by the medial line sharply angled beyond discal spot onto costa, by the obliquely straight postmedial line which is twice sharply angled near costa and by the absence of pair of dark dashes on M1 and M2 and pair of dark dashes on CuA2 and behind it as well (Text-fig. 46). The narrower uncus near apex, the valva with bent apex, the smaller aedeagus and the shape of sternum A8 in the male, and the small pouch-like, membranous bursa copulatrix without spines in the female genitalia should distinguish *E. rosmarinata* from both the compared species.

The ***tripunctaria*** species-group

Group introduced here for the first time. It is a species-poor species-group of *Eupithecia* represented in Europe by two species: *E. barteli* and *E. tripunctaria*. The following Asian species *E. tenuisquama* (WARREN, 1896), *E. toshimai* INOUE, 1980, *E. cotidiana* VOJNITS, 1979, *E. invicta* VOJNITS, 1979, *E. incohata* VOJNITS, 1979 and *E. acuta* VOJNITS, 1973 belong to the *tripunctaria* group as well.

MALE GENITALIA: Uncus biapical, short, stout, very broad near base and tapering towards apex. Papillae length about 0.6-0.7 times uncus length. Valva large, with small blunt ventral process; apex broad and rounded; sacculus sclerotized. Vinculum short, broad, with shallow medial hollow. Aedeagus length about 0.7-0.8 times valva length; about 3.5-4.0 times longer than medial width. Vesica with transverse striations; armed with one slim horn-like cornutus near apex, with length about 0.3 times aedeagus length, also with one or two smaller, irregular, flat cornuti at ductus ejaculatorius base. Sternum A8 elongated, peg-like, with broad basal part; basal hollow shallow; there are two short, blunt or pointed apical lobes; sternum length about 0.6-0.8 times valva length.

FEMALE GENITALIA: Bursa copulatrix sub-globular or round, almost completely covered with small spines except membranous area at base; and with long, narrow, cylindrical ductus bursae without or with spines, covered by numerous longitudinal striations. Ductus seminalis large, long and broad, medially broadened, with longitudinal striations also; attached to ductus bursae near colliculum on right side. Colliculum collar-like, short and narrow. Antrum membranous, short and broad. Anterior and posterior apophyses medium length and thin. Spurs of anterior apophyses well defined, evenly expanded to apices. Tergum A8 almost square. Papillae anales small, short and rounded.

REMARKS: The species of the *tripunctaria* group can usually be easily distinguished from other *Eupithecia* species by the short and very stout uncus; the large valva with blunt ventral process; the armature of vesica and the shape of sternum A8 in male; as well as by the female genitalia.

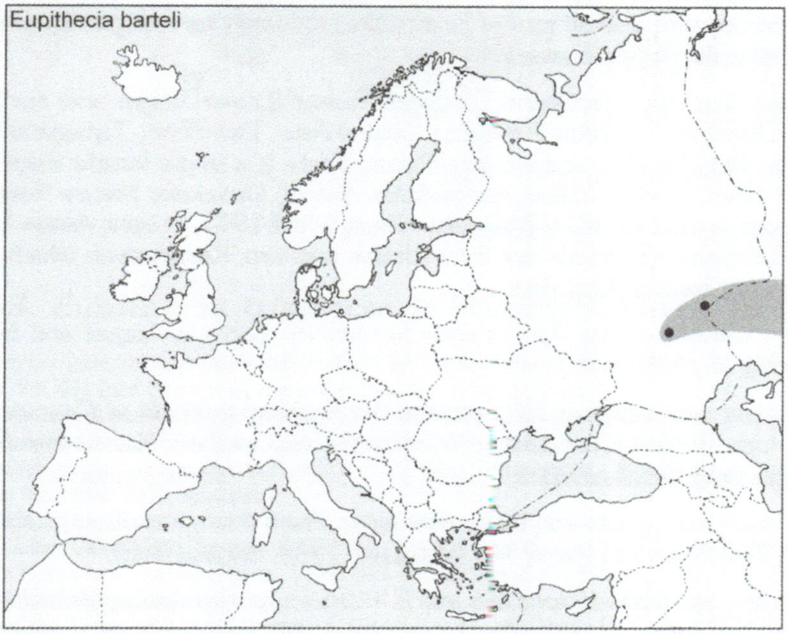

62. *Eupithecia barteli* DIETZE, 1908

Eup(ithecia) barteli DIETZE, 1908: Dt. ent. Z. Iris 21: 162, pl. 2: 13, 14 (Kazakhstan: Uralsk). Syntypes 3, including 1♂2♀ (MNHU, syntype ♀ examined).

Eupithecia artshae VIIDALEPP, 1988: Fauna pjadenitz gor Srednej Azii: 124, pl. 2: 21; figs 28: 2, 3, 6-8 (Tajikistan: SW Pamir, Khorog, botanical garden). Holotype ♂ (ZISP, examined).

DIAGNOSIS: Wingspan 17-22 mm. Labial palpi dark grey. Frons and vertex ash-grey to dark grey. Notum bicoloured: anterior part blackish grey; posterior part ash-grey. Forewing costa slightly arched; apex somewhat pointed; termen rounded; ground colour greyish fuscous with a soft ochreous tinge; costal margin slightly darker; basal area to medial line almost completely unicolorous greyish fuscous; transverse lines rather indistinct, cloudy; medial much sinuate, sharply angled from discal spot (or beyond it) towards costa; postmedial line slightly curved, twice right-angled near costa; medial area with distinct, pale ochreous cloud under discal spot; double pale ochreous stripe between medial and terminal areas usually very evident; whitish wavy line fine, multiply indented, often broken, forming a series of small triangular spots or sometimes obsolete, pale tornal dot lacking; discal spot large, very distinct, ovoid, intensely black. Hindwing slightly paler; transverse lines more or less distinct, especially evenly curved light medial line; terminal area usually darker; paler wavy line ill-defined; discal spot conspicuous, dark, circular or ovate. Fringe light grey with weak ochreous tinge, spotted with dark grey at vein ends. Abdomen ash-grey to fuscous grey, with short and narrow black lateral stripes on posterior segments.

MALE GENITALIA: Vesica armed with one horn-like cornutus and with two smaller, flat cornuti (one at spine base, the other at ductus ejaculatorius base). Sternum A8 with two very short, sclerotized, blunt apical lobes.

FEMALE GENITALIA: Bursa copulatrix round, almost completely covered by spines; cylindrical ductus bursae with a short row of small spines, strongly curved to left, curving round left side of globular corpus bursae.

DISTRIBUTION: Turanian. Southern European Russia (Lower Volga) and north-western Kazakhstan (Uralsk). – Outside Europe in Central Asia: Tien-Shan, Tarbagatai Mts, Saur Mts, Ghissar Mts, Alai Mts and western Pamir. There is a single female from European Russia with label: Saratov oblast, Krasnokutsk district, Djakovsky Nature Reserve, right side of Eruslan river, 3 km W. of Djakovka village, 9.VIII.1991, at light, ANIKIN leg. (ZISP). All other European specimens are from Uralsk (western Kazakhstan), which is on the right (European) bank of Ural river.

PHENOLOGY: Univoltine: late July to early September. Larva in August and September. Overwintering as pupa.

BIOLOGY: Larva monophagous (2). Found in the ripening seedballs of *Cuscuta monogyna* and *C. lupuliformis* (Cuscutaceae) (Kirghizstan and southern Kazakhstan (Tien-Shan Mts): original data); feeds at night.

HABITAT: Flood plain forests and bushy river sides. From 0 m up to about 2,300 m above sea-level in Tien-Shan and Pamir Mts (VIIDALEPP 1988; original data).

SIMILAR SPECIES: Similar to *E. lariciata* and *E. subfuscata*, two species with which it could be confused, but distinguished by the darker colour with pale ochreous medial blotch and double stripe beyond postmedial line, and by the less distinct transverse lines on the forewing as well. The male and female genitalia of *E. barteli* can be easily distinguished from those of all other *Eupithecia* species. The genitalia are very similar to those of East Asian *E. tenuisquama* WARREN, 1896, especially in the armature of aedeagus, the shape of valva and shape of bursa copulatrix in the female genitalia. However, *E. barteli* can usually be distinguished by the shape of sternum A8 in male.

REMARKS: A species with little variation in the coloration and markings. Some specimens have veins covered by black scales and without discal spots on the hindwings. Unfortunately, all known European specimens, which are deposited in MNHU and ZISP, are in relatively bad condition, including three type-specimens.

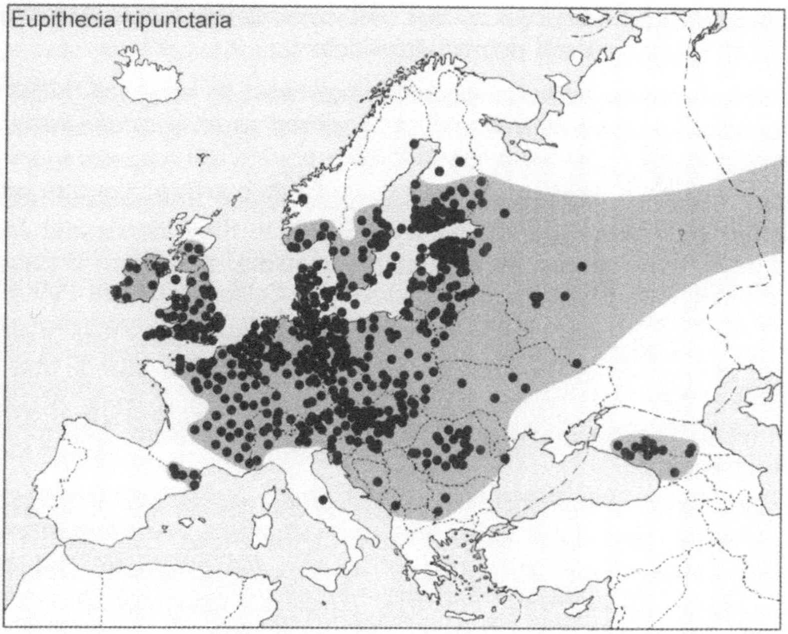

63. *Eupithecia tripunctaria* HERRICH-SCHÄFFER, 1852

Eupithecia tripunctaria HERRICH-SCHÄFFER, 1852: Syst. Bearb. Schmett. Eur. 6 (55): 77; ibidem (1851), 3 (Suppl.), pl. 75: 461 (non binominal), (Germany: Frankfurt a/M). Syntype(s).
Phalaena albipunctata HAWORTH, 1809: Lepid. Br.: 360 (United Kingdom: London). Syntype(s) (BMNH). Junior primary homonym of *Phalaena albipunctata* HUFNAGEL, 1767.
Eupithecia albipunctata var. *angelicata* BARRETT, 1877: The Entomologist's Monthly Mag. (1876-1877) 8: 278 (United Kingdom: York). Syntype(s).
Eupithecia promulgata PEARSALL, 1909: Proc. Ent. Soc. Wash., 11: 121; partim ♀; nec ♂ (U.S.A.: New York, Big Indian Valley, Catskill Mts). Syntype ♀ (AMNH).
(Eupithecia) albipunctata f. *privata* DIETZE, 1913: Biol. Eupithecien 2: 87; ibidem (1910), 1, pl. 85: 941 (Italy: Tyrol, Waidbruck). Syntypes 4, including ♂ ♀ (MNHU). Valid at subspecific rank.
Unavailable names (infrasubspecific): *tripunctata*: CREWE (1862); *angelicata*: BARRETT (1877); *aestiva*: DIETZE (1910); *intermedia*: LEMPKE (1951).

DIAGNOSIS: Wingspan 16-23 mm. Labial palpi dark grey. Frons and vertex dark grey to ash-grey. Notum dark grey with white posterior spot. Forewing rather wide; costa arched; unicolorous dark greyish fuscous with very weak darkish costal spots; all transverse lines indistinct, in some specimens only first medial and postmedial lines evident, first medial line sharply curved around discal spot; veins black, but usually chequered; white wavy line fine, multiply indented, often broken, with two or three conspicuous white spots, of which the tornal is largest; discal spot rather small, circular to ovate, black. Hindwing fuscous grey; veins chequered black and white; basal area darker along dorsum; transverse lines invisible; terminal area sometimes slightly darker; pale wavy line inconspicuous but two small white tornal spots often distinct; discal spot small, circular, dark grey. Fringe grey, slightly spotted with dark grey at vein ends. Abdomen fuscous grey, with small white dorsal crests on the posterior margin of each segment and with black lateral stripes.

MALE GENITALIA: Valva with small blunt or sharp ventral process. Vesica armed with one horn-like cornutus, and with one smaller, flat cornutus at ductus ejaculatorius base. Sternum A8 with two short and pointed apical lobes.

FEMALE GENITALIA: Corpus bursae elongated, subglobular; its basal half thickened, membranous, apical half covered dorsally with small spines, ventrally membranous, spineless; ductus bursae elongated, narrow and cylindrical.

DISTRIBUTION: Holarctic. Widespread in Palaearctic region from Ireland and Pyrenees throughout Europe, Caucasus and Siberia eastwards to Kuril Islands and Japan. Also, widespread in North America from eastern Newfoundland to western British Columbia; north to Labrador, and south to San Bernadino Mts, California (BOLTE 1990).

PHENOLOGY: Uni- or bivoltine: early May (mid-April) to early June (late June); mid-July (early July) to late August (late September); in the north one generation in May and June). Larva: mid-June (late May) to mid-July (late July); late August (early August) to mid-September (early October); in the north one generation from July to September. Overwintering as pupa.

BIOLOGY: Larva rather polyphagous (1). Found on the flowers and developing seeds of *Angelica sylvestris, Heracleum sphondylium, Cicuta virosa, Laserpitium latifolium, Peucedanum oreoselinum, P. palustre, Anthriscus sylvestris, Daucus carota, Astrantia major, Bupleurum, Falcaria vulgaris, Chaerophyllum bulbosum, Aegopodium podagraria, Ligusticum, Pastinaca sativa, Pimpinella saxifraga, Ferulago campestris* (Apiaceae) and probably occasionally on the flowers and seeds of *Solidago virgaurea, Eupatorium cannabinum* (Asteraceae), *Knautia arvensis* (Dipsacaceae), *Filipendula ulmaria* (Rosaceae); the second generation has been found on *Viburnum, Sambucus nigra, S. ebulus* (Caprifoliaceae) (United Kingdom: CREWE 1861a, 1862, 1863; BUCKLER 1899; BARRETT 1904; France: GOOSSENS 1869; LHOMME 1935; Germany: DIETZE 1874a, 1910, 1913; WEIGT 1980, 1993; WEGNER pers. comm.; Romania: CARADJA 1896; Denmark: JUUL 1948; Finland: SEPPÄNEN 1970; Russia: original data). In North America found on flower-heads of *Viburnum* (Caprifoliaceae), umbels of *Heracleum maximum (=lanatum)* (Apiaceae) (Canada: FORBES 1948; RINDGE 1952) and one stunted adult was reared on *Abies balsamea* (Pinaceae) (Canada (CNC): BOLTE 1990, probably in error). Often heavily parasitized.

PARASITOIDS: Hymenoptera, Ichneumonidae: *Endromopoda detrita, Deuterolabops eupitheciae, Platylabus iridipennis, P. pedatorius, P. pumilio, Agrypon clandestinum* - Hymenoptera, Braconidae: *Apanteles cajae, A. tibialis (=congestus), Meteorus ictericus* - Hymenoptera, Encyrtidae: *Litomastix* sp. - Hymenoptera, Eulophidae: *Pediobius flaviscapus* (THOMPSON 1945; SHENEFELT 1969, 1972; HERTING 1976).

HABITAT: Open, light forests, forest clearings and meadows, edges of forests, parks, damp areas, shady roadsides and forest paths, as well as open, uncultivated areas. From 0 m up to about 1,600 m above sea-level in Apennines (Mt. Majella: ZAHM pers. comm.), to about 1,800 m in Alps and Pyrenees (DANTART 1988).

SIMILAR SPECIES: Similar to *E. virgaureata, E. lariciata* and *E. subfuscata*. However, it is clearly distinguished from these species by the combination of white spot on the notum, the indistinct transverse lines and the very weak darkish costal spots on the forewing, the chequered veins and presence of some white spots on the wavy line, especially tornal on the fore- and hindwings. The very stout uncus; the large valva with blunt ventral process; the vesica armed with one horn-like cornutus; the unique shape of sternum A8 in male, and the shape of bursa copulatrix covered by spines only in the medial area in the female genitalia are very good features that serve to distinguish this species.

REMARKS: The typical form is a little variable in the coloration and distinctness of markings. The melanic ab. *angelicata* BARRETT, 1877 has wings completely unicolorous, sooty black with more intensely black discal spots and veins, and white subterminal spots lacking, and also with unicolorous dark fringe.

The *fletcherata* species-group

This is a species-poor group of *Eupithecia* represented in Europe by *E. virgaureata*. Three North American species: *E. casloata* (DYAR, 1904), *E. fletcherata* TAYLOR, 1907 and *E. sharronata* BOLTE, 1990 were separated to the *fletcherata* species-group by BOLTE (1990) for the first time. The *E. dealbata* INOUE, 1988 from Taiwan, can be placed in this group as well.

MALE GENITALIA: Uncus biapical, short, broadened in medial part, when viewed laterally. Papillae narrow and short; length about 0.5-0.7 times uncus length. Valva without ventral process, medially broadened, with curved ventral margin and almost straight dorsal margin; apex broad and rounded; sacculus weakly sclerotized. Vinculum short and wide. Aedeagus relatively thin, with striated and oblique apex; length about 0.7-0.8 times valva length; about 4.0-5.0 times longer than medial width. Vesica armed with one thin cornutus medially (except *E. casloata* with two cornuti) and one large and elongated flat cornutus at ductus ejaculatorius base. Sternum A8 with broad base and shallow hollow; shaft narrowed medially; apex slightly broader than shaft and terminating in two sharp horns, the right one slightly longer than left one and bent toward left side, and with shallow hollow; sternum length about 0.6-0.7 times valva length.

FEMALE GENITALIA: Bursa copulatrix diverse in size and shape: usually membranous, small, rather rounded or pouch-like, broadest near base or in middle, with diverse spinulated area (with a few scattered spines medially, or with a semi-circular bulge with spinules near apex and numerous spinules around perimeter medially or almost completely and uniformly covered by spinules). Ductus seminalis membranous, short and narrow; usually attached to apical part of corpus bursae near colliculum on left side. Colliculum collar-like, short and narrow. Antrum membranous, short and broad. Anterior and posterior apophyses long and thin, not expanded at apices. Spurs of anterior apophyses well defined, elongated, broadened to apices. Tergum A8 square or rectangular. Papillae anales usually large, rounded or slightly elongated.

REMARKS: The species of the *fletcherata* group are close to species of the *lariciata* and *tripunctaria* groups in the maculation of their wings. The cornuti on the vesica, the unusual shape of sternum A8 in the male and the areas covered by spines in the bursa copulatrix in the female are good distinguishing group characteristics.

64. *Eupithecia virgaureata* DOUBLEDAY, 1861

Eupithecia virgaureata DOUBLEDAY, 1861: The Zoologist 19: 7566 (United Kingdom). Syntypes, at least 3 (BMNH).

Eupithecia offirmata SPEYER, 1869: in GRENTZENBERG, Schr. phys.-ökon. Ges. Königsb. 10: 120 (Poland: Johannisberge near Danzig [=Gdansk]). Syntypes (MNHU).

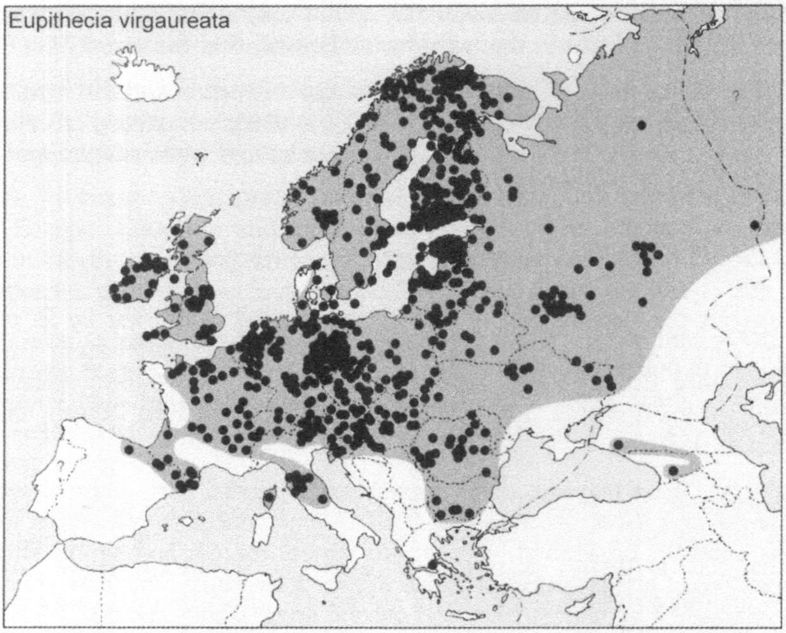

Eupithecia invisa BUTLER, 1878: Ann. Mag. nat. Hist. (5) 1: 444 (Japan: Hakodaté). Holotype ♀ (BMNH, examined: INOUE 1980).

Eupithecia altenaria STAUDINGER, 1861: Stett. Ent. Z. 22 (10-12): 401 (Norway: near Bossekop). Syntypes ♂ ♀ (MNHU).

Unavailable names (misidentification): *pimpinellata*: sensu DOUBLEDAY (1856) nec HÜBNER, 1813 (identity: *Eupithecia pimpinellata*). – (infrasubspecific): *nigronotata*: DIETZE (1910); *nigrofasciata*: DIETZE (1910); *aestiva*: DIETZE (1910); *notata*: DIETZE (1913); *nigra*: LEMPKE (1951). - (incorrect subsequent spelling): *virgaurearia*: MORRIS (1861).

DIAGNOSIS: Wingspan 16-23 mm. Labial palpi blackish grey, usually with oblique medial row of white scales. Frons, vertex and notum fuscous grey. Forewing fuscous grey; transverse lines indistinct, slightly sinuous, often reduced to series of short black dashes on the veins, especially on the stalk of Cu beyond discal spot; costa usually with five blackish grey spots; medial line usually conspicuous, sharply angled near costa beyond discal spot or touching discal spot; terminal area with pale wavy line inconspicuous, broken or often obsolete, but with small whitish tornal spot; discal spot distinct, ovate or rounded, black. Hindwing slightly paler, with darker terminal area and minute pale tornal spot; transverse lines usually ill-defined but in some specimens they are more or less distinct; discal spot small, dark, rounded. Fringe grey, slightly spotted with dark grey at vein ends. Abdomen fuscous grey with black lateral stripes; last segment paler, ash-grey in male.

MALE GENITALIA: See the *fletcherata* species-group diagnosis.

FEMALE GENITALIA: Bursa copulatrix small, broadened near base, with small patch of spinules near apex and with numerous spinules around perimeter medially. Tergum A8 rectangular.

DISTRIBUTION: Eurasiatic. Widespread in Palaearctic region from Spain throughout Europe and Siberia to Japan.

PHENOLOGY: Bivoltine: late April (mid-March) to mid-June (late June); late July (late June) to late August (early October). Larva: mid-June (mid-May) to early July (mid-July); late August (mid-July) to late September (late October). Overwintering as pupa.

BIOLOGY: Larva of the first generation is oligophagous (1). Found on the leaves of *Crataegus monogyna, Prunus padus, P. spinosa, Rubus chamaemorus* (Rosaceae). Larva of the second generation is polyphagous (1). Found on the flowers and seeds of *Solidago virgaurea, S. canadensis, S. gigantea, S. japonica, Senecio jacobaea, S. nemorensis, S. vulgaris, Centaurea, Cirsium palustre, Artemisia vulgaris, Hieracium, Eupatorium cannabinum, Taraxacum officinale, Inula conyza* (Asteraceae), *Calluna vulgaris, Erica tetralix* (Ericaceae), *Lysimachia* (Primulaceae), *Heracleum sphondylium, Anthriscus sylvestris, Chaerophyllum, Pimpinella saxifraga, Angelica sylvestris* (Apiaceae), *Gentiana* (Gentianaceae) (United Kingdom: CREWE 1862, 1863; Germany: DIETZE 1872, 1875b, 1910, 1913; WEIGT 1993; GELBRECHT and WEGNER pers. comm.; France: LHOMME 1935; Denmark: JUUL 1948; Finland: SEPPÄNEN 1970; Russia: original data). Found also on the flowers and ripening seeds of *Heracleum moellendorffi* (Apiaceae) (Far East of Russia: original data). Recorded on the flowers of *Kalimeris yomena* and *Dendranthema japonicum* (Asteraceae) (Japan: MURASE 2000).

PARASITOIDS: Hymenoptera, Ichneumonidae: *Netelia tarsata, Triclistus* sp., *Hypsicara* sp. - Hymenoptera, Braconidae: *Meteorus ictericus, M. pulchricornis, Rasivalva (=Protomicroplitis) calceata, Catesia* sp. - Diptera, Tachinidae: *Blondelia (=Lydella) nigripes* (THOMPSON 1945; HERTING 1960, 1976; SHENEFELT 1969, 1973; original data).

HABITAT: Deciduous and mixed forests, forest edges and clearings, along forest paths, warm, uncultivated slopes, gardens and parks. From 0 m up to about 1,000 m above sea-level in Apennines (TEOBALDELLI 1994), to about 1,400 m in Alps and Carpathians (VORBRODT 1931; POPESCU-GORJ 1964; FORSTER & WOHLFAHRT 1981), and to about 1,600-1,700 m in Pyrenees (MAZEL & PESLIER 1997).

SIMILAR SPECIES: This species is distinguished with difficulty from *E. subfuscata* HAWORTH, 1809. The larger, rounded discal spot and the chequered veins on the forewing can usually be used to separate adults from *E. subfuscata* (Text-fig. 84). The male and female genitalia, and the shape of the sternum A8 in male of *E. virgaureata* easily distinguished this species from all other European *Eupithecia* species.

REMARKS: Variable in the distinctness of the wing pattern; in some specimens the maculation almost is wholly obsolete, except black discal spot, dark veins along postmedial transverse line and pale tornal spot on the forewing. Specimens from northern Europe usually have similar inconspicuous, clouded markings on the wings (f. *altenaria* STAUDINGER, 1897). The melanic ab. *nigra* LEMPKE, 1951 with ground colour black occurs sporadically in many parts of Europe.

The ***tantillaria*** species-group

This species-group introduced here for the first time. Represented in Europe by two species: *E. tantillaria* and *E. liguriata*. One East Asian species that has been examined and could be placed in this species-group is *E. tantilloides* INOUE, 1958.

MALE GENITALIA: Uncus short or medium length, narrow, biapical. Papillae length about 0.6-0.9 times uncus length. Valva with one prominent, blunt, sclerotized ventral process and one smaller, blunt ventral process, with shallow hollow between them; apex of valva narrow; sacculus heavily sclerotized. Vinculum diverse in size and shape. Aedeagus slim, long and thin, widest at base; length nearly equal to valva length or slightly shorter, about 4.0-5.0 times longer than medial width. Vesica armed with one to three slim, narrow horn-like cornuti (one larger than the others); with one elongated, flat, convex, V-shaped cornutus and one small, irregular, folded cornutus at ductus ejaculatorius base. Sternum A8 small; base broadened with shallow basal hollow; apical hollow deeper with length about 0.3 times sternum length; with two narrow horns, pointed at apices; sternum length about 0.4-0.5 times valva length.

FEMALE GENITALIA: Bursa copulatrix diverse in size and shape (pear-shaped or elongated, pouch-like), membranous with more sclerotized apical half; covered with small spines in basal or apical half. Ductus seminalis short and narrow, usually evenly curved; attached to corpus bursae near colliculum or to medial part. Colliculum collar-like, usually short and narrow. Antrum membranous or sclerotized. Anterior and posterior apophyses short and thin. Spurs of anterior apophyses narrow, usually slightly expanded to apices. Tergum A8 square. Papillae anales small, short and rounded.

REMARKS: Species of the *tantillaria* group are closely related to species of the *lariciata* group. However, they have narrower uncus, valva with two ventral processes, differences in armature of vesica and the shape of sternum A8 in the male; the corpus bursae also shows good distinguishing group characteristics.

65. Eupithecia tantillaria BOISDUVAL, 1840

Eupithecia tantillaria BOISDUVAL, 1840: Gen. Index meth. Eur. Lepid.: 210 (France: Gall. mer.). Syntype(s).
Eupithecia pusillata var. *laricis* SPEYER, 1873: Stett. Ent. Z. 34 (7-9): 362 (Germany: Arolsen). Syntype(s), including ♂ (MNHU?). Valid at subspecific rank.
(Eupithecia) subumbrata f. *calabrica* DIETZE, 1913: Biol. Eupithecien 2: 156; ibidem (1910), 1, pl. 78: 729, 730 (Italy: Calabria, Aspromonte). Syntype(s) (MNHU). Valid at subspecific rank.
Eupithecia tantillaria f. *piceata* PROUT, 1914: in SEITZ, Macrolep. 4: 297, pl. 13: O (North and Central Europe; Caucasus). Syntype(s) (BMNH). Valid at subspecific rank.
Unavailable names (misidentification): *subumbrata*: sensu HÜBNER (1799) nec DENIS & SCHIFFERMÜLLER, 1775 (identity: *Eupithecia subumbrata*); *pusillata*: sensu HÜBNER (1813) nec DENIS & SCHIFFERMÜLLER 1775 (identity: *Eupithecia pusillata*); *pusillaria*: sensu HERRICH-SCHÄFFER (1848) nec HÜBNER, 1825 (identity: *Eupithecia pusillata*). – (infrasubspecific): *pallida*: LEMPKE (1951); *marginata*: LEMPKE (1969); *nigrescens*: LEMPKE (1969); *mediofasciata*: LEMPKE (1969); *approximata*: LEMPKE (1969). – (incorrect subsequent spellings): *pussilata*: SOFFNER, (1928); *tandillaria*: TOMIČ & al. (1994).

DIAGNOSIS: Wingspan 14-22 mm. Labial palpi brown or blackish brown. Frons pale brownish grey with dark brown scales placed anteriorly. Vertex and notum pale brownish grey. Forewing pale greyish brown; transverse lines darker, greyish brown, usually wide, distinct but clouded, slightly sinuous, usually broadened at costa and so forming rather ill-defined dark costal spots; two medial lines usually curved around discal spot; postmedial line evenly curved near costa; terminal area darker, greyish brown, usually with lighter apical and medial blotches; sometimes small pale tornal spot present; pale wavy line usually indistinct and broken among fuscous clouding; discal spot large, prominent, rounded or ovate, fuscous. Hindwing paler, pale grey with brownish tinge; ante- and postmedial lines conspicuous, wide and clouded; postmedial line usually right-angled or curved around discal spot; terminal area slightly darker, clouded with inconspicuous pale wavy line; discal spot conspicuous, rounded or ovate-orbicular, sometimes obsolete. Fringe pale

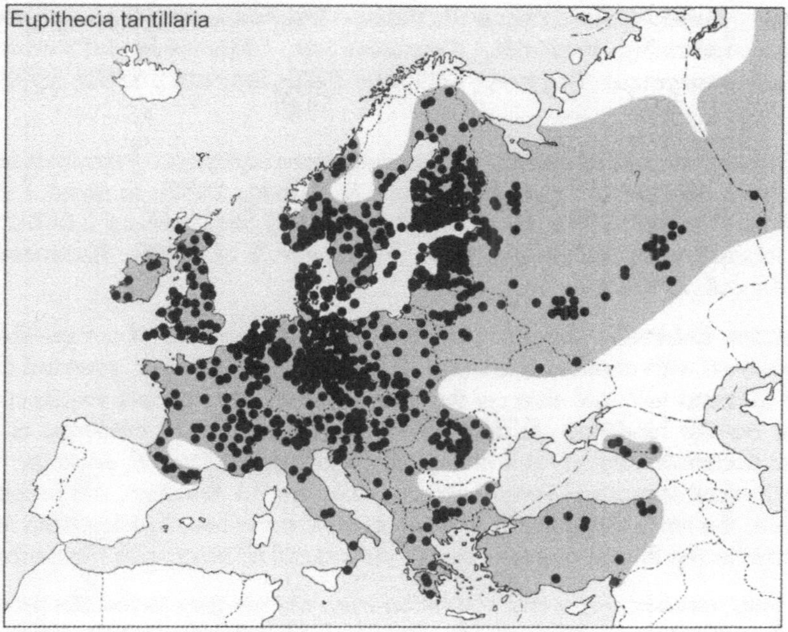

grey, spotted greyish brown at vein ends. Abdomen pale brownish grey with small, dark brown lateral spots; second segment darker, brownish.

MALE GENITALIA: Vinculum elongated, narrow, basally tapered. Vesica armed with two small and narrow horn-like cornuti (one medially and one larger apically); with one elongated, V-shaped, apically broadest cornutus with dentate apical margin; and with one small folded cornutus at ductus ejaculatorius base.

FEMALE GENITALIA: Bursa copulatrix pear-shaped, sclerotized; more than half covered with small spines at base; with one patch of small spines near colliculum. Ductus seminalis attached to medial part of corpus bursae on left side. Colliculum slightly inclined to left. Antrum heavily sclerotized. Tergum A8 with deep and narrow medial hollow in the posterior margin.

DISTRIBUTION: Euro-Siberian. Widely distributed in Europe except Iceland, Portugal and Mediterranean islands. – Outside Europe in Asia Minor, Caucasus, western Transcaucasus and eastward through Urals and western Siberia to Altai Mts.

PHENOLOGY: Uni- or bivoltine: late April (mid-April) to late June (late July). In the south a second generation in August (MAZEL & PESLIER 1997). Larva: mid-May to late July (late August) (usually mid-June to mid-July; in the north in August). Overwintering as pupa.

BIOLOGY: Larva oligophagous (2). Found on *Picea abies (=excelsa)* (United Kingdom: CREWE 1862, as *pusillata*) and also *Abies alba, A. sibirica, Pinus sylvestris, Larix sibirica, L. decidua* (Pinaceae) and *Juniperus* (Cupressaceae), preferring the young needles (Germany: DIETZE 1874a, 1910, 1913; WEIGT 1980, 1993; Denmark: JUUL 1948; Finland: SEPPÄNEN 1970). Recorded also on *Pseudotsuga menziesii* (Pinaceae) (United Kingdom: GOATER & al. 1981).

PARASITOIDS: Hymenoptera, Ichneumonidae: *Trychosis legator* - Hymenoptera, Braconidae: *Apanteles immunis*, *Apanteles* sp., *?Rasivalva (=Protomicroplitis; Microgaster) circumvecta*, *Rogas* sp. (PATOČKA 1960; SHENEFELT 1972, 1973; HERTING 1976).

HABITAT: Spruce- and mixed forests, spruce plantations and parks. From 0 m up to about 1,600 m above sea-level in Apennines (PROLA & RACHELI 1980), to about 1,900-2,000 m in Pyrenees (DANTART 1991; MAZEL & PESLIER 1997), and to about 2,000-2,100 m in Alps and mountains of Balkan peninsula (HELLMANN & al. 1999; REZBANYAI-RESER & MAIER 2000; coll. ZMUC).

SIMILAR SPECIES: Externally rather similar to *E. analoga*, which is of comparable size and is also associated with conifers. However, *E. analoga* has a larger, rounded discal spot, more liver-coloured terminal area on the forewing and more or less evenly curved postmedial line on the hindwing. *E. tantillaria* usually has lighter hindwing with sharply angled dark postmedial line. This is a very reliable difference from *E. analoga*. The shape of valva with blunt sclerotized ventral process, the tapered vinculum, the peculiar cornuti on the vesica, the small sternum A8 in male, as well as the heavily sclerotized antrum and corpus bursae in the female genitalia should distinguish *E. tantillaria* from other species.

REMARKS: Very variable in size and coloration from whitish grey to fuscous brown ground colour, with markings varying greatly in intensity (DIETZE 1910; LEMPKE 1969).

66. *Eupithecia liguriata* MILLIÈRE, 1884

Eupithecia liguriata MILLIÈRE, 1884: Revue D'Ent. 3 (1): 3 (Italy: Bordighéra; France: Cannes). Syntypes ♂ ♀ (MNHN?).

Eupithecia roederaria STANDFUSS, 1888: Berl. Ent. Z. 32 (1): 243, pl. 3: 7, 8 (France: Digne). Syntypes 3♂ 7 ♀.

(Eupithecia) bordigherata DIETZE, 1913: Biol. Eupithecien 2: 39. Unnecessary replacement name for *liguriata* MILLIÈRE, 1884.

DIAGNOSIS: Wingspan 15-18 mm. Labial palpi grey. Frons, vertex and notum pale grey. Forewing pale grey; basal area brownish grey; basal line right-angled near apex; postbasal line brownish grey, clouded and rather indistinct, broadened, darker and angled near costa; antemedial line inconspicuous, oblique, sharply angled near costa, slightly darker on the stalk of Cu and A, forming darker, brownish grey costal spot; postmedial line sinuous, evenly curved around discal spot, well indicated at costa as darker brownish grey spot, with narrow dark dashes on the veins; medial area slightly darker near costa; terminal area narrow, its inner part pale grey with ochreous tinge, outer part pale grey; pale wavy line well defined; discal spot dark grey, small and slightly elongate. Hindwing pale grey; transverse lines indistinct; terminal area usually slightly darker with inconspicuous pale wavy line; discal spot minute, weakly elongate or obsolete. Fringe pale grey, usually with faint darker spots at vein ends. Abdomen pale grey with ochreous tinge, with a series of small brown dorsal spots; first and second segments usually with small brown lateral spots.

MALE GENITALIA: Vinculum elongated and broad, with shallow medial hollow. Vesica armed with three small horn-like cornuti; with one elongated, V-shaped cornutus, and with one small folded cornutus at ductus ejaculatorius base.

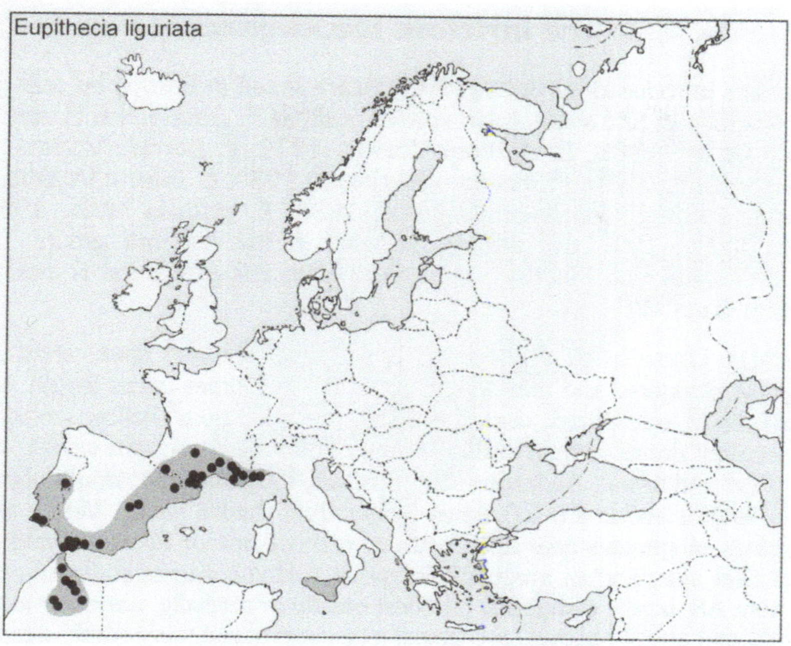

FEMALE GENITALIA: Bursa copulatrix small, irregular, rather pouch-like, membranous, with short and broad diverticulum at base on right side and with numerous striations; covered by spinules in apical half. Ductus seminalis attached to corpus bursae near colliculum on right side. Colliculum relatively broad.

DISTRIBUTION: West-Mediterranean. Portugal, southern Spain, southern France, north-western Italy and Sicily. – Outside Europe in North Africa (Morocco): subsp. *ketama* HERBULOT, 1981.

PHENOLOGY: Univoltine: early May (mid-April) to mid-June (late June). In North Africa bivoltine: from April to September. Larva in July and August. Overwintering as pupa in flimsy cocoon in the ground.

BIOLOGY: Larva monophagous (1). Found on the flowers of *Sedum dasyphyllum* (Crassulaceae) (France: DIETZE 1910, 1913).

HABITAT: Warm and dry landscapes. From 0 m up to about 1,200 m above sea-level in southern Alps (coll. ZSM), to about 1,600 m in Sierra Nevada (coll. ZMUC), and to about 1,800 m in Middle Atlas (RUNGS 1981).

SIMILAR SPECIES: This invariable species appears most similar to *E. pantellata*. Distinguished by the pale grey ground colour; the indistinct transverse lines, terminal area without small white spots, especially white tornal spot on the forewing; the hindwing with inconspicuous pattern, pale grey, the discal dot smaller, rounded, but not narrow, ovate-oblong, as in *E. pantellata*. The male genitalia, especially stout uncus, shape of valva with two blunt processes, narrow apex and sclerotized sacculus, large vinculum, cornuti on the vesica and shape of sternum A8, should distinguish this species from *E. pantellata*. Nevertheless, the female genitalia of both compared species are very similar.

The *lariciata* species-group

This group was introduced by Bolte (1990). Represented in Europe by one species: *E. lariciata*. Several east and south Asian species, such as *E. conjunctiva* HAMPSON, 1895; *E. emanata* DIETZE, 1908; *E. studiosa* VOJNITS, 1979; *E. horrida* VOJNITS, 1980; *E. stomachosa* VOJNITS, 1980; *E. maenamiella* INOUE, 1980; *E. infecta* VOJNITS, 1981; *E. sola* VOJNITS, 1983; *E. exrubicunda* INOUE, 1988; *E. stataria* INOUE, 1988 and *E. aspectabilis* INOUE, 1996 can also be included in the *lariciata* group. The North American *E. harrisonata* MACKAY, 1951 belongs in this group and is included in the species-group diagnosis.

MALE GENITALIA: Uncus biapical, short and broad; enlarged near apex, when viewed laterally. Papillae elongated and thin, length about 0.7-0.9 times uncus length. Valva large with a small, blunt, sclerotized, ventral process, preceded by a shallow wrinkled depression, with straight dorsal margin and broadly rounded apex; sacculus weakly sclerotized. Vinculum short and broad. Aedeagus slim, relatively long and thin, length about 0.6-0.8 times valva length; about 4.0-6.0 times longer than medial width. Vesica armed with patch of spines or spinules near apex; with or without one or two small and slim horn-like cornuti near apex and an irregular, elongated, folded cornutus at ductus ejaculatorius base. Sternum A8 usually elongated, broadest basally or medially, narrower to apex, with shallow basal and apical hollows; two apical lobes sclerotized, very small; sternum length about 0.5-0.6 times valva length.

FEMALE GENITALIA: Bursa copulatrix round; completely covered by small spines; with more or less long and narrow apical part with numerous longitudinal striations, without or with small patch of spines. Ductus seminalis usually bulging medially; attached to apical part of corpus bursae on right side. Colliculum usually not present or short and narrow. Antrum funnel-shaped, sclerotized. Anterior and posterior apophyses short, thin, not expanded at apices. Spurs of anterior apophyses evenly broadened at apices. Tergum A8 almost square, with rounded posterior corners. Papillae anales slim, small and rounded.

REMARKS: According to characters of the male genitalia the species of the *lariciata* group are closely related to species of the *lanceata* group. The valva with one blunt ventral process and wrinkled depression, the short vinculum, the vesica with its patch of spinules near apex, as well as good differences in bursa copulatrix, ductus seminalis, colliculum and antrum should distinguish species of the *lariciata* group.

67. Eupithecia lariciata (FREYER, 1842)

Larentia lariciata FREYER, 1842: Neuere Beitr. Schmett. 4 (61): 135, pl. 366 (Germany: Wellenburg; Leitershausen). Syntype(s).

Geometra residuata HÜBNER, 1817: Samml. Eur. Schmett. 5 Geometrae (1), pl. 91: 467 (Europe). Syntype(s) lost.

Eupithecia luteata PACKARD, 1867: Proc. Boston Soc. Nat. Hist. 11: 45 (Canada: Labrador, Cariboo Isl., Straits of Belle Isle). Lectotype ♂ (MCZ).

Tephrocystis [sic!] *luteata bifasciata* DYAR, 1904: Proc. U.S. Natn. Mus. 27 (1376): 891 (Canada: British Columbia, Kaslo). Holotype ♀ (USNM). Valid at subspecific rank.

Eupithecia catskillata PEARSALL, 1908: Ent. News 19 (5): 192 (U.S.A.: New York, Catskill Mts, Big Indian Valley). Lectotype ♂ (AMNH).

Eupithecia fasciata TAYLOR, 1910: The Can. Ent. 42 (3): 79 (Canada: Ontario, Ottawa). Holotype ♀ (USNM).

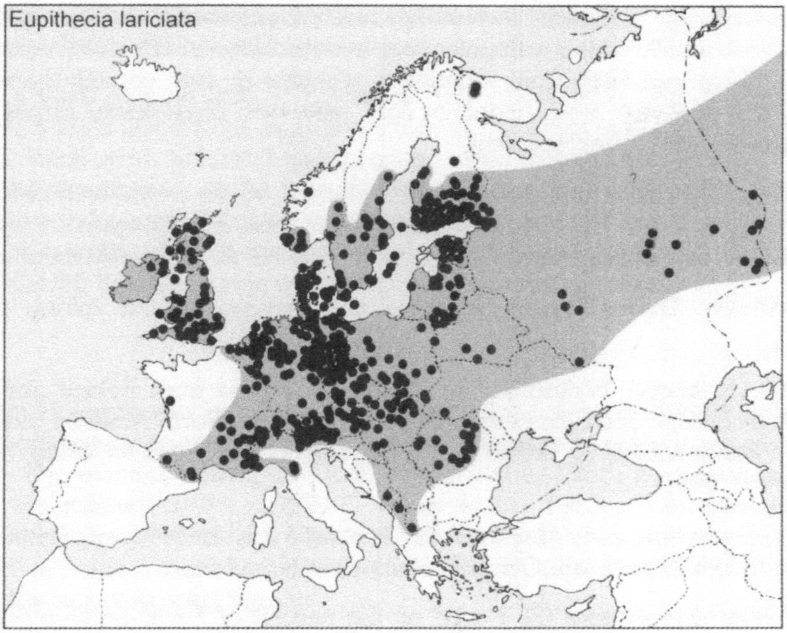

(Eupithecia) lariciata f. *uniformis* DIETZE, 1913: Biol. Eupithecien 2: 155 (Europe ?: type locality not stated) Syntype(s) (MNHU). Valid at subspecific rank.

(Eupithecia) lariciata f. *luxuriosa* DIETZE, 1913: Biol. Eupithecien 2: 155 (Europe ?: type locality not stated) Syntype(s) (MNHU). Valid at subspecific rank.

(Eupithecia) lariciata f. *strigata* DIETZE, 1913: Biol. Eupithecien 2: 155; ibidem (1910), 1, pl. 78: 734 (Germany: Frankfurt a.M.). Syntype(s) (MNHU). Valid at subspecific rank.

(Eupithecia) lariciata f. *basifasciata* DIETZE, 1913: Biol. Eupithecien 2: 155; ibidem (1910), 1, pl. 78: 736 (Germany: Sachsen). Syntype(s) (MNHU). Valid at subspecific rank.

(Eupithecia) lariciata f. *mediopallens* DIETZE, 1913: Biol. Eupithecien 2: 155; ibidem (1910), 1, pl. 78: 733 (Switzerland: Wallis; Germany: Sachsen). Syntypes (MNHU). Valid at subspecific rank.

Eupithecia lariciata mesodeicta PROUT, 1938: in SEITZ, The Macrolepidoptera of the World 4 (Suppl.): 208, pl. 18: e (Kashmir: Gulmarg; Scind Valley). Syntypes ♂ ♀ (BMNH). Valid at subspecific rank.

Eupithecia laboriosa VOJNITS, 1977: Acta Zool. Acad. Sci. Hung. 23 (3-4): 479, fig. 15 (Mongolia: Central aimak, SO of Somon Bajanzogt). Holotype ♀ (HNHM, examined; the genital preparate of *Pasiphila* sp.). New junior synonym.

Unavailable names (infrasubspecific): *impuncta*: LEMPKE (1969). - (incorrect subsequent spellings): ?*residuaria*: BOISDUVAL (1840); *lariciaria*: HERRICH-SCHÄFFER (1848).

DIAGNOSIS: Wingspan 16-25 mm. Labial palpi dark grey. Frons and vertex whitish grey. Notum light grey with whitish posterior spot. Forewing light grey; stalk of Cu chequered black and light grey; transverse lines usually distinct, dark grey, slightly sinuate and sharply angled near costa; medial area between antemedial and medial lines often slightly darker; medial line sharply angled beyond discal spot or touching it, with delicate paler double stripe beyond medial line and with pale ochreous cloud under discal spot in some specimens; postmedial line slightly curved and right-angled near costa, usually with dark dashes on the veins; pale wavy line sometimes present, fine, sometimes forming a weak light tornal dot; discal spot distinct, relatively small, obliquely ovate to circular, intensely black. Hindwing rather angulated; termen gently concave medially; ground colour slightly paler than on the forewing; transverse lines weakly marked; postmedial slightly angled medially; terminal area usually darker with pale wavy line obsolete and often with a minute, pale tornal spot; discal spot faint, from rounded to elongate or even comma-shaped. Fringe chequered light and dark grey. Abdomen light grey, with black lateral stripes or spots, usually beginning with fourth segment.

MALE GENITALIA: Vinculum with shallow medial hollow. Vesica armed with patch of spinules near apex; with two slim horn-like cornuti near apex, one elongated, irregular, folded cornutus and one small, irregular and curved cornutus at ductus ejaculatorius base.

FEMALE GENITALIA: Ductus bursae elongated and narrow, without spines. Colliculum absent.

DISTRIBUTION: Holarctic. Widespread in Palaearctic region from Ireland and northern Spain through Europe and Siberia to China and Japan. In North America widely distributed from Newfoundland to British Columbia and in the south to White Mountains, Arizona (BOLTE 1990).

PHENOLOGY: Univoltine: early May (mid-April) to early July (mid-August). Larva: mid-July (early June) to late August (late October). Overwintering as pupa.

BIOLOGY: Larva oligophagous (2). Found on the needles of *Larix decidua (=europaea)* (Germany: FREYER 1842; DIETZE 1872, 1910, 1913; WEIGT 1980, 1993; United Kingdom: CREWE 1864b, 1865; BUCKLER 1899; Denmark: JUUL 1948), also *Larix sibirica, L. kaempferi (=leptolepis)* (Pinaceae) (Finland: SEPPÄNEN 1970) and occasionally mentioned on *Picea abies* (Pinaceae) (United Kingdom: CREWE 1864b, 1865), *Juniperus* (?) (Cupressaceae) (Denmark: JUUL 1948). In North America on *Picea glauca, P. mariana, Abies balsamea* and *Larix laricina,* occasionally on *Picea rubens, Tsuga canadensis, Pinus strobus, Pseudotsuga menziesii* (Pinaceae) and *Thuja occidentalis* (Cupressaceae) (Canada: ROSS & EVANS 1956; MCGUFFIN 1958; PRENTICE 1963; BOLTE 1990).

PARASITOIDS: Hymenoptera, Ichneumonidae: *Mesochorus fulgurans* (hyperparasite), *M. tachypus*, *Netelia tarsata*, *Platylabops lariciatae*, *Agrypon clandestinum* - Hymenoptera, Braconidae: *Aleiodes (=Rogas) modestus*, *Apanteles juniperatae*, *A. vitripennis*, *Zele (=Meteorus) caligatus* - Hymenoptera, Encyrtidae: *Litomastix* sp. - Hymenoptera, Larvivoridae: *Caranthia fulvipes* var. *lichtwardtiana* - Diptera, Tachinidae: *Phryxe nemea* (BUCKLER 1899; THOMPSON 1945; HERTING 1960, 1976; SHENEFELT 1972, 1975; ZWAKHALS pers. comm.).

HABITAT: Larch forests, mature stands of larch and parks, rarely in rocky places. From 0 m up to about 2,200 m above sea-level in Alps (VORBRODT 1928; HELLMANN & al. 1999; REZBANYAI-RESER & MAIER 2000), and to about 2,300 m in Pyrenees (MAZEL & PESLIER 1997). Recorded even at 3,482 m and 3,678 m above sea-level in Alps (WEHRLI 1926; VORBRODT 1928).

SIMILAR SPECIES: *E. lariciata* can easily be confused with *E. subfuscata* and distinguished with difficulty from it. In *E. subfuscata*, the transverse lines are usually more distinct and more evenly curved, including postmedial line, discal spot usually smaller, and on the whole this species is slightly lighter with more distinct pale wavy lines on all wings than in *E. lariciata*. The presence of white thoracic spot in *E. lariciata* can be used for separation of it from *E. subfuscata* with accuracy (Text-fig. 85). The male and female genitalia are very clearly distinguished in these two species.

REMARKS: Usually varying little in coloration and maculation. The melanic ab. *nigra* PROUT, 1938 has a unicolorous more sooty tone of the wings with at least traces of transverse lines, black veins, normal discal spots and white thoracic spot as well.
After careful examination of the type-specimen of *E. laboriosa* VOJNITS, 1977, I have come to the conclusion that this species-name was applied to a specimen of *E. lariciata*. However, the genital slide of the type-specimen of *E. laboriosa* (No. 10,559, Dr. A. VOJNITS, HNHM) belongs to one of the *Pasiphila* species by mistake.

The ***lanceata*** species-group

This distinct Holarctic species-group is introduced here for the first time. Represented in Europe by only one species: *E. lanceata*. Several species from eastern and southern Asia, such as *E. daemionata* DIETZE, 1903; *E. takao* INOUE, 1955; *E. shikokuensis* INOUE, 1980; *E. emendata* VOJNITS, 1983; *E. kudoi* INOUE, 1983; *E. longipennata* INOUE, 1988; *E. habermani* VIIDALEPP & MIRONOV, 1988 and two North American *Eupithecia* species: *E. zelmira* SWETT & CASSINO, 1920 and *E. vitreotata* CASSINO, 1927 also belong to the *lanceata* species-group.

MALE GENITALIA: Uncus medium length, stout and biapical. Papillae usually elongated, length about 0.5-0.8 times uncus length. Valva large and broad, with two elongated, heavily sclerotized ventral processes (or with one ventral process in *E. shikokuensis*) and large medial gap between them; with almost straight or slightly curved dorsal margin and rounded apex; sacculus heavily sclerotized. Vinculum elongated and broad. Aedeagus stout and large, length about 0.7-0.8 times valva length, about 3.0-4.0 times longer than medial width. Vesica armed with one or two large horn-like cornuti, with some spinules (or with patch of spinules); with one large, broad, folded, H- or U-shaped cornutus and usually with one smaller curved cornutus at ductus ejaculatorius base. Sternum A8 peg-like; basal hollow shallow; apical hollow quite small or absent; sternum length about 0.4-0.6 times valva length.

FEMALE GENITALIA: Bursa copulatrix usually funnel-shaped (or pear-shaped in *E. daemionata*), with numerous spines at base and with one or two rows of large spines and usually with patch of spinules in apical part of corpus bursae. Ductus seminalis from short to long and from narrow to broad. Colliculum collar-like, wide or plate-like, elongated and narrow. Antrum membranous. Anterior and posterior apophyses short or medium length, thin, not expanded at apices. Spurs of anterior apophyses well defined, broadened to apices. Tergum A8 square. Papillae anales short, wide and rounded.

REMARKS: This group is closely related to the *interruptofasciata* and the *lariciata* species-groups according to characters of the male and female genitalia. The species of the *lanceata* group can usually be distinguished from all other *Eupithecia* species by the shape of valva, with two long, heavily sclerotized ventral processes and large medial hollow between them; the heavily sclerotized sacculus; the long and broad vinculum; the cornuti on the vesica. The female genitalia can be distinguished from those of other European species by the funnel-shaped bursa copulatrix, with numerous spines at base, with one or two rows of large spines and usually with apical patch of spinules.

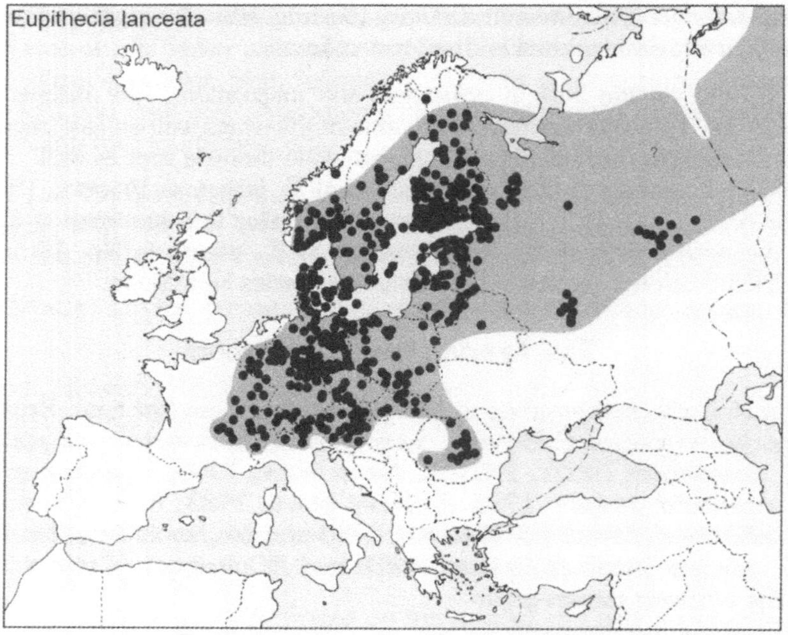

68. Eupithecia lanceata (HÜBNER, 1825)

Dyscymatoge lanceata HÜBNER, 1825: Verz. bekannt. Schmett. (21): 324 (Europe). Syntype(s) lost.
Larentia hospitata TREITSCHKE, 1828: Die Schmett. Eur. 6 (2): 120 (Austria; Hungary). Syntype(s) (HNHM).
Unavailable names (misidentification): *succenturiata*: sensu HÜBNER (1799) nec LINNAEUS, 1758 (identity: *Eupithecia succenturiata*). - (incorrect subsequent spelling): *hospitaria*: BOISDUVAL (1840).

DIAGNOSIS: Wingspan 16-22 mm. Labial palpi yellowish brown, scales with lighter greyish apices; concolorous with frons and vertex. Male antennae bipectinate; with dense and long setae. Female antennae with rather sparse and short setae. Notum brown. Forewing yellowish brown, basal area broad; basal and antemedial lines distinct, brown, slightly oblique, almost straight and sharply angled near costa; medial area narrow, yellowish brown, usually slightly darker than other areas; postmedial line distinct, brown, slightly sinuate, almost straight near costa; terminal area with brown, usually darker basal half, with yellowish brown or brown, usually lighter terminal half, sometimes with large lighter medial spot, with dark-brown M1 and M2, and with wavy line lighter, sinuate (not wavy or dentated); discal dot small, elongated, oblique, dark brown. Hindwing paler, brownish grey with brownish, slightly darker basal and terminal areas; discal dot minute, inconspicuous or usually absent. Fringe long, shiny yellowish brown, with indistinct brown spots. Abdomen yellowish brown, scales with lighter greyish apices; with dark brown lateral stripes; second segment dark brown.

MALE GENITALIA: Valva with one pointed ventral process and one apically expanded and rounded ventral process. Vinculum with shallow medial hollow. Vesica armed with one large horn-like cornutus; two small cornuti; one large, broad, curved, H-shaped cornutus and one smaller curved cornutus at ductus ejaculatorius base. Sternum A8 without apical hollow, narrowing to apex.

FEMALE GENITALIA: Ductus seminalis short; broadly attached to corpus bursae near base, narrowing to apex. Colliculum elongated, plate-like and narrowing to apex. Tergum A8 almost square.

DISTRIBUTION: Euro-Siberian (boreal). Northern and Central Europe, eastwards to Sajan Mts in southern Siberia (coll. ZISP).

PHENOLOGY: Univoltine: early April (early March) to late May (mid-June). Larva: mid-May (late March) to mid-June (early July); in the north in June and July. Overwintering as pupa in flimsy cocoon in the ground.

BIOLOGY: Larva oligophagous (2). Found on the young needles, occasionally on the female inflorescences of *Picea abies (=excelsa)* (Pinaceae) (Germany: DIETZE 1874b, 1910, 1913; WEIGT 1980, 1993) and also on *Juniperus communis* (Cupressaceae) (Finland: SEPPÄNEN 1970). Recorded on *Abies alba (=pectinata), Larix decidua (=europaea), Pinus* (Pinaceae) (France: LHOMME 1935; Denmark: JUUL 1948).

PARASITOIDS: Hymenoptera, Ichneumonidae: *Gelis gonatopinus* (THOMPSON 1945; PATOČKA 1960; doubtful according to ZWAKHALS pers. comm.).

HABITAT: Spruce forests and plantations, especially in areas with older spruce trees, rarely in fir-, larch- and mixed forests, forest edges and parks. From 0 m up to about 1,500 m above sea-level in Alps (WEIGT 1993).

SIMILAR SPECIES: *E. lanceata* can be distinguished easily from other European *Eupithecia* species by the bipectinate antennae in male and the maculation of the forewing, especially the narrow darker medial area between antemedial line which is almost straight, sharply angled near costa and slightly sinuate postmedial line. The male and female genitalia can be used for accurate determinations.

The *egenaria* species-group

Group introduced here for the first time. It is possibly one of the most species-rich and ancient groups of the genus *Eupithecia*. Represented in Europe by nine species: *E. groenblomi*, *E. selinata*, *E. quercetica*, *E. actaeata*, *E. egenaria*, *E. repentina*, *E. orana*, *E. pimpinellata* and *E. karadaghensis*. Three species from Canary Islands can be placed in this group: *E. tenerifensis* (REBEL, 1906), *E. stertzi* (REBEL, 1911) and *E. schuetzeata* PINKER, 1961 and one species from Madeira: *E. latipennata* PROUT, 1914. At least four North African *Eupithecia* species, *E. undulataria* (TURATI, 1934), *E. acutula* (TURATI & KRÜGER, 1936), *E. jefrenata* SCHÜTZE, 1959 and *E. mirei* HERBULOT, 1965 belong to the *egenaria* species-group. A number of Asian species can be placed in this group also, such as: *E. detritata* STAUDINGER, 1897, *E. assectata* DIETZE, 1903, *E. recens* DIETZE, 1903, *E. kobayashii* INOUE, 1958, *E. nepalata* SCHÜTZE, 1961, *E. jinboi* INOUE, 1976, *E. kunzi* PINKER, 1976, *E. shikokuensis* INOUE, 1980, *E. tricornuta* INOUE, 1980, *E. tsushimensis* INOUE, 1980, *E. coccinea* VOJNITS, 1981, *E. lata* VOJNITS, 1981, *E. concinna* VOJNITS, 1983, *E. jizlensis* WILTSHIRE, 1986, *E. tricerata* WILTSHIRE, 1990 and many others. Some North American species belong to the *egenaria* group also, as follows: *E. graefii* (HULST, 1896), *E. agnesata* TAYLOR, 1908, *E. sheppardata* MCDUNNOUGH, 1938 and *E. neomexicana* MCDUNNOUGH, 1945.

MALE GENITALIA: Uncus usually short and narrow, biapical. Papillae large, elongated, length about 0.7-1.2 times uncus length. Valva large, with small, usually pointed ventral process or without it; broadest medially, with almost straight or slightly curved dorsal margin and with narrowly rounded or pointed apex; sacculus often weakly sclerotized. Vinculum short and broad. Aedeagus large and broad, stout, length about 0.7-1.0 times valva length; about 3.0-5.0 times longer than medial width. Vesica armed with one to five extremely stout horn-like cornuti, diverse in size and shape; usually with one more or less elongated, flat, convex U-shaped cornutus and one smaller, irregular, curved cornutus at ductus ejaculatorius base. Sternum A8 usually elongated, broadest at base, with shallow to deep basal hollow; with two short, pointed, heavily sclerotized, symmetrical or asymmetrical, often pincer-like lobes and with shallow and broad apical hollow between them (except in *E. kobayashii*); sternum length about 0.6-0.9 times valva length.

FEMALE GENITALIA: Bursa copulatrix diverse in shape, spherical, pouch-like or irregular; covered with spines uniformly or irregularly, or armed with different curved bands or patches of spines (even differing between individuals). Ductus bursae usually sclerotized. Ductus seminalis from short and narrow to long and broad; attached to ductus bursae on right side. Colliculum varied, from collar-like to plate-like. Antrum membranous, usually broad and weakly defined. Anterior and posterior apophyses short or medium length, thin and usually not expanded at apices. Spurs of anterior apophyses broadened to apices. Tergum A8 of various shapes. Papillae anales varied also, but usually short, broad and rounded.

REMARKS: INOUE (1980) observed a similarity in the morphological characteristics of the genitalia of five Japanese *Eupithecia* species (*tricornuta-shikokuensis*) for the first time. These species were the basis for the description of a new large and variable species complex under the name *egenaria* species-group. It is a group closely related to the *interruptofasciata* species-group according to characteristics of the male genitalia. However, representatives of the *egenaria* group can easily be distinguished from other European *Eupithecia* species by their extremely stout horn-like cornuti on the vesica and the sternum A8 with two short, pointed and sclerotized apical lobes; and the female genitalia can usually be distinguished by the pouch-like or mis-shapen corpus bursae without a row of spines in ductus bursae (with few exceptions).

69. *Eupithecia groenblomi* URBAHN, 1969

Eupithecia groenblomi URBAHN, 1969: Ann. Ent. Fenn. 35 (4): 186, pl. 1a, 1b, 2, 3 (Finland: EH, Pälkäne). Holotype ♂ (ZMUH).

Eupithecia tsushimensis INOUE, 1980: Bull. Fac. domest. Sci., Otsuma Wom. Univ. 16: 167, figs 40C (nec 43B), 47D (nec 47E), 49A. (Japan: Tsushima Isl., Shirikubiyama). Holotype ♂ (BMNH). New junior synonym.

Eupithecia fujisana INOUE, 1980: Bull. Fac. domest. Sci., Otsuma Wom. Univ. 16: 204, figs 55W, 57B (Japan: Shizuoka Prefecture, Mt. Fuji, Shin 5-gomé). Holotype ♀ (BMNH). New junior synonym.

Eupithecia kunashiriensis VIIDALEPP & MIRONOV, 1988: Eesti NSV Tead. Akad. Toim. (Biol.) 37 (3): 212, figs 3 d-z (Russia: Kuril Islands, Kunashir Isl., The Pioneer Camp 'Dozornyi'. Holotype ♂ (IZBE, examined). New junior synonym.

DIAGNOSIS: Wingspan 14.5-24 mm. Labial palpi grey. Frons and vertex ash-grey. Notum grey. Forewing unicolorous smoky grey to grey; transverse lines inconspicuous, except weak, fine, dark grey medial and postmedial lines; medial line slightly sinuate, sharply angled around discal spot and forming a weak, small dark grey costal spot; postmedial line evenly curved, well marked on the veins forming a series of narrow, elongated, black dashes and a larger dark grey costal spot; terminal area faintly darker to costa with black

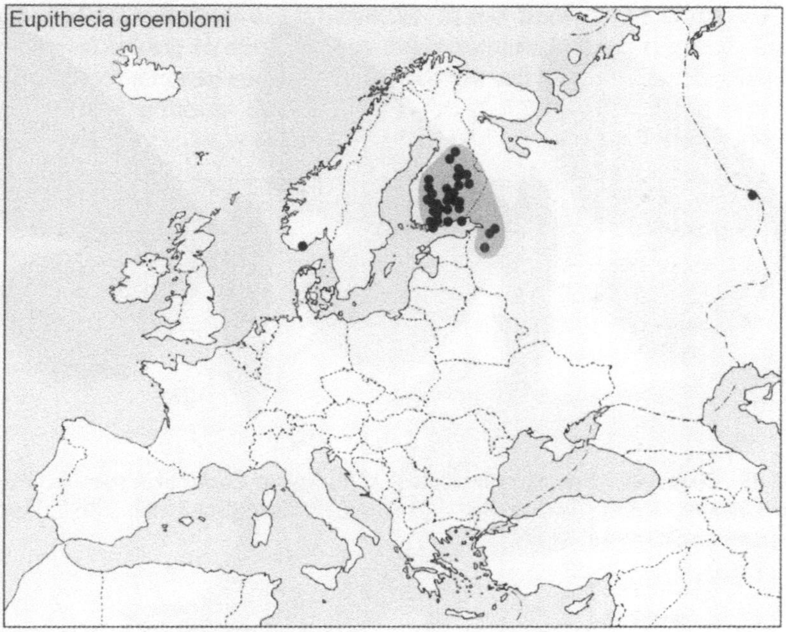

or chequered veins; whitish wavy line more or less distinct, fine, multiply indented, sometimes with minute pale tornal dot; discal spot relatively large, ovate or ovate-orbicular, intensely black. Hindwing paler; transverse lines invisible or only evenly curved postmedial line and paler stripe beyond it are evident; discal spot distinct, relatively large, circular. Fringe faintly chequered light and dark grey. Abdomen grey with narrow black lateral stripes.

MALE GENITALIA: Valva large, elongated, with weak visible broad and blunt ventral process. Vesica armed with three stout different horn-like cornuti, one elongated flat U-shaped cornutus and one smaller, irregular curved cornutus at ductus ejaculatorius base. Sternum A8 elongated with two symmetrical pincer-like horns with deep and broad apical hollow between them.

FEMALE GENITALIA: Bursa copulatrix elongated, pouch-like, with small membranous diverticulum; uniformly covered with small spines. Ductus bursae with numerous longitudinal striae, without spines. Ductus seminalis narrow, but broadened at base. Colliculum collar-like, elongated and relatively narrow. Anterior and posterior apophyses medium length, thin. Tergum A8 narrow and elongated.

DISTRIBUTION: Euro-Siberian (boreal; disjunct ?). Southern Norway, central and southern Finland, north-western Russia (St. Petersburg and Novgorod provinces) and Urals. – Outside Europe in Far East of Russia (Primorje territory and southern Kuril Islands) and Japan. – There is one specimen from Urals mountains: 1♀, Russia, central Urals, near Dvurechensk, 58 km south-east of Ekaterinburg, 30.VII.1997 (at light), G. ZAMSHINA leg. (ZISP).

PHENOLOGY: Univoltine: late July to late August. Larva: mid-August to early October. Overwintering as pupa.

Text-figs 47-49: Diagnostic characters (indicated) of European *Eupithecia* species. Text-fig. 47: *E. groenblomi* URBAHN, 1969. Text-fig. 48: *E. selinata* HERRICH-SCHÄFFER, 1861. Text-fig. 49: *E. trisignaria* HERRICH-SCHÄFFER, 1848.

BIOLOGY: Larva monophagous (1). Found on the flowers and seeds of *Solidago virgaurea* (Asteraceae) (Finland: PELTONEN & al. 1970; SEPPÄNEN 1970; MIKKOLA & al. 1989; Russia: MIRONOV 2001).

PARASITOIDS: Hymenoptera, Ichneumonidae: *Platylabus pumilio* - Hymenoptera, Braconidae: *Zele caligatus, Protapanteles* sp. - Diptera, Tachinidae: *Ceranthia lichtwardtiana* (PELTONEN & al. 1970; ANDERSEN 1996; original data; ZWAKHALS pers. comm.).

HABITAT: Old and light spruce or mixed forests and forest clearings where goldenrod grows.

SIMILAR SPECIES: This rather invariable species is very similar to *E. trisignaria* and *E. virgaureata*. In *E. trisignaria*, the discal spot is usually larger, the medial line not curved beyond discal spot, the postmedial line more angled on the forewing, the discal spot on the hindwing smaller and narrower. *E. virgaureata* has more distinct transverse lines and the veins evidently more rouge chequered; in *E. groenblomi*, the forewing has more conspicuous dark costal spots and the hindwing has medially concave termen and usually smaller discal spot. *E. groenblomi* is narrower-winged than *E. selinata*, another species with which it could be confused, with slim body (rather as in *E. virgaureata*) and with less rounded postmedial line on the forewing and with larger discal spot on the hindwing (Text-fig. 47). The elongated valva, the larger aedeagus with big cornuti on the vesica, the shape of sternum A8 in the male, and the narrower, elongated bursa copulatrix completely covered by spines with small membranous diverticulum near base, the ductus bursae without spines and the narrow and elongated tergum A8 in female should distinguish *E. groenblomi* from these compared species.

REMARKS: After carefully examining some specimens of *E. groenblomi* from Finland and north-western Russia, the types of *E. kunashiriensis*, as well as studying the literature on *E. tsushimensis* and *E. fujisana* descriptions (INOUE 1980, 2000), I have come to the conclusion that all these species are conspecific. In my opinion, the females of *E. tsushimensis* (paratypes) belong to species from the *tripunctaria* species-group. In all probability, the rare and local boreal species *E. groenblomi* will found in the Siberian taiga from Urals to Primorje in the future.

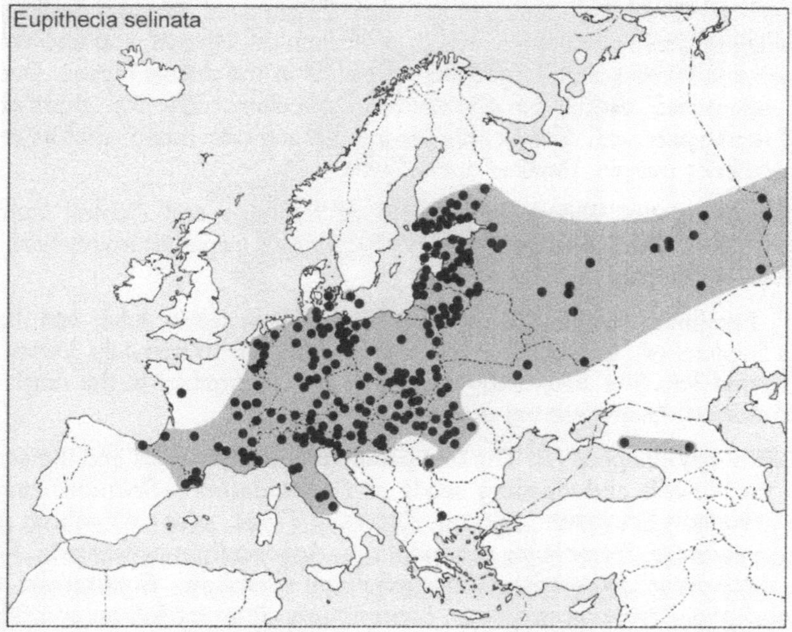

70. *Eupithecia selinata* HERRICH-SCHÄFFER, 1861

Eupithecia selinata HERRICH-SCHÄFFER, 1861: Corresp.-Bl. Sammler Insect., insbesondere Schmett. 17: 131 (Germany: Mombach). Syntype(s).

Eupithecia divina VOJNITS, 1981: Acta Zool. Acad. Sci. Hung. 25 (1-2): 205 fig. 16 (China: Chekiang Province, West Tien-mu-shan). Holotype ♂ (ZFMK, examined). New junior synonym.

Unavailable names (incorrect subsequent spelling): *silenata*: DIETZE (1871a) nec ASSMANN, 1848 (identity: *Eupithecia silenata*).

DIAGNOSIS: Wingspan 18-21 mm. Labial palpi grey. Frons and vertex pale grey to grey. Notum shiny grey. Forewing unicolorous, brownish grey, with darker costal margin; transverse lines inconspicuous; medial line much sinuate, crossing discal spot; postmedial line finely waved, evenly curved; pale wavy line ill-defined or obsolete; discal spot large, conspicuous, dark brown, ovate or sometimes rounded, placed near costa and usually touching dark medial costal spot. Hindwing only slightly paler than forewing, unicolorous brownish grey; all transverse lines rather obsolete; discal spot distinct, small, rounded. Fringe pale brownish grey, slightly spotted with brownish grey. Abdomen shiny grey, with small dark dorsal tufts and black lateral stripes.

MALE GENITALIA: Uncus enlarged near apex, when viewed laterally. Valva large and elongated, without ventral process. Vesica armed with two short, stout, slightly curved horn-like cornuti (one apical and one medial); with one irregular, folded cornutus at medial horn-like cornutus base; with numerous small spines and with one, elongated, flat, folded, V-shaped cornutus at ductus ejaculatorius base; the base of ductus ejaculatorius with two or three small, spindle-shaped cornuti. Sternum A8 peg-like, elongated; base broadened with deep medial hollow; apex slightly broadened, with two very small blunt or pointed lobes.

FEMALE GENITALIA: Bursa copulatrix elongated, pouch-like, strongly sclerotized; 1/3 covered by spines with marginal spines largest, with longitudinal, broad, heavily sclerotized row of large spines from spiniferous area to colliculum on left side, and also with numerous scattered small spines and longitudinal striations in the ductus bursae. Ductus seminalis thick, sclerotized, evenly curved ventrad. Colliculum collar-like, short and broad. Tergum A8 rectangular with convex anterior margin and with broad, shallow medial hollow in the posterior margin. Papillae anales small.

DISTRIBUTION: Eurasiatic. Widespread mainly in southern and Central Europe from Pyrenees to Urals. Outside Europe in northern Caucasus (separate population) and from Urals to China and Japan.

PHENOLOGY: Bivoltine: mid-May (early May) to mid-June (early July); late July to late August (late September). Univoltine in the north: early June to mid-July. Larva: late June (late May) to mid-July (late July); early August to late September. In the north from late June to mid-August. Overwintering as pupa.

BIOLOGY: Larva oligophagous (1). The first generation on the leaves and the second generation on the flowers and ripening seeds of *Peucedanum (=Selinum) oreoselinum* (Germany: HERRICH-SCHÄFFER 1861a; DIETZE 1971a), also *Angelica sylvestris, Peucedanum palustre, Heracleum sphondylium, Aegopodium podagraria, Pimpinella saxifraga, Anthriscus sylvestris, Chaerophyllum temulum, Bupleurum falcatum, Selinum carvifolia, Ptychotis saxifraga, Laserpitium, Sium latifolium* and *Berula erecta* (Apiaceae) (Germany: DIETZE 1872, 1910, 1913; WEIGT 1980, 1990; GELBRECHT and WEGNER pers. comm.; France: LHOMME 1935; Finland: SEPPÄNEN 1970).

HABITAT: Light, open deciduous, coniferous and mixed forests, forest edges and paths, country roads, small forest clearings, hedges, scrub and scrub-covered meadows, wet meadows, grassland near water, as well as warm and dry slopes and rocks. From 0 m up to about 1,000 m above sea-level in Pyrenees (MAZEL & PESLIER 1997), and to about 1,500 m in southern Alps (FORSTER & WOHLFAHRT 1981).

SIMILAR SPECIES: Very similar to *E. trisignaria* and can easily be confused with this species. *E. selinata* is distinguished from it by the more shiny, silken colour of wings, the smaller costal spot, the weaker wavy line, by absence of dark points on the veins of forewing, round discal spot on the hindwing (Text-fig. 48). The structure of genitalia, especially cornuti on the vesica and the shape of sternum A8 in the male; the elongated shape of bursa copulatrix with unique disposition of spines in the female genitalia should distinguish this species from others.

71. *Eupithecia actaeata* WALDERDORFF, 1869

Eupithecia actaeata WALDERDORFF, 1869: Corresp.-Bl. Zool.-Min. Ver. Regensburg 23 (6): 82 (Germany). Holotype ♂.

Eupithecia bergunensis DIETZE, 1875: Stett. Ent. Z. 36 (4-6): 248, pl. 2: 1 (Switzerland: Kanton Graubünden, Bergün). Syntypes 1♂2♀ (MNHU).

Unavailable names (incorrect subsequent spelling): *acteata*: KOSTROWICKI (1953).

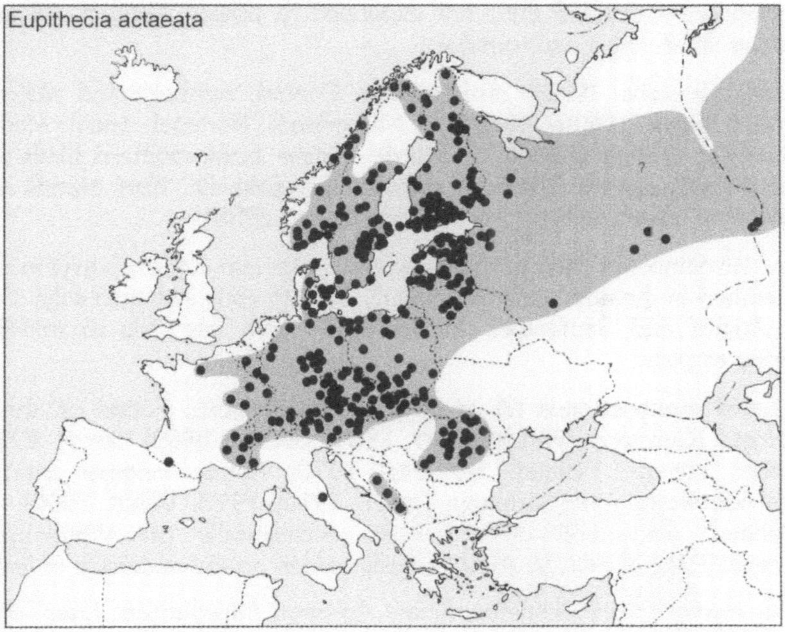

DIAGNOSIS: Wingspan 19-24 mm. Labial palpi dark grey. Frons and vertex grey. Notum grey or dark grey. Forewing dark grey with brownish tinge; transverse lines usually inconspicuous; postbasal and antemedial lines straight, oblique, angled near costa; postmedial line slightly sinuate, well defined as dark dashes on the veins; ante- and postmedial lines usually forming two dark costal spots; terminal area broad; pale multiply indented wavy line from very distinct to obsolete but white tornal spot always well defined; discal spot large, ovate-oblong or semilunar, black. Hindwing slightly paler; transverse lines usually ill-defined; terminal area darker with small white tornal spot and with obsolete pale wavy line; discal spot large, ovate-acuminate or slightly elongated, dark grey. Fringe grey or pale grey, spotted with dark grey at vein ends. Abdomen grey or dark grey with black lateral stripes and sometimes with a series of small black dorsal spots on the segment ends.

MALE GENITALIA: Uncus biapical, short and narrow. Papillae long, nearly equal to uncus length. Valva relatively short, without ventral process; with rounded ventral margin; with straight or weakly curved dorsal margin and pointed apex; sacculus slightly sclerotized. Vinculum short and very broad. Aedeagus large, stout; length nearly equal to valva length. Vesica armed with one apical, very broad, blunt, hooked, horn-like cornutus; with one large, strong, horn-like cornutus medially and with one irregular, elongated, curved cornutus at ductus ejaculatorius base. Sternum A8 large, long and broad, basally and apically broadened; basal hollow relatively deep; apical hollow short, but broad; apical horns sclerotized and tapered, with very pointed tips; sternum length about 0.8-0.9 times valva length.

FEMALE GENITALIA: Bursa copulatrix large, pouch-like; membranous; armed with a narrow band of large spines which almost encircle medial area; also armed with a number of smaller spines in medial area. Ductus seminalis short and narrow, arising from medial part of corpus bursae from right side. Colliculum broad. Antrum membranous, short and very broad. Apophyses short and thin, not expanded to apices. Tergum A8 rectangular. Papillae anales small, short and rounded.

DISTRIBUTION: Eurasiatic. Widely distributed in Central, northern and eastern Europe except Iceland, Ireland, United Kingdom, Netherlands, Portugal, Spain, Mediterranean islands, Bulgaria, Greece, Turkey. – Outside Europe from southern Urals throughout southern and north-eastern Siberia to Kamtchatka, Sakhalin, Kuril Islands and Japan, southeastward to Taiwan (subsp. *praenubilata* INOUE, 1958).

PHENOLOGY: Bivoltine: mid-May to mid-June (mid-July); late July (mid-July) to mid-August (early September) In the north univoltine: early June to early July (mid-July). Larva: June and July; August and September. In the north from late July to mid-September. Overwintering as pupa.

BIOLOGY: Larva monophagous (2) or disjunct oligophagous. Found on the leaves of *Actaea spicata* (Germany: WALDERDORFF 1869; SPEYER 1869; DIETZE 1910, 1913; WEGNER pers. comm.; Finland: SEPPÄNEN 1970; Russia: original data), also on *Thalictrum aquilegifolium* (Germany: DIETZE 1910, 1913; WEIGT 1990), *T. flavum* (Ranunculaceae) (Denmark: JUUL 1948), occasionally on *Viburnum opulus* (Caprifoliaceae) (Poland: SPEYER 1872), usually hidden on the underside of leaves.

PARASITOIDS: Hymenoptera, Ichneumonidae: *Agrypon flaveolatum*, *Lissonota albicoxis*, *Platylabus pactor*, *Agrypon anxium* - Hymenoptera, Encyrtidae: *Tyndarichus scaurus* - Diptera, Tachinidae: *Bactromyia (=Erycia) aurulenta*, *Ceranthia pallida* (THOMPSON 1945; HERTING 1960, 1976; ANDERSEN 1996; original data; ZWAKHALS pers. comm.).

HABITAT: Mature light spruce, mixed and deciduous forests with moist top soil, where baneberry grows. From 0 m up to about 1,300 m above sea-level in Apennines (Mt. Majella: ZAHM pers. comm.), to about 1,900 m in Alps (DANNEHL 1927; REZBANYAI-RESER 1988), and to about 2,000 m in mountains of Balkan peninsula (Montenegro, Durmitor: coll. FIBIGER).

SIMILAR SPECIES: Similar to *E. trisignaria* but usually larger, the wings broader, more shiny coloured, the discal spot on the forewing larger and more elongate, the pale tornal spots on all wings more distinct, the abdomen always with continuous black lateral stripes. The genitalia of *E. actaeata* very clearly distinguished from those of all other European *Eupithecia* species by the combination of short uncus, the long papillae, the shape of valva and sternum A8, the very broad vinculum, the armature on the vesica in the male and by the shape of corpus bursae with irregular disposition of spines and with short and broad colliculum in the female.

72. ***Eupithecia quercetica*** PROUT, 1938

Eupithecia quercetica PROUT, 1938: in SEITZ, Macrolep. 4 (Suppl.): 204, pl. 14: i (Lebanon: Arayah). Syntypes, at least ♀ (BMNH, 1 syntype examined).

Eupithecia buxata PINKER, 1958: Fragm. balcan. 2 (12): 92, figs 2, 3, 6-9 (Macedonia: Treska valley). Holotype ♂ (NHMW; 1 paratype examined).

DIAGNOSIS: Wingspan 19-23 mm. Labial palpi whitish grey. Frons and vertex greyish white. Notum ash-grey with dark grey or brownish grey transverse band in front.

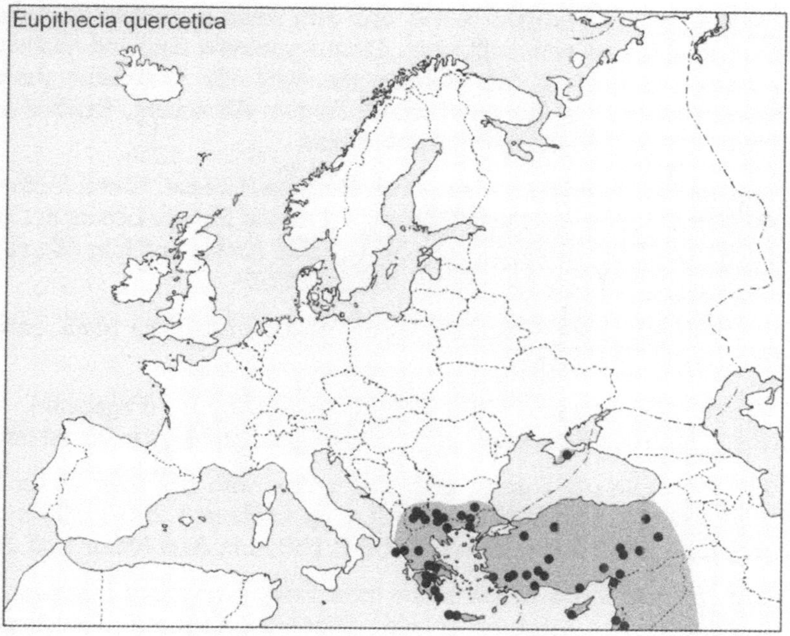

Forewing elongate, costa straight, yet arched near base, apex pointed, termen oblique, slightly rounded; ground colour ash-grey; transverse lines rather inconspicuous, forming dark costal spots; basal line oblique, sharply angled near costa; postbasal and antemedial lines slightly sinuate, oblique, not angled onto costa; the first medial line very fine, much sinuate, crossing discal spot and twice sharply angled on to costa; postmedial line curved, twice angled near costa, formed a series of black dashes on the veins; terminal area with indistinct wavy line; discal spot large, ovoid, intensely black. Hindwing rather elongated and angulated with slightly concave termen near apex; slightly paler than forewing; transverse lines indistinct; M3, CuA1 and CuA2 chequered black and white; terminal area narrow; pale wavy line ill-defined; discal spot large, ovate to ovate-acuminate (comma-shaped), dark, but always paler than that on the forewing. Fringe greyish white, distinctly spotted with grey or dark grey at vein ends. Abdomen grey, often with brownish tinge and with narrow black lateral lines, except on the last segment; second segment usually darker, greyish brown, the last segment light grey, paler than the others.

MALE GENITALIA: Uncus narrow, biapical. Papillae elongated, about 0.8-0.9 times uncus length. Valva widest at base and strongly narrowing toward apex, with broad, blunt, ventral process, and with narrow and rounded apex; sacculus sclerotized. Vinculum short and very broad. Aedeagus relatively large, stout; length about 0.8-0.9 times valva length. Vesica armed with five large, strongly sclerotized horn-like cornuti of which first (basal) cornutus largest and broadest; also with one elongated, flat cornutus, and one small, irregular, curved cornutus at ductus ejaculatorius base. Sternum A8 with base slightly broadened and basal hollow prominent; narrowed toward apex; with two pincer-like, thick, sclerotized and pointed lobes; sternum length about 0.8-0.9 times valva length.

FEMALE GENITALIA: Bursa copulatrix irregular, rather pear-shaped, sclerotized; with basal half completely covered with small spines, bordered by strongly sclerotized, narrow, transverse band with largest and longest spines, and with some large spines on the left side; with a patch of small spines near colliculum. Ductus seminalis long and relatively narrow, attached to corpus bursae near colliculum on the right side. Colliculum short and very broad. Antrum membranous, short and broad. Tergum A8 square. Papillae anales relatively elongated, narrow and tapered toward apexes.

DISTRIBUTION: East-Mediterranean. Macedonia, Bulgaria, Greece, Crete, Turkey and separate population in Ukraine (southern Crimea). – Outside Europe occurs in Cyprus, Asia Minor (Anatolia), Lebanon, Israel, Jordan, Syria, Saudi Arabia and Iraq (WILTSHIRE 1936, 1939, 1957; SCHÜTZE 1961b; HAUSMANN 1991; 1995).

PHENOLOGY: Univoltine: early April (late March) to early May (late May). Larva in April and May. Overwintering as pupa.

BIOLOGY: Larva oligophagous (3). Found on *Quercus* (Fagaceae), *Rhamnus* (Rhamnaceae), *Crataegus* (Rosaceae) and *Cistus* (Cistaceae) (Syria: WILTSHIRE 1936).

HABITAT: Light mountain oak forests, cliffs and gorges with oak trees or scrub on limestone, calcareous sandstone and lava. From 0 m up to about 1,200-1,300 m above sea-level in Greece (coll. ZSM) and to about 2,100-2,150 m in Asia Minor (coll. SOMMERER).

SIMILAR SPECIES: *E. quercetica* is one of the most easily recognized European species of the genus *Eupithecia,* and needs no detailed differential diagnosis. The shape of valva, the series of large, stout, horn-like cornuti on the vesica and the sternum A8 with two asymmetrical, heavily sclerotized, pincer-like apical lobes in the male; and the corpus bursae armed with small spines, margined by longer elegant spines, the very broad and flat colliculum and the tapered papillae anales in the female genitalia readily distinguish this species from others.

73. *Eupithecia egenaria* HERRICH-SCHÄFFER, 1848

Eupithecia egenaria HERRICH-SCHÄFFER, 1848: Syst. Bearb. Schmett. Eur. 3: 121, 132, pl. 46: 279, 280 (Germany: Gunzenhausen). Syntype(s).

Eupithecia undosata DIETZE, 1875: Stett. Ent. Z. 36 (4-6): 250, pl. 2: 2 (Latvia: Livland). Syntypes 1♂2♀ (MNHU).

Unavailable names (incorrect subsequent spellings): *egenata*: CREWE (1863); *egeneria*: MIRONOV (1988).

DIAGNOSIS: Wingspan 18-24 mm. Labial palpi, frons, vertex and notum light grey. Forewing light grey; transverse lines fine, conspicuous, dark grey; basal multiply indented; postbasal much wavy; antemedial evenly curved, thickened and darker near costa; medial line narrow, rather indistinct, sharply angled near costa and touching discal spot; postmedial line double, faintly sinuate, obtuse angled beyond discal spot, broadened and darker near costa, with broad lighter stripe, which crosses the forewing beyond postmedial line and parallel with it; terminal area broad, usually darker; pale wavy line multiply indented, sometimes broadened at tornus; discal spot distinct, obliquely elongate or ovate-oblong, black. Hindwing paler; transverse lines inconspicuous, more distinct along dorsum; pale transverse band beyond postmedial line weakly defined; terminal area faintly darker; light wavy line much indented but often inconspicuous or obsolete; discal spot minute, rather ill-defined, circular or comma-like. Fringe chequered light and dark grey. Abdomen light grey, with short black lateral stripes.

MALE GENITALIA: Valva large, with one small pointed ventral process (sometimes with one

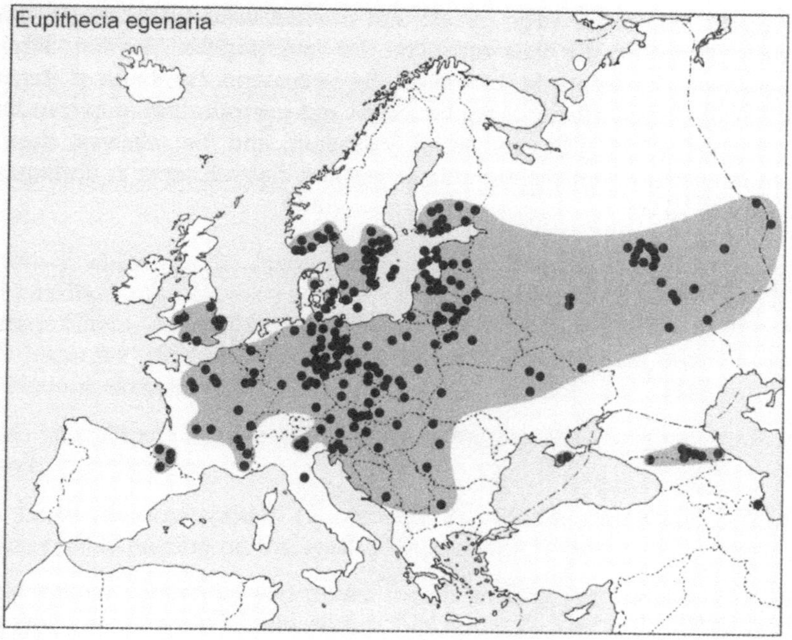

or two adjacent smaller processes). Vesica armed with one small apical hook-like cornutus, one large stout medial horn-like cornutus and broad, elongated cornutus.

FEMALE GENITALIA: Bursa copulatrix large, of irregular shape; membranous basal part variable, usually the largest and broadest part of the organ, with small patch of spinules; sclerotized medial part has broad, heavily sclerotized folds and is covered by numerous spinules; membranous apical part short, striated. Ductus seminalis short and narrow; attached to apical part of corpus bursae on the right side. Colliculum large and broad. Tergum A8 almost square with convex lateral margins.

DISTRIBUTION: Euro-Caucasian. Widespread from Spain and United Kingdom to southern Urals; northward to southern Finland, southward to about 43º latitude. – Outside Europe in Caucasus and Transcaucasus.

PHENOLOGY: Univoltine: mid-May to mid-June (early July). Larva: mid-June (early June) to early July (early August). Overwintering as pupa.

BIOLOGY: Larva monophagous (2). Found on the flowers of *Tilia vulgaris* (United Kingdom: HAGGETT 1968), *T. platyphyllos, T. cordata* (Tiliaceae) (Finland: SEPPÄNEN 1970; Germany: WEIGT 1980, 1993; United Kingdom: GOATER & al. 1981), preferring the pistils and stamens. In captivity the larva will accept the flowers of the common hybrid *Tilia* (United Kingdom: SKINNER 1998).

HABITAT: Stands of lime, broad-leaved forests and parks, forest fringes. From 0 m up to about 800 m above sea-level in Alps (WEIGT 1993).

SIMILAR SPECIES: A little variable species, easily distinguished from all other European representatives of the genus by the double postmedial line, with characteristic curve beyond discal spot on the forewing, and by the soft light grey ground colour. The combination of the short biapical uncus, the large, broad and pointed valva with one to three minute pointed ventral processes, the wide vinculum, the large papillae, the very large and very thick aedeagus with characteristic cornuti on the vesica and the shape of sternum A8 in male, as well as the extraordinary irregular shape and sclerotization of corpus bursae with two patch of small spines, the very broad colliculum, and the relatively short and thin apophyses in the female genitalia are unique features that will serve to distinguish *E. egenaria* from others.

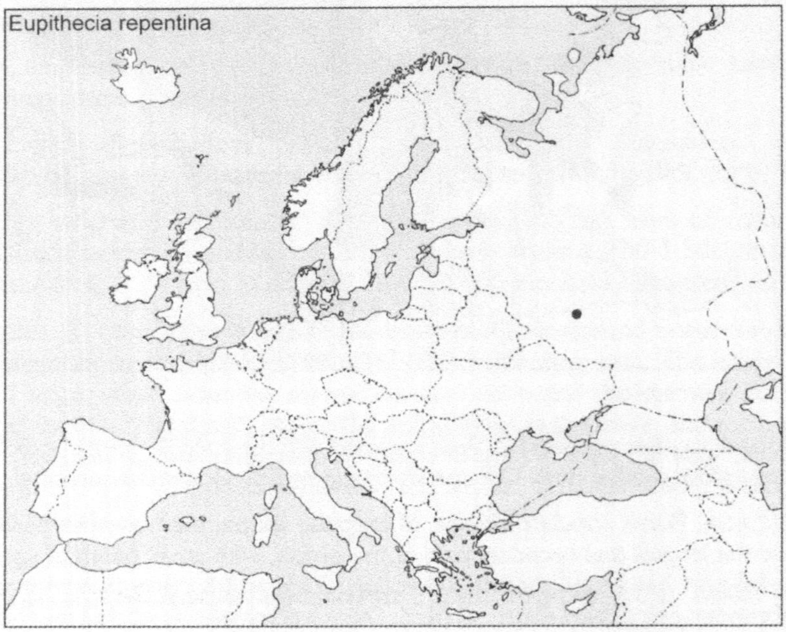

74. *Eupithecia repentina* Vojnits & de Laever, 1978

Eupithecia repentina Vojnits & de Laever, 1978: Acta zool. Acad. Sci. hung., 24 (1-2): 237, figs 15, 17 (China, Hunan Province: Hoeng-Shan). Holotype ♂ (ZFMK, examined).

Unavailable names (misidentification): *recens*: sensu Inoue (1955) nec Dietze, 1903 (identity: *Eupithecia recens*).

Diagnosis: Wingspan 19-21 mm. Labial palpi dark grey or grey. Frons, vertex and notum grey to whitish grey. Forewing narrow, elongated; costa straight, evenly curved near apex; termen obliquely rounded; apex rather pointed; ground colour grey or brownish grey; transverse lines inconspicuous, well marked only along costa by dark spots; postmedial line thin, oblique, sharply angled near costa; pale wavy line indistinct, exept small pale tornal spot; discal spot large, elongated, black. Hindwing unicolorous pale grey; dark transverse lines and pale wavy line inconspicuous; discal dot small, rounded, blackish grey. Fringe chequered whitish grey and grey. Abdomen whitish grey with black lateral lines.

Male genitalia: Valva with narrow and heavily sclerotized ventral process. Vesica armed with three small stout horn-like cornuti (two placed apically and one medially), one elongated, broad cornutus with longitudinal striations in its medial part and one small, irregular, curved cornutus at ductus ejaculatorius base. Sternum A8 broadened at base and at apex, with two small pointed apical horns and shallow, broad hollow between them.

FEMALE GENITALIA: Bursa copulatrix small, rather pouch-like, covered with small spines from base to colliculum on the left side. Ductus seminalis long and narrow; attached to corpus bursae near its base on right side. Colliculum collar-like, small (short and narrow). Anterior and posterior apophyses relatively long. Tergum A8 short and broad. Papillae anales large.

DISTRIBUTION: East-Palaearctic (nemoral). Central European Russia (one micropopulation?). – Outside Europe in the Far East of Russia (Amur and Ussuri regions), Korea, south-western China, Japan (VOJNITS & DE LAEVER 1978; VIIDALEPP & MIRONOV 1988; OH 1992; VIIDALEPP 1996). - New for the fauna of Europe: 1♂ and 9♀, Russia, Tula province, Novomoskovsk, 4.VI.2000, S. ANDREEV leg. (ZISP and ZMMU).

PHENOLOGY: Univoltine?: June. In the East Asia bivoltine: late May (mid-April) to mid-July; late August to mid-October.

BIOLOGY: Larva monophagous? (1). Found on *Mosla dianthera* (=*grosseserrata*) (Lamiaceae), mainly feeding on the seeds (Japan: INOUE 1980).

HABITAT: Mountain coniferous and mixed forests in East Asia. From 0 m up to about 1,600 m above sea-level in Ussuri region (VIIDALEPP & MIRONOV 1988), and to about 3,000 m in Tsinling Mts in central China (VOJNITS & DE LAEVER 1978).

SIMILAR SPECIES: Similar to *E. egenaria*, but smaller, forewing narrower without double postmedial line and with larger and more elongated discal spot. The narrower valva, with wide apex and one thin sclerotized ventral process, the smaller papillae, the narrower vinculum, the smaller aedeagus with different cornuti and shorter sternum A8, with shallower and broader apical hollow and shorter apical rods in the male distinguish this species from the closely related *E. egenaria*. The female genitalia of *E. repentina* are very distinct from those of *E. egenaria* also.

REMARKS: The separate micropopulation of this East Asian species was founded accidentally in a park (about 5 ha) near Novomoskovsk town (Tula province, Russia), where East Asian trees and shrubs were introduced from 1950 till 1990. The small adults of *E. repentina* were caught within this park and nearby on the trunks of old limes (BOLSHAKOV pers. comm.). European specimens are smaller than East Asian, which have wingspan 21-24 mm in the first generation and 16.5-18 mm in the second generation according to the original description.

75. *Eupithecia orana* DIETZE, 1910

Eup(ithecia) unitaria f. *orana* DIETZE, 1910: Biol. Eupithecien 1, pl. 75: 589; ibidem (1913), 2: 101 (Algeria: Oran). Holotype ♀ (MNHU, examined: SCHÜTZE 1961). Raised to species rank by PROUT (1954).

Eupithecia soricella LUCAS, 1938: Bull. Soc. ent. France 43 (13/14): 182 (Algeria: Biskra). Holotype ♀ (HERB/ZSM, examined: HERBULOT 1984).

Eupithecia maspalomae PINKER, 1961: Z. Wien. ent. Ges. 46 (5): 71, text. figs 18, 19; pl.4: 3, 4 (Canary Islands: Maspalomas). Holotype ♂ (NHMW). New junior synonym.

DIAGNOSIS: Wingspan 16-21 mm. Labial palpi yellowish grey. Frons, vertex and notum grey with yellowish suffusion. Forewing rather narrow and elongated, costa straight,

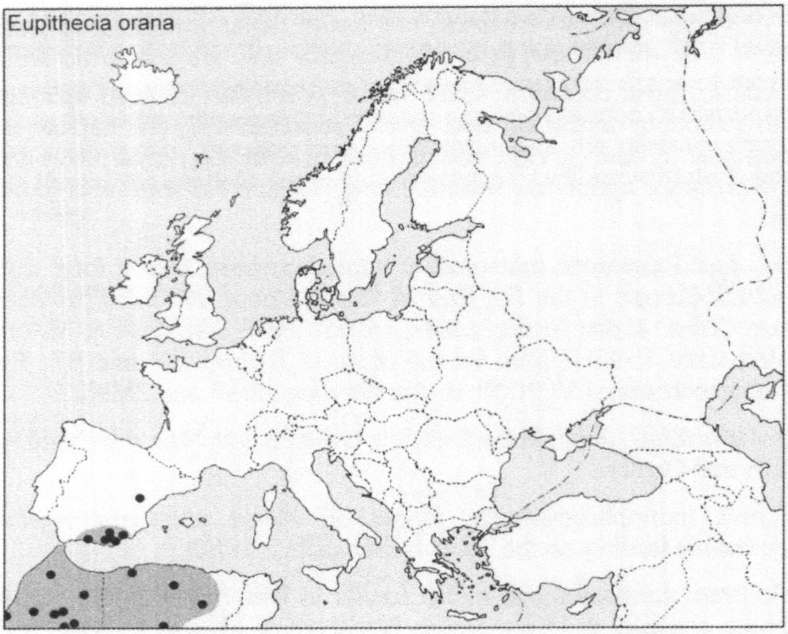

curved near apex only, apex pointed, termen oblique, slightly curved; ground colour grey with distinct yellowish tinge; pattern inconspicuous; basal, postbasal and antemedial lines faintly sinuate, angled near costa; medial area between antemedial and medial lines usually darker; medial line sharply angled on to costa beyond discal spot; postmedial line minutely wavy, obtusely curved on to costa; pale wavy line indistinct, without tornal dot; discal spot minute, obliquely elongate, blackish grey. Hindwing ovoid with evenly rounded termen; costal area pale, without pattern; anal and terminal areas darker, especially towards tornus, yellowish grey; traces of transverse lines visible along anal margin only; pale wavy line inconspicuous; discal spot absent. Fringe unicolorous, yellowish grey or sometimes slightly spotted with dark grey at vein ends. Abdomen yellowish grey.

MALE GENITALIA: Valva with pointed ventral process; sacculus heavily sclerotized. Vesica armed with three slim horn-like cornuti, one elongated, flat cornutus and one smaller, irregular, curved cornutus at ductus ejaculatorius base. Sternum A8 rather short and wide, with shallow and broad basal and apical hollows, and with two narrow, pointed, symmetrical, heavily sclerotized apical horns.

FEMALE GENITALIA: Bursa copulatrix irregularly shaped or rather pear-shaped; its membranous basal part armed with inconspicuous weakly sclerotized spines and bordered by rows of sparse large spines and wrinkles. Ductus bursae heavily sclerotized with elongated patch of very small spines at ductus seminalis base. Colliculum short, broad, inclined to right. Ductus seminalis narrow, yet broadly attached to medial part of corpus bursae. Antrum heavily sclerotized. Anterior and posterior apophyses short and thin. Tergum A8 elongated, trapeziform. Sternum A8 heavily sclerotized (as in species of the *suboxydata*-group). Papillae anales rather short and rounded.

DISTRIBUTION: West-Mediterranean. Southern Spain (Granada, Almeria, Cuenca and

Teruel). – Outside Europe in the Canary Islands (Tenerife and Lanzarote) and North Africa (Morocco and Algeria).

PHENOLOGY: Bi- or polyvoltine: early January to late March (July in Morocco); mid-October (early October) to mid-November (early December).

BIOLOGY: Larva monophagous? (1). Recorded on *Launaea arborescens* (=*Zollikoferia spinosa*) (Asteraceae) (Algeria: TAUTEL pers. comm.).

HABITAT: Deserts and other arid landscapes. From 0 m up to about 1,100 m above sea-level in southern Spain (coll. ZSM) and to about 2,600 m in Great Atlas (coll. ZSM).

SIMILAR SPECIES: Smaller than *E. unitaria*, with narrower and pointed wings and oblique termen of forewing. The wing pattern is brindled and rather indistinct in both compared species. *E. orana* also resembles *E. praealta* among other European *Eupithecia* species; yet it is easily distinguished from this latter by evidently smaller size, paler costal areas and absence of discal spots on the hindwings. The shape of valva and sternum A8, the cornuti on the vesica in the male genitalia, and the shape of bursa copulatrix with unique arrangement of spines, the heavily sclerotized ductus bursae and sternum A8, the unique shape and heavily sclerotization of antrum should distinguish *E. orana* from any other European *Eupithecia* species.

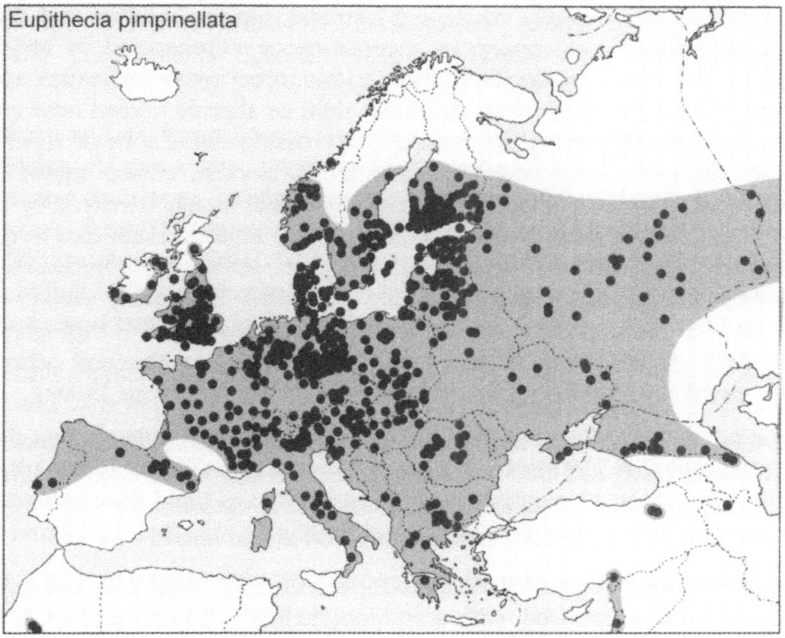

76. *Eupithecia pimpinellata* (HÜBNER, 1813)

Geometra pimpinellata HÜBNER, 1813: Samml. Eur. Schmett. 5 Geometrae (2), pl. 86: 443, 444 (Europe). Syntype(s) lost.

Eupithecia altaicata GUENÉE, 1858: in BOISDUVAL & GUENÉE, Hist. nat. Insectes (Spec. gén. Lépid.) 10 (2): 330 (Russia: Altai). Syntypes 1♂1♀ (NHMW?).

Eupithecia lantoscata MILLIÈRE, 1873: Cat. Rais. Lépid. dept. Alpes-Maritimes (2): 201 (France: Alpes-Maritimes, vallée de Lantosque). Syntype(s) (MNHN?).

Eupithecia pimpinellata var. *cinerascens* TENGSTRÖM, 1875: Notis. Sällsk. Faun. Fl. fenn. Förh. (N.S.) 11: 32 (Russia: Karelia). Holotype ♀ (ZMH). Valid at subspecific rank.

Eup(ithecia) vellicata DIETZE, 1908: Dt. ent. Z., Iris (1908) 21 (2/3): 173; ibidem (1906), 19 (3), pl. 2: 16 (China: Urumtschi). Syntypes (MNHU).

Eup(ithecia) pimpinellata f. *limbosignata* DIETZE, 1910: Biol. Eupithecien 1, pl. 72: 285; ibidem (1913), 2: 119 (China: Kuldja [Yining]; Kazakhstan: Djarkent [=Panfilov]). Syntype(s) ♀ (MNHU). Valid at subspecific rank.

(Eupithecia) pimpinellata f. *assimilis* DIETZE, 1913: Biol. Eupithecien 2: 118; ibidem (1910), 1, pl. 76: 554-556; 85: 949 (China: [Xinjiang], Juldus [Yining] region; Kazakhstan: Ili region). Syntypes 8♂ 1♀ (MNHU). Valid at subspecific rank.

(Eupithecia) pimpinellata f. *elongata* DIETZE, 1913: Biol. Eupithecien 2: 118; ibidem (1910), 1, pl. 76: 558, 559 (Italy: Pescocostanzo). Syntypes, at least 2 (MNHU). Valid at subspecific rank.

Eupithecia pimpinellata variata SCHWINGENSCHUSS, 1936: in ZERNY, Mém. Soc. Sci. nat. Phys. Maroc 42: 85 (Morocco: Great Atlas, Tachdirt; Tizi n'Tachdirt). Syntype(s), at least 1 syntype in NHRS. Valid at subspecific rank.

Eupithecia pimpinellata retheli PROUT, 1938: in SEITZ, Macrolep. 4 (Suppl.): 200 (Italy). Syntype(s) (BMNH). Valid at subspecific rank.

Unavailable names (misidentification): *denotata*: sensu DOUBLEDAY (1856) nec HÜBNER, 1813 (identity: *Eupithecia denotata*). – (infrasubspecific): *grisea*: LEMPKE (1969). - (incorrect subsequent spelling): *pimpinellaria*: BOISDUVAL (1840).

DIAGNOSIS: Wingspan 17-25 mm. Labial palpi grey. Frons, vertex and notum ash-grey. Forewing rather elongated and pointed; ash-grey with a delicate light ochreous tinge, especially in medial area, usually costal and terminal margins mottled and irrorate with white scales; transverse lines sometimes inconspicuous, represented by series of small blackish dots on the veins, forming from three to five rather weak dark costal spots; basal, postbasal and antemedial lines much indented, right or sharply angled near costa; three medial lines fine, line preceding postmedial line crossing discal spot or touching it and sharply angled towards costa; postmedial line faintly sinuate, obtuse angled near costa beyond discal spot; wavy line fine, distinct, creamy white, usually bordered by medial and tornal shades, especially on inner side; light tornal spot absent; discal spot large, oblique, ovate-oblong or ovate, intensely black. Hindwing paler, smoky-grey; transverse lines well indicated along dorsum; terminal area darker; pale wavy line faintly visible; discal spot large, very distinct, dark, circular but sometimes comma-shaped. Fringe greyish white, spotted with dark greyish brown at vein ends. Abdomen grey with a soft ochreous tinge, usually dark lateral stripes lacking; second segment darker, ochreous brown.

MALE GENITALIA: Valva without ventral process. Vinculum with shallow medial hollow. Aedeagus relatively short and thick. Vesica armed with one short, hooked horn-like apical cornutus and one small irregular cornutus at ductus ejaculatorius base. Sternum A8 with deep apical hollow and short, thin and tapered apical horns.

FEMALE GENITALIA: Bursa copulatrix elongated, pear-shaped, completely covered by small spines. Ductus bursae sclerotized, with some longitudinal striations and small sclerotized appendix. Ductus seminalis relatively long and broad, sclerotized at base. Colliculum collar-like, short and narrow. Papillae anales small.

DISTRIBUTION: West-Palaearctic. Widely distributed in Palaearctic region from Portugal eastwards throughout Europe and Siberia to Altai and Mongolia. Very isolated occurrence in Great Atlas (Morocco).

PHENOLOGY: Uni- or bivoltine: late April to mid-June; late July (mid-July) to late August (mid-September). In Central and northern Europe univoltine: mid-July (early June) to late August (mid-September). Larva: early August (late July) to late September (late October). Overwintering as pupa in insecure cocoon in the ground.

BIOLOGY: Larva rather oligophagous (1). Found on the flowers and ripening seeds of *Pimpinella saxifraga, P. major (=magna)* (Apiaceae) (United Kingdom: STAINTON 1855; CREWE 1861c; BUCKLER 1899; BARRETT 1904; Germany: DIETZE 1875a; WEGNER pers. comm.; Russia: original data), also *Solidago virgaurea* (Asteraceae) (United Kingdom: CREWE 1861c), *Bupleurum falcatum* (Apiaceae) (France: MILLIÈRE 1870), *Aegopodium podagraria, Peucedanum cervaria, P. oreoselinum, Pastinaca sativa* (Apiaceae) (Germany: DIETZE 1910, 1913; WEIGT 1980), *Angelica sylvestris* (Apiaceae), *Galium verum* (Rubiaceae), *Artemisia campestris* (Asteraceae) (Finland: SEPPÄNEN 1970), *Achillea ptarmica, Senecio vulgaris* (Asteraceae) (Switzerland: VORBRODT 1928). Recorded on *Seseli annuum* (France: LHOMME 1935) and *Bupleurum spinosum* (Apiaceae) (Morocco: RUNGS 1981). In captivity feeds also *Senecio jacobaea, S. congestus* (Asteraceae) (United Kingdom: CREWE 1861c). Often heavily parasitized.

PARASITOIDS: Hymenoptera, Ichneumonidae: *Campoplex difformis, Mesochorus brevipetiolatus, M. confusus, M. fulgurans, M. testaceus, Tycherus discoidalis, Platylabus iridipennis, P. pedatorius* - Hymenoptera, Braconidae: *Aleiodes (=Rogas) modestus, A. (=Rogas) bicolor (=tristis), Apanteles carbonarius, A. cajae (=difficilis), A. juniperatae (=sessilis), A. jucunda (=nigrinervis), A. salebrosus, A. spurius, Apanteles* sp., *Earinus delusor, Lissogaster (=Microgaster) crassicornis* - Hymenoptera, Eurytomidae: *Systole conspicua* (THOMPSON 1945; SHENEFELT 1970, 1972, 1973, 1975; HERTING 1976; ZWAKHALS pers. comm.).

HABITAT: Warm and dry forest fringes and meadows with calcareous or sandy soil, dry and sunny roadsides, old gravel pits, coasts, hedges, scrub and scrub-covered meadows, sandy fields, warm and dry slopes, heaths, dunes, sandhills, alpine meadows, rocks, screes and cultivated areas. From 0 m up to about 1,500 m above sea-level in Pyrenees (MAZEL & PESLIER 1997), to about 2,000-2,100 m in Apennines (Mt. Majella: ZAHM pers. comm.), to about 2,200-2,300 m in southern Alps (HELLMANN & al. 1999), and to about 2,100-2,500 m in Balkan peninsula (coll. ZSM).

SIMILAR SPECIES: Remotely similar to *E. absinthiata* but paler. *E. pimpinellata* is unlikely to be confused with any other European *Eupithecia* species except *E. pauxillaria* (see Text-fig. 75 and *E. pauxillaria* diagnosis). The male genitalia are distinguished by the cornuti on the vesica and the larger and broadened sternum A8. A long, sclerotized spineless ductus bursae with short heavily sclerotized appendix in its base, and narrow colliculum in the female genitalia are should distinguish this species.

REMARKS: Usually variable in coloration from light grey to light ochreous brown.

77. *Eupithecia karadaghensis* MIRONOV, 1988

Eupithecia karadaghensis MIRONOV, 1988: Vestnik zool. 4: 19 (Ukraine, Crimea: Karadagh). Holotype ♀ (ZISP, examined).

DIAGNOSIS: Wingspan 18-20 mm. Labial palpi greyish brown. Frons and vertex with mixture of grey and brown scales. Notum covered with grey and scattered brown scales.

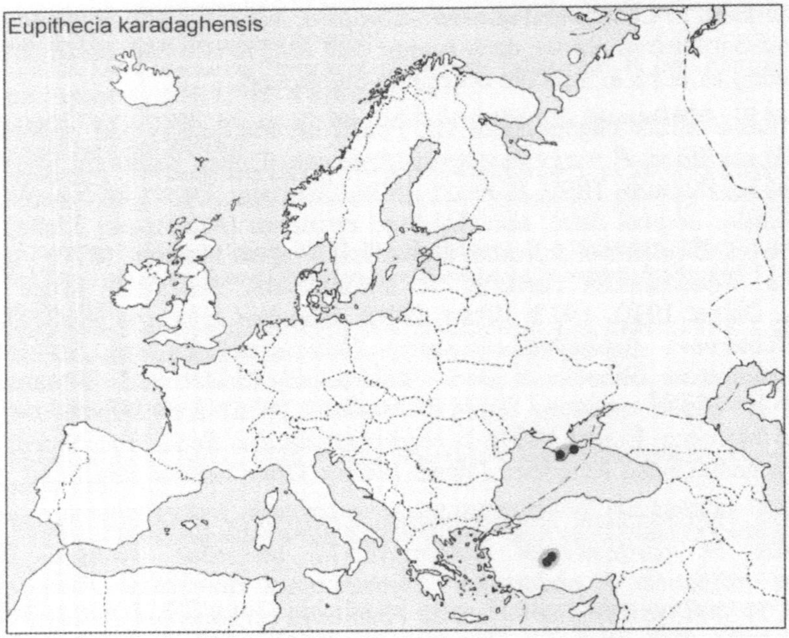

Forewing pale brownish grey; basal, antemedial and medial lines indistinct, usually well marked with dark brown costal spots; postmedial line brown, conspicuous, slightly sinuate and evenly curved near costa; terminal area darker, brownish grey, with faint, pale grey wavy line; discal spot large, oblique, oval, intensely dark brown or black. Hindwing concolorous with forewing; postmedial line broad, rather faint, dark brown; terminal area darker, brownish grey; discal spot small, rounded or oval, intensely dark brown. Fringe grey, spotted with greyish brown. Abdomen greyish brown, with dark brown or black lateral spots.

MALE GENITALIA: Valva without ventral process. Vinculum with shallow medial hollow. Vesica armed with two large and broad horn-like cornuti and with large, elongated U-shaped cornutus. Sternum A8 rather peg-like with two small and blunt apical lobes.

FEMALE GENITALIA: Bursa copulatrix large, rather pear-shaped; small spines with row of larger spines forming a circle along basal part of corpus bursae from base of ductus seminalis; ductus bursae spineless, with longitudinal striations. Ductus seminalis long and broad, unevenly curved backwards. Colliculum collar-like, short and relatively broad, slightly inclined to right. Tergum A8 broad, rectangular, with posterior corners projecting and rounded.

DISTRIBUTION: Pontic. Ukraine (southern Crimea). – Outside Crimea in Asian Turkey (Prov. Konia). - New for the fauna of Turkey: 5♂5♀, Asia min. c., Anatolia c. Akşehir Sultan Dagh, 15-1700 m, 1.-15.7.1934, 15.-30.7.1934, 15.-30.8.1934, coll. E. PFEIFFER (HNHM).

PHENOLOGY: Univoltine: mid-August (early July) to late August (early September). Overwintering possibly as pupa.

BIOLOGY: Unknown, but larvae probably on the flowers or seeds of *Seseli tortuosum (S. pauciradiatum)* or other *Seseli* species (Apiaceae) in September and October (?).

HABITAT: Warm, dry and light slopes and light Mediterranean low forests. From 0 m to about 400-500 m above sea-level in southern Crimea and to about 1,500-1,700 m in Asia Minor (coll. HNHM).

SIMILAR SPECIES: Adults of *E. karadaghensis* are difficult to separate from *E. pimpinellata*. *E. karadaghensis* is on the whole more greyish, with rounded and more conspicuous discal dots on the hindwings, but these cannot be used as reliable diagnostic characteristics. Only the male and female genitalia can be used to make determinations: the vesica with two identical, large and broad horn-like cornuti, the narrower sternum A8 with two smaller and blunt apical lobes, as well as the shape and sclerotization of corpus bursae and area covered by spines.

The ***suboxydata*** species-group

This is a species-poor Asian *Eupithecia* group, which was described as genus *Catarina* VOJNITS & DE LAEVER, 1973. One species *E. suboxydata* has a western limit of its distribution in south European Russia (Lower Volga). The other species occurs in various places from Central Asia to southern and south-eastern Asia, such as: *E. subbrunneata* DIETZE, 1913 (=*Catarina carissima* VOJNITS & DE LAEVER, 1973; *Catarina dolosa* VOJNITS, 1977, syn. n.; *Catarina necessaria* VOJNITS, 1977, syn. n.), *E. propoxydata* SCHÜTZE, 1961 and *E. formosa* (VOJNITS & DE LAEVER, 1973).

MALE GENITALIA: Uncus large and stout, biapical. Papillae usually small, about 0.5-0.6 times uncus length. Valva large, sclerotized, with or without a small, sclerotized ventral process, with broadly rounded apex and sclerotized sacculus. Vinculum short and broad. Aedeagus large, stout, length about 0.8-0.9 times valva length; about 3.0-4.0 times longer than medial width. Vesica armed with one horn-like cornutus with wrinkled base and one large and broad flat cornutus at ductus ejaculatorius base. Sternum A8 consists of elongated central part and two heavily sclerotized, concave and rounded latero-caudal sclerites connected with central part by membranous bands.

FEMALE GENITALIA: Bursa copulatrix almost spherical; completely covered with small uniform spines. Ductus bursae short, with frequent longitudinal striations and with or without one patch of sparse spines. Ductus seminalis long and thick, attached to ductus bursae near colliculum on right side. Colliculum collar-like, usually short and narrow. Antrum broad and sclerotized. Anterior and posterior apophyses short, relatively thick, whitout expanded to apices. Spurs of anterior apophyses small or normal, broadened to apices. Tergum A8 almost square. Sternum A8 large, broad and heavily sclerotized. Papillae anales short, broad and rounded.

REMARKS: The species of the *suboxydata* group are unlikely to be confused with any other *Eupithecia* species in the World because of their unique sternum A8 in the male. But a heavily sclerotized sternum A8 of female, as recorded in VOJNITS & DE LAEVER (1973), occurs also in *E. orana* DIETZE, 1910 and some Asian *Eupithecia* species, as following: *E. detritata* STAUDINGER, 1897, *E. recens* DIETZE, 1903 and *E. daemionata* DIETZE, 1903.

The relationship of this unique Asian species-group is not known. The presence of transverse vein in areole of forewing, and broad flat cornutus and one horn-like cornutus on the vesica in male, indicate a possible remote relationship to the *simpliciata* and the *sinuosaria* species-groups.

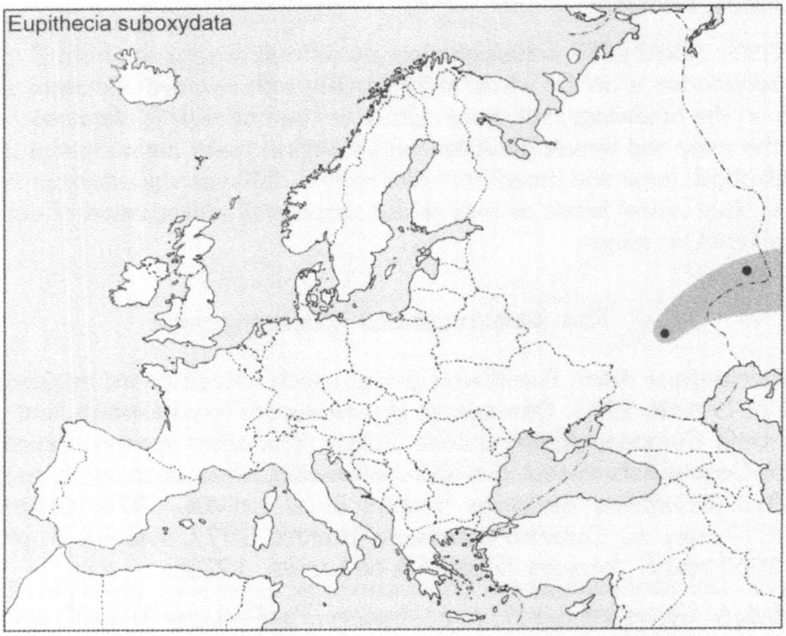

78. *Eupithecia suboxydata* STAUDINGER, 1897

Eupithecia suboxydata STAUDINGER, 1897: Dt. ent. Z. Iris 10 (1): 112, pl. 3: 74 (Russia: 'Amur', Vladivostok, Askold Isl., Suifun, Sutschan). Syntype(s) (MNHU).
Unavailable names (misidentification): *minutata:* sensu EVERSMANN (1844) nec DENIS & SCHIFFERMÜLLER, 1775 (identity: *Eupithecia absinthiata*). - (infrasubspecific): *minorata*: DIETZE (1913).

DIAGNOSIS: Wingspan 19-24 mm. Labial palpi greyish brown. Frons, vertex and notum pale greyish brown. Forewing with slightly arched costa and rounded termen; fuscous greyish brown ground colour; antemedial line obtuse angled near costa; three fine medial lines more or less conspicuous; postmedial line most distinct, almost straight as far as obtuse angle near costa beyond discal spot; medial and terminal areas separated by wide, double, light stripe; terminal area broad; whitish wavy line usually broken, forming a distinct, ovate tornal spot, edged inwardly by darker scales; discal spot large, ovate or ovate-oblong, blackish. Hindwing fuscous, grey or greyish brown, yet paler than forewing; traces of transverse lines developed along dorsum; terminal area usually slightly darker; wavy line invisible; discal spot small, rounded, considerably paler than on the forewing. Fringe whitish grey, spotted with dark greyish brown at vein ends. Abdomen brownish grey, often with fine blackish lateral lines.

MALE GENITALIA: Valva with a small, sclerotized ventral process, preceded by a shallow wrinkled impression. The central part of sternum A8 resembles the figure of a man; two latero-caudal sclerites of sternum A8 very large, ovate.

FEMALE GENITALIA: Ductus bursae without patch of spines. Ductus seminalis has a short row of spinules near its base. Antrum with sclerotized swelling. Spurs of anterior apophyses very short.

DISTRIBUTION: East-Palaearctic (sub-eremic). South-east of European Russia and southern Urals. – Outside Europe in western Siberia (Tjumen' oblast'), Kazakhstan, Central Asia (Tien-Shan Mts), Altai, southern Siberia to Ussuri region, Mongolia, China and Japan (Hokkaido).

PHENOLOGY: Bivoltine probably: late March to early August. Larva in August and September. Overwintering as pupa.

BIOLOGY: Larva monophagous? (1). Found on the flowers of *Artemisia santolinifolia* (Asteraceae) (Kirghizstan: original data).

HABITAT: Steppes and in places with abundant ruderal vegetation. From 0 m up to about 2,200 m above sea-level in Tien-Shan Mts (original data).

SIMILAR SPECIES: Not unlike a very dark *E. pimpinellata* which differs, however, in possessing a less distinct light stripe between medial and terminal areas and having a smaller pale tornal spot on the forewing, and larger discal spots on the hindwings. The broadened tip of uncus; the shape of valva with obtuse, broadly rounded apex and with an obtuse ventral process, preceded by a shallow wrinkled impression; the vesica armed with straight horn-like cornutus, and the unique sternum A8 with two large, heavily sclerotized ovate caudal discs in the male; and the large and broad corpus bursae with short ductus bursae; the small patch of minute spinules at the base of ductus seminalis; the wide colliculum; the broadened and sclerotized antrum; the broad and heavily sclerotized sternum A8 in the female can be used to identify *E. suboxydata* with accuracy.

REMARKS: There are two worn specimens of *E. suboxydata* from European Russia. One male from Orenburg Province with label 'Spask' (=Spasskoe - the former estate of Prof. E. EVERSMANN), which had been incorrectly identified as '*Larentia minutata*' by E. EVERSMANN in ZISP (EVERSMANN 1844: 413), and another recently caught female from Saratov Province.

The ***simpliciata*** species-group

Group introduced by MIRONOV (1991). Four *Eupithecia* species can be included in this group, only one occurring in Europe: *E. simpliciata*. Two Central Asian species, *Eupithecia inculta* VOJNITS, 1975 and *E. tshimkentensis* VIIDALEPP, 1988, and *E. suspiciosata* DIETZE, 1875 from North America (California) also belong to the *simpliciata* species-group.

MALE GENITALIA: Uncus biapical, short or medium length. Papillae short, length about 0.6-0.8 times uncus length. Valva without ventral process, of normal width, with curved ventral margin and rounded apex; sacculus weakly sclerotized. Vinculum short and broad. Aedeagus stout and relatively long, slightly shorter than valva length. Vesica covered with a great number of small spines with one more dense patch of the same spines, armed with one small and one broad, elongated and slightly spiralled cornutus at ductus ejaculatorius base. Sternum A8 peg-like, elongated and narrow, with two short and sclerotized horns; sternum length about 0.6-0.7 times valva length.

FEMALE GENITALIA: Bursa copulatrix large, pear-shaped, with numerous small spines at base and longitudinal striations in ductus bursae. Ductus seminalis short and broad, attached to bursa copulatrix near middle on right side. Colliculum collar-like, short and wide. Antrum membranous, short and broad. Anterior and posterior apophyses relatively short and thin, not expanded to apices. Tergum A8 almost square. Papillae anales small, short and rounded.

REMARKS: The species of the *simpliciata* group can usually be distinguished from other *Eupithecia* species by the vesica, which armed with one small and one broad, elongated and slightly spiralled cornutus at ductus ejaculatorius base, as well as being covered by a great number small spines, with one more dense patch of the same spines in the male genitalia.

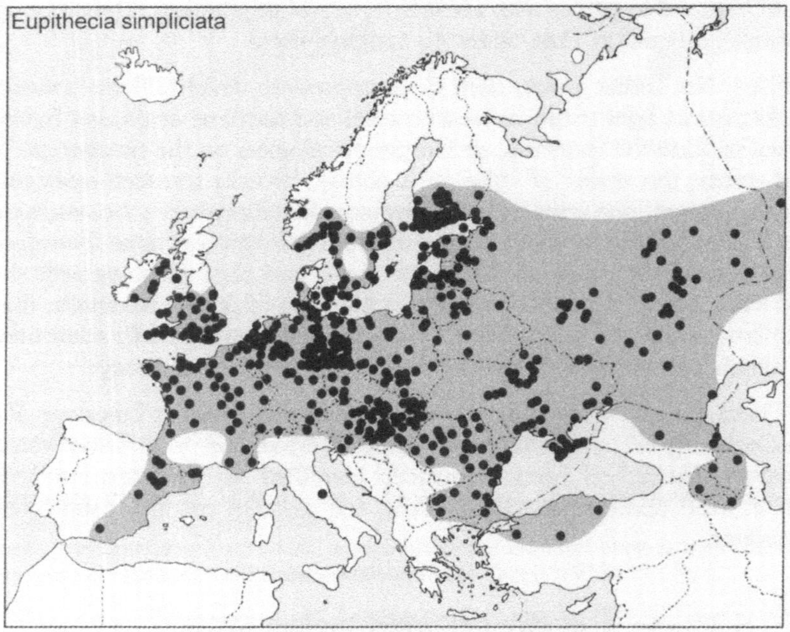

79. Eupithecia simpliciata (HAWORTH, 1809)

Phalaena simpliciata HAWORTH, 1809: Lepid. Br. (2): 359 (United Kingdom). Syntype(s) (BMNH).
Geometra subnotata HÜBNER, 1813: Samml. Eur. Schmett. 5 Geometrae (2), pl. 89: 458 (Europe). Syntype(s) lost.
Unavailable names (misidentification): *scabiosata*: sensu DUPONCHEL (1830) nec BORKHAUSEN, 1794 (identity: *Eupithecia subumbrata*). – (infrasubspecific): *brunnea*: LEMPKE (1951); *impuncta*: LEMPKE (1951). - (incorrect subsequent spellings): *subnotaria*: BOISDUVAL (1840); *subnotataria*: HERRICH-SCHÄFFER (1848).

DIAGNOSIS: Wingspan 17-25 mm. Labial palpi, frons, vertex and notum light greyish ochreous. Antennae chequered pale ochreous and brown. Forewing with oblique transverse vein in areole; wide with broadly rounded apex; smooth ochreous; costal margin covered with mixture of whitish and light brown scales; transverse lines fine, rather inconspicuous; postmedial line distinct, brownish, finely waved, slightly curved and not angled near costa; whitish line between medial and terminal areas ribbon-like, wide and distinct, rather slightly and evenly curved, more prominent towards costa: terminal area broad, pale tornal spot lacking; wavy line whitish, conspicuous, sinuate and evenly curved; discal spot small, rounded, light brown or sometimes dark brown. Hindwing light greyish ochreous with darker brownish posterior margin and with fine, indistinct transverse lines; terminal area darker, more ochreous in tornus; wavy line distinct, whitish, fine, fine-dentated; discal spot minute, rounded, sometimes absent. Fringe unicolorous brownish grey with ochreous tinge. Abdomen light greyish ochreous.

MALE AND FEMALE GENITALIA: See species-group diagnosis.

DISTRIBUTION: European-Westasiatic. Almost throughout Europe from Ireland and Spain to southern Urals (except Iceland, Portugal, former Yugoslavia, Albania, Greece and Mediterranean islands). – Outside Europe in Asia Minor, Caucasus, Transcaucasus and western Siberia.

PHENOLOGY: Univoltine: early July (early June) to early August (late September). Larva: late August (mid-August) to late September (early October). Overwintering as pupa, often twice.

BIOLOGY: Larva oligophagous (1). Found on the flowers and ripening seedheads of *Chenopodium album, C. vulvaria, C. hybridum, C. bonus-henricus, C. glaucum, C. rubrum, Atriplex patula, A. prostrata (=hastata), A. littoralis, A. rosea, A. oblongifolia* (Chenopodiaceae) (United Kingdom: CREWE 1861c; BUCKLER 1899; BARRETT 1904; France: GOOSSENS 1869; MILLIÈRE 1870; MABILLE 1872; Germany: DIETZE 1874a, 1910, 1913; WEIGT 1991; WEGNER pers. comm.; Denmark: JUUL 1948; Finland: SEPPÄNEN 1970), and occasionally (?) on *Mercurialis annua* (Euphorbiaceae) (Germany: DIETZE 1871b) and *Artemisia maritima* (Asteraceae) (Netherlands: LEMPKE 1969).

PARASITOIDS: Hymenoptera, Ichneumonidae: *Cyclolabus dubiosus, Platylabus pedatorius, P. punctifrons* (BUCKLER 1899; THOMPSON 1945; HERTING 1976).

HABITAT: Fields, gardens, along the coast and especially places with ruderal vegetation: cultivated areas, farms, cattle-yards, old construction sites and rubbish tips. From 0 m up to about 1,200 m above sea-level in Alps (VORBRODT 1931; WOLFSBERGER 1950).

SIMILAR SPECIES: *E. simpliciata* is readily recognized and can be easily distinguished from other European *Eupithecia* species by the combination of wide wings, smooth pale ochreous coloration, fine and slightly curved postmedial line and whitish outer ribbon-like line along postmedial line. The cornuti on the vesica, the shape of sternum A8 in the male, and the spiniferous area in bursa copulatrix in the female genitalia should distinguish this species.

The ***sinuosaria*** species-group

Group introduced by VOJNITS (1976). Three *Eupithecia* species can be included in this group, of which only *E. sinuosaria* occurs in Europe. Two Central Asian species: *E. pallescens* DIETZE, 1910 occuring from Tien-Shan to Mongolia and *E. ingrata* VOJNITS, 1981 which is distributed from western Tien-Shan to northern Pakistan (Prov. Swat) also belong to the *sinuosaria* species-group.

MALE GENITALIA: Uncus short, narrow, biapical. Papillae length about 0.6-0.9 times uncus length. Valva without ventral process, broadest medially; with smoothly rounded ventral margin, slightly curved dorsal margin and narrowly rounded apex; sacculus weakly sclerotized. Vinculum short and broad. Aedeagus stout, relatively long and thick, length about 0.7-0.9 times valva length; about 3.0-3.5 times longer than medial width. Vesica armed with one horn-like or two dentate cornuti at apex; with one large, elongated, twisted, V-shaped cornutus and one small, irregular cornutus at ductus ejaculatorius base. Sternum A8 long with two sclerotized, more or less long and thin apical rods, with blunt apices; basal hollow shallow; apical sinus deep; sternum length about 0.6-0.8 times valva length.

FEMALE GENITALIA: Bursa copulatrix usually elongated, pear-shaped, sclerotized, almost completely covered by small spines except right side. Ductus bursae with numerous longitudinal striations. Ductus seminalis short and narrow; attached to middle of corpus bursae on right side. Colliculum collar-like, elongated, from narrow to broad, slightly inclined to right. Antrum membranous, short and broad. Anterior and posterior apophyses short and thin. Tergum A8 rectangular. Papillae anales relatively small, rounded.

REMARKS: The shape of sternum A8 in the male separates this group from all other European *Eupithecia* species. I have placed the *sinuosaria* species-group between the *simpliciata* and the *innotata* species-groups because of their similarities, such as the large and broad, slightly spiralled cornutus on the vesica in the male genitalia and the structure of the female genitalia, which indicate their close relationship.

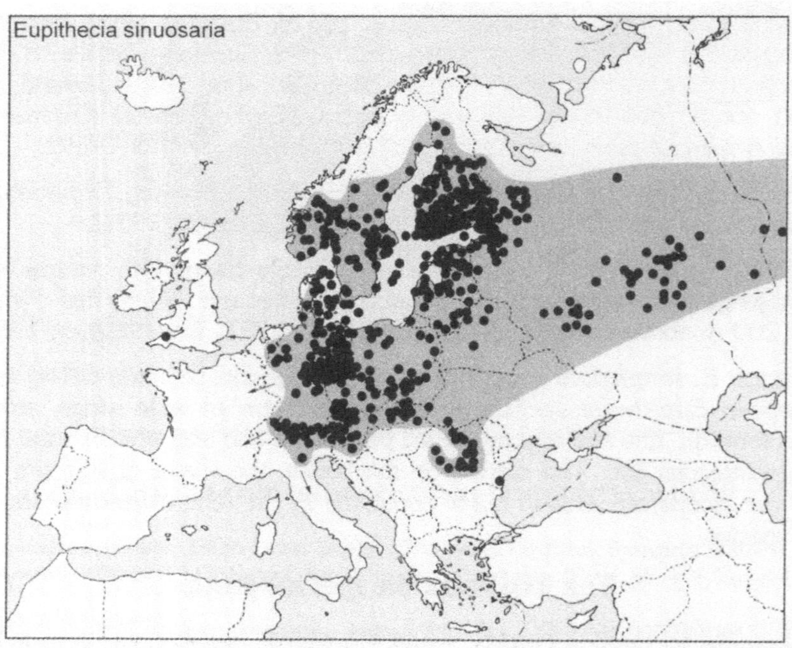

80. Eupithecia sinuosaria (EVERSMANN, 1848)

Larentia sinuosaria EVERSMANN, 1848: Bull. Soc. Imp. Nat. Moscou 21 (3): 230 (Russia: Irkutsk). Lectotype ♂, herewith designated to stabilize nomenclature (ZISP, examined).

Eupithecia sinuosaria tenella VOJNITS, 1976: Acta Zool. Acad. Sci. Hung. 22 (3-4): 417, fig. 2b (China: Prov. Shansi, Mien-shan). Holotype ♀ (ZFMK, examined). Valid at subspecific rank.

Unavailable names (incorrect subsequent spellings): *sinnosaria*: DIETZE (1913); *sinuoaria*: VALLE (1946); *sinnuosaria*: BIELEWICZ (1966); *sinousaria*: LASTUKHIN (1998).

DIAGNOSIS: Wingspan 18-26 mm. Labial palpi and frons brown. Vertex and notum paler, yellowish brown. Forewing with oblique transverse vein in areole; yellowish brown; costal margin rather ash-grey; basal area with usually pale brownish to black basal line and blackish grey, obscure to inconspicuous, thrice sharply angled postbasal line; antemedial line almost straight or slightly sinuate, right angled, black and widened near costa; medial line slightly curved and obtuse angled near discal spot, straight, black and widened near costa; basal part of medial area between antemedial and medial lines narrow, constricted in the middle, broadened towards costa, dark, reddish brown with black stalk of Cu; outside of medial area yellowish brown; postmedial line inconspicuous, sinuate and evenly curved; terminal area with large blackish brown apical blotch, sometimes with small similar tornal blotch and always with slightly sinuate pale wavy line; discal spot black, large, linear, oblique. Hindwing paler, brownish grey; dark transverse lines marked only on the posterior margin of wing; terminal area inconspicuous; discal spot usually conspicuous, rounded. Fringe pale brownish grey with yellowish tinge, spotted with brown at vein ends; paler on the hindwing. Abdomen reddish brown with narrow white ring on the first segment and usually with small dark brown or black lateral spots on each segment or with dark lateral stripes; first and second segment dark brown or black.

MALE GENITALIA: Vesica armed with one large and slim horn-like cornutus at apex and with one large, elongated, flat, basally curved, V-shaped cornutus at ductus ejaculatorius base. Sternum A8 with two apically diverging sclerotized rods.

FEMALE GENITALIA: Colliculum broad. Tergum A8 with shallow medial hollow in the posterior margin.

DISTRIBUTION: Euro-Siberian. East and Central Europe; westward to Netherlands, Germany, Switzerland and northern Italy and eastward to Urals. Gradually spreading to the west of Europe. – Outside Europe in western and southern Siberia (Burjatia, Dahuria, Amur region), Sakhalin, Mongolia, north-eastern China and Korea.

PHENOLOGY: Univoltine: mid-June (late May) to early August (late August). Larva: mid-July to late August (mid-September). Overwintering as pupa in flimsy cocoon in the ground, occasionally twice.

BIOLOGY: Larva oligophagous (1). Found on the leaves, flowers and seeds of *Chenopodium album, C. pratericola, C. hybridum, C. glaucum, C. rubrum, C. bonus-henricus, Atriplex patula, A. littoralis, A. laciniata, A. oblongifolia* (Chenopodiaceae) (Germany: DIETZE 1910, 1913; WEIGT 1990; WEGNER pers. comm.; Denmark: JUUL 1948; Finland: SEPPÄNEN 1970) and also on *Polygonum aviculare* (Polygonaceae) (Finland: SEPPÄNEN 1970).

HABITAT: Sunny roadsides, fields, gardens, dry and warm areas, along the coasts and also places with ruderal vegetation: cultivated areas, near farms, cattle-yards, old construction sites and rubbish tips. From 0 m up to about 2,000-2,060 m above sea-level in Alps (REZBANYAI-RESER 1989; REZBANYAI-RESER & al. 1998).

SIMILAR SPECIES: *E. sinuosaria* can usually be easily separated from other European *Eupithecia* species by the distinctive maculation and unique, narrow, reddish brown medial fascia on the forewing. The shape of sternum A8, the cornuti on the vesica in the male and female genitalia can be used to make exact determinations.

The ***innotata*** species-group

Group introduced by McDunnough (1949) with three species from North America: *E. gelidata gelidata* MÖSCHLER, 1860, *E. multistrigata* (HULST, 1896) and *E. perfusca* (HULST, 1898). Represented in Europe by the following species: *E. gelidata*, *E. nanata*, *E. innotata*, *E. ochridata*, *E. unedonata* and *E. chalikophila*. Numerous Central Asian species have been examined and included in the *innotata* group by DIETZE (1908) and VOJNITS (1982, 1988). *Eupithecia emittens* INOUE, 1996 from Pakistan can also be included in the *innotata* group. Three species of this group from North Africa are known: *E. marmaricata* (TURATI, 1922); *E. fioriata* SCHÜTZE, 1959; *E. tripolitaniata* SCHÜTZE, 1959; and also *E. incommoda* HERBULOT, 1983 from Ethiopia.

MALE GENITALIA: Uncus medium length, narrow, biapical. Papillae large, length nearly equal to uncus length. Valva with or without small spine-like or wrinkled ventral process; broadened medially; with nearly straight dorsal margin and narrowly rounded or pointed apex; sacculus lightly sclerotized. Vinculum short and broad. Aedeagus large, stout and long, length about 0.8-1.0 times valva length; about 3.0-4.0 times longer than medial width. Vesica armed with two small, horn-like cornuti; with large elongated patch of spinules; with one large, broad, folded and one smaller curved cornutus at ductus ejaculatorius base. Sternum A8 peg-like with base broadened, narrowing to apex; basal hollow shallow; apical hollow absent or very small; sternum length about 0.6-0.7 times valva length.

FEMALE GENITALIA: Bursa copulatrix pear-shaped; almost completely covered with small spines; ductus bursae usually sclerotized and striated on right side and membranous and striated on left side. Ductus seminalis medium length, evenly curved, broadly attached to middle of corpus bursae on dorsal side. Colliculum collar-like, short and narrow. Antrum large, lightly sclerotized with medial width usually greater than height. Anterior and posterior apophyses thin, short or medium length, not expanded at apices. Spurs of anterior apophyses well defined, slightly broadened at apices, but nevertheless relatively narrow. Tergum A8 almost square or rectangular with medial width about to 1.5 times longer than height. Papillae anales rounded, short and wide.

REMARKS: The species of the *innotata* group can usually be distinguished from other European species by the large papillae, the slight ventral process of valva and the characteristic cornuti on the vesica in the male genitalia. The female genitalia can be distinguished by the pear-shaped, weakly sclerotized bursa copulatrix which is covered by small spines at base, with longitudinal striations in ductus bursae and with broad sclerotized antrum. Both male and female genitalia of all species in this group appear very similar. Only the adult external features can usually be used for reliable determinations. Therefore, they have been separated by various authors into questionable species and subspecies, especially the Central Asian representatives of the *innotata* species-group.

81. *Eupithecia gelidata hyperboreata* STAUDINGER, 1861

Eupithecia hyperboreata STAUDINGER, 1861: Stett. Ent. Z. 22 (10-12): 400 (Norway: Bossekop). Syntypes ♂ ♀ (MNHU).

(Eupithecia) nanata f. *transversa* DIETZE, 1913: Biol. Eupithecien 2: 122; ibidem (1910), 1, pl. 85: 1001 (Germany: Berlin). Syntypes 2 (MNHU). Valid at subspecific rank.

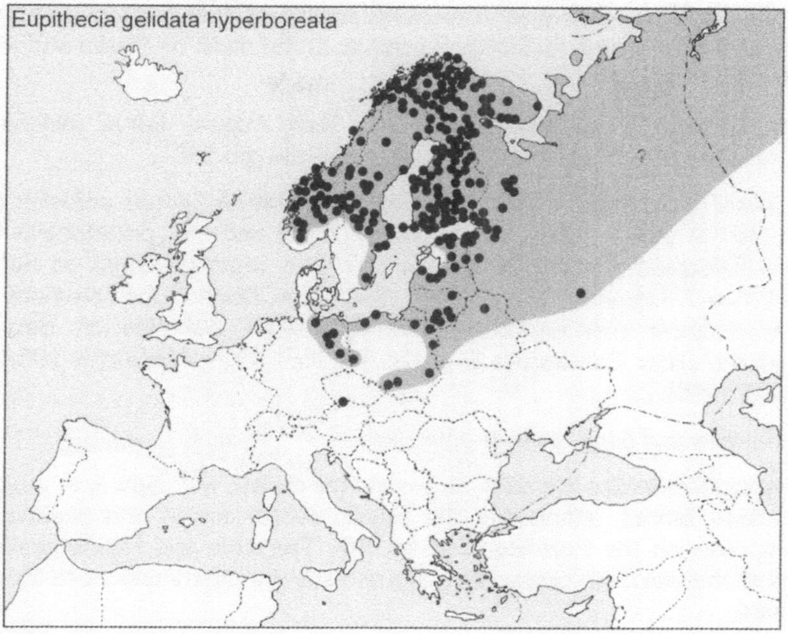

DIAGNOSIS: Wingspan 16-22 mm. Labial palpi grey. Frons and vertex greyish white. Antennae chequered. Notum ash-grey to dark grey. Forewing narrow, rather elongated and pointed; costa slightly arched; termen oblique and evenly curved; ground colour greyish fuscous, often mottled and irrorate with whitish scales; transverse lines inconspicuous, except ante- and postmedial; antemedial line sinuate, evenly curved or obtuse angled near costa; medial line right or sharply angled beyond discal spot; postmedial line multiply waved, obtuse angled near costa; whitish wavy line fine, sometimes broken or obsolete, forming more or less distinct, small, triangular white tornal spot; discal spot small, circular to narrow and elongate, black. Hindwing slightly paler; transverse lines usually distinct along dorsum; postmedial weakly defined, angled, with paler stripe beyond it; terminal area slightly darker; pale wavy line ill-defined, multiply indented, often broken; discal spot minute, circular to comma-shaped, sometimes obsolete. Fringe distinctly chequered greyish white and dark grey, especially on the hindwings. Abdomen ash-grey to dark grey, often with narrow black lateral stripes; second segment usually slightly darker than others.

MALE GENITALIA: See species-group diagnosis.

FEMALE GENITALIA: Tergum A8 rectangular, with small tooth in the middle of shallow hollow in posterior margin.

DISTRIBUTION: Subsp. *hyperboreata* is Euro-Siberian (arctic), but the species as a whole is Holarctic. Widespread in northern Europe. In Central Europe restricted to Poland, northeastern Germany and southern Czech Republic. One record from Durmitor in Montenegro from 1,350 m above sea-level (TOMIČ & al. 1990) doubtful. – Outside Europe in Greenland and northern Asia from Polar Urals through Siberia to Magadan province, Kamchatka peninsula and northern Kuril Islands. The nominate subsp. *gelida-*

ta MÖSCHLER, 1860 was described from North America (Okak, Labrador); it is widely distributed from Newfoundland to British Columbia, in the north to Alaska and in the south to New Mexico (BOLTE 1990).

PHENOLOGY: Univoltine: late May to early July (early August). Larva: mid-June to mid-August. Overwintering as pupa in flimsy cocoon in the ground.

BIOLOGY: Larva oligophagous (2). Found on the leaves of *Ledum palustre* (Ericaceae) (Germany: DIETZE 1910, 1913; WEIGT 1980, 1993) and also occasionally *Vaccinium uliginosum* (Ericaceae) (Finland: SEPPÄNEN 1970). In captivity reared on *Salix cinerea* (Salicaceae) (Czech Republic: SPITZER & al. 1991). The larvae of the nominate subspecies on *Ledum palustre* ssp. *groenlandicum* (Ericaceae) and also on *Betula*, *Alnus* (Betulaceae) and *Salix* (Salicaceae) (Canada: MACKAY 1951; MCGUFFIN 1958; PRENTICE 1963; BOLTE 1990).

HABITAT: Marshes and bogs, paludal pine forests.

SIMILAR SPECIES: Resembles the allied *E. nanata* but darker, with light lines much reduced, maculation less distinct, antemedial line usually evenly curved and postmedial obtuse angled near costa on the forewing (Text-fig. 50). The male and female genitalia typical for species of the *innotata* group; they are no clear-cut differences from those of other near relatives.

82. Eupithecia nanata (HÜBNER, 1813)

Geometra nanata HÜBNER, 1813: Samml. Eur. Schmett. 5 Geometrae (1), pl. 75: 387 (Europe). Syntype(s) lost.

Phalaena angustata HAWORTH, 1809: Lepid. Br. (2): 362 (United Kingdom: Westerham). Syntype(s) (BMNH).

(Eupithecia) nanata f. *nigrofasciata*: DIETZE, 1913: Biol. Eupithecien 2: 124; ibidem (1910), 1, pl. 74: 473 (Germany: Taurus, near Frankfurt a.M.). Syntypes 2 (MNHU). Valid at subspecific rank.

Eupithecia nanata zebrata WOLFF, 1929: in SPAERCK, The Zoology of the Faroes 39, 2 (1): 14, fig. 6 (Faroe Islands: Thorshavn (Strømø) and Klaksvig (Bordø). Syntypes 5 (ZMUC, examined). Valid at subspecific rank.

Eupithecia nanata angusta PROUT, 1938: in SEITZ, Macrolep. 4 (Suppl.): 201, pl. 17: k. Unnecessary replacement name for *angustata* HAWORTH, 1809. Valid at subspecific rank.

Eupithecia nanata gelidatoides WARNECKE, 1951: Flora og Fauna 57 (1/2): 43 (Iceland). Syntype(s) (ZMUH). Valid at subspecific rank.

Unavailable names (misidentification): *pauxillaria*: sensu STAUDINGER & REBEL (1901) nec BOISDUVAL, 1840 (identity: *Eupithecia pauxillaria*); *gelidata*: sensu LINDROTH (1928) nec MÖSCHLER, 1860 (identity: *Eupithecia gelidata*). – (infrasubspecific): *mediofaciata*: DIETZE (1910); *bistrigata*: LEMPKE (1951); *bicolor*: LEMPKE (1951); *obscura*: LEMPKE (1969); *unicolor*: LEMPKE (1969). - (incorrect subsequent spellings): *nanaria*: BOISDUVAL (1840); *gelatoides*: HARDY & KENNARD (1960).

DIAGNOSIS: Wingspan 15-23 mm. Labial palpi dark grey. Frons and vertex ash-grey. Notum ash-grey with brownish tinge, with dark frontal transverse stripe. Forewing narrow and pointed; termen slightly curved, oblique; ground colour whitish grey with brownish tinge, on the whole rather brindled; transverse lines distinct, very oblique, sharply angled near costa; basal, postbasal and antemedial lines much indented; medial area between antemedial line and discal spot usually lighter; medial line sharply curved beyond discal spot; postmedial line twice angled near costa; white faintly wavy line distinct, sometimes broken, relatively broad, thickened at tornus forming a conspicuous triangular spot; discal spot small, intensely black, from narrow to ovate or circular. Hindwing paler; dark transverse lines well marked along dorsum; light transverse medial stripe usually also distinct; terminal area broad, darker; whitish multiply wavy line usually wide, almost coin-

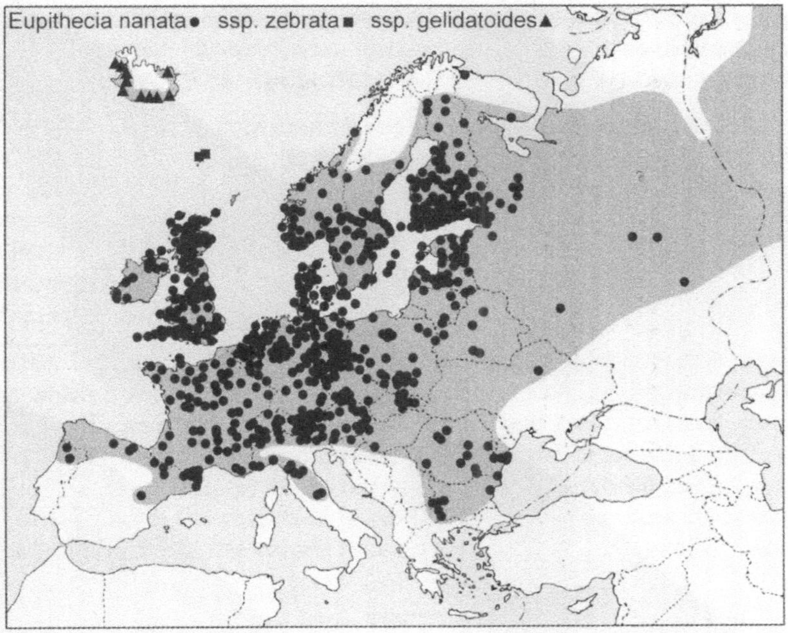

ciding with termen; discal spot minute, dark, usually circular. Fringe distinctly chequered smoky-white and dark brownish grey. Abdomen ash-grey with brownish tinge; lateral series of black dots present; second segment usually darker.

The subsp. ***zebrata*** WOLFF, 1929 is distinguished from nominate subspecies by the narrower and more pointed forewing, with broad whitish basal area between basal and antemedial lines, the more oblique postmedial line, the broader whitish area between postmedial and subterminal lines on both fore- and hindwings, and the broader whitish, not clear wavy lines in terminal areas of wings as well.

The subsp. ***gelidatoides*** WARNECKE, 1951 is distinguished by the clear whitish line between postmedial and subterminal lines and also the conspicuous whitish wavy line in terminal area on the forewing; the hindwing with definite white wavy line along terminal margin. The adults of this subspecies are rather similar to f. *angusta* from British Isles.

MALE GENITALIA: See species-group diagnosis.

FEMALE GENITALIA: Tergum A8 almost rectangular, with broadly rounded and prominent posterior corners and with small tooth in the middle of shallow hollow in posterior margin.

DISTRIBUTION: European. Nominate subspecies widely distributed in Europe except south of Iberian and Balkan peninsulas, Mediterranean islands and south of Ukraine and Russia. In the Faroe Islands replaced by subsp. *zebrata*, in Iceland by subsp. *gelidatoides*. Probably occurs eastward to Enisej river in central Siberia, which marks the eastern limit of *Calluna vulgaris*.

PHENOLOGY: Uni- or bivoltine: early June to mid-July (late July). In the south bivoltine: mid-April (early March) to late June (early July); late July (late June) to mid-August (mid-

Text-figs 50-53: Diagnostic characters (indicated) of European *Eupithecia* species. Text-fig. 50: *E. gelidata hyperboreata* STAUDINGER, 1861. Text-fig. 51: *E. nanata* (HÜBNER, 1813). Text-fig. 52: *E. innotata* (HUFNAGEL, 1767). Text-fig. 53: *E. ochridata* SCHÜTZE & PINKER, 1968.

October). Larva: early May to late July (early August); mid-August (late July) to mid-September (early October). Overwintering as pupa in flimsy cocoon in the ground.

BIOLOGY: Larva oligophagous (1). Found on the flowers and leaves of *Calluna vulgaris* (United Kingdom: CREWE 1861c; BUCKLER 1899; BARRETT 1904; France: MILLIÈRE 1870; MABILLE 1872; Germany: DIETZE 1910, 1913; WEIGT 1980, 1993; WEGNER pers. comm.; Denmark: JUUL 1948; Finland: SEPPÄNEN 1970), *Erica herbacea (=carnea)* (Alps: DIETZE 1874a, 1913) and *Erica tetralix* (Ericaceae) (Germany: WEGNER pers. comm.).

PARASITOIDS: Hymenoptera, Ichneumonidae: *Dysona juvenilis*, *Chorinaeus funebris* -

Hymenoptera, Braconidae: *Aleiodes modestus*, *Apanteles juniperatae*, *A. praepotens*, *Meteorus pulchricornis* - Hymenoptera, Sphecidae: *Ammophila pubescens* (THOMPSON 1945; SHENEFELT 1969, 1972, 1975; HERTING 1976; ZWAKHALS pers. comm.).

HABITAT: Heaths, moorlands, light pine forests and sandy edges of forests. The subsp. *gelidatoides* is found in dry localities, preferably on heaths and sea coasts, but even in perfectly biotopes with no plants (WOLFF 1971). From 0 m up to about 1,000 m above sealevel in Apennines (PROLA & RACHELI 1980), to about 2,000-2,200 m in Alps (VORBRODT & MÜLLER-RUTZ 1914; WOLFSBERGER 1950; 1955; BURMANN 1957; HELLMANN & al. 1999; REZBANYAI-RESER & MAIER 2000), and to about 2,400 m in Pyrenees (MAZEL & PESLIER 1997; coll. FIBIGER).

SIMILAR SPECIES: Usually smaller than *E. innotata* and especially than *E. ochridata*, two species with which it could be confused, with more distinctly marked forewing with darker medial fascia. *E. gelidata*, the other very similar and closely related species, has a cloudy and inconspicuous maculation, usually evenly curved antemedial and less angulated postmedial line near costa on the forewing (Text-fig. 51). No significant structural differences can be found in the male and female genitalia of *E. nanata* from others in the *innotata* species-group, yet the ventral process of valva is usually smaller than in *E. gelidata*.

REMARKS: Rather slightly variable species. The f. *angusta* PROUT, 1938 with very distinct markings appears occasionally throughout the range of the species. A melanic ab. *oliveri* PROUT, 1938 occurs in the Midlands and occasionally elsewhere in the British Isles.

83. *Eupithecia innotata* (HUFNAGEL, 1767)

Phalaena innotata HUFNAGEL, 1767: Berl. Mag. 4 (6): 616 (Germany: Berlin region). Syntype(s) lost.

Phalaena innotata THUNBERG, 1788: D. D. Mus. nat. Acad. Upsal. 6: 74 (Europe). Lectotype ♀ (DEUU). Junior primary homonym and junior synonym of *innotata* HUFNAGEL, 1767.

Larentia tamarisciata FREYER, 1836: Neuere Beitr. Schmettkde 2 (32): 153, pl. 192: 1a-e (Germany: Wolfratshausen). Syntype(s).

Eupithecia fraxinata CREWE, 1863: The Entomologist's Annual 1863: 116 (United Kingdom: England). Syntypes (BMNH?).

Eupithecia innotata f. *meridionalis* MABILLE, 1868: Annls Soc. ent. France (4) 9: 73 (Corsica). Syntype(s). Valid at subspecific rank.

Eupithecia suspectata DIETZE, 1871: Stett. Ent. Z. 32 (4-6): 210 (France: Paris; Germany: Warthe; Wiesbaden; Frankfurt a.M.). Syntypes (MNHU).

Eupithecia innotata var. *grisescens* PETERSEN, 1910: Dt. ent. Z. Iris (1909) 22 (4): 271 (Estland). Syntypes (ZMTU).

Eupithecia alexandriana WARDIKJAN, 1972: Dokl. Acad. Nauk Arm. SSR 55 (4): 253, figs 1, 2 (Armenia: Razdan region, Tzakhkadzor). Holotype ♂ (IZAR, examined; the slide of holotype included genitalia of 3 females, 5 armature of male genitalia, 3 aedeaguses and 1 sternum A8).

Eupithecia sergiana WARDIKJAN, 1972: Dokl. Acad. Nauk Arm. SSR 55 (4): 255, figs 3, 4 (Armenia: Erivan). Holotype ♀ (IZAR, examined).

Unavailable names (infrasubspecific): *rotundata*: BASTELBERGER (1908); *melanotica*: WARNECKE (1962). - (incorrect subsequent spellings): *innotaria*: BOISDUVAL (1840); *innonata*: KOSTROWICKI (1953).

DIAGNOSIS: Wingspan 18-25 mm. Labial palpi greyish fuscous. Frons and vertex ash-grey. Notum brownish grey with dark grey transverse band in front. Forewing narrow, elongate; costa almost straight, gently curved near apex; apex pointed; termen oblique; ground colour brownish grey; veins a little darker than ground colour; transverse lines inconspicuous, straight, oblique, sharply angled and slightly widened near costa; medial

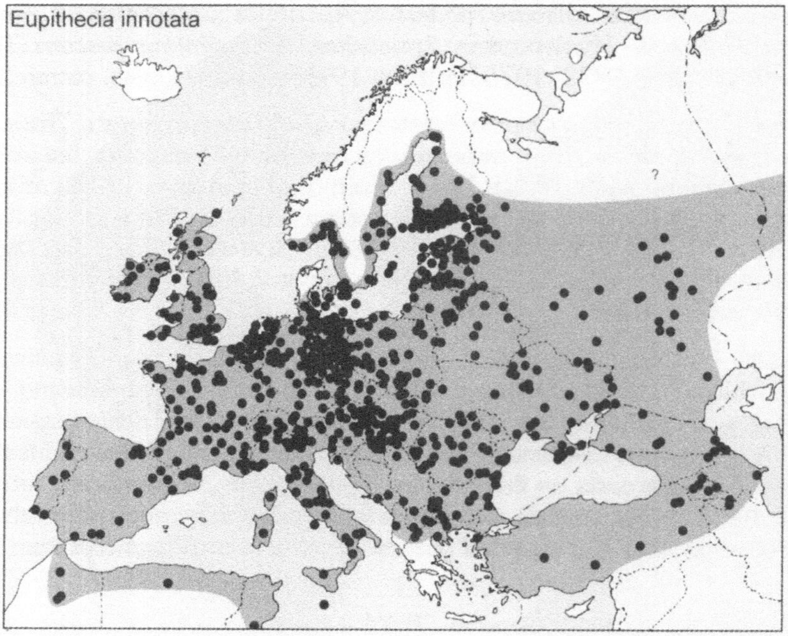

line sharply angulated beyond discal spot; postmedial line usually more evident, twice angled near costa (the first time sharply angled and the second time obtuse angled); whitish faintly wavy line fine, sometimes indistinct, broken and slightly broadened at tornus; lighter and narrow apical streak sometimes visible; discal spot narrow, obliquely elongated, intensely black but in some specimens more or less rounded. Hindwing paler than forewing; terminal area and dorsum darker than ground colour; transverse lines ill-defined; pale wavy line fine, multiply indented, usually indistinct, very close to terminal line; discal spot minute, circular. Fringe pale grey, spotted with grey at vein ends. Abdomen brownish grey with narrow black lateral stripes.

MALE GENITALIA: See species-group diagnosis.

FEMALE GENITALIA: Tergum A8 almost square.

DISTRIBUTION: West-Palaearctic. Widely distributed in Europe except Iceland, northern Scandinavia, Balearics, southern Balkan peninsula and Crete. – Outside Europe in North Africa (Morocco, Algeria and Tunisia), Asia Minor, Caucasus, Transcaucasus, Urals and Kazakhstan, eastward to south-western Siberia.

PHENOLOGY: Bivoltine: mid-April (early March) to late May (late June); mid-July (early July) to early September (mid-October). Larva: mid-June (late May) to early July (late August); September and October. Overwintering as pupa, sometimes for a few years.

BIOLOGY: Larva polyphagous (1). The first generation found on the leaves of *Prunus spinosa, Crataegus monogyna, Rosa canina* (Rosaceae), *Olea, Fraxinus excelsior* (Oleaceae), *Coriaria myrtifolia* (Coriariaceae), *Sambucus* (Caprifoliaceae), *Myricaria germanica*, *Tamarix gallica* (Tamaricaceae) and *Hippophae rhamnoides* (Elaeagnaceae) (United Kingdom: CREWE 1863, 1865; BUCKLER 1899; BARRETT 1904; France: MILLIÈRE

1870; MABILLE 1872; LHOMME 1935; Switzerland: PÜNGELER 1889; Germany and Austria: DIETZE 1871b, 1872, 1874a, 1875a, 1910, 1913; WEIGT 1993; Denmark: JUUL 1948); larva of the second generation on the flowers and seeds of *Artemisia campestris, A. absinthium, A. vulgaris, A. caerulescens, A. maritima, A. alba (=camphorata)* (Asteraceae) and *Suaeda maritima* (Chenopodiaceae) (United Kingdom: CREWE 1862, 1863; BUCKLER 1899; BARRETT 1904; Corsica: MABILLE 1868; France: MILLIÈRE 1870; MABILLE 1872; LHOMME 1935; Germany: DIETZE 1871b, 1910, 1913; WEIGT 1993; WEGNER pers. comm.; Finland: SEPPÄNEN 1970; Denmark: JUUL 1948; Russia: original data).

PARASITOIDS: Hymenoptera, Ichneumonidae: *Angitia* sp., *Mesochorus brevipetiolatus, Netelia thomsoni, Platylabus pedatorius* - Hymenoptera, Braconidae: *Apanteles tibialis (=congestus)* - Hymenoptera, Pteromalidae: *Habrocytus semotus* (hyperparasite) - Hymenoptera, Encyrtidae: *Litomastix truncatulus, Tyndarichus scaurus* - Diptera, Tachinidae: *Blondelia (=Lydella) nigripes, Phryxe vulgaris* (THOMPSON 1945; BAKKE 1955; HERTING 1960, 1976; SHENEFELT 1972).

HABITAT: In most types of biotopes, except moors and spruce forests. From 0 m up to about 1,000 m above sea-level in Pyrenees (MAZEL & PESLIER 1997), to about 1,300 m in Apennines (PROLA & RACHELI 1980), to about 1,500 m in mountains of Balkan peninsula (coll. FIBIGER), to about 1,800 m in southern Spain (coll. ZSM), and to about 2,000 m in Alps (WOLFSBERGER 1955; FORSTER & WOHLFAHRT 1981).

SIMILAR SPECIES: Often *E. innotata* can be difficult to separate from *E. ochridata*, however they do tend to be smaller and greyer than *E. ochridata* and the light medial areas and more or less distinct light apical streaks on the forewings are always absent in *E. innotata* (Text-fig. 52). The male and female genitalia very similar to others in the *innotata* species-group.

REMARKS: It is variable in size, depth of coloration and distinctness of markings. In some melanic individuals the wings are darker, of sootier tone and showing at least traces of transverse lines, and with black discal spots evident (f. *unicolor* PROUT, 1938).
In my opinion the widespread continental *E. innotata* and the British *E. fraxinata* are conspecific. There are no significant morphological differences between adults, caterpillars and host preferences of these populations.

84. *Eupithecia ochridata* SCHÜTZE & PINKER, 1968

Eupithecia ochridata SCHÜTZE & PINKER, 1968: in PINKER: Posebno Izd. prirodon. Muz. Skopje 4: 17 (Macedonia: Ochrid). Holotype ♂ (NHMW; 3 paratypes examined).
Eupithecia szelenyii VOJNITS, 1969: Acta Zool. Acad. Sci. Hung. 15 (3-4): 463, figs 1a-c, 3a (Hungary: Mátra Mts, Galyatetö). Holotype ♂ (HNHM, examined).

DIAGNOSIS: Wingspan 22-26 mm (first generation) and 18-21 mm (second generation). Labial palpi grey to greyish fuscous. Frons and vertex from greyish white to dark grey. Notum whitish grey with blackish grey transverse band in front. Forewing narrow, elongated; costa straight, slightly curved near apex; apex pointed; termen obliquely rounded; ground colour from grey to greyish fuscous with delicate ochreous suffusion; transverse lines oblique, rather indented; basal sharply angled near costa; postbasal and antemedial lines right-angled onto costa; medial area between angle of antemedial line and discal spot

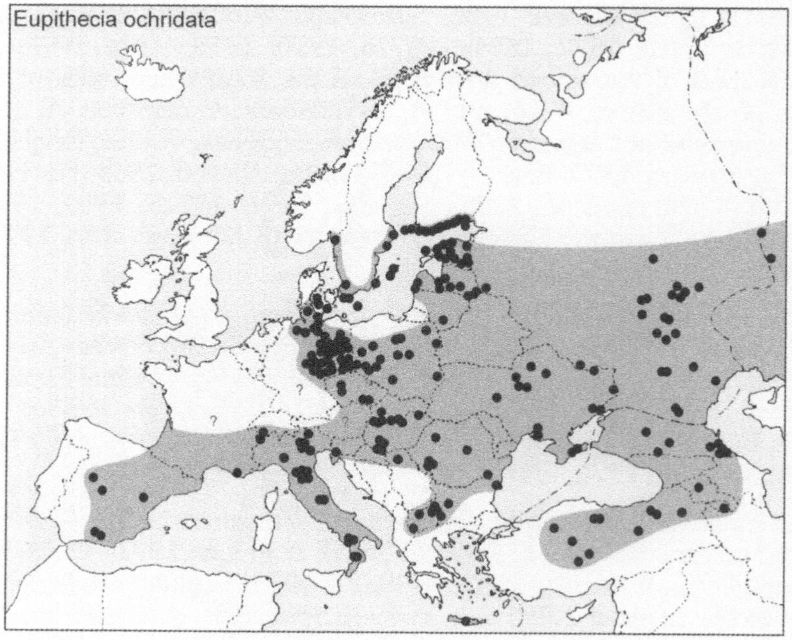

whitish grey; first medial line oblique, sharply angled beyond discal spot onto costa; postmedial line twice right-angled near costa; terminal area with light apical streak; whitish minutely wavy line sharply toothed near costa and forming a sharp inward angle near tornus; discal spot obliquely elongate to ovoid, intensely black. Hindwing ovate, with evenly rounded margins; paler, smoky-grey; transverse lines well marked along dorsum to stalk of Cu; terminal area grey with weak ochreous tinge, darker than the rest of wing; whitish wavy line fine, much indented, usually ill-defined; discal spot minute, circular or obsolete. Fringe greyish white, spotted with dark grey at vein ends. Abdomen pale grey to greyish fuscous, sometimes with brownish tinge; fine black lateral lines present.

MALE GENITALIA: See the *innotata* species-group diagnosis.

FEMALE GENITALIA: Tergum A8 rectangular, with shallow medial hollow in posterior margin.

DISTRIBUTION: West-Palaearctic. A generally southern species distributed from Spain to south-western Siberia and Tibet. Also, one specimen was reported from 'Amur' by VOJNITS (1982).

PHENOLOGY: Bivoltine: mid-May (late April) to early June (late June); late July to late September (early October). Larva: late June to late July; mid-September to mid-October. Overwintering as pupa.

BIOLOGY: Larva monophagous (2). Found on the leaves, flowers and seeds of *Artemisia alba (=camphorata)* (Macedonia: PINKER 1968), also *A. campestris* (Germany: GELBRECHT and WEGNER pers. comm.), *A. scoparia, A. schmidtiana* (Asteraceae) (Austria and Italy: WEIGT 1993).

HABITAT: Hedges, scrub and scrub-covered meadows, warm and dry slopes, especially on sandy soil, rocks and stony slopes. From 0 m up to about 1,200 m above sea-level in southern Alps (coll. ZSM), to about 1,300 m in Apennines (Sirino Mt.: coll. ZSM), and to about 1,500-1,600 m in southern Spain and Asia Minor (coll. ZSM; coll. FIBIGER).

SIMILAR SPECIES: The moths of the first generation are larger, their forewings usually wider, with weak ochreous tinge, lighter medial areas and more or less conspicuous light apical streaks, which are absent in *E. innotata*, the only species with which it could be easily confused (Text-fig. 53). The adults of the second generation have no significant external differences from those of *E. innotata* (KAILA 1989; MIKKOLA & al. 1989). No significant structural differences can be found in the male and female genitalia from those of other species in the *innotata* group. However, the ventral process of the valva is usually larger than in *E. innotata*.

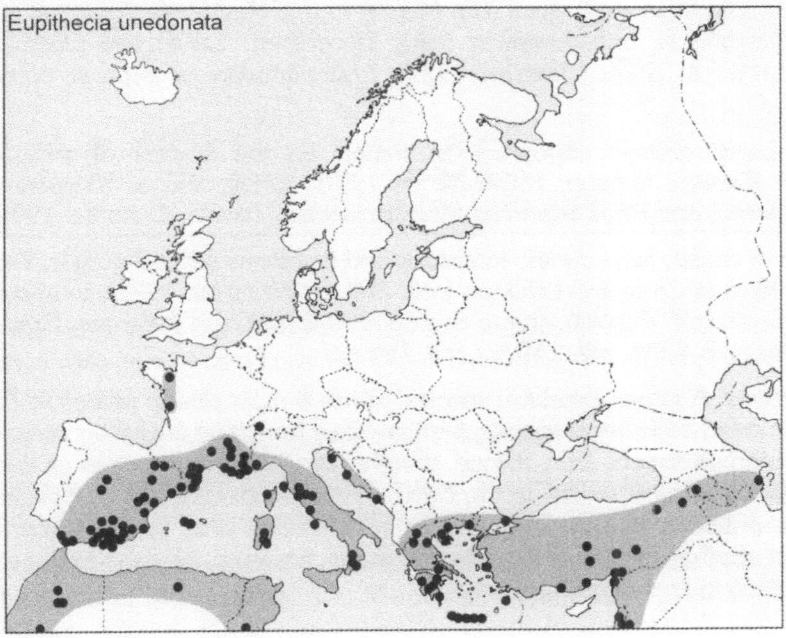

85. *Eupithecia unedonata* MABILLE, 1868

Eupithecia unedonata MABILLE, 1868: Annls Soc. ent. France 7 (4): 649, pl. 14: 3a-c (Corsica: Bastia; Porto-Vecchio). Syntype(s).
Unavailable names (infrasubspecific): *autumnalis*: DIETZE (1910); *hybrida*: DIETZE (1913).

DIAGNOSIS: Wingspan 17-24 mm. Labial palpi grey. Frons and vertex ash-grey. Notum ash-grey with dark grey transverse band in front. Forewing narrow, elongate, pointed; termen very oblique; ground colour light grey; sometimes the veins covered by black scales, especially stalk of Cu; transverse lines usually fine, conspicuous, oblique, straight, sharply angled near costa, dark grey; basal, postbasal and antemedial much indented; three medial lines fine, sharply angled beyond discal spot, parallel to and close to postmedial line; postmedial line twice sharply angled near costa; pale wavy line fine, unbroken, sharply toothed near apex and often slightly thickened at the tornus; sometimes a lighter oblique apical streak visible; discal spot narrow, elongate, oblique, sometimes ovoid, intensely black. Hindwing ovate; ground colour paler, smoky-white; numerous dark grey transverse lines well marked only along dorsum; pale transverse stripe between medial and terminal areas often evident; terminal area usually darker; discal speck minute, dark, circular. Fringe chequered greyish white and grey, especially distinct on the hindwing. Abdomen ash-grey to grey with slightly darker second segment.

MALE AND FEMALE GENITALIA: See diagnosis of species-group.

DISTRIBUTION: Mediterranean. Widely distributed in southern Europe from southern Spain to European Turkey, Crete and Rhodos Island. Recorded for northern Hungary (VOJNITS 1973) probably erroneously. – Outside Europe in North Africa (Morocco to Egypt), Lebanon and Israel (ANDERS 1913; ZERNY 1933; WILTSHIRE 1939, 1949; SCHÜTZE 1959; HAUSMANN 1997), Turkey and Transcaucasus.

PHENOLOGY: Bivoltine: early April (late March) to mid-May (early June); mid-September (early September) to mid-November (early December). Larva: late October to early December (DIETZE 1913; CHRÉTIEN 1917). Overwintering as pupa in cocoon in the ground.

BIOLOGY: Larva disjunct oligophagous. Found on the flowers of *Arbutus unedo* (Ericaceae) (Corsica: MABILLE 1868; DIETZE 1910, 1913), also on *Thymelaea hirsuta* (Thymelaeaceae) and *Rhus tripartita* (Anacardiaceae) (Tunisia: CHRÉTIEN 1917).

HABITAT: Sea coasts, sand dunes, sandstone and limestone hill and gorges. From 0 m to about 1,000 m in Crete and Lebanon (coll. ZMUC; WILTSHIRE 1939), to about 1,800 m in central Spain (coll. FIBIGER), and to about 1,900-2,000 m in Pyrenees, Great Atlas and Middle Atlas (coll. ZSM; coll. ZMUC; coll. FIBIGER).

SIMILAR SPECIES: A rather invariable species, which is most closely related to *E. innotata* among European *Eupithecia* species. Distinguished from it by the lighter colour and presence of numerous distinct, fine, oblique, straight, parallel transverse lines on the forewing. Its appeared to be most similar to the Central Asian *E. parallelaria*, but evidently smaller with less distinct markings. The genitalia in both sexes of *E. unedonata* are very similar to those of other species of the *innotata* group, however, the valva has poorly marked ventral projection in the male genitalia.

86. Eupithecia chalikophila WEHRLI, 1926

Eupithecia chalikophila WEHRLI, 1926: Schweiz. Ent. Anz. 5 (11): 2 (Spain: Andalusia, Sierra Nevada, Peñon di S.Francisco). Syntypes 14♂ 29 ♀ (ZFMK, examined).
Unavailable names (incorrect subsequent spelling): *chalckophila*: WEIDNER (1974).

DIAGNOSIS: Wingspan 16-23 mm. Labial palpi grey. Frons and vertex greyish white. Notum light grey. Forewing narrow, elongate and pointed; termen slightly rounded, oblique; ground colour whitish grey; veins often chequered dark and light grey, especially stalk of Cu; transverse lines fine, oblique, straight, rather inconspicuous, right or obtuse angled, broadened and darker near costa; the first medial line sharply angled beyond discal spot; postmedial line twice obtuse angled near costa; pale wavy line fine, inconspicuous, multiply indented but sharply toothed near apex of forewing; discal spot small, black, from circular to narrow and elongate. Hindwing paler; transverse lines indistinct, cloudy; pale medial stripe broad, more or less evident; terminal area darker with a series of dark

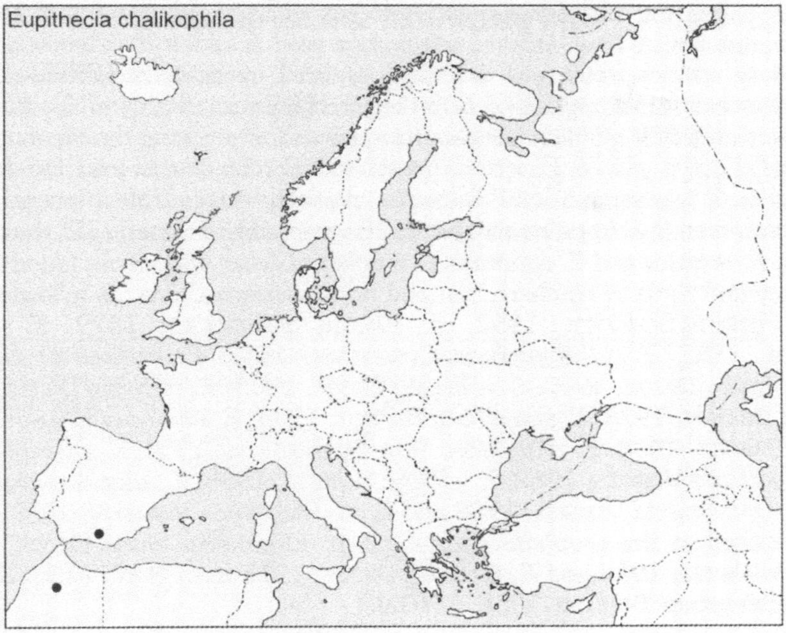

dashes on the veins; wavy line obsolete; discal spot minute, circular, dark. Fringe greyish white, spotted with grey at vein ends. Abdomen light grey with slightly darker, grey second segment.

MALE GENITALIA: See the *innotata* species-group diagnosis.

FEMALE GENITALIA: Tergum A8 almost square, with broadly rounded and prominent posterior corners and with shallow medial hollow in posterior margin.

DISTRIBUTION: West-Mediterranean. Southern Spain and North Africa (Middle Atlas). – New for the fauna of North Africa: 1♀, Marokko, M(oyen)-Atlas, Mischliffen, 2000 m, 29.07.1993, leg. STENGEL (ZSM).

PHENOLOGY: Univoltine: mid-June (late May) to mid-July (late July).

BIOLOGY: Unknown.

HABITAT: Unshaded calcareous rocks. From 1,500 m up to about 2,300 m above sea-level in Sierra Nevada Mts (WEHRLI 1927).

SIMILAR SPECIES: On the whole the most brindled and light-coloured species among the European representatives of the *innotata* species-group, which is only remotely similar to *E. unedonata*. Distinguished from it by the less distinct markings, the obtuse angled ante- and postmedial lines near the costa and usually smaller and more rounded discal spots on the forewings, and more distinctly chequered fringe. The male and female genitalia are typical for species of the *innotata* group, but appear to be distinguished by the more pointed ventral process of valva, and shorter and sparser brush-like patch of minute spines on the vesica in the male genitalia, and by the presence of medial patch of spines in corpus bursae in the female genitalia.

The *graphata* species-group

The *graphata* species-group was described in detail by SCHÜTZE (1958) as subgenus *Petersenia* including 12 species of *Eupithecia*. However, after carefully examining all these species and their genitalic preparations, as well as studying the literature on their descriptions, I have placed *E. spissilineata* and *E. cuculliaria* into the *breviculata* species-group. It is a large West-Palaearctic, probably Ancient Mediterranean species-group, represented in Europe by six species: *E. graphata*, *E. riparia*, *E. thurnerata*, *E. gemellata*, *E. mystica* and *E. cooptata*. A number of Asian species are found in the area from Turkey and Syria to southern Iran and north-western China, as following: *E. subsequaria* HERRICH-SCHÄFFER, 1852; *E. furcata* STAUDINGER, 1879; *E. saisanaria* STAUDINGER, 1882; *E. lithographata* CHRISTOPH, 1887; *E. brunneata* STAUDINGER, 1900; *E. novata* DIETZE, 1903; *E. relictata* DIETZE, 1903; *E. farinosa* DIETZE, 1913; *E. aradjouna* BRANDT, 1938; *E. convallata* BRANDT, 1938; *E. harenosa* BRANDT, 1938; *E. tesserata* BRANDT, 1938; *E. undulifera* SCHWINGENSCHUSS, 1939; *E. adjemica* BRANDT, 1941; *E. vaticina* VOJNITS, 1982; *E. elbursi* VOJNITS, 1988; *E. edaphopteryx* VOJNITS, 1988; *E. xanthomixta* VOJNITS, 1988 and *E. kopetdaghica* MIRONOV, 1989 which can also be included in the *graphata* species-group. *Eupithecia elissa* DIETZE, 1910; *E. desertorum* DIETZE, 1910 and *E. farinosa* DIETZE, 1913 from North Africa also belong to this species-group (SCHÜTZE 1958; 1961a).

MALE GENITALIA: Uncus short, narrow, biapical. Papillae usually short and broad, length about 0.4-0.8 times uncus length. Valva without ventral process, with evenly curved ventral margin, usually with dorsal margin convex medially and with very narrow apex; sacculus weakly sclerotized. Vinculum short and rather narrow. Aedeagus stout, large and broad, length about 0.8-0.9 times valva length; about 2.5-4.0 times longer than medial width. Vesica armed with elongated patch of spinules; without or with one or two weakly defined dentate cornuti and always with one elongated, flat, folded, U- or V-shaped cornutus at ductus ejaculatorius base. Sternum A8 peg-like, usually short; base broad with shallow to deep basal hollow; apical hollow shallow or not present; sternum length about 0.5-0.7 times valva length.

FEMALE GENITALIA: Bursa copulatrix usually elongated, pouch-like; 0.5-0.8 covered with uniformly sized spines at base and obliquely up left side; with or without row of more or less large spines between ductus seminalis base and colliculum on right side; ductus bursae more sclerotized with several or numerous longitudinal striations. Ductus seminalis of medium length, from narrow to broad, always evenly curved backwards; attached to middle part of corpus bursae on right side. Colliculum collar-like, usually short, narrow and inclined to right. Antrum usually funnel-shaped, sclerotized. Anterior and posterior apophyses short or very short, usually thin and not expanded at apices (or thick and broadened at apices in *E. wehrlii, E. cooptata* and *E. subsequaria*). The spurs of anterior apophyses broadest apically or narrow along their entire length. Tergum A8 broad, rectangular or trapeziform. Papillae anales usually large, broad and rounded (in some species, such as *E. wehrlii, E. cooptata* and *E. subsequaria*, they are large, elongated, pointed to apex and sclerotized).

REMARKS: The *graphata* species-group appears to be remotely related to the *satyrata* or possibly the *distinctaria* species-groups. The male genitalia can be distinguished from those of other European species by the valva with medially convex dorsal margin and very narrow apex and the vesica with its large, elongated, V-shaped cornutus and with numerous spinules; and the female genitalia can be distinguished by the corpus bursae which is diagonally covered with spines at base and on the left side, with ductus seminalis evenly curved backwards and sclerotized antrum.

I think that the *graphata*, *satyrata*, *russeliata*, *semigraphata* and allied smaller groups include many 'young' species, which are very variable, have many subspecies and colour forms and are closely related to each other. Moreover, these 'young' species are usually associated with herbaceous plants, especially with Asteraceae. Therefore, these species-groups have been placed at the end of the genus *Eupithecia*.

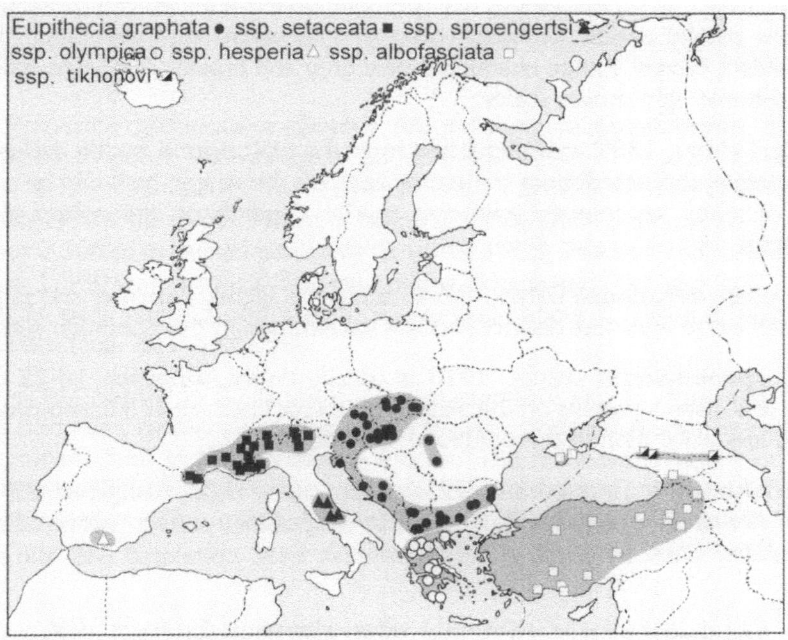

87. Eupithecia graphata (TREITSCHKE, 1828)

Cidaria graphata TREITSCHKE, 1828: Die Schmett. Eur. 6 (2): 144 (Hungary: (near Budapest), Ofen). Lectotype ♂ (HNHM, examined).

Acidalia corticulata FREYER, 1840: Neuere Beitr. Schmettkde 4 (54): 53, pl. 323: 1 (Slovenia: Laybach [=Ljubljana]). Syntype(s).

Eupithecia mayeri MANN, 1852: Verh. zool.-bot. Ver. Wien 2 (Abh.): 134 (Austria Vienna district). Syntype(s) (NHMW?).

Eupithecia albofasciata STAUDINGER, 1879: Horae Soc. Ent. Ross. 14: 479 (Turkey: Amasia). Holotype ♀ (MNHU).

Eupithecia graphata var. *setaceata* DIETZE, 1904: Dt. ent. Z. Iris 16 (2): 374, pl. 4: 33 (France: Digne?). Syntype(s) (MNHU). Valid at subspecific rank.

Eup(ithecia) graphata f. *sproengertsi* DIETZE, 1910: Biol. Eupithecien 1, pl. 77: 636; ibidem (1913), 2: 114 (Italy: Roccoraso). Syntype(s) (MNHU). Valid at subspecific rank.

Eupithecia graphata f. *hesperia* WEHRLI, 1927: Int. Ent. Z. 20 (38): 349 (Spain: Sierra Nevada). Holotype ♀ (ZFMK, examined). Valid at subspecific rank.

Eupithecia olympica TULESCHKOV, 1951: Bull. Inst. Zool. Acad. Bulg. Sci. 1: 327, fig. 3 (Greece: Mt Olympe). Lectotype ♂, herewith designated for stabilize nomenclature (NMNH, Sofia, examined). New junior synonym.

Eupithecia graphata graeciata SCHÜTZE, 1958: Ent. Z., Frankf. a.M. 68 (5): 54, pl. 1: 31 (Greece: Peloponnes, Achaia Mons, Mt. Chelmos). Holotype ♀ (ZSM, examined). Valid at subspecific rank.
Unavailable names (incorrect subsequent spellings): *grapharia*: BOISDUVAL (1840); *mayerata*: GUENÉE, 1858.

DIAGNOSIS: Wingspan 15-21 mm. Labial palpi mainly light brown. Frons and vertex covered with mixture of light brown and pale grey scales. Antennae ringed pale grey and light brown. Notum grey with numerous light brown scales. Forewing on the whole brightly coloured, pale grey or pale ash-grey; all transverse lines dark grey or brownish grey, slightly sinuous, usually prominent, fine-dentated and weakly angled near costa; postmedial line light grey, narrow and fine-dentated; wavy line inconspicuous; discal dot minute, rounded, dark brown. Hindwing pale grey, rather brightly coloured; all transverse lines greyish brown, indistinct except whitish, definite and fine-dentated postmedial line; terminal area usually darker, brownish grey with inconspicuous wavy line; discal dot minute, rounded, brown. Fringe chequered pale grey and brown. Abdomen pale grey or grey with scattered light brown scales.

The f. *mayeri* MANN, 1852 is distinguished from the typical form by the darker, brownish ground colour and less distinct transverse lines on the wings. It should be noted that the colour of many saxicolous *Eupithecia* species depends on the colour of the surrounding rocks.

The alpine subsp. ***setaceata*** DIETZE, 1904 usually has slightly narrower and more pointed wings, transverse lines less distinct and less waved than in the nominate subspecies.

The subsp. ***sproengertsi*** DIETZE, 1910 is usually larger (wingspan 18-22 mm) and lighter, with less distinct pattern on the wings; transverse lines rather inconspicuous, paler, clouded and less wavy than in the nominate subspecies.

The subsp. ***olympica*** TULESCHKOV, 1951 (**stat. n.**) is larger (wingspan 22-26 mm); wings more elongated, pointed and narrower in proportion; pattern less brindled on the whole; transverse lines indistinct, especially whitish lines, compared with the nominate subspecies.

The subsp. ***hesperia*** WEHRLI, 1927 has wings narrower and more elongated in proportion; forewing with more obliquely rounded termen, antemedial and first medial lines more evenly curved, postmedial less waved; hindwing with less waved inner margin of terminal area than in nominate subspecies.

The subsp. ***albofasciata*** STAUDINGER, 1879 is variable and the smallest subspecies (wingspan 15-18 mm). The wings are narrower, forewing with more pointed apex, transverse lines usually less distinct than in nominate subspecies.

MALE GENITALIA: See species-group diagnosis. The shapes of valva and sternum A8 diverse (sternum often with small apical lobes) (SCHÜTZE 1958).

FEMALE GENITALIA: Ductus bursae with a row of spines on right side; spiniferous area reaching to colliculum.

DISTRIBUTION: Sub-Mediterranean. Nominate subspecies in mountains of Austria, Hungary, Czech Republic, Romania and Balkan peninsula. In the Alps (southern France, northern Italy and Switzerland) replaced by subsp. *setaceata*, in Apennines (Italy) by subsp. *sproengertsi*, in the south of Balkan peninsula (Greece) by subsp. *olympica*, in southern Spain (Andalusia) by subsp. *hesperia* and in southern Crimea (Ukraine) by subsp. *albofasciata*. – Outside Europe in Asia Minor and Transcaucasus (subsp. *albofasciata*), and northern Caucasus (subsp. *tikhonovi* MIRONOV, 2001).

PHENOLOGY: Probably bivoltine: mid-May (late April) to late July (mid-August). Larva: mid-July (early July) to late August (early September). Overwintering as pupa, in flimsy cocoon in the ground, occasionally twice.

BIOLOGY: Larva oligophagous (1). Found on the flowers and seeds of *Minuartia* (=*Alsine*) *verna* (France: MILLIÈRE 1870), *M. setacea, Gypsophila repens* (France and Switzerland: DIETZE 1910, 1913), *Minuartia austriaca* (Austria: DIETZE 1910, 1913), *M. mucronata* (Switzerland: VORBRODT 1928), *M. mutabilis, M. capillacea, M. rupestris, M. recurva* (Caryophyllaceae) (France and Italy: WEIGT 1991) and *Drypis spinosa* (Caryophyllaceae) (Croatia: STADIE pers. comm.).

HABITAT: Preferring calcareous or siliceous, dry and warm mountain slopes, steppe slopes and rocks. From about 900 m up to about 2,000-2,100 m above sea-level in Alps (VORBRODT 1928; coll. ZMUC; TAUTEL pers. comm.), from 300 m up to about 2,300 m in Apennines (SCHÜTZE 1958; PROLA & RACHELI 1980; ZAHM pers. comm.), from 500 m up to about 2,300 m in Balkan peninsula (SCHÜTZE 1958; PINKER 1968; coll. ZMUC; coll. ZSM), from 1,500 up to about 2,300 m in Sierra Nevada and Pyrenees (coll. ZMUC; coll. ZSM; TAUTEL pers. comm.), and from 1,300 m up to about 2,700 m in Asia Minor and Transcaucasus (coll. ZISP and ZSM).

SIMILAR SPECIES: Similar to *E. satyrata, E. semigraphata* and *E. impurata* but smaller; the wings narrower, usually paler; discal dots smaller; postmedial lines fine-dentated and more prominent than in the compared species. The male genitalia are on the whole smaller; vesica without dentate cornuti; sternum A8 smaller and of different shape, as well as bursa copulatrix smaller and of different shape to those of similar species.

88. Eupithecia riparia HERRICH-SCHÄFFER, 1851

Eupithecia riparia HERRICH-SCHÄFFER, 1851: Syst. Bearb. Schmett. Eur. 3 (Suppl.), pl. 85: 518, 519 (non binominal); ibidem (1852), 6 (55): 76 (Croatia: Fjume [Rijeka]). Holotype lost.
Eupithecia drupisaria PETERSEN, 1910: Dt. ent. Z. Iris (1909) 22 (4): 261, pl. 18: 74 (Italy: Majella Mts). Syntypes 1♂1♀.
Unavailable names (incorrect subsequent spelling): *drypidaria*: DIETZE (1910).

DIAGNOSIS: Wingspan 14-21 mm. Labial palpi, frons and vertex smoky white. Notum greyish white. Forewing narrow and elongate, costa straight, apex pointed, termen obliquely rounded; ground colour pale, greyish white; transverse lines inconspicuous, slightly sinuate, broader, darker and not sharply angled near costa; medial line weak, crossing discal spot, right or obtuse angled onto costa; postmedial line minutely wavy, usually forming a series of small and faint dark dashes on the veins; terminal area narrow, sometimes gently darker, with trace of pale wavy line; discal spot small, rounded or ovoid, relatively pale. Hindwing ovate, termen slightly concave medially; ground colour paler, smoky-white; transverse lines indistinct, only postmedial line more or less conspicuous; terminal area narrow, sometimes slightly darker, with wavy inner margin, wavy line not visible; discal spot small, circular to elongate, comma-shaped, relatively pale. Fringe

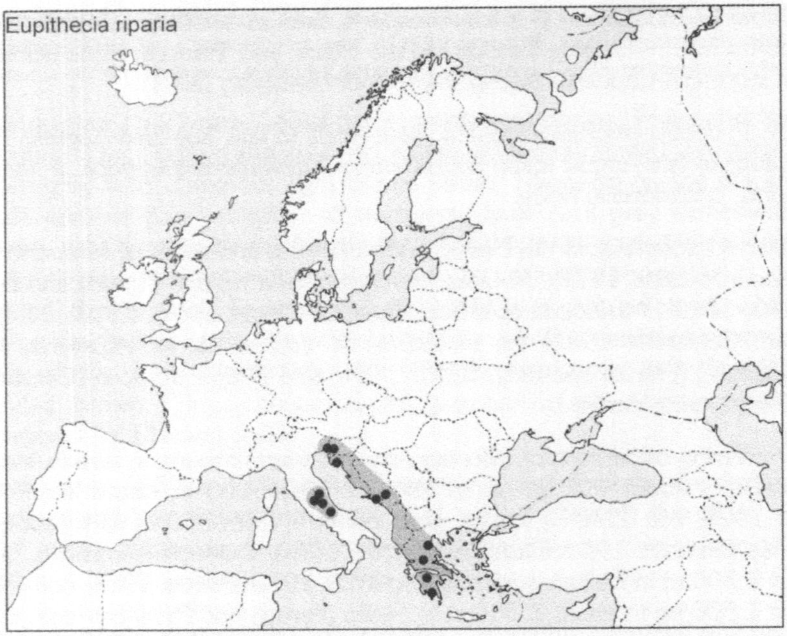

white, delicately spotted with pale grey at vein ends. Abdomen greyish white, sometimes with very narrow, short, black lateral stripes.

MALE GENITALIA: See species-group diagnosis.

FEMALE GENITALIA: Ductus bursae with a row of spines on right side; spiniferous area not reaching colliculum.

DISTRIBUTION: East-Mediterranean. Appenines and mountains of Balkan peninsula.

PHENOLOGY: Bivoltine: late May (mid-May) to mid-July (early August); September. Larva: late August to early September (late September). Overwintering as pupa.

BIOLOGY: Larva monophagous (1). Recorded on *Drypis spinosa* (Caryophyllaceae) (Italy: DIETZE 1910, 1913; SCHÜTZE 1958).

HABITAT: Warm exposed extensive pastures with cliffs and exposed rocks (GOMBOC pers. comm.). From 1,400 m up to about 2,300 m above sea-level in Apennines (SCHÜTZE 1958), and to about 2,500 m in Greece (coll. HNHM).

SIMILAR SPECIES: Closely related and similar to *E. graphata*, especially to *E. graphata setaceata*. Distinguished from it by the smaller size, the slightly paler, whitish colour and the less brindled wings, especially hindwings, the less dark pattern, the less wavy transverse lines and usually smaller discal spots. The genitalia of both male and female of *E. riparia* show no morphological differences from those of *E. graphata* specimens.

REMARKS: A little variable species in the depth of colour and distinctness of markings. No distinct differences in size, wing coloration and maculation, and structure of genitalia between adults of Balkan and Apennine populations, which was described as *E. drupisaria* PETERSEN and reported as subspecies of *E. riparia* later (SCHÜTZE 1958).

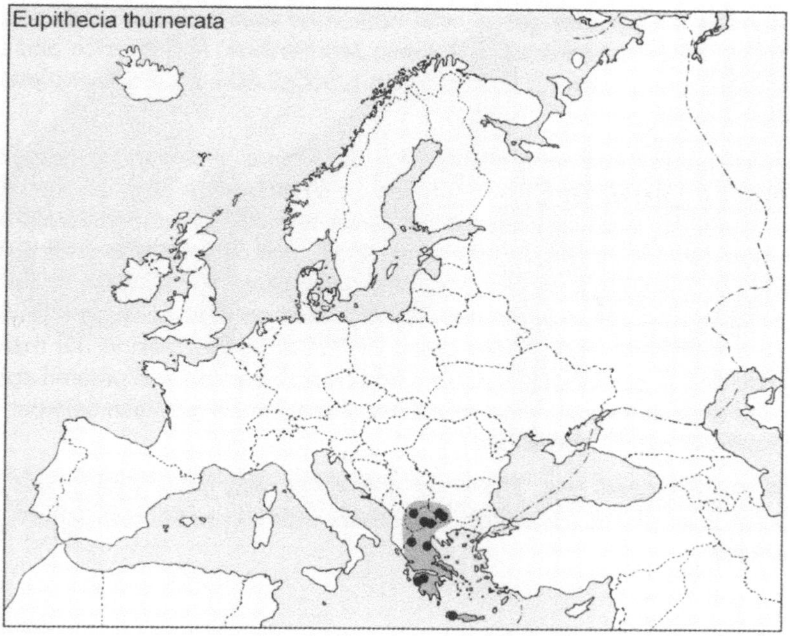

89. Eupithecia thurnerata SCHÜTZE, 1958

Eupithecia thurnerata SCHÜTZE, 1958: Ent. Z., Frankf. a.M. 68 (5): 62, pl. 1: 21-30 (Macedonia: Stari-Dojran, Drenovo near Kavadar). Holotype ♂ (ZSM; examined).

DIAGNOSIS: Wingspan 15-20 mm. Labial palpi grey. Frons, vertex and notum greyish white. Forewing grey; transverse lines numerous, fine, rather indistinct, dark grey or brownish grey; basal, postbasal and antemedial lines oblique, sharply angled onto costa; the first medial line less oblique, sinuate, touching discal spot and right or obtuse angled near costa; postmedial line minutely wavy, slightly sinuous, obtuse angled onto costa; terminal area brownish grey, sometimes mottled and irrorate with whitish scales at apex; pale grey wavy line usually distinct, fine, much indented, not forming a pale tornal spot; discal spot minute, circular to ovoid, but sometimes obsolete. Hindwing of the same colour or a little paler than forewing; grey with slightly darker terminal area, more or less weak indications of transverse lines; discal spot minute, rounded, dark. Fringe greyish white, spotted with grey at vein ends. Abdomen pale grey.

MALE GENITALIA: See species-group diagnosis.

FEMALE GENITALIA: Ductus bursae has an oblique row of minute spines on right side; spiniferous area reaching colliculum; marginal and posterior spines longer and more sclerotized than others.

DISTRIBUTION: East-Mediterranean. South of Balkan peninsula (south-western Bulgaria, Macedonia, Greece and Crete).

PHENOLOGY: Univoltine or probably bivoltine: mid-May (early May) to mid-June (late July).

BIOLOGY: Unknown.

HABITAT: Found in a deep river gorge, with numerous stones on the sandy soil, and with growing trees of *Platanus orientalis*, *Pistacia terebinthus*, *Ficus carica* and *Juniperus excelsa* (BESHKOV pers. comm.). Up to about 1,500-1,600 m above sea-level in Crete (coll. FIBIGER).

SIMILAR SPECIES: Similar to *E. graphata* and *E. gemellata*. However, *E. graphata* is to a considerable extent lighter and more brindled, with transverse lines on the wings more distinct and wavy. The other similar species *E. gemellata* has a more curved postmedial line on the forewing, larger discal spots on all wings, and can easily be distinguished from *E. thurnerata* by the distinct pattern and larger, elongated discal spots on the underside of all wings. Valva of *E. thurnerata* usually with more pointed apex; vesica with smaller elongated U-shaped plate and smaller group of minute spines than in the male genitalia of *E. graphata* and *E. gemellata*. Sternum A8 has pronounced and pointed apical horns. No differences were apparent in the structures of the female genitalia between *E. thurnerata*, *E. graphata* and *E. gemellata*.

REMARKS: A variable small pug. There are specimens with more distinct and dark ante- and postmedial lines on the forewings, and others with inconspicuous pattern, especially on the hindwings.

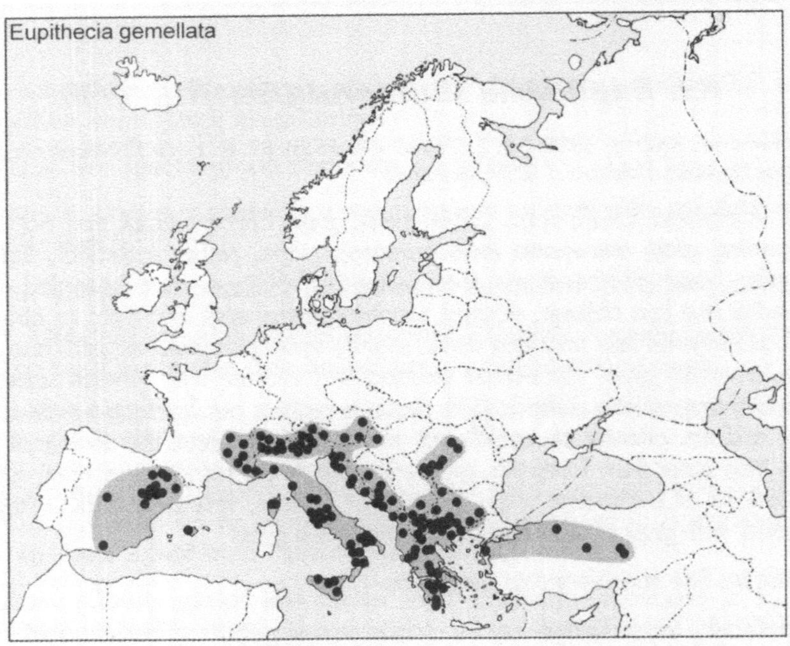

90. *Eupithecia gemellata* HERRICH-SCHÄFFER, 1861

Eupithecia gemellata HERRICH-SCHÄFFER, 1861: Corresp.-Bl. Samml. Insect., insbesondere Schmett. 2 (17): 131 (Switzerland: Engadin; Austria: Tyrol). Syntypes lost.

Eupithecia schmidii DIETZE, 1872: Stett. Ent. Z. 33 (4-6): 190, pl. 1: 2 (Italy: Tyrol, Botzen [=Bozen]). Syntype(s) (MNHU).

Unavailable names (incorrect subsequent spellings): *gemmelata*: GANEV (1984); *gemellatata*: WEIGT (1991).

Text-figs 54-56: Diagnostic characters (indicated) of European *Eupithecia* species. Text-fig. 54: *E. gemellata* HERRICH-SCHÄFFER, 1861. Text-fig. 55: *E. mystica* DIETZE, 1910. Text-fig. 56: *E. cooptata* DIETZE, 1904.

DIAGNOSIS: Wingspan 14-20 mm. Labial palpi brownish grey. Frons and vertex greyish white. Notum pale brownish grey. Forewing pale brownish grey; basal, postbasal and antemedial lines gently sinuate, oblique, sharply angled onto costa; the first medial line much sinuate, crossing discal spot and afterwards right-angled near costa; two other medial lines usually fine, well defined, parallel with first medial; postmedial line much sinuate, minutely wavy, evenly curved onto costa; terminal area often with weak pale apical and medial blotches; pale wavy line usually fine, much indented, tornal spot lacking; discal spot small, ovate or triangular, black. Hindwing slightly paler; transverse lines minutely wavy, usually well defined or evident along dorsum; terminal area slightly darker with wavy inner margin; pale wavy line weak; discal spot minute, dark, circular, ovate or even comma-shaped. Fringe greyish white, spotted with brownish grey at vein ends. Abdomen pale brownish grey, sometimes with a row of small dark dorsal tufts; second segment slightly darker, brownish grey.

MALE GENITALIA: Papillae large, length nearly equal to length of uncus. Vesica with longitudinal striations and without dentate cornuti. Sternum A8 small; its basal hollow shallow and apical hollow not present.

FEMALE GENITALIA: Bursa copulatrix with basal half completely covered by spines, with a patch of spinules near colliculum and also with a row of small spines between colliculum and ductus seminalis base. Tergum A8 rectangular with lightly sclerotized, shallow medial hollow in posterior margin.

DISTRIBUTION: Mediterranean. Spain, Balearics, south-eastern France, Corsica, Sicily, Italy, Switzerland, Austria, Hungary, Romania, former Yugoslavia, Bulgaria, Albania and Greece. – Outside Europe in Asian Turkey.

PHENOLOGY: Bi- or probably polyvoltine: early May (early April) to mid-June (mid-July); late July (early July) to late August (mid-September). In southern Spain, Italy and Greece probably a third generation from late September until late October. Larva: mid-June to mid-July (late July); late September (mid-August) to mid-October (early November). Overwintering as pupa.

BIOLOGY: Larva monophagous (2). Found on the flowers and seed capsules of *Petrorhagia (=Tunica) saxifraga* (Caryophyllaceae) (Italy: DIETZE 1872; Austria: DIETZE 1910, 1913; France and Austria: WEIGT 1991) and *Gypsophila struthium* (Caryophyllaceae) (Spain: KING pers. comm.).

HABITAT: Rocks and screes. From 0 m up to about 1,700 m above sea-level in Alps (VORBRODT 1928), to about 2,000 m in Apennines and Balkan peninsula (coll. ZSM), and to about 2,200 m in Peloponnese (THURNER 1967).

SIMILAR SPECIES: Similar to two following species, however, distinguished by the fine detail of the wing pattern (Text-fig. 54). The male and female genitalia are typical for the *graphata* species-group. For the external and genital differentiation of this species see below under the descriptions of *E. cooptata* and *E. wehrlii* and in the publication of SCHÜTZE (1958).

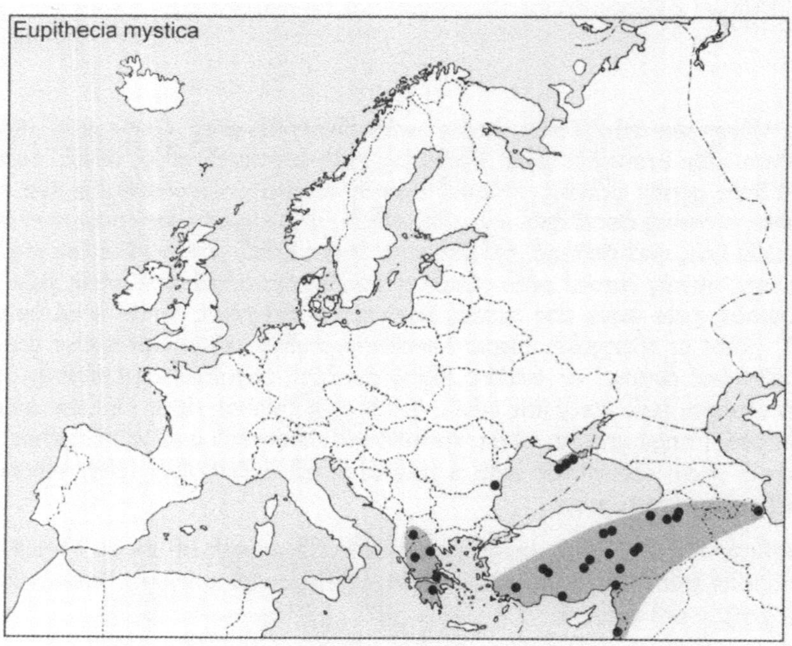

91. *Eupithecia mystica* DIETZE, 1910

Eup(ithecia) gemellata f. *mystica* DIETZE, 1910: Biol. Eupithecien 1, pl. 76: 617; ibidem (1913), 2: 112 (Syria: Akbès). Holotype ♂ (MNHU; examined). Raised to species rank by SCOBLE (1999).
Eupithecia wehrlii WAGNER, 1931: Intern. Ent. Z. (Guben) 25: 369 (Turkey: Marasch). Syntype(s) (NHMW). New junior synonym.

DIAGNOSIS: Wingspan 17-22 mm. Labial palpi brownish grey. Frons, vertex and notum ash-grey with delicate brownish suffusion. Forewing rather narrow and pointed with straight costa and obliquely rounded termen; brownish grey ground colour; basal, post-basal and antemedial lines oblique, slightly sinuate, sharply angled onto costa; the first medial line touching antemedial on the stalk of Cu, where a small, dark, blackish grey or brownish grey spot is usually formed; ante- and postmedial lines always broader and darker near costa; postmedial line much sinuate, obtuse angled near costa, often with tip of angle pointed outward; terminal area narrow; pale wavy line fine, minutely indented, rather indistinct, without pale tornal spot; discal spot large, ovate or ovate-oblong, intensely black. Hindwing paler, smoky-grey, with inconspicuous transverse lines; terminal area slightly darker; pale wavy line much indented, ill-defined; discal spot large, ovate, or triangular, or sometimes comma-shaped, intensely black. Fringe greyish white, distinctly spotted with brownish grey at vein ends. Abdomen pale brownish grey with narrow black band between first and second segments and with black lateral stripes; sometimes with a row of small dark dorsal dots.

MALE GENITALIA: Aedeagus large, length 1.5 mm or slightly longer. Vesica without dentate cornuti, with elongated V-shaped cornutus. Sternum A8 with slightly convex lateral margins; its basal hollow deep, apical hollow very shallow.

FEMALE GENITALIA: Bursa copulatrix elongated, rather pear-shaped, with larger marginal spines in spiniferous area; without spines between ductus seminalis base and colliculum. Colliculum straight, not inclined. Anterior and posterior apophyses short, thick and broadened at apices. Spurs of anterior apophyses narrow along their entire length. Tergum A8 with convex lateral margins, rounded posterior angles and with broad and shallow medial hollow in the posterior margin. Papillae anales heavily sclerotized, elongated and pointed.

DISTRIBUTION: East-Mediterranean. Romania (Dobrogea), Macedonia, Greece, Ukraine (southern Crimea). – Outside Europe in Asia Minor and Lebanon, eastward to Azerbaijan (Talysh Mts). – New for the faunas of Romania, Macedonia and Greece: 1♀, Romania, Dobrogea, Hagieni, 18.VIII.1983, leg. L. SZEKELY (HNHM); 1♂2♀, Macedonia, Ohrid, 20.VIII-8.IX.1936, leg. WOLFSCHLÄGER (ZSM); 1♂1♀, Greece, Joaninnia-Mitsikeli, 1000 m, 10.VIII.1977, leg. GROSS (ZSM); 1♂, Graecia, Parnassos, ndl. Arakhova, 1900 m, 24.VII.1984 leg. M. u. E. ARENBERGER (ZSM); 1♂2♀, Griechenland, Parnassos Oros, Umg. Arachova, 4.VIII.1985, Li, 800-1100 m, leg. G. BAISCH (ZSM).

PHENOLOGY: Univoltine: early August (late July) to mid-September (early October). Reported for southern Turkey as bivoltine: June; August and September (WEHRLI, 1934).

BIOLOGY: Unknown.

HABITAT: Mediterranean thorn forests, dry and light scrub, cliffs, rocks and mountain slopes. From 0 m up to about 1,900 m above sea-level in Greece (coll. GELBRECHT), and to about 1,700-2,000 m in Asia Minor (coll. ZSM, coll. HNHM; coll. FIBIGER).

SIMILAR SPECIES: Rather variable only in size; it is similar to *E. gemellata*, the only species with which it could really be confused. Distinguished from *E. gemellata* by the usually larger size, the relatively more elongated and pointed wings; the less distinct transverse lines, except ante- and postmedial, the more sharply angled postmedial line on the forewing, the larger discal spots on all wings, and by the presence of a narrow black transverse band between first and second abdominal segments and also the black longitudinal lateral stripes (Text-fig. 55). On the whole the male and female genitalia are larger; valva broader medially and narrower at apex, with convex costa, sternum A8 of male with convex lateral margins; and ductus bursae without rows of small spines, apophyses broadened to apexes, papillae anales narrower, more elongated, acute and sclerotized in the female genitalia than those in *E. gemellata*. The elongate and narrow tuft of scales at the end of the abdomen in the male and the sharply pointed papillae anales, projecting beyond abdomen in female can usually be used to distinguish *E. mystica* successfully from other European *Eupithecia* species.

REMARKS: According to the 'International Code of Zoological Nomenclature' (article 45g) the name *Eupithecia gemellata* f. *mystica* DIETZE, 1910 is available, and it has subspecific (not infrasubspecific) rank. *Eupithecia mystica* VOJNITS, 1988 (*innotata* species-group) from the Pamir Mountains is a primary homonym of this name (see also SCOBLE 1999).

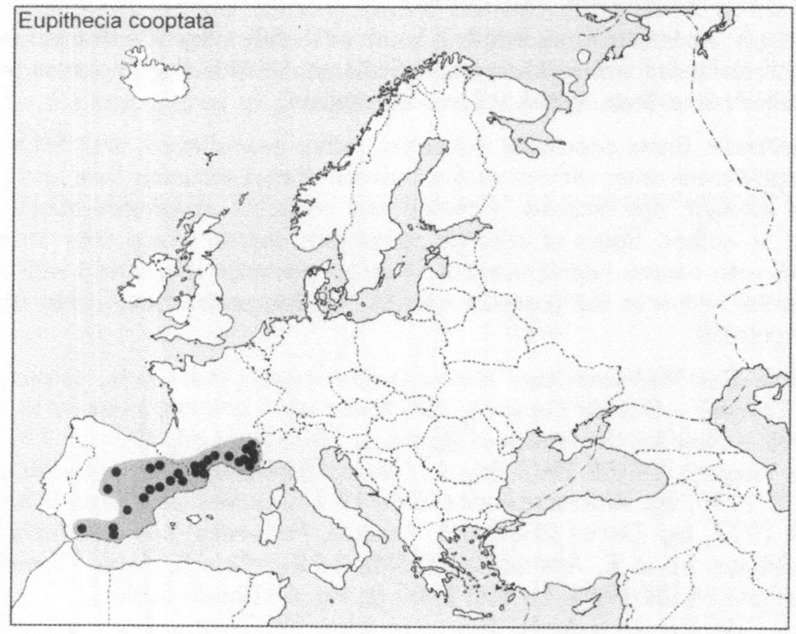

92. *Eupithecia cooptata* DIETZE, 1904

Eup(ithecia) cooptata DIETZE, 1904: Dt. ent. Z. Iris (1903) 16 (2): 335, pl. 4: 24-27 (France: Digne). Syntypes ♂ ♀ (MNHU).

DIAGNOSIS: Wingspan 16-22 mm. Labial palpi, frons and vertex light brownish grey. Notum pale brownish grey. Forewing with straight costa, rather pointed apex and obliquely rounded termen; pale brownish grey with weak ochreous tinge; transverse lines numerous, fine, oblique, greyish brown; basal, postbasal and antemedial lines sharply angled on to costa; the first medial line oblique, slightly sinuate, close to antemedial line, crossing discal spot, afterwards sharply angled and separated from antemedial line near costa; postmedial line very sinuate, evenly curved on to costa, shaded on inner side; pale double stripe between medial and terminal areas broadened between R5 and M2; terminal area narrow, widened near costa; pale wavy line fine, much indented, formed of small pale tornal spot with dark inner shade; discal spot narrow, obliquely elongate, merging with first medial line. Hindwing paler, smoky-grey; transverse lines fine, conspicuous only along dorsum; terminal area slightly darker, narrow, with rather inconspicuous, minutely indented pale wavy line; discal spot minute, circular. Fringe chequered greyish white and brownish grey. Abdomen pale brownish grey, with narrow black transverse stripe between first and second segments; sometimes with a series of small dark dorsal dots.

MALE GENITALIA: Papillae narrow and elongated, nearly equal to uncus length.

FEMALE GENITALIA: Ductus bursae with a short row of stout spines on right side between ductus seminalis base and colliculum; spiniferous area not reaching colliculum; marginal spines larger than others. Anterior and posterior apophyses thick and expanded at their apices. Papillae anales narrow and pointed, sclerotized, with sparse setae.

DISTRIBUTION: West-Mediterranean. Spain, Southern France.

PHENOLOGY: Bivoltine: mid-June (early June) to mid-July (early August); mid-August (mid-July) to mid-September (late September). Overwintering as pupa.

BIOLOGY: Larva oligophagous (2). Found on *Artemisia alba (=camphorata)* (Asteraceae) (France: LHOMME 1935; TAUTEL pers. comm.).

HABITAT: Very xerothermic rocky places and mountain slopes. From 250 m up to about 1,300 m above sea-level in south-western Alps (coll. ZSM), and to about 1,200 m in central Spain (coll. ZSM; coll. FIBIGER).

SIMILAR SPECIES: Very similar to *E. gemellata* but usually larger and lighter, with postmedial line less strongly curved near costa of forewing (Text-fig. 56). Valva broader medially and narrower towards apex, sternum A8 of male larger than in *E. gemellata*. The female genitalia are distinguished from those of *E. gemellata* by the broadened and more rounded shape of corpus bursae with larger marginal spines, the patch of robust spines in ductus bursae, the broadened apices of anterior and posterior apophyses, and narrower, tapered papillae anales (see also SCHÜTZE 1958).

The ***breviculata*** species-group

Newly introduced species-group of probably Mediterranean origin. At least five species can be included in the *breviculata* group: *E. spissilineata*, *E. cuculliaria*, *E. unitaria*, *E. breviculata* and *E. cerussaria*.

MALE GENITALIA: Uncus short, biapical. Papillae usually narrow and relatively elongated, length about 0.7-1.0 times uncus length. Valva without ventral process; sacculus weakly sclerotized. Vinculum short, normal or wide. Aedeagus large, long and thick, broader at base; length about 0.8-1.0 times valva length, about 3.0-4.0 times longer than its medial width. Vesica armed with two dentate cornuti, one or two elongated patches of denticles and two flat cornuti (one elongated, U-shaped, the other smaller, irregular) at ductus ejaculatorius base. Sternum A8 peg-like, elongate, usually without apical lobes; basal hollow shallow; sternum length about 0.6-0.8 times valva length.

FEMALE GENITALIA: Bursa copulatrix pear-shaped, spherical or ovate; almost completely covered with numerous small spines. Ductus bursae short, usually with longitudinal striations and with a small patch of spines or with short rows of small spines. Ductus seminalis short, broadened at base, curved backwards, usually attached to corpus bursae near colliculum on right side. Colliculum collar-like, relatively short and narrow. Antrum membranous. Anterior and posterior apophyses short and thin, not expanded at apices. Spurs of anterior apophyses thin at base but abruptly broadened at apices. Tergum A8 small, almost square. Papillae anales small, usually rounded.

REMARKS: According to the structure of the male and female genitalia, the *breviculata* species-group is closely related to the *graphata* species-group. The presence of two dentate cornuti on the vesica in the male genitalia should readily distinguish the species of the *breviculata* group from species of the *graphata* group.

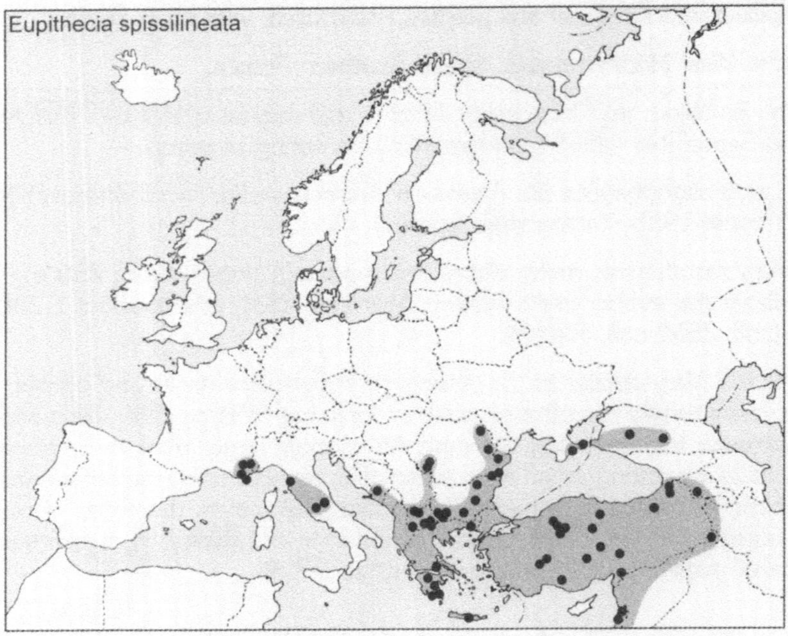

93. Eupithecia spissilineata (METZNER, 1846)

Larentia (Eupithecia) spissilineata METZNER, 1846: Stett. Ent. Z. 7 (8): 241 (Turkey). Syntype(s).
Eupithecia multilineata MANN, 1866: Verh. zool.-bot. Ges. Wien 16 (Abh.): 339, pl. 1: 1 (Romania: Dobruja, Tultscha). Syntypes 2♂ (NHMW?).
Unavailable names (incorrect subsequent spellings): *spissilinearia*: HEYDENREICH (1851); *spinilineata*: DIETZE (1913).

DIAGNOSIS: Wingspan 15-21 mm. Labial palpi covered with light ochreous scales with brownish apices. Frons, vertex and notum light ochreous. Antennae whitish ringed light brown. Forewing bright, light ochreous ground colour; transverse lines rather wide, slightly sinuate and weakly angled near costa, unicolorous light ochreous, alternating with fine, broken, whitish lines; postmedial line finely waved, slightly sinuate and evenly curved near costa; sometimes the veins chequered with light brown and white; wavy line distinct, unbroken, fine-dentated or fine-waved, whitish, sometimes with small white tornal spot; discal spot absent. Hindwing paler, rather brightly coloured, with alternating fine, conspicuous, light ochreous and whitish transverse lines; sometimes whitish transverse lines broken between veins; whitish wavy line fine, distinct, unbroken, fine-dentated; discal spot obsolete or usually absent. Fringe with narrow whitish and wide ochreous chequers at vein ends. Abdomen light ochreous.

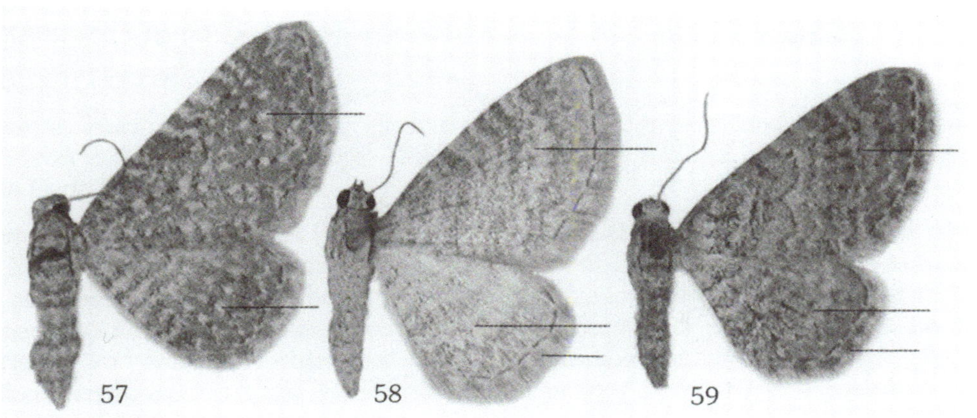

Text-figs 57-59: Diagnostic characters (indicated) of European *Eupithecia* species. Text-fig. 57: *E. spissilineata* (METZNER, 1846). Text-fig. 58: *E. plumbeolata lutosaria* BOHATSCH, 1893. Text-fig. 59: *E. cuculliaria* (REBEL, 1901).

MALE GENITALIA: Valva tapered to apex. Vinculum wide. Vesica armed with two dentate cornuti of almost equal size. Sternum A8 elongated, with two obtuse united apical lobes.

FEMALE GENITALIA: Bursa copulatrix pear-shaped. Ductus bursae with some rows of large spines.

DISTRIBUTION: East-Mediterranean. South-eastern France, Italy, Herzegovina, Macedonia, Romania, Bulgaria, Greece, Crete, probably European Turkey, Ukraine (south-eastern Crimea) and southern Russia (Krasnodar and Stavropol prov.). – Outside Europe in Asia Minor, Lebanon and Armenia (ZERNY 1933; WILTSHIRE 1939), whence subsp. *libanotica* SCHWINGENSCHUSS, 1955 was described.

PHENOLOGY: Uni- or probably bivoltine: mid-May (early April) to mid-July (early August).

BIOLOGY: Unknown.

HABITAT: Limestone, rocks, screes and gorges. From 0 m up to about 1,100 m above sea-level in Apennines (Mt. Majella: ZAHM pers. comm.), to about 1,700 m in Balkan peninsula (coll. ZMUC), and to about 2,000 m in Asia Minor (RIEMIS pers. comm.).

SIMILAR SPECIES: *E. spissilineata* can usually be easily separated from other *Eupithecia* species by the wings with alternating light ochreous and whitish transverse lines, discal spots lacking (Text-fig. 57). The male and female genitalia are typical for the *breviculata* species-group, and most resemble the genitalia of *E. cuculliaria* (SCHÜTZE 1958), which has larger denticles and dentate cornuti in the vesica, broader sternum A8 in male, and smaller spines in ductus bursae between ductus seminalis base and colliculum in female genitalia.

REMARKS: Adults from central Greece have more pointed wings with darker, grey, rather inconspicuous and less wavy transverse lines. I have named these here as a new form *helassata*.

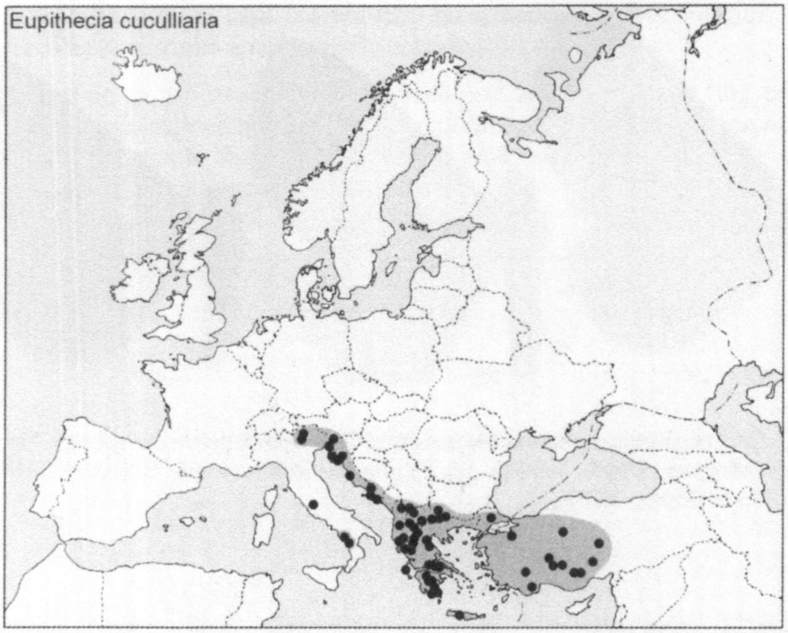

94. *Eupithecia cuculliaria* (REBEL, 1901)

Tephroclystia cuculliaria REBEL, 1901: in STAUDINGER & REBEL, Cat. Lepid. Palaearct. Faunengeb. 1: 316 (Slovenia and NE. Italy: Carniolia; Croatia, Dalmatia; Hercegovina; Turkey: ? Bithynia; Konia). Syntypes (NHMW).

Eupithecia undata BOHATSCH, 1893: Dt. ent. Z. Iris 6: 28 (Croatia: Dalmatia). Syntypes including 4♂1♀ (NHMW). Junior homonym of *undata* FREYER, 1842.

Unavailable names (incorrect subsequent spellings): *cuculiaria* SCHWINGENSCHUSS & WAGNER (1927); *cucculiaria*: RIEMIS (1994).

DIAGNOSIS: Wingspan 14-19 mm. Labial palpi, frons, vertex and notum light yellowish brown. Antennae yellowish grey ringed brown. Forewing pale yellowish brown; transverse lines slightly sinuate, weakly angled near costa, yellowish brown, alternating with fine, inconspicuous pale grey lines; postmedial line weakly curved near costa, with dark brown dashes on the veins on its inner side; wavy line pale grey, more or less distinct; discal spot usually absent or minute, rounded or slightly elongated. Hindwing concolorous; transverse lines conspicuous but obscure, alternating with narrow pale grey lines; wavy line indistinct; discal spot minute, slightly elongated, inconspicuous or obsolete. Fringe with narrow pale grey and wide brown chequers. Abdomen light yellowish brown with weak greyish tinge.

MALE GENITALIA: Valva with tapered apex. Two dentate cornuti of almost equal size on the vesica. Sternum A8 narrow, usually with small membranous blotch near apex.

FEMALE GENITALIA: Bursa copulatrix ovate.

DISTRIBUTION: East-Mediterranean. Italy, Slovenia, Croatia, Herzegovina, Albania, Macedonia, south-western Bulgaria, Greece, Crete and European Turkey. – Outside Europe in western and central Asian Turkey.

PHENOLOGY: Univoltine: early May (mid-April) to late June (late July).

BIOLOGY: Unknown.

HABITAT: Warm and dry slopes, rocks and screes. From 0 m up to about 1,400 m above sea-level in Balkan peninsula (coll. ZMUC) and to about 2,000-2,100 m in Asia Minor (Ala Dagh Mts: coll. ZSM).

SIMILAR SPECIES: This species (Text-fig. 59) is closely related to *E. spissilineata* but adults of *E. cuculliaria* are most similar to light specimens of *E. pygmaeata* and also to typical *E. plumbeolata*. However, the forewing of *E. pygmaeata* has more rounded apex, the ground colour usually darker, chocolate-brown, with distinct pale tornal spot on the forewing and with more chequered fringe. *E. plumbeolata* has more greyish ground colour, less conspicuous and less sinuate transverse lines on the wings, especially on the hindwing, and unicolorous grey fringe. The male and female genitalia of *E. cuculliaria* can be difficult to distinguish from those of *E. spissilineata* (SCHÜTZE 1958).

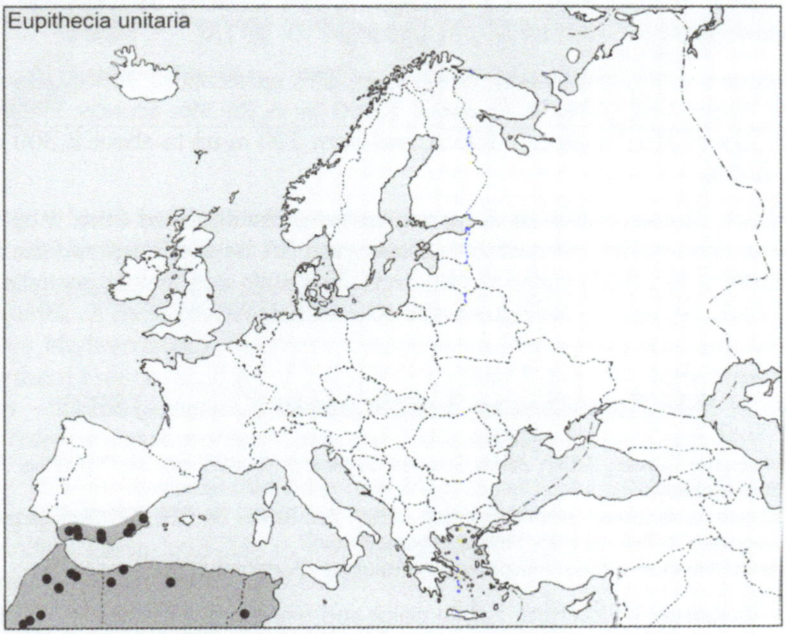

95. *Eupithecia unitaria* HERRICH-SCHÄFFER, 1851

Eupithecia unitaria HERRICH-SCHÄFFER, 1851: Syst. Bearb. Schmett. Eur. 3 (Suppl.), pl. 85: 524, 525 (non binominal); ibidem (1852), 6 (55): 77 (Spain: Ronda). Syntype(s).

Cidaria roseocinnamomaria ROTHSCHILD, 1914: Novit. zool. 21: 346 (Algeria: Guelt-Es-Stel). Holotype ♂ (BMNH).

DIAGNOSIS: Wingspan 20-25 mm. Labial palpi pale grey. Frons, vertex and notum whitish grey with yellowish tinge. Forewing rather elongate and pointed, costa arched, termen rounded. Ground colour whitish grey, tinged yellowish; transverse lines moderately inconspicuous, narrow, wavy, usually darker along costa; postmedial marked by a series of closely set brownish dashes beside it; terminal area brownish grey; pale wavy line forming two light blotches (one near apex of wing, the other between M3 and CuA1); discal dot scarcely visible, narrow and oblique. Hindwing with rather conspicuous transverse lines, especially postmedial line which is marked by a series of small brownish specks on the veins; terminal area light brownish grey with ill-defined pale wavy line and pale blotch between M3 and CuA1; discal spot similar to that on the forewing. Fringe dirty yellowish white, narrowly spotted with brownish at vein ends. Abdomen pale yellowish grey.

MALE GENITALIA: Sternum A8 with two narrow apical horns and narrow, deep membranous hollow between them.

FEMALE GENITALIA: Bursa copulatrix rather spherical.

DISTRIBUTION: West-Mediterranean. Southern Spain. – Outside Europe in Northern Africa (Morocco, Algeria, Tunisia and Tibesti Mts) (DIETZE, 1913, HERBULOT, 1965).

PHENOLOGY: Bivoltine: mid-April (early April) to mid-June (late June); December. Larva in April.

BIOLOGY: Larva monophagous (1). Found on the flowers of *Diplotaxis pendula* (Brassicaceae) (Tunisia: CHRÉTIEN 1910; DIETZE 1910, 1913).

HABITAT: Warm and dry slopes and rocks. From 0 m up to about 900 m above sea-level in southern Spain (coll. ZMUC), to about 1,600 m in Rif Mts (RUNGS 1981), to about 2,200 m in Tibesti Mts (HERBULOT 1965), and from 700 m up to about 2,300 m in Great Atlas (coll. ZSM).

SIMILAR SPECIES: Rather similar to *E. semigraphata,* distinguished from it by the lighter ground colour, the smaller and narrower discal spots on the forewings and the fringe with narrower spots of brownish scales at vein ends. The male and female genitalia of *E. unitaria* can usually be used to distinguish this species successfully from *E. semigraphata*.

96. Eupithecia breviculata (DONZEL, 1837)

Melanthia breviculata DONZEL, 1837: Annls Soc. ent. France 6 (4): 478, pl. 18: 7 (France: Basses-Alpes, Hières, Digne). Syntypes.

Eupithecia breviculata anatolica SCHWINGENSCHUSS, 1939: Ent. Rdsch. 56 (23): 251 (Turkey: Inner Anatolia, A(kshehir)). Syntypes 1♂3♀ (NLHW). Valid at subspecific rank.

Unavailable names (incorrect subsequent spelling): *brevicularia*: BOISDUVAL (1840).

DIAGNOSIS: Wingspan 15-20 mm. Labial palpi and frons yellowish brown. Vertex yellowish brown, usually with whitish marginal scales. Notum with mixture of yellowish brown and whitish scales. Forewing with basal area and basal half of medial area to discal dot yellowish brown with four whitish costal spots; the other half of medial area white; postmedial line double, dark brown, very winding; terminal area white with large, brown or yellowish brown apical spot and narrow, whitish, wavy medial line; tornal spot smaller, dark-brown; discal dot small, oval, blackish-brown; fringe yellowish brown from apex to M3; white from M3 to CuA2 and pale whitish brown from CuA2 to tornus. Hindwing with brownish basal area; medial and terminal areas white with pale, inconspicuous brownish lines, especially near inner and terminal margins; discal dot minute, dark brown; fringe white with four brownish spots between R4 and M3 and with similar spot at tornus. Abdomen white; first and second segments yellowish brown; fourth segment with small, brown dorsal spot.

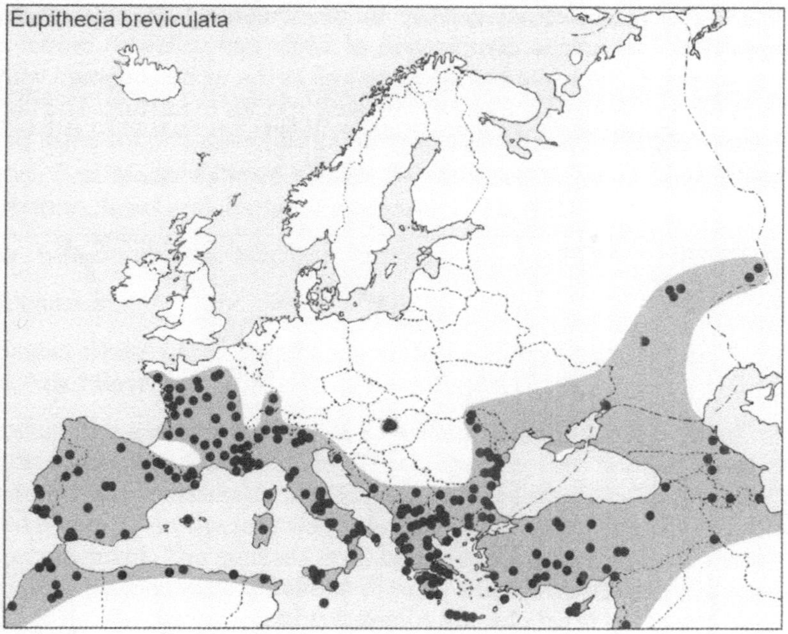

MALE GENITALIA: Valva with more or less rounded apex. One dentate cornutus larger than the other in the vesica. Sternum A8 has undulate lateral margins.

FEMALE GENITALIA: Bursa copulatrix spherical.

DISTRIBUTION: Mediterranean. Southern Europe from Portugal to southern Urals, northward to northern France, southward to Crete. – Outside Europe in North Africa (Morocco and Algeria - subsp. *georgica* VOJNITS, 1977), Cyprus, Asian Turkey, Israel, Jordan, Caucasus, Transcaucasus, northern Iran and Turkmenistan (Kopet-Dagh Mts).

PHENOLOGY: Uni- or bivoltine: mid-June (late May) to mid-July (late August). In the south from mid-April (mid-March) to late June (early July) and from early June to early August (mid-September). Larva: mid-July to mid-August (early October). Overwintering as pupa in flimsy cocoon in the ground, sometimes twice.

BIOLOGY: Larva oligophagous (1). Found on the flowers and seeds of *Ptychotis heterophylla (=Carum bunias)* (France: CHRÉTIEN 1904; DIETZE 1910, 1913), *Heracleum sphondylium, Peucedanum oreoselinum, P. palustre, Pimpinella peregrina* (?), *Pimpinella saxifraga* (Apiaceae) (Germany: SCHÜTZE 1957; WEIGT 1990) and probably occasionally on *Clematis* (Ranunculaceae) (Corsica: MILLIÈRE 1870).

HABITAT: Forest fringes and open deciduous forests, hedges, scrub and scrub-covered meadows, warm and dry slopes, rocks and screes. From 0 m up to about 1,100-1,200 m above sea-level in Alps (VORBRODT 1931; SCHÜTZE 1957; HELLMANN & al. 1999; coll. ZMUC), to about 1,300 m in Pyrenees (MAZEL & PESLIER 1997; coll. ZMUC), to about 1,400 m in Apennines (coll. ZSM), to about 1,500 m in mountains of Balkan peninsula (PINKER 1968), to about 1,600 m in Middle Atlas (RUNGS 1981), and to about 1,600-1,800 m in Asia Minor (coll. ZSM).

SIMILAR SPECIES: *E. breviculata* is unlikely to be confused with any other European *Eupithecia* species. The unique combination of white and yellowish brown colour and maculation of forewing, the white hindwing, as well as the white abdomen with yellowish brown first and second segments, make this an easy species to identify. The narrow valva, the cornuti, especially two small dentate cornuti on the vesica and the small peg-like sternum A8 in male can be used to separate this species accurately.

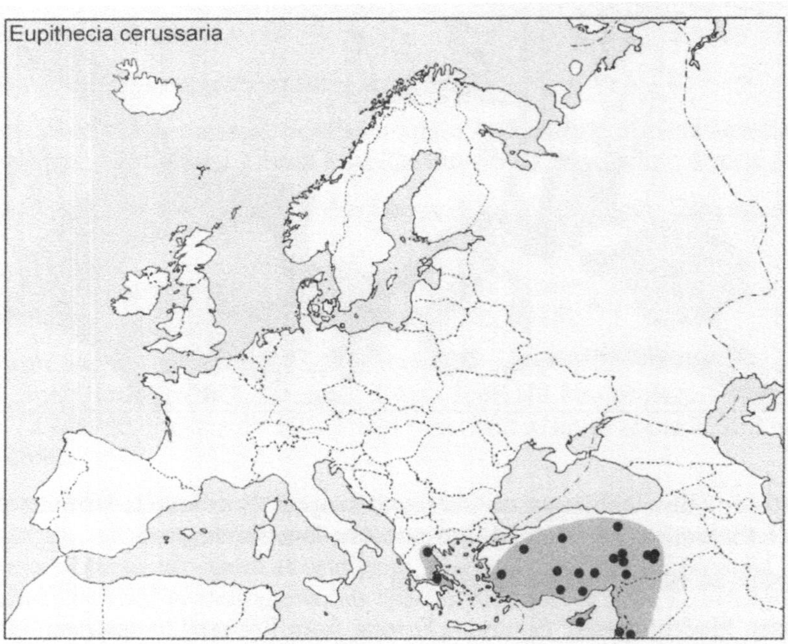

97. Eupithecia cerussaria (LEDERER, 1855)

Cidaria cerussaria LEDERER, 1855: Verh. zool.-bot. Ver. Wien 5 (Abh.): 215, pl. 3: 9 (Lebanon: Beirut). Syntype(s) (NHMW).
Unavailable names (incorrect subsequent spelling): *cerusaria*: REBEL (1939).

DIAGNOSIS: Wingspan 18-22 mm. Labial palpi pale brown with scarce white scales. Frons light ochreous. Vertex white or creamy white. Notum usually light ochreous or white with two light ochreous spots posteriorly. Forewing unicolorous white with five light yellow, slightly sinuate, delicate and rather inconspicuous transverse lines: basal, antemedial, medial, postmedial and wavy, which are distinctly marked on the veins as small, brown dots (sometimes the wavy line is absent, and/or the brown dots on the veins are absent but they are always present on the postmedial line); costal margin slightly darker with six to eight ochreous brown spots between costal and radial veins; discal spot minute, circular, brownish. Hindwing unicolorous white with two to four transverse lines, which are distinctly marked on the veins as small brown dots (sometimes some lines absent but postmedial line always present); discal spot minute, similar to that on forewing. Fringe white, slightly spotted with light brown at vein ends or unicolorous white. Abdomen yellowish white with small, dark brown dorsal spot on each segment.

MALE GENITALIA: Valva with angular ventral margin. Vesica armed with two elongated and dense patches of relatively large denticles. Sternum A8 asymmetrical, with digitiform apex.

FEMALE GENITALIA: Bursa copulatrix rather pouch-like; with larger marginal spines. The posterior margin of tergum A8 with medial membranous hollow.

DISTRIBUTION: East-Mediterranean. Greece. – Outside Europe occurs in Cyprus, Asian Turkey, Lebanon, Israel and Jordan (HAUSMANN 1991, 1997).

PHENOLOGY: Univoltine: early May (late April) to early June (late June).

BIOLOGY: Unknown.

HABITAT: Slopes where garlic plants abound. From 200 m up to about 1,300 m above sea-level in Asia Minor (coll. ZSM, RIEMIS pers. comm.).

SIMILAR SPECIES: The external features of *E. cerussaria* can usually be used to distinguish it successfully from other European *Eupithecia* species. This species is rather similar to *E. fenestrata* but it is considerably smaller, brown costal spots and dots on the veins more distinct, costal margin on the forewing more yellowish and abdomen with brown dorsal spots on each segment. The genitalia of *E. cerussaria* can easily be separated from those of other *Eupithecia* species by the shape of valva and sternum A8, the unique cornuti on the vesica in male, the disposition of spiniferous areas in bursa copulatrix, and poriferous posterior part of tergum A8 in female.

The *irriguata* species-group

Group introduced by MIRONOV (2001). Four European *Eupithecia* species can be included to the *irriguata* group: *E. irriguata*, *E. antalica*, *E. indigata* and *E. conterminata*. *E. marasa* WEHRLI, 1934 from Asian Turkey belongs in this group also.

MALE GENITALIA: Uncus biapical, short and narrow. Papillae relatively narrow and elongated; length about 0.9-1.2 times length of uncus. Valva without ventral process, broadest medially, with evenly curved ventral margin and rounded apex; sacculus slightly sclerotized. Vinculum short and broad. Aedeagus length about 0.6-0.9 times length of valva; about 3-4 times longer than medial width. Vesica covered by numerous very small spinules and armed with one or three dentate cornuti, which are not connected at base, and with one thin, flat V-shaped cornutus at ductus ejaculatorius base. Sternum A8 peg-like, usually small and short, with base broadened, narrowing to apex; basal and apical hollows shallow; accordingly with two small apical lobes; sternum length about 0.4-0.6 times length of valva.

FEMALE GENITALIA: Bursa copulatrix pear-shaped, half or more covered with small spines, sometimes with longer marginal spines. Ductus bursae with patch of spines or without it. Ductus seminalis narrow; attached near colliculum or to middle part of corpus bursae on right side. Colliculum collar-like, relatively short and narrow. Anterior and posterior apophyses short and thin. Spurs of anterior apophyses broadened to apices. Tergum A8 usually short and broad or square, with rounded apical corners. Papillae anales rather small and rounded.

REMARKS: Species of the *irriguata* group are closely related to species from the *satyrata* group. Distinguished from these by the vesica, which is covered with numerous spinules, and the shape of irregular plate in the vesica in the male genitalia; ductus seminalis mainly narrow but not evenly curved posteriorly in the female genitalia as well.

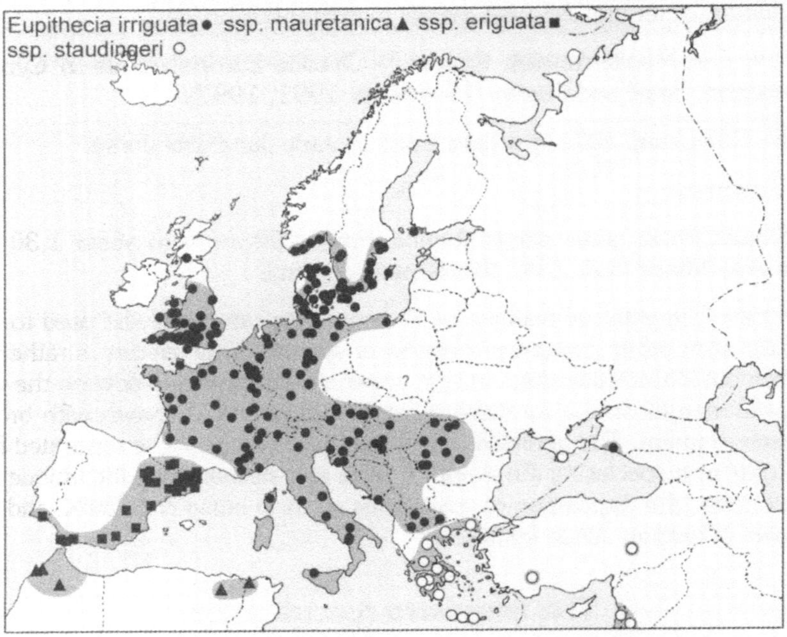

98. *Eupithecia irriguata* (HÜBNER, 1813)

Geometra irriguata HÜBNER, 1813: Samml. Eur. Schmett. 5 Geometrae (1), pl. 77: 397 (Europe). Syntype(s) lost.

Phalaena variegata HAWORTH, 1809: Lepid. Br. (2): 362 (United Kingdom: London). Syntype(s) (ZMUC). Junior primary homonym of *Phalaena variegata* SCOPOLI, 1763.

Eupithecia irriguata var. ? *eriguata* STAUDINGER, 1871: Cat. Lepid. Eur. Faunengeb. 1: 193 (Spain, Andalusia: Granada). Syntype(s). Valid at subspecific rank.

Eupithecia staudingeri BOHATSCH, 1893: Dt. ent. Z. Iris 6: 11 (Lebanon: Beirut, Cremona). Lectotype ♂ (MNHU, examined). Valid at subspecific rank.

Eup(ithecia) irriguata f. *franconica* DIETZE, 1910: Biol. Eupithecien 1, pl. 78: 803, 804; ibidem (1913), 2: 41 (Germany: Frankfurt a.M.). Syntypes 4, ♂ ♀ (MNHU). Valid at subspecific rank.

Eup(ithecia) irriguata f. *mauretanica* DIETZE, 1910: Biol. Eupithecien 1, pl. 78: 805; ibidem (1913), 2: 42 (Algeria: Mauretania, Batna). Syntypes (MNHU). Valid at subspecific rank.

Eupithecia irriguata kurdica PROUT, 1938: in SEITZ, Macrolep. 4 (Suppl.): 186, fig. 14: f (Turkey: Kurdistan, Malatya). Syntype(s) (BMNH; two syntypes examined). Valid at subspecific rank.

Unavailable names (incorrect subsequent spellings): *irriguaria*: BOISDUVAL (1840); *erriguata*: PROUT (1914).

DIAGNOSIS: Wingspan 18-23 mm. Labial palpi brownish grey. Frons, vertex and notum brownish white. Forewing elongated, triangular; costa gently arched; apex pointed; termen oblique, almost straight; ground colour whitish with elongated, blackish brown costal spots; transverse lines ill-defined, except three fine, brown, closely placed medial lines; inner and outer parts of basal area ochreous brown, broadened and blackish to costa; antemedial line invisible, forming only a blackish brown costal spot; first and second medial lines slightly sinuate, sharply angled near costa and forming dark costal spot; postmedial line well marked only near dorsum, with some elongated black dashes on the veins and broad blackish brown costal spot; terminal area narrow, abruptly broadened to costa, ochreous brown, with three slightly darker clouds (costal, medial and tornal); pale wavy line broad, multiply indented, often broken; discal spot conspicuous, large, oblique, ovate, ovate-oblong or ovate-acuminate, intensely black. Hindwing rather narrow, angular; termen slightly concave medially; ground colour whitish, paler than forewing; basal area blackish brown along dorsum; transverse lines inconspicuous; terminal area narrow, brownish, gently darker; pale wavy line indistinct; discal spot evident, circular or ovate, dark. Fringe brownish white, distinctly spotted with dark brownish grey at vein ends. Abdomen light brownish grey, with black lateral spots.

The Iberian subsp. ***eriguata*** STAUDINGER, 1871 is darker, grey or whitish grey, with costal spots less visible, three narrow medial lines indistinct on the forewing, all discal spots larger, ovate or rounded, more intensely black than in nominate subspecies.

The south-eastern subsp. ***staudingeri*** BOHATSCH, 1893 is darker, usually almost unicolorous grey or pale grey, costal spots less distinct and three narrow medial lines rather inconspicuous on the forewing, discal spots paler on all wings than in nominate subspecies.

MALE GENITALIA: Vesica armed with three small, dentate cornuti: two apical more or less elongated and one shorter near ductus ejaculatorius base.

FEMALE GENITALIA: Without the patch of spinules in ductus bursae.

DISTRIBUTION: West-Palaearctic (?). Widely distributed in western, Central and southern Europe except Iceland, Ireland, Balearics and Malta. Separate populations found in western Caucasus. In Iberian peninsula (Spain, Andorra and south-western France) replaced by subsp. *eriguata*, in Greece, Crete and southern Ukraine (Crimea) by subsp. *staudingeri*. – Outside Europe in North Africa (Morocco, Algeria, Tunisia: subsp. *mauretanica* DIETZE, 1910); Asian Turkey and Levant: northern Jordan and Lebanon (subsp. *staudingeri*). – New for the fauna of Ukraine: 1♀, (Ukraine), Crimea, Yalta, (the date not marked), leg. Bekendorff (ZISP). Recorded for Mongolia (VOJNITS 1977), probably in error.

PHENOLOGY: Univoltine: mid-April (late March) to mid-June (early July). Larva: early May (late April) to late May (late July). In the north in June-July (until August). Overwintering as pupa; about 25% of pupae overwintering twice (WEIGT 1988).

BIOLOGY: Larva monophagous (3). Found on the flowers and young leaves of *Quercus*, rarely on *Fagus* (Fagaceae) (Germany: DIETZE 1870; United Kingdom: BARRETT 1904), more exactly on *Quercus robur (=pedunculata)* (United Kingdom: BUCKLER 1899; BARRETT 1904; Denmark: JUUL 1948), *Q. petraea (=sessiliflora)* (Germany: DIETZE 1910, 1913; WEIGT 1980, 1988), *Q. ilex* (Alps: WEIGT 1988) and *Q. suber* (Mediterranean: DIETZE 1913; WEIGT 1988).

PARASITOIDS: Hymenoptera, Braconidae: *Apanteles vitripennis* (SHENEFELT 1972).

HABITAT: Oak and oak-beech forests, oak plantations and edges of forests, scrub. From 0 m up to about 900 m above sea-level in Apennines (PROLA & RACHELI 1980), to about 1,000 m in Pyrenees (MAZEL & PESLIER 1997), to about 1,500-1,600 m in southern Spain (Sierra Guillimona: coll. ZSM), to about 1,300 m in Greece (coll. ZSM), to about 1,500 m in Asia Minor (coll. ZSM), and to about 2,000 m in Alps (VORBRODT & MÜLLER-RUTZ 1914).

SIMILAR SPECIES: An unmistakable species, which is unlikely be confused with any other European *Eupithecia* species. *E. irriguata* can usually be easily distinguished by the shape, coloration and unique maculation of the wings. The short and thick aedeagus with vesica armed with three small dentate cornuti, the small, short, triangular sternum A8 in male, and the bursa copulatrix without spines between ductus seminalis base and colliculum in the female genitalia should distinguish this species.

REMARKS: The species name *eriguata* (RAMBUR 1866; pl. 21: 7) is not binominal. It is valid in STAUDINGER (1871) as *Eupithecia irriguata* var. *eriguata*. The type specimens were not traced in materials of P. RAMBUR (HERBULOT 1987a).

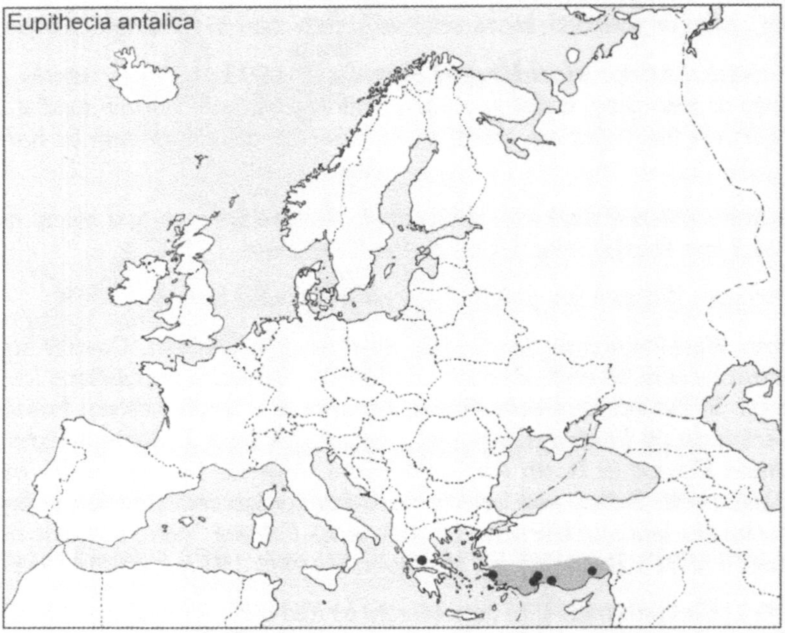

99. *Eupithecia antalica* MIRONOV, 2001

Eupithecia antalica MIRONOV, 2001: Trudy Zool. Inst. RAN 291: 113, figs 1-4, 14 (Greece: Thessalia, Agios Nikolaos). Holotype ♀ (HNHM, examined).

DIAGNOSIS: Wingspan 16-19 mm. Labial palpi brownish grey. Frons, vertex and notum whitish grey. Forewing rather narrow and elongated, costa slightly arched, termen evenly curved, apex pointed; ground colour pale grey; transverse lines inconspicuous, but well marked by dark costal spots; two medial lines faintly sinuate and sharply angled from discal spot onto costa; postmedial line marked with a series of small black dashes on the veins, twice angled near costa and forming triangular dark grey costal spot; terminal area dark grey, usually with whitish grey wavy line; discal spot oblique, narrow, ovate-oblong or ovate-acuminate, black. Hindwing rather narrow and elongated with more or less pointed apex and slightly concave termen; ground colour whitish grey; transverse lines inconspicuous, well marked along dorsum only; discal spot distinct, rounded, ovate-acuminate or triangular, black. Fringe greyish white, spotted by dark grey opposite vein ends. Abdomen whitish grey with black lateral stripes; second abdominal segment darker, greyish brown or ochreous brown.

MALE GENITALIA: Vinculum trapeziform. Vesica covered with numerous small spinules, armed with one dentate cornutus and one thin V-shaped cornutus. Sternum A8 peg-like, elongated, with two short and blunt pincer-like apical lobes.

FEMALE GENITALIA: Bursa copulatrix oblong pear-shaped, almost completely covered by small spines; marginal spines heavily sclerotized with a series of small wrinkles. Ductus bursae heavily sclerotized, without spines. Ductus seminalis bent backwards, attached to middle of corpus bursae.

DISTRIBUTION: East-Mediterranean. Greece (Thessalia). – Outside Europe in south of Asian Turkey (Prov. Antalya and Adana).

PHENOLOGY: Univoltine: late March to mid-May.

BIOLOGY AND HABITAT: Unknown. Up to about 500 m above sea-level in Greece (coll. HNHM).

SIMILAR SPECIES: There are no reliable and constant external differences between *E. antalica* sp. n. and *E. irriguata staudingeri*. However, the male and female genitalia of the two compared species are very clearly distinguished. The trapeziform vinculum, one dentate cornutus on the vesica, more elongated sternum A8 in male, and elongated pear-shaped bursa copulatrix covered with smaller spines, the heavily sclerotized small marginal spines and ductus bursae should distinguish the new species from the closely related *E. irriguata staudingeri* and *E. marasa* as well.

100. Eupithecia indigata (HÜBNER, 1813)

Geometra indigata HÜBNER, 1813: Samml. Eur. Schmett. 5 Geometrae (1), pl. 77: 399 (Europe). Syntype(s) lost.

Eupithecia albifronsata DE GRASLIN, 1863: Annls Soc. Ent. France 3 (4): 328, pl. 8: 13 (France: Pyrénées Orientales, Cabanasse). Syntypes.

Tephroclystia indigata var. *turfosata* DRAUDT, 1903: Beitr. Naturk. Preuss. 9: 72 (Prussia). Holotype ♀. Valid at subspecific rank.

Eupithecia pliniata STAUDER, 1929: Ent. Anz. 9: 191 (Italy: Gebirge von Sorrento, Monte Faito, Boscoreale). Holotype ♀. New junior synonym (?).

Unavailable names (infrasubspecific): *limbofasciata*: DIETZE (1913); *uniformata*: LEMPKE (1969); *nigrescens*: LEMPKE (1969). - (incorrect subsequent spellings): *indigaria*: BOISDUVAL (1840); *indignata*: LAINE (1977).

DIAGNOSIS: Wingspan 15-21 mm. Labial palpi and notum ochreous grey. Frons and vertex light ochreous grey with dark transverse band between them. Forewing rather triangular, narrow; apex pointed; termen very oblique; light ochreous grey of a very soft and delicate tinge; transverse lines weak, oblique, slightly darker, often ill-defined; in some specimens only medial and postmedial lines more distinct, slightly angled near costa; medial line touching discal spot; terminal area narrow, usually darker; pale wavy line absent; discal spot ovate or ovate-orbicular, relatively large and always conspicuous. Hindwing paler than forewing, smoky-grey with delicate ochreous tinge, unmarked or with weak transverse lines, especially with more distinct, darker, medially angled postmedial line; terminal area clouded, slightly darker; discal spot minute, circular or elongated. Fringe unicolorous, light ochreous grey or slightly spotted with ochreous grey at vein ends. Abdomen light grey with delicate ochreous tinge.

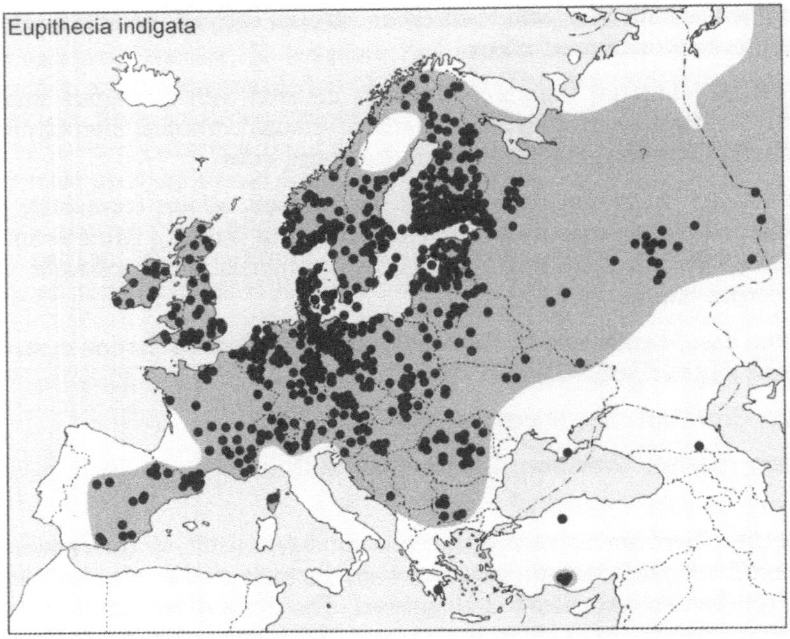

MALE GENITALIA: Vesica armed with dense patch of small dentate cornuti at ductus ejaculatorius base.

FEMALE GENITALIA: Ductus bursae with one patch of spines on the right side between ductus seminalis base and colliculum.

DISTRIBUTION: Eurasiatic. Widespread in Europe westward to Ireland, eastward to Urals, northward to northern Scandinavia and southward to southern Spain and Peloponnese. New for the fauna of Greece: 1♀, Greece, Palaeohorion, Parnon, 800 m, 21.-24.IV.1981, leg. Schepler (ZMUC). – Outside Europe in Asian Turkey, Caucasus, northern Kazakhstan, Jacutia, Altai, southern Siberia; eastward to Ussuri.

PHENOLOGY: Univoltine: late April (late March) to mid-June (late July). Larva: late June (early May) to early July (mid-September). Overwintering as pupa in dense cocoon in the ground.

BIOLOGY: Larva oligophagous (2). Found on the inflorescences and needles, mainly on the young twigs of *Pinus sylvestris* (Pinaceae) (United Kingdom: CREWE 1862), also *Juniperus* and *Cupressus* (Cupressaceae) (United Kingdom: CREWE 1863; BUCKLER 1899; BARRETT 1904; Germany: DIETZE 1910, 1913), and rarely on *Larix decidua (=europaea)* (France: LHOMME 1935; Denmark: JUUL 1948; Finland: SEPPÄNEN 1970) and *Picea abies* (Pinaceae) (Switzerland: VORBRODT 1928; Germany: KOCH 1984; WEIGT 1991). Larvae of this species have been recorded eating aphids on the needles of *Pinus sylvestris* (Scotland: LEVERTON 1998).

PARASITOIDS: Hymenoptera, Ichneumonidae: *Diadegma majale* - Diptera, Tachinidae: *Campylochaeta inepta*, *Frivaldzkia distincta* (THOMPSON 1945; HERTING 1960; ZWAKHALS pers. comm.).

HABITAT: Pine and mixed forests, plantations. From 0 m up to about 2,200-2,300 m above sea-level in Alps (VORBRODT & MÜLLER-RUTZ, 1914; VORBRODT, 1928), to about 2,300-2,400 m in Pyrenees (MAZEL & PESLIER 1997), and to about 2,500 m in Sierra Nevada (coll. ZMUC).

SIMILAR SPECIES: A rather invariable species, which is similar to *E. conterminata*. The differences between these species are given below under *E. conterminata*. In comparison with other similar small *Eupithecia*, *E. distinctaria* has ground colour much greyer, discal spot more elongated and dark costal spots on the forewing conspicuous; *E. plumbeolata*, *E. pygmaeata* and *E. valerianata* lack dark discal spots on the forewings. The shape of the valva and sternum A8, the cornuti in the vesica in the male genitalia, the shape of the bursa copulatrix and disposition of spines in the female genitalia should distinguish this species.

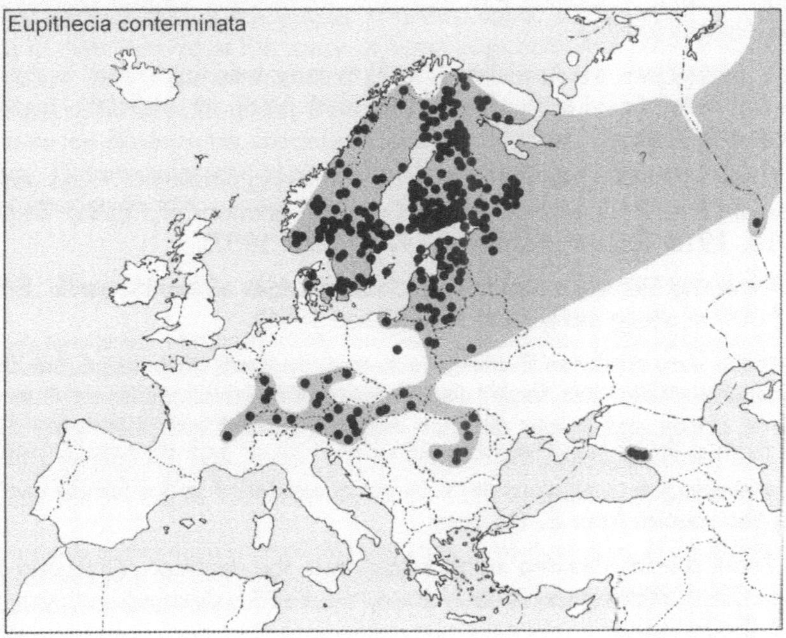

101. *Eupithecia conterminata* (LIENIG & ZELLER, 1846)

Larentia conterminata LIENIG & ZELLER, 1846: Isis 3: 197 (Latvia: ?Koknese [=Kokenhusen]). Syntypes 1♂ 1♀ (BMNH?).

Eupithecia manniaria HERRICH-SCHÄFFER, 1848: Syst. Bearb. Schmett. Eur. 3 (32): 122, 136; ibidem (1846), pl. 25: 156 (non binominal), (Czech Republic, Böhmen). Syntype(s).

Eupithecia idiopusillata INOUE, 1979: Bull. Fac. domest. Sci., Otsuma Wom. Univ. 15: 213, figs 22M, N; fig. 23K; fig. 31B; fig. 33A (Japan: Toyama Prefecture, Kanetsuri Spa, Kurobe). Holotype ♂ (BMNH).

Unavailable names (incorrect subsequent spellings): *maniaria*: SIEVERS (1852); *conterminanta*: GARTHE (1976).

DIAGNOSIS: Wingspan 14-18 mm. Labial palpi grey. Frons greyish white with group of grey scales placed anteriorly. Vertex and notum greyish white. Forewing smoky grey; transverse lines rather oblique, inconspicuous, little darker than ground colour; basal, ante- and postmedial lines usually forming dark grey costal spots; medial and postmedial lines sometimes more distinct, oblique and slightly sinuate, right-angled near costa; medial line crossing or curved around discal spot; terminal area narrow, sometimes slightly darker; pale wavy line absent; discal spot large, ovate or ovate-oblong, black or dark grey, placed near costal margin. Hindwing slightly paler than forewing, smoky grey, slightly darker towards termen, unmarked except for dark, minute, narrow and elongated discal spot and sometimes invisible, evenly curved, slightly darker postmedial transverse line. Fringe unicolorous smoky-grey or slightly chequered whitish grey and grey. Abdomen greyish white, with a series of dark grey dorsal dots.

MALE GENITALIA: Vesica armed with three dentate cornuti: two apical elongated and one shorter near base of ductus ejaculatorius.

FEMALE GENITALIA: Ductus bursae with two patches of spines.

DISTRIBUTION: Eurasiatic. Central and eastern Europe and Scandinavia. – Outside Europe in the west of Great Caucasus Mts (separate population), Siberia eastward to Ussuri region and Japan (Honshu).

PHENOLOGY: Univoltine: late April (late March) to early June (mid-June). In the north from mid-May to mid-June. Larva: late June (mid-June) to mid-July (early August). Overwintering as pupa.

BIOLOGY: Larva monophagous (1). Found on the young needles of *Picea abies (=excelsa)* (Pinaceae) ('East Prussia' [now isolated western territory of Russia]: DRAUDT 1900; DIETZE 1910, 1913; Finland: SEPPÄNEN 1970; WEIGT 1993).

HABITAT: Preferring old, spruce and mixed forests, edges of forests, parks. From 0 m up to about 1,000 m above sea-level in Alps (WEIGT 1993).

SIMILAR SPECIES: Very similar to *E. indigata* in colouring and wing shape, but distinguished from it by usually smaller size, larger discal spot, more distinct transverse lines, larger and darker costal clouds and longer distance between medial and postmedial lines on the forewing. The narrower valva, the cornuti on the vesica, and the smaller sternum A8 in the male, and also the spiniferous areas in bursa copulatrix in the female can be used to distinguish this species from *E. indigata*.

REMARKS: Small species showing a little variation in the depth of colour and distinctness of markings. The f. *manniaria* HERRICH-SCHÄFFER, 1848 is distinguished by very large discal spot on the forewing.

The ***distinctaria*** species-group

This is a species-poor Mediterranean group introduced here for the first time. Only two species, *E. distinctaria* and *E. senorita* sp. n. occur in Europe. In addition five species of this group have been described from Caucasus and Transcaucasus: *E. subdeverrata* VOJNITS, 1975; *E. dimidia* VOJNITS, 1982; *E. aliena* VOJNITS, 1982; *E. decrepita* VOJNITS, 1982; *E. transacta* VOJNITS, 1982, and also *E. elbursiata* SCHÜTZE, 1960 from Elburs Mts (Iran).

MALE GENITALIA: Uncus short or medium length, with elongated basal part, biapical. Papillae short, length about 0.5-0.7 times uncus length. Valvae sometimes slightly asymmetrical, with short, broad and blunt ventral process near base, preceded by shallow medial hollow; with straight or slightly curved dorsal margin, tapered to apex; sacculus heavily sclerotized. Vinculum usually short, broad or medium width. Aedeagus relatively long, slim, length nearly equal to valva length, about 4-5 times longer than medial width. Vesica armed with numerous minute spines, one small, broad, flat apical cornutus and with one elongated, broad, twisted cornutus at ductus ejaculatorius base. Sternum A8 peg-like, slightly broadened at base; basal hollow shallow; with one more or less elongate digitiform, sclerotized apical lobe; sternum length about 0.6-0.7 times valva length.

FEMALE GENITALIA: Bursa copulatrix elongated, pouch-like, with almost straight right side and prominent, rounded left side; left half covered by small spines from base to colliculum. Ductus bursae short, weakly sclerotized. Ductus seminalis attached to corpus bursae between base and middle on right side. Colliculum collar-like, narrow, inclined to right. Antrum broadened and sclerotized. Anterior and posterior apophyses short and thin. Spurs of anterior apophyses short and thin. Tergum A8 almost square. Papillae anales small, slim, rounded.

REMARKS: The combination of the shapes of uncus, valva, sternum A8, and the cornuti on the vesica in male, as well as the shape of bursa copulatrix, the area covered by spines in corpus bursae, sclerotized antrum and short spurs of anterior apophyses in female should distinguish this species-group from others. From the structure of the male and female genitalia the *distinctaria* species-group appears to be related to the *centaureata* species-group. The unique shape of sternum A8 in the male is the main reason for separating these species-groups.

102. Eupithecia distinctaria HERRICH-SCHÄFFER, 1848

Eupithecia distinctaria HERRICH-SCHÄFFER, 1848: Syst. Bearb. Schmett. Eur. 3 (32): 121, 131; ibidem (1846), pl. 26: 162, 163 (non binominal), (Germany: Regensburg). Syntype(s) lost.

Eupithecia constrictata GUENÉE, 1858: in BOISDUVAL & GUENÉE, Hist. nat. Insectes (Spec. gén. Lépid.) 10 (2): 334 (United Kingdom: England). Syntypes 9.

Eupithecia sextiata DARDOIN & MILLIÈRE, 1867: in MILLIÈRE, Iconogr. Descr. Chenilles Lépid. inédits 2 (20): 370, pl. 89: 5 (France: Bouches-du-Rhône, near Aix). Syntypes 2♂1♀ (MNHN).

Eupithecia heydenaria STAUDINGER, 1870: Berl. Ent. Z. 14: 128 (Switzerland: Rigi; St.Moritz). Syntypes 2♂ 2♀ (MNHU).

Eupithecia distinctaria piemonticola SCHWINGENSCHUSS, 1954: Z. Wien. ent. Ges. 39 (6): 219 (Italy: Piemont, Col di Sestriere). Syntype(s) (NLHW). Valid at subspecific rank.

Unavailable names (incorrect subsequent spellings): *constrictaria*: MORRIS (1861); *destinctaria*: RAKOSY & WEBER (1986).

DIAGNOSIS: Wingspan 14-21 mm. Labial palpi grey. Frons, vertex and notum greyish white. Forewing with oblique transverse vein in areole; rather narrow, with straight costa, pointed apex and weakly curved termen; ground colour light grey; transverse lines fine, rather obscurely developed, forming dark costal spots; basal line oblique, evenly curved near costa; postbasal line inconspicuous; antemedial line straight, right or obtuse angled onto costa; first medial line slightly sinuate, crossing discal spot and sharply angled towards costa; postmedial line minutely waved, curved, usually formed a series of small dark dashes on the veins; terminal area often faintly darker; pale wavy line distinct, sometimes forming a weak pale tornal dot; discal spot large, obliquely elongate, black. Hindwing with shallow concave termen near apex; slightly paler; transverse lines indistinct except postmedial line; terminal area usually slightly darker; pale wavy line fine, minutely indented, sometimes obsolete; discal spot small, narrow and elongate, dark but paler than on forewing. Fringe whitish grey, spotted with grey or dark grey at vein ends.

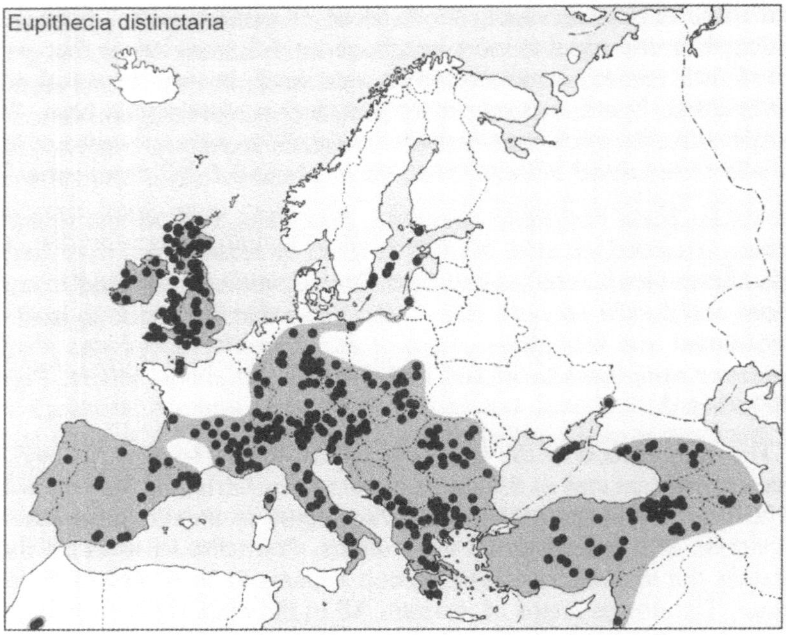

Abdomen light grey with narrow black lateral lines or with rows of small black lateral dots; second segment usually darker, brownish.

MALE AND FEMALE GENITALIA: See species-group diagnosis.

DISTRIBUTION: West-Palaearctic. Sporadically in Central and southern Europe: westward to western coast of Ireland and Portugal; eastward to south of European Russia; northward to Finland; southward to southern Spain and Greece. – Outside Europe in Asia Minor, Lebanon, Caucasus and Transcaucasus. Recorded also for Egypt (ANDERS 1913) and Iran (SCHÜTZE 1960).

PHENOLOGY: Uni- or bivoltine: (early April) early June to mid-July (early August). In the south bivoltine: mid-May (early May) to mid-June (mid-July); early August (mid-July) to early September (mid-September). Larva in July and August; in the south May-June and from mid-July to late August (early September); its suffers heavily from parasitization. Overwintering as pupa in flimsy cocoon in the ground.

BIOLOGY: Larva rather oligophagous (2). Found on the flowers and seeds of *Thymus serpyllum* (United Kingdom: CREWE 1862; BUCKLER 1899; BARRETT 1904; Germany: DIETZE 1875a, 1910, 1913; WEIGT 1991), also *T. vulgaris* (France: MILLIÈRE 1869; MABILLE 1872; DIETZE 1913) and *T. praecox (=drucei)* (United Kingdom: GOATER & al. 1981), *Odontites lutea* (Corsica: MABILLE 1869), *Satureja montana* (Lamiaceae) (Italy: CARRARA 1928), *Cytisus purgans* (Fabaceae), *Nepeta nepetella* (Lamiaceae), *Laserpitium gallicum, Pastinaca sativa, Peucedanum oreoselinum* (Apiaceae), *Solidago* (Asteraceae) (France: LHOMME 1935), *Hypericum perforatum* (Clusiaceae), *Galium pumilum* (Rubiaceae), *Leucanthemum atratum* (=*Chrysanthemum coronopifolium)* (Asteraceae) (DIETZE 1877; SPULER 1910), *Origanum vulgare* (Lamiaceae) (Germany: KOCH 1984; WEIGT 1991).

Text-figs 60-61: Diagnostic characters (indicated) of European *Eupithecia* species. Text-fig. 60: *E. distinctaria* HERRICH-SCHÄFFER, 1848. Text-fig. 61: *E. senorita* MIRONOV, sp. n.

HABITAT: Warm and dry landscapes, stony slopes and rocks, rocky coasts and limestone cliffs, preferring a southern aspect, forest clearings, fringes and 'alvars'. From 0 m up to about 1,600-1,700 m above sea-level in Apennines (Mt. Sirino, Mt. Pollino: coll. ZSM), to about 2,000 m in Balkan peninsula (coll. ZSM), to about 2,300-2,400 m in Alps (DANNEHL 1927; VORBRODT 1928), to about 2,400-2,600 m in Pyrenees (HAUSMANN pers. comm.; coll. FIBIGER) and to about 2,200-2,900 m in Sierra Nevada (WEHRLI 1926, 1927; WEIGT 1991).

SIMILAR SPECIES: *E. distinctaria* can be easily confused with *E. extraversaria* because of presence of distinct obliquely elongate and intensly black discal spots on the forewings in both species. However, *E. distinctaria* lacks the warm brownish suffusion; in this species on the forewing the antemedial line is always angled near costa, the postmedial line usually more waved and not angled onto costa; and second abdominal segment usually darker than the others (Text-fig. 60). The presence of an obtusely rounded ventral process, preceded by a deep notch of valva; the armature on the vesica and the sternum A8 with one digitiform apical lobe in male; and the shape of bursa copulatrix and disposition of spines and ductus seminalis in the female genitalia should distinguish this species.

REMARKS: A variable species. No significant differences can be found in colour and maculation between British and continental populations of *E. distinctaria*.

103. *Eupithecia senorita* MIRONOV, sp. n.

Unavailable names (misidentification): *distinctaria*: sensu DIETZE (1910, pl. 76: 608) nec HERRICH-SCHÄFFER, 1848 (identity: *Eupithecia distinctaria*); *distinctaria*: sensu WEHRLI (1926) nec HERRICH-SCHÄFFER, 1848 (identity: *Eupithecia distinctaria*); *deverrata*: sensu MIRONOV (2001) nec DIETZE, 1910 (identity: *Eupithecia deverrata*).

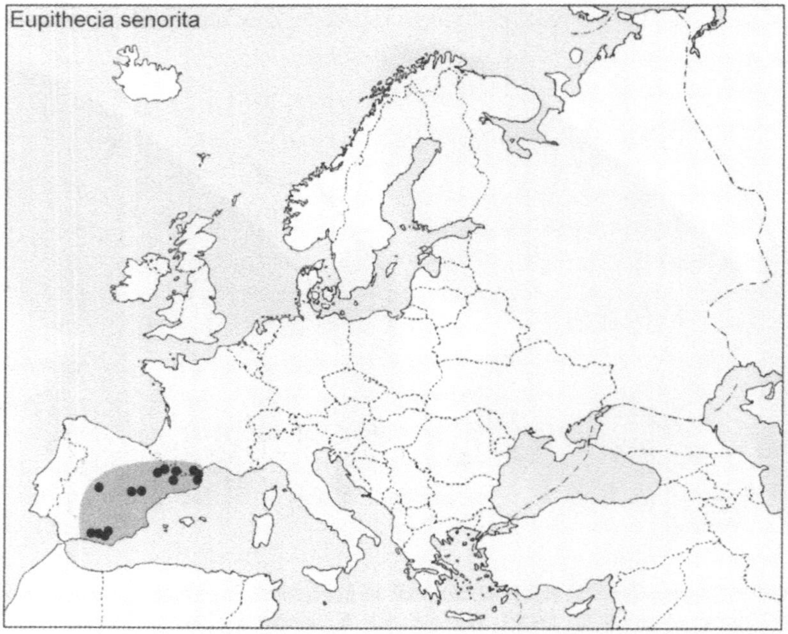

HOLOTYPE: ♂, Süd-Frankreich, Corbières, Albas, 400 m, 13.V.(19)70, L. (LUKASCH leg.), coll. ZSM.

PARATYPES: 1♂, Spanien, Aragon, Albarracin, 21.V.1955 (v. BUDDENBROCK coll.); 1♂1♀, Hispania, Sierra Nevada, Carr. Veleta, 1600 m, 8.VII.1971, KLIMESCH leg.; 1♀, Nordspanien, Prov. Gerona, Llansa, E.IV.(19)85, W. PAVLAS leg.; 1♀, Hispania süd, Sierra Nevada, Routa de Veleta, 2700 m, 17.VII.1991, G. BAISCH leg.; 1♂, Hispania, Andalucia, Sierra Nevada, El Parador Nat., 2500 m, 21.VII.1980, E. TRAUGOTT-OLSEN leg.; 1♀, Hispania, Cataluna, Espot, 1300 m, 9.VII.1984, F. SCHEPLER leg.; 2♂1♀, Hispania, Andalucia, Sierra Nevada, Cam. de Veleta, 2250 m, 31.VII.1984, 18.VII.1985, E. TRAUGOTT-OLSEN leg.; 3♀, the same localities, 1600 m, 13.VI, 2.-3.VII.1986, E. TRAUGOTT-OLSEN leg.; 1♂, Hispania, Granada, 25 km NO Granada, 1400 m, 7.V.1989, leg. F. SCHEPLER (coll. FIBIGER); 1♀, Spain, Prov. Teruel, Albarracin, 1200 m, 8.VII.1992, leg. B. SKULE; 1♂1♀, Spain, Prov. Lerida, Tremp Valley, Aransis, 8.VII.1993, 700 m, leg. P. SKOU; 2♀, Spain, Prov. Teruel, Albarracin, Val de Vecar, 1100 m, 22.V.1998, 10.VI.1999, leg. P. SKOU; 2♂3♀, Spanien, Sierra Nevada, Pte. De Ragua, 1800 m, 5-7.7.(19)78, L, leg. LUKASCH.

Paratypes are partially deposited in ZSM, ZFMK, ZMUC and private collections of M. FIBIGER (Sorø, Denmark) and P. SKOU (Stenstrup, Denmark).

DIAGNOSIS: Wingspan 14-19 mm. Labial palpi grey or yellowish grey. Frons, vertex and notum pale grey, often with yellowish tinge. Forewing rather narrow and elongate, costa slightly arched, apex pointed, termen obliquely straight; ground colour pale yellowish grey; transverse lines brownish, slightly broadened and darker near costa; basal line dentated, forming right-angle near costa; antemedial line almost straight, obtuse angled near costa; first medial line slightly sinuate, crossing discal spot and sharply angled toward costa; postmedial line fine, faintly sinuate and waved, perpendicularly curved to costa; terminal area relatively narrow; pale wavy line distinct, indented, with brownish shade on the innerside; discal spot large, obliquely elongate, rather ovate-oblong, intensely black. Hindwing with shallowly concave termen near apex, of the same colour, mottled and irrorate with dark brown scales; transverse lines indistinct, except basal and postmedial lines; terminal area narrow; pale wavy line conspicuous, indented, crossing close to termen; discal spot distinct, relatively large, elongate, paler than on forewing. Fringe whitish yellow, slightly spotted brownish at vein ends. Abdomen pale yellowish grey.

MALE AND FEMALE GENITALIA: See diagnosis of species-group and 'similar species' below.

DISTRIBUTION: West-Mediterranean. South-western France (Pyrenees-Orientales) and Spain (Gerona, Lerida, Teruel, Madrid, Almeria, Granada).

PHENOLOGY: Bivoltine: mid-April to late May; early July to mid-August.

BIOLOGY: Unknown.

HABITAT: In very different habitat types, e.g. warm and dry slopes and rocks with sparse vegetation, scrub or open woodland, uncultivated areas in farmland or hot lowlands close to the coast (SKOU, pers. comm.), possibly simply occurring where the so far unknown food-plant grows. From 0 m up to about 1,900 m above sea-level in Pyrenees and central Spain and 2,700-2,800 m in Sierra Nevada.

SIMILAR SPECIES: According to the structure of the male genitalia, *E. senorita* is closely related to *E. distinctaria*. There are no differences in structure of the male genitalia between these species. However, sternum A8 of male *E. senorita* has shorter, broader and blunter apical projection than in *E. distinctaria*. No differences in the female genitalia between *E. senorita* and *E. distinctaria* could be found. Adults of *E. senorita* rather similar to *E. liguriata*, but can be distinguished by the less arched costa near apex, more pointed apex and less rounded termen of forewing, and by the yellowish suffusion (Text-fig. 61). The costal spots on the forewing are smaller, the antemedial line obtuse angled near costa, the postmedial line less curved, the pale wavy line more distinct and more indented, shaded on the inner side; the postmedial line less angled on the hindwing than in *E. liguriata*.

REMARKS: Adults of the second generation of this species from Sierra Nevada are more greyish, without yellowish suffusion.

The ***centaureata*** species-group

Species-group introduced by VOJNITS (1977). This group is represented mainly by North African and Central Asian species. Four species which are found in Europe can be included in the *centaureata* group, as following: *E. extraversaria*, *E. centaureata*, *E. limbata* and *E. nachadira*. Four North African species *E. desertorum* DIETZE, 1910; *E. deverrata* DIETZE, 1910; *E. lecerfi* PROUT, 1928; *E. schwingenschussi* ZERNY, 1934 and some Central Asian species such as *E. irritaria* STAUDINGER, 1892; *E. subtilis* DIETZE, 1906; *E. decipiens* PETERSEN, 1909; *E. repetita* VOJNITS, 1981; *E. tonu* VIIDALEPP, 1988 and two East Asian species *E. chinae* VOJNITS, 1977 and *E. exquisita* VOJNITS, 1979 belong in this group and are included in the species-group diagnosis.

MALE GENITALIA: Uncus biapical, long or medium length. Papillae length about 0.4-0.7 times uncus length. Valva large and broad, with sclerotized, broad and blunt ventral process, and nearly straight dorsal margin; apex narrow and rounded; sacculus heavily sclerotized. Vinculum usually large, broad and elongated. Aedeagus relatively thin, length about 0.7-0.8 times valva length; about 3.5-5.0 times longer than medial width. Vesica armed with one small, irregular, flat cornutus at apex, one patch of minute spines and one elongated, flat, V-shaped cornutus at ductus ejaculatorius base. Sternum A8 with more or less prominent basal hollow; medially slightly narrower; its apical half broadened with two sclerotized, broad, slightly incurved and apically blunt lobes; sternum length about 0.6 times valva length.

FEMALE GENITALIA: Bursa copulatrix pear-shaped; completely covered with small spines in broadened basal part; with narrower, spineless, usually sclerotized ductus bursae, without longitudinal striations. Ductus seminalis medium length; broadly attached to apical part of corpus bursae on right side. Colliculum collar-like, usually short and narrow. Antrum large, broad, funnel-shaped, strongly sclerotized, with two long and broad ventral lobes. Anterior and posterior apophyses short and thin, not expanded at apices. Spurs of anterior apophyses short and thin. The shape of tergum A8 is specific for each species. Papillae anales small, slim and rounded.

REMARKS: This distinctive species-group can usually be distinguished from most other European *Eupithecia* species-groups by the large and broad valva with blunt ventral process, with heavily sclerotized sacculus, and also by the armature of vesica and shape of sternum A8 in the male. In the female genitalia the characteristic features are the large, funnel-shaped, strongly sclerotized antrum with two long and broad ventral lobes and the very short spurs of anterior apophyses.

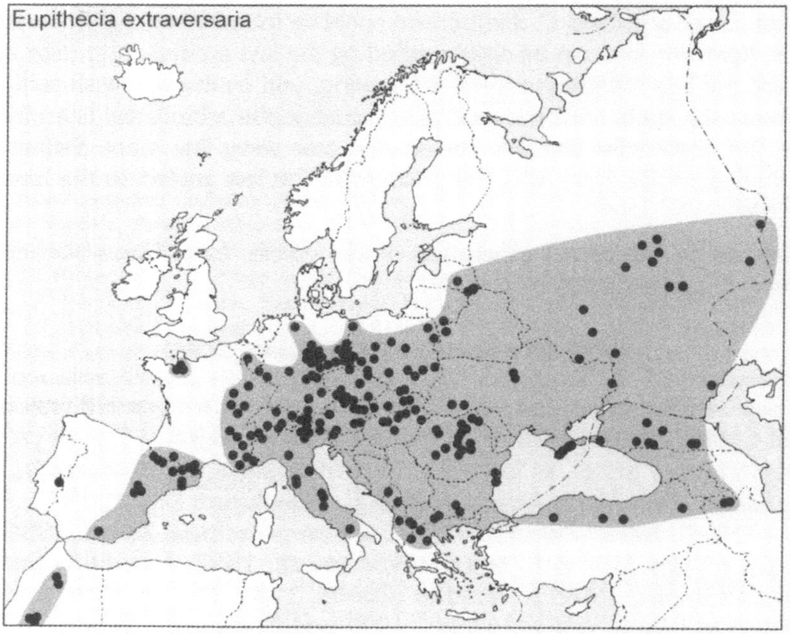

104. *Eupithecia extraversaria* HERRICH-SCHÄFFER, 1852

Eupithecia extraversaria HERRICH-SCHÄFFER, 1852: Syst. Bearb. Schmett. Eur. 6 (55): 77; ibidem (1853), 3 (Suppl.), pl. 88: 543 (non binominal), (France). Syntype(s).

Eupithecia libanotidata GUENÉE, 1858: in BOISDUVAL & GUENÉE, Hist. nat. Insectes (Spec. gén. Lépid.) 10 (2): 333 (Poland: Silesia). Syntypes 3♀.

Eupithecia extraversaria atlanticaria Schwingenschuss, 1955: Z. Wien. Ent. G. 40 (8): 229 (Morocco: Great Atlas). Syntype(s) (NHMW). Valid at subspecific rank.

DIAGNOSIS: Wingspan 16-20 mm. Labial palpi brownish grey. Frons, vertex and notum pale greyish ochreous. Forewing unicolorous pale greyish ochreous; transverse lines fine, rather inconspicuous, not sinuate, brownish grey, slightly broadened at costa; basal and antemedial lines slightly curved near costa, three obscurely developed medial lines sharply angled near discal spot; postmedial line evenly curved near costa; pale indented wavy line more or less distinct; discal dot large, narrow, black. Hindwing slightly paler; transverse lines weakly indicated; terminal area and pale wavy line indistinct; discal spot minute, rounded or elongated. Fringe pale grey or pale brownish grey, slightly spotted with darker brownish grey, more distinctly chequered on the hindwing. Abdomen pale brownish grey, usually with ochreous tinge; with dark brown lateral stripes.

MALE GENITALIA: Valva with weak, broad and blunt ventral process. Vinculum broad but short. In the vesica the elongated, V-shaped cornutus has two apically tapered lobes at ductus ejaculatorius base. Sternum A8 with two relatively narrow apical lobes.

FEMALE GENITALIA: Tergum A8 triangular, with straight anterior margin, and with small, pointed medial tip on the posterior margin.

DISTRIBUTION: West-Palaearctic. In Europe from Portugal to Urals, northward to Latvia, southward to Granada (Spain). – Outside Europe in North Africa (Morocco: ? subsp. *atlanticaria* SCHWINGENSCHUSS, 1955), north of Asian Turkey, Caucasus, Transcaucasus and northern Iran. Found in south-western Siberia and Altai Mts (MIRONOV 1991; VIIDALEPP 1996). – New for the fauna of Portugal: 1 ♂ Portugal, Alto Alentejo, Portalegre, Marvão, 3.VI.1997, M.F.V. CORLEY leg. (Coll. Corley).

PHENOLOGY: Uni- or bivoltine: mid-June (early June) to mid-July (early August); in the south bivoltine: from late May to late August (early September). Larva: early July to early September (mid-September). Overwintering as pupa, occasionally twice.

BIOLOGY: Larva oligophagous (1). Found on the flowers of *Peucedanum oreoselinum, Angelica sylvestris* (Germany: DIETZE 1872), also *Pimpinella saxifraga, Bupleurum falcatum* (Switzerland: FAVRE 1905), *Laserpitium latifolium, Pimpinella major (=magna), Seseli libanotis, Daucus carota* (Germany: DIETZE 1910, 1913), *Athamanta turbith (=mathioli)* (Apiaceae) (Alps: DIETZE 1913), *Eupatorium* (Asteraceae) (France: LHOMME 1935) and *Pastinaca sativa* (Apiaceae) (Germany and France: WEIGT 1990).

HABITAT: Warm and dry slopes, rocks and screes, often in mountains. From 0 m up to about 1,400 m above sea-level in Apennines (coll. ZSM), to about 1,800-1,900 m in Balkan peninsula and Asia Minor (TOMIČ & al. 1990; coll. SOMMERER), to about 2,000 m in southern Alps (HELLMANN & al. 1999), and to about 2,200-2,300 m in Pyrenees (MAZEL & PESLIER 1997).

SIMILAR SPECIES: *E. extraversaria* differs from the following species by brownish grey colour of forewing, with elongated discal spots. In the male genitalia the ventral process of the valva is shorter; elongated cornutus on the vesica narrower, and apical rods of sternum A8 longer and narrower than in *E. centaureata*. The female genitalia are very similar to those in the following species except the shape of tergum A8.

105. *Eupithecia centaureata* (DENIS & SCHIFFERMÜLLER, 1775)

Geometra centaureata DENIS & SCHIFFERMÜLLER, 1775: Syst. Verz. Schmett. Wien.: 114 (Austria: Vienna district). Syntype(s) lost.

? *Phalaena signata* SCOPOLI, 1763: Ent. Carniol., Ordo 3: 231 (Slovenia and NE. Italy: Carniolia). Syntype(s) lost. Senior synonym ?

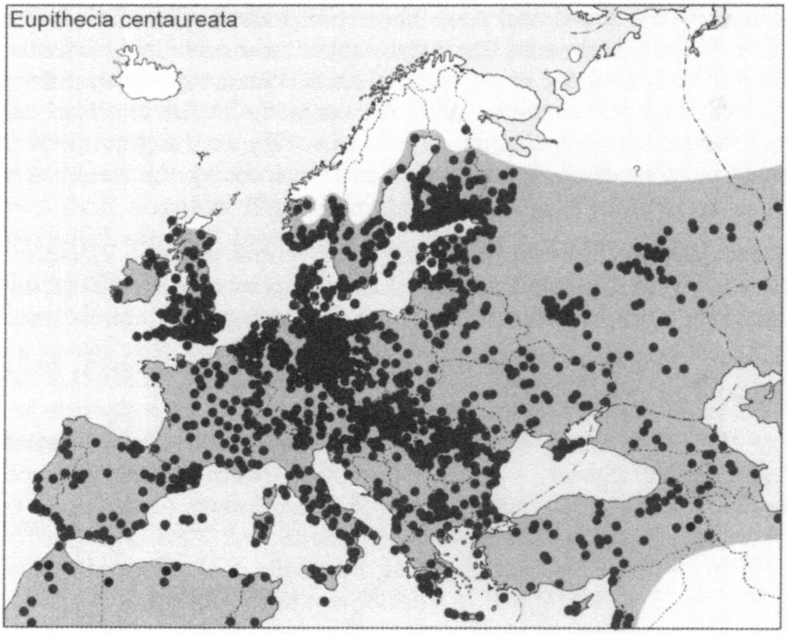

Geometra oblongata THUNBERG, 1784: Diss. ent. sistens Insecta Suecica: 1: 14, fig. 1 (Sweden). Lectotype ♂ (DEUU).

Phalaena boloniensis FOURCROY, 1785: Ent. Parisiensis 2: 282 (France: Paris). Syntype(s) in Mus. Autun?

Eupithecia oblongata var. *centralisata* STAUDINGER, 1892: Dt. ent. Z. Iris 5: 250 (Central Asia: Uzbekistan: Margelan; Namangan; Samarkand; Kirghizstan: Alexander Mts; Kazakhstan: Saisan). Syntypes 13 (MNHU). Valid at subspecific rank.

Eup(ithecia) oblongata f. *obscura* DIETZE, 1910: Biol. Eupithecien 1, pl. 70: 132; ibidem (1913), 2: 63 (Kirghizstan: Alai Mts; Kazakhstan: Uralsk; Italy: southern Tyrol). Syntypes 6 (MNHU). Valid at subspecific rank.

Eupithecia centaureata dagestani VOJNITS, 1977: Acta Zool. Acad. Sci. Hung. 23 (1-2): 229, fig. 2 (Russia: Dag(h)estan, Petrovsk-port [=Makhachkala]). Holotype ♂ (ZFMK). Valid at subspecific rank.

Eupithecia centaureata dsharkendi VOJNITS, 1977: Acta Zool. Acad. Sci. Hung. 23 (1-2): 230, fig. 3 (Kazakhstan: Dsharkend [=Panfilov], Semiretschensk). Holotype ♀ (ZFMK). Valid at subspecific rank.

Unavailable names (misidentification): *succenturiata*: sensu HUFNAGEL (1767) nec LINNAEUS, 1758 (identity: *Eupithecia succenturiata*). – (infrasubspecific): *albidior*: HEINRICH (1916); *bistrigata*: LEMPKE (1969). - (incorrect subsequent spellings): *centaurearia*: BOISDUVAL (1840); *centauerata* TOMIČ & al. (1990).

DIAGNOSIS: Wingspan 15-24 mm. Labial palpi, frons and vertex brown or whitish brown. Notum white. Forewing white; basal area with three blackish costal spots; basal and antemedial lines often inconspicuous, black, narrow and angled; postmedial line black, narrow, sinuate, sharply angled near costa; medial area with a large, blackish costal spot between discal dot and postmedial line and with smaller whitish brown marginal spot; inner half of terminal area ochreous, its outside half brownish white; discal dot obliquely elongate, usually slightly curved, intensely black. Hindwing white with whitish brown spots along inner margin; discal dot minute, brownish. Fringe white, spotted with pale brown. Abdomen ochreous brown, with small white dorsal spots; with large black lateral spots at least on the first and second segments; the last segment white.

MALE GENITALIA: Valva with more or less long and blunt ventral process (VOJNITS 1977).

Vesica armed with one elongated, V-shaped cornutus with two apically expanded lobes at ductus ejaculatorius base. Sternum A8 with deep basal hollow; apical lobes slightly tapered toward apexes.

FEMALE GENITALIA: Tergum A8 almost rectangular with shallow medial hollow on the anterior margin, also with rounded corners and small, pointed medial tip on the posterior margin.

DISTRIBUTION: Palaearctic. All Europe, except Iceland and northern Scandinavia. – Outside Europe widespread in North Africa and Asia (from Turkey, Lebanon, Israel and Syria eastward to Mongolia and Amur region), where occurs the lighter f. *centralisata* STAUDINGER, 1892.

PHENOLOGY: Bi- or polyvoltine: early May (mid-April) to late June (mid-July); mid-July to mid-September (late September). In extreme south on the wing all year round. Larva: mid-June to late July; late July to late September (late October). Overwintering as pupa.

BIOLOGY: Larva polyphagous (1). Found on the flowers, seeds and leaves of various herbaceous plants, such as *Rumex acetosa, R. acetosella* (Polygonaceae), *Silene vulgaris* (Caryophyllaceae), *Clematis vitalba* (Ranunculaceae), *Plantago major* (Plantaginaceae), *Genista triacanthos, Vicia cracca, Ononis spinosa, Lotus corniculatus, Trifolium pratense* (Fabaceae), *Silaum silaus, Pimpinella major (=magna), P. saxifraga, Angelica sylvestris, Peucedanum oreoselinum, P. palustre, Selinum carvifolia, Daucus carota, Pastinaca sativa, Heracleum sphondylium, Seseli annuum (=coloratum), Bunium bulbocastanum, Torilis japonica, Cicuta virosa, Eryngium, Ferulago* (Apiaceae), *Erica tetralix, Calluna vulgaris* (Ericaceae), *Galium album, G verum, G. mollugo* (Rubiaceae), *Lysimachia vulgaris* (Primulaceae), *Linaria vulgaris* (Scrophulariaceae), *Origanum* (Lamiaceae), *Succisa pratensis, Scabiosa columbaria, Knautia arvensis* (Dipsacaceae), *Atriplex littoralis, Chenopodium* (Chenopodiaceae), *Campanula glomerata, C. rotundifolia* (Campanulaceae), *Filipendula ulmaria* (Rosaceae), *Dittrichia viscosa, Solidago virgaurea, Tanacetum vulgare, Hieracium umbellatum, Bidens tripartita, Eupatorium cannabinum, Aster tripolium, Centaurea, Senecio jacobaea, S. vulgaris, S. erucifolius, S. aquaticus, S. bicolor, Achillea millefolium, A. ligustica, Anthemis, Chrysanthemum coronarium, Matricaria, Calendula arvensis, Gnaphalium, Leucanthemum vulgare, Matricaria perforata (Tripleurospermum maritimum* var. *inodorum), Artemisia vulgaris, A. campestris, Arctium* (Asteraceae) (United Kingdom: CREWE 1861c; BUCKLER 1899; BARRETT 1904; Corsica: MABILLE 1868, 1869; France: GOOSSENS 1871a; Germany: DIETZE 1871b, 1910, 1913; WEIGT 1990; WEGNER pers. comm.; Italy: CARRARA 1928; Denmark: JUUL 1948; Finland: SEPPÄNEN 1970; Malta: VALLETTA 1973). Found also on the flowers of *Achillea biebersteinii* (Asteraceae) (Tadjzhikistan: original data) and *Ziziphora clinopodioides* (Lamiaceae) (Kirghizstan: original data).

PARASITOIDS: Hymenoptera, Ichneumonidae: *Dusona* sp., *Bathythrix aerea, Mesochorus pectoralis, Platylabus iridipennis* - Hymenoptera, Braconidae: *Apanteles cajae (=difficilis)* - Hymenoptera, Encyrtidae: *Litomastix chalconotus, L. truncatellus* (THOMPSON 1945; SHENEFELT 1972; HERTING 1976; original data).

HABITAT: Many habitats with open terrain, gardens, also found at ragwort blossoms. From 0 m up to about 1,000 m above sea-level in Pyrenees (MAZEL & PESLIER 1997), to about 1,500 m in Balkan peninsula (PINKER 1968), to about 2,000-2,100 m in Alps and

Apennines (WOLFSBERGER 1981; HELLMANN & al. 1999; REZBANYAI-RESER & MAIER 2000), and to about 2,300 m in Asia Minor (coll. FIBIGER).

SIMILAR SPECIES: This is the best known *Eupithecia* species and can easily be recognised and distinguished from other *Eupithecia* by its white forewing with large blackish costal spot between discal dot and postmedial line, and other maculation on the wings, together with the marked abdomen. The male and female genitalia are similar to those of *E. extraversaria* but its has good differential characteristics from those in other *Eupithecia* species.

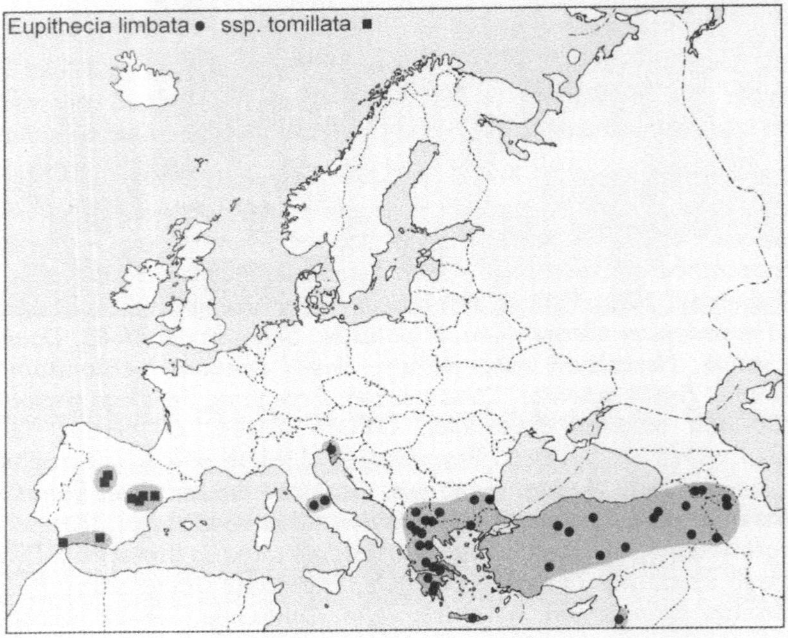

106. Eupithecia limbata STAUDINGER, 1879

Eupithecia limbata STAUDINGER, 1879: Horae Soc. ent. Ross. 14 (4): 476 (Turkey: Asia Minor, Kerasdere). Syntypes (MNHU).

Tephroclystia tomillata CHRÉTIEN, 1904: Bull. Soc. Ent. France: 133 (Spain: Segovia, San Ildefonso). Lectotype ♀ (coll. HERBULOT in ZSM, examined).

Eupithecia limbata var. *occidens* WEHRLI, 1927: Int. Ent. Z. 20 (37): 343 (Spain: Sierra Nevada). Syntypes 1♂3♀ (ZFMK, examined). Valid at subspecific rank.

DIAGNOSIS: Wingspan 13-18 mm. Labial palpi brownish grey. Frons, vertex and notum pale ash-grey. Forewing pale ash-grey; transverse lines brownish grey, slightly sinuate, inconspicuous but well marked as small costal spots; postmedial line evenly curved near costa and usually well indicated on the veins by minute brownish grey spots; terminal area ochreous or greyish ochreous, greyer at apex; pale wavy line usually distinct but sometimes broken; discal spot large, ovate, intensely black. Hindwing pale ash-grey; transverse lines conspicuous, brownish grey; terminal area ochreous or greyish ochreous; pale wavy line fine-dentated, indistinct; discal spot large, ovate, dark brown. Fringe pale grey, distinctly spotted with greyish ochreous at vein ends. Abdomen pale ash-grey with ochreous tinge and usually with small black lateral spots on each segment; second segment ochreous.

The subsp. ***tomillata*** CHRÉTIEN, 1904 with discal spots on all wings larger, ovate or circular; terminal areas usually darker, rusty brownish or rusty ochreous, and postmedial line on the hindwing often less angled than in nominate subspecies.

MALE GENITALIA: See species-group diagnosis.

FEMALE GENITALIA: Ductus bursae slightly sclerotized. Ductus seminalis not strongly broadened at base. Tergum A8 rectangular, elongated and relatively narrow.

DISTRIBUTION: Mediterranean. Nominate subspecies in Macedonia, Bulgaria, Greece and Crete; in Iberian peninsula replaced by subsp. *tomillata*. Intermediate separate populations found in Italy. – Outside Europe in Asian Turkey, Armenia, Azerbaijan and Lebanon (nominate subspecies).

PHENOLOGY: Bi- or probably polyvoltine: late April to late May; late June (mid-June) to mid-July (early August). In Lebanon found in September and October. Iberian subsp. *tomillata* probably bivoltine: mid-June (late May) to late July (early September). Larva: late August to September.

BIOLOGY: Larva monophagous (1). Found on the flowers of *Eryngium amethystinum* (Italy: REBEL 1923b, 1924; CARRARA 1928) and *Eryngium* sp. (Spain: SCHWINGENSCHUSS 1962).

HABITAT: Limestone and calcareous sandstone. From 0 m up to about 1,900-2,000 m above sea-level in Asia Minor (coll. FIBIGER), and to about 2,000-2,300 m in Greece (coll. ZSM). The subsp. *tomillata* up to about 1,800-1,900 m above sea-level in central Spain (coll. ZSM).

SIMILAR SPECIES: Unlikely to be confused with any other European *Eupithecia* species. *E. limbata* can be easily recognized by the ash-grey or yellowish grey ground colour with dark ochreous terminal areas and large, ovate-oblong, black discal spots on all wings. The male and female genitalia can be reliably used to separate *E. limbata* from other species, including other representatives of the *centaureata* species-group. The larger papillae, smaller ventral process of valva, broader and shorter apical lobes of the male sternum A8, and tergum A8 in the female are of different shape to those in both the preceding species.

107. Eupithecia nachadira BRANDT, 1941

Eupithecia nachadira BRANDT, 1941: Mitt. Münch. Ent. Ges. 31 (3): 877, pl. 29: 23 (Iran: Belutschistan, Kouh i Taftan). Lectotype ♂, herewith designated to stabilize nomenclature (NHRS, examined).
Eupithecia mandarinca MIRONOV, 2001: Nota lepid. 24 (1/2): 27, fig. 3, 15 (Ukraine: Crimea, Karadagh). Holotype ♀ (ZISP, examined). New junior synonym.

DIAGNOSIS: Wingspan 16.0 mm. Labial palpi, frons, vertex and notum whitish grey. Forewing with slightly arched costa, pointed apex and oblique faintly rounded termen; ground colour whitish grey; transverse lines light brownish grey; basal, ante- and postmedial lines forming broad dark costal spots; basal and antemedial lines evenly curved; postbasal line inconspicuous; postmedial line slightly sinuate, three times obtusely angled

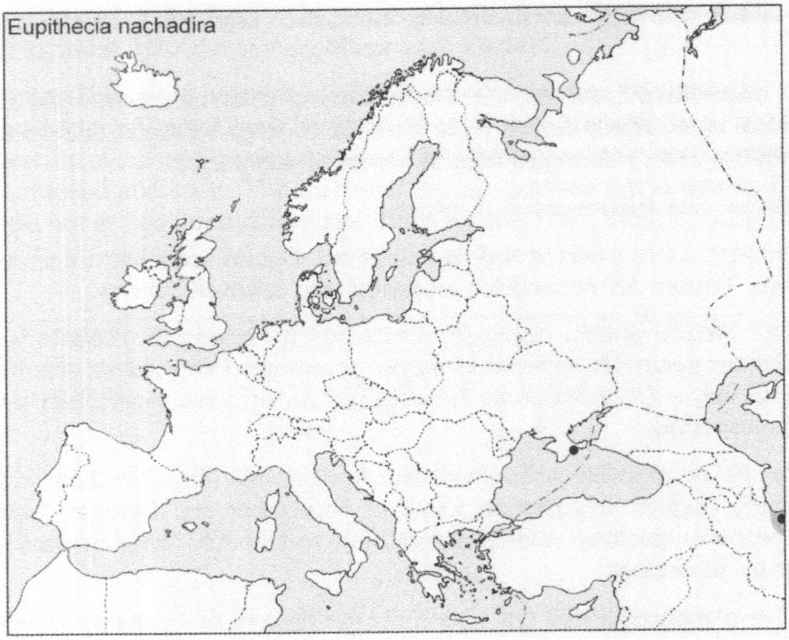

toward costa; terminal area broad, darker, brownish grey; whitish wavy line wide, broken; interrupted by veins; discal spot large, elongate, rather ovate-oblong, brownish grey. Hindwing with weakly concave termen; slightly paler, whitish grey; basal and medial areas mottled and irrorate with brownish grey scales; terminal area also broad, darker, brownish grey as on the forewing; pale wavy line rather indistinct, indented, fine and broken; discal spot large, narrow and elongate, dark brownish grey. Fringe smoky white, spotted with brownish grey at vein ends.

MALE GENITALIA: Uncus elongated. Valvae asymmetrical: the left valva with two broad and blunt sclerotized processes; the right valva with one longer and rather pointed sclerotized process. Sternum A8 with pointed lateral basal lobes and blunt incurved apical lobes.

FEMALE GENITALIA: See species-group diagnosis.

DISTRIBUTION: Turanian. Ukraine (southern Crimea). – Outside Europe in Iran (Elburs Mts, Belutschistan).

PHENOLOGY: Univoltine: late April to late June.

BIOLOGY AND HABITAT: Unknown. From 2,500 m up to about 2,800 m above sea-level in southern Iran (BRANDT 1941).

SIMILAR SPECIES: *E. limbata*, the species with which it could be confused. Externally distinguished from nominate subspecies of *E. limbata* by paler ground colour, more curved antemedial line, postmedial line curved more closely around discal spot; short, black dashes on the veins of forewing lacking; transverse lines on the hindwing indistinct, terminal areas brownish grey without rusty tinge, and lighter, larger and more elongated discal spots on all wings. The uncus narrower at base, the pointed vinculum, the wider and asymmetrical valvae with narrower apices and longer ventral processes, especially on the right valva, the longer aedeagus in the male genitalia and the different shape of sternum A8 in male, can be successfully used to separate *E. nachadira* from *E. limbata*.

REMARKS: Three adults of *E. nachadira*, including lectotype and paralectotype, were examined and compared with holotype of *E. mandarinca* from Crimea. There were no morphological differences or differences in wing colour and maculation between both these species.

The ***despectaria*** species-group

Species group introduced here for the first time. It is a species-poor group including mainly species from North Africa, Middle East and Central Asia, as following: *E. despectaria* LEDERER, 1853; *E. tenellata* DIETZE, 1906; *E. graciliata* DIETZE, 1908; *E. mekrana* BRANDT, 1941; *E. relativa* VOJNITS, 1982 and *E. peridonea* MIRONOV, 1989. Although not yet known from Europe, there is a strong possibility that *E. mekrana* might be discovered in the far south-east of European Russia.

MALE GENITALIA: Uncus medium length, narrow, biapical. Papillae usually large, length nearly equal to uncus length. Valva without ventral process, with broad and blunt apex; sacculus weakly sclerotized. Vinculum short and broad or very broad, sometimes with small medial hollow. Aedeagus short or medium length and rather stout, about 0.7-1.0 times valva length. Vesica armed with two small cornuti (one elongated, the other irregular) at ductus ejaculatorius base and numerous minute spinules. Sternite A8 usually with parallel lateral margins, narrow and deep basal and especially apical hollows, two elongated and narrow apical rods with apices rounded or blunt; sternum A8 about 0.7-0.8 times valva length.

FEMALE GENITALIA: Bursa copulatrix pear-shaped; almost completely covered with small spines; ductus bursae usually sclerotized. Ductus seminalis short or medium length, sharply curved, attached to middle of corpus bursae or ductus bursae on right side. Colliculum collar-like, short and narrow. Antrum membranous. Anterior and posterior apophyses thin and short, not expanded at their apices. Spurs of anterior apophyses broadened at apices. Tergite A8 usually trapeziform. Papillae anales rounded, short and rather wide.

REMARKS: I have placed this species-group near the *centaureata* species-group based on external similarity, as well as relatively similar shape of sternite A8 in male and cornuti on the vesica in the male genitalia.

108. Eupithecia mekrana BRANDT, 1941

Eupithecia mekrana BRANDT, 1941: Mitt. Münch. Ent. Ges. 31 (3): 877, pl. 29: 24 (Iran, Bender Tchahbahar). Lectotype ♂, herewith designated to stabilize nomenclature (NHRS, examined).

Eupithecia mekrana khorassana BRANDT, 1941, Mitt. Münch. Ent. Ges. 31 (3): 877 (Iran: Binaloud Mts). Lectotype ♂, herewith designated to stabilize nomenclature (NHRS, examined). Valid at subspecific rank.

Eupithecia idonea VOJNITS, 1982: Acta Zool. Acad. Sci. Hung. 28 (3-4): 412, figs 12, 13 (Armenia, Megri on Araks river). Holotype ♂ (HNHM, examined).

Eupithecia stulta VOJNITS, 1982: Acta Zool. Acad. Sci. Hung. 28 (3-4): 413, figs 14 (Armenia, St. Negram on Araks river). Holotype ♂ (HNHM, examined).

Eupithecia commenticia VOJNITS, 1982: Acta Zool. Acad. Sci. Hung. 28 (3-4): 414, figs 15, 17 (Armenia, Megri on Araks river). Holotype ♂ (HNHM, examined).

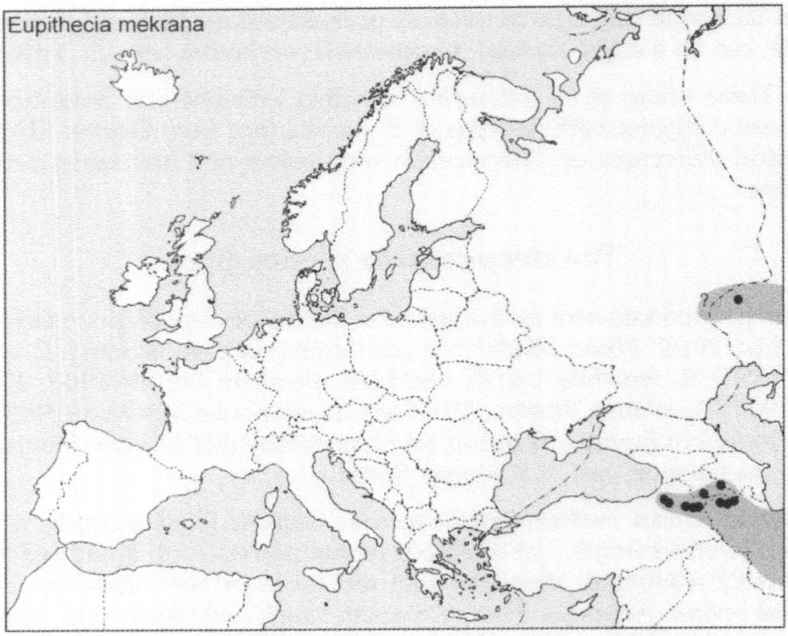

DIAGNOSIS: Wingspan 11-16 mm. Labial palpi pale whitish brown. Frons, vertex and notum white with yellowish tinge. Forewing narrow and elongated, with straight costa and oblique, slightly curved termen; apex pointed; ground colour pale yellowish white; transverse lines rather faint, pale brown; basal and antemedial lines more conspicuous, slightly curved or almost straight, well marked on the veins and near costa; two thin medial lines sinuate and rather indistinct; postmedial line evenly curved, minutely wavy, sometimes broken, well marked by a series of brown dots on the veins; terminal area darker, pale yellowish brown, narrow, but broadened on to apex; wavy line whitish; discal spot narrow, elongated, black or dark brown. Hindwing pale yellowish white; three brownish transverse lines evident: basal, ante- and postmedial; terminal area narrow, darker, pale brown with weak defined whitish wavy line; discal spot very small, usually linear or sometimes rounded. Fringe yellowish white, slightly spotted with pale brown at vein ends. Abdomen white with a soft yellowish tinge; segment A3 light yellowish brown.

MALE GENITALIA: See species-group diagnosis, including the following: valva with wide and blunt ventral extension. Vinculum without medial hollow.

FEMALE GENITALIA: See species-group diagnosis.

DISTRIBUTION: Turanian. Armenia, Azerbaijan (Nakhichevan region), Turkey (Prov. Agri and Kars), Iran, Saudi Arabia (*subsp. amiralis* WILTSHIRE, 1986), Turkmenia (Kopet-Dagh Mts), northward to Asian part of Orenburg province in Russia (BRANDT 1941; WILTSHIRE 1986; MIRONOV 1991; VIIDALEPP 1996). – New for the fauna of Russia: about 40 exx. (including 1♀ only), Russia, S-Urals, Orenburg oblast, (50°40-45'N, 54°26-28'E), 170-230 m, Pokrovka vill. 20 km S, Schibendy valley, 4.-7.VI, 17.-18.VII.1998; 21.-23.VI.1999; 31.VII-1.VIII, 30.-31.VIII.2000; 10.VI.2001, T. & K. NUPPONEN leg. (coll. K.NUPPONEN and ZISP).

PHENOLOGY: Bivoltine: early June (mid-May) to late June (mid-July); mid-July to late August (late September).

BIOLOGY: Unknown.

HABITAT: Whitish limestone rocks (chalk hills) surrounding the flat and dry *Artemisia* steppe. From 0 m up to about 230 m above sea-level in Orenburg province (K. NUPPONEN pers. comm.), and to about 2,500 m in Iran (BRANDT 1941).

SIMILAR SPECIES: *E. mekrana* is most similar to very light forms of *E. distinctaria* than to other European *Eupithecia* species. Distinguished by the small size, light whitish yellow ground colour of wings and usually by more pointed forewings. The shapes of uncus, valva, sternum A8 and cornuti in the vesica in the male genitalia, as well as the shape of bursa copulatrix in the female genitalia can be used to distinguish *E. mekrana* from *E. distinctaria* with accuracy, as described and illustrated.

REMARKS: The population of this species was found about 240 km to the south of the Ural river, of which the right bank marks the south-east border of the territory of Europe. This locality lies in the extensive eastern European steppe zone. Possibly, *E. mekrana* will be found on the same chalk hills to the west or northwest (for example in Samara or Saratov provinces) in the future.

The ***variostrigata*** species-group

Introduced here for the first time. Represented in Europe by one species *E. variostrigata*. At least seven African *Eupithecia* species can also be included in the *variostrigata* species-group: *E. falkneri* VOJNITS, 1978 from Algeria, *E. ochralba* HERBULOT, 1983 and *E. dinshoensis* HERBULOT, 1983 from Ethiopia, *E. astales* FLETCHER, 1978, *E. isopsoliodes* FLETCHER, 1978, *E. nigrataenia* FLETCHER, 1978 and *E. cretosa* VOJNITS, 1994 from Tanzania. There are also two species from Iran in this group: *E. siata* BRANDT, 1938 and *E. keredjana* BRANDT, 1938.

MALE GENITALIA: Uncus relatively long, biapical. Papillae narrow and elongate, length about 0.8 times uncus length. Valva without ventral process, elongate and narrow, broadened at base, with slightly curved, almost parallel ventral and dorsal margins; apex rather angular; sacculus weakly sclerotized. Vinculum small, short and narrow. Aedeagus slim, small: short and thin; length about 0.5-0.6 times valva length and about 4.0 times longer than medial width. Vesica armed with five small dentate cornuti, one elongate, flat cornutus and one small, irregular, curved cornutus at ductus ejaculatorius base. Sternum A8 with two narrow symmetrical rods narrowly joined at base, distally sclerotized and tapering with apex incurved, slightly expanded and obtuse; rods with deep sinus between them; basal hollow very shallow; sternum length about 0.5 times valva length.

FEMALE GENITALIA: Bursa copulatrix membranous, small, spherical or pouch-like; about 1/2-1/3 covered by small spines at base, with small patch of minute spines in ductus bursae between ductus seminalis and colliculum. Ductus seminalis short and narrow. Colliculum collar-like, short, broad and slightly sclerotized. Antrum membranous, short and broad. Anterior and posterior apophyses relatively long and thin. Spurs of anterior apophyses long, thin, gradually expanded to apices. Tergum A8 rectangular (height shorter than medial width). Papillae anales relatively large, broad and rounded.

REMARKS: Representatives of the *variostrigata* species-group are closely related to those of the *satyrata* species-group on the basis of male and female genitalic characteristics. However, the longer and narrower valva; the several small dentate cornuti on the vesica; sternum A8 with two long rods narrowly joined at base; and the size and shape of corpus bursae and area covered by spines are clear differences.

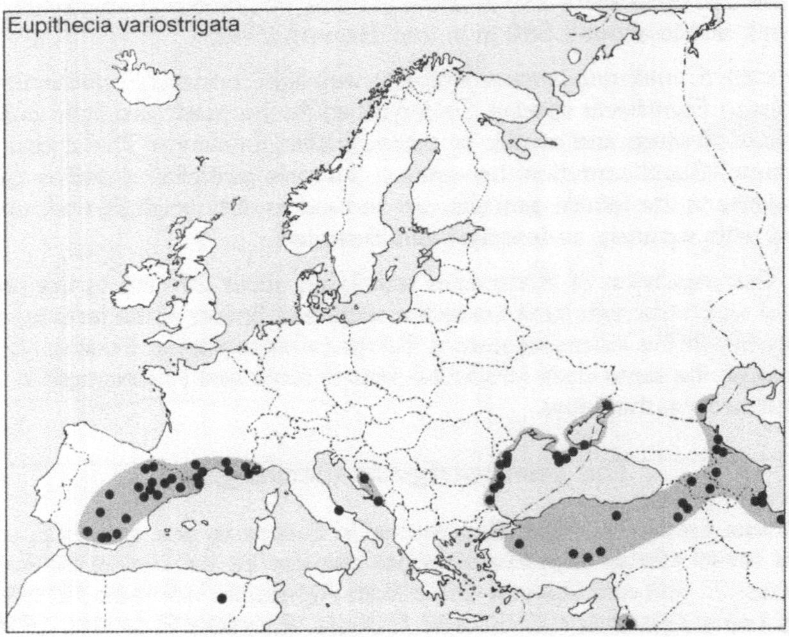

109. *Eupithecia variostrigata* ALPHÉRAKY, 1876

Eupithecia variostrigata ALPHÉRAKY, 1876: Horae Soc. ent. Ross. 10: 40 (Russia: Taganrog). Holotype ♂ (one topotype ♀ in ZISP, examined).

Eupithecia artemisiata CONSTANT, 1884: Annls Soc. ent. France (6) 4 (3): 203, pl. 9: 3 (France: Alpes Maritimes, littoral region). Syntypes, at least 20, (MNHN?). Valid at subspecific rank.

Eupithecia littorata CONSTANT, 1884: Annls Soc. ent. France (6) 4 (3): 205, pl. 9: 4 (France: Alpes Maritimes, littoral region). Syntypes 8 (MNHN?).

?*Eupithecia artemisiata* var. *constantina* BETHUNE-BAKER, 1885: The Entomologist's Mon. Mag. 21: 244 (Algeria Mauritania: Lambessa). Syntypes 4, ♂ ♀ (BMNH). Valid at subspecific rank.

?*Eupithecia albosparsata* JOANNIS, 1891: Bull. Soc. ent. France 6: 81 (Turkey: Césarée). Syntypes 2 (MNHN).

(Eupithecia) variostrigata f. *designata* DIETZE, 1913: Biol. Eupithecien 2: 49; ibidem (1910), 1, pl. 73: 339, 340 (as *variostrigata*) (Caucasus). Syntypes, including ♂ (MNHU). Valid at subspecific rank.

Unavailable names (incorrect subsequent spelling): *artemesiata* SCOBLE (1999).

DIAGNOSIS: Wingspan 18-23 mm. Labial palpi brownish ash-grey. Frons, vertex and notum ash-grey usually with scattered brownish scales. Forewing with extremely straight costa, termen obliquely rounded and apex narrow, rather pointed; ground colour mainly ash-grey; basal area, stalks of R and Cu and A, and area between M3 and CuA1 pale reddish brown; transverse lines blackish brown, oblique, almost straight or slightly sinuate, sharply angled near costa, especially distinct medial line with darker inner shade and with outer white broad or finer transverse line; postmedial line slightly sinuate, right or obtuse angled near costa, with inner darker shade; pale wavy line conspicuous, slightly sinuate, usually also with darker inner shade; discal spot absent. Hindwing pale ash-grey with slightly dark terminal area, weak indications of transverse lines, and small dark, usually indistinct discal spot. Fringe pale grey, distinctly spotted with dark brown at vein ends. Abdomen ash-grey, usually with light reddish brown tinge.

MALE AND FEMALE GENITALIA: See species-group diagnosis.

DISTRIBUTION: Mediterranian-Turanian. In Europe: Spain, southern France, Adriatic coasts of Italy and former Yugoslavia, coastal areas of Black and Caspian Sea. – Outside Europe in Algeria, Asian Turkey, Jordan, eastern Transcaucasus (Armenia, Azerbaijan), northern and western Iran, Turkmenistan (Kopet-Dagh Mts), Uzbekistan and Tadjikistan (south-western Pamirs).

PHENOLOGY: Univoltine: mid-September (early September) to mid-October (late October). However, recorded for Gravosa (Croatia) on 25 April (SCHWINGENSCHUSS & WAGNER 1927). Larva: August until November. Overwintering as pupa.

BIOLOGY: Larva oligophagous (1). Found on the flowers and seeds of *Artemisia caerulescens* subsp. *gallica* (France: DIETZE 1910, 1913) and *A. herba-alba* (Asteraceae) (Spain: ZERNY 1927), recorded also on *Camphorosma monspeliaca* and *Solidago nudiflora* (Asteraceae) (France: CONSTANT 1884; LHOMME 1935).

HABITAT: Steppes, including arid low mountains and plateaus, dunes and sandy places on beaches or by salt lakes. From 0 m up to about 1,200 m above sea-level in central Spain (coll. ZSM), to about 2,300 m in Central Asia (Pamirs: coll. ZISP).

SIMILAR SPECIES: Unlikely to be confused with any other European *Eupithecia* species except *E. extensaria*. However, *E. variostrigata* can usually be easily distinguished by the darker forewing with pale reddish brown stalks of R, Cu and A and M3-CuA1 area and the absence of oblique, broad whitish lines, separating basal, medial and terminal areas. The male and female genitalia, especially the shapes of valva and sternum A8, the small corpus bursae and the area covered by spines, can be used to separate *E. variostrigata* from other European species successfully.

REMARKS: According to the last data, the population of a very small form *artemisiata* CONSTANT, 1884 is certainly extinct. All suitable littoral sites on the east coast of Rhône river (Hyères and Cannes) are now destroyed (TAUTEL pers. comm.). This form has the white oblique lines, including the wavy line, less distinct on the forewing than those in the nominate subspecies; the medial transverse whitish band on the hindwing is inconspicuous as well.

The *nevadata* species-group

The *nevadata* species-group was introduced by MCDUNNOUGH (1949), consisting of some North American species, as follows: *E. nevadata* PACKARD, 1871; *E. ravocostaliata* PACKARD, 1876; *E. cestata* (HULST, 1896); *E. implorata* (HULST, 1896) and *E. cestatoides* MCDUNNOUGH, 1949. Represented in Europe by one species, *E. insigniata*. The East Asian species *E. insignioides* WEHRLI, 1923, also belongs to the *nevadata* species-group and *E. costimacularia* LEECH, 1897 from Japan can probably be included in this species-group also.

MALE GENITALIA: Uncus usually medium length, biapical. Papillae broad, length about 0.6-0.7 times uncus length. Valva with a large, sclerotized ventral process, usually with apex blunt but also occasionally pointed; with nearly straight dorsal margin and rounded apex; sacculus sclerotized. Vinculum short and broad. Aedeagus thick and short, length about 0.7-0.8 times valva length. Vesica armed with two medial dentate cornuti, apically with numerous spinules and one or two large spines, and with one large, elongated and folded cornutus at ductus ejaculatorius base. Sternum A8 short and wide, with very shallow basal hollow and with two short lateral lobes usually over 0.4-0.5 sternum length; sternum length about 0.3-0.5 times valva length.

FEMALE GENITALIA: Bursa copulatrix rounded or elongated, completely spined; ductus bursae with small membranous area without spines on left side. Ductus seminalis long, broadly attached to corpus bursae. Colliculum collar-like, stout. Antrum membranous. Anterior and posterior apophyses short and thin. Papillae anales short, wide and rounded.

REMARKS: The species of this group can usually be easily distinguished from other *Eupithecia* species by the distinctive purplish brown costal blotches on the forewings; valva with prominent ventral process and specific shape of sternum A8 in male (MCDUNNOUGH 1949; BOLTE 1990).

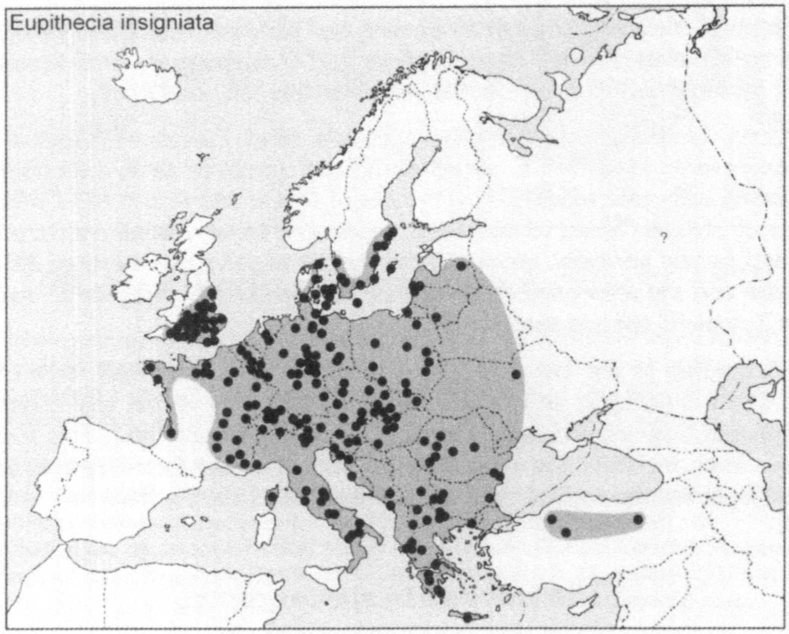

110. Eupithecia insigniata (HÜBNER, 1790)

Ph(alaena) Geom(etra) insigniata HÜBNER, 1790: Beitr. Gesch. Schmett. 2 (4): 97, pl. 3: U (Europe). Syntype(s) ♀ lost.
Phalaena (Geometra) consignata BORKHAUSEN, 1794: Naturgesch. Eur. Schmett. 5: 315 (Europe). Syntype(s).
Unavailable names (incorrect subsequent spellings): *consignaria*: BOISDUVAL (1840); *insignata*: MILLIÈRE (1870).

DIAGNOSIS: Wingspan 18-24 mm. Labial palpi brown. Frons and vertex pale ash-grey. Notum purplish brown with wide light ash-grey longitudinal medial stripe. Forewing elongate and rather pointed, of a delicate pale ash-grey with four large, purplish brown costal blotches (basal, postbasal, medial and triangular postmedial); transverse lines brownish, fine, distinct, sinuate, especially postbasal, antemedial and double medial; medial area with fine, black dashes on the veins; postmedial line evenly curved near costa; terminal area narrow, pale wavy line lacking; discal spot large, intensely black, elongate and narrow, contiguous with medial purplish brown costal blotch. Hindwing slightly paler, greyish; two basal transverse lines oblique, indicated only near dorsum; discal spot small, conspicuous, round or slightly elongated. Fringe pale grey, spotted with grey or brownish grey. Abdomen purplish brown with black lateral stripes; last segment usually paler, brownish grey or ash-grey.

MALE GENITALIA: Papillae broad, slightly sclerotized, apically covered by short setae. Valva with a large, sclerotized ventral process; these processes usually asymmetric: blunt on left valva and prominently produced on right valva.

FEMALE GENITALIA: Bursa copulatrix elongated. Tergum A8 width about 1.5 times longer than its length.

DISTRIBUTION: European-Westasiatic. Mainly in Central and southern Europe; northward to Sweden (along coast of Baltic Sea); southward to Crete. Recorded for Kazan (Russia) by KRULIKOWSKY (1908). However, I have not seen a single specimen from Russia in collections so far. – Outside Europe in Asia Minor.

PHENOLOGY: Univoltine: mid-April (late March) to mid-June (late July). Larva: early April to mid-May (in the south), usually from late May (mid-May) to mid-June (early July). Recorded in June to August in United Kingdom. Overwintering as pupa, sometimes twice or even three times.

BIOLOGY: Larva oligophagous (1). Found on the leaves of *Malus domestica* (United Kingdom: BUCKLER 1899) and *Prunus spinosa, Crataegus monogyna* (Germany: DIETZE 1910, 1913), *Malus sylvestris* (Germany: WEIGT 1988), *Pyrus amygdaliformis* (Corsica: MABILLE 1868, 1869); recorded also on leaves of *Prunus dulcis (P. amygdalus)* and *Rubus idaeus* (France: LHOMME 1935), *Pyrus communis, Amelanchier ovalis, Prunus padus, P. avium, P. domestica* and *P. cerasus* (Rosaceae) (Europe: WEIGT 1980, 1988).

PARASITOIDS: Hymenoptera, Ichneumonidae: *Cyclolabus pactor* (THOMPSON 1945; HERTING 1976; ZWAKHALS pers. comm.).

HABITAT: Gardens, orchards and apple plantations, scrub and hedges, forest edges, warm and dry slopes, rocks and screes. From 0 m up to about 1,500-1,700 m above sea-level in Balkan peninsula (coll. ZMUC), to about 1,300-1,500 m in Apennines (Mt. Sirino: coll. ZSM), to about 1,600 m in southern Alps (WEIGT 1988), and to about 2,100-2,200 m in central Turkey (coll. SOMMERER).

SIMILAR SPECIES: A handsome, rather elegant and invariable species, which impossible to confuse with other European *Eupithecia* species. *E. insigniata* is easily recognized by the delicate pale ash-grey ground colour with large, purplish brown costal blotches on the forewings. The sclerotized papillae, the asymmetrical valva with large blunt ventral process, the cornuti on the vesica, the shape of sternum A8 in male, as well as the shape of corpus bursae with spiniferous area in the female genitalia should distinguish this species.

The *gueneata* species-group

The *gueneata* species-group was introduced by SCHÜTZE (1961) as subgenus *Delaeveria*. This is a western and central Palaearctic species-group which is represented in Europe by four species: *E. trisignaria*, *E. gueneata*, *E. gratiosata* and *E. scalptata*. In Asia there are eleven additional species: *E. syriacata* STAUDINGER, 1879; *E. subpulchrata* ALPHÉRAKY, 1882; *E. cingulata* CHRISTOPH, 1885; *E. accurata* STAUDINGER, 1892; *E. conviva* DIETZE, 1903; *E. gluptata* DIETZE, 1903; *E. albertiata* SCHÜTZE, 1961; *E. kruusi* VIIDALEPP, 1988; *E. remmi* VIIDALEPP, 1988; *E. dechkanata* MIRONOV, 1989 and *E. turkmena* MIRONOV, 1989 (SCHÜTZE 1961; VOJNITS 1978; VIIDALEPP 1988; MIRONOV 1989, 1991).

MALE GENITALIA: Uncus biapical, short or medium length. Papillae narrow and rather short; length about 0.5-0.8 times uncus length. Valva without ventral process; broadest medially; with angulated or evenly curved (as in *E. trisignaria, E. scalptata* and *E. accurata*) ventral margin and with rather narrow apex; sacculus sclerotized. Vinculum short and usually broad. Aedeagus large and stout, length about 0.7-0.9 times valva length, about 2.5-4.0 times longer than medial width. Vesica armed with two more or less large dentate cornuti, dense elongated patch of spinules, one elongated, narrow, U-shaped cornutus and one small, irregular, flat cornutus at ductus ejaculatorius base. Sternum A8 peg-like with broad base, narrowing to apex; basal hollow shallow; apical hollow about 0.03-0.1 or 0.4-0.6 times sternum length (or absent in *E. kruusi*); accordingly with two short or long and narrow apical lobes; sternum length about 0.5-0.6 times valva length.

FEMALE GENITALIA: Bursa copulatrix rounded, rather pear-shaped, usually heavily sclerotized, especially ductus bursae, which has a patch of small spines near colliculum; corpus bursae almost completely covered by small spines, usually with larger marginal spines; in the central part of corpus bursae there are often several large horn-like or drawing-pin-like spines. Ductus seminalis short and narrow; attached to corpus bursae near colliculum or to middle on right side. Colliculum collar-like, broad and short. Antrum membranous. Anterior and posterior apophyses thin, short or of medium length. Spurs of anterior apophyses broadened to apices. Tergum A8 square or rectangular (in some species with shallow medial hollow in the posterior margin). Papillae anales rather small and rounded.

REMARKS: This species-group can usually be separated from other European *Eupithecia* species by the usually rich colour and contrasting pattern of the forewings with darker medial area (exept *E. trisignaria*). The valva is broader medially, often with angulated ventral margin; the large and stout aedeagus; the vesica armed with two dentate cornuti, dense patch of spinules, and one small, irregular, flat cornutus at ductus ejaculatorius base; the sclerotized corpus bursae with large horn-like or drawing-pin-like spines in central part should distinguish this species-group.

111. *Eupithecia trisignaria* HERRICH-SCHÄFFER, 1848

Eupithecia trisignaria HERRICH-SCHÄFFER, 1848: Syst. Bearb. Schmett. 3 (31): 120, 131; ibidem (1846), pl. 28: 175, 176 (non binominal), (Europe). Syntype(s).
Unavailable names (incorrect subsequent spelling): *trisignata*: HERRICH-SCHÄFFER (1861).

DIAGNOSIS: Wingspan 15-25 mm. Labial palpi blackish grey. Frons and vertex ash-grey to dark grey. Notum grey. Forewing dark greyish fuscous or brownish fuscous; costal

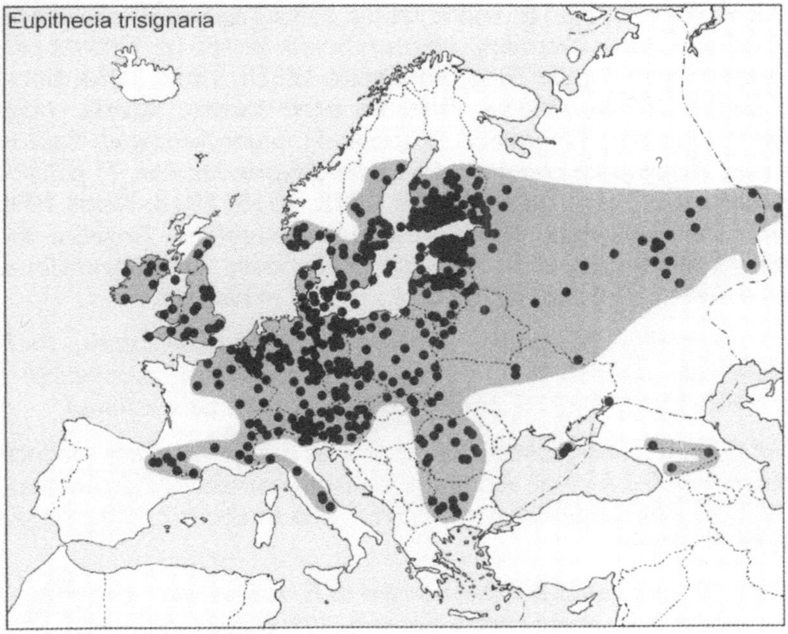

margin usually mottled and irrorate with white scales with two dark costal spots (ante- and postmedial) and sometimes with a third dark costal spot near apex; transverse lines inconspicuous or obsolete; antemedial line straight, oblique, with a hook-like curve near costa; medial line much sinuate, crossing discal spot and perpendicularly angled to costa; postmedial line sinuate, rather straight and curved only near costa; the veins usually with dark dashes in place of postmedial line and in terminal area; pale wavy line often clouded and weakly waved, without tornal spot; discal spot black, rather small, obliquely elongate or sometimes circular. Hindwing greyish fuscous but paler than forewing; transverse lines indistinct except weak-defined, medially angled postmedial line and faintly paler band beyond it; pale wavy line absent; discal spot minute, narrow, elongate. Fringe pale grey, spotted with dark grey at vein ends. Abdomen grey with a dorsal row of small black dots and black lateral spots on two or three anterior segments.

MALE GENITALIA: Uncus short. Valva with evenly curved ventral margin; sacculus weakly sclerotized. Sternum A8 with a deep apical hollow.

FEMALE GENITALIA: Bursa copulatrix rather pear-shaped, sclerotized, without large horn-like or drawing-pin-like spines in the middle. Ductus seminalis attached to middle of corpus bursae.

DISTRIBUTION: Euro-Siberian. Widespread in Europe from Ireland and Pyrenees to Urals; northward to central Scandinavia; southward to south-western Bulgaria. – Outside Europe in Caucasus and western Transcaucasus (Georgia); found also in south-western and southern Siberia.

PHENOLOGY: Uni- or bivoltine: mid-May (early April) to late June; mid-August (late July) to early September (late September). In the north univoltine: early June to early August. Larva: early August (late July) to mid-September (late October). Overwintering as pupa.

BIOLOGY: Larva oligophagous (1). Found on the flowers and seeds of *Heracleum sphondylium, Pastinaca sativa* (Germany: HERRICH-SCHÄFFER 1861a; WEGNER pers. comm.), also *Angelica sylvestris* (United Kingdom: CREWE 1861b, 1862, 1863; BUCKLER 1899; Germany: DIETZE 1874a, 1875a; WEGNER pers. comm.; Russia: original data), *Aegopodium podagraria, Pimpinella saxifraga, P. major (=magna), Laserpitium latifolium, L. siler, Athamanta cretensis, Peucedanum oreoselinum, P. palustre, P. officinale, P. cervaria* (Germany: DIETZE 1871a, 1872, 1910, 1913; WEIGT 1990; WEGNER pers. comm.; Denmark: JUUL 1948), *Anethum graveolens, Angelica archangelica* subsp. *litoralis* (Apiaceae) and probably occasionally on *Artemisia campestris* (Asteraceae) (Finland: SEPPÄNEN 1970). Often heavily parasitized.

PARASITOIDS: Hymenoptera, Ichneumonidae: *Agrypon flaveolatum*, *Endromopoda detrita*, *Platylabus iridipennis* - Hymenoptera, Encyrtidae: *Litomastix quercicola* (THOMPSON 1945; HERTING 1976; original data; ZWAKHALS pers. comm.).

HABITAT: Light, open forests, forest edges, clearings and paths, shady roadsides. From 0 m up to about 1,300-1,400 m above sea-level in Pyrenees and Apennines (MAZEL & PESLIER 1997; Mt. Majella: ZAHM pers. comm.), and to about 2,000 m in Alps (DIETZE 1913; VORBRODT 1925).

SIMILAR SPECIES: *E. trisignaria* is a most similar to *E. subfuscata*, yet darker, with inconspicuous transverse lines, exept postmedial line, with distinct dark costal spots and evident pale wavy line on the forewing (Text-fig. 49). The characteristic cornuti on the vesica, the narrow and elongated sternum A8 with two narrow digitiform apical rods in male, and the sclerotized base of ductus seminalis in the female genitalia allow reliable separation of *E. trisignaria* from all other European *Eupithecia* species.

REMARKS: In my opinion *E. trisignaria* belongs to the *gueneata* group. This placement is based mainly on genitalic characteristics: the vesica with two dentate cornuti, elongate patch of spinules and elongated flat cornutus in male genitalia, and heavily sclerotized corpus bursae in female genitalia.

112. *Eupithecia gueneata* MILLIÈRE, 1862

Eupithecia gueneata MILLIÈRE, 1862: Iconogr. Descr. Chenilles Lépid. inédits 1 (6): 257, 279, pl. 1: 1, 2 (France: Ardéche, Celles-les-Bains). Syntypes ♂ ♀ (MNHN?).

Eupithecia separata STAUDINGER, 1879: Horae Soc. Ent. Ross. 14 (4): 476 (Turkey: Asia Minor, Amasia). Holotype ♀ (MNHU).

Eupithecia gueneata var. *busambraria* RAGUSA, 1889: Il Naturalista Siziliano 8 (10/11): 229, pl. 3: 3-4 (Sicily: Busambra; Ficuzza). Syntype(s). Valid at subspecific rank.

Eupithecia saueri VOJNITS, 1978: Acta Zool. Acad. Sci. Hung. 24 (3-4): 432, figs 1, 2 (Armenia: Yerevan [=Erivan]). Holotype ♂ (ZFMK; examined).

Unavailable names (incorrect subsequent spellings): *gueneeata*: SPULER (1910); *guencata*: CARRARA (1928); *guenata*: GARTHE (1976); *gueniata*: LASTUKHIN (1998).

DIAGNOSIS: Wingspan 18-24 mm. Labial palpi brownish grey. Frons and vertex whitish grey, with brownish tinge. Notum pale brick-red, darker, blackish grey in front. Forewing from bright to pale brick-red ground colour; basal area along costa blackish grey to ash-grey, mottled and irrorate with white scales, edged outwardly with white costal spot; antemedial line oblique, slightly sinuate evenly curved or obtuse angled towards costa; one medial line fine, whitish, angled beyond discal spot; postmedial line less oblique, faintly sinuate, obtuse angled near costa, broken between M3 and CuA2; medial area broadened

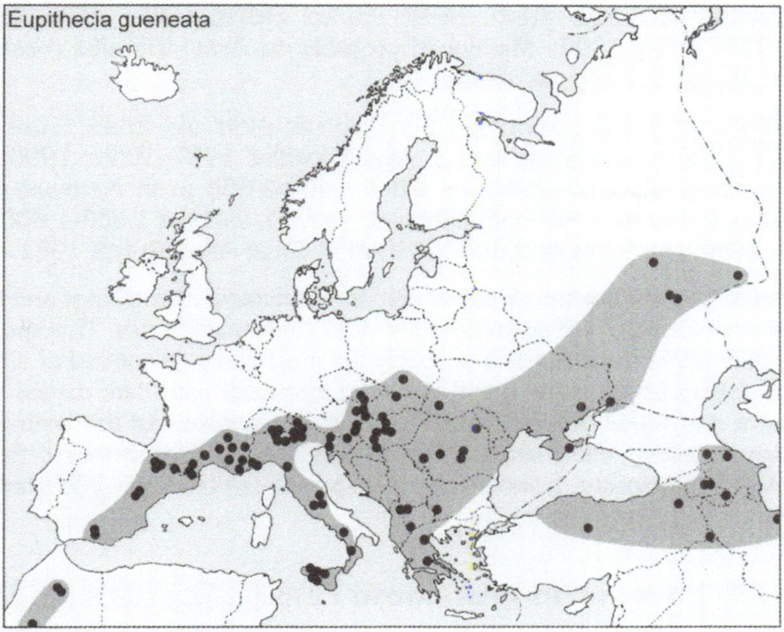

to costa and along dorsum, interrupted by triangular brick-red blotch under stalk of Cu, blackish grey, mottled and irrorate with white scales, edged on the inner and outer sides by white lines, which are sometimes faintly black-margined; whitish wavy line relatively wide, evenly curved, parallel to termen, scarcely waved or dentated, shaded on both sides with blackish grey; discal spot rounded or ovoid, blackish but usually inconspicuous against the dark background of medial area. Hindwing grey, darker blackish grey along dorsum and blackish grey with brick-red suffusion along termen; transverse lines invisible except a weak double pale stripe beyond medial area and pale, fine, minutely waved line; discal spot conspicuous, rounded, blackish. Fringe dark, slightly chequered whitish grey and blackish grey. Abdomen brick-red of various intensity; the first, second segments and abdominal tip are covered with blackish grey scales.

MALE GENITALIA: Sternum A8 with shallow apical hollow between two small apical lobes.

FEMALE GENITALIA: Bursa copulatrix with large medial spines, joined into a longitudinal and sclerotized row, broadened at base. Colliculum slightly inclined to right.

DISTRIBUTION: West-Palaearctic. Mainly in southern Europe, except Mediterranean islands (Balearics, Corsica, Sardinia, Malta and Crete). – Outside Europe separate population in North Africa (Morocco: subsp. *plantei* HERBULOT, 1981), eastwards through Asia Minor, Caucasus and Transcaucasus to Turkmenistan (Kopet-Dagh Mts).

PHENOLOGY: Univoltine: mid-July (late June) to late August (mid-September). In the south has a partial second generation until mid-October. Larva: late August (mid-July) to early October (late October). Overwintering as pupa in flimsy cocoon in the ground, sometimes several years.

BIOLOGY: Larva monophagous (1). Found on the flowers and seeds of *Pimpinella sax-*

ifraga (Apiaceae) (Hungary: VIERTL 1898; France: DIETZE 1901, 1910, 1913; Italy: CARRARA 1928; WEIGT 1990). Mentioned probably on *Artemisia alba (=camphorata)* (Asteraceae) (Romania: CARADJA 1896).

HABITAT: Steppe landscapes, warm and dry meadows, rocks and screes. From 0 m up to about 1,200-1,300 m above sea-level in Alps (DANNEHL 1927; WEIGT 1990), to about 1,400 m in Apennines (coll. ZSM), to about 1,500-1,600 m in Pyrenees and Sierra Nevada (MAZEL & PESLIER 1997; coll. ZSM; coll. ZMUC), to about 1,600-1,800 m in Asia Minor (coll. ZSM), and to about 2,600-2,700 m in Great Atlas (RUNGS 1981).

SIMILAR SPECIES: Variable in size, depth of colour, distinctness of markings and breadth of dark medial area. Usually wild specimens are paler than reared ones. This species needs no detailed differential diagnosis; it is probably the most easily recognized of all European *Eupithecia*, distinguished by the warm brick-red forewings with dark medial band. The shape of valva and shape of sternum A8 in male are variable, but the ventral angle on valva is always present. The male and female genitalia are very similar to those in the closely related *E. gratiosata*, however, the two species can be easily separated on external features.

113. *Eupithecia gratiosata* HERRICH-SCHÄFFER, 1861

Eupithecia gratiosata HERRICH-SCHÄFFER, 1861: Neue Schmett. Eur. angr. Ländern 3: 27, pl. 135 (Europe?). Syntypes ♂ ♀ lost.

Eupithecia gueneata var. *subseparata* CHRISTOPH, 1885: in ROMANOFF, Mém. Lépid. 2: 134, pl. 7: 3 (Turkmenistan: Kisil-Arwat). Holotype ♀.

Eupithecia gratiosata var. *lavata* FUCHS, 1901: Stett. Ent. Z. 62 (1-2): 124 (Sicily). Holotype ♀ (NSMW?). Valid at subspecific rank.

Eup(ithecia) gratiosata f. *pallida* DIETZE, 1910: Biol. Eupithecien 1, pl. 70: 127; ibidem (1913), 2: 67 (Russia: Orenburg). Holotype (MNHU). Valid at subspecific rank.

DIAGNOSIS: Wingspan 19-27 mm. Labial palpi brown or pale brown. Frons white with brown band anteriorly and brown blotches near antennal bases, or sometimes entirely brown. Vertex white, often with brownish transverse band. Notum snowy white or with weak ochreous tinge. Forewing snowy white; inner part of basal area ash-grey or light brownish grey at costa; outer part of basal area pale greyish ochreous, darker near costa; antemedial line usually three times sharply angled; medial area pale ash-grey, sometimes with brownish tinge, darker towards costa, usually with white or whitish ochreous area between M3 and CuA2; medial line fine, sinuate, right or sharply angled near costa, white; postmedial line sinuate with black dashes or points on the veins; inner part of terminal area pale greyish ochreous, outer part more or less pale ash-grey, with large white apical blotch and with smaller white blotch between M3 and CuA1; white wavy line broad; discal dot small, elongated, dark grey, rather inconspicuous. Hindwing snowy white; ash-grey transverse lines well indicated near dorsum; postmedial line indistinct but well marked with a series of blackish dots on the veins; terminal area pale ash-grey, well marked with two large blotches (one near tornus, the other between M1 and M3 or sometimes absent); white wavy line broad, indented; discal spot greyish, rounded or slightly elongated. Fringe white, distinctly spotted with more or less pale ash-grey at vein ends. Abdomen snowy white, sometimes with weak ochreous tinge; first and second segments with small black lateral dots; last segment blackish grey at end.

MALE GENITALIA: Sternum A8 with shallow apical hollow.

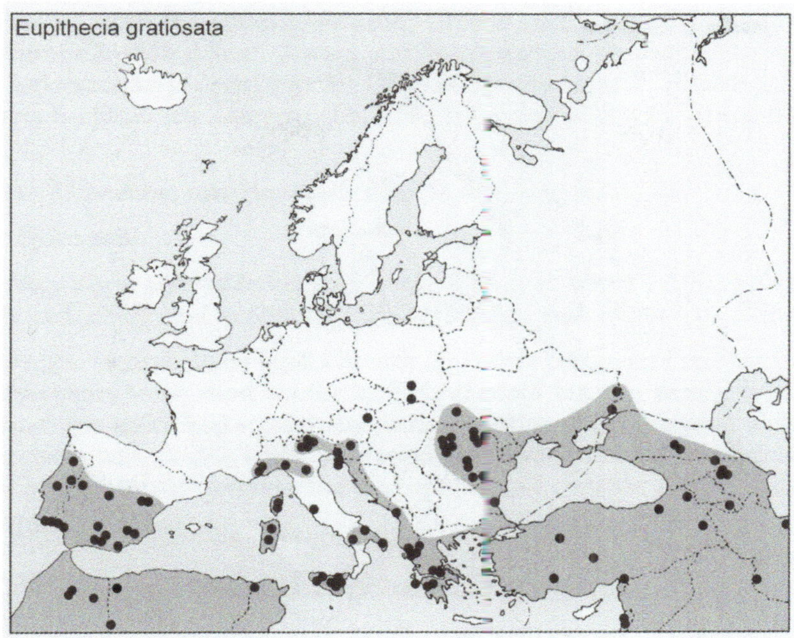

FEMALE GENITALIA: Bursa copulatrix with numerous large spines in the more sclerotized central part.

DISTRIBUTION: West-Palaearctic. Southern Europe. – Outside Europe in North Africa, Asia Minor, Caucasus, Transcaucasus, Lebanon, Syria, Iraq, Iran, Turkmenistan (Kopet-Dagh Mts) and north-western China (Kuldja).

PHENOLOGY: Univoltine: early May (mid-April) to late June (mid-July). Larva: mid-June to late July (mid-August). Overwintering as pupa, sometimes up to four times.

BIOLOGY: Larva oligophagous (1). Found on the flowers and seeds of *Ferulago campestris (=galbanifera), Heracleum sphondylium* (Romania: CARADJA 1895, 1896; DIETZE 1910, 1913), *Anethum* and *Pastinaca* (Apiaceae) (Greece: THURNER 1967). In captivity has been fed on *Foeniculum vulgare (=officinale)* (Romania: CARADJA 1895).

PARASITOIDS: Hymenoptera, Ichneumonidae: *Phobocampe* sp. (THOMPSON 1945).

HABITAT: Steppes, warm and dry low mountains, oases at middle altitudes. From 0 m up to about 1,000-1,300 m above sea-level in Alps and Balkan peninsula (ZANGHERI 1975; coll. ZSM), to about 1,600 m in Middle Atlas (RUNGS 1981), and to about 2,200-2,300 m in Asia Minor (coll. ZSM).

SIMILAR SPECIES: *E. gratiosata* is readily recognized and can be easily distinguished from other European *Eupithecia* species by the large size, white wings and body, broad ash-grey medial area with fine, white medial line on the forewing and other distinctive maculation. The shapes of valva and sternum A8, cornuti on the vesica, shape of bursa copulatrix and the area covered by spines should distinguish this species.

REMARKS: A variable species; in f. *pallida* DIETZE, 1910 the white on the forewing greatly predominates.

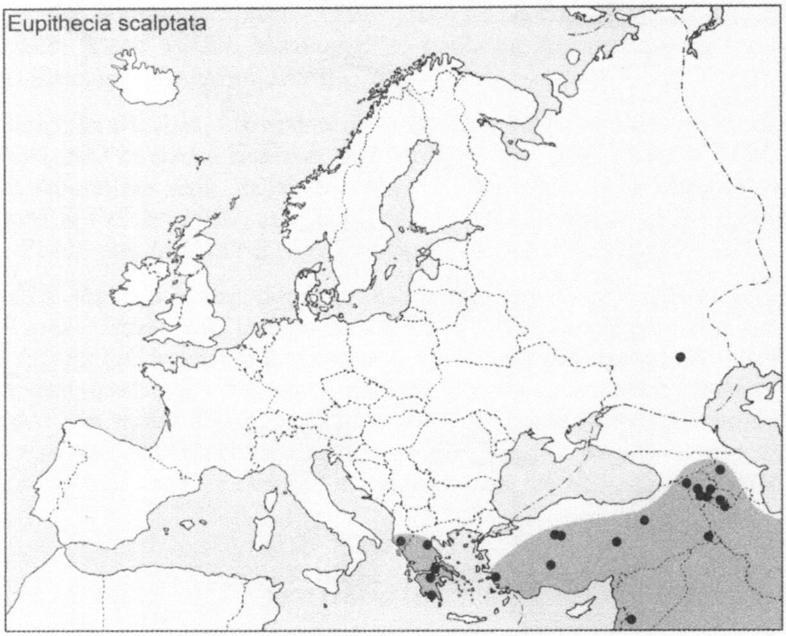

114. *Eupithecia scalptata* CHRISTOPH, 1885

Eupithecia scalptata CHRISTOPH, 1885: in ROMANOFF: Mém. Lépid. 2: 134, pl. 7: 4 (Turkmenistan: Askhabad; Azerbaijan: Valley of Araks, near Ordubad). Lectotype ♂ (ZISP, examined).

DIAGNOSIS: Wingspan 16-22 mm. Labial palpi whitish grey, sometimes with scattered pale brownish scales. Frons, vertex and notum white with greyish tinge. Antennae whitish ringed light brown. Forewing with ground colour whitish; inner part of basal area ash-grey, usually darker near costa; outer part of basal area pale greyish ochreous, likewise darker and greyer near costa; postbasal and antemedial lines always straight and oblique, sharply angled at costa; medial area pale blackish grey with fine, white, evenly sinuate medial line which is sharply angled or strongly curved near costa, and also with lighter longitudinal streak between M3 and CuA1; postmedial line evenly sinuate; terminal area pale greyish ochreous with more or less distinct whitish apical streak; white wavy line fine, straight or slightly sinuate, usually fine-dentated near apex; discal dot small, rounded, black, usually inconspicuous. Hindwing whitish with distinct pale grey transverse lines, darker near dorsum; terminal area slightly darker, greyish ochreous near tornus, with more or less conspicuous, fine, dentated white wavy line; discal dot minute, darkish, rounded. Fringe white, distinctly spotted with grey or dark grey at vein ends. Abdomen white with greyish tinge; first and second segments slightly darker, brownish.

MALE GENITALIA: Valva with evenly curved ventral margin. Vinculum rather narrow. Sternum A8 with deep apical hollow about 0.4-0.5 times sternum length; with two rather long and narrow apical rods.

FEMALE GENITALIA: Bursa copulatrix with several large spines in heavily sclerotized medial part on dorsal side; appendix bursae well defined, sclerotized. Colliculum slightly inclined to right.

DISTRIBUTION: East-Mediterranean. Albania, Greece and south of European Russia. – Outside Europe in Asia Minor, Jordan and Transcaucasus (Armenia, Azerbaijan), eastward to Turkmenistan (Kopet-Dagh Mts). – New for the fauna of Russia: 2♂, European p. of Russia S., Elton salt lake env., 18.-23.V.2001, leg. KARALIUS V. and MIATLEUSKI J. (coll. SKOU).

PHENOLOGY: Univoltine: mid-May (early May) to late June (mid-July).

BIOLOGY: Unknown.

HABITAT: Very warm and dry landscapes. From 0 m up to about 1,100 m above sea-level in Greece (coll. ZSM), and to about 1,600-1,800 m in Asia Minor (coll. ZSM).

SIMILAR SPECIES: Unlikely to be confused with any other European *Eupithecia* species. *E. scalptata* appears to be most similar to *E. gratiosata* but can usually be easily distinguished from this species by smaller size, ochreous tinge of the forewing, finer and more delicate pattern with double white transverse lines between basal, medial and terminal areas and other external features. The male and female genitalia are easily differentiated in the two compared species.

The *cretaceata* species-group

Introduced by BOLTE (1990) with four North American species: *E. cretaceata* (PACKARD, 1874), *E. zygadeniata* PACKARD, 1876, *E. nimbosa* (HULST, 1896) and *E. indistincta* TAYLOR, 1910. The *cretaceata* species-group consists mainly of northern and alpine *Eupithecia*. In Europe there are only two species: *E. veratraria* and *E. cretaceata* (as subsp. *fenestrata*). *E. subfenestrata* STAUDINGER, 1892 (Caucasus, Transcaucasus and eastern Turkey) and two species from eastern Asia, *E. mandschurica* STAUDINGER, 1897 and *E. lvovskyi* MIRONOV, 1988 also belong to the *cretaceata* group.

MALE GENITALIA: Uncus from medium length to short, stout, biapical. Papillae large, length about 0.5-0.8 times uncus length. Valva without ventral process; medially broadened, with smoothly curved ventral margin, usually with slightly prominent dorsal margin and with narrowly rounded apex; sacculus weakly sclerotized. Vinculum short and broad. Aedeagus relatively long and stout, length about 0.7-0.8 times valva length; about 3.0-4.0 times longer than medial width. Vesica armed with three elongated and narrow multiply dentate cornuti, large elongated patch of spinules in medial part as well as elongated, curved, V-shaped cornutus (usually with a smaller irregular one at its base) at ductus ejaculatorius base. Sternum A8 peg-like; base broad with shallow hollow; apical hollow deeper, narrow; two apical rods narrow with apices tapered or blunt; sternum length about 0.5-0.6 times valva length.

FEMALE GENITALIA: Bursa copulatrix elongated, pear-shaped; mostly covered with large, uniform sized spines; left side usually with a longitudinal membranous gap, of variable width, separating the spines from medial part of corpus bursae to its base. Ductus bursae with one or two patches of spines and with longitudinal striations. Ductus seminalis with broadened base, evenly curved; attached to corpus bursae near colliculum on right side. Colliculum collar-like, relatively short and narrow. Antrum membranous. Anterior and posterior apophyses short or medium length, thin. Spurs of anterior apophyses quite broad at apices. Tergum A8 usually wide. Papillae anales rather small, rounded.

REMARKS: According to the structure of the genitalia, the *cretaceata* species-group is closely related and similar to the *satyrata* species-group. The vesica with two or three dentate cornuti, together with the prominent, elongated patch of spinules at apex, and the apical hollow of the sternum A8 in male, and the longitudinal membranous gap separating the spines in the corpus bursae in the female genitalia are good characteristics separating this species-group from the *satyrata* group. The other characteristics reported by BOLTE (1990), are not applicable to the Palaearctic representatives of the *cretaceata* species-group.

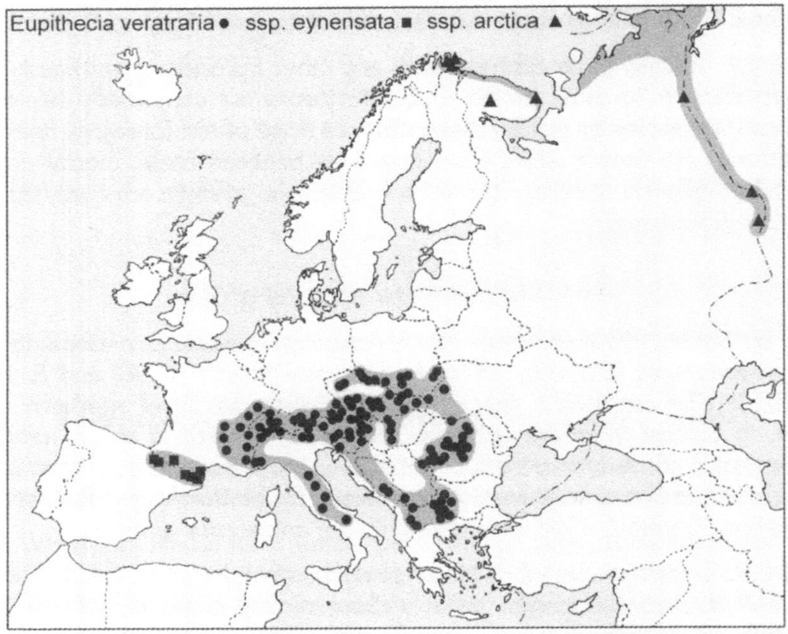

115. *Eupithecia veratraria* HERRICH-SCHÄFFER, 1848

Eupithecia veratraria HERRICH-SCHÄFFER, 1848: Syst. Bearb. Schmett. Eur. 3 (31): 120, 130; ibidem (1846), pl. 25: 152 (non binominal), (Europe, Alps). Syntype(s) lost.

Eupithecia eynensata DE GRASLIN, 1863: Annls Soc. ent. France (4) 3 (2): 326, pl. 8: 3 (France: Pyrénées-Orientales, d'Eyna valley). Holotype ♀.

Eupithecia magnata MILLIÈRE, 1873: Revue Mag. Zool. (3) 1: 2 (Switzerland: Mt. Muottas, near Célérina). Holotype ♀ (MNHN?).

Eupithecia veratraria arctica VIIDALEPP, 1974: Tartu Riikliku ulikooli Toime., Zool.-Alaseid toid 8 (327): 88 (Russia: Polar Urals, Krasny Kamen). Holotype ♂ (ZMTU). Valid at subspecific rank.

Eupithecia certa VOJNITS, 1981: Acta Zool. Acad. Sci. Hung. 27 (3-4): 420, figs. 33-35 (China: Shansi Province, Mien Shan). Paratype ♂ only, nec holotype ♀ (ZFMK, examined). New junior synonym.

DIAGNOSIS: Wingspan 24-28 mm. Labial palpi grey. Frons, vertex and notum whitish grey to grey. Forewing light grey, sometimes with delicate brownish tinge; costal margin mottled and irrorate with whitish scales with rather weak blackish grey costal spots at the ends of the transverse lines; transverse lines inconspicuous, evenly curved on to costa, forming series of small blackish grey dots on the veins; wavy line white, clouded, with large white tornal spot; discal spot relatively small, ovoid or rounded, intensely black. Hindwing slightly paler; transverse lines indistinct, marked with dark dots on the veins along dorsum; terminal area usually faintly darker; whitish wavy line is fine, much indented, scarcely visible, forming a minute white tornal speck; discal spot, minute, rounded, blackish grey. Fringe chequered with whitish grey and dark grey. Abdomen whitish grey with, or sometimes without, a series of small blackish grey dorsal tufts.

Subsp. ***eynensata*** DE GRASLIN, 1863 is usually larger (wingspan 25-30 mm) and lighter than nominate subspecies, with whitish grey ground colour.

Subsp. ***arctica*** VIIDALEPP, 1974 is smaller (wingspan 17-24 mm) and also lighter, with pale-grey colour and faintly marked dark dashes on the veins of forewings, and with smaller discal dots on all wings than in nominate subspecies.

MALE GENITALIA: See diagnosis of species-group.

FEMALE GENITALIA: Ductus bursae with two patches of spines. Posterior apophyses with short spurs.

DISTRIBUTION: Eurasiatic (arctic-alpine), with separate populations often forming subspecies. Nominate subspecies in the mountains of Central and southern Europe. In Pyrenees (France, Andorra, Spain) replaced by subsp. *eynensata*; in northern Europe (arctic Norway and Russia: Kola peninsula and Polar Urals) by subsp. *arctica*. – Outside Europe several mountain subspecies are distributed in East Asia, such as: subsp. *homophaea* DJAKONOV, 1926 (from Altai to Ussuri region), subsp. *geiserata* MIRONOV, 1988 (Kamchatka) and subsp. *perpaupera* INOUE, 1965 (Japan).

PHENOLOGY: Univoltine: late June (early June) to early August (late August). Larva: mid-July (early July) to late September (early October). Overwintering as pupa, often twice or even three times.

BIOLOGY: Larva monophagous (2). Found on the inflorescences and especially the seed capsules of *Veratrum nigrum* (Austria: HERRICH-SCHÄFFER 1848), *V. album* (France: MILLIÈRE 1870; Germany: DIETZE 1874a, 1877, 1910, 1913; WEIGT 1990) and *V. lobelianum* (Liliaceae) (Norway: JUUL 1948; TAMMARU & al. 1995, 1996; Germany: WEIGT 1990).

PARASITOIDS: Hymenoptera, Ichneumonidae: *Agrypon clandestinum* (HERTING 1976); *Campoletis rectangulata* (TAMMARU & al. 1996).

HABITAT: Alpine meadows. The subsp. *arctica* has been recorded from luxuriant mountain birch forests, river meadows, sheltered pastures where the food-plant grows along streams and river banks around a coastal settlement. Nominate subspecies from 500 m to about 1,000 m above sea-level in Apennines (PROLA & RACHELI 1980), to about 1,700 m in mountains of Balkan peninsula (PINKER 1968), and from 900 m to about 2,500 m in Alps (WEIGT 1990). The subsp. *eynensata* from 850 m up to about 2,400 m above sea-level in Pyrenees (MAZEL & PESLIER 1997); subsp. *arctica* up to 50 m above sea-level in Lapland (TAMMARU & al. 1996).

SIMILAR SPECIES: Rather invariable species similar to *E. satyrata*, but easily distinguished from it by their considerably larger size and usually proportionately narrower and more elongated wings. The male and female genitalia can be easily distinguished from those of *E. satyrata*.

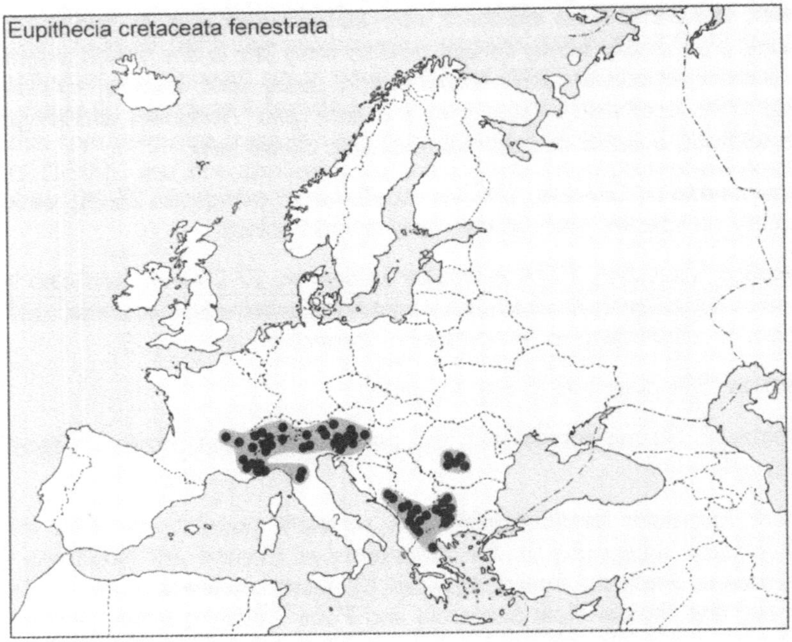

116. Eupithecia cretaceata fenestrata MILLIÈRE, 1874

Eupithecia ? fenestrata MILLIÈRE, 1874: Iconogr. Descr. Chenilles Lépid. inédits 3 (35): 431, pl. 153: 14, 15 (France: Alpes-Maritimes: Fenestre; Saint-Martin-Lantosque). Syntype(s) ♂ (MNHN?).

Eup(ithecia) fenestrata f. *mediopunctata* DIETZE, 1910: Biol. Eupithecien 1, pl. 71: 224; ibidem (1913), 2: 81 (France: Fenestre). Syntypes 6, ♂ ♀ (MNHU). Valid at subspecific rank.

(Eupithecia) fenestrata f. *inanis* DIETZE, 1913: Biol. Eupithecien 2: 81; ibidem (1910), 1, pl. 71: 225 (France: Allos). Syntypes 5, ♂ ♀ (MNHU). Valid at subspecific rank.

DIAGNOSIS: Wingspan 22-32 mm. Labial palpi whitish brown. Frons, vertex and notum snowy white, sometimes with a soft yellowish tinge. Forewing white; transverse lines slender and faint, minutely wavy, pale yellowish grey or pale yellowish brown, broadened and darker along costal margin; basal and postbasal lines right or obtuse angulated; antemedial usually evenly curved; two medial lines (one before, the other beyond discal spot) are well marked near costa; postmedial line faintly curved, minutely waved, with one sharp inward projection near costa, often well marked on the veins with a series of dark dots; double minutely wavy transverse line edges terminal area inwardly; the outer margin of terminal area narrow, pale yellowish brown; discal spot minute, rounded, brown but frequently obsolete. Hindwing white; three medial lines weakly defined; terminal area edged by minutely waved, slender and faint line, which crosses parallel with termen; discal spot absent. Fringe white, minutely spotted with pale yellowish brown at vein ends. Abdomen unicolorous snowy-white.

MALE GENITALIA: See diagnosis of species-group.

FEMALE GENITALIA: Ductus bursae with two small patches of spines. Posterior apophyses with small spurs.

DISTRIBUTION: Subspecies *fenestrata* is European (alpine), but the species as a whole is

Holarctic (alpine). The Alps (France, Italy, Switzerland, Austria, Slovenia), the southern Carpathians (Romania) and the mountains of Balkan peninsula (Herzegovina, Montenegro, Serbia (Kosovo), Macedonia, Albania, Bulgaria and Greece). Reputedly in Molise (southern central Italy: SCIARETTA & RAINERI 2001). – Outside Europe in North America (nominate subspecies). In western Caucasus and Transcaucasus (Georgia) and in north-eastern Turkey occurs the separate but closely related species *E. subfenestrata* STAUDINGER, 1892 (ALBERTI 1965; GELBRECHT pers. comm.).

PHENOLOGY: Univoltine: early July (mid-June) to early August (mid-August). Larva: late June to late August (late September). Overwintering as pupa, sometimes two or even up to four years.

BIOLOGY: Larva monophagous (2). Found on the flowers and seeds of *Veratrum album* (France: DIETZE 1910, 1913) and *V. lobelianum* (Liliaceae) (France and Austria: WEIGT 1990). Larvae of the nominate American subspecies was found on the flowers, seeds and leaves of *Veratrum viride* (Liliaceae) (Canada: TAYLOR 1907; MCGUFFIN 1958; BOLTE 1990).

HABITAT: Alpine meadows. From 650 m up to 1,900-2,200 m above sea-level in mountains of Balkan peninsula (PINKER 1968; TOMIČ & al. 1990), and from 1,000 m up to about 2,600 m in Alps (coll. ZSM; WEIGT 1990).

SIMILAR SPECIES: *E. cretaceata fenestrata* is a readily recognized species, which can be easily distinguished from any other European relatives by its large size, the white head, body and wings, with weak light yellowish grey or yellowish brown transverse lines. The European subspecies is distinguished from nominate North American subspecies by the broader terminal area on the forewing and less evident transverse lines especially on the hindwing. The male and female genitalia are variable, especially the shape of valva, the dentate cornuti on the vesica, the sternum A8 of male and the spiniferous area of bursa copulatrix in female.

The ***satyrata*** species-group

Introduced by MCDUNNOUGH (1949). It is a large *Eupithecia* species-group in Europe and Palaearctic region. The *satyrata* group is represented in Europe by nine species: *E. praealta, E. intricata, E. satyrata, E. pernotata, E. cauchiata, E. extensaria, E. absinthiata, E. expallidata, E. valerianata.* A number of Asian species found from Turkey to Japan and from northern Siberia to southern China can also be placed in this group (INOUE 1979; MIRONOV 1991). In North America this group is represented by at least eight species (MCDUNNOUGH 1949; BOLTE 1990).

MALE GENITALIA: Uncus short, narrow to broad, biapical in all species except *E. intricata* in which it is uniapical. Papillae about 0.5-1.0 times uncus length. Valva without ventral process; broadest medially; usually with evenly curved ventral margin, straight dorsal margin and with narrowly rounded apex; sacculus weakly sclerotized. Vinculum short, usually medium width, rounded. Aedeagus stout, relatively long and thick, length about 0.7-1.0 times valva length; about 2.5-4.0 times longer than medial width. Vesica armed with two to four prominent dentate cornuti, which are not joined at base (in some species with one or two horn-like cornuti replacing the corresponding dentate cornuti), of which one or two smaller near apex and usually one or two larger and broader in medial part of aedeagus; also usually with one elongated, convex, flat, U-shaped cornutus and with one smaller irregular, curved cornutus at base and accordingly at ductus ejaculatorius base. Sternum A8 peg-like with shallow basal hollow; apical hollow, if present, quite small; sternum length about 0.5-0.8 times valva length.

FEMALE GENITALIA: Bursa copulatrix usually sclerotized, pear-shaped (in some species subglobular); usually 0.5-0.8 covered with spines at base; left side often with a longitudinal, membranous gap, of variable width, separating the spines from apex to at least the base; ductus bursae with patch of numerous larger spines, usually with a few spines or with a row of large spines near ductus seminalis base and always with longitudinal striations. Ductus seminalis short or medium length, broadest at base, usually evenly curved backwards; attached near apex or to middle of corpus bursae on right side. Colliculum collar-like, usually short from narrow to broad. Antrum membranous (some species have large shagreened dorsal patch). Anterior and posterior apophyses short or medium length, thin, pointed at apices. Spurs of anterior apophyses with apices usually broadened. Tergum A8 variable in size and shape: square or rectangular. Papillae anales usually large, relatively broad, rounded.

REMARKS: The representatives of the *satyrata* species-group can usually be distinguished from other *Eupithecia* by vesica with two to four dentate cornuti, separate each from other and elongated, curved U-shaped cornutus at ductus ejaculatorius base, as well as, by peg-like sternum A8 in male, and pear-shaped bursa copulatrix with evenly curved ductus seminalis and membranous strip separating the spiniferous area in corpus bursae in the female genitalia.

117. Eupithecia praealta WEHRLI, 1926

Eupithecia praealta WEHRLI, 1926: Schweiz. Ent. Anz. 5 (11): 1 (Spain: Sierra Nevada). Syntypes 7 ♂9 ♀ (ZFMK, examined).
Unavailable names (incorrect subsequent spellings): *praealata*: SCHMIDLIN (1964); *prealta*: EXPOSITO HERMOSA (1978).

DIAGNOSIS: Wingspan 23-26 mm. Labial palpi pale brownish grey. Frons, vertex and notum pale brownish grey. Notum pale brownish grey. Forewing elongate and pointed at apex, costa straight, termen oblique, rounded; brindled pale brownish grey; transverse lines rather inconspicuous, minutely waved, slightly darker towards costa; basal evenly curved; postbasal broad, almost straight; medial area usually darker between antemedial and first medial; first and second medial lines right-angled near costa; postmedial line minutely waved and more or less evenly curved; pale wavy line usually invisible; discal spot large, ovoid, dark brownish grey but inconspicuous against background of wing. Hindwing paler, brownish grey with inconspicuous transverse lines, darker terminal area, pale wavy line lacking and with small, rounded or ovoid dark discal spot. Fringe greyish white, some brownish grey spots at vein ends. Abdomen whitish grey with brownish tinge.

MALE GENITALIA: Uncus short, stout, biapical. Papillae length about 0.7-0.8 uncus length. Valva large, without ventral process, broadened medially and sharply tapered towards apex; sacculus weakly sclerotized. Vinculum medium length and width. Aedeagus stout, long and thick, length about 0.7 times valva length and about 3.0 times longer than medial width. Vesica armed with two apical dentate cornuti, one dense patch of scaly cornuti, one elongated U-shaped flat cornutus and smaller, irregular, curved cornutus at ductus

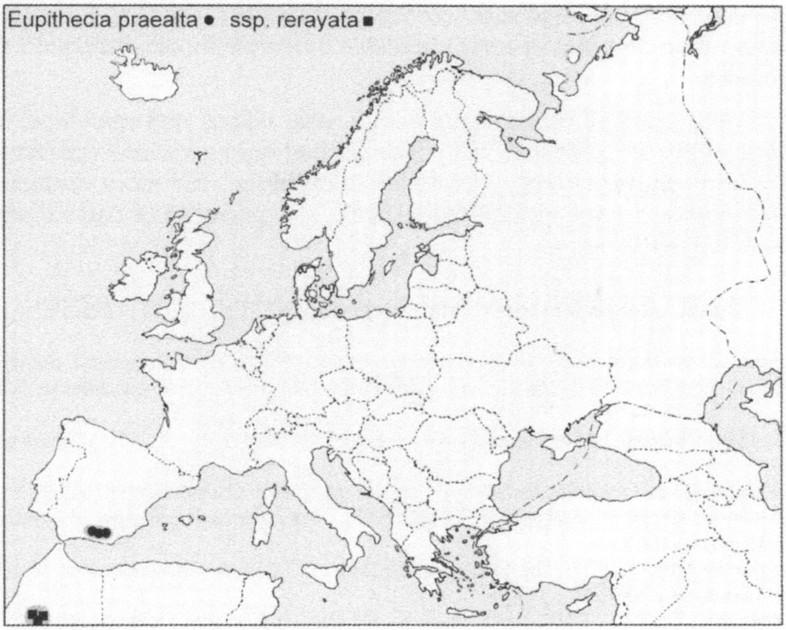

ejaculatorius base. Sternum A8 almost triangular; its basal hollow shallow and apical hollow short and narrow; sternum length about 0.7 times valva length.

FEMALE GENITALIA: Bursa copulatrix elongated, almost pear-shaped, with long membranous gap, completely separating the spiniferous area on dorsal side; also with sclerotized area around ductus seminalis base. Ductus bursae with longitudinal striations and two small patches of spines. Ductus seminalis short, broadened basally, attached to middle of corpus bursae on right side. Colliculum collar-like, relatively short and narrow. Anterior and posterior apophyses thin, apically tapered. Spurs of anterior apophyses broadened apically. Tergum A8 broad, rectangular with small, ovate membranous hollow on posterior margin. Papillae anales small, slim, rounded.

DISTRIBUTION: West-Mediterranean. Southern Spain (Sierra Nevada). Occurs also in Morocco (subsp. *rerayata* REISSER, 1933).

PHENOLOGY: Univoltine: July.

BIOLOGY: Unknown.

HABITAT: High mountain species; among rocks, screes, alpine meadows and alpine stony deserts. From 1,800 m up to about 2,800-2,900 m above sea-level in Sierra Nevada (WEHRLI 1926, 1927; coll. ZMUC), and from 1,600 m up to about 3,400 m in Great Atlas (REISSER 1933; RUNGS 1981).

SIMILAR SPECIES: *E. praealta* can usually be distinguished from all other European *Eupithecia* species by the large size and elongated wings with numerous transverse lines. It is rather remotely similar to *E. impurata* but distinguished from it by the characteristics enumerated above. From the structure of the male and female genitalia *E. praealta* appears to be remotely related to *E. intricata*. However, the biapical uncus, two dentate cornuti and dense patch of large scaly cornuti on the vesica, the shape of sternum A8 in male, and disposition of spines in bursa copulatrix in female should distinguish this species from *E. intricata*.

REMARKS: A large species, rather invariable in size, colour and markings. The North African subsp. *rerayata* REISSER, 1933 is distinguished from nominate Spanish subspecies by the more brindled colour with pale ochreous suffusion, the more distinct and much more waved transverse lines, including pale wavy line, presence of narrow, almost complete, blackish terminal line.

118. Eupithecia intricata (ZETTERSTEDT, 1839)

Larentia intricata ZETTERSTEDT, 1839: Insecta Lapponica 4 (6): 962 (Sweden: Lapponia). Syntype(s) (MZLU).
Eupithecia helveticaria BOISDUVAL, 1840: Gen. Index meth. Eur. Lepid.: 208 (Switzerland: Wallis, Valesia). Syntype(s).
Larentia arceuthata FREYER, 1842: Neuere Beitr. Schmett. 4 (62): 145, pl. 372 (Germany: Augsburg). Syntype(s).
Eupithecia helveticaria var. *anglicata* MILLIÈRE, 1870: Iconogr. Descr. Chenilles Lépid. inédits 3 (25): 102, pl. 110: 20, 21 (United Kingdom). Syntypes 8♂♀, (MNHN?). Junior primary homonym of *Eupithecia anglicata* HERRICH-SCHÄFFER, 1863.
Eupithecia taylorata SWETT, 1907: The Can. Ent. 39 (11): 377 (U.S.A.: Massachusetts, Sherborn; Maine, Monmouth). Lectotype ♂ (MCZ).
Eupithecia gibsonata TAYLOR, 1910: The Can. Ent. 42 (3): 78 (Canada: Ontario, Ottawa). Holotype ♀ (USNM).
Eup(ithecia) helveticaria f. *septemtrionalis* DIETZE, 1910: Biol. Eupithecien 1, pl. 74: 423, 424, 434, 435; ibidem (1913), 2: 75 (Finland: Kuusamo). Syntypes 13, ♂♀ (MNHU). Valid at subspecific rank.
Eupithecia chagnoni SWETT, 1911: The Can. Ent. 43 (7): 255 (Canada: Quebec, Montreal). Holotype ♂ (MCZ).
(Eupithecia) helveticaria f. *suffusa* DIETZE, 1913: Biol. Eupithecien 2: 76 (Germany: Mittelrhein; Austria: Stainz). Syntypes 5, ♂♀ (MNHU). Valid at subspecific rank.
(Eupithecia) helveticaria f. *mediofasciata* DIETZE, 1913: Biol. Eupithecien 2: 76 (Finland; Germany: Erfurt). Syntypes 3♂ (MNHU). Valid at subspecific rank.
Eupithecia helveticaria millieraria WNUKOWSKY, 1929: Zool. Anz. 83 (9/10): 223. Replacement name for *anglicata* MILLIÈRE, 1870. Valid at subspecific rank.
Eurithecia [sic!] *intricata hibernica* MERE, 1964: Entomologist's Gaz. 15 (2): 73, pl. 3: 2 (Ireland: Co(unty) Clare, The Burren, Doughbranneen). Holotype ♂ (BMNH). Valid at subspecific rank.
Unavailable names (incorrect subsequent spellings): *helveticata*: DUPONCHEL (1842); *arceutharia*: HERRICH-SCHÄFFER (1848).

DIAGNOSIS: Wingspan 18-26 mm. Labial palpi grey. Frons, vertex and notum whitish grey. Forewing rather elongate and pointed: costa slightly arched, apex more or less pointed, termen oblique, rounded; pale grey; veins slightly chequered blackish grey and whitish grey; transverse lines very weakly defined, faintly waved, right or obtuse angled onto costa; pale wavy line almost invisible; discal spot distinct, small and ovate, black. Hindwing paler, smoky grey; transverse lines inconspicuous; discal spot minute, circular or slightly elongate. Fringe distinctly chequered greyish white and grey or dark grey. Abdomen covered with whitish grey scales.

Subsp. ***hibernica*** MERE, 1964 is distinguished by the small size, the narrow and pointed forewings, the delicate, bluish grey ground colour of the head, body and wings and very distinct blackish transverse lines.

MALE GENITALIA: Uncus narrow, uniapical. Valva slightly angled medio-ventrally, tapering in apical half to narrow apex. Vesica armed with four dentate cornuti, one elongated U-shaped cornutus and one smaller, irregular cornutus. Sternum A8 with slightly convex

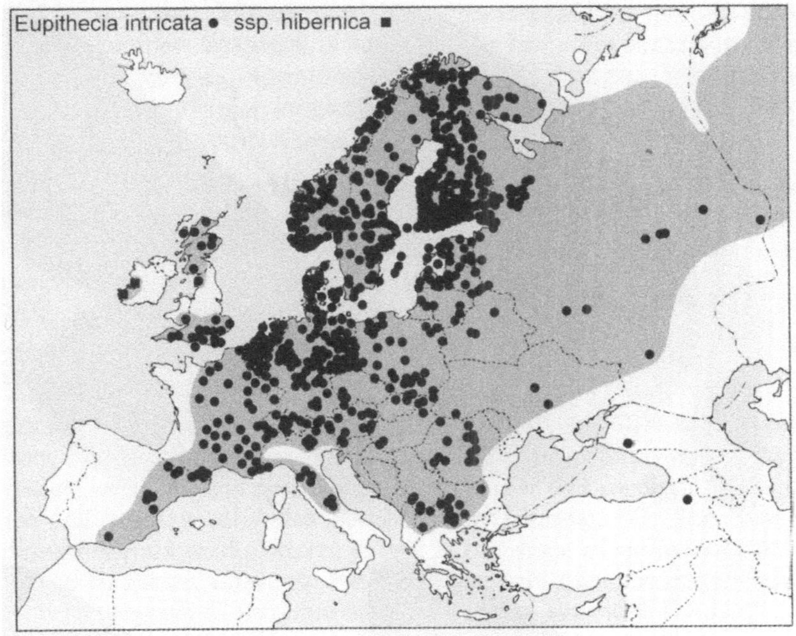

and apically sclerotized lateral margins, without apical hollow. The males of subsp. *hibernica* have smaller genitalia than specimens of nominate subspecies: juxta rather slim, narrower with slightly concave posterior margin and tapered sides; aedeagus shorter, with cornuti about half the size of those in nominate subspecies. Sternum A8 short, almost triangular, evenly tapered to apex and heavily sclerotized near it.

FEMALE GENITALIA: Bursa copulatrix with spines separated by wide, distinct, membranous gap on left side and with larger marginal spines. Antrum with elongated shagreened dorsal patch. Tergum A8 rectangular with slightly convex anterior margin. In females of subsp. *hibernica* the bursa copulatrix is smaller, the marginal spines of general spiniferous area are not larger than the others, and the spines in the ductus bursae and around the base of ductus seminalis are larger than in females of the nominate subspecies.

DISTRIBUTION: Nominate subspecies Euro-Siberian (boreal), but species as a whole Holarctic. Almost all Europe from Spain to Urals except Iceland, Portugal, Mediterranean islands, south of Apennines, Balkan peninsula and European Turkey. The subsp. *hibernica* is endemic to western Ireland (The Burren, County Clare). – Outside Europe found in western Caucasus, north-eastern Turkey and southern Siberia. Recorded by VIIDALEPP & MIRONOV (1988) for Amur region on the basis of one male. Replaced in North America by subsp. *taylorata* SWETT (1907).

PHENOLOGY: Univoltine: mid-May (late April) to mid-July (late July), in some years a second generation from mid-August to mid-September. Larva: late July (early July) to late September (early November). Overwintering as pupa.

BIOLOGY: Larva monophagous (3). Found on the needles of *Juniperus communis* (United Kingdom: CREWE 1861c, 1862; BUCKLER 1899; BARRETT 1904; France: MILLIÈRE 1870; Germany: DIETZE 1910, 1913; WEIGT 1990; Denmark: JUUL 1948; Finland: SEPPÄNEN

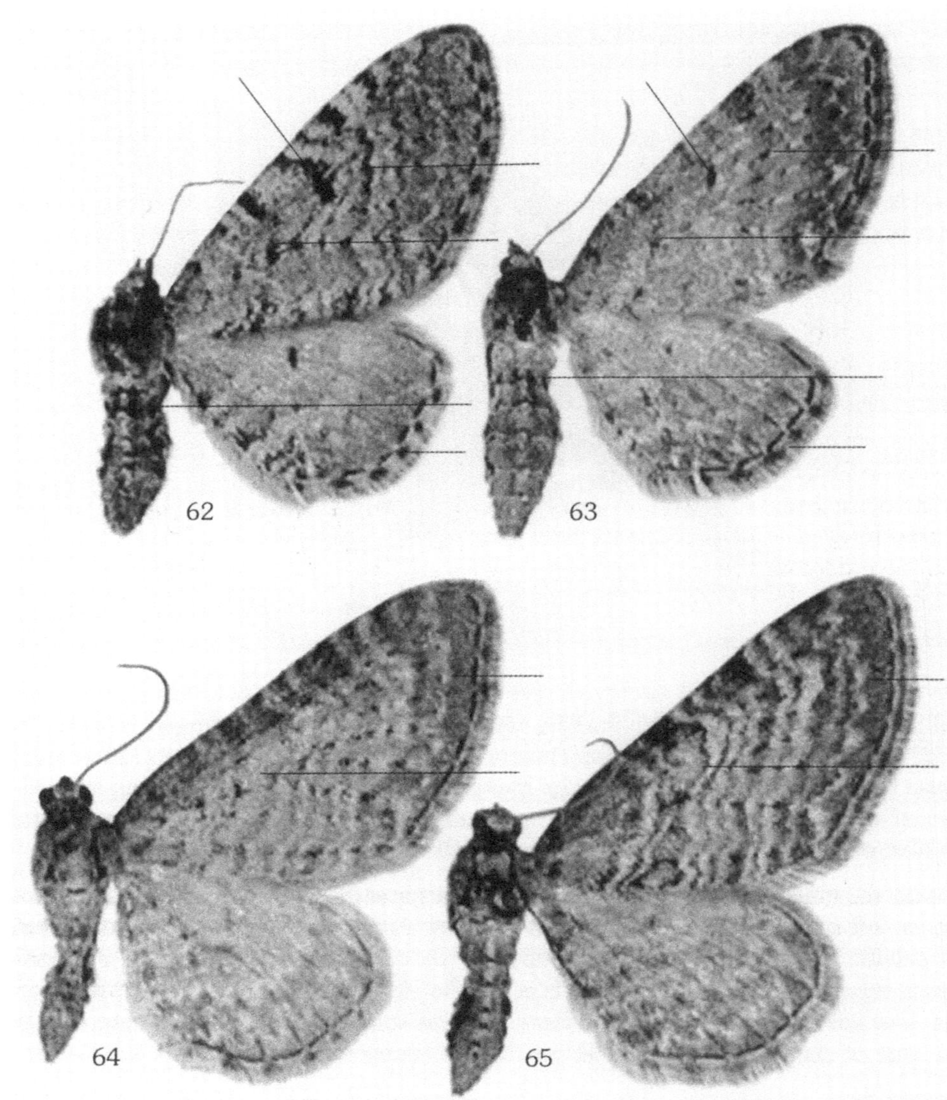

Text-figs 62-65: Diagnostic characters (indicated) of European *Eupithecia* species. Text-fig. 62: *E. intricata* (ZETTERSTEDT, 1839). Text-fig. 63: *E. satyrata* (HÜBNER, 1813). Text-fig. 64: *E. cauchiata* (DUPONCHEL, 1830). Text-fig. 65: *E. pernotata* GUENÉE, 1858.

1970), *J. sabina* (Switzerland: BOISDUVAL 1840), *J. communis* subsp. *alpina* (Switzerland: PÜNGELER 1889; DIETZE 1910, 1913), also on introduced *Juniperus chinensis, J. rigida, J. drupacea, J. squamata, J. recurva* and *Cupressus* (Cupressaceae) (United Kingdom: GOATER & al. 1981; Germany: WEIGT 1990). Larvae of subsp. *hibernica* found on the needles of *Juniperus communis* subsp. *alpina* (Ireland: MERE 1964; HAGGETT 1968). The North American subspecies was found on *Juniperus scopulorum* (Canada: RINDGE 1952; MCGUFFIN 1958).

PARASITOIDS: Hymenoptera, Ichneumonidae: *Agrypon clandestinum*, *Cratichneumon fugitivus*, *Virgichneumon tergenus*, *Platylabus pactor* - Hymenoptera, Braconidae: *Agathis (=Bassus) cingulipes*, *Apanteles salebrosus*, *A. spurius* - Diptera, Tachinidae: *Campylochaeta inepta*, *Phryxe longicauda* (THOMPSON 1945; HERTING 1960, 1976; SHENEFELT 1970, 1972; ZWAKHALS pers. comm.).

HABITAT: Pine forests, heaths, moors, gardens and also along coasts. The subsp. *hibernica* inhabits limestone pavement of Mount Doughbranneen (MERE 1964; GOATER & al. 1981; SKINNER 1984). Nominate subspecies from 0 m up to about 1,500 m above sea-level in Apennines (PROLA & RACHELI 1980), to about 1,600 m in Balkan peninsula (PINKER 1968), to about 2,400-2,500 m in Pyrenees (MAZEL & PESLIER 1997; coll. FIBIGER), and to about 2,500-2,600 m in Alps (WEHRLI 1926; WOLFSBERGER 1955; FEICHTENBERGER 1962; REZBANYAI-RESER 1988; WEIGT 1990).

SIMILAR SPECIES: This species is closely related and similar to *E. satyrata*. Usually distinguishable from it by darker ante- and postmedial lines on the forewings, veins more significantly marked with light and dark scales, faintly larger oblique discal spots, more chequered fringe and darker second abdominal segment (Text-fig. 62). The uniapical uncus and three dentate cornuti on the vesica, as well as the shape of sternum A8 in male, and the area covered by spines, especially the larger marginal spines in corpus bursae in the female genitalia distinguish *E. intricata* from *E. satyrata* very well.

REMARKS: Very variable in coloration and distinctness of markings. There are two general forms in Europe, which are distinguished from the more pale northern typical form: f. *arceuthata* (FREYER, 1842) which has the widest distribution in Europe. It is usually larger, darker, light grey or light greyish brown in colour, sometimes with delicate mauvish tinge in fresh specimens, larger discal spots, more distinct transverse lines and chequered veins in the forewings than in the northern typical form. The f. *helveticaria* BOISDUVAL, 1840, from the Alps, is very similar to f. *millieraria* WNUKOWSKY, 1929 from Scotland. Both forms have head, body and wings brown; transverse lines more distinct than those in typical form but they are less conspicuous than in f. *arceuthata*. No significant differences can be found in colour and maculation between alpine f. *helveticaria* and dark Scottish f. *millieraria*. I have come to the conclusion that these two forms should be treated as synonymous. A single high mountain Caucasian specimen of *E. intricata* (ZISP) is similar to the moths of f. *helveticaria*.

119. *Eupithecia satyrata* (HÜBNER, 1813)

Geometra satyrata HÜBNER, 1813: Samml. Eur. Schmett. 5 Geometrae (1), pl. 85: 439 (Europe). Syntype(s) lost.

Eupithecia grammaria BOISDUVAL, 1840: Gen. Index meth. Eur. Lepid.: 208 (France: Alp. Gall.). Syntype(s).

Eupithecia callunaria DOUBLEDAY, 1850: The Zoologist 8 (App., Art. 8): 55 (United Kingdom: Scotland; northern England). Syntypes (BMNH).

Eupithecia italicata GUENÉE, 1858: in BOISDUVAL & GUENÉE, Hist. nat. Insectes (Spec. gén. Lépid.) 10 (2): 316 (Italy: Lombardie, Domo-Dossola). Holotype ♂.

Eupithecia satyrata ab./var. *subatrata* STAUDINGER, 1871: in STAUDINGER & WOCKE, Cat. Lepid. eur. Faunengeb. 1: 197 (Central Asia: Ferghana; Alai Mts; Ili region). Syntype(s) (MNHU). Valid at subspecific rank.

Eupithecia rivosulata DIETZE, 1875: Stett. ent. Ztg. 36 (4-6): 254, pl. 2: 6 (Russia: Altai Mts). Syntypes 1♂ 1♀ (MNHU).

Eupithecia curzoni GREGSON, 1884: The Entomologist 17 (257): 230 (United Kingdom: Shetland Islands, Balta Sound). Syntype(s) (BMNH?).

Eupithecia satyrata var. fagicoloria ROBSON & GARDNER, 1886: Young Naturalist 7 (Suppl.): (part 84) 7, 41 (type locality not stated). Syntype(s) not traced in BMNH (L. PITKIN pers. comm.). Valid at subspecific rank.

Eupithecia dodata TAYLOR, 1906: The Can. Ent. 38 (3): 103 (Canada: Calgary, Head of Pine Creek). Lectotype ♀ (USNM).

Eupithecia slocanata TAYLOR, 1908: The Can. Ent. 40 (2): 58 (Canada: British Columbia, Kaslo). Lectotype ♂ (USNM).

Eupithecia terminata TAYLOR, 1908: The Can. Ent. 40 (2): 58 (Canada: British Columbia, Kaslo). Lectotype ♂ (USNM).

Eupithecia satyrata intimata PEARSALL, 1908: Jl. N. Y. ent. Soc. 16 (2): 100 (U.S.A.: North Carolina, Black Mts). Lectotype ♀ (AMNH).

Eupithecia fumata TAYLOR, 1910: The Can. Ent. 42 (3): 82 (Canada: Ontario, Ottawa). Holotype ♀ (USNM).

Eup(ithecia) satyrata f. *juldusi* DIETZE, 1910: Biol. Eupithecien 1, pl. 71: 196, 197 (Kazakhstan: Ili region; China: [Xinjiang], Juldus [Yining] Mountains). Syntypes 6♂ 1♀ (MNHU). Valid at subspecific rank.

Eup(ithecia) satyrata f. *medionotata* DIETZE, 1910: Biol. Eupithecien 1, pl. 71: 206, 207; ibidem (1913), 2: 85 (Russia: Altai Mts; Sajan Mts). Syntypes, at least 2, ♂ ♀ (MNHU). Valid at subspecific rank.

(Eupithecia) satyrata f. *concolor* DIETZE, 1913: Biol. Eupithecien 2: 85. Unnecessary replacement name for *juldusi* DIETZE.

Eupithecia divinula CASSINO & SWETT, 1924: Lepidopterist 4 (4): 26 (Canada: Alberta, Nordegg). Holotype ♂ (MCZ).

Eupithecia mackieata CASSINO & SWETT, 1925: Lepidopterist 4 (6-7): 45 (Canada: Alberta, Edmonton). Holotype ♂ (MCZ).

Eupithecia satyrata f. *zermattensis* WEHRLI, 1928: in VORBRODT, Dt. ent. Z. Iris 42 (2): 85 (Switzerland: Zermatt). Syntype(s). Valid at subspecific rank.

Eupithecia submelanochroa VOJNITS, 1973: Folia ent. hung. 26 (1): 211; figs 1c, d; 2b; 3c, d (Kirghizstan: Turkestan Oriental, Province Semirechgensee, Fort Narine). Holotype ♂, not traceable in ZFMK.

Unavailable names (misidentification): *innotata*: sensu STANESCU (1997) nec HUFNAGEL, 1767 (identity: *Eupithecia innotata*), - (infrasubspecific): *strandi*: FUCHS (1901); *limbopunctata*: DIETZE (1910; *caeca*: DIETZE (1913); *nigrofasciata*: DIETZE (1913); *transversa*: DIETZE (1913); *bistrigata*: DIETZE (1913); *marginata*: LEMPKE (1969). - (incorrect subsequent spellings): *satirata*: ALEXINSCHI, 1942; *satyraria*: BOISDUVAL (1840).

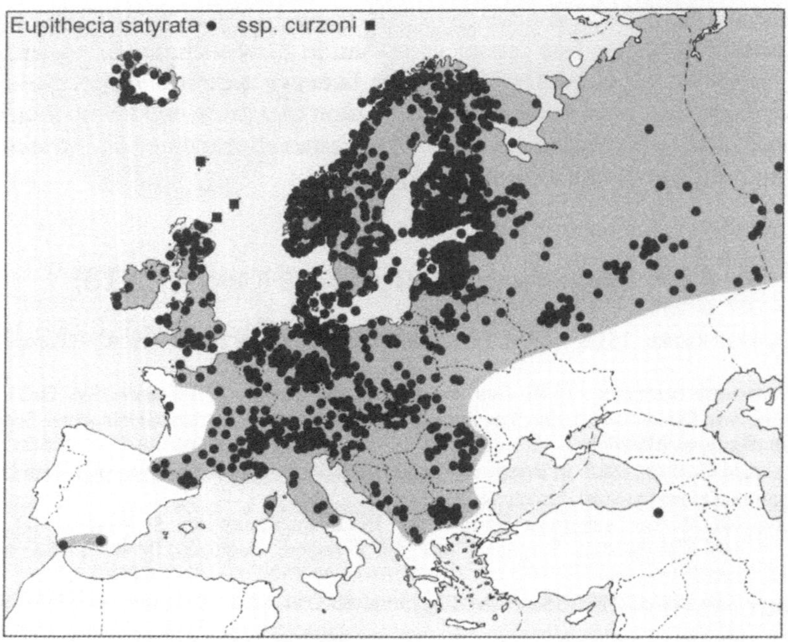

DIAGNOSIS: Wingspan 15-25 mm. Labial palpi, frons, vertex and notum whitish grey or grey. Forewing whitish grey to dark grey; transverse lines fine, numerous, rather indistinct and broken, faintly broadened near costa and slightly sinuate, dark grey; basal line from evenly curved to sharply angled near costa; postbasal and antemedial lines straight, curved towards costa; first medial line right or sharply angled near costa, touching discal spot; postmedial line usually marked on the veins by a series of minute wedge-shaped dark dots; terminal area sometimes darker; white wavy line evident, indented, often with several small triangular white dots and with similar or rounded larger tornal spot (or sometimes obsolete); discal spot small, rounded, black, sometimes finely white-edged, but obsolete in some specimens. Hindwing usually slightly paler; transverse lines inconspicuous; terminal area sometimes darker with fine, indistinct and indented pale wavy line and with small white tornal spot; discal spot small, circular to linear or obsolete. Fringe chequered pale grey and grey or dark grey. Abdomen whitish grey to dark grey.

Subsp. ***curzoni*** GREGSON, 1884 is usually smaller (wingspan 17-22 mm), with narrower, pointed and brindled wings compared with nominate subspecies. It has mainly pale grey ground colour of forewing, sometimes with brownish tinge; transverse lines numerous, oblique, fuscous, often forming three distinct dark bands (antemedial, postmedial and before pale wavy line); terminal area narrow with conspicuous, wide, indented, white wavy line; discal spot minute, black, further reduced in some specimens; hindwing usually with narrow white transverse stripe between medial and terminal areas.

MALE GENITALIA: Vesica armed with four dentate cornuti, one elongated U-shaped cornutus and one smaller, irregular cornutus. Sternum A8 with slightly concave lateral margins; without apical hollow.

FEMALE GENITALIA: Bursa copulatrix with narrow membranous gap, separating spiniferous area on left side. Tergum A8 almost square with shallow medial hollow in posterior margin.

DISTRIBUTION: Holarctic, nominate subspecies Euro-Siberian. Widely distributed in Europe except Portugal and Mediterranean islands, but present in Corsica. In the Shetland and Faroe Islands replaced by subsp. *curzoni*. – Outside Europe the dark subsp. *subatrata* STAUDINGER, 1871 occurs in north-east Asian Turkey and mountains of Central Asia; eastward to Transbaikalia another form occurs (? subsp. *rivosulata* DIETZE, 1875). In North America subsp. *dodata* TAYLOR, 1906 represents the species. – New for the fauna of Turkey: 1♂, NE-Türkey, Provinz Erzurum, Dog. Karadeniz Daglari: Korga Dagi, Umg. Köprüköy bei Ispir, 2000 m, 07.VI.2000, leg. J. GELBRECHT, T. DRECHSEL, R. BUSSE & E. SCHWABE (coll. GELBRECHT).

PHENOLOGY: Univoltine: early May (mid-April) to mid-July (late August); subsp. *curzoni* univoltine: late May (mid-May) to late June. Larva: mid-July (mid-June) to late August (late September). Overwintering as pupa in flimsy cocoon in the ground.

BIOLOGY: Larva polyphagous (1). Found on the flowers and seeds of various, mainly herbaceous plants such as: *Centaurea jacea, C. nigra, Cirsium palustre, Senecio jacobaea, S. nemorensis, Achillea millefolium, Hieracium, Tragopogon, Solidago virgaurea, Anthemis, Matricaria, Aster, Inula, Chrysanthemum, Tanacetum vulgare, Leucanthemum vulgare, Artemisia vulgaris, A. campestris, Eupatorium cannabinum, Leontodon autumnalis, L. hispidus, Arnica montana* (Asteraceae), *Gentiana pneumonanthe, Gentianella ciliata, G. amarella, G. campestris* (Gentianaceae),

Campanula, *Jasione montana* (Campanulaceae), *Galium mollugo*, *G. verum*, *G. sylvaticum* (Rubiaceae), *Erodium cicutarium*, *Geranium sylvaticum* (Geraniaceae), *Helianthemum nummularium (Cistaceae)*, *Valeriana officinalis* (Valerianaceae), *Clematis vitalba*, *Aconitum*, *Thalictrum flavum* (Ranunculaceae), *Hypericum maculatum*, *H. perforatum* (Clusiaceae), *Veronica longifolia*, *Rhinanthus minor*, *Linaria vulgaris*, *Scrophularia nodosa*, *Digitalis purpurea*, *Verbascum thapsus* (Scrophulariaceae), *Polygonum viviparum*, *P. bistorta*, *Rumex acetosella* (Polygonaceae), *Minuartia*, *Silene vulgaris*, *S. dioica*, *S. latifolia (Melandrium album)*, *Lychnis flos-cuculi* (Caryophyllaceae), *Chamaespartium sagittale*, *Trifolium pratense*, *Ononis spinosa*, *Lotus corniculatus* (Fabaceae), *Knautia arvensis*, *Scabiosa columbaria*, *Succisa pratensis* (Dipsacaceae), *Meum athamanticum*, *Pimpinella saxifraga*, *Peucedanum palustre*, *Angelica sylvestris* (Apiaceae), *Prunella vulgaris*, *Thymus vulgaris*, *T. serpyllum*, *Origanum vulgare*, *Galeopsis* (Lamiaceae), *Epilobium angustifolium* (Onagraceae), *Veratrum* (Liliaceae), *Filipendula ulmaria*, *Rubus idaeus* (Rosaceae), *Myricaria* (Tamaricaceae), *Calluna vulgaris*, *Erica tetralix*, *Vaccinium*, *Arctostaphylos uva-ursi* (Ericaceae), *Armeria maritima* (Plumbaginaceae), *Juncus conglomeratus*, *J. effusus* (Juncaceae), *Parnassia palustris* (Parnassiaceae), *Ribes rubrum* (Grossulariaceae), *Humulus lupulus* (Cannabaceae), *Betula pendula*, *B. pubescens* (Betulaceae) and *Salix phylicifolia* (Salicaceae) (United Kingdom: CREWE 1861c; BUCKLER 1899; BARRETT 1904; France: MILLIÈRE 1870; LHOMME 1935; Germany: DIETZE 1871b, 1872, 1874a, 1875b, 1910, 1913; SÄLZL 1935; WEIGT 1991; WEGNER pers. comm.; Iceland: LINDROTH 1931; WOLFF 1971; Denmark: JUUL 1948; Finland: SEPPÄNEN 1970; Russia: original data). Larvae of subsp. *curzoni* found on the leaves of *Calluna vulgaris* (Ericaceae) and flower heads of *Succisa pratensis* (Dipsacaceae) (Shetland and Faroe Islands: GREGSON 1884; DIETZE 1913; WOLFF 1929a; GOATER pers. comm.). Found in North America on *Aralia* (Araliaceae), *Thalictrum*, *Clematis virginiana* (Ranunculaceae), *Scrophularia* (Scrophulariaceae), *Epilobium* (Onagraceae), *Rubus*, *Potentilla fruticosa* (Rosaceae), *Aster* (Asteraceae), *Salix* (Salicaceae) and *Alnus* (Betulaceae), fruits of *Viburnum cassinoides* (Caprifoliaceae), flowers of *Chelone glabra* (Scrophulariaceae) and blossoms of *Spiraea* (Rosaceae) (Canada: FORBES 1948; MCDUNNOUGH 1949; RINDGE 1952; MCGUFFIN 1958; BOLTE 1990).

PARASITOIDS: Hymenoptera, Ichneumonidae: *Phobocampe crassiuscula*, *Platylabus iridipennis* (THOMPSON 1945; HERTING 1976; ZWAKHALS pers. comm.).

HABITAT: Various types of forest, forest clearings, plantations, gardens, fields and meadows, heaths, moors and also among rocks. Subsp. *curzoni* on damp moorland and roadsides. From 0 m up to about 1,300 m above sea-level in Apennines (PROLA & RACHELI 1980), to about 1,800-1,900 m in Pyrenees (MAZEL & PESLIER 1997), and to about 2,200 m in Alps (VORBRODT & MÜLLER-RUTZ 1914).

SIMILAR SPECIES: In *E. vulgata*, one species of several with which it could be confused, the costa is arched and wings narrower and longer in proportion, transverse lines are less distinct, postmedial line evidently obtuse angled near costa, and veins are not chequered. *E. virgaureata* has veins more chequered, transverse lines darker and broader near costa, first medial line always sharply angled beyond discal spot, and forewing discal spot larger and blacker. The genitalia of *E. satyrata* show very good differences from those of these two compared species. See also Text-fig. 63.

The small and very variable subsp. *curzoni* is more likely to be confused with *E. nanata* than with the nominate subspecies of *E. satyrata*. However, *E. nanata* has transverse lines which are more oblique, with the basal, postbasal and antemedial sharply angled near costa. The valva without small ventral process, the dentate cornuti on the vesica, the shape of sternum A8 in male, and disposition of spines in bursa copulatrix in the female genitalia can be used to distinguish *E. satyrata curzoni* from *E. nanata* with accuracy.

REMARKS: Very variable species in the depth of colour from pale whitish grey to dark grey or brownish fuscous.

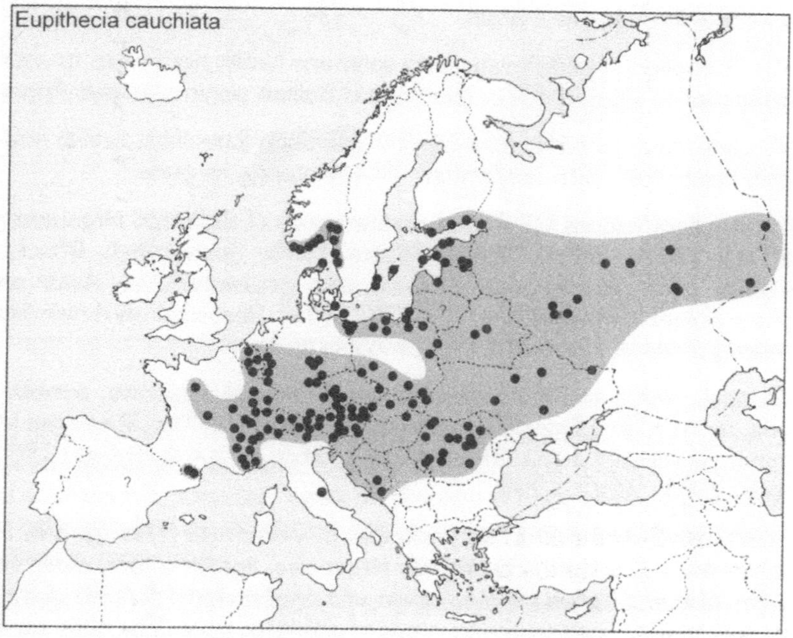

120. *Eupithecia cauchiata* (DUPONCHEL, 1830)

Larentia cauchiata DUPONCHEL, 1830: in GODART & DUPONCHEL, Hist. Nat. Lepid. Papillons France 8 (5; 1): 440, pl. 202: 1 (France: near Bourbonne-les-Bains). Syntype(s) (MNHN?).

Eupithecia griseimarginata DE LA HARPE, 1855: Neue Denkschr. allgem. Schweiz. Ges. ges. Naturwiss. 13 (Suppl. 2): 11 (France: Doubs). Syntypes (at least 1♂1♀).

Eup(ithecia) cauchiata f. *klosi* DIETZE, 1910: Biol. Eupithecien 1, pl. 71: 215, 216; ibidem (1913), 2: 82 (Austria: Stainz). Syntypes 5♂ (MNHU). Valid at subspecific rank.

Unavailable names (incorrect subsequent spellings): *cauchyaria*: BOISDUVAL (1840); *cauchyata*: STAUDINGER (1871).

DIAGNOSIS: Wingspan 18-22 mm. Labial palpi covered by pale grey scales with dark tips. Frons, vertex and notum whitish grey. Forewing pale grey; veins chequered whitish grey and dark grey; transverse lines numerous, distinct, fine, faintly broadened near costa and slightly sinuate, grey; basal line gently angled on to costa; postbasal and antemedial evenly curved; two medial lines right-angled near costa (and first medial usually touching discal spot); postmedial line marked with a series of wedge-shaped dark dots; terminal area wide, grey, darker; pale wavy line fine, with small whitish tornal spot; discal spot small, circular or often imperceptible. Hindwing of similar colour to forewing or slightly paler;

transverse lines more weakly marked, fine and grey; terminal area broad, grey, darker as well; pale wavy line fine, indented, sometimes with small whitish tornal spot; discal spot tiny, circular or absent. Fringe pale grey, spotted with grey or dark grey at vein ends. Abdomen pale grey.

MALE GENITALIA: Vesica armed with three dentate cornuti, one elongated U-shaped cornutus and one irregular, curved cornutus. Sternum A8 elongated, slightly broadened at apex; apical hollow shallow and broad.

FEMALE GENITALIA: Tergum A8 broad, rectangular, with convex anterior margin and shallow medial hollow in posterior margin.

DISTRIBUTION: European. From Pyrenees to southern Urals; northward to south coast of Finland; southward to latitude 42° in Iberian and Balkan peninsulas and Appenines.

PHENOLOGY: Univoltine: late May (mid-May) to late June (late July). Larva: mid-July (mid-June) to early September (late September). Overwintering as pupa.

BIOLOGY: Larva oligophagous (1). Found on the leaves of *Solidago virgaurea* (Germany: DIETZE 1871b, 1871c, 1910, 1913; WEIGT 1990; Switzerland: PÜNGELER 1889; Denmark: JUUL 1948; Russia: original data), also occasionally on *Aster amellus, A. linosyris* (Asteraceae) (Germany: DIETZE 1872, 1913). Recorded on *Artemisia absinthium* (Asteraceae) (France: LHOMME 1935) probably in error.

HABITAT: Steppes, warm and dry meadows, scrub-covered meadows, scrubby heathland on sandy soil, forest edges. From 0 m up to about 1,500-1,600 m above sea-level in Alps and Apennines (VORBRODT 1931; WOLFSBERGER 1965; PROLA & RACHELI 1980; FORSTER & WOHLFAHRT 1981; WEIGT 1990).

SIMILAR SPECIES: Very similar to *E. satyrata* and *E. pernotata* (Text-fig. 64). However, it is distinguished from *E. satyrata* by slightly larger size, lighter colour, more visible transverse lines, broader and darker terminal areas and smaller discal dots on all wings. *E. pernotata* is usually smaller, with narrower wings of yellowish tinge. The male genitalia, especially the dentate cornuti on the vesica and shape of sternum A8 can be use for determination with accuracy.

121. Eupithecia pernotata GUENÉE, 1858

Eupithecia pernotata GUENÉE, 1858: in BOISDUVAL & GUENÉE, Hist. nat. Insectes (Spec. gén. Lépid.) 10 (2): 316 (Italy: Mont-Rose; Mont-Ossolano). Syntypes 2♂ 1♀.

Eupithecia antaggregata INOUE, 1977: Bull. Fac. domest. Sci., Otsuma Wom. Univ. 13: 275, 321, fig. 39 (Japan, Hokkaido: Tokachi, Nukabira). Holotype ♂ (BMNH). New junior synonym.

Eupithecia pernotata enictata PELLMYR & MIKKOLA, 1984: Notulae Ent. 64 (2): 54, fig. 2 (Finland: E-K., Hamina). Holotype ♂ (ZMUH). Valid at subspecific rank.

Unavailable names (incorrect subsequent spelling): *pernotaria*: MORRIS (1861).

DIAGNOSIS: Wingspan 20-23 mm. Labial palpi light ochreous brown. Frons, vertex and notum whitish grey with ochreous tinge. Forewing rather narrow, with straight costa and oblique, rounded termen; ground colour whitish grey with delicate light ochreous suffusion; transverse lines numerous, fine, alternately dark and pale, slightly sinuate, rather oblique, right-angled, slightly broader and darker near costa; whitish wavy line sometimes broken; small whitish tornal spot present; discal spot minute, rounded or ovoid, frequently

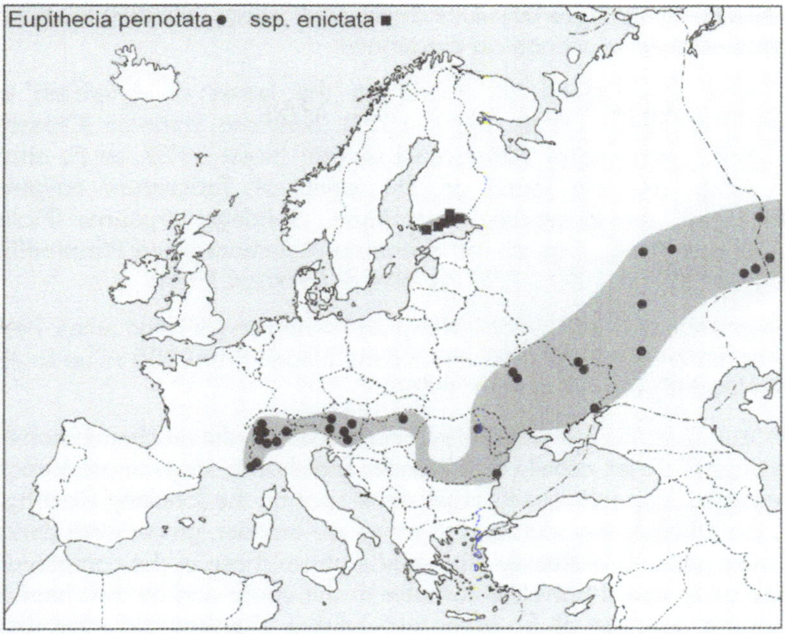

ill-defined. Hindwing paler with rather inconspicuous transverse lines except double light stripe between medial and terminal areas; pale wavy line much indented, slender and faint; discal spot minute, rounded, weak or obsolete. Fringe chequered whitish ochreous brown and ochreous brown. Abdomen light ochreous grey, sometimes with a series of small dark brown dorsal dots.

Subsp. ***enictata*** PELLMYR & MIKKOLA, 1984 is usually smaller (wingspan 17-20.5 mm), less brindled than nominate subspecies; at least the alternate dark and light transverse lines on all wings are less well developed.

MALE GENITALIA: Vesica armed with three dentate cornuti (one larger than others), with one apical, short and broad horn-like cornutus, one elongated U-shaped cornutus and one small, irregular cornutus. The males of subsp. *enictata* have similar genitalia, but the apical horn-like cornutus and largest dentate cornutus on the vesica are broader, sternum A8 is narrow toward apex, almost triangular.

FEMALE GENITALIA: Ductus bursae with prominent appendix on left side. Tergum A8 with straight posterior margin.

DISTRIBUTION: Euro-Siberian. The Alps (France, Italy, Switzerland, Austria and Slovenia) and steppe zone (Hungary, Romania, Ukraine and southern Russia). Along the southern coast of Finland replaced by subsp. *enictata*. – Outside Europe occurs eastward through Urals, northern Kazakhstan (Kokchetav) and southern Siberia (Altai, Minussinsk region, Transbaikalia, Amur and Ussuri regions) to Korea and Japan (Hokkaido, Honshu). – New for the fauna of Hungary: 1♂, (Hungary), Bükk hg., BNP, Oszlai kut. ház, 3.VI.1981, leg. L. RONKAY & SINK (HNHM).

PHENOLOGY: Univoltine: late May (early May) to late June (early August). The subsp. *enic-*

tata: mid-June (early June) to late June (early July). Larva: late July to mid-September. Overwintering as pupa in cocoon on the ground.

BIOLOGY: Larva polyphagous (1). Found on the leaves of *Artemisia absinthium* (Asteraceae) (Switzerland: DIETZE 1910, 1913), *Scabiosa japonica* (Dipsacaceae) and *Serratula wolffii (=coronata)* (Asteraceae) (Japan: INOUE 1979, as *E. antaggregata*). Larvae of subsp. *enictata* found on the leaves of *Tanacetum vulgare* (Finland: KONTUNIEMI 1941), also *Artemisia absinthium, Solidago virgaurea* (Finland: VALLE 1946), *A. campestris, A. vulgaris* var. *coarctata* (Asteraceae) and *Pimpinella saxifraga* (Apiaceae) (Finland: SEPPÄNEN 1970; PELLMYR & MIKKOLA 1984).

HABITAT: Steppes and ruderal places, warm and sandy uncultivated areas, beach fringes. The subsp. *enictata* in coastal areas and ruderal places. From 900 m up to about 2,400 m above sea-level in Alps (VORBRODT 1928).

SIMILAR SPECIES: *E. pernotata* can be easily confused with the much commoner *E. satyrata*, but the latter is greyer, usually with broader and shorter wings in proportion, with less oblique transverse lines and usually larger discal spot on the forewing (Text-fig. 65). In *E. cauchiata*, the other similar species, the wings are broader, greyer, with darker terminal area. The male genitalia are easily distinguished from those in the compared species by the presence of an apical horn-like cornutus in the vesica and by the shape of sternum A8. The female genitalia of *E. pernotata* have a characteristic sclerotized rounded swelling on the left side of corpus bursae beyond spiniferous area.

REMARKS: I have compared the moths from southern Siberia (Altai, Dahuria, Amur and Ussuri regions) with European specimens of *E. pernotata* from Switzerland, Hungary, Romania, Ukraine and southern Russia on the one hand, and with Japanese *E. antaggregata* on the other hand. The eastern moths have clearer yellowish ochreous ground colour; male genitalia with slightly longer uncus and papillae, with larger (longer and broader) valva, especially near apex, with two larger and broader dentate cornuti in the vesica, and with more parallel margins of sternum A8. In my opinion, *E. pernotata* and *E. antaggregata* are conspecific.

122. *Eupithecia extensaria* (FREYER, 1845)

Acidalia extensaria FREYER, 1845: Neuere Beitr. Schmettkde 5 (73): 96, pl. 438: 3 (Type locality not stated). Syntype(s).

Larentia prolongata LIENIG & ZELLER, 1846: in LIENIG & ZELLER, Isis 3: 198 (Latvia: ?Koknese [=Kokenhusen]). Holotype (BMNH?).

Eupithecia sydyi STAUDINGER, 1885: Stett. ent. Zeitung 46 (7-9): 349-352 (China: Kuldja district). Syntypes 9 ♂♀ (MNHU).

Eup(ithecia) extensaria f. *leuca* DIETZE, 1910: Biol. Eupithecien 1, pl. 74: 467; ibidem (1913), 2: 121 (Russia: Ussuri, Kasakewitsch). Holotype (MNHU). Valid at subspecific rank.

Eup(ithecia) extensaria f. *prolongata* DIETZE, 1910: Biol. Eupithecien 1, pl. 74: 468, 469; ibidem (1913), 2: 121. Junior secondary homonym of *Larentia prolongata* LIENIG & ZELLER, 1846.

Eupithecia extensaria occidua PROUT, 1914: in SEITZ, Macrolep. 4: 293, pl. 13: O (United Kingdom: England, Yorkshire, Norfolk). Syntype(s), at least 1♂1♀ (MNHU). Valid at subspecific rank. [Replacement name for *prolongata* DIETZE, 1910 (nec LIENIG & ZELLER, 1846)].

DIAGNOSIS: Wingspan 22-26 mm. Labial palpi ash-grey or brownish. Frons, vertex and notum ash-grey. Forewing with extremely straight costa, oblique, rounded termen and with narrow but rounded, scarcely pointed apex; ground colour ash-grey; transverse lines

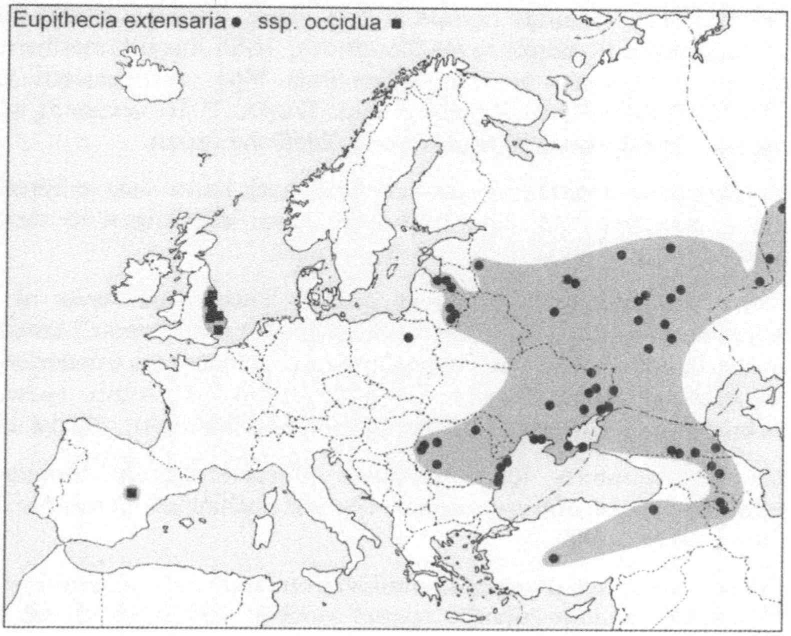

dark grey, oblique, almost straight, sharply angled near costa; three conspicuous broad oblique, chalk-white transverse lines (one between basal and medial areas, one medial and a broader one between medial and terminal areas); whitish wavy line broad, distinct, almost straight but curved only near apex; discal spot absent. Hindwing pale grey with slightly darker terminal area, weak indications of transverse lines, usually with conspicuous whitish transverse band and without discal spot. Fringe pale whitish grey, distinctly spotted with dark grey at vein ends. Abdomen ash-grey with scattered brown scales.

Subsp. ***occidua*** Prout, 1914 has brighter more contrasting brownish ochreous colour on the wings and body. Forewing also with small brown or dark brown costal spots and with dark brown dashes on the veins.

Male genitalia: Vesica armed with two small identical horn-like cornuti near apex, with two small identical dentate cornuti in medial part of aedeagus, with one elongated U-shaped cornutus and one small, irregular cornutus at ductus ejaculatorius base. Sternum A8 peg-like, broad, almost triangular, without apical hollow.

Female genitalia: Bursa copulatrix with wide and deep membranous gap, separating the spines on left side; the spiniferous area relatively small. Spurs attached to anterior apophyses near their apices. Tergum A8 rectangular with medially convex anterior margin and with medially concave posterior margin.

Distribution: Eurasiatic. In Europe nominate subspecies in Poland, Latvia, Lithuania, Romania, Ukraine, central and southern European Russia. In eastern England (Essex, north of Norfolk, Lincolnshire, south-eastern Yorkshire) and central Spain (Prov. Teruel) replaced by subsp. *occidua*. - New for the fauna of Spain: 1♂, Hisp. centr., (Prov. Teruel), Noguera, 1600 m, 7.6.1951, leg. J. R. Caron (ZSM). A single Spanish specimen, which was found in ZSM, shows no difference from English specimens of subsp.

occidua PROUT, 1914. – Outside Europe in Asia Minor, Caucasus and Transcaucasus. The Central Asian form (? subsp.) *sydyi* STAUDINGER, 1885 found in northern and eastern Kazakhstan, Kirghizstan (northern Tien-Shan Mts), north-eastern China and Mongolia. The East Asian form (? subsp.) *leuca* DIETZE, 1910 occurs in western and southern Siberia, Altai, Dahuria, Ussuri region, Korea and Japan.

PHENOLOGY: Univoltine: mid-May to mid-June (late June). Larva: July to September. The subsp. *occidua* univoltine also: June and July. Larva: late August to early October. Overwintering as pupa in dense cocoon in the ground.

BIOLOGY: Larva monophagous (2). Found on the flowers and leaves of *Artemisia campestris* (Asteraceae) (Poland: BUSZKO and GELBRECHT pers. comm.); larvae of subsp. *occidua* recorded on the flowers and leaves of *Artemisia maritima* (Asteraceae) growing amongst tall grasses (United Kingdom: BARRETT 1904), and subsp. *sydyi* found on *Artemisia santolinifolia* (Asteraceae) (Kirghizstan (Tien-Shan Mts): original data).

PARASITOIDS: Hymenoptera, Ichneumonidae: *Mesochorus* sp. (original data) - Hymenoptera, Braconidae: *Microgaster* sp. - Diptera, Tachinidae: *Ceranthia tristella* (all original data for subsp. *sydyi*).

HABITAT: Steppes, warm and dry slopes, preferably on sandy soil, old gravel pits and gravel fields. The subsp. *occidua* inhabits coastal sandhills and edges of salt marshes in England. From 0 m up to about 1,600 m above sea-level in central Spain (coll. ZSM), and to about 2,200-2,300 m in Tien-Shan Mts in Central Asia (original data).

SIMILAR SPECIES: One of the most easily recognized species among the European representatives of the genus. The distinctive maculation and colour can be used with reasonable accuracy. The broad triangular sternum A8, the cornuti on the vesica, and the shape of bursa copulatrix should distinguish this species.

123. *Eupithecia absinthiata* (CLERCK, 1759)

(Phalaena) absinthiata CLERCK, 1759: Icon. Insect. rariorum 1, pl. 8: 9 (Sweden). Lectotype ♂ (NHRS, examined: MIKKOLA (1985).

Geometra minutata DENIS & SCHIFFERMÜLLER, 1775: Syst. Verz. Schmett. Wien. (2): 110 (Austria: Vienna distr.). Syntype(s) lost.

Phalaena elongata HAWORTH, 1809: Lepid. Br. (2): 358 (United Kingdom). Syntype(s) (BMNH).

Eupithecia notata STEPHENS, 1831: Illustr. Br. Ent. (Haustellata) 3: 286 (United Kingdom: England, near London). Syntype(s) (BMNH).

Eupithecia absinthiata var. *callunae* SPEYER, 1867: Verh. naturh. Ver. preuss. Rheinl. 24: 249 (Germany: Westfallen, Waldeck). Syntype(s). Valid at subspecific rank.

Eupithecia goossensiata MABILLE, 1869: Annls Soc. ent. France (4) 9 (1): 78 (France). Lectotype ♀ (coll. HERBULOT in ZSM, examined).

Eupithecia knautiata GREGSON, 1874: The Entomologist 7 (135): 255 (United Kingdom: Lancashire, Bolton, Ball Hill Lodges). Syntype(s) (BMNH?).

Eupithecia catharinae VOJNITS, 1969: Acta Zool. Acad. Sci. Hung. 15 (1-2): 232, figs 1a, 2a (Hungary: Kunfehértó). Holotype ♂ (HNHM, examined). New junior synonym.

Unavailable names (misidentification): *innotata*: sensu WOOD (1835) nec HUFNAGEL, 1767 (identity: *Eupithecia innotata*). – (infrasubspecific): *obscura*: DIETZE (1910); *obscura*: LEMPKE (1969); *nigrescens*: LEMPKE (1969). - (incorrect subsequent spellings): *absynthiata*: FISCHER VON RÖSLERSTAMM (1829); *minutaria*: BOISDUVAL (1840); *absinthiaria*: HERRICH-SCHÄFFER (1848); *elongaria*: DOUBLEDAY (1849); *absynthiaria*: DE LA HARPE (1855); *goosensiata*: STERNECK (1929); *goossentiata*: SCHMIDLIN (1964); *gooseniata*: BIELEWICZ (1966); *goosentiata*: GARTHE (1976).

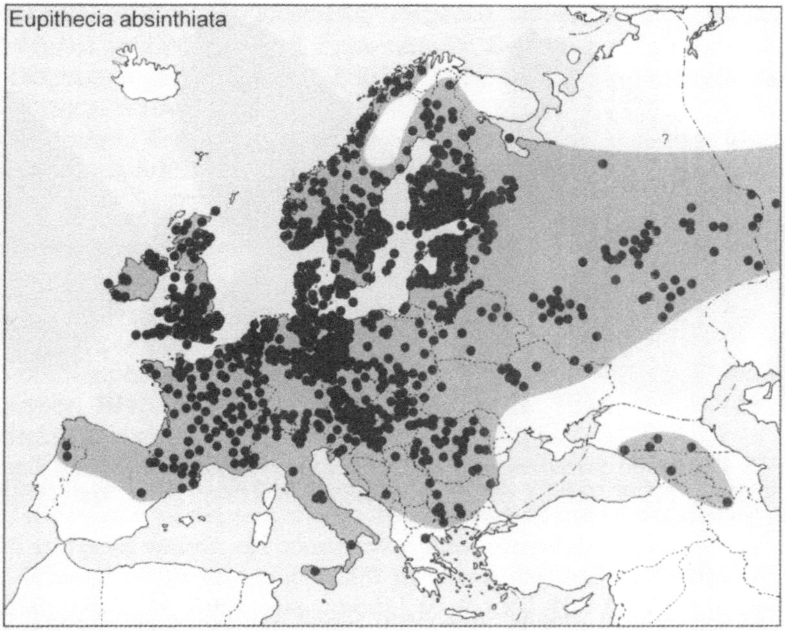

DIAGNOSIS: Wingspan 15-30 mm. Labial palpi blackish grey. Frons and vertex pale ash-grey. Notum reddish brown. Forewing elongated; reddish brown or reddish grey brown; costal margin mottled and irrorate with white scales; transverse lines inconspicuous, more or less well marked on the veins as blackish dots or dashes and also always as larger black costal spots, especially ante- and postmedial costal spots; two weak medial lines cross close to and parallel with postmedial line, curved around discal spot; a series of dark dots or dashes in place of postmedial line usually margined by white scales outside; terminal area sometimes darker and reddish; whitish wavy line ill-defined, but usually broken, forming a series of minute white dots and always with larger white tornal spot, which is margined by dark scales on inner side; discal spot ovate-oblong, more or less elongated, intensely black. Hindwing brownish grey without reddish tinge; transverse lines rather inconspicuous, darker and more distinct along dorsum; in some specimens paler medial transverse band well defined; terminal area usually slightly darker; pale wavy line invisible except small whitish tornal spot; discal spot minute, rounded, rather indistinct. Fringe greyish brown, slightly spotted with dark greyish brown at vein ends. Abdomen reddish or greyish brown, usually with pale ash-grey first segment, with dark brown second segment, with black lateral spots on the three anterior segments and with black scales at posterior margin of last segment.

MALE GENITALIA: Vesica armed with four dentate cornuti, one elongated U-shaped cornutus and one smaller, irregular cornutus. Sternum A8 elongated with quite small apical hollow.

FEMALE GENITALIA: Bursa copulatrix with prominent medial appendix on left side. Tergum A8 almost square.

DISTRIBUTION: Euro-Siberian. All Europe except Iceland, south of Iberian and Balkan

Text-figs 66-68: Diagnostic characters (indicated) of European *Eupithecia* species. Text-fig. 66: *E. absinthiata* (CLERCK, 1759). Text-fig. 67: *E. expallidata* DOUBLEDAY, 1856. Text-fig. 68: *E. assimilata* DOUBLEDAY, 1856.

peninsulas, Mediterranean islands (present in Sicily) and European Turkey. Widely distributed in Palaearctic region from Spain and Ireland to Sakhalin and Japan; separate population in Caucasus and Transcaucasus.

PHENOLOGY: Uni- or bivoltine: late June (early June) to mid-August (late September); in some years a partial first generation from late April to mid-May. Larva: early August (late June) to mid-October (early November). Overwintering as pupa.

BIOLOGY: Larva polyphagous (1). Found on the flowers and seeds of various herbaceous plants, such as *Solidago virgaurea, S. gigantea, S. canadensis, Centaurea nigra, Eupatorium cannabinum, Achillea millefolium, Senecio jacobaea, S. erucifolius, S. nemorensis, S. aquaticus, S. cordatus, S. vulgaris, Artemisia absinthium, A. vulgaris, A. campestris, A. maritima, A. alba (=camphorata), Aster tripolium, Matricaria, Tanacetum vulgare, Cirsium arvense, Chrysanthemum, Conyza canadensis, Adenostyles alpina* (Asteraceae), *Campanula, Phyteuma* (Campanulaceae), *Origanum vulgare, Lavandula angustifolia, Ballota nigra* (Lamiaceae), *Scabiosa, Knautia arven-*

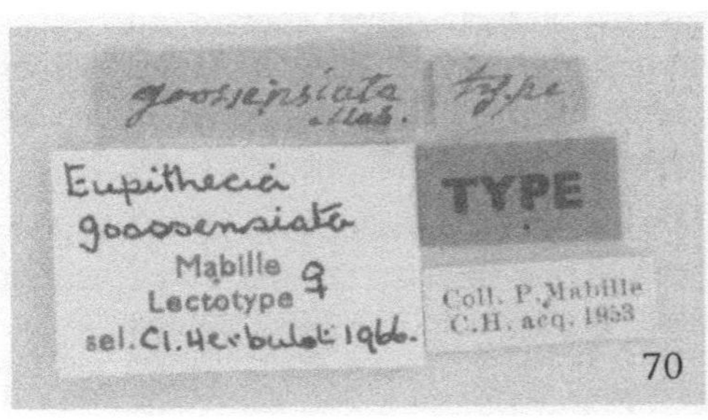

Text-figs 69-70: Lectotype of *Eupithecia goossensiata* MABILLE, 1869 (female, wingspan 22 mm) and original labels of Pierre MABILLE and Claude HERBULOT (Photo Sir Anthony GALSWORTHY).

sis (Dipsacaceae), *Anthriscus sylvestris, Angelica sylvestris, Pimpinella saxifraga* (Apiaceae), *Myrica gale* (Myricaceae), *Erica tetralix, Calluna vulgaris* (Ericaceae) (United Kingdom: DOUBLEDAY 1861; CREWE 1861c, 1862; BUCKLER 1899; BARRETT 1904; Germany: HERRICH-SCHÄFFER 1861a; DIETZE 1872, 1874a, 1910, 1913; WEIGT 1991; WEGNER pers. comm.; France: MABILLE 1869; GOOSSENS 1869, 1871a; MILLIÈRE 1870; LHOMME 1935; Denmark: JUUL 1948; Finland: SEPPÄNEN 1970; Russia: original data).

PARASITOIDS: Hymenoptera, Ichneumonidae: *Dusona pugillator, Casinaria morionella, Mesochorus tachypus, M. temporalis, M. vittator, Netelia tarsata, Platylabus iridipennis, Agrypon clandestinum* - Hymenoptera, Braconidae: *Aleiodes (=Rogas) modestus, Apanteles tibialis (=gracilis), Ascogaster quadridentata, Zele (=Meteorus) caligatus, Meteorus pulchricornis (=striatus), Microplitis* sp. - Hymenoptera, Encyrtidae: *Copidosoma coleophorae* - Diptera, Tachinidae: *Blondelia nigripes, Clausicella turmalis* (THOMPSON 1945; SHENEFELT 1969, 1972, 1973a, 1975; HERTING 1976; original data; ZWAKHALS pers. comm.).

HABITAT: Open, light forests, forest paths and clearings, edges of forest, meadows, open, uncultivated areas such as steppes, moors, heaths, disused gravel pits, karstic slopes with sparse scrub and grass swards, old construction sites, plantations, gardens, parks and near the coasts. From 0 m up to about 1,300 m above sea-level in Apennines (PROLA & RACHELI 1980; coll. ZSM; ZAHM pers. comm.), to about 1,400-1,500 m in Balkan peninsula (TOMIČ & al. 1990), to about 1,900-2,000 m in Alps (DANNEHL 1927; REZBANYAI-RESER 1988; REZBANYAI-RESER & MAIER 2000), and to about 2,000-2,100 m in Pyrenees (coll. FIBIGER).

SIMILAR SPECIES: Variable in the depth of colour from pale ochreous grey to reddish brown and in the distinctness of markings on the wings. It is very similar to species such as *E. denotata, E. assimilata, E. expallidata* and *E. moecha* (see Text-fig. 66). The differences between these species and *E. absinthiata* are given under the relevant species diagnosis.

REMARKS: There have been many publications about the status of *Eupithecia goossensiata* MABILLE, 1869 (DIETZE 1913; SCHÜTZE 1954; VOJNITS 1969; FIBIGER 1979; KAABER 1980; MIKKOLA 1982; SKOU 1986; WEIGT 1991). The lectotype of *E. goossensiata* is deposited in the collection of HERBULOT (ZSM). In my opinion the answer is clear. The specimen is certainly conspecific with *E. absinthiata* (Text-figs. 69-70).
After carefully examining many specimens of *E. absinthiata* from West Europe to the Far East of Russia and the holotype and some paratypes of *E. catharinae*, as well as studying the original description of *E. catharinae*, I have come to the conclusion that *E. absinthiata* and *E. catharinae* are conspecific. VOJNITS (1969) distinguished *E. catharinae* from *E. absinthiata* mainly by its maculation, greyer colour, size and shape of wings. However, to my surprise, no differences in external and morphological characteristics could be found between the specimens of *E. absinthiata* and the holotype of *E. catharinae*.

124. *Eupithecia expallidata* DOUBLEDAY, 1856

Eupithecia expallidata DOUBLEDAY, 1856: The Zoologist 14: 5140, 5142 (United Kingdom). Syntype(s) (BMNH).
Unavailable names (incorrect subsequent spelling): *expallidaria*: MORRIS (1861).

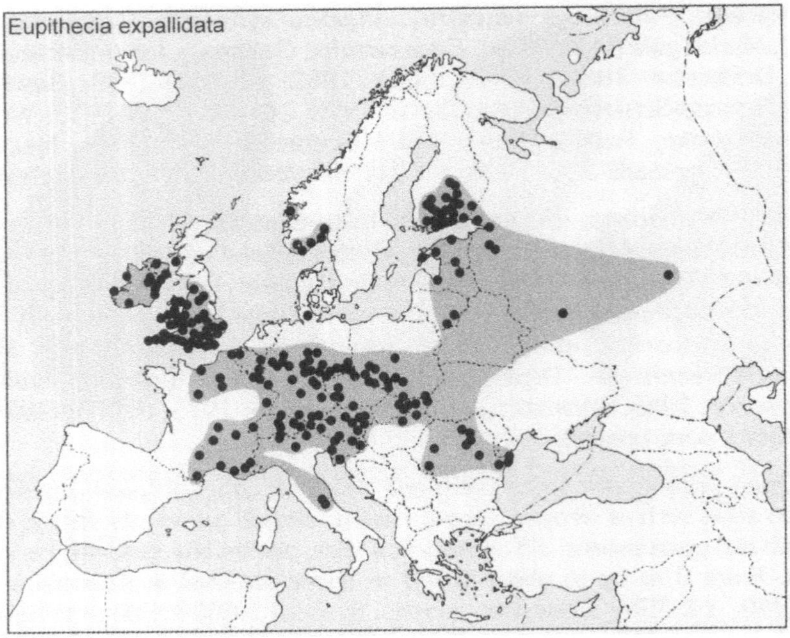

DIAGNOSIS: Wingspan 20-28 mm. Labial palpi blackish grey. Frons, vertex and notum pale greyish ochreous. Forewing pale greyish ochreous of a very soft and delicate tinge; costal and sometimes terminal areas mottled and irrorate with white scales; transverse lines inconspicuous, well marked only with series of black or dark brown dots on the veins and also with four larger costal spots (basal, antemedial, medial and postmedial); terminal area usually slightly darker; pale wavy line weakly defined or obsolete, forming a small and indistinct pale tornal spot; discal spot large, obliquely elongate or ovate-oblong, intensely black. Hindwing paler, weakly marked with fine transverse lines, postmedial line darker and more conspicuous; terminal area usually slightly darker; pale wavy line ill-defined or absent; discal spot minute, rounded or elongated, dark. Fringe light greyish ochreous, spotted with dark greyish ochreous at vein ends. Abdomen pale greyish ochreous with black lateral stripes from third to last segments.

MALE GENITALIA: Similar to those of *E. absinthiata*.

FEMALE GENITALIA: Similar to those of *E. absinthiata* but corpus bursae without medial appendix on the left side.

DISTRIBUTION: European. Occurs in Ireland, United Kingdom, Norway, Finland, European Russia, Estonia, Latvia, Lithuania, Poland, Czech Republic, Slovakia, Germany, Netherlands, Belgium, Luxemburg, France, Italy, Switzerland, Austria, Slovenia, Croatia, Hungary and Romania.

PHENOLOGY: Univoltine: early July (mid-May) to mid-August (late August); in the south probably bivoltine. Larva: late September (early August) to early October (early November). Overwintering as pupa, sometimes two years.

BIOLOGY: Larva oligophagous (1). Found on the flowers and seeds of *Solidago virgaurea*

(Asteraceae) (United Kingdom: CREWE 1861c, 1862; BUCKLER 1899; Germany: HERRICH-SCHÄFFER 1861a; DIETZE 1910, 1913; WEIGT 1991; Russia: original data) and also on *Senecio jacobaea* (United Kingdom: BUCKLER 1899), *S. nemorensis* (Asteraceae) (Germany: KOCH 1984; WEIGT 1991).

PARASITOIDS: Hymenoptera, Ichneumonidae: *Dusona disclusa*, *Casinaria morionella*, *Cyclolabus dubiosus* - Hymenoptera, Braconidae: *Apanteles praepotens*, *Zele (=Meteorus) albiditarsus*, *Z. caligatus* - Hymenoptera, Encyrtidae: *Litomastix* sp. (THOMPSON 1945; SHENEFELT 1969, 1972; HERTING 1976; original data; ZWAKHALS pers. comm.).

HABITAT: Old deciduous, spruce-, and mixed forests, forest clearings, paths and edges, shady roadsides. From 0 m up to about 1,300 m above sea-level in Apennines (Mt. Majella: ZAHM pers. comm.), to about 1,800 m in Alps (WEIGT 1991).

SIMILAR SPECIES: A little varying species, which is similar to *E. absinthiata* and can be easily confused with it. Distinguished by the paler, greyer ochreous ground colour often with weak violet tinge compared with the rich reddish brown *E. absinthiata*, by more intensely black, larger and more obliquely elongate discal spot, the less conspicuous pale wavy line and tornal spot on the forewing, and more distinct transverse series of dark dots on the veins, as well (Text-fig. 67). The male and female genitalia of *E. expallidata* and *E. absinthiata* are difficult to distinguish. The males can be separated by comparing the shape of sternum A8. The female genitalia of *E. expallidata* are distinguished by the smaller corpus bursae with longer spines and by absence of the diverticulum on the left side of ductus bursae, which is present in *E. absinthiata*.

125. *Eupithecia valerianata* (HÜBNER, 1813)

Geometra valerianata HÜBNER, 1813: Samml. Eur. Schmett. 5 Geometrae (1), pl. 76: 395 (Europe). Syntype(s) lost.

Eupithecia laevilignata BRUAND, 1850: Mém. Soc. libre d'emulation Doubs 3 (3) (5/6): 65 (France: Besancon). Syntype(s).

Eupithecia viminata DOUBLEDAY, 1858: The Zoologist 17: 6103 (United Kingdom). Syntype(s) (BMNH).

Unavailable names (infrasubspecific): *grisea*: LEMPKE (1969). - (incorrect subsequent spellings): *valerianaria*: BOISDUVAL (1840); *viminaria*: MORRIS (1861).

DIAGNOSIS: Wingspan 14-22 mm. Labial palpi covered with light brown or brown scales with greyish apices, concolorous with frons, vertex and notum. Forewing almost unicolorous greyish ochreous; transverse lines inconspicuous, sometimes markings with faint greyish brown costal spots; pale narrow transverse stripe between medial and terminal areas often conspicuous; pale wavy line and small pale tornal spot distinct; discal dot minute, rounded, inconspicuous, brown or brownish grey. Hindwing slightly paler, greyish ochreous; transverse lines indistinct except pale narrow medial transverse line or pale wavy line; pale tornal spot distinct; discal dot absent. Fringe unicolorous brownish grey or slightly spotted with brown at vein ends. Abdomen unicolorous brownish grey.

MALE GENITALIA: Uncus broad, when viewed laterally. Papillae usually very narrow. Vesica armed with three dentate cornuti (one apical and two medial) and one elongated, U-shaped cornutus at ductus ejaculatorius base. Sternum A8 peg-like, elongated with well defined apical hollow and two small, pointed apical lobes.

FEMALE GENITALIA: Bursa copulatrix with a large, broad, transverse spiniferous band along

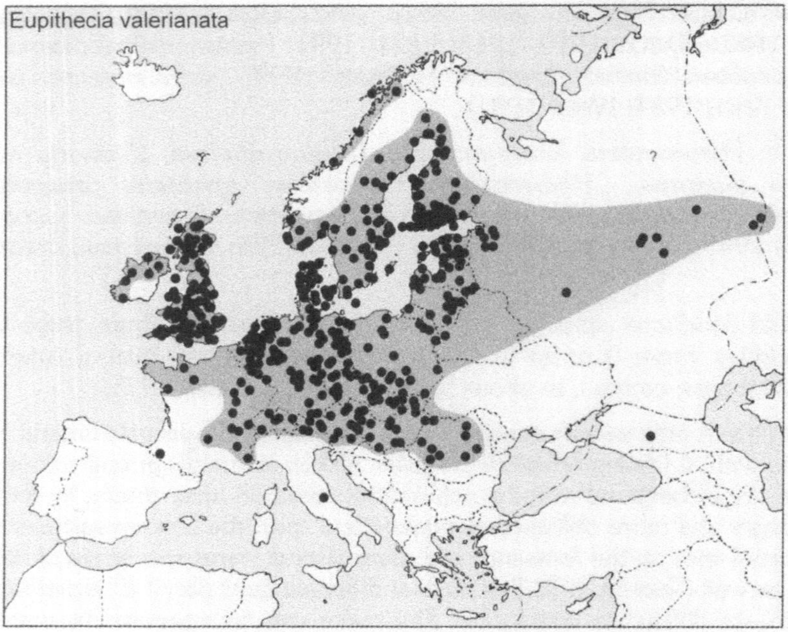

its base; this band evenly curved in medial part of corpus bursae. Tergum A8 almost square with shallow medial hollow in posterior margin.

DISTRIBUTION: Euro-Caucasian. Widely distributed in Europe from Ireland and France to Urals (except Iceland, Portugal, Spain, Mediterranean islands and Balkan peninsula). Outside Europe found in the northern Caucasus (MIRONOV 1991; VIIDALEPP 1996).

PHENOLOGY: Univoltine: mid-June (mid-May) to mid-July (early August). Larva: late July (mid-June) to late August (mid-September). Overwintering as pupa, sometimes two years.

BIOLOGY: Larva monophagous (2). Found on the flowers and ripening seeds of *Valeriana officinalis* (Valerianaceae) (United Kingdom: CREWE 1863; BARRETT 1904; Germany: DIETZE 1910, 1913; WEIGT 1990; Denmark: JUUL 1948; Russia: original data) and also on *Valeriana pyrenaica* in the south (WEIGT 1990). Larvae often suffers heavily from parasitization.

PARASITOIDS: Hymenoptera, Ichneumonidae: *Agrypon flaveolatum*, *Gelis monozonius* (hyperparasite), *Agrypon clandestinum* - Hymenoptera, Braconidae: *Apanteles cajae (=difficilis)*, *A. salebrosus*, *A. praepotens (=sericeus)* (THOMPSON 1945; SHENEFELT 1972; HERTING 1976; original data; ZWAKHALS pers. comm.).

HABITAT: Wet meadows, scrub, banks, shady roadsides, ditches, edges of forests, fens, open and marshy places, also in limestone quarries at Portland and Dorset in southern England (GOATER & al. 1981). From 0 m up to about 1,100 m above sea-level in Apennines (Mt. Majella: ZAHM pers. comm.), and to about 1,500-1,600 m in Alps (FORSTER & WOHLFAHRT 1981; WEIGT 1990).

SIMILAR SPECIES: The adults of *E. valerianata* can easily be distinguished from *E. plumbeolata* by the larger pale tornal spots on forewing and hindwing and usually with more distinct pale wavy line on the forewing (Text-fig. 20). The other similar species, *E. vulgata* has postmedial line sharply angled near costa on the forewing. The papillae, the cornuti on the vesica and the shape of sternum A8 in the male; and also the shape of corpus bursae and position of spiniferous area in bursa copulatrix in the female genitalia are quite unlike those in other European *Eupithecia* species.

The ***russeliata*** species-group

Species-group introduced by BOLTE (1990) as a segregate from the *satyrata* species-group (see MCDUNNOUGH 1949). Afterwards, the similar Palaearctic *vulgata* species-group was introduced by MIRONOV (1991). Represented in Europe by five species: *E. assimilata*, *E. vulgata*, *E. biornata*, *E. sutiliata* and *E. immundata*. Several Asian species could be placed in this group as follows: *E. nobilitata* STAUDINGER, 1882; *E. rubeni* VIIDALEPP, 1976; *E. subvulgata* VOJNITS, 1982, *E. pseudassimilata* VIIDALEPP & MIRONOV, 1988 and *E. acosmos* MIRONOV, 1989. Four species from North America belong to this group, as following: *E. assimilata* DOUBLEDAY, 1856; *E. cimicifugata* PEARSALL, 1908; *E. russeliata* SWETT, 1908; *E. ammonata* MCDUNNOUGH, 1929.

MALE GENITALIA: Uncus usually short, slim to stout, biapical. Papillae short, length about 0.4-0.6 times uncus length. Valva diverse in size and shape, without ventral process, usually broadened at base, with smoothly curved ventral margin, straight or slightly curved dorsal margin and narrowly rounded apex; sacculus weakly sclerotized. Vinculum short and broad. Aedeagus relatively large, usually narrowing anteriorly; length about 0.6-1.0 times valva length, about 3.2-4.0 times longer than medial width. Vesica armed with two or three more or less large dentate cornuti near apex connected at the base, and with one large, elongated, folded, usually V-shaped cornutus at ductus ejaculatorius base. Sternum A8 peg-like with base broadened, narrowing to apex; basal hollow shallow to deep; apical hollow usually absent (but in *E. nobilitata, E. biornata* and *E. immundata* this apical hollow small and narrow); sternum length about 0.6-0.8 times valva length.

FEMALE GENITALIA: Bursa copulatrix usually rounded (in some species its pear-shaped); almost completely covered with small spines (except in *E. immundata* and *E. rubeni*), and with small patch of spinules near colliculum (except in *E. biornata* and *E. pseudassimilata*); ductus bursae with longitudinal striations. Ductus seminalis long and broad; usually attached to corpus bursae near colliculum on right side. Colliculum collar-like, short and narrow. Antrum membranous. Anterior and posterior apophyses short and thin, usually not expanded at apexes. Spurs of anterior apophyses narrow along their entire length. Tergum A8 usually broad, rectangular, with more or less deep medial hollow in the posterior margin. Papillae anales slim, small and rounded (only in *E. nobilitata* the papillae anales large and elongated; covered by numerous small spines).

126. *Eupithecia assimilata* DOUBLEDAY, 1856

Eupithecia assimilata DOUBLEDAY, 1856: The Zoologist 14: 5140, 5142 (United Kingdom). Syntype(s) (BMNH).

Tephroclystis fumosa HULST, 1896: Trans. Amer. ent. Soc. 23 (3): 269 (U.S.A.: Massachusetts, South Abington). Lectotype ♂ (AMNH).

Unavailable names (misidentification): *minutata*: sensu HÜBNER (1813) nec DENIS & SCHIFFERMÜLLER, 1775 (identity: *Eupithecia absinthiata*); *absynthiata*: sensu PACKARD (1876) nec FISCHER VON RÖSLERSTAMM, 1829

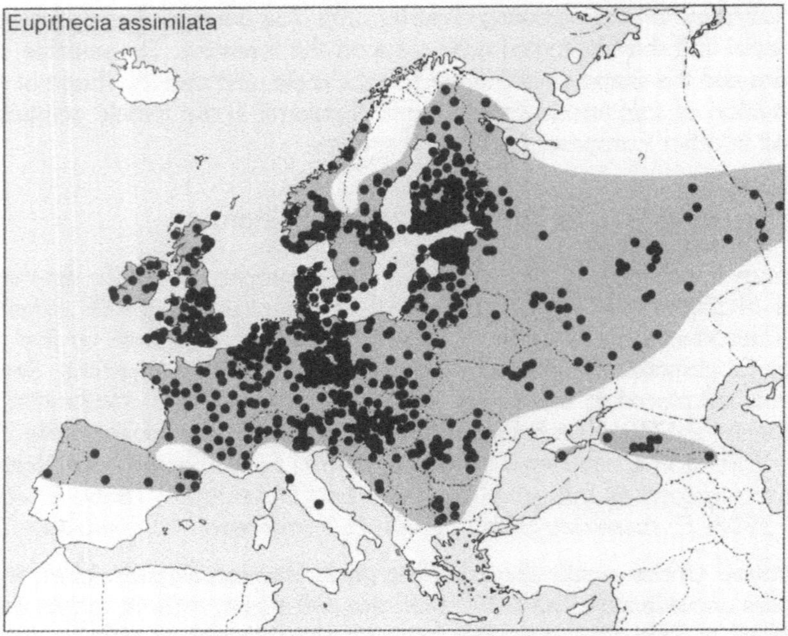

(identity: *Eupithecia absinthiata*); *coagulata*: sensu McDunnough (1949) nec Guenée, 1858 (identity: *Eupithecia coagulata*). - (incorrect subsequent spelling): *assimilaria*: Morris (1861).

Diagnosis: Wingspan 17-24 mm. Labial palpi brownish grey. Frons and vertex pale ash-grey or brownish grey. Notum greyer brown, sometimes with reddish tinge. Forewing greyer brown, often with reddish tinge; costal margin mottled and irrorate with white scales, with two to five darkish costal spots in place of transverse lines; transverse lines inconspicuous, except postmedial line, which is right-angled beyond discal spot and well marked on the veins with a series of dark dashes; medial line fine, weak, angled around discal spot; wavy line broken, forming a series of small, usually triangular white dots and a larger, often double, white tornal spot, shaded on inner margin; discal spot narrow, elongate, oblique, intensely black. Hindwing paler, greyer brown; transverse lines weak, indicated along dorsum; terminal area sometimes darker; pale wavy line ill-defined; a small white tornal spot present; discal spot small, rounded or slightly elongate. Fringe light greyish brown, spotted with dark greyish brown at vein ends. Abdomen greyish brown, with narrow whitish dorsal stripe.

Male genitalia: Uncus slim, narrow. Valva narrow, broadest at base, with concave dorsal margin. Vesica armed with three dentate cornuti and one elongated, narrow, folded cornutus. Sternum A8 with deep basal hollow.

Female genitalia: Bursa copulatrix pear-shaped, with dense patch of longitudinal striations in its medial part. Tergum A8 with shallow medial hollow in the posterior margin, about 1/3 tergum length.

Distribution: Euro-Siberian. Widely distributed in Europe from Ireland and Portugal to Urals except Iceland, Mediterranean islands, south of Appenines and Iberian and Balkan peninsulas. Outside Europe in Caucasus; eastward from Urals to Ussuri region and Sakhalin Island. Recorded for North America by Bolte (1990).

PHENOLOGY: Bivoltine: mid-April (early April) to late June (early July); mid-July (late June) to late August (mid-September). Larva: late May to mid-July (late July); early September (mid-August) to early October (late October). Overwintering as pupa.

BIOLOGY: Larva disjunct oligophagous. Found on the leaves of *Ribes nigrum* (Grossulariaceae) (Germany: HERRICH-SCHÄFFER 1861a) and also *Humulus lupulus* (Cannabaceae) (United Kingdom: BUCKLER 1899; France: GOOSSENS 1869; Germany: DIETZE 1910, 1913; WEIGT 1991; Denmark: JUUL 1948; Finland: SEPPÄNEN 1970), occasionally on *Ribes rubrum* (United Kingdom: CREWE 1861c) and *Urtica dioica* (Urticaceae) (Finland: SEPPÄNEN 1970). In North America has been found on cultivated *Ribes rubrum* (Grossulariaceae) (Canada: BOLTE 1990), and also recorded on *Picea glauca* (Pinaceae) (Canada: PRENTICE 1963), probably in error.

PARASITOIDS: Hymenoptera, Braconidae: *Apanteles lateralis* - Diptera, Tachinidae: *Bessa selecta* (THOMPSON 1945; SHENEFELT 1972; HERTING 1976; ZWAKHALS pers. comm.).

HABITAT: Forest edges, scrub, field hedges, gardens and parks. From 0 m up to about 1,500 m above sea-level in Pyrenees (MAZEL & PESLIER 1997), and to about 1,800 m in Alps (FORSTER & WOHLFAHRT, 1981; WEIGT 1991; HELLMANN & al. 1999).

SIMILAR SPECIES: Similar to *E. absinthiata* but distinguished by the shorter and broader forewing, with costa more arched, by discal spot usually more obliquely elongate, the series of white subterminal dots and white tornal spot on the forewing of *E. assimilata* more prominent than in *E. absinthiata*, the postmedial line rather angled (not evenly curved) on the hindwing, the fringe more chequered than in other closely related species (Text-fig. 68). However, valva small and narrow, aedeagus narrowed toward base, vesica with three dentate cornuti connected with each other at base, sternum A8 short, peg-like in the male and corpus bursae without longitudinal row of small spines near ductus seminalis base in the female genitalia should distinguish this species.

127. *Eupithecia vulgata* (HAWORTH, 1809)

Phalaena vulgata HAWORTH, 1809: Lepid. Br. (2): 359 (United Kingdom). Syntype(s) (BMNH).
?*Phalaena (Geometra) pygmeata* BORKHAUSEN, 1794: Naturgesch. Eur. Schmett. 5: 334 (Europe: type locality not stated). Syntypes.
Geometra clusterata HÜBNER, 1813: Samml. Eur. Schmett. 5 Geometrae (2), pl. 89: 457 (Europe). Syntype(s) lost.
Tarachia austerata HÜBNER, 1825: Verz. bek. Schmett. (21): 324 (Europe). Syntype(s) lost.
Eupithecia lepsaria STAUDINGER, 1882: Stett. Ent. Z. 43 (1-3): 74 (Kazakhstan: Lepsa). Syntypes 2♂ (MNHU).
Eup(ithecia) vulgata f. *montium* DIETZE, 1910: Biol. Eupithecien 1, pl. 72: 299. 300; ibidem (1913), 2: 91 (Switzerland: Zermatt). Syntypes, at least 2♂ (MNHU).
Eup(ithecia) vulgata f. *atropicta* DIETZE, 1910: Biol. Eupithecien 1, pl. 72: 301. 302; ibidem (1913), 2: 91 (Austria: Tyrol). Syntypes (MNHU).
Eupithecia vulgata var. *cyrneata* SCHAWERDA, 1933: Z. Öst. Ent. Ver. 18 (6): 48, pl. 8: 1, 4, 7, 8 (Corsica: Vizzavona, Ponte de Fulminato). Syntypes (NHMW). Valid at subspecific rank.
Eupithecia vulgata scotica COCKAYNE, 1951: The Entomologist 84 (1058): 154 (United Kingdom: Scotland, Aviemore). Holotype ♂ (BMNH). Valid at subspecific rank.
Eupithecia vulgata clarensis HUGGINS, 1962: The Entomologist 95 (1185): 45 (Ireland: Co(unty) Clare, the Burren, Ballyvalackan). Holotype ♀ (BMNH). Valid at subspecific rank.
Unavailable names (infrasubspecific): *nigrofasciata*: DIETZE (1913); *unicolor*: LEMPKE (1951); *impuncta*: LEMPKE (1951). - (incorrect subsequent spellings): *austeraria*: BOISDUVAL (1840); *vulgaria*: MORRIS (1861)

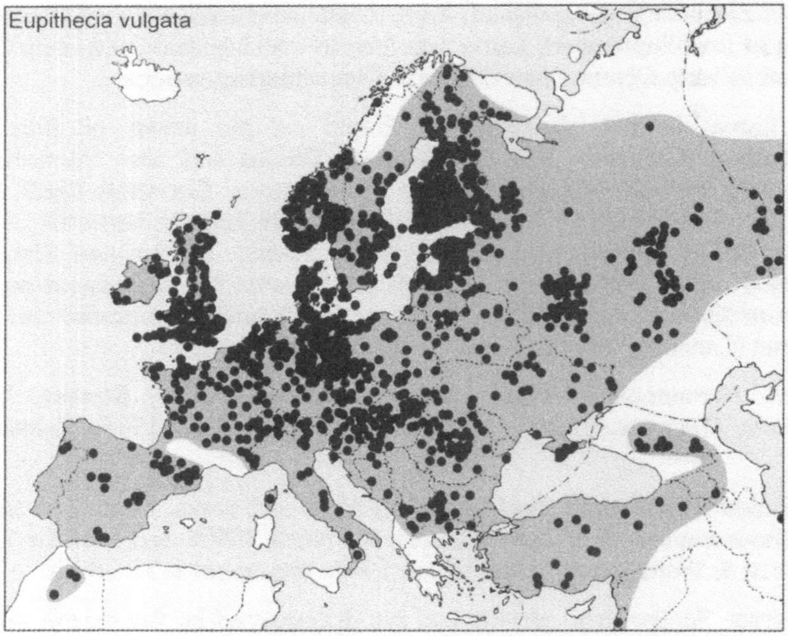

DIAGNOSIS: Wingspan 16-23 mm. Labial palpi brown. Frons and vertex pale greyish brown. Notum pale greyish brown, often with rusty tinge. Forewing rather narrow, with costa arched, termen rounded and with more or less pointed apex, usually more pointed in male. Ground colour greyish brown, often with a rusty tinge; veins often chequered dark brown and white. Transverse lines slightly sinuate, sharply or right-angled, broadened and slightly darker near costa. Basal, postbasal and antemedial lines rather inconspicuous; postmedial line usually straight, minutely waved, twice sharply or right-angled onto costa; pale wavy line fine, sometimes broken, shaded by dark scales from within, forming a distinct, whitish tornal spot. Discal spot minute, circular to ovate, black or blackish brown, still more reduced in some specimens. Hindwing slightly paler than forewing, greyish brown, darker towards termen; transverse lines weakly marked; pale wavy line conspicuous, fine, much indented but sometimes ill-defined; discal spot minute, rounded or obsolete. Fringe distinctly chequered brownish white and brown or dark brown at vein ends. Abdomen greyish brown with a series of small, black lateral spots, and sometimes with a row of small dark brown or white dorsal tufts.
The Corsican subsp. ***cyrneata*** SCHAWERDA, 1933 is darker and greyer, without brown but with less conspicuous transverse lines on the wings than in the nominate subspecies.

MALE GENITALIA: Uncus stout, thick, when viewed dorsoventrally. Valva narrow, broadest at base, with concave dorsal margin. Vesica armed with three dentate cornuti and one elongated, folded cornutus. Sternum A8 with deep basal hollow and shallow apical hollow.

FEMALE GENITALIA: Tergum A8 with convex lateral margins and deep medial hollow in the posterior margin, about 3/4 tergum heighth.

DISTRIBUTION: Palaearctic. All over Europe except Iceland, Sardinia, Sicily, Malta and

Crete. On Corsica replaced by subsp. *cyrneata*. – Outside Europe in North Africa (Morocco), Asia Minor, Lebanon, Caucasus, Transcaucasus and northern Iran. The darker f. *lepsaria* STAUDINGER, 1882 occurs in Central Asia (Tien-Shan, Ghissar Mts to western Pamirs), eastern Kazakhstan, western and southern Siberia, Amur region, Mongolia and Korea. – New for the fauna of Greece: 1♂, Grecia, V. Kala(m)baka, Kal(l)ithea, 24.VI.1982, leg. F. SCHEPLER (ZMUC); 1♀, Hellas, Dráma, Falakro(n), 1700 m, 10.VII.2001, leg. M. MØLGAARD (coll. FIBIGER).

PHENOLOGY: Uni- or bivoltine: mid-May (mid-April) to mid-July (early August); sometimes a second generation from mid-August (early August) to late September. Larva: early July (mid-May) to mid-August (early September). Overwintering as pupa.

BIOLOGY: Larva polyphagous (1). Found on the leaves of numerous mainly herbaceous plants: *Thalictrum foetidum, Clematis vitalba* (Ranunculaceae), *Chaerophyllum temulum, Pimpinella saxifraga, Anthriscus, Daucus* (Apiaceae), *Gallium mollugo* (Rubiaceae), *Polygonum* (Polygonaceae), *Campanula rotundifolia* (Campanulaceae), *Succisa pratensis* (Dipsacaceae), *Solidago virgaurea, Achillea millefolium, Artemisia campestris, Aster, Centaurea jacea, Senecio jacobaea* (Asteraceae), *Cucubalus, Silene vulgaris* (Caryophyllaceae), *Sedum telephium* subsp. *maximum* (Crassulaceae), *Hedera helix* (Araliaceae), *Lonicera* (Caprifoliaceae), *Rubus idaeus, Crataegus monogyna, Prunus, Mespilus* (Rosaceae), *Calluna vulgaris, Ledum palustre, Vaccinium myrtillus* (Ericaceae) (United Kingdom: CREWE 1861c, 1862; BUCKLER 1899; France: MILLIÈRE 1870; MABILLE 1872; LHOMME 1935; Germany: DIETZE 1871b, 1910, 1913; WEIGT 1991; WEGNER pers. comm.; Switzerland: VORBRODT 1931; Finland: SEPPÄNEN 1970).

PARASITOIDS: Hymenoptera, Ichneumonidae: *Netelia tarsata* - Hymenoptera, Braconidae: *Meteorus gyrator* (THOMPSON 1945; SHENEFELT 1969; HERTING 1976).

HABITAT: All types of habitats from old wet forests to dry open areas, plantations, gardens and parks, cultivated areas and fallow fields. From 0 m up to about 1,800 m above sea-level in Apennines (Mt. Pollino: coll. ZSM), to about 1,900-2,000 m in Alps and southern Spain (REZBANYAI-RESER 1988; REZBANYAI-RESER & MAIER 2000; coll. ZMUC), and to about 2,300-2,400 m in Pyrenees (MAZEL & PESLIER 1997).

SIMILAR SPECIES: In spite of its variability, *E. vulgata* can be easily distinguished from any other European *Eupithecia* species by the combination of the characteristic wingshape, the postmedial line twice sharply angled near forewing costa, the usually conspicuous pale wavy line with pale tornal spot and the distinctly chequered fringe. Aedeagus tapered anteriorly, vesica armed with three dentate cornuti, connected at base in the male genitalia; the rounded corpus bursae 2/3 covered by minute spinules with longer marginal spines and with small spiniferous area nearly colliculum, absence of broad spurs of anterior apophyses, and presence of deep membranous medial hollow in tergum A8 in female should distinguish this species.

128. *Eupithecia biornata* CHRISTOPH, 1867

Eupithecia biornata CHRISTOPH, 1867: Stett. Ent. Z. 28 (4-6): 238 (Russia: Sarepta [=Krasnoarmeysk]). Syntype(s) ♀ not traceable in ZISP.

DIAGNOSIS: Wingspan 19-24 mm. Labial palpi and notum pale yellowish grey. Frons and vertex white. Forewing elongated with straight costa, oblique and weakly curved termen

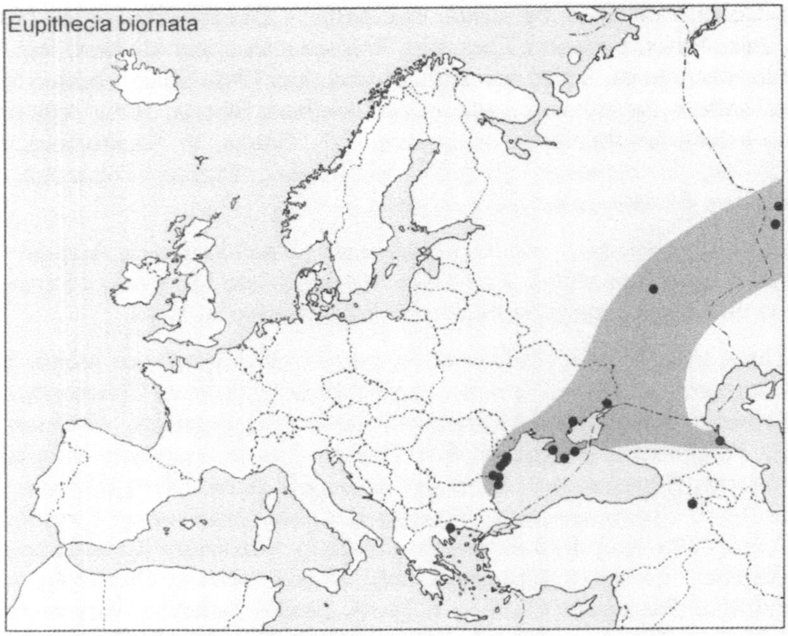

and narrowly rounded apex; delicate pale ash-grey with yellowish tinge and with broad light yellow, longitudinal medial stripe, broadened at base; transverse lines oblique, inconspicuous or usually absent; discal dot minute, dark, rounded. Hindwing paler, light grey; transverse lines inconspicuous, weakly indicated on the posterior margin; discal spot absent. Fringe unicolorous, pale grey. Abdomen pale yellowish grey with scattered brown scales, and always with broad white dorsal stripe and with minute brown lateral spots on each segment.

MALE GENITALIA: Uncus slim, rather long and narrow. Valva large, elongated, length more than 2 mm, with almost parallel ventral and dorsal margins. Vinculum relatively narrow. Vesica armed with three large dentate cornuti and one elongated, narrow, folded cornutus. Sternum A8 elongated with broad, shallow basal hollow and narrow, shallow apical hollow.

FEMALE GENITALIA: Bursa copulatrix without separate patch of spinules near colliculum. Colliculum inclined to left. Tergum A8 square, with deep, narrow medial hollow in the posterior margin.

DISTRIBUTION: Euro-Siberian. In Europe found in Greece, Bulgaria, Romania, southern Ukraine and southern European Russia. – Outside Europe in Asia Minor, northern Kazakhstan, south-western and southern Siberia (including Dahurian steppes in Transbaikalia), north-western and north-eastern China and Mongolia.

PHENOLOGY: Univoltine: late June to mid-August (mid-September).

BIOLOGY: Larva found on *Artemisia* (Asteraceae) (Russia: CHRISTOPH 1867).

HABITAT: Steppes. Up to about 900-1,000 m above sea-level in eastern Turkey (coll. FIBIGER).

SIMILAR SPECIES: *E. biornata* can be easily distinguished from other European *Eupithecia* species by the distinctive maculation and presence of light yellow longitudinal medial stripe from base to apex on the forewing. The shapes of valva and sternum A8, the armature of vesica, and the shape of corpus bursae almost completely covered by small spines should distinguish this species.

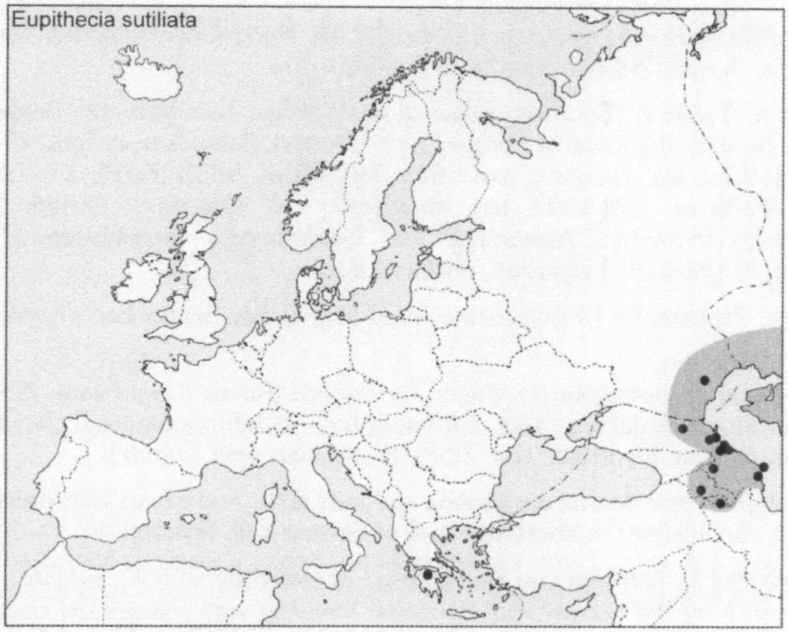

129. *Eupithecia sutiliata* CHRISTOPH, 1876

Eupithecia sutiliata CHRISTOPH, 1876: Horae Soc. Ent. Ross. 12: 262, pl. 7: 34 (Iran: Schahrud). Lectotype ♂ (ZISP; without abdomen, examined).

Eupithecia aequata STAUDINGER, 1892: Dt. ent Z. Iris 5: 256 (Kirghizstan: Transalai; Uzbekistan: Margelan). Syntypes 3♀ (MNHU, 1 syntype examined).

Eupithecia scopariata var.? *orientata* STAUDINGER, 1901: in STAUDINGER & REBEL, Cat. Lepid. Palaearct. Faunengeb. (Edn 3) 1: 314 (Arm(enia); Tura). Lectotype ♂, herewith designated to stabilize nomenclature (ZISP, examined).

Eupithecia rjabovi VOJNITS, 1975: Acta zool. Acad. Sci. Hung. 21 (1-2): 237, fig. 3 (Russia: Daghestan, district Petrovsk [=Makhachkala]). Holotype ♂ (HNHM, examined).

Eupithecia kasahstanica VIIDALEPP, 1988: Fauna pjadenitz gor Srednej Azii: 120, pl. 2: 17; text-pl. 26: 2-4 (Kazakhstan: Kzyl-Orda). Holotype ♂ (ZISP, examined).

Unavailable names (misidentification): *satyrata*: sensu LVOVSKY (1971) nec HÜBNER, 1813 (identity: *Eupithecia satyrata*).

DIAGNOSIS: Wingspan 16-23.5 mm. Labial palpi, frons, vertex and notum pale grey with delicate ochreous tinge. Forewing narrow, elongated, with straight or weakly arched costa, rather pointed tip and oblique, rounded termen; ground colour pale greyish ochreous; numerous closely placed, fine and oblique transverse lines, sharply angled near costa; postmedial line often more distinct than others; pale wavy line usually indistinct; discal dot very small, rounded, blackish. Hindwing narrow and elongated, paler than forewing, with terminal shade; numerous transverse lines well marked along dorsum; discal dot inconspicuous or absent. Fringe weakly chequered greyish white and greyish brown. Abdomen pale ochreous grey.

MALE GENITALIA: Uncus stout. Vinculum narrow. Vesica armed with one small patch of granulation and one elongated, thin, slightly folded V-shaped cornutus at ductus ejaculatorius base; dentate cornuti absent. Sternum A8 with deep basal hollow and sclerotized apex, with deep hollow.

FEMALE GENITALIA: Bursa copulatrix rounded, lightly elongated, completely covered with small spines. Tergum A8 with very deep medial hollow.

DISTRIBUTION: Turanian. Southern Russia (Lower Volga, Kalmykia and Daghestan) and Greece (!). New for the fauna of Europe: 2♀♀, (Russia), Astrakhan, at light, 27.V. (19)70 and 7.VII.(19)70, leg. LVOVSKY; gen. prep. No. 9851, 9852 (ZISP); 1♀, Greece, Mt. Chelmos, 1750 m, -.VII.1997, leg. M. FIBIGER (coll. FIBIGER). – Outside Europe in Transcaucasus (Armenia, Azerbaijan) and Central Asia (Kazakhstan, Turkmenia, Uzbekistan, Kirghizstan, Tadjikistan, northern Iran).

PHENOLOGY: Probably bi- or polyvoltine: early May to late September. Overwintering as pupa.

BIOLOGY: Larva oligophagous (1). Found on *Suaeda* (Russia (Daghestan): ZISP, original data), *Kalidium foliatum* and *Climacoptera* (Chenopodiaceae) (south-eastern Kazakhstan (Ili river, Mynbulak, coll. ZISP): FALKOVITSH pers. comm.).

HABITAT: Steppes and deserts on alkaline and salty soils, and places with ruderal vegetation. Up to about 1,800 m above sea-level in Greece (coll. FIBIGER).

SIMILAR SPECIES: *E. sutiliata* can be confused occasionally with *E. vulgata*, but distinguished from it by the narrow and elongated forewing with less arched costa, the soft ochreous tinge of coloration, the numerous fine oblique transverse lines, the minute circular discal dot, the less waved and weakly visible pale wavy line and absence of pale tornal spot on the forewing; the more elongated hindwing, and the less chequered fringe. The larger and more elongate uncus and valva; the narrower vinculum, absence of dentate cornuti on the vesica in the male genitalia, and the small marginal spines in bursa copulatrix in the female genitalia all assist separation of *E. sutiliata* from *E. vulgata*.

130. *Eupithecia immundata* (LIENIG & ZELLER, 1846)

Larentia immundata LIENIG & ZELLER, 1846: Isis 3: 194 (Latvia: ?Koknese [=Kokenhusen]). Syntypes 2♂3♀ (BMNH?).

Eupithecia argillacearia HERRICH-SCHÄFFER, 1848: Syst. Bearb. Schmett. Eur. 3 (32): 122, 136; ibidem (1846), pl. 23: 143, 144 (non binominal), (Czech Republic, Böhmens). Syntype(s).

DIAGNOSIS: Wingspan 17-21 mm. Labial palpi yellowish brown. Frons, vertex and notum pale yellowish brown. Forewing pale yellowish brown with a silky sheen; transverse lines indistinct, slightly sinuate, evenly curved near costa; three paler lines usually more or less conspicuous (antemedial, medial and more waved postmedial); pale wavy line indistinct or obsolete; discal spot absent. Hindwing concolorous or slightly paler, transverse lines lacking or only with inconspicuous postmedial line; discal spot absent. Fringe unicolorous, pale yellowish brown with a silky sheen, sometimes slightly spotted with dark. Abdomen pale yellowish brown.

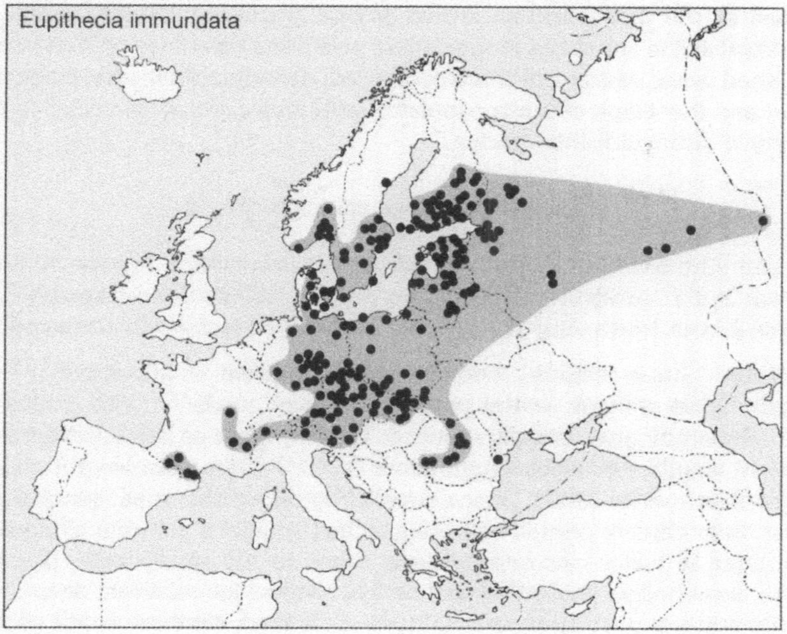

MALE GENITALIA: Valva broad with apex broadly rounded. Vesica armed with three large dentate cornuti, one elongated, folded cornutus and one small, irregular cornutus at ductus ejaculatorius base. Sternum A8 elongated with shallow basal hollow and with deep and very narrow apical hollow.

FEMALE GENITALIA: Bursa copulatrix rather pouch-like, with waved longitudinal striations; 1/3-1/4 covered by uniform-sized small spines at base and with large patch of spinules in ductus bursae. Colliculum short and broad. Anterior and posterior apophyses very short. Tergum A8 square, without medial hollow in the posterior margin.

DISTRIBUTION: European. Mainly in northern and Central Europe. In southern Europe occurs in the mountain ranges (Pyrenees, Alps and Carpathians).

PHENOLOGY: Univoltine: late May (mid-May) to mid-July (late July). Larva: mid-July (mid-June) to early August (late August). Overwintering as pupa.

BIOLOGY: Larva monophagous (1). Found in unripe berries of *Actaea spicata* (Ranunculaceae), feeding on the seeds (Switzerland: FAVRE 1905; Germany: DIETZE 1910, 1913; WEIGT 1988; Denmark: JUUL 1948; Finland: SEPPÄNEN 1970; Russia: original data).

PARASITOIDS: Hymenoptera, Braconidae: *Lissogaster (=Microgaster) globata*, *Microgaster rufipes* (SHENEFELT 1973; HERTING 1976).

HABITAT: Shady deciduous, mixed and spruce forests with fertile, moist soil, and in old neglected parks. From 0 m up to about 1,600 m above sea-level in Pyrenees (DANTART 1988), and to about 1,800 m in southern Alps (WEIGT 1988).

SIMILAR SPECIES: Rather remotely similar to *E. plumbeolata* and *E. thalictrata*.

Distinguished by the pale yellowish brown ground colour with a silky sheen, the more curved postmedial line and three more distinct pale transverse lines on the forewing. The short and broad valva, vesica with three connected, dentate cornuti, the shape of sternum A8 in male, and the shape of bursa copulatrix with area covered by spines in the female genitalia should distinguish this species.

The ***addictata*** species-group

Species-group introduced by KRAMPL (1994) and represented in Europe by two species: *E. thalictrata* and *E. addictata*. *E. caliginea* BUTLER, 1878 from Japan and *E. anticaria* WALKER, 1862 from North America can also be included in the *addictata* species-group.

MALE GENITALIA: Uncus biapical, medium length. Papillae length about 0.7-0.8 times uncus length. Valva without ventral process, broadest medially; with angled or evenly curved ventral margin; apex broadly rounded; sacculus more or less sclerotized. Vinculum short, medium length. Aedeagus length about 0.7-0.9 times valva length; about 3.0-4.0 times longer than medial width. Vesica covered by numerous small spinules and armed with one or three dentate cornuti with longitudinal striations and one elongated, irregular, flat cornutus at ductus ejaculatorius base. Sternum A8 usually wide, slightly narrowing to apex; basal hollow shallow; apical hollow shallow (or relatively deep in *E. thalictrata*); lateral margins near apex usually more sclerotized; sternum length about 0.5-0.7 times valva length.

FEMALE GENITALIA: Bursa copulatrix spherical or slightly elongated, almost completely or more than half covered with small spines. Ductus bursae short, more sclerotized than corpus bursae; without spines or only with separate spines or sparse row of very small spines. Ductus seminalis usually long and broad; attached to bursa copulatrix near colliculum on right side. Colliculum collar-like, short and not very broad. Anterior and posterior apophyses short or medium length. Spurs of anterior apophyses narrow at base and broadened to apices. Tergum A8 broad, usually wider than long; with rounded apical corners. Papillae anales short, rounded or slightly tapered.

REMARKS: I have placed the *addictata* species-group close to the *druentiata* and *russeliata* species-groups because their genitalia show similarities, especially the cornuti on the vesica in the male genitalia and disposition of spines in the female genitalia. One dentate cornutus larger than the other, longitudinal striations on the vesica and broad sternum A8 in male should distinguish representatives of the *addictata* species-group from others.

131. Eupithecia thalictrata (PÜNGELER, 1902)

Tephroclystis thalictrata PÜNGELER, 1902: Dt. ent. Z. Iris 15: 137, pl. 6: 20, 21 (Switzerland: Wallis). Syntypes ♂ ♀ (MNHU).

Eupithecia ijimai INOUE, 1963: Tinea 6 (1/2): 33, text-fig. 5; pl. 7: 11, 12 (Japan, Hokkaido: Kushiro, Shibecha). Holotype ♂ (BMNH).

Eupithecia indecora VOJNITS, 1981: Acta Zool. Acad. Sci. Hung. 27 (3-4): 429, fig. 52 (China: Shansi Province, Mien Shan). Paratype ♀ only, nec holotype ♂ (ZFMK, examined). New junior synonym.

DIAGNOSIS: Wingspan 16-20 mm. Labial palpi, frons, vertex and notum grey or brownish grey. Forewing unicolorous grey or brownish grey, usually with paler band between medial and terminal areas, well indicated only near costa; transverse lines indistinct, obscure, weakly sinuate, slightly angled and darker near costa; postmedial line evenly curved near

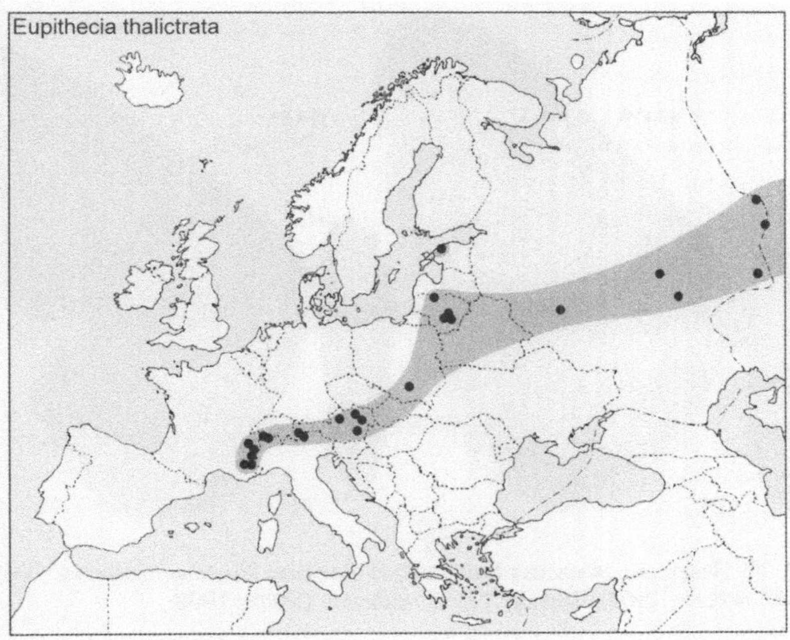

costa; pale wavy line indistinct or absent; discal spot brownish, weakly discernible or absent. Hindwing paler, grey or brownish grey; transverse lines indistinct, more or less marked only at posterior margin; slightly darker terminal area and pale wavy line inconspicuous; discal spot absent. Fringe brownish grey, weakly spotted with dark, more strongly spotted on the hindwing. Abdomen grey or brownish grey.

MALE GENITALIA: Valva with broadly rounded apex. Vesica granulated; armed with three dentate cornuti (two small at apex and one larger, longitudinally wrinkled in medial part of aedeagus), also with elongated, flat, curved cornutus at ductus ejaculatorius base. Sternum A8 short and broad, with two relatively broad and blunt apical rods, and with deep medial hollow between them, about 0.5 times sternum length.

FEMALE GENITALIA: Bursa copulatrix subglobular, with transverse row of small sparse spines in ductus bursae. Tergum A8 rectangular.

DISTRIBUTION: Eurasiatic. In Europe has been found in south-eastern France, Switzerland, northern Italy, Austria, Poland, Lithuania, Estonia and central European Russia to Urals. – Outside Europe in northern Kazakhstan (Kokchetav province), Altai, southern Siberia, Amur and Ussuri regions, China, Sakhalin, Kuril Islands (Kunashir Island) and Japan (Hokkaido).

PHENOLOGY: Univoltine: early May (late April) to late June. Larva: mid-June to mid-August. Overwintering as pupa in cocoon usually amongst moss near base of host-plant, sometimes twice.

BIOLOGY: Larva monophagous (2). Found on the leaves, flowers and young seeds of *Thalictrum foetidum* (Switzerland: PÜNGELER 1902; FAVRE 1905; DIETZE 1913) and also *T. minus* (Ranunculaceae) (Italy: WEIGT 1990).

Text-figs 71-72: Diagnostic characters (indicated) of European *Eupithecia* species. Text-fig. 71: *E. thalictrata* (PÜNGELER, 1902). Text-fig. 72: *E. addictata* DIETZE, 1908.

HABITAT: Deciduous and mixed forests, scrub and scrub-covered meadows, rocks and stony slopes. From 0 m up to about 1,800 m above sea-level in Alps (VORBRODT 1928).

SIMILAR SPECIES: Most similar to *E. plumbeolata* and *E. valerianata*. However, *E. plumbeolata* has more rounded apex and more convex terminal margin of forewing; the transverse lines are usually more waved and evenly curved, and the pale band outside the postmedial line less conspicuous than in *E. thalictrata* (Text-fig. 71). *E. valerianata* has greyish ochreous ground colour, less conspicuous transverse lines and well defined pale tornal spot on the forewing. The cornuti on the vesica and the shape of sternum A8 in the male, as well as the shape of bursa copulatrix and disposition of the spiniferous area in corpus bursae in the female genitalia can be used to distinguish *E. thalictrata* from other extremely similar European *Eupithecia* species (see also KRAMPL 1994).

132. *Eupithecia addictata* DIETZE, 1908

Eup(ithecia) ? selinata f. *addictata* DIETZE, 1908: Dt. ent. Z. Iris 21: 194, pl. 2: 5, 6 (Russia: Ussuri, Kasakewitsch). Lectotype ♂ (MNHU, examined by KRAMPL 1994). Raised to species rank by PROUT (1930).

Eupithecia danielata SCHÜTZE, 1959: Mitt. Münchn. Ent. Ges. 49: 35, pl. 1: 1, 2, 5, 6 (Greece: Olympos Mt., Enipefs-Tal, Dionyssios). Holotype ♀ (ZSM, examined).

Eupithecia rudniki VOJNITS, 1973: Annales hist.-nat. Mus. Nat. Hung. 65: 253, fig. 3b (Russia: Ussuri region, Sutschanski Rudnik). Holotype ♀ (ZFMK; examined).

Eupithecia pseudoplumbeolata VOJNITS, 1973: Annales hist.-nat. Mus. Nat. Hung. 65: 254 (Russia: southern Siberia, near Minussinsk). Holotype ♂ (ZFMK, 5 paratypes examined).

Eupithecia falkovitshi VIIDALEPP, 1976: Nasekomye Mongolii 4: 386, figs 15, 19-23 (Russia: Khakassia, Sajan Mts, valley of Sartaban river). Holotype ♂ (IZBE, examined).

Eupithecia koreaica VOJNITS, 1978: in VOJNITS & DE LAEVER, Acta Zool. Acad. Sci. Hung. 24 (1-2): 225, figs 1, 2 (Korea: Province Ryang-gang, Hyesan). Holotype ♂ (HNHM, examined).

Unavailable names (misidentification): *selinata*: sensu INOUE (1953) nec HERRICH-SCHÄFFER, 1861 (identity: *Eupithecia selinata*). - (incorrect subsequent spelling): *addictaria*: PROUT (1914).

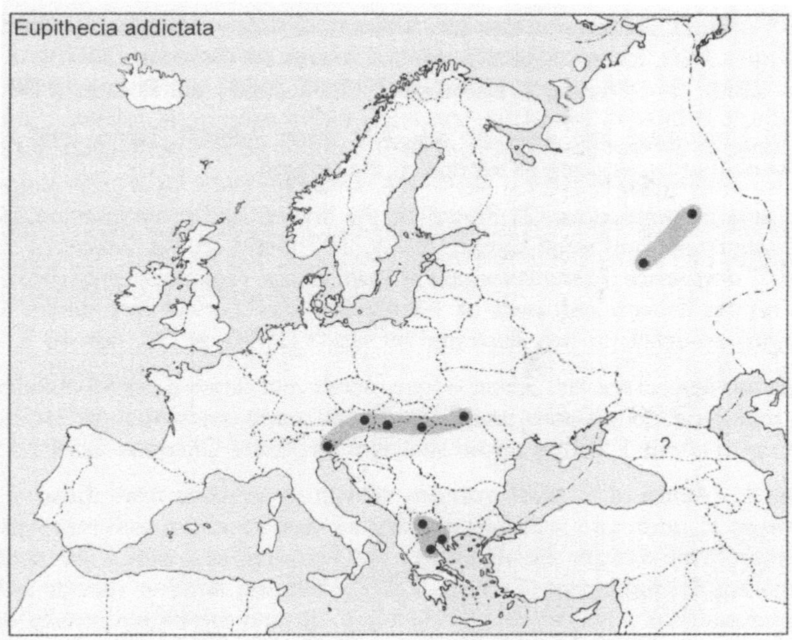

DIAGNOSIS: Wingspan 16-21 mm. Labial palpi, frons, vertex and notum brownish grey. Forewing unicolorous, brownish grey; transverse lines rather indistinct, usually obtuse angled and slightly broadened near costa; medial line with two pronounced sinuate curves, crossing discal spot and obtuse angled towards costa; postmedial line multiply waved, straight and obtuse angled near costa or evenly curved; pale double stripe beyond postmedial line often evident; wavy line inconspicuous, tornal spot lacking; discal spot variable from large, ovate-orbicular, dark brownish grey to very small or obsolete. Hindwing faintly paler, with inconspicuous transverse lines, other than paler, evenly curved stripe between medial and terminal areas; discal spot minute, very narrow, linear or absent. Fringe unicolorous, light brownish grey, only sometimes faintly dark-spotted at vein ends, especially on the hindwing. Abdomen brownish grey, with a series of small dark dorsal dots and short black lateral stripes.

MALE GENITALIA: Valva with obtuse angled ventral margin and broadly rounded apex. Vesica granulated, with one elongated, longitudinally wrinkled dentate cornutus at apex and one elongated, U-shaped cornutus at ductus ejaculatorius base. Sternum A8 peg-like, with two very small sclerotized apical lobes and shallow hollow between them.

FEMALE GENITALIA: Bursa copulatrix almost completely covered by small spines, without separate patch of spines in ductus bursae. Tergum A8 almost square with rounded posterior corners.

DISTRIBUTION: Eurasiatic. In Europe very sporadic and local in north-eastern Italy, Austria, Slovakia, Hungary, Macedonia, Greece, western Ukraine and central European Russia. – Outside Europe in western Siberia (Tjumen province), Altai, southern Siberia, Amur and Ussuri regions, southern Kuril Islands, China, Korea and Japan. Recorded for northern Caucasus (Mamyx and Mashur [= Mt. Mashuk near Pjatigorsk]) by VOJNITS (1973; as paratypes ♀♀ of *E. pseudoplumbeolata* VOJNITS, 1973). – New for the faunas of Italy and Hungary: 1♂1♀, Italien, Udine, Lago di Cavazzo bei Gemona, 300 m, 1.VII.1971, leg. GROSS (ZSM); 1♀, (Hungary), Bátorliget, 3-5.VII.1988, leg. RONKAY (HNHM).

PHENOLOGY: Univoltine: late June to late July (early August). Larva: mid-July to late August. Overwintering as pupa in cocoon in the ground.

BIOLOGY: Larva monophagous (2). Found on the flowers and sometimes on the ripening seeds of *Thalictrum thunbergii* (Japan: INOUE, 1979) and on the flowers of *Thalictrum minus* and *T. amurense* (Ranunculaceae) (Russia, Ussuri region: original data). Probably, in Europe on the flowers and seeds of *Thalictrum* also. Larvae very similar to those of *E. thalictrata*, especially to two illustrated by WEIGT (1990; pl. 36, figs 40.7, 40.8).

HABITAT: Broad-leaved forests, forest edges, scrub and scrub-covered meadows, shady roadsides, rocks and stony slopes; usually in dry and warm landscapes (in Far East Russia). From 0 m up to about 1,200 m above sea-level on Mount Olympus (SCHÜTZE 1959).

SIMILAR SPECIES: Adults of *E. addictata* are difficult to separate from those of *E. thalictrata*, however, *E. addictata* is darker, rather dark grey in colour, with more distinct postmedial line and with a larger discal spot on the forewing, and with a less spotted fringe of wings than in *E. thalictrata* (Text-fig. 72). *E. selinata*, another species with which it could be confused, has a lighter, shiny, silken grey ground colour and usually larger discal spots, especially on the hindwings. The genitalia of *E. addictata* are very easily distinguished from those of all other similar European *Eupithecia* species by the combination of the ventrally obtuse angled valva, the presence of only one elongated dentate cornutus on the vesica, the characteristic shape of sternum A8 in the male and by the rather pear-shaped bursa copulatrix, completely covered with uniformed small spines, without patch of spines in ductus bursae in female.

The *druentiata* species-group

Introduced here for the first time. This is a species-poor Palaearctic group which is represented in Europe by three species, as following: *E. exiguata, E. druentiata* and *E. moecha*. Two East Asian species *E. bella* STAUDINGER, 1897 and *E. subexiguata* VOJNITS, 1974 can also be included in the *druentiata* species-group. *E. denotata* and *E. pseudocastigata* Pinker, 1976 from Asia Minor show characters both of the *druentiata* species-group and the *millefoliata* species-group. *E. denotata* is treated here along with the *druentiata* group species.

MALE GENITALIA: Uncus medium length, narrow, biapical (uniapical in *E. exiguata*). Papillae various, length about 0.5-0.7 times uncus length. Valva various in size and shape, from rather narrow and elongated to very broad, with or without wide, blunt ventral process. Vinculum large, short and broad. Aedeagus relatively large and stout, length about 0.7-0.9 times valva length, about 3.0-4.0 times longer than medial width. Vesica armed with two large, heavily sclerotized dentate cornuti (usually one apically and one medially) and one large, elongated, slightly folded, V-shaped cornutus, with two broad apices at ductus ejaculatorius base. Sternum A8 peg-like with base broadened, narrowing to apex; basal hollow deep; apical hollow absent; sternum length about 0.6-0.8 times valva length.

FEMALE GENITALIA: Bursa copulatrix sclerotized, pear-shaped, almost half covered by small spines at base and with numerous longitudinal striations in ductus bursae. Ductus seminalis long and broad, attached to middle part of corpus bursae on right side. Colliculum collar-like, short and rather broad, slightly inclined to right. Antrum lightly sclerotized. Anterior and posterior apophyses rather short and thick, not expanded at apices. Spurs of anterior apophyses narrow throughout their length or only weakly expanded towards apices. Tergum A8 rather square. Papillae anales slim, small and rounded.

REMARKS: The presence of a large, elongated V-shaped cornutus with expanded apices at ductus ejaculatorius base and two large dentate cornuti in the vesica can be used for separating the *druentiata* species-group from any others.

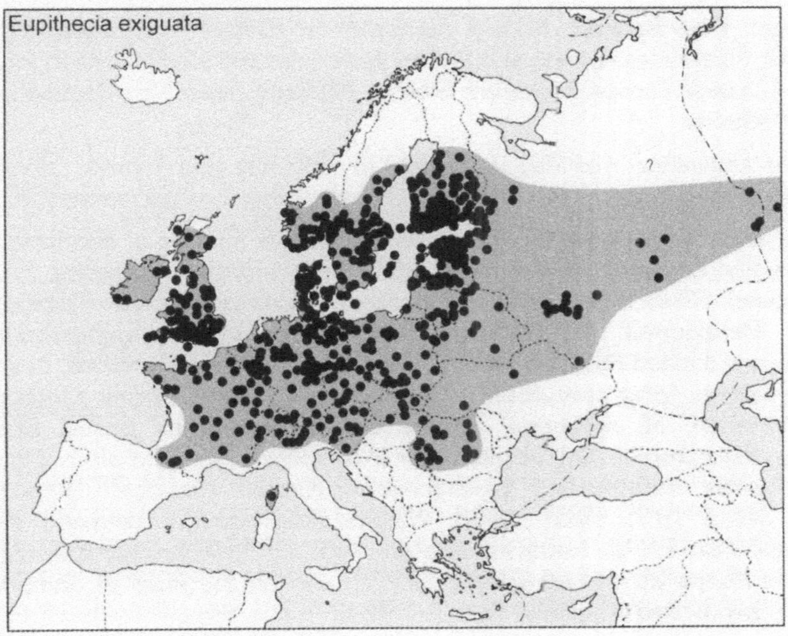

133. *Eupithecia exiguata* (HÜBNER, 1813)

Geometra exiguata HÜBNER, 1813: Samml. Eur. Schmett. 5 Geometrae (2), pl. 73: 379 (Europe). Syntype(s) lost.

Phalaena trimaculata HAWORTH, 1809: Lepid. Br. (2): 362 (United Kingdom). Syntype(s) (BMNH). Junior primary homonym of *Phalaena Geometra trimaculata* DE VILLERS, 1789 (identity: *Stegania trimaculata*, Ennominae).

Eupithecia ochreata STEPHENS, 1831: Illustr. Br. Ent. (Haustellata) 3: 283 (United Kingdom: London). Syntype(s) (BMNH).

Eupithecia lanceolaria WOOD, 1854: Index Ent. (Edn 2): 261 (United Kingdom: England, Sudbury, Suffolk; Isle of Wight). Syntypes not traced in BMNH (L. PITKIN pers. comm.).

Unavailable names (incorrect subsequent spellings): *exiguaria*: BOISDUVAL (1840); *exiquata*: IVINSKIS (1993).

DIAGNOSIS: Wingspan 17-24 mm. Labial palpi light brown. Frons, vertex and notum brownish grey. Forewing pale brownish grey; transverse lines light brown, obscure, almost straight or slightly sinuate, widened and right-angled near costa; postmedial line angulated near costa, with brownish shade and with small black wedge-shaped marks on the veins; terminal area brownish with slightly sinuate pale wavy line and with a more or less quadrate dark blotch between M1 and M3 and weaker blotches on costa and in tornus; discal spot conspicuous, black, elongated or ovate (or sometimes rounded). Hindwing pale brownish grey and almost unmarked; dark transverse lines marked only along posterior margin; discal spot brownish, rounded or slightly elongated. Fringe pale brownish grey, spotted with brown at vein ends. Abdomen brownish grey; first and second segments usually darker, brown.

MALE GENITALIA: Uncus uniapical. Valva with very large, broad, blunt ventral process, with medially convex dorsal margin and very narrow apex; sacculus strongly sclerotized. Vinculum tapered to apex. Sternum A8 elongated, with small apical hollow.

FEMALE GENITALIA: Ductus bursae with numerous longitudinal striations.

DISTRIBUTION: Euro-Siberian. Widely distributed in Europe except Iceland, northern Scandinavia, Appenines, Iberian and Balkan peninsulas and Mediterranean islands (exept Corsica). – Outside Europe in western Siberia, Altai and Sajan Mts, Transbaikalia, Amur and Ussuri regions.

PHENOLOGY: Univoltine: mid-May (early May) to early July (mid-August). Larva: late June (early June) to late August (early October). Overwintering as pupa, sometimes twice.

BIOLOGY: Larva polyphagous (1). Found on leaves of a number of deciduous trees and shrubs including *Berberis vulgaris* (Berberidaceae), *Crataegus monogyna, Prunus spinosa* (Rosaceae), *Ribes nigrum* (Grossulariaceae), *Fraxinus excelsior* (Oleaceae), *Alnus glutinosa* (Betulaceae), *Cornus sanguinea* (Cornaceae), *Symphoricarpos albus* (Caprifoliaceae) (United Kingdom: CREWE 1861c), also *Lonicera tatarica, L. xylosteum, L. periclymenum, Viburnum opulus* (Caprifoliaceae), *Ribes rubrum* (Grossulariaceae), *Malus domestica, M. sylvestris, Sorbus aucuparia, Prunus padus, Rosa canina* (Rosaceae), *Salix fragilis* (Salicaceae), *Acer* (Aceraceae), *Frangula alnus* (Rhamnaceae) and *Hippophae rhamnoides* (Elaeagnaceae) (United Kingdom: BUCKLER 1899; Germany: DIETZE 1913; WEIGT 1988; WEGNER pers. comm.; Denmark: JUUL 1948; Finland: SEPPÄNEN 1970). Recorded also on *Angelica* (!) (Apiaceae) (Germany: DIETZE 1875a) and *Rubus* (Rosaceae) (Austria: 1971), and on the fruits of *Sambucus nigra* (Caprifoliaceae) (United Kingdom: CORLEY 1984).

PARASITOIDS: Hymenoptera, Ichneumonidae: *Netelia tarsata*, *Phobocampe crassiuscula*, *Platylabus pedatorius* - Hymenoptera, Braconidae: *Aleiodes (=Rogas) circumscriptus*, *A. modestus*, *Apanteles juniperatae*, *A. tetricus*, *A. xanthostigmus*, *Meteorus gyrator (=scutellator)*, *M. versicolor*, *Lissogaster (=Microgaster) tibialis*, *Microplitis spectabilis* (?), *M. tuberculifer*, *M. varipes* (THOMPSON 1945; SHENEFELT 1969, 1972, 1973, 1975).

HABITAT: Deciduous, mixed and old spruce forests, forest edges, bushy thickets, field hedges, gardens and parks. From 0 m up to about 1,800 m above sea-level in Alps (VORBRODT 1931).

SIMILAR SPECIES: *E. exiguata* can usually be easily separated from species like *E. pusillata* by the greyer ground colour, the medial line more strongly angled near costa, larger discal spots, less curved postmedial line, absence of dark blotches and small pale tornal spot in terminal area and lighter brownish grey abdomen, lacking dark brown or black lateral stripes. The uniapical uncus, broader valva with larger blunt process, the cornuti on the vesica and the shape of sternum A8 in the male, and the shorter bursa copulatrix without row of large spines on the left side in the female should distinguish this species from *E. pusillata* and other European *Eupithecia* species.

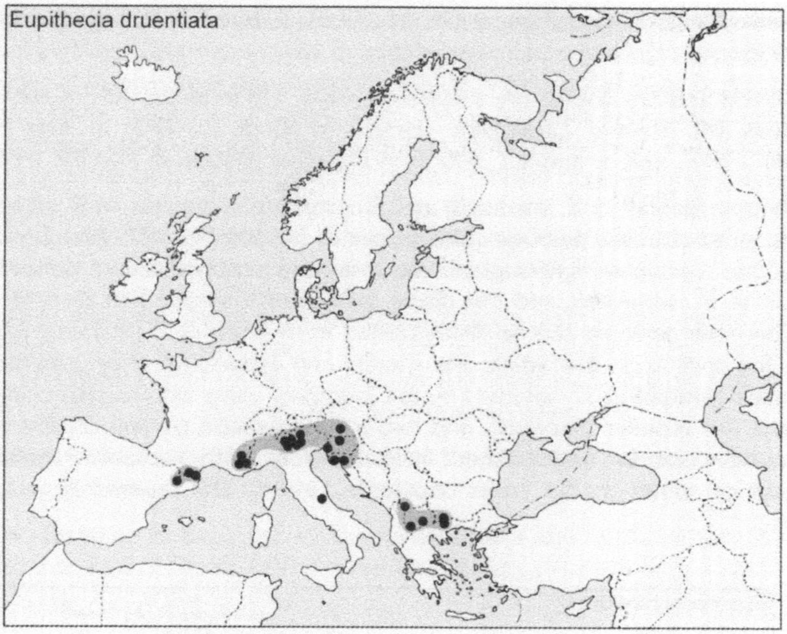

134. *Eupithecia druentiata* DIETZE, 1902

Eupithecia druentiata DIETZE, 1902: Dt. ent. Z. Iris (1901) 14 (2): 254, pl. 4 (France: Digne). Syntype(s) (MNHU).

DIAGNOSIS: Wingspan 18-24 mm. Labial palpi grey. Frons, vertex and notum ash-grey. Forewing whitish grey; transverse lines numerous fine, slightly sinuate, and sharply or right-angled, broadened and darker near costa, especially ante- and postmedial lines; postmedial rather straight and oblique, minutely waved, angled near costa, usually with black dashes on the veins; pale minutely wavy line fine, sometimes faintly broadened at tornus; discal spot large, ovoid or ovate-oblong, intensely black. Hindwing slightly paler with weak transverse lines and darker terminal suffusion; pale wavy line obsolete; discal spot evident, small and rounded. Fringe chequered smoky-white and grey or dark grey. Abdomen ash-grey with dark lateral stripes.

MALE GENITALIA: Uncus with slightly sclerotized ventral tip. Valva with weakly developed blunt ventral process, with medially convex dorsal margin and narrow apex; sacculus slightly sclerotized. Vinculum very broad, trapeziform, with wide and shallow medial hollow.

FEMALE GENITALIA: Ductus bursae with small patch of spinules near base of ductus seminalis. Tergum A8 with shallow medial hollow in posterior margin.

DISTRIBUTION: European (mountains). The Pyrenees, Alps, Appenines and mountains of Balkan peninsula.

PHENOLOGY: Univoltine: early May (late April) to mid-June (late June). Larva: late June to early October (early November). Overwintering as pupa in flimsy cocoon in the ground.

BIOLOGY: Larva monophagous (1). Found on the flowers and seeds of *Artemisia alba*

(=*camphorata*) (Asteraceae) (France: DIETZE 1902, 1910, 1913; Italy: CARRARA 1928; WEIGT 1991).

HABITAT: Warm and dry slopes, dry calcareous rocks. Up to about 900 m above sea-level in Apennines (Mt. Majella: ZAHM pers. comm), to about 1,100 m in Alps (FORSTER & WOHLFAHRT 1981), and to about 1,300 m in Pyrenees (MAZEL & PESLIER 1997).

SIMILAR SPECIES: Similar to *E. intricata* and *E. satyrata*. However, in *E. druentiata* the forewing is more pointed, narrower and longer in proportion, with straighter costa, the antemedial line sharply or right-angled near costa, the postmedial line more oblique (and darker than in *E. satyrata*), and the discal spot usually larger than in both compared species. The male genitalia can be distinguished from those of *E. intricata* by the biapical uncus, the cornuti on the vesica, the shorter and apically narrower sternum A8; and from the male genitalia of *E. satyrata* by the 'humped' valva with convex costal and ventral margins, the broader vinculum, and two longer dentate cornuti on the vesica. The female genitalia can be distinguished from those of both compared species by the arrangement of spines in the bursa copulatrix, and by the tapered spurs of anterior apophyses.

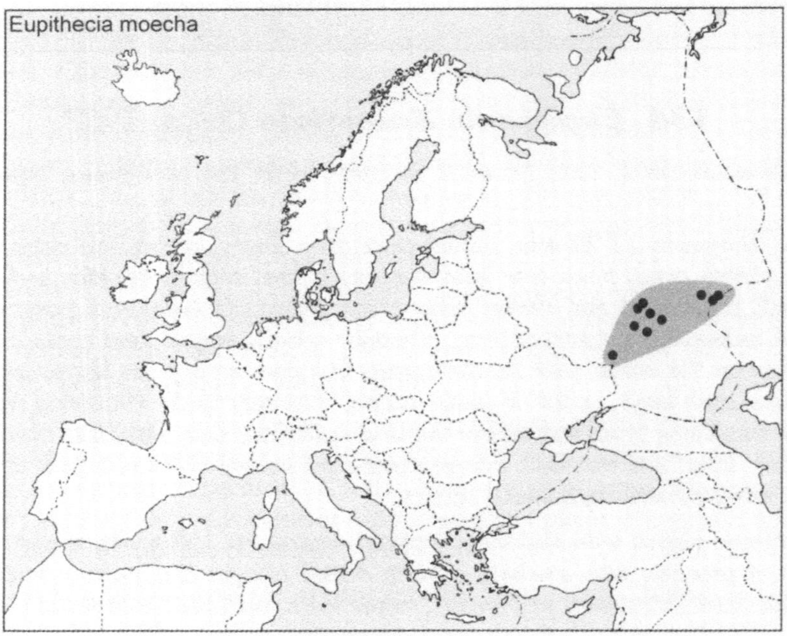

135. *Eupithecia moecha* DIETZE, 1904

Eup(ithecia) bella var. *moecha* DIETZE, 1904: Dt. ent. Z. Iris (1903) 16 (2): 373, pl. 3: 25 (Russia: Sarepta [=Krasnoarmeysk]). Syntypes, at least 3, ♂♀ (MNHU). Valid at subspecific rank. Raised to species rank by VOJNITS & DE LAEVER (1973).

Eupithecia magnifica VOJNITS & DE LAEVER, 1973: Acta zool. Acad. Sci. Hung. 19 (3-4): 438, figs 3b, 7b, 10b (Kazakhstan: Uralsk). Holotype ♂ (MNHU).

Unavailable names (incorrect subsequent spellings): *moeche*: DIETZE (1913); *moechaa*: VOJNITS & DE LAEVER (1973).

DIAGNOSIS: Wingspan 21-22 mm. Labial palpi, frons, vertex and notum reddish brown. Forewing with oblique transverse vein in areole; unicolorous reddish brown; transverse lines inconspicuous; postbasal line straight, oblique, right-angled near costa; postmedial line straight, obtuse angled and darker near costa; terminal area with inconspicuous, slightly sinuate pale wavy line and always with distinct whitish tornal spot; discal dot large, black, oval. Hindwing usually paler, greyish brown, reddish tinge lacking; transverse lines slightly marked only on the posterior margin; discal dot usually absent or indistinct, minute and rounded. Fringe greyish brown, weakly spotted with brown at vein ends. Abdomen reddish brown; first and second segments darker, brown.

MALE GENITALIA: Valva elongated, without ventral process; with evenly curved ventral margin; with slightly concave dorsal margin and with broadly rounded apex; sacculus weakly sclerotized. Vinculum evenly rounded. Sternum A8 broadened at base and in its apical half, tapered towards apex and with blunt and rounded apex.

FEMALE GENITALIA: Ductus bursae with numerous longitudinal striations between spiniferous area and colliculum.

DISTRIBUTION: Endemic to the south of European Russia and north-western Kazakhstan. Recorded for Far East of Russia (Manchuria) erroneously (VOJNITS & DE LAEVER 1973).

PHENOLOGY: Univoltine: mid-June to mid-July.

BIOLOGY: Unknown.

HABITAT: Flood plain forests and bushy riverine places in steppe zone.

SIMILAR SPECIES: Similar to *E. absinthiata*. However, the wings are narrower and lighter with visible reddish or reddish ochreous tinge, inconspicuous transverse lines not marked with dark costal spots and pale wavy line without the dark shade on inner side, that *absinthiata* usually has. The male and female genitalia can be used to separate *E. moecha* from *E. absinthiata* and other species with accuracy.

136. *Eupithecia denotata* (HÜBNER, 1813)

Geometra denotata HÜBNER, 1813: Samml. Eur. Schmett. 5 Geometrae (1), pl. 38: 455 (Europe). Syntype(s) lost.

Eupithecia atraria HERRICH-SCHÄFFER, 1848: Syst. Bearb. Schmett. Eur. 3 (32): 121, 132; ibidem (1846), pl. 25: 154, 155 (non binominal), (Austria: Schneeberg Mt.). Syntype(s).

Eupithecia campanulata HERRICH-SCHÄFFER, 1861: Corresp.-Bl. Sammler Insect., insbesondere Schmett. 17: 130 (Germany: Frankfurt; Braunschweig [=Brunswick]). Syntypes.

Eupithecia primulata MILLIÈRE, 1874: Revue Mag. d'Zool. (3) 2: 244 (Switzerland: Célérina). Syntype(s) (MNHN?).

Eupithecia jasioneata CREWE, 1881: The Entomologist 14 (220): 198; pl. 1: 4, 5 (United Kingdom: North Devon). Syntype(s) (BMNH).

Eupithecia denotata var. ? *ferreata* FUCHS, 1901: in PAGENSTECHER, Jahrb. nassau. Ver. Naturkde 54: 57 (Italy: Trentino, Trafoi). Syntype(s) (NSMW?). Valid at subspecific rank.

Eup(ithecia) ? denotata f. *livida* DIETZE, 1910: Biol. Eupithecien 1, pl. 72: 278, 279; ibidem (1913), 2: 92 (China: (Xinjiang), Juldus (Yining) region; Ili). Syntypes 7, including ♂ ♀ (MNHU). Valid at subspecific rank.

Eup(ithecia) pimpinellata f. *limbosignata* DIETZE, 1910: Biol. Eupithecien 1, pl. 72: 285; ibidem (1913), 2: 119 (China: Kuldja [Yining (Gulja)]; Kazakhstan: Djarkent [=Panfilov]). Syntype(s) ♀ (MNHU). Valid at subspecific rank.

Eup(ithecia) denotata f. *difficilis* DIETZE, 1910: Biol. Eupithecien 1, pl. 85: 905-907; ibidem (1913), 2: 92 (Russia: Saratov; Altai; Ispajran). Syntypes 3, including ♂ ♀ (MNHU). Valid at subspecific rank.

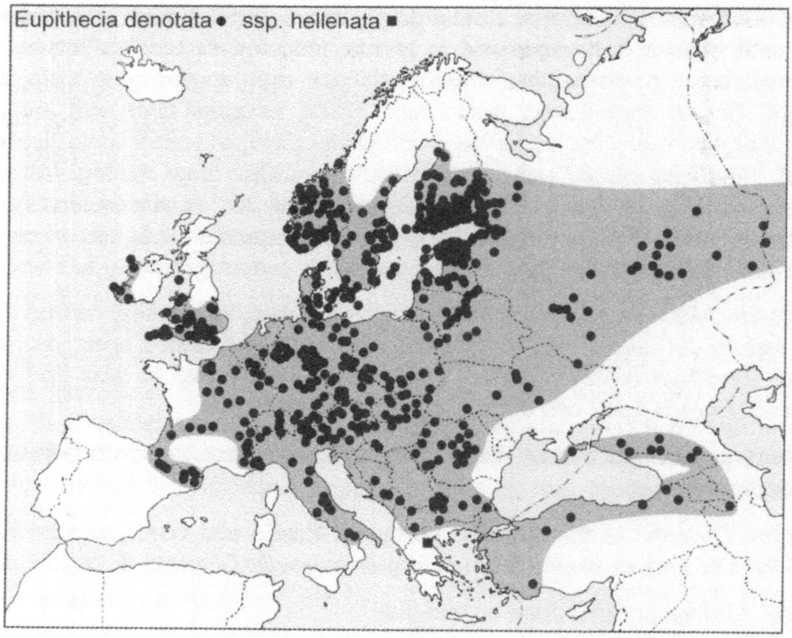

Eupithecia denotata hellenata SCHÜTZE, 1959: Mitt. Münchn. Ent. Ges. 49: 37, pl. 1: 3, 4 (Greece: Olympus Mt., Enipefs-Tal, Kataphygion). Holotype ♂ (ZSM, examined). Valid at subspecific rank.
Unavailable names (infrasubspecific): *solidaginis*: FUCHS (1902). - (incorrect subsequent spellings): *denotaria*: BOISDUVAL (1840); *campannlata*: REISEN (1891).

DIAGNOSIS: Wingspan 14-24 mm. Labial palpi brownish grey. Frons and vertex pale brownish grey. Notum brownish grey. Forewing brownish grey to light brown; transverse lines fine, indistinct or sometimes evident, especially postmedial line, which is slightly sinuate, right-angled near costa, with weak dark dashes on the veins; dark costal spots weakly marked; pale wavy line usually unbroken, multiply indented, with dark shade on inner side, tornal spot small, or ill-defined, or even absent; discal spot small, circular or ovate-orbicular, intensely black. Hindwing paler and greyer than forewing; dark transverse lines conspicuous only along dorsum except postmedial line, which is sharply angled near to costa; pale double medial stripe between medial and terminal areas sometimes evident; terminal area usually darker; pale wavy line fine, much indented with small light tornal spot, or obsolete; dark discal spot well defined, minute, rounded or slightly ovate. Fringe chequered pale and dark brownish grey. Abdomen brownish grey with a dorsal row of small black dots and lateral black stripes or with lateral rows of small black spots, or in some specimens without these.

Subsp. ***hellenata*** SCHÜTZE, 1959 is distinguished by lighter, greyish ground colour, inconspicuous transverse lines and smaller, faintly visible rounded discal dots on all wings. Usually smaller than nominate subspecies.

MALE GENITALIA: Uncus short and stout, biapical. Papillae rather normal, about half uncus length. Valva large, broadened medially, with broad and blunt ventral process; sacculus heavily sclerotized. Vinculum short and wide. Aedeagus large: long and stout. Vesica armed with one slightly curved horn-like cornutus, one elongated, U-shaped flat cornutus and one irregular, curved cornutus at ductus ejaculatorius base. Sternum A8 elongated peg-like, broadened at base, with two small, blunt apical lobes and with shallow and narrow hollow between them.

FEMALE GENITALIA: Bursa copulatrix elongated pouch-like; small area at its base and along left side covered with small spines. Ductus bursae with longitudinal striations and separate patch of small spines. Ductus seminalis relatively broad, long and evenly curved; attached to middle of corpus bursae. Colliculum collar-like, rather small and narrow. Anterior and posterior apophyses medium length, thin and tapered: posterior apophyses with minute spurs near papillae anales. Spurs of anterior apophyses broadened to their apices. Papillae anales short and rounded.

DISTRIBUTION: Euro-Siberian. Widespread in Europe from Ireland and Pyrenees to Urals; northward to central Finland; southward to about 40º latitude in Apennines and Balkan peninsula. On Mount Olympus (Greece) replaced by subsp. *hellenata*. – Outside Europe eastward through southern Urals in north and through Asia Minor, Caucasus, Transcaucasus and Central Asian mountains (Tien-Shan Mts) in south to southern Siberia and Mongolia.

PHENOLOGY: Univoltine, but probably bivoltine in the south (?): mid-June (early April) to early August (late August). Larva: late July (mid-June) to late September (late October). Overwintering as pupa, often twice.

BIOLOGY: Larva oligophagous (3). Found upon the unripe seeds and seed capsules, and rarely on the flowers of *Campanula trachelium* (Germany: HERRICH-SCHÄFFER 1861a; DIETZE 1874a, 1910, 1913; WEIGT 1991; WEGNER pers. comm.; United Kingdom: CREWE 1864c; 1865; BUCKLER 1899; BARRETT 1904; France: MILLIÈRE 1870), also *C. rotundifolia* (Germany: DIETZE 1872), *C. barbata* (Switzerland: PÜNGELER 1889; FAVRE 1905), *C. persicifolia, C. medium, C. latifolia (=urticaefolia), C. collina, C. patula, C. carpatica, C. rapunculus, C. rapunculoides, Phyteuma campanuloides* (United Kingdom: BARRETT 1904), *Campanula glomerata* (Germany: DIETZE 1913; WEIGT 1991), *Phyteuma ovatum* (Alps: DIETZE 1913), flower heads of *Jasione montana* (Campanulaceae) (United Kingdom: BUCKLER 1899; BARRETT 1904, as *E. jasioneata*), and probably occasionally on *Ballota nigra* (Lamiaceae) (France: MABILLE 1872), *Primula integrifolia* (Primulaceae) (Switzerland: FAVRE 1905; VORBRODT 1928) and *Vaccinium uliginosum* (Ericaceae) (Finland: SEPPÄNEN 1970). Found in the ripening seed capsules of *Codonopsis clematidea* (Campanulaceae) (Kirghizstan (Tien-Shan Mts): original data) in August.

PARASITOIDS: Hymenoptera, Ichneumonidae: *Campoletis incisa, C. raptor, Mesochorus pectoralis* (hyperparasite) - Hymenoptera, Braconidae: *Apanteles juniperatae, A. nanus, A. spurius, Lissogaster (=Microgaster) crassicornis, L. globatus, L. tibialis, Microplitis strenuus* - Diptera, Tachinidae: *Actia bicolor, Ceranthia pallida* (BUCKLER 1899; THOMPSON 1945; HERTING 1960, 1976; SHENEFELT 1972, 1973; ANDERSEN 1996).

HABITAT: Open, light forests, forest clearings, edges of forests, rocks. From 0 m up to about 1,500 m above sea-level in Pyrenees (MAZEL & PESLIER 1997), to about 1,600 m in Apennines (coll. ZSM), and to about 2,200 m in Alps (VORBRODT 1928). Subsp. *hellenata* is found from 2,000 m up to about 2,500 m above sea-level on Mount Olympus (SCHÜTZE 1959).

SIMILAR SPECIES: A variable species in size, colour (from grey to light brown), and distinctness of pattern on the wings. Similar to *E. absinthiata* and also to other relatives such as *E. expallidata* and *E. assimilata*. However, *E. denotata* can usually be recognized from these species by the paler, less reddish and ochreous ground colour, the smaller and more rounded discal spot, the weakly defined dark costal spots and indistinct, very small pale tornal spot on the forewing, and the sharply angled dark postmedial line on the hindwing. *E. denotata* is also closely related to species of the *millefoliata* group. The male genitalia can be easily distinguished from those of other allied species by the broader valva, with broad and blunt ventral process, by the presence of one slightly curved horn-like cornutus in the vesica and by the shape of sternum A8. The smaller area covered with spines in the corpus bursae will distinguish the female genitalia of *E. denotata* from those of *E. millefoliata*.

REMARKS: Some English specialists distinguished *E. denotata jasioneata* CREWE, 1881 from the typical subspecies mainly by the smaller size and darker fuscous grey colour; and they also mentioned that wild adults of subsp. *jasioneata* were hardly ever encountered. Thus, all known specimens of subsp. *jasioneata* were described and determined on the basis of reared adults, which have a tendency to produce specimens with colour and size variations differing from wild specimens. I have come to the conclusion that *E. denotata denotata* and *E. denotata jasioneata* are synonymous.

The ***millefoliata*** species-group

Introduced here for the first time. It is a small Palaearctic species-group probably of ancient Mediterranean origin. Represented in Europe by five species: *E. weissi*, *E. pauxillaria*, *E. santolinata*, *E. millefoliata* and *E. spadiceata*. Two Central Asian species, *E. exactata* STAUDINGER, 1882 and *E. nephelata* STAUDINGER, 1896 also belong to the *millefoliata* species-group.

MALE GENITALIA: Uncus biapical, short, thick and blunt, when viewed dorso-ventrally, but thin, when viewed laterally. Papillae large, often elongated and thick, length about 0.6-1.2 times uncus length. Valva usually large, broadened medially; without ventral process; sacculus lightly sclerotized. Vinculum rather short and wide. Aedeagus large and stout, broader at base; length about 0.8-1.0 times valva length; and about 3.5-5.0 times longer than medial width. Vesica armed with one to four dentate or short horn-like cornuti, and always with one elongate and narrow tubular cornutus, and one smaller, irregular, usually curved cornutus at ductus ejaculatorius base. Sternum A8 rather peg-like, broadened at base; basal and apical hollows shallow; two apical lobes small, usually blunt and heavily sclerotized; sternum about 0.6-0.7 times valva length.

FEMALE GENITALIA: Bursa copulatrix elongated, pouch-like or pear-shaped; basal 0.5-0.8 covered with uniform spines. Ductus bursae with a large patch of spines or with sclerotized row of spines, and usually with numerous longitudinal striae. Ductus seminalis usually long and thick, broadly attached to corpus bursae. Colliculum collar-like, relatively short and narrow, usually inclined to the right. Antrum membranous. Anterior and posterior apophyses short to long, thin, usually slightly expanded at apices. Spurs of anterior apophyses broadened to their apices. Tergum A8 almost square. Papillae anales normal (medium length and width) and rounded, but elongated and pointed in some species.

REMARKS: The genital characteristics of this species-group can usually be used to distinguish them successfully from other European *Eupithecia* species. The blunt and thick uncus, vesica with elongated, tubular cornutus and shape of sternum A8 in male clearly indicate this group.

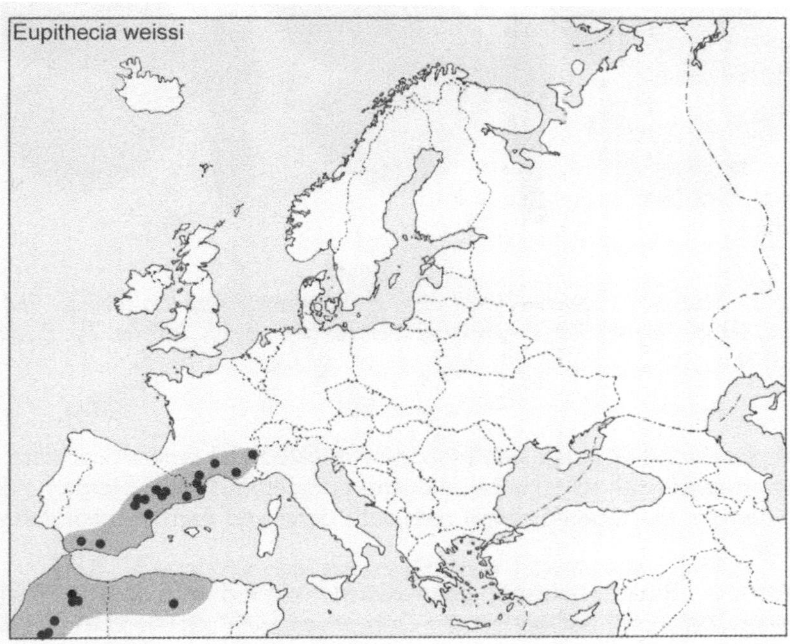

137. *Eupithecia weissi* PROUT, 1938, stat. n.

Eupithecia euphrasiata weissi PROUT, 1938: in SEITZ, Macrolep. World 4 (Suppl.): 200, pl. 17: k (Spain: Albarracin). Syntypes (BMNH, syntype 1♂ and some topotypes examined).

Eupithecia barbaria SCHWINGENSCHUSS, 1955: Z. Wien. ent. Ges. 40 (8): 229 (Morocco: Great Atlas). Holotype ♀ (NLHW). New junior synonym.

Unavailable names (misidentification): *santolinata* var.?: sensu ZERNY (1934) nec MABILLE, 1871 (identity: *Eupithecia santolinata*).

DIAGNOSIS: Wingspan 17-22 mm. Labial palpi greyish brown. Frons whitish grey with blackish stripe anteriorly. Vertex whitish grey. Notum whitish grey or pale brownish grey with black transverse stripe in front. Forewing elongate and rather pointed, costa straight, termen oblique, slightly arched; ground colour pale brownish grey; transverse lines conspicuous, dark brownish grey; basal, antemedial and medial lines oblique, straight, sharply angled near costa; postmedial line oblique, almost straight, weakly curved and perpendicular to costa; pale wavy line sinuate with brownish shade on inner side; black discal spot large, oblique and narrowly elongate, or sometimes ovate. Hindwing paler, whitish grey with weak brownish tinge; transverse lines evenly curved, distinct along dorsum, especially postmedial line; pale wavy line indistinct; discal spot small, rounded, paler than on forewing. Fringe chequered whitish grey and brownish grey. Abdomen whitish grey, usually with black lateral spots; tergum A2 lighter, whitish with posterior black transverse band; tergum A3 ochreous brown, darker than others.

MALE GENITALIA: Papillae large, broadened to apices. Vesica armed with two horn-like cor-

Text-figs 73-75: Diagnostic characters (indicated) of European *Eupithecia* species. Text-fig. 73: *E. weissi* PROUT, 1938. Text-fig. 74: *E. pauxillaria* BOISDUVAL, 1840. Text-fig. 75: *E. pimpinellata* (HÜBNER, 1813).

nuti (one large, the other much smaller), one elongated and tapered cornutus, and with one smaller, irregular, curved cornutus at ductus ejaculatorius base. Sternum A8 with two elongated, narrow and tapered apical rods with deep and narrow membranous hollow between them.

FEMALE GENITALIA: Bursa copulatrix pear-shaped, covered with large uniform spines. Ductus bursae sclerotized, without spines. Ductus seminalis attached to corpus bursae in middle. Anterior and posterior apophyses tapered; posterior apophyses have small spurs near papillae anales. Papillae anales small, elongated and rather pointed.

DISTRIBUTION: West-Mediterranean. Southern France and Spain. – Outside Europe in North Africa (Morocco and Algeria). – New for the fauna of France: 1♂, (France), Hérault, St. Guilhem-le-Désert, 30.VII.1952, Coll. D. LUCAS (HERB); 1♀, (France), Vernet, Juillet, prep. N: 1417 (HERB); 1♂, Süd-Frankreich, Corbières, Albas, 400 m, 27.7.(19)69, L, leg. LUKASCH (ZSM); 2♂21♀, Süd-Frankreich, Bouches du Rhône, Aureille, 275 m, 4.-6.8.(19)78, L, leg. LUKASCH (ZSM).

PHENOLOGY: Univoltine: late July (mid-July) to early September (early October).

BIOLOGY: Larva monophagous? (1). Found on *Ormenis scariosa* (Asteraceae) (Morocco: ZERNY 1935).

HABITAT: Unknown. From 0 m up to about 400 m above sea-level in southern France (coll. ZSM), to about 1,600 m in Sierra Nevada (coll. ZMUC), and from 2,000 m to about 2,900 m in Great Atlas (coll. ZSM).

SIMILAR SPECIES: *E. weissi* can be distinguished with difficulty from the closely related species *E. pauxillaria* by the more distinct and broader dark transverse lines and by larger discal spots on the forewings only (Text-fig. 73). The two horn-like cornuti (one larger than the other) on the vesica in the male genitalia will distinguish *E. weissi* from *E. pauxillaria* with accuracy. The female genitalia of *E. weissi* are similar to the genitalia of *E. pauxillaria*, but can be easily distinguished by the absence of the broad and heavily sclerotized band armed with numerous spines in the corpus bursae.

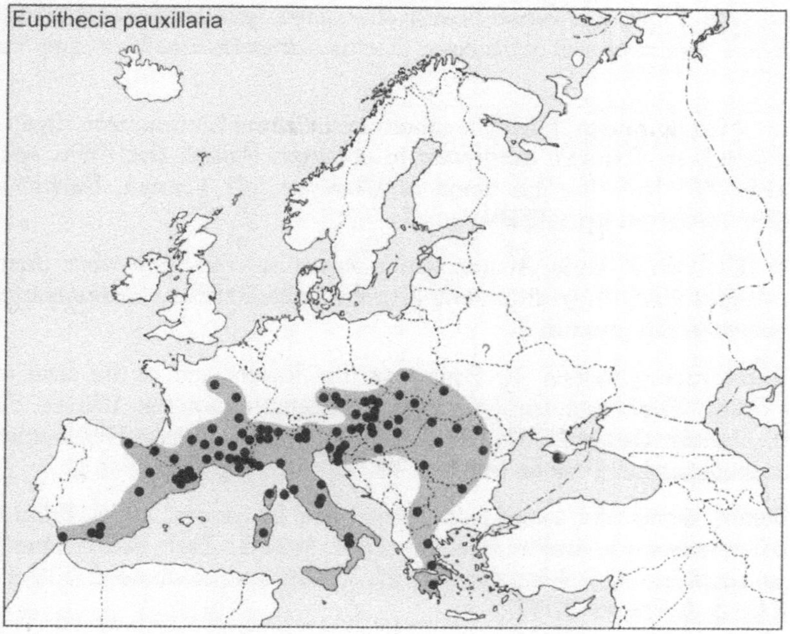

138. *Eupithecia pauxillaria* BOISDUVAL, 1840

Eupithecia pauxillaria BOISDUVAL, 1840: Genera Index meth. eur. Lepid.: 210 (France Gall. mer.). Holotype ♀ (coll. HERBULOT in ZSM).

Eupithecia euphrasiata HERRICH-SCHÄFFER, 1861: Corresp.-Bl. Sammler Ins., insbesondere Schmett. 1861 (17): 130 (Germany: Mombach). Syntypes.

Unavailable names (misidentification): *constrictata*: sensu MILLIÈRE (1870) nec GUENÉE, 1858 (identity: *Eupithecia distinctaria*).

DIAGNOSIS: Wingspan 17-23 mm. Labial palpi grey. Frons and vertex pale ash-grey. Notum ash-grey with narrow greyish fuscous transverse stripe in front. Forewing of a soft ash-grey ground colour; transverse lines fine, distinct, blackish grey; basal, postbasal and antemedial oblique, straight, then sharply angled, faintly broadened and darker near costa; medial line obliquely sinuate, touching discal spot and sharply curved beyond it; postmedial line multiply waved, more or less straight, obtusely curved to costa just beyond discal spot; pale wavy line fine; discal spot large, oblique, ovate-acuminate or ovate-oblong, black. Hindwing paler; transverse lines well marked only along dorsum; pale stripe beyond postmedial line is broad and evenly curved, with waved outer margin; terminal area darker; pale wavy line rather indistinct, fine, multiply indented; discal spot relatively large, visible, circular. Fringe greyish white, spotted with grey or dark grey at vein ends. Abdomen ash-grey with narrow black transverse band at posterior margin of first segment; sometimes with narrow black lateral stripes or with series of black lateral dots.

MALE GENITALIA: Valva rather narrow, with slightly angulated ventral margin. Vinculum relatively narrow. Vesica armed with three small and broad horn-like cornuti; one long tubular cornutus and one irregular curved cornutus at ductus ejaculatorius base. Sternum A8 with two apical lobes with shallow hollow between them.

FEMALE GENITALIA: Corpus bursae elongated, almost egg shaped, covered with large uniform spines, with a long and broad heavily sclerotized spiniferous band along right side between ductus seminalis and colliculum. Ductus seminalis usually sharply curved ventrad.

DISTRIBUTION: Mediterranean. Mainly southern and Central Europe from Spain eastwards to Ukraine (southern Crimea), northward to southern Poland and Paris, southward to about latitude 36º. New for the fauna of Ukraine: 1♂, Crimea, Bakhchisaraj, 11.-12.VIII.1993, VASILENKO leg. (ZISP).

PHENOLOGY: Univoltine: early August (early June!) to mid-September (mid-October). Larva: late August (early August) to early October (mid-October). Overwintering as pupa in flimsy cocoon in the ground.

BIOLOGY: Larva monophagous (1). Found on the flowers and in the seed capsules of *Odontites lutea, Euphrasia* spp. (Germany: HERRICH-SCHÄFFER 1861a; Switzerland: FAVRE 1905; France: DIETZE 1910, 1913; Italy and Austria: WEIGT 1993) and *Odontites verna* (Scrophulariaceae) (France: MABILLE 1872)

HABITAT: Dunes, sandy hills, sandy fields, warm and dry slopes, rocks. From 0 m up to about 1,300 m above sea-level in Apennines (Mt. Majella: ZAHM pers. comm.), to about 1,500 m in southern Alps (HELLMANN & al. 1999), and to about 1,700-1,800 m in Pyrenees (MAZEL & PESLIER 1997).

SIMILAR SPECIES: Usually lighter and greyer than *E. pimpinellata*, the species with which it most likely to be confused, with straighter and more oblique basal, postbasal and antemedial lines, usually smaller discal spot on the forewing and black band on tergum A1 (Text-fig. 74). The male genitalia can be distinguished from those of other species by the vesica with three small horn-like cornuti combined with very long tubular cornutus; and the female genitalia can be distinguished by the shape of the corpus bursae which is armed with large spines and sclerotized, spiniferous band on right side.

139. *Eupithecia santolinata* MABILLE, 1871

Eupithecia santolinata MABILLE, 1871: Petites Nouv. Ent. 1 (42): 168 (France: southern Pyrenees, Carcasson). Holotype ♀ (coll. HERBULOT in ZSM, examined).

DIAGNOSIS: Wingspan 19-24 mm. Labial palpi brownish grey. Frons, vertex and notum whitish grey. Forewing whitish grey; transverse lines distinct, fine, oblique, brown-grey; basal, postbasal and antemedial almost straight, sharply angled near costa; three medial lines rather evident; the first medial sharply angled beyond discal spot and usually separated from second medial line by narrow whitish stripe; postmedial line with shade on inner side, slightly sinuate, less oblique than other lines, right or obtuse angled near costa, but this angle with pointed apex; pale stripe between medial and terminal areas is broadened medially, with oblique streak from angle of postmedial line; inner part of terminal area forming three dark, brown-grey blotches (costal, medial and tornal); pale wavy line faintly broadened near tornus; discal spot large, circular or ovate-orbicular, black. Hindwing evidently paler, smoky brownish grey; transverse lines numerous, dark, but visible only along dorsum except postmedial line; terminal area darker, brownish grey, with inner edge wavy; pale wavy line much dentate, very fine, frequently inconspicuous; discal spot large, rounded, but smaller and paler than on the forewing. Fringe whitish grey, distinctly spotted with dark brownish grey at vein ends. Abdomen grey to ash-grey with black

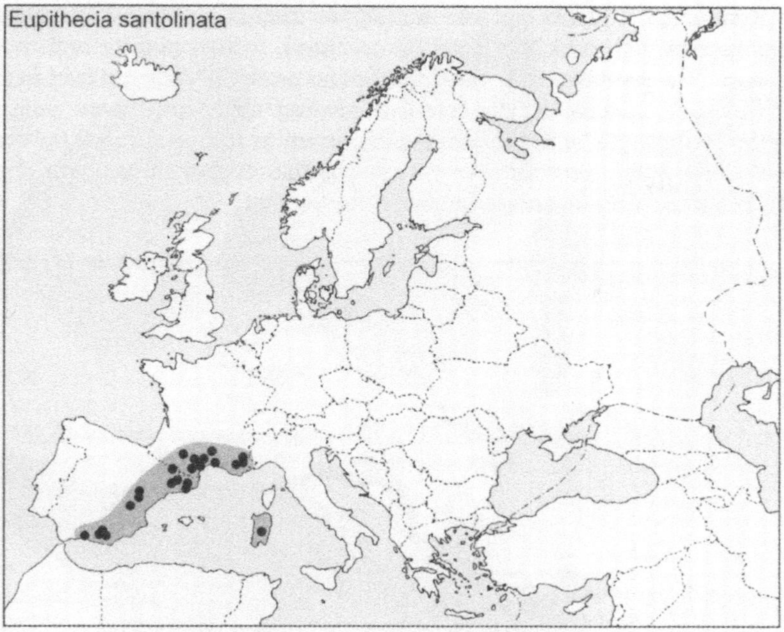

lateral stripes and sometimes with a series of small black dorsal dots; the third segment darker than others.

MALE GENITALIA: Vesica armed with one apical small, broad horn-like cornutus; one medial dentate cornutus; with patch of sclerotized striae and teeth near long tubular cornutus base and with one large, irregular, curved cornutus at ductus ejaculatorius base. Sternum A8 broad, medially slightly narrowed; apical hollow broad.

FEMALE GENITALIA: Bursa copulatrix covered by small spines in basal part, with two large, elongated and broad areas of larger spines on dorsal and ventral sides, with a narrow, longitudinal membranous gap separating the spines on right side, and with heavily sclerotized left side. Ductus seminalis with a wide, heavily sclerotized and striated base, attached to corpus bursae near base on left side. Colliculum slightly inclined to right. Tergum A8 with small, rounded medial hollow on the posterior margin.

DISTRIBUTION: West-Mediterranean. Spain, southern France and Sardinia.

PHENOLOGY: Univoltine: late June (mid-June) to mid-July (early August). Larva in August and September. Overwintering as pupa.

BIOLOGY: Larva monophagous (1). Found on the flowers and seeds of *Santolina chamaecyparissus* (France: MABILLE 1872; CHRÉTIEN 1905; DIETZE 1913; TAUTEL 2000).

HABITAT: Very dry and warm mountain slopes, terraces, rocky and stony places where the food-plant is present. From 0 m up to about 600 m above sea-level in Pyrenees (TAUTEL 2000), and to about 1,800 m in Sierra Nevada (coll. ZSM).

SIMILAR SPECIES: An unmistakable species, which is easily distinguished from any other European *Eupithecia* species because of the characteristic forewing postmedial line with angle with pointed tip. It is less variable and lighter than *E. icterata* f. *cognata*, the only form of that species with which it could be confused, with straighter and more oblique transverse lines. The genitalia of E. *santolinata* can easily be distinguished from those of all other *Eupithecia* species by the papillae covered with short setae only near their apices, the large aedeagus with characteristic cornuti in the vesica and by the shape of sternum A8 in the male, and the elongate pouch-like corpus bursae with characteristic dispositions of spines and ductus seminalis in the female.

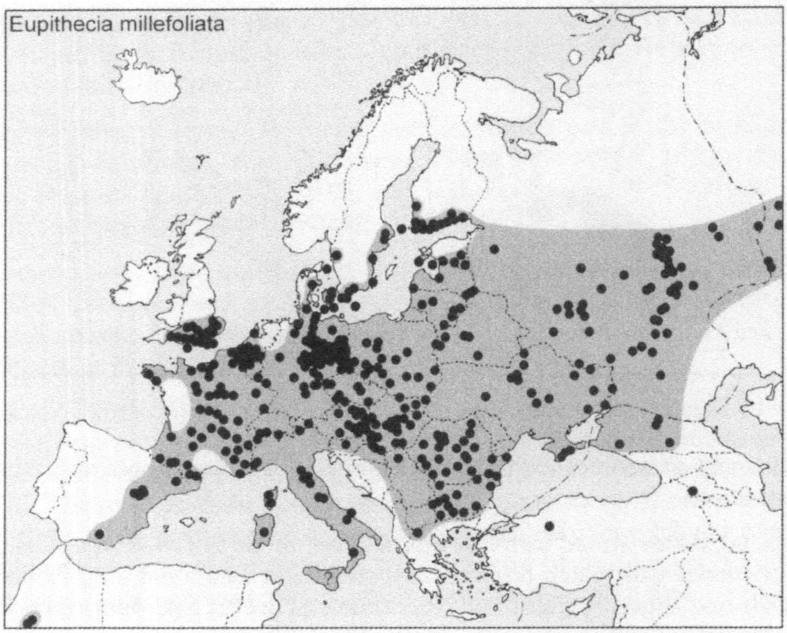

140. *Eupithecia millefoliata* RÖSSLER, 1866

Eupithecia millefoliata RÖSSLER, 1866: Jb. nassau. Ver. Naturwiss. 19/20: 153 (Germany: Nassau). Syntypes ♂♀.

Eupithecia achilleata MABILLE, 1869: Annls Soc. ent. France (4) 9 (1): 70, pl. 2: 5 (Corsica: Bastia). Lectotype ♀ (coll. HERBULOT in ZSM).

Eupithecia millefoliata var. *maeoticaria* BOHATSCH, 1893: Dt. ent. Z. Iris 6: 19 (Russia: Taganrog). Lectotype ♂, herewith designated to stabilize nomenclature (ZISP, examined). Valid at subspecific rank.

Eup(ithecia) subfulvata f. *flavescens* DIETZE, 1910: Biol. Eupithecien 1, pl. 73: 324; ibidem (1913), 2: 96 (Kirghizstan: Issyk-Kul). Holotype (MNHU). Valid at subspecific rank.

Eupithecia danilevskyi WARDIKJAN, 1961: Dokl. Acad. Nauk Arm.SSR 32 (1): 61, pl. 1; only ♂; nec ♀ (Armenia: Erivan). Holotype ♂ (IZAR, examined).

Eupithecia wettsteini VOJNITS, 1974: Folia Ent. Hung. 27 (2): 237, fig. 1 (Hungary: Lake Balaton). Holotype ♀, not traceable in HNHM (the slide examined). New junior synonym.

DIAGNOSIS: Wingspan 19-26 mm. Labial palpi brownish grey. Frons, vertex and notum whitish grey with brownish tinge. Forewing with oblique transverse vein in areole; ground colour whitish grey with warm yellowish brown suffusion; transverse lines numerous, fine, oblique, slightly broadened and darker along costa; basal, postbasal, antemedial and three medial lines are much dentate or much waved, sharply angled near costa; postmedial line less oblique than others, multiply indented and twice obtuse angled near costa; pale stripe

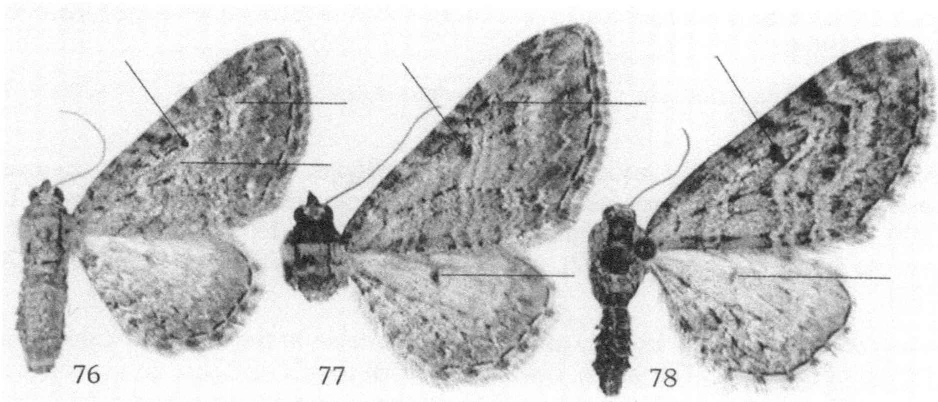

Text-figs 76-78: Diagnostic characters (indicated) of European *Eupithecia* species. Text-fig. 76: *E. millefoliata* RÖSSLER, 1866. Text-fig. 77: *E. spadiceata* ZERNY, 1933. Text-fig. 78: *E. icterata* (DE VILLERS, 1789).

beyond postmedial line is widest at the angle near costa and forms an oblique light streak; pale wavy line much indented, sometimes broken with narrow pale tornal dot; discal spot small, rounded, or ovate, or even longitudinally triangular, dark, sometimes inconspicuous but in some specimens intensely black. Hindwing paler; dark transverse lines are usually visible along dorsum; terminal area always darker, with distinct, multiply waved inner edge; pale wavy line much dentate, weak, often rather inconspicuous; discal spot minute, rounded. Fringe whitish grey, spotted with brownish grey. Abdomen whitish grey with weak brownish tinge; a series of very small white dorsal crests and sometimes dark lateral dots are present.

MALE GENITALIA: Vesica armed with four dentate cornuti (one larger and broader, two smaller and one very small, inconspicuous), also with one long tubular cornutus and one large, irregular, curved cornutus at ductus ejaculatorius base. Sternum A8 broad, peg-like with relatively broad apical hollow.

FEMALE GENITALIA: Bursa copulatrix covered with spines in basal part and on left side; the separate patch of spines also on the left side. Ductus seminalis attached to middle of corpus bursae on right side. Colliculum slightly inclined to right.

DISTRIBUTION: Palaearctic. Widespread in Europe but not found in Iceland, Ireland, Portugal, Balearics, Malta, Crete, western and southern parts of Balkan peninsula. – Outside Europe in Asia Minor, Caucasus, Transcaucasus, southern Urals, Central Asia (Tien-Shan Mts: southern Kazakhstan, Kirghizstan, Uzbekistan; Tadjikistan: Darvaz Mts), Altai, south-western and southern Siberia, southern Yakutia.

PHENOLOGY: Univoltine: mid-June (early May) to late July (mid-August). Larva: late August (early August) to early October (late October). Overwintering as pupa, occasionally twice.

BIOLOGY: Larva monophagous (1). Found on the flowers and especially on the ripening seeds of *Achillea millefolium* (Asteraceae) (Germany: RÖSSLER 1866; FUCHS 1875; DIETZE 1910, 1913; WEIGT 1991; Corsica: GOOSSENS 1869; Finland: SEPPÄNEN 1970; Russia: original data), several specimens found on *Achillea filipendulina* (Germany:

WEIGT 1991). Found occasionally (?) on *Linaria vulgaris* (Scrophulariaceae) (Netherlands: AARTSEN 1964).

PARASITOIDS: Hymenoptera, Braconidae: *Apanteles praepotens* (SHENEFELT 1972; HERTING 1976).

HABITAT: Dry fields and meadows, dry sunny roadsides, dry and warm areas near the coasts, beach ridges, fringes, slopes and among rocks. From 0 m up to about 1,500-1,600 m above sea-level in southern Alps (TEOBALDELLI 1979), to about 1,900 m in Pyrenees (MAZEL & PESLIER 1997), and to about 2,600-2,700 m in Great Atlas (RUNGS 1981).

SIMILAR SPECIES: Variable in size; usually a little variable in the depth of colour and distinctness of markings. The warm, delicate yellowish brown suffusion on the wings distinguishes this species from *E. egenaria*, *E. santolinata* and all colour forms of *E. icterata* (Text-fig. 76). The light ochreous *E. simpliciata* has broader and more rounded wings with evenly curved postmedial line on the forewing. The combination of the broad and short uncus, the elongate and narrow valva, the wide and short papillae, the cornuti in the vesica and the shape of sternum A8 in the male, and the elongated pouch-like corpus bursae with ductus seminalis arising from its middle in the female genitalia should distinguish *E. millefoliata* from other relatives.

REMARKS: After carefully examining the genitalia slide of the holotype of *E. wettsteini*, as well as studying the type description with very vague remarks and illustration of *E. wettsteini* (FORSTER & WOHLFAHRT 1981), I have come to the conclusion that *E. millefoliata* and *E. wettsteini* are conspecific. The female of the following species has short medial spurs on the posterior apophyses which are absent in the female of *E. millefoliata* and in the holotype of *E. wettsteini* as well.

141. *Eupithecia spadiceata* ZERNY, 1933

Eupithecia spadiceata ZERNY, 1933: Dt. ent. Z. Iris 47: 95, pl. 1: 22 (Lebanon: Beiruth). Syntypes ♂ ♀ (NHMW; two syntypes in ZISP, examined).

Unavailable names (misidentification): *maeoticaria*: sensu WEHRLI (1934) nec BOHATSCH, 1893 (identity: *Eupithecia millefoliata*).

DIAGNOSIS: Wingspan 20-24 mm. Labial palpi dark grey or brownish grey. Frons and vertex pale grey. The frontal part of notum dark grey or brownish grey, the other part ash-grey. Forewing grey with slightly yellowish tinge; transverse lines brownish grey; basal and antemedial lines oblique, angled near costa; the first medial line slightly sinuate, crossing discal spot and sharply angled to costa; the second medial line thinner, slightly sinuate and angled near costa also; postmedial line brownish grey with shade on inner side, evenly curved and sharply angled to costa, forming a sharp angle on M1; terminal area darker, yellowish brown; pale wavy line usually conspicuous, sometimes with very small whitish tornal spot; discal spot large, rounded, dark brown or black. Hindwing slightly paler; transverse lines inconspicuous; terminal area darker; pale wavy line thin, dentate; discal spot distinct, relatively large, circular to comma-shaped, dark, brownish grey, but smaller and paler than on forewing. Fringe pale grey, broadly spotted with brownish grey at vein ends. Abdomen grey with weak yellowish tinge.

MALE GENITALIA: Papillae blunt with sclerotized apices. Vesica armed with one large den-

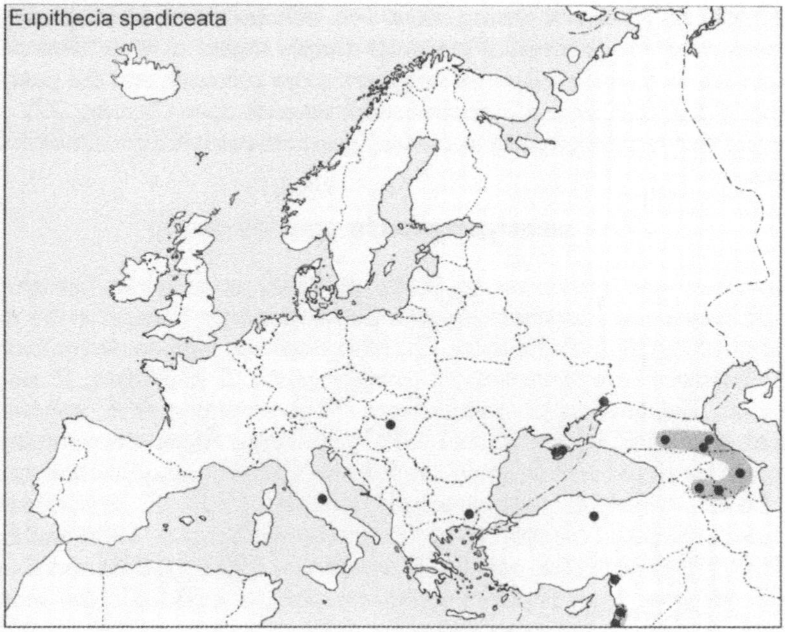

tate cornutus, one elongated and tapered cornutus, and with smaller, irregular curved cornutus at ductus ejaculatorius base. Sternum A8 elongated with two small blunt and heavily sclerotized apical lobes and with shallow and narrow hollow between them.

FEMALE GENITALIA: Bursa copulatrix elongated pear-shaped. Ductus bursae with a large patch of spines, reaching base of ductus seminalis. Posterior apophyses with small medial spurs.

DISTRIBUTION: East-Mediterranean. Sporadic and very local in southern Europe: central Italy, Hungary, Bulgaria, southern Ukraine (Crimea) and south of European Russia. – Outside Europe in Asian Turkey, Lebanon, Transcaucasus (Armenia and Azerbaijan) and northern Caucasus (Mt. Mashuk near Pjatigorsk; Daghestan). – New for the faunas of Hungary, Bulgaria, Italy and European Russia: 1♀, Hungaria centr., Fót, 3.VI.1950, leg. Dr. ISSEKUTZ (HNHM); 1♀, (Hungaria), Fót-Csomad, 23.V.1953, leg. Dr. LENGYEL (HNHM); 1♀, Bulgaria, East Rhodopi Mts, Siv Kladenetz v., Ivailov grad region, 21.-22.VI.1993, leg. BESHKOV (ZISP); 1♂ 1♀, Italien, 20 km W v Castel d. leri Abruzzen, 1000 m, 18.VI.1966, leg. E. u. A. VARTIAN (HNHM); 1♀, Italien, Abruzzen, M.S.Franko, 1400 m, Gran Sasso, 2.VIII.1970, leg. GROSS (ZSM); 1♂, (Russia), Taganrog, 29.V.(18)75, (leg.) Alph(éraky), v. *Oxydata*, Coll. Vel. Knjazja NIKOLAJA MIKHAJLOVICHA (ZISP).

PHENOLOGY: Probably univoltine: early June (late May) to early July (early August).

BIOLOGY: Unknown.

HABITAT: Dry limestone and calcareous sandstone slopes and gorges. From 0 m up to about 1,200 m above sea-level in Lebanon (WILTSHIRE 1939), and to about 1,400 m in central Apennines and Asia Minor (coll. ZSM; coll. SOMMERER).

SIMILAR SPECIES: This species is distinguished with difficulty from *E. icterata* f. *cognata*: the lighter colour and the postmedial line more sharply angled at costa towards apex provide useful guides; in *E. millefoliata* all wings are more yellowish and the postmedial line on the forewing does not form a V-shaped mark towards apex (Text-fig. 77). In cases of doubt, the male and female genitalia show very good differences from those in both compared species.

The ***semigraphata*** species-group

This species-group was introduced by SCHÜTZE (1956) as subgenus *Dietzea*. SCHÜTZE included eight palaearctic *Eupithecia* species (six species from Europe) in this group. This is one of the most species-rich Holarctic *Eupithecia* groups represented in Europe by ten species: *E. icterata*, *E. succenturiata*, *E. semigraphata*, *E. impurata*, *E. poecilata*, *E. fuscicostata*, *E. denticulata*, *E. subumbrata*, *E. orphnata* and *E. subfuscata*. One North African species *E. tarfata* (LUCAS, 1907) and some Asian species that have been examined could be placed in this group, as follows: *E. marginata* STAUDINGER, 1892; *E. lacteolata* DIETZE, 1906; *E. limbofasciata* SCHÜTZE, 1956; *E. forsterata* SCHÜTZE, 1960; *E. pseudoicterata* SCHÜTZE, 1960; *E. verecunda* VOJNITS, 1980 and *E. vitiosata* MIRONOV, 2001. Two additional species, *E. latipennis* (HULST, 1898) and *E. sewardata* BOLTE, 1978 described from North America, can also be included in the *semigraphata* species-group.

MALE GENITALIA: Uncus relatively short or medium length, biapical. Papillae length about 0.6-0.8 times uncus length. Valva without ventral process, large, usually with evenly curved ventral margin, with straight or slightly convex dorsal margin and usually pointed at apex; sacculus slightly sclerotized. Vinculum rounded, short and broad. Aedeagus large, stout, length about 0.7-1.0 times valva length; about 3.0-4.5 times longer than medial width. Vesica armed with one to three large, broad dentate cornuti (except in some species), with one large, elongated, flat, folded U- or V-shaped cornutus and one large, irregular, curved cornutus at ductus ejaculatorius base. Sternum A8 variable in size and shape: usually elongated, broad peg-like or almost rectangular; basal hollow shallow to deep; apical hollow shallow (except in some species); with two short, blunt apical lobes and heavily sclerotized lateral margins in apical part of sternum; sternum length about 0.6-0.8 times valva length.

FEMALE GENITALIA: Bursa copulatrix usually pear-shaped (sub-globular in some species); from half to almost completely covered by a numerous spines, and with a separate group of spines between colliculum and ductus seminalis base, and usually also with short longitudinal striations near colliculum. Ductus seminalis very characteristic: usually thick, very long and evenly curved around corpus bursae towards its base, attached to ductus bursae on right side. Colliculum collar-like, usually short, narrow, slightly tapered to apex. Antrum membranous. Anterior and posterior apophyses thin, normal or short pointed or slightly expanded at apices. Spurs of anterior apophyses usually broadened or very broad at apices. Tergum A8 square or rectangular. Papillae anales rather slim: relatively short, narrow and rounded.

REMARKS: The representatives of this species-group can usually be distinguished from other *Eupithecia* species by the large valva with usually pointed apex, cornuti on the vesica, broad and elongated shape of sternum A8 in the male, as well as by the thick and very long ductus seminalis, which is evenly curved around corpus bursae towards its base in the female genitalia. However, it is necessary to take into account that the shape of sternum A8 in the males of each species are variable (SCHÜTZE 1956).

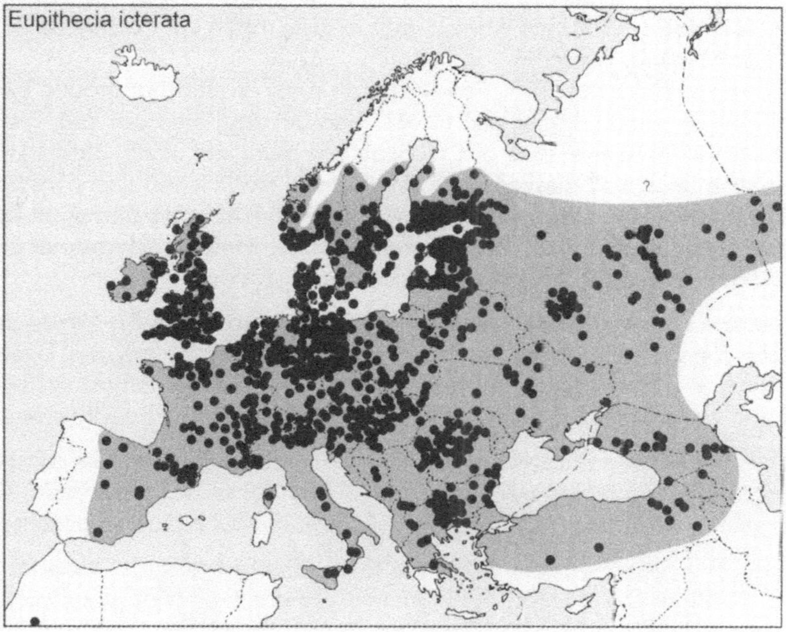

142. *Eupithecia icterata* (DE VILLERS, 1789)

Phal(aena) (Geom(etra)) icterata DE VILLERS, 1789: Carol. Linn. ent. 2: 382 (France Gallia). Syntype(s).
Geometra disparata HÜBNER, 1799: Samml. Eur. Schmett. 5 Geometrae (2), pl. 47: 247 (nec 246) (Europe). Syntype(s) lost.
Phalaena subfulvata HAWORTH, 1809: Lepid. Br. (2): 357 (United Kingdom). Syntype(s) (BMNH).
Larentia oxydata TREITSCHKE, 1828: Die Schmett. Eur. 6 (2): 114 (Austria: Vienna district). Holotype ♀ (HNHM, examined).
Eupithecia cognata STEPHENS, 1831: Illustr. Br. Ent. (Haustellata) 3: 288 (United Kingdom, Scotland: near Edinburgh). Syntypes, at least 3 (BMNH).
Larentia ligustigata DONZEL, 1838: Annls Soc. ent. France 7 (3): 431, pl. 12: 5 (France: Pyrénées-Orientales, Cabanasse). Syntype(s).
Eupithecia ferruginata DUPONCHEL, 1842: in GODART & DUPONCHEL, Hist. nat. Lépid. Papillons France (Suppl. 4): 50, pl. 54: 8 (Europe?). Holotype (MNHN?).
Unavailable names (infrasubspecific): *intermedia*: DIETZE (1910); *excelsa*: DIETZE (1910); *obscura*: LEMPKE (1969). - (incorrect subsequent spellings): *ligusticaria*: BOISDUVAL (1840); *oxydaria*: BOISDUVAL (1840); *subfulvaria*: DOUBLEDAY (1849); *ligusticata*: DIETZE (1910).

DIAGNOSIS: Wingspan 20-25 mm. Labial palpi dark grey. Frons and vertex whitish grey to dark grey. Notum whitish grey with wide, blackish grey transverse band in front.
Occurring in two general colour forms:
Form *cognata* STEPHENS, 1831. Forewing whitish grey to deep leaden grey, often with a rich rusty or chestnut triangular blotch between bases of M3 and CuA1 or even CuA2; stalk of Cu usually chequered black and white; basal, postbasal and antemedial lines indented, sharply or right-angled near costa; first medial line much waved, sharply angled where it touches discal spot or beyond it towards costa; postmedial line much sinuate,

right or obtuse angled on to costa; whitish wavy line usually distinct; discal spot large, ovoid, intensely black. Hindwing fuscous, lighter towards costa, slightly marked with dark transverse lines along dorsum; terminal area usually darker; wavy line fine, much dentated, whitish; discal spot minute, rounded dark. Fringe greyish white, spotted with grey or dark grey at vein ends. Abdomen whitish grey to grey, often with ochreous brown suffusion; black lateral stripes present.
Form *subfulvata* HAWORTH, 1809. Forewing rich rusty or chestnut ground colour; basal area, costal margin to discal spot and to M1, and terminal area are dark grey, mottled and irrorate with white scales; stalk of Cu chequered black and white; dark and light alternating transverse lines well marked only along costa; white wavy line conspicuous; terminal line black; discal spot large, ovoid, intensely black. Hindwing similar or slightly more 'brindled' than in preceding form. Fringe greyish white, spotted with grey or dark grey at vein ends. Abdomen reddish brown with black lateral stripes.

MALE GENITALIA: Uncus broadened near apex, when viewed laterally. Vesica armed with three dentate cornuti; one large, narrow, U-shaped cornutus and one large irregular, curved cornutus at ductus ejaculatorius base. Sternum A8 large, almost rectangular with deep basal and apical hollows; two apical lobes broad, blunt and heavily sclerotized.

FEMALE GENITALIA: Bursa copulatrix rather elongated, pouch-like, almost completely covered by small spines with a small patch of spines near ductus seminalis base. Tergum A8 rectangular with convex anterior margin and shallow medial hollow in posterior margin.

DISTRIBUTION: West-Palaearctic. Widespread in Europe except Iceland, northern Scandinavia, Portugal and Mediterranean islands (present in Corsica and Sicily). – Outside Europe in Asian Turkey, Caucasus, Transcaucasus, northern Iran, northern Kazakhstan, western Siberia to Altai. Recorded in North Africa (Morocco: Great Atlas) by ZERNY (1935).

PHENOLOGY: Univoltine: mid-June (late May) to late August (early September). Larva: early September (mid-July) to early October (late October). Overwintering as pupa.

BIOLOGY: Larva oligophagous (1). Found on flowers, seeds and occasionally on the leaflets of *Achillea millefolium* (United Kingdom: CREWE 1861c, 1862; BUCKLER 1899; BARRETT 1904; Germany: HERRICH-SCHÄFFER 1861a; DIETZE 1871b, 1910, 1913; WEIGT 1991; WEGNER pers. comm.; France: GOOSSENS 1869; MILLIÈRE 1870; Denmark: JUUL 1948), also on flowers of *Tanacetum vulgare* (United Kingdom: BARRETT 1904; Germany: DIETZE 1913; Denmark: JUUL 1948; Finland: SEPPÄNEN 1970) and rarely on *Artemisia vulgaris, A. alba (=camphorata), A. absinthium, A. campestris* (Germany: DIETZE 1913; Denmark: JUUL 1948; Finland: SEPPÄNEN 1970), *Achillea tomentosa, A. filipendulina* (Asteraceae) (Germany: WEIGT 1991).

PARASITOIDS: Hymenoptera, Ichneumonidae: *Platylabus pumilio* - Hymenoptera, Braconidae: *Microgaster dorsalis*, *Microplitis mediator*, *M. tuberculifer* - Hymenoptera, Encyrtidae: *Tyndarichus scaurus* (BUCKLER 1899; THOMPSON 1945; SHENEFELT 1973; HERTING 1976).

HABITAT: Dry fields and meadows, dry sunny roadsides, forest edges and parks. From 0 m up to about 1,900 m above sea-level in Balkan peninsula (TOMIČ & al. 1990; coll. GELBRECHT), to about 2,000 m in Apennines (PROLA & RACHELI, 1980), to about 2,300-2,400 m in Alps and Pyrenees (VORBRODT 1928; MAZEL & PESLIER 1997; HELLMANN & al. 1999; coll. FIBIGER), to about 2,400-2,800 m in Great Atlas (coll. ZSM), and to about

3,000 m in Sierra Nevada (coll. ZMUC). Recorded in southern Tirol (Piz Umbrail) at about 2,800 m above sea-level (DANNEHL 1927).

SIMILAR SPECIES: The f. *subfulvata* is one of the most easily recognized of all European *Eupithecia*. The f. *cognata* can occasionally be confused with *E. millefoliata*, but is distinguished from it by absence of delicate light brown suffusion and by the larger, intensely black discal spot on the forewing (Text-fig. 78). The large, pointed valva, three dentate cornuti on the vesica and very broad, characteristic shape of sternum A8 in the male, as well as the large pouch-like corpus bursae almost competely covered by small spines in the female genitalia should distinguish this species.

REMARKS: A large variable species. There are many individuals intermediate between the two general forms, with more or less predominance of rusty colour.

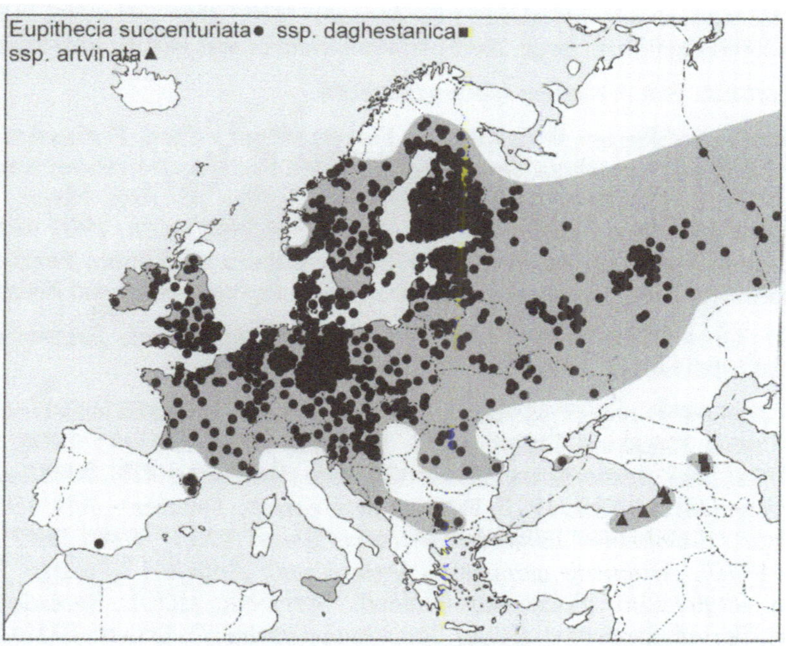

143. Eupithecia succenturiata (LINNAEUS, 1758)

Phalaena (Geometra) succenturiata LINNAEUS, 1758: Syst. Nat. (Ed. 10) 1: 528 (Sweden). Syntype(s) (LSL).
Geometra disparata HÜBNER, 1799: Samml. Eur. Schmett. 5 Geometrae (2), pl. 47: 246 (nec 247) (Europe). Syntypes lost.
Tephroclystia succenturiata var./ab. *exalbidata* STAUDINGER, 1901: in STAUDINGER & REBEL, Cat. Lepid. palaearct. Faunengeb. (Edn 3) 1: 315 (Russia: Altai; Kazakhstan: Ala Tau Mts, Ili region; Kirghizstan: Issyk-Kul; Germany). Syntype(s) (MNHU). Valid at subspecific rank.
(Eupithecia) succenturiata f. *extrema* DIETZE, 1913: Biol. Eupithecien 2: 98; ibidem (1910), 1, pl. 73: 337 (China: Kuldja). Holotype (MNHU). Valid at subspecific rank.
Unavailable names (infrasubspecific): *obscurata*: LEMPKE (1951); *bistrigata*: LEMPKE (1951). - (incorrect subsequent spellings): *succenturiaria*: BOISDUVAL (1840); *subccenturiata* VALLHONRAT (1989).

DIAGNOSIS: Wingspan 20-27 mm. Labial palpi brown with a row of white scales. Frons and vertex dark brown or covered with mixture of dark brown and white scales. Notum chalky white except dark brown pronotum. Forewing chalky white, along costa brown usually with reddish tinge; transverse lines indistinct, brownish, well marked on costa with large dark brown spots; medial area more or less brownish with reddish tinge along posterior margin; terminal area dark brown on inner side, and reddish brown with short dark brown dashes between veins towards margin, usually with whitish apical streak; wavy line white, distinct, with more or less large white tornal spot; discal dot intensely black, large, oval. Hindwing chalky white with darker shade and weak indications of transverse lines on the posterior margin; terminal area distinctly darker with dentated white wavy line, which separates terminal area into brown inner side and reddish brown outer side; discal dot minute, more or less conspicuous, dark brown, rounded. Abdomen dark reddish-brown with black lateral spots; first abdominal segment white; last segment also white but only in male.

MALE GENITALIA: Uncus and vesica as those in preceding species. Sternum A8 large, almost rectangular, slightly broadened near base and apex; basal and apical hollows deep; two apical lobes digitiform, large, blunt, tapered to apex and heavily sclerotized.

FEMALE GENITALIA: See preceding species diagnosis.

DISTRIBUTION: Euro-Siberian. Widespread in Europe except Iceland, Portugal and Albania. In southern Europe populations very isolated. Absent from Mediterranean islands except Sicily (GRILLO & PARENZAN 1995). – Outside Europe in Asia Minor, Caucasus, Transcaucasus and Urals. The more light f. *exalbidata* STAUDINGER, 1901 widely distributed in Central Asia (south-eastern Kazakhstan, northern Tien-Shan, Ferghana and Ili basin in north-western China) and Siberia from Altai to Amur basin and Kamchatka.

PHENOLOGY: Univoltine: mid-June (late May) to mid-August (early September). Larva: early August (mid-July) to early October. Overwintering as pupa.

BIOLOGY: Larva rather oligophagous (1). Found on flowers, fruits and leaflets of *Artemisia vulgaris* (United Kingdom: CREWE 1862; BUCKLER 1899; BARRETT 1904; Germany: DIETZE 1874a), also *A. maritima* (United Kingdom: BARRETT 1904), *Achillea millefolium* (Germany: DIETZE 1910, 1913; WEGNER pers. comm.; Denmark: JUUL 1948; Russia: original data), *Tanacetum vulgare* (Germany: SÄLZL 1935; WEIGT 1991; Finland: SEPPÄNEN 1970), *Artemisia campestris* (Asteraceae), *Angelica sylvestris* (Apiaceae), *Vaccinium uliginosum* (Ericaceae) (Finland: SEPPÄNEN 1970). Recorded also on *Verbascum* (Scrophulariaceae), *Rubus* (Rosaceae) (Germany?: SPULER 1910), *Artemisia absinthium* (Asteraceae) (France: LHOMME 1935) and *Calluna vulgaris* (Ericaceae) (Germany: SÄLZL 1935).

PARASITOIDS: Hymenoptera, Ichneumonidae: *Casinaria morionella, Mesochorus brevipetiolatus*, *Netelia tarsata* - Hymenoptera, Braconidae: *Apanteles falcatus*, *A. melanoscelus (=solitarius)*, *Microplitis spectabilis* (?), *M. tuberculifer* - Hymenoptera, Encyrtidae: *Litomastix quercicola* - Diptera, Tachinidae: *Eurysthaea scutellaris*, *Siphona setosa*, *S. cristata*, *Discochaeta hyponomeutae* (THOMPSON 1945; HERTING 1960, 1976; SHENEFELT 1972, 1973; ANDERSEN 1982, 1996; ZWAKHALS pers. comm.).

HABITAT: Most types of open areas, gardens, parks, and it has been found in spruce and mixed forests in north-western European Russia. From 0 m up to about 1,000 m above sea-level in Alps (WEIGT 1991), to about 1,300 m in Sicily (GRILLO & PARENZAN 1995), to about 1,200-1,500 m in Carpathians (MESZAROS & VOJNITS 1966), and to about 1,800 m in Pyrenees (MAZEL & PESLIER 1997).

SIMILAR SPECIES: The external features, such as white ground colour and distinctive maculation of wings and body of typical *E. succenturiata* can usually be used to distinguish it from other European *Eupithecia* species with accuracy. One cannot easily separate *E. succenturiata* from the closely related *E. icterata* by the male and female genitalia. However, the shape of sternum A8 in these two species is very distinct.

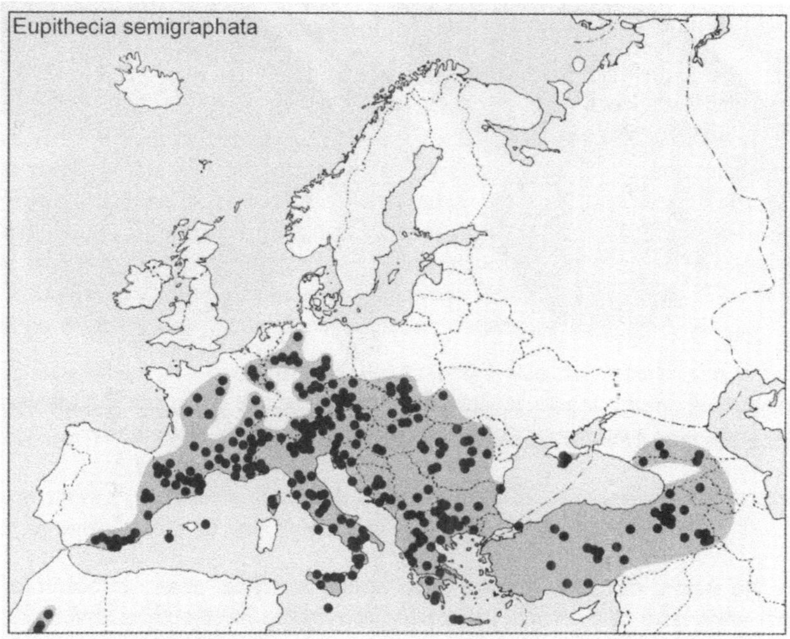

144. *Eupithecia semigraphata* BRUAND, 1851

Eup(ithecia) semigraphata BRUAND, 1851: Mém. Soc. libre d'emulation Doubs 3 (3): 64 (France: Maison-Rouge, Douds, pres Saint Vit). Syntype(s).
Eupithecia nepetata MABILLE, 1869: Annls Soc. ent. France (4) 9 (1): 68, pl. 2: 4 (Corsica: Bastia). Lectotype ♀ (coll. HERBULOT in ZSM, examined).
Eup(ithecia) semigraphata f. *canariensis* DIETZE, 1910: Biol. Eupithecien 1, pl. 76: 601; ibidem (1913), 2: 105 (Canary Islands). Syntypes 3 (MNHU). Valid at subspecific rank.
Eup(ithecia) semigraphata f. *valida* DIETZE, 1910: Biol. Eupithecien 1, pl. 76: 602; ibidem (1913), 2: 105 (Germany: Mittelrhein). Syntypes (MNHU). Valid at subspecific rank.
Eup(ithecia) semigraphata f. *arida* DIETZE, 1910: Biol. Eupithecien 1, pl. 76: 603; ibidem (1913), 2: 105 (Turkey: Taurus, Zeitun). Syntypes 9 (MNHU). Valid at subspecific rank.
Eupithecia semigraphata gravosata SCHÜTZE, 1956: Z. wien. ent. Ges. 41 (11): 309, pl. 33A: 1-10; text-figs 1E, F (Croatia, Dalmatia: Gravosa; Orebic). Syntypes 30, including 25♂5♀ (ZSM). Valid at subspecific rank.
Unavailable names (misidentification): *modicaria*: sensu HERRICH-SCHÄFFER (1848) nec BOISDUVAL, 1840 (identity: *Eupithecia impurata*); *impurata*: sensu REDONDO (1990) nec HÜBNER, 1813 (identity: *Eupithecia impurata*). - (incorrect subsequent spelling): *semigrapharia*: HERRICH-SCHÄFFER (1852).

DIAGNOSIS: Wingspan 20-21 mm. Labial palpi grey. Frons, vertex and notum whitish grey. Forewing with oblique transverse vein in areole; pale grey; transverse lines numerous, fine, rather inconspicuous, much indented, sharply angled and slightly broadened and darker near costa; medial area between antemedial and first medial lines usually faintly darker; postmedial line much waved, often with a series of short, wedge-shaped dark

Text-figs 79-80: Diagnostic characters (indicated) of European *Eupithecia* species. Text-fig. 79: *E. semigraphata* BRUAND, 1851. Text-fig. 80: *E. impurata* (HÜBNER, 1813).

dashes on the veins, obtusely angled onto costa; terminal area concolorous or slightly darker; pale wavy line fine, much indented, forming a minute pale tornal dot but sometimes this is obsolete; discal spot more or less large, ovate, blackish grey. Hindwing the same colour as forewing; transverse lines rather indistinct; postmedial line and the inner edge of terminal area usually marked with short, wedge-shaped dark dashes on the veins; terminal area narrow; whitish wavy line fine, much indented but sometimes invisible; discal spot small, ovate or rounded, blackish grey. Fringe whitish grey, distinctly spotted with dark grey at vein ends. Abdomen whitish grey.

MALE GENITALIA: Vesica armed with two dentate cornuti, one large, elongated, U-shaped cornutus and one large, irregular, curved cornutus at ductus ejaculatorius base. Sternum A8 diverse in size and shape (see SCHÜTZE, 1956).

FEMALE GENITALIA: See species-group diagnosis.

DISTRIBUTION: Mediterranean. Central and southern Europe (except Portugal and Sardinia). – Outside Europe in North Africa (Morocco; f. *porphyrata* ZERNY, 1934), Asia Minor (f. *arida* DIETZE, 1910), Caucasus and Transcaucasus.

PHENOLOGY: Uni- or bivoltine: early July (late June) to early August (mid-August). In the south bivoltine: late June to mid-October (late October). Larva: late August (mid-July) to early October (late October). Overwintering as pupa, sometimes two years.

BIOLOGY: Larva oligophagous (1). Found on the flowers and seeds of *Calamintha nepeta* (Corsica: MABILLE 1869; GOOSSENS 1869), *Acinos alpinus, Hyssopus officinalis* (Italy: DIETZE 1872), *Thymus serpyllum, Satureja montana* (Germany: DIETZE 1975a, 1910,

1913), *Thymus vulgaris* (Switzerland: VORBRODT 1928) and *Origanum vulgare* (Lamiaceae) (Germany: KOCH 1984; WEIGT 1991). Recorded also on *Nepeta cataria* (France: LHOMME 1935). Larva often suffers heavily from parasitization.

HABITAT: Warm and dry, preferably stony slopes, rocks and screes. From 0 m up to about 2,000 m above sea-level in Apennines and Balkan peninsula (PROLA & RACHELI 1980; coll. FIBIGER), to about 2,300 m in Alps (VORBRODT 1928), to about 2,300-2,400 m in Pyrenees (MAZEL & PESLIER 1997), to about 2,600-2,700 m in Great Atlas (ZERNY 1934; RUNGS 1981), and to about 3,000 m in Asia Minor (coll. FIBIGER).

SIMILAR SPECIES: Usually smaller than *E. impurata*, the only species with which it could be confused, with less whitish colouration, more sharply angled basal, postbasal and antemedial lines and often smaller discal spots on the hindwings. *E. impurata* is more broadly winged, more brindled over all, often with darker terminal areas on all wings (Text-fig. 79). The absence of small apical dentate cornutus on the vesica and usually narrower sternum A8 of male, despite the variability of this structure in the two compared species (SCHÜTZE, 1956), will distinguish *E. semigraphata* from *E. impurata*. The female genitalia cannot be distinguished from those of other closely related species.

REMARKS: *E. semigraphata nepetata* MABILLE, 1869 was described from a single specimen, distinguished from other Corsican adults by larger discal spots, darker costal spots and especially terminal area. I cannot to accept the moths from Corsica as a valid subspecies.
There is one reddish ochreous Spanish specimen of *E. semigraphata* in collection of ZSM, which is very similar to the Moroccan f. *porphyrata* ZERNY, 1934.

145. *Eupithecia impurata* (HÜBNER, 1813)

Geometra impurata HÜBNER, 1813: Samml. Eur. Schmett. 5 Geometrae (1), pl. 67: 347 (Europe). Syntype(s) lost.
Geometra modicata HÜBNER, 1813: Samml. Eur. Schmett. 5 Geometrae (1), pl. 70: 361 (Europe). Syntype(s) lost.
Larentia proluaria FREYER, 1852: Neuere Beitr. Schmettkde 6 (99): 182, pl. 593: 1 (Germany: Füssen). Syntype(s).
Eupithecia impurata gremmingerata SCHÜTZE, 1952: Z. Wien. Ent. Ges. 36 (10/12): 164, pl. 13: 1-10 (Germany: Schwarzwald, Wutachgebiet). Holotype ♂ (ZSM, examined). Valid at subspecific rank.
Eupithecia impurata badeniata SCHÜTZE, 1952: Z. Wien. Ent. Ges. 36 (10/12): 165, pl. 13: 11-20 (Germany: Baden, Kaiserstuhl). Holotype ♂ (ZSM, examined). Valid at subspecific rank.
Eupithecia impurata langeata SCHÜTZE, 1952: Z. Wien. Ent. Ges. 36 (10/12): 166, pl. 13: 21-30 (Germany: Saxonia, Zschopautal). Holotype ♂ (ZSM, examined). Valid at subspecific rank.
Eupithecia impurata germanicata SCHÜTZE, 1952: Z. wien. ent. Ges. 36 (10/12): 167, pl. 13: 31-40 (Germany: Edersee/Waldeck). Holotype ♂ (ZSM, examined). Valid at subspecific rank.
Eupithecia impurata thueringiata SCHÜTZE, 1956: Z. Wien. Ent. Ges. 41 (11): 312, pl. 34 A: 1-10 (Germany: Thuringia, Blankenstein). Holotype ♂ (ZSM, examined). Valid at subspecific rank.
Eupithecia impurata franconiata SCHÜTZE, 1956: Z. Wien. Ent. Ges. 41 (11): 313, pl. 34 B: 11-20; text-figs 4a-c (France, Jura: Pottenstein). Holotype ♂ (ZSM, examined). Valid at subspecific rank.
Eupithecia impurata westfalicata WEIGT, 1982: Dortm. Beitr. Landeskde 16: 4, figs 1-14 (Germany: Marsberg). Holotype ♂ in coll. WEIGT? (not information). Valid at subspecific rank.
Eupithecia impurata canigouata WEIGT, 1989: Dortm. Beitr. Landeskde 23: 39, figs 1-4 (France: Pyrenees, Nyer near Olette). Holotype ♂ in coll. WEIGT? (not information). Valid at subspecific rank.
Unavailable names (misidentification): *minoraria*: sensu DUPONCHEL (1842) nec BOISDUVAL, 1840 (identity: *Perizoma minorata*). - (incorrect subsequent spellings): *modicaria*: BOISDUVAL (1840); *impuraria*: BOISDUVAL (1840).

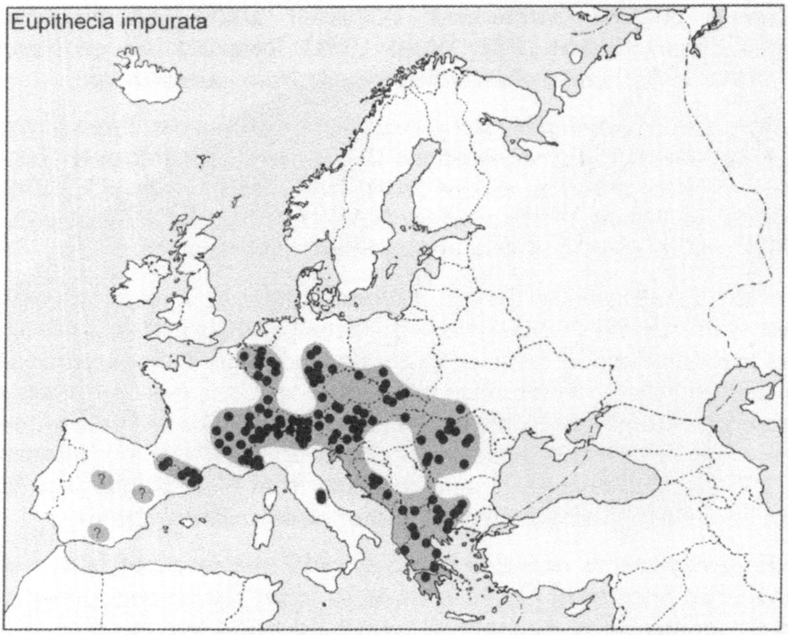

DIAGNOSIS: Wingspan 19-24 mm. Labial palpi grey. Frons, vertex and notum whitish grey. Forewing wide with slightly arcuate costa; evenly rounded termen and rounded apex. Ground colour of wings whitish grey or ash-grey; transverse lines on the forewing sinuate, usually forming dark grey costal spots; medial line sharply angled near costa and curved round outside of discal spot; postmedial line often forming a series of blackish wedge-shaped dashes on the veins; the terminal area sometimes darker, with more or less thin distinct pale wavy line; discal spot usually clear, dark grey, rounded or ovoid. Hindwing wide and rounded with rather inconspicuous transverse lines, with the postmedial line sometimes forming a series of blackish wedge-shaped dashes on the veins; terminal area sometimes darker; discal spot frequently clear, dark grey, rounded. Fringe greyish white spotted with grey or dark grey. Abdomen whitish grey, often with delicate ochreous tinge.

MALE GENITALIA: Vesica armed with three dentate cornuti (one apical very small), one large, elongated, U-shaped cornutus and one large, irregular, curved cornutus at ductus ejaculatorius base. Sternum A8 diverse in size and shape (see SCHÜTZE, 1956).

FEMALE GENITALIA: See species-group diagnosis.

DISTRIBUTION: European. Mountain ranges of Central and southern Europe.

PHENOLOGY: Bivoltine: mid-June (late May) to mid-July (late August); early September to mid-September (late October). Larva: late July (mid-June) to mid-August (mid-October). Overwintering as pupa.

BIOLOGY: Larva monophagous (2). Found on the flowers and seeds of *Campanula rotundifolia* (Germany: RÖSSLER 1866; DIETZE 1875b; Switzerland: PÜNGELER 1889), also *C.*

cochlearifolia (Austria: DIETZE 1910, 1913; Germany: WEIGT 1982, 1991); recorded also on *Thymus* (Lamiaceae) (Austria: HABELER 1971) probably in error.

HABITAT: Warm and dry, preferably stony slopes, rocks and screes. From 0 m up to about 1,900-2,200 m above sea-level in Balkan peninsula (THURNER 1967; PINKER 1968; coll. ZMUC; coll. FIBIGER), and to about 2,400 m in Alps and Pyrenees (VORBRODT 1928; coll. FIBIGER).

SIMILAR SPECIES: *E. impurata* can be confused with *E. semigraphata*, but distinguished from it by the larger size, the weakly elongated forewings, the rather lighter ground colour, the well marked but soft pattern on the wings and the rounded discal spots on all wings. Usually, the projection of the postmedial line between M2 and M3 on the forewing is shorter than in *E. semigraphata* (Text-fig. 80). In the male genitalia the two large dentate cornuti on the vesica are usually broader than those in the aedeagus of *E. semigraphata*.

REMARKS: Many subspecies of *E. impurata* have been described from Europe, mainly from Germany and France, based on reared specimens: subsp. *gremmingerata* SCHÜTZE, 1951, subsp. *badeniata* SCHÜTZE, 1951, subsp. *langeata* SCHÜTZE, 1951, subsp. *thueringiata* SCHÜTZE, 1956, subsp. *franconiata* SCHÜTZE, 1956, subsp. *westfalicata* WEIGT, 1982, subsp. *canigouata* WEIGT, 1989. In my opinion these have little value and cannot be compared with the nominate subspecies of *E. impurata* since its origin and location are unknown.

146. *Eupithecia poecilata* PÜNGELER, 1888

Eupithecia poecilata PÜNGELER, 1888: Stett. ent. Z. 49 (10-12): 349 (Corsica; Sardinia). Syntypes ♂ ♀ (MNHU).

DIAGNOSIS: Wingspan 18-21 mm. Labial palpi blackish grey. Frons and vertex yellowish white with scattered black scales. Notum pale yellowish grey. Forewing rather narrow, elongate; costa straight and slightly curved to near apex; termen oblique, slightly rounded; tip of wing pointed; ground colour brindled, delicate yellowish or creamy with dark grey transverse lines broadest at costa; basal line dentate; antemedial line usually smoothly curved, medial line sinuate and sharply angled near discal spot, postmedial line multiply waved and forming a series of dark dots on the veins, sharply angled near costa; medial area between antemedial and medial lines creamy grey; terminal area grey with two paler blotches (one medial, the other near apex) and darker costal blotch; pale wavy line narrow, sometimes broken, not forming a pale tornal dot; discal spot conspicuous, small, ovate to narrow ovate-oblong, black. Hindwing brindled, of the same colour as the forewing, lighter near dorsum; transverse lines distinct, dark grey; postmedial line represented by a series of black closely set dashes on the veins; terminal area dark grey with lighter blotch between M3 and CuA1; discal spot conspicuous, black, as on the forewing. Fringe creamy white, distinctly spotted with dark grey at vein ends. Abdomen yellowish grey, with some minute black lateral dots.

MALE GENITALIA: Vesica armed with two dentate cornuti, one large, elongated, flat, folded U-shaped cornutus and one large, irregular, curved cornutus at ductus ejaculatorius base. Sternum A8 elongated; basal hollow rather deep; apical hollow shallow; with two short, blunt apical lobes and heavily sclerotized lateral margins in apical part of sternum.

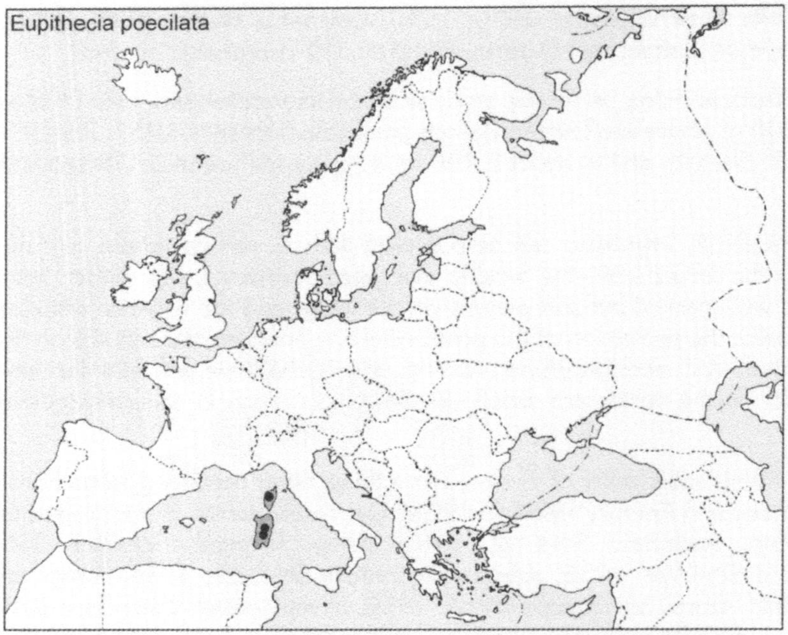

FEMALE GENITALIA: Bursa copulatrix pear-shaped, heavily sclerotized in medial part. Anterior and posterior apophyses thin and pointed.

DISTRIBUTION: Tyrrhenic. Corsica and Sardinia.

PHENOLOGY: Probably bivoltine: early June (early April) to mid-July (early August).

BIOLOGY: Unknown.

HABITAT: Warm and dry slopes, rocks. From 0 m up to about 1,300 m above sea-level in Sardinia (coll. HNHM, ZSM).

SIMILAR SPECIES: Most similar to *E. semigraphata* among European *Eupithecia* species, but distinguished by narrower wings with yellowish or creamy tinge, more brindled coloration and transverse lines more sharply angled near costa. On the whole the male and female genitalia are smaller than in *E. semigraphata*: valva narrower and sternum A8 of male narrower near apex.

REMARKS: *E. poecilata* is a good species, and not a subspecies of the Transcaucasian *E. lithographata* (see BOHATSCH 1893). This is amply proved in detail by SCHÜTZE (1956). I also confirm, after careful examination of the type-specimen of *E. lithographata* described by CHRISTOPH (ZISP) and many other specimens, that *E. poecilata* and *E. lithographata* are not conspecific (see also the classic work by DIETZE 1910: Taf. 80, fig. 890). In *E. lithographata*, which really is closely related to it, the head, body and wings are of a rich ochreous colour, the postmedial line not forming a series of dark dashes on the veins; the wavy line ochreous, broad, cloudy and broken; the dark terminal line obsolete; the discal specks usually larger; the hindwing without a series of dark dashes on the veins and pale wavy line lacking, but with very narrow terminal area.

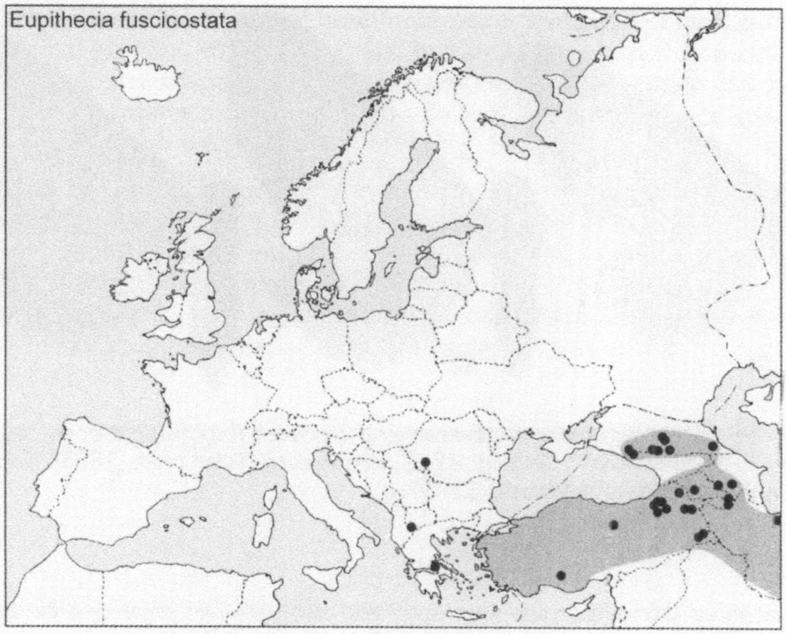

147. *Eupithecia fuscicostata* CHRISTOPH, 1887

Eupithecia fuscicostata CHRISTOPH, 1887: in ROMANOFF, Mém. Lépid. 3: 11 pl. 1: 6 (Azerbaijan: Karabagh, Guéroussi). Holotype ♂ (ZISP, examined).
Eupithecia salami BRANDT, 1938: Ent. Rdsch. 55 (50): 587, fig. 289 (Iran: Fars, Strasse Chiraz-Kazeroun, Fort Sine-Sefid). Holotype ♂ (NHRS, examined). New junior synonym.
Eupithecia domogledana VOJNITS & SZABÓ, 1988: Nota lepid. 11 (2): 160, figs 2-6 (Romania: Dom. [=Mt. Domogled]). Holotype ♂ (HNHM, examined). New junior synonym.
Unavailable names (incorrect subsequent spelling): *fuscostigma*: ALPHÉRAKY (1892).

DIAGNOSIS: Wingspan 17-23 mm. Labial palpi pale brownish yellow. Frons, vertex and notum yellowish white. Forewing narrow and elongated with pointed apex; costa straight; termen oblique and almost straight; ground colour white or creamy white; transverse lines brownish yellow, usually darker greyish brown and sharply angled near costa; medial area between antemedial and medial lines usually darker, covered with brownish yellow scales; postmedial line often additionally marked by a series of brown specks on the veins; terminal area completely brownish yellow; white wavy line conspicuous, sometimes shaded on inner side; discal spot small, circular, brown or sometimes absent. Hindwing concolorous with forewing; transverse lines brownish yellow covered with scattered brown scales; terminal area broad, brownish yellow with very distinct white wavy line; discal spot small, round, brown. Fringe white, spotted with yellowish brown opposite vein ends. Abdomen yellowish white, with dorsal row of brown spots.

MALE GENITALIA: Uncus broadened near apex, when viewed laterally. Valva with narrowly rounded apex. Vesica armed with three small dentate cornuti, with one elongated, U-shaped cornutus and one irregular, curved, rather S-shaped cornutus at ductus ejaculatorius base. Sternum A8 elongated, slightly broadened at base and tapered to apex, with

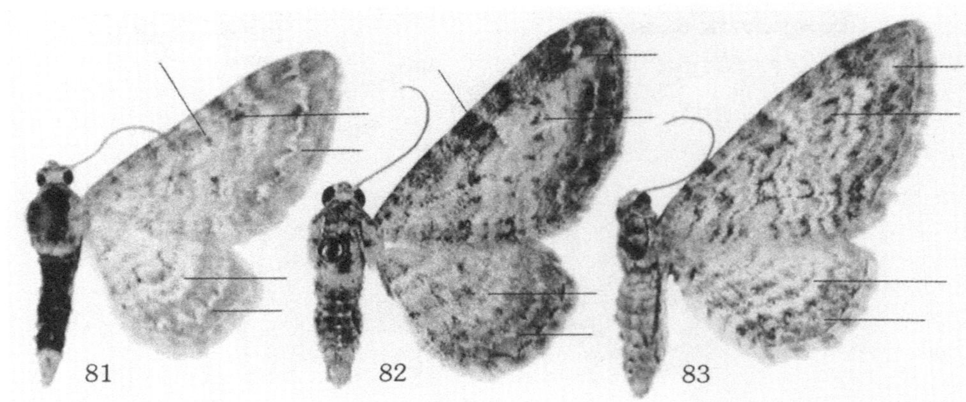

Text-figs 81-83: Diagnostic characters (indicated) of European *Eupithecia* species. Text-fig. 81: *E. fuscicostata* CHRISTOPH, 1887. Text-fig. 82: *E. denticulata* (TREITSCHKE, 1828). Text-fig. 83: *E. subumbrata* (DENIS & SCHIFFERMÜLLER, 1775).

heavily sclerotized lateral margins near apex and also with two small clublike apical lobes; basal hollow shallow; apical hollow small and shallow.

FEMALE GENITALIA: Bursa copulatrix more than half covered by spines at base, with a large patch of spines in ductus bursae. Tergum A8 square.

DISTRIBUTION: East-Mediterranean. Apparently isolated populations have been discovered in Romania, Macedonia and Greece. – Outside Europe in Asian Turkey (Prov. Tokat, Sivas, Erzurum and Agri), Caucasus, Transcaucasus and Iran. – New for the faunas of Macedonia, Greece and Turkey: 1♂, Macedonia, Ohrid, Rasmun, 1.-31.7.1935, leg. WOLFSCHLÄGER (coll. KLIMESCH, ZSM); 2♂, Hellas, Volotia, 1400 m, 8 km NW. Arachova, Mts Parnassos, 28.VI.1985, leg. P. SKOU & B. SKULE (coll. SKOU); 1♂ and 1♀, Kleinasien, Prov. Tokat-Sivas, 1700 m, Çamlibel-Pass, 01.7.-10.7.(19)78, leg. de FREINA (HNHM); 1♀, Türkey, Prov. Agri, 7 km W of Aydintepe, 1800 m, 42'30'E, 39'49'N, 05.-06.08.1988, leg. GYULAI, HREBLAY, RONKAY & Ronkay (ZMKU); 1♂, Türkey, Prov. Erzurum, 8 km SW of Askale, 40'40'E, 39'52'N, 13.VII.1989, leg. P. GYULAI & M. HREBLAY (ZMKU); 1♀, Türkei, Erzurum, 20 km nördl. Askale, Kop Geç, 2300 m, 28.7.1992, Li, leg. G. BAISCH (ZSM).

PHENOLOGY: Univoltine: mid-July (late June) to mid-August (early September).

BIOLOGY: Unknown.

HABITAT: Limestone slopes and rocks. Up to about 1,400 m above sea-level in Greece (coll. SKOU), to about 2,300 m in northern Iran (Elburs Mts: coll. ZSM), and to about 2,500-2,600 m in Caucasus and eastern Turkey (coll. ZISP; coll. GELBRECHT).

SIMILAR SPECIES: *E. fuscicostata* is most like *E. subumbrata*, but that species never has yellowish wing colour. The yellowish transverse lines are also present in *E. pernotata*, but that species has broader wings, the apex of forewing more rounded, the transverse lines less distinct, especially white wavy line, and small pale tornal spot always present on the forewing (Text-fig. 81). The combination of a few features, such as: the relatively short valva with obtuse apex, the short and narrow vinculum, the different cornuti in the vesica and the shorter sternum A8 with small heavily sclerotized posterior apices in the male, and the elongated pear-shaped bursa copulatrix in the female genitalia can be used to separate *E. fuscicostata* from other representatives of the *semigraphata* species-group and from *E. pernotata* as well.

REMARKS: According to the collection of HNHM, *E. satyrata rivosulata* DIETZE, 1875 was recorded for Mongolia by VOJNITS (1977), not *E. fuscicostata* CHRISTOPH, 1887.
After carefully examining adults and their genitalia of *E. fuscicostata, E. salami* and *E. domogledana* from Romania, Macedonia, Greece, Asian Turkey, south of European Russia, Georgia, Armenia, Azerbaijan and Iran, as well as studying the type descriptions, I have come to the conclusion that all of these species are conspecific.

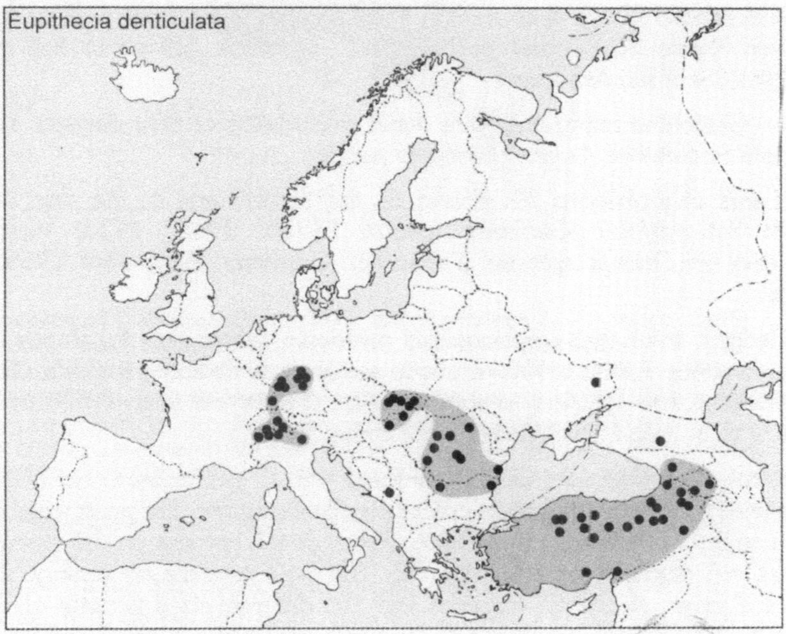

148. Eupithecia denticulata (TREITSCHKE, 1828)

Larentia denticulata TREITSCHKE, 1828: Die Schmett. Eur. 6 (2): 132 (Hungary: Ofen). Lectotype ♂ (HNHM, examined).
Unavailable names (incorrect subsequent spelling): *denticularia*: BOISDUVAL (1840).

DIAGNOSIS: Wingspan 19-23 mm. Labial palpi pale brown, sometimes brown or dark brown. Frons dirty white with pale brown band anteriorly and two pale brown blotches near antennal bases. Vertex dirty white. Notum dirty white with pale brown transverse band in front. Forewing chalky white or yellowish white; transverse lines slightly sinuate, inconspicuous but distinctly indicated with large, broad, brownish grey costal spots; ante- and postmedial lines well marked by a series of brown dots on the veins; terminal area brownish grey with broad chalky white apical streak between R5 and M1 or M2; whitish wavy line conspicuous, dentate, without tornal spot; discal dot large, rounded or slightly elongated, dark brown or black. Hindwing chalky white; transverse lines brownish, wavy, strongly indicated near dorsum; terminal area darker, brownish grey with distinct and waved inner margin; whitish wavy line conspicuous, fine, indented; discal dot small, usually rounded, dark brown. Fringe chalky white, broadly spotted with light brown. Abdomen dirty white with soft creamy tinge; a row of small pale brown dorsal tufts with white tips and lateral rows of small pale brown specks are present.

MALE GENITALIA: Vesica armed with two dentate cornuti, one large, elongated, U-shaped cornutus and one large, irregular, curved cornutus at ductus ejaculatorius base. Sternum A8 diverse in size and shape as in closely related species (SCHÜTZE 1956).

FEMALE GENITALIA: See species-group diagnosis.

DISTRIBUTION: Euro-Caucasian. Germany, Switzerland, north-western Italy, Slovakia, Hungary, Romania and Bulgaria. Isolated occurrence in southern Italy (SCALERCIO pers. comm.), Montenegro (TOMIČ & al. 1990), eastern Ukraine (Donetzk region: coll. ZMKU) and southern Russia (Kislovodsk: coll. ZMKU). – Outside Europe in Asia Minor and Transcaucasus (Georgia, Armenia).

PHENOLOGY: Univoltine: early July (late June) to mid-August (late August). Larva: mid-August to late September. Overwintering as pupa.

BIOLOGY: Larva oligophagous (2). Found on the flowers and in the seed capsules of *Campanula rotundifolia* (Germany: DIETZE 1875b, 1910, 1913; WEIGT 1991). Recorded also on *Ononis spinosa* (Fabaceae) (Germany: BERGMANN 1955) probably erroneously.

HABITAT: Hedges, scrub and scrub-covered meadows, warm and dry slopes, rocks and screes. Up to about 1,300-1,400 m above sea-level in Balkan peninsula (TOMIČ & al. 1990), to about 1,700-1,800 m in southern Alps (HELLMANN & al. 1999), and to about 2,100-2,400 m in Asia Minor (RIEMIS pers. comm.).

SIMILAR SPECIES: *E. denticulata* can be confused with *E. subumbrata* but is usually larger, with pointed forewings, the white colour usually dominant, the postmedial line represented by a series of dark dots and a white apical streak present on the forewing, these being absent in *E. subumbrata* (Text-fig. 82). The male genitalia are distinguished by the presence of a large, elongated cornutus and two different-sized dentate cornuti in the vesica, and by the broader sternum A8; the bursa copulatrix is larger and broader, with different arrangement of spines from that in the female genitalia of *E. subumbrata*.

149. *Eupithecia subumbrata* (DENIS & SCHIFFERMÜLLER, 1775)

Geometra subumbrata DENIS & SCHIFFERMÜLLER, 1775: Syst. Verz. Schmett. Wien.: 110 (Austria: Vienna district). Syntype(s) lost.

Phalaena (Geometra) scabiosata BORKHAUSEN, 1794: Natur. Eur. Schmett. 5: 336 (Europe). Syntype(s).

Eupithecia piperata STEPHENS, 1831: Illustr. Br. Ent. (Haustellata) 3: 288 (United Kingdom: England, Surrey, Riddlesdown, near Croydon). Syntypes (BMNH).

Eupithecia obrutaria HERRICH-SCHÄFFER, 1848: Syst. Bearb. Schmett. Eur. 3 (31): 119, 126, pl. 24: 145, 146 (Germany: Regensburg). Syntypes.

Eupithecia majoraria DE LA HARPE, 1855: Fauna Suisse. Lépidoptères. 4: 149 (Switzerland). Syntypes 2♀.

Eup(ithecia) scabiosata f. *limbofasciata* DIETZE, 1910: Biol. Eupithecien 1, pl. 76: 570, 571; ibidem (1913), 2: 103 (Kirghizstan: North Tibet, Alexander Mts). Lectotype ♂ (MNHU, examined: SCHÜTZE, 1956). Valid at subspecific rank.

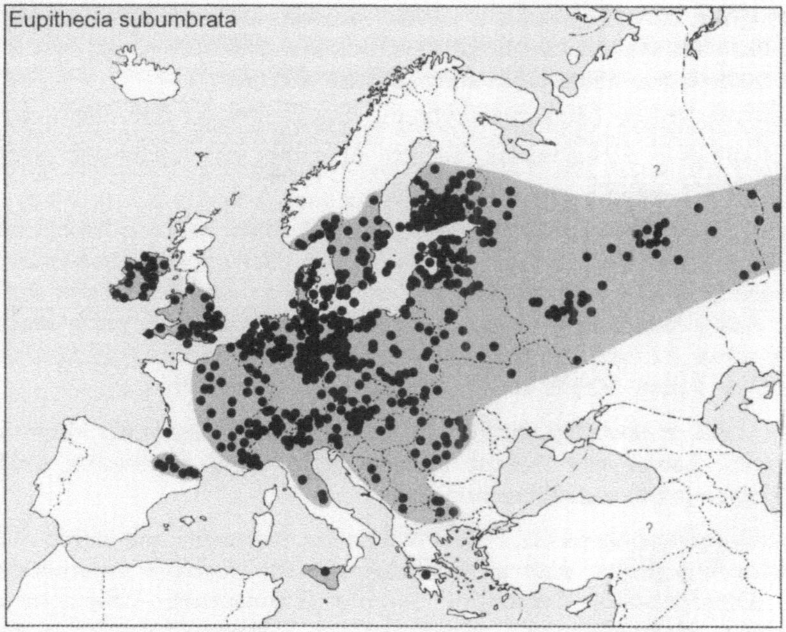

(*Eupithecia*) *scabiosata* f. *bistrigata* DIETZE, 1913: Biol. Eupithecien 2: 103; ibidem (1910), 1, pl. 76: 578; 85: 993 (Austria: Styria, Stainz). Syntype(s) (MNHU). Valid at subspecific rank.

Unavailable names (misidentification): *residuaria*: sensu HERRICH-SCHÄFFER (1856) nec BOISDUVAL, 1840 (identity: *Eupithecia lariciata*). – (infrasubspecific): *aequistrigata*: STAUDINGER (1871); *obscurata*: LEMPKE (1951); *impuncta*: LEMPKE (1951). - (incorrect subsequent spellings): *piperitata*: STEPHENS (1831): *subumbraria*: BOISDUVAL (1840); *piperaria*: DOUBLEDAY (1849); *subumrata*: BIELEWICZ (1973); *subbumbrata*: VIVES MORENO (1994).

DIAGNOSIS: Wingspan 15-23 mm. Labial palpi grey. Frons and vertex greyish white. Notum greyish white with grey transverse band in front. Forewing with oblique transverse vein in areole; rather narrow, costa straight, termen oblique, rounded; ground colour greyish white, with darker suffusion, grey along costa and termen; stalks of R and Cu often ochreous; transverse lines numerous, fine, oblique, more or less waved, and broadened, darker, right or sharply angled onto costa; first medial line crossing discal spot; postmedial line shaded on inner side and usually well marked by a series of dark grey dashes on the veins, twice angled near costa; terminal area darker, grey, with ochreous tinge in some specimens and sometimes with black marginal dashes between veins; whitish wavy line more or less distinct; pale tornal spot absent; discal spot minute, rounded, dark, or sometimes obsolete. Hindwing paler, smoky white; transverse lines inconspicuous or well marked along dorsum; terminal area grey, distinctly darker; whitish wavy line much dentated, close to termen; discal spot more or less small, circular and dark. Fringe greyish white, spotted with grey or dark grey at vein endings. Abdomen greyish white with pale ochreous brown suffusion; the second segment ochreous brown; usually a series of minute dark dorsal tufts bordered by white and sometimes black lateral dots are present.

MALE GENITALIA: Uncus narrow. Vesica armed with two almost identical dentate cornuti and one irregular, curved cornutus at ductus ejaculatorius base. Sternum A8 elongated, rather peg-like with heavily sclerotized lateral margins near apex; basal hollow deep; apical hollow shallow, two apical lobes short, narrow and blunt.

FEMALE GENITALIA: Bursa copulatrix covered by spines in basal half, with a large triangular patch of spines in ductus bursae. Tergum A8 rather rectangular with shallow medial hollow in posterior margin.

DISTRIBUTION: Euro-Siberian. Widespread in Europe from Spain to Urals; northward to central Finland; southward to about latitude 37-38º (Sicily, Iberian and Balkan peninsulas). – Outside Europe occurs eastward through Urals to southern Siberia and Mongolia. In Central Asia (Tien-Shan Mts, south-eastern Kazakhstan and north-western China) replaced by subsp. (or form ?) *iliata* SCHÜTZE, 1956. Mentioned in a provisional list of the Geometridae of Turkey (RIEMIS 1994), but identification uncertain.

PHENOLOGY: Uni- or probably bivoltine in the south: mid-May (early May) to late June (early August). Larva: late August (mid-June) to early September (mid-October). Overwintering as pupa, sometimes two years.

BIOLOGY: Larva polyphagous (1). Found on the flowers, seeds and leaves of numerous, mainly herbaceous plants, such as *Knautia arvensis*, *Scabiosa columbaria*, *Succisa pratensis* (Dipsacaceae), *Origanum vulgare* (Lamiaceae), *Valeriana officinalis* (Valerianaceae), *Silene vulgaris* (Caryophyllaceae), *Pimpinella saxifraga, Bupleurum, Heracleum sphondylium, Angelica, Pastinaca* (Apiaceae), *Galium mollugo, G. verum, G. boreale* (Rubiaceae), *Thalictrum foetidum, T. flavum* (Ranunculaceae), *Hypericum maculatum* (Clusiaceae), *Epilobium angustifolium* (Onagraceae), *Rosa centifolia* (Rosaceae), *Euphrasia, Rhinanthus minor* (Scrophulariaceae), *Globularia* (Globulariaceae), *Asparagus officinalis* (Liliaceae), *Gentiana pneumonanthe* (Gentianaceae), *Campanula rotundifolia, Jasione* (Campanulaceae), *Ononis spinosa, Medicago* (Fabaceae), *Lysimachia* (Primulaceae), *Eupatorium*, *Aster*, *Tanacetum vulgare, Achillea*, *Leontodon hispidus, Crepis vesicaria*, *Solidago virgaurea, Centaurea jacea, C. nigra, Senecio jacobaea, Artemisia campestris* (Asteraceae), *Erica tetralix, Calluna vulgaris* (Ericaceae) (United Kingdom: CREWE 1861c; BARRETT 1904; Germany: DIETZE 1874a, 1875a, 1875b, 1910, 1913; SÄLZL 1935; WEIGT 1991; WEGNER pers. comm.; France: MILLIÈRE 1870), LHOMME 1935; Denmark: JUUL 1948; Finland: SEPPÄNEN 1970; Russia: original data); also found on the flowers of *Scabiosa songarica* (Dipsacaceae), *Gypsophila altissima* (Caryophyllaceae), *Gentiana turkestanorum* (Gentianaceae), *Codonopsis clematidea* (Campanulaceae), *Ziziphora clinopodioides* (Lamiaceae), *Echinops tjanschanicus* and *Artemisia* (Asteraceae) (Kirghizstan (Tien-Shan Mts): original data).

PARASITOIDS: Hymenoptera, Ichneumonidae: *Dusona juvenilis, Mesochorus pectoralis* - Hymenoptera, Braconidae: *Apanteles cajae (=difficilis)* - Hymenoptera, Encyrtidae: *Copidosoma chalconotum* (THOMPSON 1945; ZWAKHALS pers. comm.).

HABITAT: Meadows, edges of forests, roadsides, beach ridges, marshes, parks. From 0 m up to about 1,600 m above sea-level in Apennines (Mt. Majella: ZAHM pers. comm.), to about 1,400-1,700 m in mountains of Balkan peninsula (REBEL 1917), to about 1,900-2,000 m in Alps (VORBRODT 1928; WEIGT 1991; REZBANYAI-RESER & MAIER 2000), and to about 2,400 m in Pyrenees (coll. FIBIGER).

SIMILAR SPECIES: A very variable species from white or pale greyish white with distinct transverse lines to darker, grey ground colour with inconspicuous markings on the forewings. In spite of its variability, usually the pale greyish white coloration with characteristic distinct dark border along costa and termen on the forewing and along termen on the hindwing, distinguish this species from *E. orphnata*, the only species with which it could really be confused and from any other European congeners (Text-fig. 83). The two dentate cornuti and one flat, irregular, curved cornutus in the vesica, the shape of sternum A8 in male, and the pear-shaped corpus bursae with characteristic disposition of spines in the female genitalia, make this an easy species to identify.

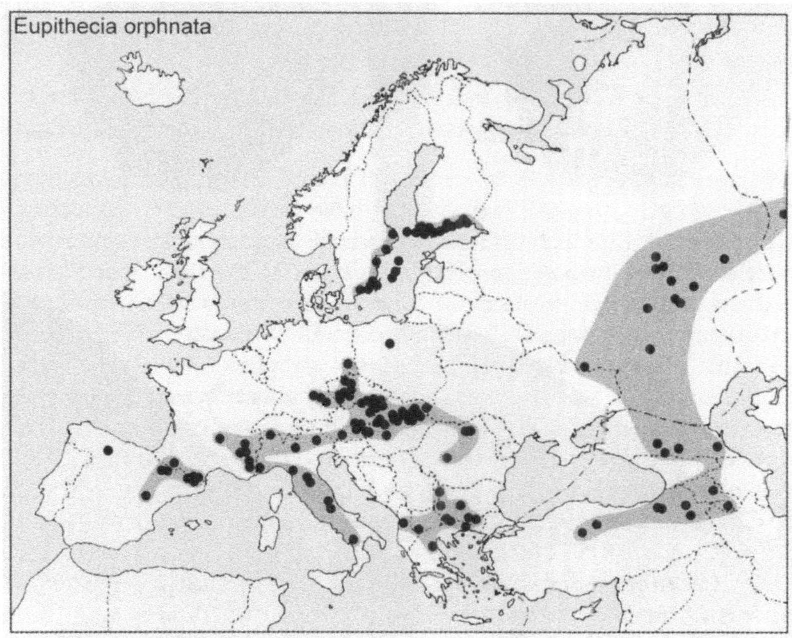

150. Eupithecia orphnata PETERSEN, 1909

Eupithecia orphnata PETERSEN, 1909: Dt. ent. Z. Iris 22 (4): 257, pl. 16: 67 ('Austro-Hungary'). Syntype(s).
Unavailable names (infrasubspecific): *orphnata* BOHATSCH (1883, as aberration of *Eupithecia scabiosata* BORKHAUSEN, 1794). - (incorrect subsequent spelling): *orphanata*: GANEV (1985).

DIAGNOSIS: Wingspan 16-21 mm. Labial palpi, frons, vertex and notum grey. Forewing with oblique transverse vein in areole; grey with reddish stalk of R1+5 vein and stalk of Cu (near M3-CuA1 base); transverse lines dark grey, slightly sinuate and indistinct, more strongly indicated at costa; postmedial line inconspicuous, usually right-angled near costa; pale wavy line rather broken and indistinct; discal spot large, black, rounded or ovate. Hindwing pale grey with posterior margin and terminal area darker; transverse lines inconspicuous, usually indicated on the veins; pale wavy line fine, sometimes indistinct; discal spot large, slightly smaller than on the forewing. Fringe grey, weakly spotted with dark grey at vein ends. Abdomen grey, usually with intensely reddish brown tinge; black lateral stripes often present.

MALE GENITALIA: Uncus narrow. Vesica armed with three large and three to four smaller

Text-figs 84-87: Diagnostic characters (indicated) of European *Eupithecia* species. Text-fig. 84: *E. virgaureata* DOUBLEDAY, 1861. Text-fig. 85: *E. lariciata* (FREYER, 1842). Text-fig. 86: *E. orphnata* PETERSEN, 1909. Text-fig. 87: *E. subfuscata* (HAWORTH, 1809).

straight horn-like cornuti connected at base and with one large, irregular, curved cornutus at ductus ejaculatorius base. Sternum A8 broadened near base, with two long, narrow apical rods with blunt, recurved apices; basal hollow shallow; apical hollow deep, length about 0.5-0.7 times sternite length.

FEMALE GENITALIA: Bursa copulatrix more than half covered by spines at base, with large, elongated, transverse patch of spines in ductus bursae; the main spiniferous area with characteristic, deep, V-shaped membranous gap on ventral side. Tergum A8 almost square.

DISTRIBUTION: West-Palaearctic. Sporadic and very local in Europe: Sweden, southern Finland, Poland, Czech Republic, Slovakia, Germany, Spain, France, Italy, Switzerland, Austria, Hungary, Romania, Bulgaria, Macedonia, Albania, Greece and south of European Russia. – Outside Europe in Asian Turkey, Caucasus, Transcaucasus, northern Kazakhstan and western Siberia (Kurghan province). The Central Asian subsp. *ferghanata* SCHÜTZE, 1956 occurs in the mountains: Tien-Shan Mts, Ghissar Mts, Darvaz Mts, Alai Mts, Ferghansky Mts, Dzhungarsky Alatau Mts).

PHENOLOGY: Univoltine: late June (early June) to mid-July (late July). Larva: early August (late July) to early September (early October). Overwintering as pupa.

BIOLOGY: Larva polyphagous (1). Found on leaves of *Solidago virgaurea,* flowers and seeds of *Centaurea, Picris, Artemisia* (Asteraceae) (Austria: DIETZE 1910, 1913), *Lavandula angustifolia* (Lamiaceae) (France (Alps): DIETZE 1913), *Centarium erythraea* (Gentianaceae) (France: LHOMME 1935), *Echium* (Boraginaceae) (Austria: WAGNER 1935), *Campanula rotundifolia* (Campanulaceae), *Tanacetum corymbosum* (Asteraceae) (Germany: SÄLZL 1935), leaves of *Quercus robur* (!) (Fagaceae), *Tanacetum vulgare, Artemisia campestris, A. vulgaris* var. *coarctata* (Asteraceae), *Silene vulgaris* (Caryophyllaceae), *Rumex crispus* (Polygonaceae), flowers of *Sonchus maritimus* (Asteraceae), seeds of *Veronica longifolia* (Scrophulariaceae), *Angelica archangelica* subsp. *litoralis* (Apiaceae) (Finland: KONTUNIEMI 1941), *Polygonum* (Polygonaceae), *Hypericum perforatum* (Clusiaceae), *Lysimachia* (Primulaceae), *Galium verum* (Rubiaceae), *Pimpinella* (Apiaceae), *Hieracium* (Asteraceae) (Finland: VALLE 1946), *Valeriana officinalis* (Valerianaceae), *Sonchus arvensis* (Asteraceae) (Finland: SEPPÄNEN 1970), *Knautia* (Dipsacaceae), *Senecio jacobaea* (Asteraceae) (Germany: WEIGT 1980), *Rumex acetosa, R. acetosella* (Polygonaceae), *Potentilla argentea* (Rosaceae), *Achillea millefolium* (Asteraceae) (Finland: LAASONEN & FIBIGER 1982), flowers of *Centranthus* (Valerianaceae), *Heracleum sphondylium* (Apiaceae), seeds of *Silene italica* (Caryophyllaceae) (France: WEIGT 1991).

HABITAT: Dry fringes and 'alvars', edge of forests, near the coast and on rocky coastal cliffs. From 0 m up to about 1,500-1,600 m above sea-level in Balkan peninsula (coll. ZISP; coll. BESHKOV), to about 1,600 m in Apennines (Mt. Majella: ZAHM pers. comm.), to about 1,700 m in Alps (VORBRODT 1928), and to about 1,800 m in Asia Minor (coll. SOMMERER).

SIMILAR SPECIES: This species is distinguished with difficulty from *E. subfuscata*: the narrower and more elongated wings, the larger, rounded discal spots, especially on the hindwings and the reddish brown tergum A2 provide useful guides (Text-fig. 86). The narrower and elongated valva, the tapered papillae on the ends of labides, presence of group of large, horn-like, basally united cornuti in the vesica in the male genitalia, and the elongated, rather pear-shaped corpus bursae with unique arrangement of spines in the female genitalia can be used to distinguish this species from *E. subfuscata* with accuracy.

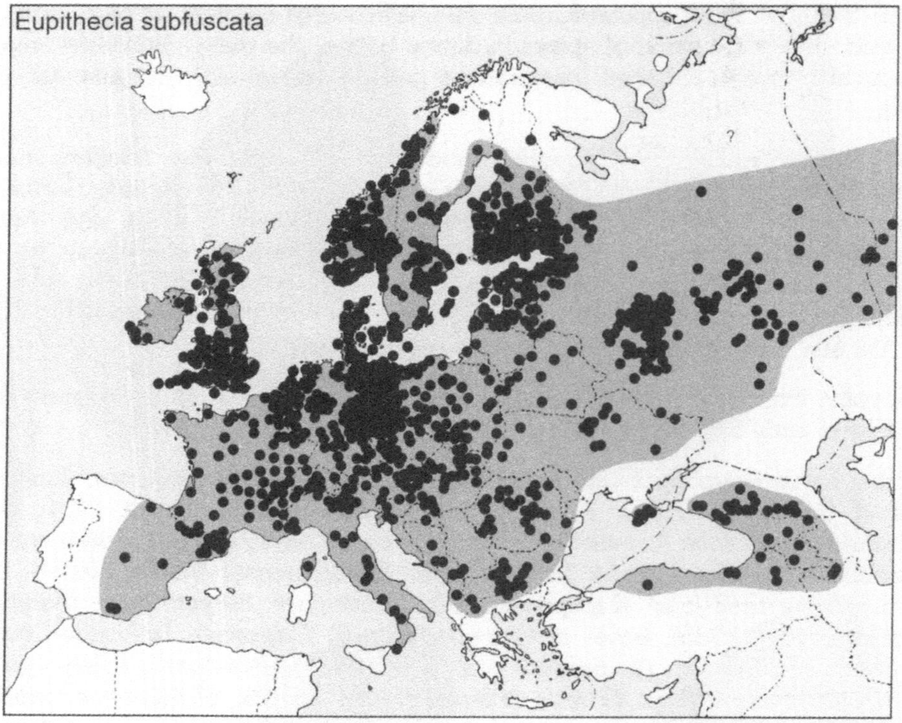

151. Eupithecia subfuscata (HAWORTH, 1809)

Phalaena subfuscata HAWORTH, 1809: Lepid. Br. (2): 360 (United Kingdom). Syntype(s) (BMNH).
Phalaena singulariata HAWORTH, 1809: Lepid. Br. (2): 360 (United Kingdom). Syntype(s) (BMNH).
Geometra castigata HÜBNER, 1813: Samml. Eur. Schmett. 5 Geometrae (2), pl. 89: 456 (Europe). Syntype(s) lost.
Eupithecia compressata GUENÉE, 1858: in BOISDUVAL & GUENÉE, Hist. nat. Insectes (Spec. gén. Lépid.) 10 (2): 324 (Type locality not stated, Europe ?). Holotype ♀.
Eupithecia implicata WALKER, 1862: List Specimens lepid. Insects. Colln. Br. Mus. 24: 1241 (Canada: Hudson's Bay, Albany River, St Martins Falls; Nova Scotia). Syntypes 8 ♀ (BMNH).
Eupithecia blancheata COOKE, 1881: The Entomologist 14 (213): 43 (United Kingdom: Isle of Man). Holotype ♀ (BMNH?).
Tephroclystis latipennis HULST, 1898: The Can. Ent. 30 (5): 114 (Canada: Quebec, Quebec City). Lectotype ♂ (AMNH).
Eup(ithecia) subfuscata sp./f./ab. *ussuriensis* DIETZE, 1910: Biol. Eupithecien 1, pl. 72: 314-316; ibidem (1913), 2: 94 (Russia: Ussuri, Kasakewitsch). Syntypes 21, including ♂ ♀ (MNHU). Valid at subspecific rank.
Eupithecia latimarginata MATSUMURA, 1925: J. Coll. Agric. Hokkaido Imp. Univ. 15 (3): 171, pl. 10: 17 (Japan: South-Saghalien, (Ichinosawa). Holotype ♀ (EIHU).
Eupithecia ichinosawana MATSUMURA, 1925: J. Coll. Agric. Hokkaido Imp. Univ. 15 (3): 172, pl. 11: 23 (Japan: South-Saghalien, (Ichinosawa). Holotype ♀ (EIHU).
Unavailable names (misidentification): *residuata*: sensu EVERSMANN (1844) nec HÜBNER, 1817 (identity: *Eupithecia lariciata*). – (infrasubspecific): *obscura*: DIETZE (1910); *obscurissima*: PROUT (1914). - (incorrect subsequent spellings): *castigaria*: BOISDUVAL (1840); *eastigata*: FASSNIDGE (1926).

DIAGNOSIS: Wingspan 16-24 mm. Labial palpi dark grey. Frons, vertex and notum pale grey to dark grey. Forewing grey; transverse lines variably expressed, usually indistinct, slightly sinuate, broadened near costa, dark grey; medial line often more conspicuous, sharply angled near costa beyond discal spot or just touching discal spot; postmedial line usually also well defined, slightly sinuate and evenly curved near costa, sometimes with dark dashes on the veins; terminal area with ill-defined, unbroken, pale wavy line and with a weak pale tornal spot; discal spot small but distinct, rounded or ovate, black. Hindwing slightly paler, with more or less conspicuous dark and light transverse lines, especially dark postmedial line and lighter double stripe beyond it; terminal area always darker with more or less distinct, fine, pale wavy line and with minute pale tornal spot near end of CuA2; discal spot minute, rounded, dark. Fringe chequered pale grey and dark grey. Abdomen grey or greyish fuscous, often with weak reddish tinge and with black lateral spots or stripes; the last segment usually paler than the others in male.

MALE GENITALIA: Uncus narrow. Vesica armed with one small, elongated and folded U-shaped cornutus and one irregular, curved cornutus at ductus ejaculatorius base, without dentate cornuti. Sternum A8 with deep basal and apical hollows. two narrow apical rods with apices blunt.

FEMALE GENITALIA: Bursa copulatrix spherical, completely covered with spines. Ductus bursae short and narrow, with a small patch of spines.

DISTRIBUTION: Holarctic. Widely distributed in Europe except Iceland, Portugal, Mediterranean islands (present in Corsica) and south of Balkan peninsula. – Outside Europe in north of Asian Turkey, Caucasus, Transcaucasus, northern and south-eastern Kazakhstan, Central Asia (northern Tien-Shan Mts), Altai, western and southern Siberia, north-western China, Korea and Japan. Widespread in North America (BOLTE 1990).

PHENOLOGY: Uni- or probably bivoltine in the south: mid-May (early April) to late July (late September). Larva: mid-July (mid-June) to late August (late October). Overwintering as pupa in cocoon in the ground.

BIOLOGY: Larva polyphagous (2). Found on the flowers, seeds and leaves of various mainly herbaceous plants, such as *Centaurea jacea, Solidago virgaurea, Achillea millefolium, A. ptarmica, Carlina acaulis, Aster, Eupatorium cannabinum, Senecio cordatus, Cirsium arvense, Tanacetum vulgare, Artemisia vulgaris, A. campestris* (Asteraceae), *Valeriana officinalis* (Valerianaceae), *Angelica sylvestris, Heracleum, Aegopodium podagraria, Pimpinella saxifraga* (Apiaceae), *Campanula rotundifolia, Phyteuma* (Campanulaceae), *Ononis* (Fabaceae), *Gentiana lutea, G. vulgaris, G. pneumonanthe* (Gentianaceae), *Galium mollugo, G. sylvaticum* (Rubiaceae). *Crataegus, Rubus idaeus, Filipendula ulmaria* (Rosaceae), *Clematis vitalba, Aconitum lycoctonum* (Ranunculaceae), *Dianthus* (Caryophyllaceae), *Ononis* (Fabaceae), *Hypericum maculatum* (Clusiaceae), *Epilobium angustifolium* (Onagraceae), *Knautia, Scabiosa, Succisa pratensis* (Dipsacaceae), *Hyssopus, Stachys officinalis, Origanum vulgare* (Lamiaceae), *Melampyrum sylvaticum, Euphrasia* spp. (Scrophulariaceae), *Lysimachia vulgaris* (Primulaceae), *Urtica dioica* (Urticaceae), *Ledum palustre, Calluna vulgaris* (Ericaceae), *Frangula alnus* (Rhamnaceae), *Alnus incana, Betula* (Betulaceae), *Salix phylicifolia* (Salicaceae), *Pteridium aquilinum* (Polypodiaceae), *Juncus conglomeratus* (Juncaceae) and *Dictamnus* (Rutaceae) (Germany: DIETZE 1871b, 1872, 1874a, 1875a, 1910, 1913; WEIGT 1991; WEGNER pers. comm.; France: LHOMME 1935; Denmark: JUUL 1948; Finland: SEPPÄNEN 1970; Austria: HABELER 1971; Russia: original data). Found also on *Picea* (Pinaceae), *Alnus rubra* (Betulaceae), *Populus balsamifera, Salix* (Salicaceae) and *Epilobium* (Onagraceae) (Canada: JONES 1951; FERGUSON 1955; ROSS & EVANS 1956; MCGUFFIN 1958).

PARASITOIDS: Hymenoptera, Ichneumonidae: *Mesochorus semirufus* (hyperparasite on *Agrypon clandestinum*), *Netelia tarsata*, *Agrypon clandestinum* - Hymenoptera, Braconidae: *Aleiodes (=Rogas) circumscriptus*, *Apanteles juniperatae*, *A. placidus*, *Microplitis tuberculifer* - Hymenoptera, Encyrtidae: *Tyndarichus scaurus* (THOMPSON 1945; SHENEFELT 1972, 1973, 1975; ZWAKHALS pers. comm.).

HABITAT: Nearly all types of biotopes. From 0 m up to about 1,600 m above sea-level in Apennines (Mt. Majella: ZAHM pers. comm.), to about 1,700-1,800 m in Balkan peninsula and Asia Minor (REBEL 1917; coll. SOMMERER; coll. GELBRECHT), to about 2,000 m in Alps (WEIGT 1991), and to about 2,000-2,100 m in Pyrenees (coll. FIBIGER).

SIMILAR SPECIES: This variable species could be confused with other *Eupithecia* species, such as *E. lariciata, E. orphnata* and *E. subumbrata*. However, *E. lariciata* has a darker area between antemedial and medial lines, more sharply angled postmedial line, larger discal spot on the forewing and usually a distinct white dot on the end of notum; *E. orphnata* has more elongated and pointed forewings, more sharply angled postmedial line on the forewing and larger discal spots on the hindwings; lastly *E. subumbrata* has more elongated and pointed forewings usually with whitish coloration and with stalks of R and Cu covered with reddish scales (Text-fig. 87). The wide valva with pointed apex, the characteristic cornuti on the vesica and the shape of sternum A8 in the male; the spherical bursa copulatrix completely covered with spines, very short and narrow ductus bursae with one small patch of spines, very long and relatively thick, evenly curved ductus seminalis in the female genitalia should distinguish this species.

Colour plates

PLATE 1 (all specimens natural size)

1a. ***Pseudobaptria bogumilaria* Rbl.**, ♂, 'N. Bosnia' (ZISP)
1b. *P. bogumilaria* Rbl. f. *eurytaenia* Rbl., ♂, Suha, Bosnia (ZMKU)
1c. *P. bogumilaria* Rbl. f. *eurytaenia* Rbl., ♀, [Gjalica e] Lumës, Albania (coll. SKOU)
1d. *P. bogumilaria* Rbl. f. *montenegrica* nova, ♂, Durmitor, Montenegro (ZSM)
1e. *P. bogumilaria* Rbl. f. *montenegrica* nova, ♀, Durmitor, Montenegro (ZSM)
2a. ***Mesotype didymata* L.**, ♂, Tåsinge, Fyn, Denmark (coll. SKOU)
2b. *M. didymata* L., ♂, St. Petersburg obl., NW. European Russia (ZISP)
2c. *M. didymata* L., ♂, Södermanland, C. Sweden (coll. SKOU)
2d. *M. didymata* L., ♀, St. Petersburg obl., NW. European Russia (ZISP)
2e. *M. didymata hethlandica* Rbl., ♂, Norðskáli, Faroe Islands (ZMUC)
2f. *M. didymata hethlandica* Rbl., ♀, Norðskáli, Faroe Islands (ZMUC)
3a. ***Mesotype parallelolineata* Retz.**, ♂, St. Petersburg obl., NW. European Russia (ZISP)
3b. *M. parallelolineata* Retz., ♂, St. Petersburg obl., NW. European Russia (ZISP)
3c. *M. parallelolineata* Retz., ♂, Göteborg, S. Sweden (ZSM)
3d. *M. parallelolineata* Retz., ♀, St. Petersburg obl., NW. European Russia (ZISP)
3e. *M. parallelolineata* Retz., ♀, Flies, N. Tirol, Austria (coll. SKOU)
4a. ***Mesotype verberata* Scop.**, ♂, Col de la Madeleine, Hautes Alpes, SE. France (ZSM)
4b. *M. verberata* Scop., ♂, Gènova, Liguria, N. Italy (coll. SKOU)
4c. *M. verberata* Scop., ♂, Nauders, W. Tirol, Austria (ZMUC)
4d. *M. verberata* Scop., ♀, Lienz, E. Tirol, Austria (ZMUC)
4e. *M. verberata* Scop., ♀, Schnalstal, Trentino-Alto Adige, NE. Italy (coll. SKOU)
5a. ***Perizoma affinitata* Stph.**, ♂, Hamrafjellet, Härjedalen, C. Sweden (ZMUC)
5b. *P. affinitata* Stph., ♀, Hamrafjellet, Härjedalen, C. Sweden (ZMUC)
5c. *P. affinitata* Stph., ♀, Snarup, Fyn, Denmark (coll. SKOU)
5d. *P. affinitata* Stph., ♀, Mecklenburg, N. Germany (coll. GELBRECHT), reared
5e. *P. affinitata* Stph., ♀, Melsomvik, Vestfold, S. Norway (coll. SKOU)
6a. ***Perizoma alchemillata* L.**, ♀, St. Petersburg obl., NW. European Russia (ZISP)
6b. *P. alchemillata* L., ♀, St. Petersburg obl., NW. European Russia (ZISP)
6c. *P. alchemillata* L., ♀, St. Petersburg obl., NW. European Russia (ZISP)
6d. *P. alchemillata* L., ♀, Ræveskov, Fyn, Denmark (coll. SKOU)
6e. *P. alchemillata* L., ♀, Brandenburg, E. Germany (coll. GELBRECHT), reared
6f. *P. alchemillata* L., ♀, Brandenburg, E. Germany (coll. GELBRECHT)
7a. ***Perizoma hydrata* Tr.**, ♀, Glocknergruppe, E. Tirol, Austria (ZMUC)
7b. *P. hydrata* Tr., ♀, Kauns, N. Tirol, Austria (ZSM)
7c. *P. hydrata* Tr., ♀, Bessans, Haute Savoie, E. France (coll. SKOU)
7d. *P. hydrata* Tr., ♀, Freiburg i Br., SW. Germany (ZSM)
7e. *P. hydrata* Tr., ♂, Wallersberg, Jura, S. Germany (ZSM)
7f. *P. hydrata* Tr., ♂, St. Petersburg obl., NW. European Russia (ZISP)
8a. ***Perizoma lugdunaria* H.-S.**, ♂, Balaton – Szentgyörgy, Hungary (ZSM)
8b. *P. lugdunaria* H.-S., ♀, Lozoya, Madrid, C. Spain (coll. SKOU)
8c. *P. lugdunaria* H.-S., ♀, Magdeburg, E. Germany (coll. GELBRECHT), reared
8d. *P. lugdunaria* H.-S., ♂, Bamberg, C. Germany (coll. SKOU), reared
8e. *P. lugdunaria* H.-S., ♂, Bamberg, C. Germany (coll. GELBRECHT), reared
8f. *P. lugdunaria* H.-S., ♀, Vienna, Austria (ZSM), reared
9a. ***Perizoma bifaciata* Hw.**, ♂, Voronezh obl., S. European Russia (ZISP)
9b. *P. bifaciata* Hw., ♀, Lübeck, Schleswig-Holstein, N. Germany (coll. WEGNER), reared
9c. *P. bifaciata* Hw., ♀, Kirkeby Sand, Fyn, Denmark (coll. SKOU)
9d. *P. bifaciata* Hw., ♂, Skallingen, Jylland, Denmark (coll. SKOU)
9e. *P. bifaciata* Hw., ♂, Albarracin, Teruel, C. Spain (ZISP)
9f. *P. bifaciata* Hw., ♂, Ispir, Erzurum, NE. Turkey (coll. GELBRECHT)
10a. ***Perizoma flavosparsata* Wagn.**, ♂, Albarracin, Teruel, C. Spain (coll. SKOU)
10b. *P. flavosparsata* Wagn., ♂, Tramacastilla, Teruel, C. Spain (coll. SKOU)
10c. *P. flavosparsata* Wagn., ♂, Albarracin, Teruel, C. Spain (coll. SKOU)
10d. *P. flavosparsata* Wagn., ♂, Albarracin, Teruel, C. Spain (coll. SKOU)
10e. *P. flavosparsata uniformata* Metzn., ♂, Skopje, Macedonia (coll. SKOU), paratype
10f. *P. flavosparsata uniformata* Metzn., ♂, Skopje, Macedonia (coll. SKOU), paratype
11a. ***Perizoma minorata* Tr.**, ♀, Briançon, Hautes Alpes, SE. France (coll. FIBIGER)
11b. *P. minorata* Tr., ♂, Vorochta, E. Carpathians, W. Ukraine (ZISP)
11c. *P. minorata norvegica* Prt., ♀, Kongsvold, Sør-Trøndelag, C. Norway (ZMUC)
11d. *P. minorata norvegica* Prt., ♀, Utsjoki, Inarin Lappi, N. Finland (ZMH)
11e. *P. minorata ericetata* Stph., ♂, 'Scotland', United Kingdom (ZISP)
11f. *P. minorata ericetata* Stph., ♀, 'Scotland', United Kingdom (ZISP)
12a. ***Perizoma blandiata* Den. & Schiff.**, ♀, Vipiteno, S. Tirol, N. Italy (ZMUC)
12b. *P. blandiata* Den. & Schiff., ♂, Vipiteno, S. Tirol, N. Italy (ZMUC)
12c. *P. blandiata* Den. & Schiff., ♂, Svenskehavn, Bornholm, Denmark (coll. SKOU)
12d. *P. blandiata* Den. & Schiff., ♀, Lodochnyi, Kola Peninsula, N. European Russia (ZISP)
12e. *P. blandiata perfasciata* Prt., ♂, Skaftafell, S. Iceland (ZMUC)
12f. *P. blandiata perfasciata* Prt., ♀, Skaftafell, S. Iceland (ZMUC)

1a
1b
1c
1d
1e
2a
2b
2c
2d
2e
2f
3a
3b
3c
3d
3e
4a
4b
4c
4d
4e
5a
5b
5c
5d
5e
6a
6b
6c
6d
6e
6f
7a
7b
7c
7d
7e
7f
8a
8b
8c
8d
8e
8f
9a
9b
9c
9d
9e
9f
10a
10b
10c
10d
10e
10f
11a
11b
11c
11d
11e
11f
12a
12b
12c
12d
12e
12f

PLATE 2 (13-18 in natural size, 19-23 in 1.5 x natural size)

13a. ***Perizoma albulata* Den. & Schiff.,** ♀, Colmars, Haute Alpes, SE. France (coll. GELBRECHT)
13b. *P. albulata* Den. & Schiff., ♂, Falakron Oros, N. Greece (coll. SKOU)
13c. *P. albulata* Den. & Schiff., ♂, St. Petersburg obl., NW. European Russia (ZISP)
13d. *P. albulata* Den. & Schiff., ♂, Storsätern-Slätten, Dalarne, C. Sweden (ZMUC)
13e. *P. albulata thules* Weir, ♀, Eiði, Faroe Islands (ZMUC)
13f. *P. albulata thules* Weir, ♀, Eiði, Faroe Islands (ZMUC)
14a. ***Perizoma flavofasciata* Thunb.,** ♀, Olonetz, Karelia, NW. European Russia (ZISP)
14b. *P. flavofasciata* Thunb., ♂, St. Petersburg obl., NW. European Russia (ZISP)
14c. *P. flavofasciata* Thunb., ♂, Østjylland, Denmark (coll. SKOU)
14d. *P. flavofasciata* Thunb., ♀, Brandenburg, E. Germany (coll. GELBRECHT)
14e. *P. flavofasciata* Thunb., ♂, Brandenburg, E. Germany (coll. GELBRECHT)
15a. ***Perizoma obsoletata* H.-S.,** ♂, Prebichl, Austria (ZSM)
15b. *P. obsoletata* H.-S., Lowerazl, N. Tirol, Austria (ZMUC)
15c. *P. obsoletata* H.-S., ♀, Aosta-Tal, Graische Alpen, NW. Italy (coll. GELBRECHT)
15d. *P. obsoletata* H.-S., ♀, Aosta-Tal, Graische Alpen, NW. Italy (coll. GELBRECHT)
15e. *P. obsoletata* H.-S., ♀, Aosta-Tal, Graische Alpen, NW. Italy (coll. GELBRECHT)
16a. ***Perizoma incultaria* H.-S.,** ♀, Colmars, Haute Alpes, SE. France (coll. GELBRECHT)
16b. *P. incultaria* H.-S., ♀, Colmars, Haute Alpes, SE. France (coll. GELBRECHT)
16c. *P. incultaria* H.-S., ♀, Colmars, Haute Alpes, SE. France (coll. GELBRECHT)
16d. *P. incultaria* H.-S., ♀, Passo Misurina, Friuli, NE. Italy (coll. SKOU)
16e. *P. incultaria* H.-S., ♀, Paznaun, Voralberg, Austria (ZSM)
16f. *P. incultaria* H.-S., ♀, Engadin, Switzerland (ZSM)
17a. ***Martania taeniata* Stph.,** ♂, Lalm, Norway (ZMUC)
17b. *M. taeniata* Stph., ♂, Ålesund, C. Norway (ZMUC)
17c. *M. taeniata* Stph., ♂, Nås, Dalarna, C. Sweden (ZMUC)
17d. *M. taeniata* Stph., ♀, Erben, Uppland, C. Sweden (coll. SKOU)
17e. *M. taeniata* Stph., ♀, Østjylland, Denmark (coll. SKOU)
17f. *M. taeniata* Stph., ♀, St. Petersburg obl., NW. European Russia (ZISP)
18a. ***Gagitodes sagittata* F.,** ♂, Brandenburg, E. Germany (coll. GELBRECHT), reared
18b. *G. sagittata* F., ♂, Novgorod obl., NW. European Russia (ZISP), reared
18c. *G. sagittata* F., ♀, Brandenburg, E. Germany (coll. GELBRECHT), reared
18d. *G. sagittata* F., ♀, Lüchow-Dannenberg, N. Germany (coll. WEGNER), reared
18e. *G. sagittata* F., ♀, St. Petersburg obl., NW. European Russia (ZISP)
19a. ***Gymnoscelis rufifasciata* Hw.,** ♀, Tivoli, C. Italy (ZSM)
19b. *G. rufifasciata* Hw., ♀, Istria, Croatia (ZSM), reared
19c. *G. rufifasciata* Hw., ♀, Torö, Södermanland, C. Sweden (coll. SKOU)
19d. *G. rufifasciata* Hw., ♂, Mti. Simbruini, C. Italy (ZSM)
19e. *G. rufifasciata* Hw., ♀, Samara obl., SE. European Russia (ZISP)
20a. ***Chloroclystis v-ata* Hw.,** ♀, Kiev obl., N. Ukraine (ZMKU)
20b. *C. v-ata* Hw., ♀, Aachen, W. Germany (ZSM)
20c. *C. v-ata* Hw., ♂, Kassel, C. Germany (ZSM), reared
20d. *C. v-ata* Hw., ♀, Kassel, C. Germany (ZSM), reared
20e. *C. v-ata* Hw., ♀, Kirkeby Sand, Fyn, Denmark (coll. SKOU)
21a. ***Pasiphila chloerata* Mab.,** ♀, Stettin (Szczecin), NW. Poland (ZISP)
21b. *P. chloerata* Mab., ♀, St. Petersburg obl., NW. European Russia (ZISP)
21c. *P. chloerata* Mab., ♀, Kassel, C. Germany (ZSM), reared
21d. *P. chloerata* Mab., ♂, Kassel, C. Germany (ZSM), reared
21e. *P. chloerata* Mab., ♀, Lüneburg, N. Germany (coll. WEGNER), reared
22a. ***Pasiphila rectangulata* L.,** ♀, 'Uppland', C. Sweden (coll. SKOU)
22b. *P. rectangulata* L., ♀, Kelheim, Bavaria, S. Germany (ZSM)
22c. *P. rectangulata* L., ♀, Säftekär, Öland, S. Sweden (coll. SKOU)
22d. *P. rectangulata* L., ♂, Kelheim, Bavaria, S. Germany (ZSM)
22e. *P. rectangulata* L. f. *anthrax* Dtze., ♀, New Forest, England, United Kingdom (ZSM)
23a. ***Pasiphila debiliata* Hbn.,** ♀, Södermanland, C. Sweden (coll. SKOU)
23b. *P. debiliata* Hbn., ♀, St. Petersburg obl., NW. European Russia (ZISP)
23c. *P. debiliata* Hbn., ♀, St. Petersburg obl., NW. European Russia (ZISP)
23d. *P. debiliata* Hbn., ♂, Løvenholm, Østjylland, Denmark (coll. SKOU), reared
23e. *P. debiliata* Hbn., ♀, St. Petersburg obl., NW. European Russia (ZISP), reared

13a
13b
13c
13d
13e
13f
14a
14b
14c
14d
14e
15a
15b
15c
15d
15e
16a
16b
16c
16d
16e
16f
17a
17b
17c
17d
17e
17f
18a
18b
18c
18d
18e
19a
19b
19c
19d
19e
20a
20b
20c
20d
20e
21a
21b
21c
21d
21e
22a
22b
22c
22d
22e
23a
23b
23c
23d
23e

PLATE 3 (all specimens 1.5 x natural size)

24a. ***Eupithecia haworthiata* Dbld.,** ♀, Brandenburg, E. Germany (coll. GELBRECHT), reared
24b. *E. haworthiata* Dbld., ♀, Brandenburg, E. Germany (coll. GELBRECHT), reared
24c. *E. haworthiata* Dbld., ♀, Yalta, Crimea, S. Ukraine (ZISP)
24d. *E. haworthiata* Dbld., ♀, Hagaparken, Öland, S. Sweden (coll. SKOU), reared
24e. *E. haworthiata* Dbld., ♀, Hagaparken, Öland, S. Sweden (coll. SKOU), reared
25a. ***Eupithecia tenuiata* Hbn.,** ♀, Stengade, Funen, Denmark (coll. SKOU)
25b. *E. tenuiata* Hbn., ♀, 'Bavaria', S. Germany (ZISP)
25c. *E. tenuiata* Hbn., ♂, Chernobyl, Kiev obl., N. Ukraine (ZMKU)
25d. *E. tenuiata* Hbn., ♀, Rilagebirge, W. Bulgaria (coll. FRIEDRICH)
25e. *E. tenuiata* Hbn., ♀, Shimshak, Elburs Mts., N. Iran (HNHM)
26a. ***Eupithecia inturbata* Hbn.,** ♂, Regensburg, SE. Germany (ZSM)
26b. *E. inturbata* Hbn., ♀, Regensburg, SE. Germany (ZSM), reared
26c. *E. inturbata* Hbn., ♀, Västerstad, Öland, S. Sweden (coll. SKOU)
26d. *E. inturbata* Hbn., ♀, Svedmyra, Södermanland, C. Sweden (coll. SKOU)
26e. *E. inturbata* Hbn., ♀, Stengade, Fyn, Denmark (coll. SKOU)
27a. ***Eupithecia abietaria* Goeze,** ♀, Moscow obl., C. European Russia (ZISP)
27b. *E. abietaria* Goeze, ♂, St. Petersburg obl., NW. European Russia (ZISP)
27c. *E. abietaria* Goeze, ♀, Norrby, Uppland, C. Sweden (coll. SKOU)
27d. *E. abietaria* Goeze, ♀, Bialowieza, E. Poland (coll. SKOU)
28a. ***Eupithecia analoga* Djak.,** ♀, St. Petersburg obl., NW. European Russia (ZISP)
28b. *E. analoga* Djak., ♀, Furnvik, Gästrikland, C. Sweden (coll. SKOU)
28c. *E. analoga* Djak., ♂, Västerås, Västmandland, C. Sweden (coll. SKOU)
28d. *E. analoga* Djak., ♀, Moscow obl., C. European Russia (ZISP)
28e. *E. analoga* Djak., ♀, Vorochta, E. Carpathians, W. Ukraine (ZISP)
29a. ***Eupithecia linariata* Den. & Schiff.,** ♀, Åsgårdstrand, Vestfold, S. Norway (coll. SKOU)
29b. *E. linariata* Den. & Schiff., ♀, Kiev obl., N. Ukraine (ZMKU)
29c. *E. linariata* Den. & Schiff., ♀, Midskov, Fyn, Denmark (coll. SKOU), reared
29d. *E. linariata* Den. & Schiff., ♂, Brandenburg, E. Germany (coll. GELBRECHT), reared
29e. *E. linariata* Den. & Schiff., ♀, Tver' obl., C. European Russia (ZISP), reared
30a. ***Eupithecia pulchellata* Stph.,** ♀, 'Scotland', United Kingdom (ZSM), reared
30b. *E. pulchellata* Stph., ♂, 'Scotland', United Kingdom (ZSM), reared
30c. *E. pulchellata intermedia* Dtze., ♀, Amfreville, Eure, N. France (ZSM/HERBULOT)
30d. *E. pulchellata intermedia* Dtze., ♂, Kassel, C. Germany (ZSM), reared
30e. *E. pulchellata intermedia* Dtze., f. *iberica* Dtze., ♀, Albarracin, Teruel, C. Spain (ZSM)
31a. ***Eupithecia pyreneata* Mab.,** ♀, Gitomir, C. Ukraine (ZMKU)
31b. *E. pyreneata* Mab., ♀, Schwarzwald, Wutach-Gebiet, SW. Germany (ZSM), reared
31c. *E. pyreneata* Mab., ♀, Varna, E. Bulgaria (coll. GELBRECHT), reared
31d. *E. pyreneata* Mab., f. *granadensis* Bub., ♂, 'Granada', S. Spain (ZSM), cotype
31e. *E. pyreneata* Mab., f. *granadensis* Bub., ♀, 'Granada', S. Spain (ZSM), cotype
32a. ***Eupithecia laquaearia* H.-S.,** ♀, Albarracin, Teruel, C. Spain (coll. SKOU)
32b. *E. laquaearia* H.-S., ♀, Albarracin, Teruel, C. Spain (coll. SKOU)
32c. *E. laquaearia* H.-S., ♀, Moscow obl., C. European Russia (ZISP)
32d. *E. laquaearia* H.-S., ♂, Heidelberg, W. Germany (ZSM), reared
32e. *E. laquaearia* H.-S., ♀, Kassel, C. Germany (ZSM), reared
33a. ***Eupithecia ultimaria* Bsd.,** ♀, Bilbao, N. Spain (ZSM)
33b. *E. ultimaria* Bsd., ♀, Rosas, Catalonia, N. Spain (ZSM)
33c. *E. ultimaria* Bsd., ♀, Puyvert, Vaucluse, SE. France (ZSM/HERBULOT)
33d. *E. ultimaria* Bsd., ♀, Puyvert, Vaucluse, SE. France (ZSM/HERBULOT)
33e. *E. ultimaria* Bsd., ♀, Rhodes, E. Greece (ZMUC)

24a
24b
24c
24d
24e
25a
25b
25c
25d
25e
26a
26b
26c
26d
26e
27a
27b
27c
27d
28a
28b
28c
28d
28e
29a
29b
29c
29d
29e
30a
30b
30c
30d
30e
31a
31b
31c
31d
31e
32a
32b
32c
32d
32e
33a
33b
33c
33d
33e

PLATE 4 (all specimens 1.5 x natural size)

34a. ***Eupithecia usbeca* Viid.**, ♂, Astrakhan' obl., SE. European Russia (ZISP)
34b. *E. usbeca* Viid., ♀, Astrakhan' obl., SE. European Russia (ZISP)
34c. *E. usbeca* Viid., ♀, Repetek Nature Reserve, E. Turkmenistan (ZISP)
34d. *E. usbeca* Viid., ♀, Kushka, S. Turkmenistan (ZISP)
34e. *E. usbeca* Viid., ♀, Kushka, S. Turkmenistan (ZISP), paratype
35a. ***Eupithecia minusculata* Alph.**, ♂, Astrakhan' obl., SE. European Russia (ZISP)
35b. *E. minusculata* Alph., ♂, Kalmykia, SE. European Russia (ZISP)
35c. *E. minusculata* Alph., ♀, Tabernas, Almeria, S. Spain (coll. SKOU)
35d. *E. minusculata* Alph., ♀, Orgiva, Andalusia, S. Spain (coll. SKOU), [*herrenschmidti* MENTZNER & MOBERG, paratype]
35e. *E. minusculata* Alph., ♀, El Pozo del Esparto, Almeria, S. Spain (coll. SKOU)
36a. ***Eupithecia pantellata* Mill.**, f. *andalusica* Whrli., ♂, Ronda, Andalusia, S. Spain (ZSM)
36b. *E. pantellata* Mill., f. *luteostrigata* Stgr., ♂, S. Martino d. Scale, Sicily (ZSM)
36c. *E. pantellata* Mill., f. *illuminata* Joan., ♀, Skikda (=Philippeville), NE. Algeria (ZSM), reared
36d. *E. pantellata* Mill., f. *illuminata* Joan., ♀, Skikda (=Philippeville), NE. Algeria (ZMKU)
36e. *E. pantellata* Mill., f. *illuminata* Joan., ♀, Skikda (=Philippeville), NE. Algeria (ZSM), reared
37a. ***Eupithecia plumbeolata* Hw.**, ♂, St. Petersburg obl., NW. European Russia (ZISP)
37b. *E. plumbeolata* Hw., ♂, St. Petersburg obl., NW. European Russia (ZISP)
37c. *E. plumbeolata* Hw., ♀, St. Petersburg obl., NW. European Russia (ZISP)
37d. *E. plumbeolata* Hw., ♀, Novgorod obl., NW. European Russia (ZISP)
37e. *E. plumbeolata* Hw., ♂, Bialowieza, E. Poland (coll. SKOU)
37f. *E. plumbeolata reikjavikaria* Stgr., ♂, Skaftafell, S. Iceland (ZMUC)
37g. *E. plumbeolata reikjavikaria* Stgr., ♀, Skaftafell, S. Iceland (ZMUC)
37h. *E. plumbeolata lutosaria* Boh., ♀, Kislovodsk, N. Caucasus, Russia (ZMKU)
37i. *E. plumbeolata lutosaria* Boh., ♂, Daghestan, N. Caucasus, Russia (ZISP)
37j. *E. plumbeolata lutosaria* Boh., ♂, Kislovodsk, N. Caucasus, Russia (ZMKU)
38a. ***Eupithecia pygmaeata* Hbn.**, ♀, Novgorod obl., NW. European Russia (ZISP)
38b. *E. pygmaeata* Hbn., ♂, St. Petersburg obl., NW. European Russia (ZISP)
38c. *E. pygmaeata* Hbn., ♀, Brandenburg, E. Germany (coll. GELBRECHT)
38d. *E. pygmaeata* Hbn., ♂, Mannersdorf, Germany (ZSM)
38e. *E. pygmaeata* Hbn., ♀, Mannersdorf, Germany (ZSM), reared
39a. ***Eupithecia undata* Frr.**, ♂, Graubünden, Switzerland (ZSM)
39b. *E. undata* Frr., ♂, Graubünden, Switzerland (ZSM)
39c. *E. undata* Frr., ♂, Pitztal, N. Tirol, Austria (ZSM/HERBULOT)
39d. *E. undata* Frr., ♀, Pitztal, N. Tirol, Austria (ZSM/HERBULOT)
39e. *E. undata* Frr., ♀, Mt. Triglav, Julische Alpen, Slovenia (ZSM), reared
40a. ***Eupithecia fennoscandica* Knb.**, ♂, Kilpisjärvi, Saana, N. Finland (ZMH)
40b. *E. fennoscandica* Knb., ♂, Reg. alp., Saana, N. Finland (ZMH)
40c. *E. fennoscandica* Knb., ♀, Snorretjåkko, Torne Lappmark, N. Sweden (coll. SKOU)
40d. *E. fennoscandica* Knb., ♂, Annjalonji, EnL, N. Finland (ZMUC)
40e. *E. fennoscandica* Knb., ♂, Kilpisjärvi, Saana, N. Finland (coll. SKOU)
41a. ***Eupithecia silenata* Assm.**, ♂, Rappensee, Allgäu, S. Germany (ZSM)
41b. *E. silenata* Assm., ♂, Riesengebirge, E. Germany (ZSM), reared
41c. *E. silenata* Assm., ♀, Riesengebirge, E. Germany (ZSM), reared
41d. *E. silenata* Assm., ♂, Witoscha near Sofia, W. Bulgaria (coll. GELBRECHT), reared
41e. *E. silenata* Assm., ♀, Witoscha near Sofia, W. Bulgaria (coll. GELBRECHT), reared
42a. ***Eupithecia carpophagata* Stgr.**, ♂, Mt. Paradiso, S. Italy (ZSM)
42b. *E. carpophagata* Stgr., ♀, Pescocostanzo, Abruzzi, C. Italy (ZSM), reared
42c. *E. carpophagata* Stgr., ♀, Casteil, Pyrénées-Orientales, S. France (ZSM/HERBULOT)
42d. *E. carpophagata* Stgr., ♀, Matka, Treska Valley, Macedonia (ZSM)
42e. *E. carpophagata* Stgr., ♂, Spanopole, Pirin (Macedonia), SW. Bulgaria (ZSM)

34a
34b
34c
34d
34e
35a
35b
35c
35d
35e
36a
36b
36c
36d
36e
37a
37b
37c
37d
37e
37f
37g
37h
37i
37j
38a
38b
38c
38d
38e
39a
39b
39c
39d
39e
40a
40b
40c
40d
40e
41a
41b
41c
41d
41e
42a
42b
42c
42d
42e

PLATE 5 (all specimens 1.5 x natural size)

43a. ***Eupithecia venosata* F.,** ♀, Brandenburg, E. Germany (coll. GELBRECHT)
43b. *E. venosata* F., ♂, Romalt, Jutland, Denmark (coll. SKOU)
43c. *E. venosata* F., ♀, Schleswig-Holstein, N. Germany (coll. WEGNER), reared
43d. *E. venosata* F., ♀, Chemnitz, Sachsen, E. Germany (ZSM), reared
43e. *E. venosata ochracae* Greg., ♂, Hoy Isl., Orkney Islands, United Kingdom (ZISP)
43f. *E. venosata fumosae* Greg., ♀, Unst Isl., Shetland Islands, United Kingdom (ZSM)
43g. *E. venosata hebridensis* Curt., ♀, Lewis Isl., Outer Hebrides, United Kingdom (ZISP)
43h. *E. venosata plumbea* Hugg., ♀, Slea Head, Co. Kerry, SW. Ireland (ZISP)
44a. ***Eupithecia schiefereri* Boh.,** ♂, Ioannina, Epirus, NW. Greece (coll. FRIEDRICH)
44b. *E. schiefereri* Boh., ♀, Arachova, Parnassos Oros, C. Greece (coll. FRIEDRICH)
44c. *E. schiefereri* Boh., ♂, Digne, Prov. Mezel, SE. France (coll. GELBRECHT)
44d. *E. schiefereri* Boh., ♀, Karadagh, Crimea, S. Ukraine (ZISP)
45a. ***Eupithecia silenicolata* Mab.,** ♀, Mt. Sirente, C. Italy (ZSM)
45b. *E. silenicolata* Mab., ♂, Graz, Styria, Austria (ZSM)
45c. *E. silenicolata* Mab., ♀, Graz, Styria, Austria (ZSM), reared
45d. *E. silenicolata* Mab., ♂, Cheljabinsk obl., S. Ural, Russia (ZISP)
46a. ***Eupithecia alliaria* Stgr.,** ♀, Dürnstein a. d. D., Austria (ZSM), reared
46b. *E. alliaria* Stgr., ♀, Unterloiben-Mentalgr., Austria (coll. SKOU), reared
46c. *E. alliaria* Stgr., ♂, Varna, E. Bulgaria (coll. GELBRECHT), reared
46d. *E. alliaria* Stgr., ♂, Digne, Basses Alpes, SE. France (ZSM)
46e. *E. alliaria* Stgr., ♂, Digne, Basses Alpes, SE. France (ZSM)
47a. ***Eupithecia dissertata* Püng.,** ♀, Val de Rheme, Graische Alpen, N. Italy (coll. GELBRECHT)
47b. *E. dissertata* Püng., ♀, Val d'Isere, Savoie, E. France (ZSM/HERBULOT)
47c. *E. dissertata* Püng., ♀, Gard, Roquedur, Ganges, S. France (ZMUC)
47d. *E. dissertata* Püng., ♀, Velka Fatra, Slovakia (NMNH)
48a. ***Eupithecia cocciferata* Mill.,** ♂, Vendée, Olonne, W. France (ZSM)
48b. *E. cocciferata* Mill., ♀, St. Laurent du Pape, La Voulte, Ardèche, S. France (coll. SKOU)
48c. *E. cocciferata* Mill., ♀, Albarracin, Teruel, C. Spain (ZMUC)
48d. *E. cocciferata* Mill., ♀, Azrou, Foret de Cèdres, Moyen Atlas, Morocco (ZMUC)
49a. ***Eupithecia abbreviata* Stph.,** ♂, Krasnodarskij kraj, S. European Russia (ZISP)
49b. *E. abbreviata* Stph., ♀, Krasnodarskij kraj, S. European Russia (ZISP)
49c. *E. abbreviata* Stph., ♂, Kaiserslautern, Pfalz, W. Germany (ZSM)
49d. *E. abbreviata* Stph., ♀, Ifrane, Moyen Atlas, Morocco (coll. GELBRECHT)
50a. ***Eupithecia lentiscata* Mab.,** ♂, Bonifacio, Corsica (ZSM/HERBULOT), holotype
50b. *E. lentiscata* Mab., ♂, Monemvasia, Peloponnes, S. Greece (ZMUC)
51a. ***Eupithecia dodoneata* Gn.,** ♀, Svanninge, Fyn, Denmark (coll. SKOU)
51b. *E. dodoneata* Gn., ♀, Puyvert, Vaucluse, SE. France (ZSM/HERBULOT)
51c. *E. dodoneata* Gn., ♀, Zaharo, Peloponnes, S. Greece (coll. FRIEDRICH)
51d. *E. dodoneata* Gn., ♀, Mt. Meden, Rhodopi Mts., S. Bulgaria (ZISP)
51e. *E. dodoneata* Gn., ♀, Kiev obl., N. Ukraine (ZMKU), aberration

43a
43b
43c
43d
43e
43f
43g
43h
44a
44b
44c
44d
45a
45b
45c
45d
46a
46b
46c
46d
46e
47a
47b
47c
47d
48a
48b
48c
48d
49a
49b
49c
49d
50a
50b
51a
51b
51c
51d
51e

PLATE 6 (all specimens 1.5 x natural size)

52a. ***Eupithecia reisserata* Pinker,** ♀, Astros, Arkadia, S. Greece (ZMUC)
52b. *E. reisserata* Pinker, ♀, Argos, Arkadia, S. Greece (ZMUC)
52c. *E. reisserata* Pinker, ♀, Astros, Arkadia, S. Greece (ZMUC)
52d. *E. reisserata* Pinker, ♀, Delphi, Parnassos Oros, C. Greece (ZSM)
52e. *E. reisserata* Pinker, ♀, Akamas Peninsula, Paphos, Cyprus (coll. MÜLLER)
53a. ***Eupithecia massiliata* Mill.,** ♂, Albarracin, Teruel, C. Spain (ZMUC)
53b. *E. massiliata* Mill., ♀, St. Roque, Algeciras, S. Spain (ZMUC)
53c. *E. massiliata* Mill., ♀, Albarracin, Teruel, C. Spain (coll. SKOU)
53d. *E. massiliata* Mill., ♀, Valverde Cam., Andalucia, S. Spain (coll. SKOU)
53e. *E. massiliata* Mill., ♀, Oraison, Provence, SE. France (ZSM)
54a. ***Eupithecia extremata* F.,** ♂, Albarracin, Teruel, C. Spain (ZSM)
54b. *E. extremata* F., ♀, Albarracin, Teruel, C. Spain (ZSM)
54c. *E. extremata* F., ♀, Kelheim, Bavaria, S. Germany (ZSM)
54d. *E. extremata* F., ♂, Nipsa, Thraki, NE. Greece (ZMUC)
54e. *E. extremata* F., ♂, Kovada Gölu, Egridir, SW. Turkey (ZSM)
55a. ***Eupithecia scopariata* Rmbr.,** ♂, Puerto de Navacerrada [near Madrid], C. Spain (ZSM)
55b. *E. scopariata* Rmbr., ♀, Puerto de Navacerrada [near Madrid], C. Spain (ZSM)
55c. *E. scopariata* Rmbr., ♀, Pigmans, Provence, SE. France (ZSM)
55d. *E. scopariata* Rmbr., ♂, Corte, Corsica (ZSM)
55e. *E. scopariata* Rmbr., ♀, St. Roque, Algeciras, S. Spain (ZMUC)
56a. ***Eupithecia pusillata* Den. & Schiff.,** ♀, Villaron near Bessans, Haute Savoie, SE. France (coll. SKOU)
56b. *E. pusillata* Den. & Schiff., ♀, Kataphygion, Mt. Olympos, C. Greece (ZSM)
56c. *E. pusillata* Den. & Schiff., ♀, Pskov obl., W. European Russia (ZISP)
56d. *E. pusillata* Den. & Schiff., ♀, Domodossola, Piemont, N. Italy (coll. EMEM in ZSM)
56e. *E. pusillata scoriata* Stgr., ♀, Hreðavatn, Iceland (ZMUC)
57a. ***Eupithecia sardoa* Dtze.,** ♂, Porto Vecchio, Corsica (ZSM/HERBULOT)
57b. *E. sardoa* Dtze., ♂, Tabarka area, NW. Tunisia (ZMUC)
57c. *E. sardoa* Dtze., ♂, Tabarka area, NW. Tunisia (ZMUC)
57d. *E. sardoa* Dtze., ♀, Tabarka area, NW. Tunisia (ZMUC)
58a. ***Eupithecia ericeata* Rmbr.,** ♀, Rovinj, Istria, Croatia (ZSM)
58b. *E. ericeata* Rmbr., ♂, Kalavrita, Peloponnes, S. Greece (coll. FRIEDRICH)
58c. *E. ericeata* Rmbr., ♂, Nikita, Crimea, S. Ukraine (ZISP)
58d. *E. ericeata* Rmbr., ♀, Sestri Levante, Liguria, N. Italy (coll. EMEM in ZSM)
58e. *E. ericeata* Rmbr., ♀, Askyphou, Crete, S. Greece (ZSM)
59a. ***Eupithecia phoeniceata* Rmbr.,** ♂, Puyvert, Vaucluse, SE. France (ZSM/HERBULOT)
59b. *E. phoeniceata* Rmbr., ♀, Puyvert, Vaucluse, SE. France (ZSM/HERBULOT)
59c. *E. phoeniceata* Rmbr., ♀, Albarracin, Teruel, C. Spain (ZSM)
59d. *E. phoeniceata* Rmbr., ♀, 'Sicily' (ZSM)
59e. *E. phoeniceata* Rmbr., ♀, Pafos/Akamas, Cyprus (coll. FRIEDRICH)
60a. ***Eupithecia oxycedrata* Rmbr.,** ♀, Uña, Cuenca, C. Spain, (coll. SKOU)
60b. *E. oxycedrata* Rmbr., ♂, Gravosa-Zaton, Dalmatia, S. Croatia (ZSM)
60c. *E. oxycedrata* Rmbr., ♂, Kosmas, Parnon Oros, Peloponnes, S. Greece (ZSM)
60d. *E. oxycedrata* Rmbr., ♀, Azrou/Ifrane area, Morocco (ZMUC)
60e. *E. oxycedrata* Rmbr., ♀, Puyvert, Vaucluse, SE. France (ZSM/HERBULOT)
61a. ***Eupithecia rosmarinata* Mill.,** ♂, Nerja, Malaga, S. Spain (ZSM)
61b. *E. rosmarinata* Mill., ♂, Orgiva, Granada, S. Spain (ZSM/HERBULOT)
61c. *E. rosmarinata* Mill., ♂, Aureille, Bouches-du-Rhône, S. France (ZSM)
61d. *E. rosmarinata* Mill., ♀, Talairan, Corbières, Pyrénées-Orientales, S. France (ZSM)

52a
52b
52c
52d
52e
53a
53b
53c
53d
53e
54a
54b
54c
54d
54e
55a
55b
55c
55d
55e
56a
56b
56c
56d
56e
57a
57b
57c
57e
58a
58b
58c
58d
58e
59a
59b
59c
59d
59e
60a
60b
60c
60d
60e
61a
61b
61c
61d
61e

PLATE 7 (all specimens 1.5 x natural size)

62a. ***Eupithecia barteli* Dtze.,** ♀, Saratov obl., SE. European Russia (ZISP)
62b. *E. barteli* Dtze., ♂, Uralsk, NW. Kazakhstan (ZISP), topotype
62c. *E. barteli* Dtze., ♂, Naryn, Kirghizstan (ZISP), reared
62d. *E. barteli* Dtze., ♀, Naryn, Kirghizstan (ZISP), reared
62e. *E. barteli* Dtze., ♀, Naryn, Kirghizstan (ZISP), reared
63a. ***Eupithecia tripunctaria* H.-S.,** ♀, Kassel, C. Germany (ZSM), reared
63b. *E. tripunctaria* H.-S., ♂, Kassel, C. Germany (ZSM), reared
63c. *E. tripunctaria* H.-S., ♀, Snarup, Fyn, Denmark (coll. SKOU)
63d. *E. tripunctaria* H.-S., ♀, St. Petersburg obl., NW. European Russia (ZISP), reared
63e. *E. tripunctaria* H.-S., ab. *angelicata* Barr., ♀, Puisieux-et-Clanlieu, Aisne, N. France (ZSM/HERBULOT)
64a. ***Eupithecia virgaureata* Dbld.,** ♀, Elnjun, Kola Peninsula, N. European Russia (ZISP)
64b. *E. virgaureata* Dbld., ♀, Jena, Thüringen, C. Germany (coll. FRIEDRICH)
64c. *E. virgaureata* Dbld., ♀, Jena, Thüringen, C. Germany (coll. FRIEDRICH)
64d. *E. virgaureata* Dbld., ♀, Kiev obl., N. Ukraine (ZMKU)
64e. *E. virgaureata* Dbld., ♀, Novgorod obl., NW. European Russia (ZISP), reared
65a. ***Eupithecia tantillaria* Bsd.,** ♀, Snarup, Fyn, Denmark (coll. SKOU)
65b. *E. tantillaria* Bsd., ♂, St. Petersburg obl., NW. European Russia (ZISP)
65c. *E. tantillaria* Bsd., ♀, St. Petersburg obl., NW. European Russia (ZISP)
65d. *E. tantillaria* Bsd., ♀, Mts. Taygetos, Lakonia, S. Greece (ZMUC)
65e. *E. tantillaria* Bsd., f. *calabrica* Dtze., ♂, Mt. Chelmos, Peloponnes, S. Greece (ZSM)
66a. ***Eupithecia liguriata* Mill.,** ♂, Ronda, Malaga, S. Spain (ZSM)
66b. *E. liguriata* Mill., ♀, Sierra de Alfacar, Andalucia, S. Spain (ZSM)
66c. *E. liguriata* Mill., ♀, Digne, Basses Alpes, SE. France (ZSM)
66d. *E. liguriata* Mill., ♀, Digne, Basses Alpes, SE. France (ZSM), reared
66e. *E. liguriata* Mill., ♀, Digne, Basses Alpes, SE. France (ZSM), reared
67a. ***Eupithecia lariciata* Frr.,** ♀, Freiberg, Sachsen, E. Germany (ZSM)
67b. *E. lariciata* Frr., ♂, Karwendelgebirge, Tirol, Austria (coll. EMEM in ZSM)
67c. *E. lariciata* Frr., ♂, Kappel, N. Estonia (ZSM)
67d. *E. lariciata* Frr., ♂, Kirkeby Sand, Fyn, Denmark (coll. SKOU)
67e. *E. lariciata* Frr., ♀, Valle de la Romanche, Dept. Isère, SE. France (coll. SKOU)
68a. ***Eupithecia lanceata* Hbn.,** ♂, Hamburg, N. Germany (ZSM)
68b. *E. lanceata* Hbn., ♂, Hamburg, N. Germany (ZSM)
68c. *E. lanceata* Hbn., ♀, Stengade, Langeland, Denmark (coll. SKOU)
68d. *E. lanceata* Hbn., ♀, Stengade, Langeland, Denmark (coll. SKOU)
68e. *E. lanceata* Hbn., ♂, St. Petersburg obl., NW. European Russia (ZISP)
69a. ***Eupithecia groenblomi* Urb.,** ♂, Lejvonmäki, Etelä-Savo, SE. Finland (ZMH), reared
69b. *E. groenblomi* Urb., ♂, Novgorod obl., NW. European Russia (ZISP), reared
69c. *E. groenblomi* Urb., ♂, Novgorod obl., NW. European Russia (ZISP), reared
69d. *E. groenblomi* Urb., ♂, St. Petersburg obl., NW. European Russia (coll. SKOU), reared
69e. *E. groenblomi* Urb., ♀, Novgorod obl., NW. European Russia (ZISP), reared
70a. ***Eupithecia selinata* H.-S.,** ♀, Gross Köris, Brandenburg, E. Germany (coll. GELBRECHT)
70b. *E. selinata* H.-S., ♀, Vidova, Slovakia (ZMUC)
70c. *E. selinata* H.-S., ♀, Kiev obl., N. Ukraine (ZMKU)
70d. *E. selinata* H.-S., ♂, Novgorod obl., NW. European Russia (ZISP)
70e. *E. selinata* H.-S., ♀, Novgorod obl., NW. European Russia (ZISP)
71a. ***Eupithecia actaeata* Wald.,** ♀, Lütjenburg, Schleswig-Holstein, N. Germany (coll. WEGNER), reared
71b. *E. actaeata* Wald., ♂, St. Petersburg obl., NW. European Russia (ZISP)
71c. *E. actaeata* Wald., ♀, Novgorod obl., NW. European Russia (ZISP), reared
71d. *E. actaeata* Wald., ♀, St. Petersburg obl., NW. European Russia (ZISP), reared

62a
62b
62c
62d
62e
63a
63b
63c
63d
63e
64a
64b
64c
64d
64e
65a
65b
65c
65d
65e
66a
66b
66c
66d
66e
67a
67b
67c
67d
67e
68a
68b
68c
68d
68e
69a
69b
69c
69d
69e
70a
70b
70c
70d
70e
71a
71b
71c
71d

PLATE 8 (all specimens 1.5 x natural size)

72a. ***Eupithecia quercetica* Prt.**, ♀, Asfaka, Ioannina, NW. Greece (coll. SKOU)
72b. *E. quercetica* Prt., ♀, Yalta, Crimea, S. Ukraine (ZISP)
72c. *E. quercetica* Prt., ♀, Yalta, Crimea, S. Ukraine (ZISP)
72d. *E. quercetica* Prt., ♂, Kovada Gölu, Egridir, SW. Turkey (ZSM)
73a. ***Eupithecia egenaria* H.-S.**, ♂, Lödaritztal near Erkner, E. Germany (coll. GELBRECHT)
73b. *E. egenaria* H.-S., ♂, Am Landgrafen, Jena, C. Germany (ZSM)
73c. *E. egenaria* H.-S., ♀, Vindeholme, Lolland, Denmark (coll. SKOU)
73d. *E. egenaria* H.-S., ♀, Uljanovsk obl., E. European Russia (ZISP)
73e. *E. egenaria* H.-S., ♀, Uljanovsk obl., E. European Russia (ZISP)
74a. ***Eupithecia repentina* Vojnits & Laever,** ♀, Tula obl., W. European Russia (ZISP)
74b. *E. repentina* Vojnits & Laever, ♀, Tula obl., W. European Russia (ZISP)
74c. *E. repentina* Vojnits & Laever, ♀, Andreevka, S. Primorje, Far East Russia (ZISP)
74d. *E. repentina* Vojnits & Laever, ♂, Rjazanovka, S. Primorje, Far East Russia (coll. SKOU)
74e. *E. repentina* Vojnits & Laever, ♀, Rjazanovka, S. Primorje, Far East Russia (ZISP)
75a. ***Eupithecia orana* Dtze.**, ♀, 'Mini Hollywood', Almeria, S. Spain (ZMUC)
75b. *E. orana* Dtze., ♀, 'Mini Hollywood', Almeria, S. Spain (coll. SKOU)
75c. *E. orana* Dtze., ♀, 'Mini Hollywood', Almeria, S. Spain (coll. SKOU)
75d. *E. orana* Dtze., ♀, 'Mini Hollywood', Almeria, S. Spain (ZISP)
75e. *E. orana* Dtze., ♂, Puerto de Carmen, Lanzarote Isl., Canary Islands (ZMUC)
76a. ***Eupithecia pimpinellata* Hbn.**, ♂, Stengade, Langeland, Denmark (coll. SKOU)
76b. *E. pimpinellata* Hbn., ♀, La Grave, Haute Alpes, SE. France (coll. SKOU)
76c. *E. pimpinellata* Hbn., ♀, Karabi-Jaila, Crimea, S. Ukraine (ZMKU)
76d. *E. pimpinellata* Hbn., ♀, Bleckede, Lüneburg, N. Germany (coll. WEGNER)
77a. ***Eupithecia karadaghensis* Mir.**, ♀, Karadagh, Crimea, S. Ukraine (ZISP), holotype
77b. *E. karadaghensis* Mir., ♂, Karadagh, Crimea, S. Ukraine (ZISP), paratype
77c. *E. karadaghensis* Mir., ♀, Karadagh, Crimea, S. Ukraine (ZISP)
77d. *E. karadaghensis* Mir., ♀, Karadagh, Crimea, S. Ukraine (ZISP)
77e. *E. karadaghensis* Mir., ♂, Karadagh, Crimea, S. Ukraine (coll. SKOU)
78a. ***Eupithecia suboxydata* Stgr.**, ♂, Orenburg, SE. European Russia (ZISP)
78b. *E. suboxydata* Stgr., ♀, Saratov, E. European Russia (ZISP)
78c. *E. suboxydata* Stgr., ♀, Minusinsk, S. Siberia, Russia (ZISP)
78d. *E. suboxydata* Stgr., ♀, near Vladivostok, S. Primorje, Far East Russia (ZISP)
78e. *E. suboxydata* Stgr., ♂, Rjazanovka, S. Primorje, Far East Russia (ZISP)
79a. ***Eupithecia simpliciata* Hw.**, ♀, Svenskehavn, Bornholm, Denmark (coll. SKOU)
79b. *E. simpliciata* Hw., ♀, Candasnos, Prov. Huesca, NE. Spain (coll. SKOU)
79c. *E. simpliciata* Hw., ♀, Voronezh obl., S. European Russia (ZISP)
79d. *E. simpliciata* Hw., ♀, Brandenburg, E. Germany (coll. GELBRECHT), reared
79e. *E. simpliciata* Hw., ♂, Kassel, C. Germany (ZSM), reared
80a. ***Eupithecia sinuosaria* Ev.**, ♂, St. Petersburg obl., NW. European Russia (ZISP)
80b. *E. sinuosaria* Ev., ♀, St. Petersburg obl., NW. European Russia (ZISP)
80c. *E. sinuosaria* Ev., ♀, Brandenburg, E. Germany (coll. GELBRECHT), reared
80d. *E. sinuosaria* Ev., ♂, Brandenburg, E. Germany (coll. GELBRECHT), reared
80e. *E. sinuosaria* Ev., ♀, Boltersen, Lüneburg, N. Germany (ZSM)
81a. ***Eupithecia gelidata* Möschler**, ♂, 'S. Greenland' (ZSM/HERBULOT)
81b. *E. gelidata hyperboreata* Stgr., ♂, Furnvik, Gästrikland, C. Sweden (coll. SKOU)
81c. *E. gelidata hyperboreata* Stgr., ♀, Kilpisjärvi (Le), Lappland, N. Finland (ZMH)
81d. *E. gelidata hyperboreata* Stgr., ♂, St. Petersburg obl., NW. European Russia (ZISP)
81e. *E. gelidata hyperboreata* Stgr., ♀, Brandenburg, E. Germany (coll. GELBRECHT), reared

72a
72b
72c
72d
73a
73b
73c
73d
73e
74a
74b
74c
74d
74e
75a
75b
75c
75d
75e
76a
76b
76c
76d
77a
77b
77c
77d
77e
78a
78b
78c
78d
78e
79a
79b
79c
79d
79e
80a
80b
80c
80d
80e
81a
81b
81c
81d
81e

PLATE 9 (all specimens 1.5 x natural size)

82a. ***Eupithecia nanata* Hbn.,** ♂, Boot, Cumberland, England, United Kingdom (ZISP)
82b. *E. nanata* Hbn., ♀, Boot, Cumberland, England, United Kingdom (ZISP)
82c. *E. nanata* Hbn., ♀, Massow near Baruth, Brandenburg, E. Germany (coll. GELBRECHT)
82d. *E. nanata* Hbn., ♀, Luckenwalde, Brandenburg, E. Germany (coll. GELBRECHT)
82e. *E. nanata* Hbn., ♂, Storelung, Fyn, Denmark (coll. SKOU)
82f. *E. nanata zebrata* Wolff, ♂, Klaksvig, Faroe Islands (ZMUC)
82g. *E. nanata zebrata* Wolff, ♂, Klaksvig, Faroe Islands (ZMUC)
82h. *E. nanata gelidatoides* Warnecke, ♀, Vifilsstaðir, Iceland (ZMUC)
82i. *E. nanata gelidatoides* Warnecke, ♀, Vifilsstaðir, Iceland (ZMUC)
82j. *E. nanata gelidatoides* Warnecke, ♀, Vifilsstaðir, Iceland (ZMUC)
83a. ***Eupithecia innotata* Hufn.,** ♂, Feering, Essex, England, United Kingdom (ZISP)
83b. *E. innotata* Hufn., ♂, Nürnberg, SE. Germany (ZSM)
83c. *E. innotata* Hufn., ♀, Urup Mose, Fyn, Denmark (coll. SKOU)
83d. *E. innotata* Hufn., ♂, Pietramurata, Trentino, N. Italy (ZSM)
83e. *E. innotata* Hufn., ♂, Saratov obl., E. European Russia (ZISP)
84a. ***Eupithecia ochridata* Schütze & Pinker,** ♀, Albarracin, Aragon, C. Spain (ZSM)
84b. *E. ochridata* Schütze & Pinker, ♂, Candasnos, Prov. Huesca, NE. Spain (coll. SKOU)
84c. *E. ochridata* Schütze & Pinker, ♂, Milanovo, Iskar Gorge Region, W. Bulgaria (coll. SKOU)
84d. *E. ochridata* Schütze & Pinker, ♀, Zunda-Tolga, Kalmykia, S. European Russia (ZISP)
85a. ***Eupithecia unedonata* Mab.,** ♀, Aureille, Bouches-du-Rhône, S. France (ZSM)
85b. *E. unedonata* Mab., ♀, Kosmas, Peloponnes, S. Greece (coll. FRIEDRICH)
85c. *E. unedonata* Mab., ♂, Evisa, Corsica (ZSM)
85d. *E. unedonata* Mab., ♀, Kagizman, Kars, E. Turkey (ZMUC)
86a. ***Eupithecia chalikophila* Whli.,** ♀, Veleta, Sierra Nevada, Granada, S. Spain (coll. SKOU)
86b. *E. chalikophila* Whli., ♀, Geronimo, Sierra Nevada, Granada, S. Spain (ZSM)
86c. *E. chalikophila* Whli., ♂, Veleta, Sierra Nevada, Granada, S. Spain (ZSM)
86d. *E. chalikophila* Whli., ♀, Veleta, Sierra Nevada, Granada, S. Spain (coll. SKOU)
87a. ***Eupithecia graphata* Tr.,** ♀, Ofen, Hungary (ZSM), topotype
87b. *E. graphata* Tr., ♀, 'Hungary' (ZISP)
87c. *E. graphata* Tr., ♂, Varna district, North Black Sea Coast, E. Bulgaria (ZISP)
87d. *E. graphata* Tr., f. *mayeri* Mann, ♀, Voslau, Austria (ZISP)
87e. *E. graphata* Tr., f. *mayeri* Mann, ♀, Voslau, Austria (ZISP)
87f. *E. graphata setaceata* Dtze., ♂, Cèreste, Vaucluse, S. France (coll. SKOU)
87g. *E. graphata setaceata* Dtze., ♀, L'Argentière, Hautes Alpes, SE. France (ZSM/HERBULOT)
87h. *E. graphata setaceata* Dtze., f. *mayeri* Mann, ♀, Valle de la Romanche, Dept. Isère, SE. France (coll. SKOU)
87i. *E. graphata sproengertsi* Dtze., ♂, Mtgna Grande, C. Apennines, C. Italy (ZSM)
87j. *E. graphata sproengertsi* Dtze., ♀, Mtgna Grande, C. Apennines, C. Italy (ZSM)
87k. *E. graphata olympica* Tul., ♀, Parnassos Oros, C. Greece (coll. GELBRECHT), I gen.
87l. *E. graphata olympica* Tul., ♂, Parnassos Oros, C. Greece (ZMUC), I gen.
87m. *E. graphata olympica* Tul., ♂, Parnassos Oros, C. Greece (ZMUC), I gen.
87n. *E. graphata olympica* Tul., ♂, Mt. Chelmos, Peloponnes, S. Greece (ZSM), I gen.
87o. *E. graphata olympica* Tul., ♀, Arnaia, Chalkidiki, N. Greece (ZSM), II gen.
87p. *E. graphata hesperia* Whli., ♂, Veleta, Sierra Nevada, Granada, S. Spain (coll. SKOU)
87q. *E. graphata hesperia* Whli., ♀, Veleta, Sierra Nevada, Granada, S. Spain (coll. SKOU)
87r. *E. graphata albofasciata* Stgr., ♀, Karadagh, Crimea, S. Ukraine (ZSM)
87s. *E. graphata albofasciata* Stgr., ♂, Karadagh, Crimea, S. Ukraine (ZISP)
87t. *E. graphata albofasciata* Stgr., ♀, Karadagh, Crimea, S. Ukraine (ZISP)
88a. ***Eupithecia riparia* H.-S.,** ♂, Mt. Paradiso, S. Apennines, S. Italy (ZSM)
88b. *E. riparia* H.-S., ♀, Mt. Paradiso, S. Apennines, S. Italy (ZSM)
88c. *E. riparia* H.-S., ♀, Mte. Rotella, S. Apennines, S. Italy (ZSM)
88d. *E. riparia* H.-S., ♂, Parnassos Oros, C. Greece (ZSM)
88e. *E. riparia* H.-S., ♀, Parnassos Oros, C. Greece (ZSM)

82a
82b
82c
82d
82e
82f
82g
82h
82i
82j
83a
83b
83c
83d
83e
84a
84b
84c
84d
85a
85b
85c
85d
86a
86b
86c
86d
87a
87b
87c
87d
87e
87f
87g
87h
87i
87j
87k
87l
87m
87n
87o
87p
87q
87r
87s
87t
88a
88b
88c
88d
88e

PLATE 10 (all specimens 1.5 x natural size)

89a. ***Eupithecia thurnerata* Schütze,** ♂, Treska Schlucht, Macedonia (ZSM)
89b. *E. thurnerata* Schütze, ♀, Omalos, Crete, S. Greece (coll. SKOU)
89c. *E. thurnerata* Schütze, ♀, Kallergi Mts., Crete, S. Greece (coll. SKOU)
89d. *E. thurnerata* Schütze, ♂, Omalos, Crete, S. Greece (coll. SKOU)
89e. *E. thurnerata* Schütze, ♀, Kallergi Mts., Crete, S. Greece (coll. FIBIGER)
89f. *E. thurnerata* Schütze, ♀, Kallergi Mts., Crete, S. Greece (coll. FIBIGER)
90a. ***Eupithecia gemellata* H.-S.,** ♂, Kalavrita, Aroania Oros, Peloponnes, S. Greece (coll. GELBRECHT)
90b. *E. gemellata* H.-S., ♂, Mt. Chelmos, Aroania Oros, Peloponnes, S. Greece (coll. GELBRECHT)
90c. *E. gemellata* H.-S., ♂, Rovinj, Istria, Croatia (ZSM)
90d. *E. gemellata* H.-S., ♀, Albarracin, Teruel, C. Spain (ZSM)
90e. *E. gemellata* H.-S., ♀, Bujaraloz, Zaragoza, NE. Spain (ZSM/HERBULOT)
90f. *E. gemellata* H.-S., ♀, Candasnos, Huesca, N. Spain (coll. SKOU)
91a. ***Eupithecia mystica* Dtze.,** ♂, Ochrid, Macedonia (ZSM)
91b. *E. mystica* Dtze., ♀, Ochrid, Macedonia (ZSM)
91c. *E. mystica* Dtze., ♀, Arachova, Parnassos Oros, C. Greece (ZSM)
91d. *E. mystica* Dtze., ♀, Arachova, Parnassos Oros, C. Greece (ZSM)
91e. *E. mystica* Dtze., ♀, Alushta, Crimea, S. Ukraine (ZISP)
92a. ***Eupithecia cooptata* Dtze.,** ♂, Gorges de la Meouge, Hautes Alpes, SE. France (coll. SKOU)
92b. *E. cooptata* Dtze., ♀, Gorges de la Meouge, Hautes Alpes, SE. France (coll. SKOU)
92c. *E. cooptata* Dtze., ♀, La Besée, Hautes Alpes, SE. France (ZSM/HERBULOT)
92d. *E. cooptata* Dtze., ♀, Albarracin, Teruel, C. Spain (ZSM)
92e. *E. cooptata* Dtze., ♀, Tosos, Zaragoza, NE. Spain (coll. SKOU)
93a. ***Eupithecia spissilineata* Metzn.,** ♂, Mti. Sabini, C. Italy (ZSM)
93b. *E. spissilineata* Metzn., ♀, Mti. Sabini, C. Italy (ZSM)
93c. *E. spissilineata* Metzn., ♀, Mti. Sabini, C. Italy (ZSM)
93d. *E. spissilineata* Metzn., ♀, Inaklju, Mt. Alagez, Armenia (ZISP)
93e. *E. spissilineata* Metzn., ♀, Antalya, Anatolia, S. Turkey (ZISP)
93f. *E. spissilineata* Metzn., ♀, Ispir, Erzurum, NE. Turkey (coll. GELBRECHT)
93g. *E. spissilineata* Metzn., f. *hellasata* nova, ♀, Arachova, Parnassos Oros, C. Greece (coll. GELBRECHT)
93h. *E. spissilineata* Metzn., f. *hellasata* nova, ♀, Arachova, Parnassos Oros, C. Greece (coll. FRIEDRICH)
93i. *E. spissilineata* Metzn., f. *hellasata* nova, ♀, Arachova, Parnassos Oros, C. Greece (coll. GELBRECHT)
93j. *E. spissilineata* Metzn., f. *hellasata* nova, ♀, Zachlorou, Peloponnes, S. Greece (ZSM)
94a. ***Eupithecia cuculliaria* Rbl.,** ♀, Drenovo – Kavadar, Macedonia (ZSM)
94b. *E. cuculliaria* Rbl., ♂, Lithochoron, N. Greece (ZSM)
94c. *E. cuculliaria* Rbl., ♀, Astros, Arkadia, S. Greece (ZMUC)
94d. *E. cuculliaria* Rbl., ♀, Stara, Krk Isl., Croatia (coll. SKOU)
94e. *E. cuculliaria* Rbl., ♀, Malmašuta, Krk Isl., Croatia (coll. SKOU)
95a. ***Eupithecia unitaria* H.-S.,** ♂, Bolnuevo, Murcia, S. Spain (coll. SKOU)
95b. *E. unitaria* H.-S., ♀, Bolnuevo, Murcia, S. Spain (coll. SKOU)
95c. *E. unitaria* H.-S., ♀, Camino de Istan., Andalucia, Spain (ZMUC)
95d. *E. unitaria* H.-S., ♀, Benaojan, Malaga, S. Spain (ZMUC)
95e. *E. unitaria* H.-S., ♀, Tizi-n-Test, Haute Atlas, Morocco (coll. GELBRECHT)
96a. ***Eupithecia breviculata* Donz.,** ♂, Gard Roquedur Ganges, S. France (ZMUC)
96b. *E. breviculata* Donz., ♀, Cravant, Yonne, C. France (ZSM/HERBULOT)
96c. *E. breviculata* Donz., ♂, S. Martino delle Scale, Sicily (ZSM)
96d. *E. breviculata* Donz., ♀, Arachova, Parnassos Oros, C. Greece (coll. GELBRECHT)
96e. *E. breviculata* Donz., ♂, Uljanovsk obl., E. European Russia (ZISP)
97a. ***Eupithecia cerussaria* Led.,** ♀, Lebadeia [Levadia], C. Greece (ZSM/HERBULOT)
97b. *E. cerussaria* Led., ♀, Lithochoron, N. Greece (ZSM)
97c. *E. cerussaria* Led., ♀, Lebadeia [Levadia], C. Greece (ZSM/HERBULOT)
97d. *E. cerussaria* Led., ♀, Aladag Göksutal, Taurus Mts., S. Turkey (ZSM)
97e. *E. cerussaria* Led., ♂, Achyr Dagh, Taurus Mts., S. Turkey (ZSM)

89a
89b
89c
89d
89e
89f
90a
90b
90c
90d
90e
90f
91a
91b
91c
91d
91e
92a
92b
92c
92d
92e
93a
93b
93c
93d
93e
93f
93g
93h
93i
93j
94a
94b
94c
94d
94e
95a
95b
95c
95d
95e
96a
96b
96c
96d
96e
97a
97b
97c
97d
97e

PLATE 11 (all specimens 1.5 x natural size)

98a. ***Eupithecia irriguata* Hbn.,** ♂, Labaci, Carniolia, Austria (ZSM)
98b. *E. irriguata* Hbn., ♀, Dürnstein, Austria (ZSM)
98c. *E. irriguata* Hbn., ♀, Vegeholm, Skåne, S. Sweden (coll. SKOU)
98d. *E. irriguata mauretanica* Dtze., ♂, Azrou, Moyen Atlas, Morocco (coll. GELBRECHT)
98e. *E. irriguata mauretanica* Dtze., ♀, Azrou, Moyen Atlas, Morocco (coll. GELBRECHT)
98f. *E. irriguata eriguata* Stgr., ♀, Jumilla, Murcia, S. Spain (ZSM)
98g. *E. irriguata eriguata* Stgr., ♀, Jumilla, Murcia, S. Spain (ZSM)
98h. *E. irriguata eriguata* Stgr., ♂, Marbella-Casa y Campo, Andalucia, S. Spain (ZMUC)
98i. *E. irruguata staudingeri* Boh., ♂, Arachova, Parnassos Oros, C. Greece (ZSM)
98j. *E. irruguata staudingeri* Boh., ♀, Arachova, Parnassos Oros, C. Greece (ZSM)
99a. ***Eupithecia antalica* Mir.,** ♀, Agios Nikolaos, Thessalia, C. Greece (HNHM), holotype
99b. *E. antalica* Mir., ♂, Antalya, Anatolia, S. Turkey (ZISP), paratype
99c. *E. antalica* Mir., ♂, Antalya, Anatolia, S. Turkey (ZISP), paratype
99d. *E. antalica* Mir., ♀, Antalya, Anatolia, S. Turkey (ZISP), paratype
100a. ***Eupithecia indigata* Hbn.,** ♀, Svanninge, Fyn, Denmark (coll. SKOU)
100b. *E. indigata* Hbn., ♀, Breslau (Wrocław), SW. Poland (ZSM)
100c. *E. indigata* Hbn., ♀, Kiev obl., N. Ukraine (ZMKU)
100d. *E. indigata* Hbn., ♀, Peñones d. S. Francisco, Sierra Nevada, Granada, S. Spain (ZMUC)
100e. *E. indigata* Hbn., ♀, Albarracin, Teruel, C. Spain (coll. SKOU)
101a. ***Eupithecia conterminata* Lien. & Zell.,** ♂, Dorpat, Estonia (ZSM)
101b. *E. conterminata* Lien. & Zell., ♀, Dorpat, Estonia (ZSM)
101c. *E. conterminata* Lien. & Zell., ♀, Grängshammar, Dalarne, C. Sweden (coll. SKOU)
101d. *E. conterminata* Lien. & Zell., ♂, St. Petersburg obl., NW. European Russia (ZISP)
101e. *E. conterminata* Lien. & Zell., ♂, Monchegorsk, Kola Peninsula, N. European Russia (ZISP)
102a. ***Eupithecia distinctaria* H.-S.,** ♀, Herte v. Burs, Gotland Isl., S. Sweden (ZMUC)
102b. *E. distinctaria* H.-S., ♂, Långe Erik, Öland Isl., S. Sweden (ZMUC)
102c. *E. distinctaria* H.-S., ♀, Schwarzwald, Wutach-Gebiet, SW. Germany (ZSM), reared
102d. *E. distinctaria* H.-S., ♀, Bude, N. Cornwall, England, United Kingdom (ZISP)
102e. *E. distinctaria* H.-S., ♂, Kermen, Crimea, S. Ukraine (ZISP)
103a. ***Eupithecia senorita* Mir.,** sp. n., ♂, Albas, Corbières, SW. France (ZSM), holotype
103b. *E. senorita* Mir., sp. n., ♀, Veleta, Sierra Nevada, Granada, S. Spain (ZSM), paratype
103c. *E. senorita* Mir., sp. n., ♀, Pto. de la Ragua, Sierra Nevada, Granada, S. Spain (coll. FIBIGER), paratype
103d. *E. senorita* Mir., sp. n., ♀, Albarracin, Teruel, C. Spain (coll. SKOU), paratype
103e. *E. senorita* Mir., sp. n., ♀, Albarracin, Teruel, C. Spain (coll. SKOU), paratype
104a. ***Eupithecia extraversaria* H.-S.,** ♀, Regensburg, SE. Germany (ZSM)
104b. *E. extraversaria* H.-S., ♀, Regensburg, SE. Germany (ZSM)
104c. *E. extraversaria* H.-S., ♀, Moscow, C. European Russia (ZISP)
104d. *E. extraversaria* H.-S., ♀, Brandenburg, E. Germany (coll. GELBRECHT), reared
104e. *E. extraversaria* H.-S., ♀, Valle de la Romanche, Dept. Isère, SE. France (coll. SKOU)
105a. ***Eupithecia centaureata* Den. & Schiff.,** ♀, La Antilla, Huelva, SW. Spain (coll. SKOU)
105b. *E. centaureata* Den. & Schiff., ♀, Stengade, Langeland, Denmark (coll. SKOU)
105c. *E. centaureata* Den. & Schiff., ♂, Stengade, Langeland, Denmark (coll. SKOU)
105d. *E. centaureata* Den. & Schiff., ♂, Meskla-Zourva, Crete, S. Greece (coll. SKOU)
105e. *E. centaureata* Den. & Schiff., ♂, St. Petersburg obl., NW. European Russia (ZISP)
106a. ***Eupithecia limbata* Stgr.,** ♀, Diakofto, Aroania Ori, Peloponnes, S. Greece (coll. GELBRECHT)
106b. *E. limbata* Stgr., ♀, Diakofto, Aroania Ori, Peloponnes, S. Greece (coll. GELBRECHT)
106c. *E. limbata* Stgr., ♂, Mt. Chelmos, Aroania Ori, Peloponnes, S. Greece (coll. ZMUC)
106d. *E. limbata tomillata* Chrét., ♀, San Ildefonso, Segovia, C. Spain (ZSM/HERBULOT), lectotype
106e. *E. limbata tomillata* Chrét., ♀, Noguera, Teruel, C. Spain (ZSM)
106f. *E. limbata tomillata* Chrét., ♀, Albarracin, Teruel, C. Spain (ZSM)
107a. ***Eupithecia nachadira* Brandt,** ♂, Kouch i Taftan, S. Iran (NHRS), lectotype
107b. *E. nachadira* Brandt, ♂, Kouch i Taftan, S. Iran (NHRS), paralectotype
107c. *E. nachadira* Brandt, ♂, Keredj – Chalus, N. Iran (HNHM)
107d. *E. nachadira* Brandt, ♀, Karadagh, Crimea, S. Ukraine (ZISP)

98a
98b
98c
98d
98e
98f
98g
98h
98i
98j
99a
99b
99c
99d
100a
100b
100c
100d
100e
101a
101b
101c
101d
101e
102a
102b
102c
102d
102e
103a
103b
103c
103d
103e
104a
104b
104c
104d
104e
105a
105b
105c
105d
105e
106a
106b
106c
106d
106e
106f
107a
107b
107c
107d

PLATE 12 (all specimens 1.5 x natural size)

108a. ***Eupithecia mekrana* Brandt,** ♂, Orenburg obl., SW. Asian Russia (ZISP)
108b. *E. mekrana* Brandt, ♂, Orenburg obl., SW. Asian Russia (ZISP)
108c. *E. mekrana* Brandt, ♂, Orenburg obl., SW. Asian Russia (coll. K. NUPPONEN)
108d. *E. mekrana* Brandt, ♂, Orenburg obl., SW. Asian Russia (coll. K. NUPPONEN)
108e. *E. mekrana* Brandt, ♀, Orenburg obl., SW. Asian Russia (ZISP)
109a. ***Eupithecia variostrigata* Alph.,** ♂, Karadagh, Crimea, S. Ukraine (ZISP)
109b. *E. variostrigata* Alph., ♀, Sudak, Crimea, S. Ukraine (ZISP)
109c. *E. variostrigata* Alph., ♀, Black-Sea Coast, E. Bulgaria (coll. GELBRECHT)
109d. *E. variostrigata* Alph., ♀, Albarracin, Teruel, C. Spain (ZSM)
109e. *E. variostrigata* Alph., ♂, 'Alpes Maritimes', SE. France (ZSM/HERBULOT)
110a. ***Eupithecia insigniata* Hbn.,** ♀, Tåsinge, Lundby, Fyn, Denmark (coll. SKOU)
110b. *E. insigniata* Hbn., ♀, Naumburg, Thüringen, C. Germany (ZSM), reared
110c. *E. insigniata* Hbn., ♂, Naumburg, Thüringen, C. Germany (ZSM), reared
110d. *E. insigniata* Hbn., ♂, Sondershausen, Thüringen, C. Germany (coll. GELBRECHT), reared
110e. *E. insigniata* Hbn., ♀, Sondershausen, Thüringen, C. Germany (coll. GELBRECHT), reared
111a. ***Eupithecia trisignaria* H.-S.,** ♂, Kirkeby Sand, Fyn, Denmark (coll. SKOU), reared
111b. *E. trisignaria* H.-S., ♀, Kirkeby Sand, Fyn, Denmark (coll. SKOU), reared
111c. *E. trisignaria* H.-S., ♀, Jena, C. Germany (coll. FRIEDRICH)
111d. *E. trisignaria* H.-S., ♀, Jena, C. Germany (coll. FRIEDRICH)
111e. *E. trisignaria* H.-S., ♂, Novgorod obl., NW. European Russia (ZISP), reared
112a. ***Eupithecia gueneata* Mill.,** ♀, Busambra, Sicily (ZSM)
112b. *E. gueneata* Mill., ♀, Busambra, Sicily (ZSM)
112c. *E. gueneata* Mill., ♀, Le Madonie, Sicily (ZSM)
112d. *E. gueneata* Mill., ♂, Veleta, Sierra Nevada, Granada, S. Spain (coll. SKOU)
112e. *E. gueneata* Mill., ♀, Uljanovsk obl., E. European Russia (ZISP)
113a. ***Eupithecia gratiosata* H.-S.,** ♂, Sierra Albamilla, Almeria, S. Spain (coll. SKOU)
113b. *E. gratiosata* H.-S., ♀, Sierra Alhamilla, Almeria, S. Spain (coll. SKOU)
113c. *E. gratiosata* H.-S., ♀, Aritzo, Sardinia (ZSM), reared
113d. *E. gratiosata* H.-S., ♀, 'Romania' (ZSM)
114a. ***Eupithecia scalptata* Chr.,** ♂, Egio, Peloponnes, S. Greece (coll. GELBRECHT)
114b. *E. scalptata* Chr., ♀, Egio, Peloponnes, S. Greece (coll. GELBRECHT)
114c. *E. scalptata* Chr., ♀, Thermopilae [near Lamia], C. Greece (ZSM)
114d. *E. scalptata* Chr., ♀, Delphi, C. Greece (ZSM)
114e. *E. scalptata* Chr., ♂, near Elton Lake, SE. European Russia (coll. SKOU)
115a. ***Eupithecia veratraria* H.-S.,** ♀, Würzeralm, Oberösterreich, Austria (coll. SKOU)
115b. *E. veratraria eynensata* Grasl., ♀, Col d'Aubisque, Pyrénées-Atlantiques, S. France (ZSM/HERBULOT)
115c. *E. veratraria eynensata* Grasl., ♀, Col d'Aubisque, Pyrénées-Atlantiques, S. France (ZSM/HERBULOT)
115d. *E. veratraria arctica* Viid., ♀, Kirovsk, Kola Peninsula, N. European Russia (ZISP)
116a. ***Eupithecia cretaceata fenestrata* Mill**., ♀, Ploecken, Carinthia, Austria (ZSM), reared
116b. *E. cretaceata fenestrata* Mill., ♀, Le Lautarel, Hautes Alpes, SE. France (ZSM/HERBULOT)
116c. *E. cretaceata fenestrata* Mill., ♀, Begovica, Piringebirge, SW. Bulgaria (coll. GELBRECHT)

108a
108b
108c
108d
108e
109a
109b
109c
109d
109e
110a
110b
110c
110d
110e
111a
111b
111c
111d
111e
112a
112b
112c
112d
112e
113a
113b
113c
113d
114a
114b
114c
114d
114e
115a
115b
115c
115d
116a
116b
116c

PLATE 13 (all specimens 1.5 x natural size)

117a. ***Eupithecia praealta* Whli.,** ♂, Sierra Nevada, Granada, S. Spain (ZSM)
117b. *E. praealta* Whli., ♀, Sierra Nevada, Granada, S. Spain (ZSM)
117c. *E. praealta* Whli., ♂, Pt. del Lobo, Sierra Nevada, Granada, S. Spain (coll. SKOU)
117d. *E. praealta rerayata* Reisser, ♂, Tachdirt, Haute Atlas, Morocco (ZISP)
118a. ***Eupithecia intricata* Zett.,** f. *arceuthata* Frr., ♀, Brandenburg, E. Germany (coll. GELBRECHT), reared
118b. *E. intricata* Zett., f. *helveticaria* Boisd., ♀, Mt. Goverla, E. Carpathians, W. Ukraine (ZISP)
118c. *E. intricata* Zett., f. *millieraria* Wnuk., ♀, Aviemore, Inverness-shire, Scotland, United Kingdom (ZISP)
118d. *E. intricata* Zett., f. *longipterata* nova, ♂, Sra Javalambre, Teruel, C. Spain (ZSM)
118e. *E. intricata hibernica* Mere, ♀, Burren, Co. Clare, SW. Ireland (ZISP)
119a. ***Eupithecia satyrata* Hbn.,** ♀, Gdansk-Stoki, N. Poland (coll. GELBRECHT)
119b. *E. satyrata* Hbn., ♀, St. Petersburg obl., NW. European Russia (ZISP)
119c. *E. satyrata* Hbn., ♀, Monte Baldo, N. Italy (coll. FRIEDRICH)
119d. *E. satyrata* Hbn., ♀, Čaven, 'Yugoslavia' (ZSM)
119e. *E. satyrata* Hbn., ♂, Kirkeby Sand, Fyn, Denmark (coll. SKOU)
119f. *E. satyrata curzoni* Greg., ♂, 'Shetland Islands', United Kingdom (ZISP)
119g. *E. satyrata curzoni* Greg., ♂, 'Shetland Islands', United Kingdom (ZSM)
119h. *E. satyrata curzoni* Greg., ♀, 'Shetland Islands', United Kingdom (ZISP), reared
119i. *E. satyrata curzoni* Greg., ♂, 'Faroe Islands' (ZMUC)
119j. *E. satyrata curzoni* Greg., ♀, 'Faroe Islands' (ZMUC)
120a. ***Eupithecia cauchiata* Dup.,** ♂, Flohberg, Styria, Austria (ZSM), reared
120b. *E. cauchiata* Dup., ♂, Hulsig, Jylland, Denmark (coll. SKOU), reared
120c. *E. cauchiata* Dup., ♀, 'Silesia', S. Poland (ZISP)
120d. *E. cauchiata* Dup., ♀, Moscow, C. European Russia (ZISP)
121a. ***Eupithecia pernotata* Gn.,** ♂, Lungau, Austria (ZSM)
121b. *E. pernotata* Gn., ♀, Martigny, Wallis, Switzerland (ZSM), reared
121c. *E. pernotata* Gn., ♀, Belgorod obl., S. European Russia (ZISP)
121d. *E. pernotata* Gn., ♂, Cheljabinsk obl., S. Ural, Russia (ZISP)
121e. *E. pernotata enictata* Pellmyr & Mikkola, ♂, Pernaja, S. Finland (ZMH)
122a. ***Eupithecia extensaria* Frr.,** ♀, Wlk, Gleboczek, Poland (coll. SKOU)
122b. *E. extensaria* Frr., ♂, Lugansk, E. Ukraine (ZISP)
122c. *E. extensaria occidua* Prt., ♂, Norfolk, England, United Kingdom (ZSM)
122d. *E. extensaria occidua* Prt., ♂, Noguera, Teruel, C. Spain (ZSM)
123a. ***Eupithecia absinthiata* Cl.,** ♀, Novgorod obl., NW. European Russia (ZISP), I gen.
123b. *E. absinthiata* Cl., ♂, Schrossberg, Germany (ZSM), I. gen.
123c. *E. absinthiata* Cl., ♂, Hoplax, Helsinki, S. Finland (coll. SKOU), II gen.
123d. *E. absinthiata* Cl., ♂, Vemmenæs, Tåsinge, Denmark (coll. SKOU), II gen.
123e. *E. absinthiata* Cl., ♀, St. Petersburg obl., NW. European Russia (ZISP), reared
124a. ***Eupithecia expallidata* Dbld.,** ♂, Kassel, C. Germany (ZSM), reared
124b. *E. expallidata* Dbld., ♀, Frankenwald, Nordbayern, S. Germany (ZSM)
124c. *E. expallidata* Dbld., ♀, Sonderhausen, Thüringen, C. Germany (coll. GELBRECHT), reared
124d. *E. expallidata* Dbld., ♂, Novgorod obl., NW. European Russia (ZISP), reared
124e. *E. expallidata* Dbld., ♀, St. Petersburg obl., NW. European Russia (ZISP), reared
125a. ***Eupithecia valerianata* Hbn.,** ♀, Tåsinge, Fyn, Denmark (coll. SKOU)
125b. *E. valerianata* Hbn., ♂, Wolsig, Brandenburg, E. Germany (coll. GELBRECHT), reared
125c. *E. valerianata* Hbn., ♂, St. Petersburg obl., NW. European Russia (ZISP), reared
125d. *E. valerianata* Hbn., ♀, St. Petersburg obl., NW. European Russia (ZISP), reared
125e. *E. valerianata* Hbn., ♀, Region Riva, Bossecca, N. Italy (ZSM)

117a
117b
117c
117d
118a
118b
118c
118d
118e
119a
119b
119c
119d
119e
119f
119g
119h
119i
119j
120a
120b
120c
120d
121a
121b
121c
121d
121e
122a
122b
122c
122d
123a
123b
123c
123d
123e
124a
124b
124c
124d
124e
125a
125b
125c
125d
125e

PLATE 14 (all specimens 1.5 x natural size)

126a. ***Eupithecia assimilata* Dbld.,** ♀, St. Petersburg obl., NW. European Russia (ZISP)
126b. *E. assimilata* Dbld., ♂, Moscow obl., C. European Russia (ZISP),
126c. *E. assimilata* Dbld., ♀, Puyvert, Vaucluse, SE. France (ZSM/HERBULOT)
126d. *E. assimilata* Dbld., ♂, Kirkeby Sand, Fyn, Denmark (coll. SKOU)
126e. *E. assimilata* Dbld., ♀, Tåsinge, Fyn, Denmark (coll. SKOU)
127a. ***Eupithecia vulgata* Hw.,** ♂, Branksome, Dorset, England, United Kingdom (ZISP)
127b. *E. vulgata* Hw., ♂, Kirkeby Sand, Fyn, Denmark (coll. SKOU)
127c. *E. vulgata* Hw., ♀, Puisieux-et-Clanlieu, Aisne, N. France (ZSM/HERBULOT)
127d. *E. vulgata* Hw., ♀, St. Petersburg obl., NW. European Russia (ZISP)
127e. *E. vulgata cyrneata* Schaw., ♂, Vizzavonna, Corsica (ZSM/HERBULOT)
128a. ***Eupithecia biornata* Chr.,** ♀, Stepanovskaja kosa, Zaporozhje, SE. Ukraine (ZMKU)
128b. *E. biornata* Chr., ♂, Uljanovsk obl., E. European Russia (ZISP)
128c. *E. biornata* Chr., ♂, Astrakhan' obl., SE. European Russia (ZISP)
128d. *E. biornata* Chr., ♂, Nea Kerdylia, Strymon-Delta, N. Greece (HNHM)
128e. *E. biornata* Chr., ♀, Strymon-Delta, N. Greece (coll. FIBIGER)
129a. ***Eupithecia sutiliata* Chr.,** ♀, Astrakhan' obl., SE. European Russia (ZISP)
129b. *E. sutiliata* Chr., ♀, Astrakhan' obl., SE. European Russia (ZISP)
129c. *E. sutiliata* Chr., ♀, Mt. Chelmos, Peloponnes, S. Greece (coll. FIBIGER)
129d. *E. sutiliata* Chr., ♀, Daghestan, N. Caucasus, S. Russia (ZISP)
129e. *E. sutiliata* Chr., ♀, Daghestan, N. Caucasus, S. Russia (ZISP)
130a. ***Eupithecia immundata* Lien. & Zell.,** ♀, Novgorod obl., NW. European Russia (ZISP), reared
130b. *E. immundata* Lien. & Zell., ♀, St. Petersburg obl., NW. European Russia (ZISP), reared
130c. *E. immundata* Lien. & Zell., ♂, Kassel, C. Germany (ZSM)
130d. *E. immundata* Lien. & Zell., ♀, Kassel, C. Germany (ZSM)
130e. *E. immundata* Lien. & Zell., ♀, Sonderhausen, Thüringen, C. Germany (coll. GELBRECHT), reared
131a. ***Eupithecia thalictrata* Püng.,** ♀, Ukmerge, Lithuania (ZISP)
131b. *E. thalictrata* Püng., ♂, Traunsteingebiet, Oberösterreich, Austria (ZSM)
131c. *E. thalictrata* Püng., ♀, Schnalstal, S. Tirol, NE. Italy (ZSM), reared
131d. *E. thalictrata* Püng., ♀, Burgeis, S. Tirol, NE. Italy (ZSM), reared
131e. *E. thalictrata* Püng., ♂, Offenseegebiet, Oberösterreich, Austria (ZSM), reared
132a. ***Eupithecia addictata* Dtze.,** ♂, Rjazanovka, S. Primorje, Far East Russia (ZISP)
132b. *E. addictata* Dtze., ♀, Mt. Olymp, N. Greece (ZMUC)
132c. *E. addictata* Dtze., ♂, Mt. Olymp, Thessalia, N. Greece (coll. EMEM in ZSM)
132d. *E. addictata* Dtze., ♀, Bátorliget, Hungary (HNHM)
132e. *E. addictata* Dtze., ♂, Lago di Cavazzo, Udine, N. Italy (coll. EMEM in ZSM)
133a. ***Eupithecia exiguata* Hbn.,** ♂, Kiev obl., N. Ukraine (ZMKU)
133b. *E. exiguata* Hbn., ♀, Dorpat, Estonia (ZSM)
133c. *E. exiguata* Hbn., ♂, St. Petersburg obl., NW. European Russia (ZISP)
133d. *E. exiguata* Hbn., ♂, Tranekær, Langeland, Denmark (coll. SKOU)
133e. *E. exiguata* Hbn., f. *muricolor* Prt., ♀, Gighr, E. Aberdeenshire, Scotland, United Kingdom (ZISP)
134a. ***Eupithecia druentiata* Dtze.,** ♂, Garda, Veneto, N. Italy (ZSM)
134b. *E. druentiata* Dtze., ♂, Torbole, Trentino, N. Italy (ZSM)
134c. *E. druentiata* Dtze., ♂, Ochrid, Macedonia (ZSM)
134d. *E. druentiata* Dtze., ♀, Ochrid, Macedonia (ZSM)
135a. ***Eupithecia moecha* Dtze.,** ♂, Yanvartzevo, NW. Kazakhstan (ZISP)
135b. *E. moecha* Dtze., ♂, Samara, E. European Russia (ZISP)
135c. *E. moecha* Dtze., ♂, Yanvartzevo, NW. Kazakhstan (ZISP)
135d. *E. moecha* Dtze., ♀, Burlinsk, NW. Kazakhstan (ZISP)
136a. ***Eupithecia denotata* Hbn.,** ♀, Novgorod obl., NW. European Russia (ZISP)
136b. *E. denotata* Hbn., ♀, St. Petersburg obl., NW. European Russia (ZISP)
136c. *E. denotata* Hbn., f. *jasioneata* Crewe, ♂, N. Cornwall, England, United Kingdom (ZISP), reared
136d. *E. denotata hellenata* Schütze, ♂, Enipets-Tal, Olympos, N. Greece (ZSM), holotype
136e. *E. denotata hellenata* Schütze, ♀, Enipets-Tal, Olympos, N. Greece (ZSM), paratype

126a
126b
126c
126d
126e
127a
127b
127c
127d
127e
128a
128b
128c
128d
128e
129a
129b
129c
129d
129e
130a
130b
130c
130d
130e
131a
131b
131c
131d
131e
132a
132b
132c
132d
132e
133a
133b
133c
133d
133e
134a
134b
134c
134d
135a
135b
135c
135d
136a
136b
136c
136d
136e

PLATE 15 (all specimens 1.5 x natural size)

137a. ***Eupithecia weissi* Prt.**, ♂, Albarracin, Aragon, C. Spain (ZSM)
137b. *E. weissi* Prt., ♀, Gurb, Barcelona, NE. Spain (coll. SKOU)
137c. *E. weissi* Prt., ♂, Tosos, Zaragoza, NE. Spain (coll. SKOU)
137d. *E. weissi* Prt., ♂, Aureille, Bouches-du-Rhône, S. France (ZSM)
137e. *E. weissi* Prt., ♂, Aureille, Bouches-du-Rhône, S. France (ZSM)
137f. *E. weissi* Prt., ♀, Albarracin, Aragon, C. Spain (BMNH)
137g. *E. weissi* Prt., ♀, Aureille, Bouches-du-Rhône, S. France (ZSM)
137h. *E. weissi* Prt., ♀, Aureille, Bouches-du-Rhône, S. France (ZSM)
137i. *E. weissi* Prt., ♀, Sierra Nevada, Granada, S. Spain (ZMUC)
137j. *E. weissi* Prt., ♀, Tizi-n Tichka, Haute Atlas, Morocco (ZSM)
138a. ***Eupithecia pauxillaria* Bsd.**, ♀, Leonidon, Peloponnes, S. Greece (ZSM)
138b. *E. pauxillaria* Bsd., ♀, Puyvert, Vaucluse, SE. France (ZSM/HERBULOT)
138c. *E. pauxillaria* Bsd., ♀, Evisa, Corsica (ZSM)
138d. *E. pauxillaria* Bsd., ♂, Mödling, Austria (ZSM), reared
138e. *E. pauxillaria* Bsd., ♀, Budaörs, Budapest, Hungary (coll. GELBRECHT), reared
139a. ***Eupithecia santolinata* Mab.**, ♂, Granada, Andalucia, S. Spain (ZSM)
139b. *E. santolinata* Mab., ♀, Albarracin, Teruel, C. Spain (coll. SKOU)
139c. *E. santolinata* Mab., ♀, Veleta, Sierra Nevada, Granada, S. Spain (ZSM)
139d. *E. santolinata* Mab., ♀, Veleta, Sierra Nevada, Granada, S. Spain (ZSM)
140a. ***Eupithecia millefoliata* Rössl.**, ♀, Gerasdorf, Austria (ZSM), reared
140b. *E. millefoliata* Rössl., ♂, Floritzdorf, Austria (ZSM), reared
140c. *E. millefoliata* Rössl., ♀, St. Petersburg obl., NW. European Russia (ZISP), reared
140d. *E. millefoliata* Rössl., ♀, Novgorod obl., NW. European Russia (ZISP), reared
141a. ***Eupithecia spadiceata* Zerny**, ♂, Castel d. Ieri, Abruzzen, C. Italy (HNHM)
141b. *E. spadiceata* Zerny, ♀, Ivalovgrad, E. Rhodopi Mts., S. Bulgaria (ZISP)
141c. *E. spadiceata* Zerny, ♀, Fót, Hungary (HNHM)
141d. *E. spadiceata* Zerny, ♀, Kisvaszar, Hungary (HNHM)
141e. *E. spadiceata* Zerny, ♀, Castel d. Ieri, Abruzzen, C. Italy (HNHM)
141f. *E. spadiceata* Zerny, ♂, V. Kazanishe, Daghestan, N. Caucasus, S. Russia (ZISP)
141g. *E. spadiceata* Zerny, ♀, Valley of Arpa river, Armenia (ZISP)
141h. *E. spadiceata* Zerny, ♂, Ürgüp, Kappadocien, Nevşehir, C. Turkey (ZSM)
142a. ***Eupithecia icterata* Vill.**, f. *cognata* Stph., ♀, Col de la Cayolle, Hautes Alpes, SE. France (ZSM)
142b. *E. icterata* Vill., f. *cognata* Stph., ♀, Col d'Allos, Hautes Alpes, SE. France (ZMUC)
142c. *E. icterata* Vill., f. *cognata* Stph., ♀, Le Lautaret, Hautes Alpes, SE. France (ZSM)
142d. *E. icterata* Vill., f. *cognata* Stph., ♀, P. Yavorov Chalet, Pirin Mts., SW. Bulgaria (ZISP)
142e. *E. icterata* Vill., f. *subfulvata* Hw., ♀, St. Petersburg obl., NW. European Russia (ZISP)
142f. *E. icterata* Vill., f. *subfulvata* Hw., ♀, St. Petersburg obl., NW. European Russia (ZISP)
142g. *E. icterata* Vill., f. *subfulvata* Hw., ♀, Amfreville, Eure, NW. France (ZSM)
142h. *E. icterata* Vill., f. *subfulvata* Hw., ♀, Tring, Herts, England, United Kingdom (ZISP), reared
143a. ***Eupithecia succenturiata* L.**, ♀, Fürstenwalde, Brandenburg, E. Germany (coll. GELBRECHT)
143b. *E. succenturiata* L., ♀, Kiev obl., N. Ukraine (ZMKU)
143c. *E. succenturiata* L., ♀, St. Petersburg obl., NW. European Russia (ZISP), reared
143d. *E. succenturiata* L., ♀, Sagar/Neisse, Brandenburg, E. Germany (coll. GELBRECHT)

137a
137b
137c
137d
137e
137f
137g
137h
137i
137j
138a
138b
138c
138d
138e
139a
139b
139c
139d
140a
140b
140c
140d
141a
141b
141c
141d
141e
141f
141g
141h
142a
142b
142c
142d
142e
142f
142g
142h
143a
143b
143c
143d

PLATE 16 (all specimens 1.5 x natural size)

144a. ***Eupithecia semigraphata* Bruand**, ♂, Digne, Basses Alpes, SE. France (coll. GELBRECHT)
144b. *E. semigraphata* Bruand, ♂, Ostarije, Croatia (coll. SKOU)
144c. *E. semigraphata* Bruand, ♂, Habichtswald, Kassel, C. Germany (ZSM)
144d. *E. semigraphata* Bruand, ♀, Evisa, Corsica (ZSM)
144e. *E. semigraphata* Bruand, ♀, Agia Triada, Prevesa, W. Greece (coll. FRIEDRICH)
145a. ***Eupithecia impurata* Hbn.**, f. *germanicata* Schütze, ♀, Edersee, Kassel, C. Germany (ZSM), paratype, reared.
145b. *E. impurata* Hbn., f. *badeniata* Schütze, ♀, Baden, Kaiserstuhl, SW. Germany (ZSM), paratype, reared
145c. *E. impurata* Hbn., f. *franconiata* Schütze, ♀, Pottenstein, Jura, E. France (ZSM), paratype, reared
145d. *E. impurata* Hbn., f. *langeata* Schütze, ♀, Zschopautal, Sachsen, N. Germany (ZSM), paratype, reared
145e. *E. impurata* Hbn., f. *thueringiata* Schütze, ♀, Blankenstein, Thüringen, C. Germany (ZSM), paratype, reared
145f. *E. impurata* Hbn., f. *gremmingerata* Schütze, ♀, Schwarzwald, Wutach-Gebiet, SW. Germany (ZSM), paratype, reared
145g. *E. impurata* Hbn., ♀, St. Vallier de Thiey, Alpes Maritimes, SE. France (coll. SKOU)
145h. *E. impurata* Hbn., ♀, Alcale de la Selva, Aragon, Spain (ZSM/HERBULOT)
146a. ***Eupithecia poecilata* Püng.**, ♂, Col de Vizzavona, Corsica (ZSM)
146b. *E. poecilata* Püng., ♀, Aritzo, Sardinia (HNHM)
146c. *E. poecilata* Püng., ♀, Aritzo, Sardinia (HNHM)
146d. *E. poecilata* Püng., ♀, Gennargentu, Sardinia (ZSM)
146e. *E. poecilata* Püng., ♀, Gennargentu, Sardinia (ZSM)
147a. ***Eupithecia fuscicostata* Chr.**, ♂, Domogled Mt., SW. Romania (HNHM), [*domogledana* Vojnits & Szabó, paratype]
147b. *E. fuscicostata* Chr., ♂, Domogled Mt., SW. Romania (HNHM), [*domogledana* Vojnits & Szabó, paratype]
147c. *E. fuscicostata* Chr., ♀, Domogled Mt., SW. Romania (HNHM), [*domogledana* Vojnits & Szabó, paratype]
147d. *E. fuscicostata* Chr., ♂, Ochrid, Macedonia (ZSM)
147e. *E. fuscicostata* Chr., ♂, Parnassos Oros, Volotia, C. Greece (coll. SKOU)
147f. *E. fuscicostata* Chr., ♀, Gedzhanan, Armenia (ZISP)
147g. *E. fuscicostata* Chr., ♀, Gedzhanan, Armenia (ZISP)
147h. *E. fuscicostata* Chr., ♀, Kackar Daglari, Artvin, NE. Turkey (coll. GELBRECHT)
147i. *E. fuscicostata* Chr., ♀, Mescit Daglari, NE. Turkey (coll. GELBRECHT)
147j. *E. fuscicostata* Chr., ♀, Shimshak, Elburs Mts., N. Iran (HNHM)
148a. ***Eupithecia denticulata* Tr.**, ♂, Baden, SW. Germany (ZSM), reared
148b. *E. denticulata* Tr., ♂, Bensheim-Gronau, Hessen, W. Germany (coll. GELBRECHT), reared
148c. *E. denticulata* Tr., ♂, Bensheim-Gronau, Hessen, W. Germany (coll. GELBRECHT), reared
148d. *E. denticulata* Tr., ♀, Aydintepe, Agri, E. Turkey (ZISP)
148e. *E. denticulata* Tr., ♀, Camlidere, Ankara, C. Turkey (ZISP)
149a. ***Eupithecia subumbrata* Den. & Schiff.,** ♀, Cèdre, Hautes-Pyrénées, SW. France (ZSM)
149b. *E. subumbrata* Den. & Schiff., ♀, St. Petersburg obl., NW. European Russia (ZISP)
149c. *E. subumbrata* Den. & Schiff., ♀, St. Petersburg obl., NW. European Russia (ZISP)
149d. *E. subumbrata* Den. & Schiff., ♀, Cheljabinsk obl., S. Ural, Russia (ZISP)
149e. *E. subumbrata* Den. & Schiff., ♀, Cheljabinsk obl., S. Ural, Russia (ZISP)
150a. ***Eupithecia orphnata* Pet.**, ♀, Leon, NW. Spain (ZMUC)
150b. *E. orphnata* Pet., ♀, Majastres, Hautes Alpes, SE. France (ZMUC)
150c. *E. orphnata* Pet., ♀, Ochrid, Macedonia (ZSM)
150d. *E. orphnata* Pet., ♀, Pernaja, S. Finland (ZMH)
150e. *E. orphnata* Pet., ♀, Uljanovsk obl., E. European Russia (ZISP)
151a. ***Eupithecia subfuscata* Hw.**, ♀, Stensund, Södermanland, C. Sweden (coll. SKOU)
151b. *E. subfuscata* Hw., ♀, St. Petersburg obl., NW. European Russia (ZISP)
151c. *E. subfuscata* Hw., ♀, Peshtera, Rhodopi Mts., S. Bulgaria (coll. SKOU)
151d. *E. subfuscata* Hw., ♂, Novgorod obl., NW. European Russia (ZISP), reared
151e. *E. subfuscata* Hw., ♀, Novgorod obl., NW. European Russia (ZISP), reared

144a
144b
144c
144d
144e
145a
145b
145c
145d
145e
145f
145g
145h
146a
146b
146c
146d
146e
147a
147b
147c
147d
147e
147f
147g
147h
147i
147j
148a
148b
148c
148d
148e
149a
149b
149c
149d
149e
150a
150b
150c
150d
150e
151a
151b
151c
151d
151e

Genitalia plates

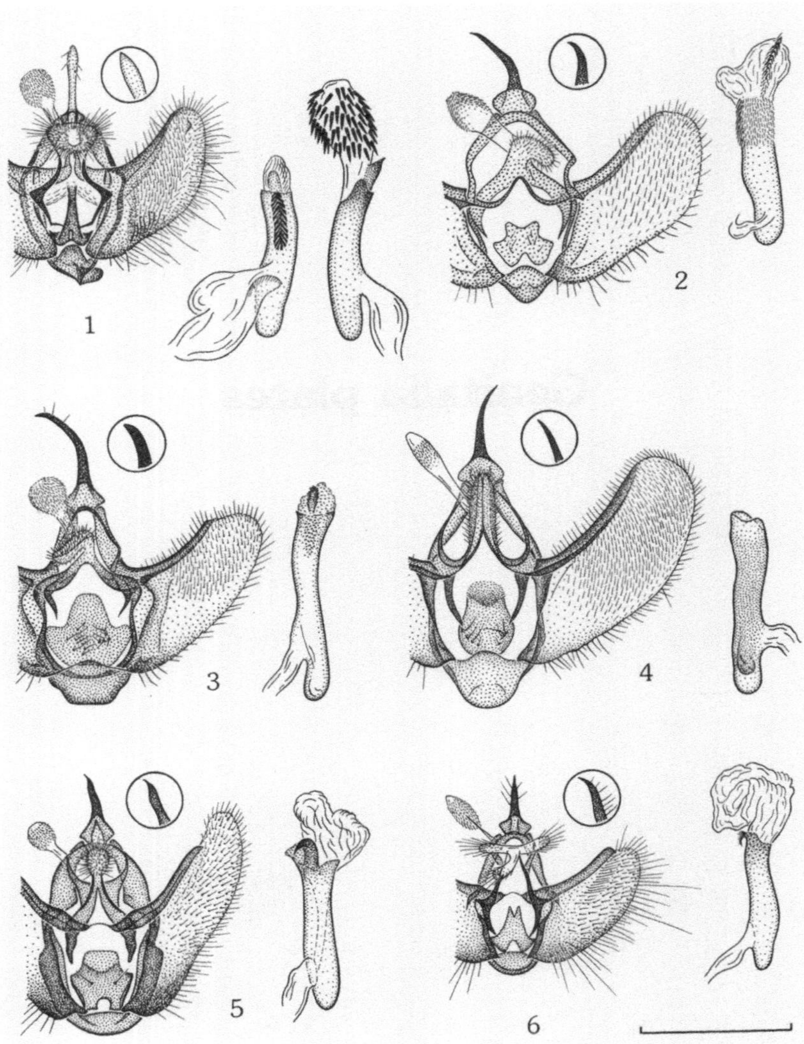

MALE GENITALIA (scale bar = 1 mm; lateral view of uncus enlarged)

1. *Pseudobaptria bogumilaria* Rbl., Bosnia [aedeagus dorsal and lateral, with vesica uneverted and everted]
2. *Mesotype didymata* L., NW. European Russia [aedeagus dorsal, with vesica everted]
3. *Mesotype parallelolineata* Retz., NW. European Russia [aedeagus lateral, with vesica semi-everted]
4. *Mesotype verberata* Scop., S. Germany [aedeagus lateral, with vesica uneverted]
5. *Perizoma affinitata* Steph., NW. European Russia [aedeagus lateral, with vesica everted]
6. *Perizoma alchemillata* L., NW. European Russia [aedeagus lateral, with vesica everted]

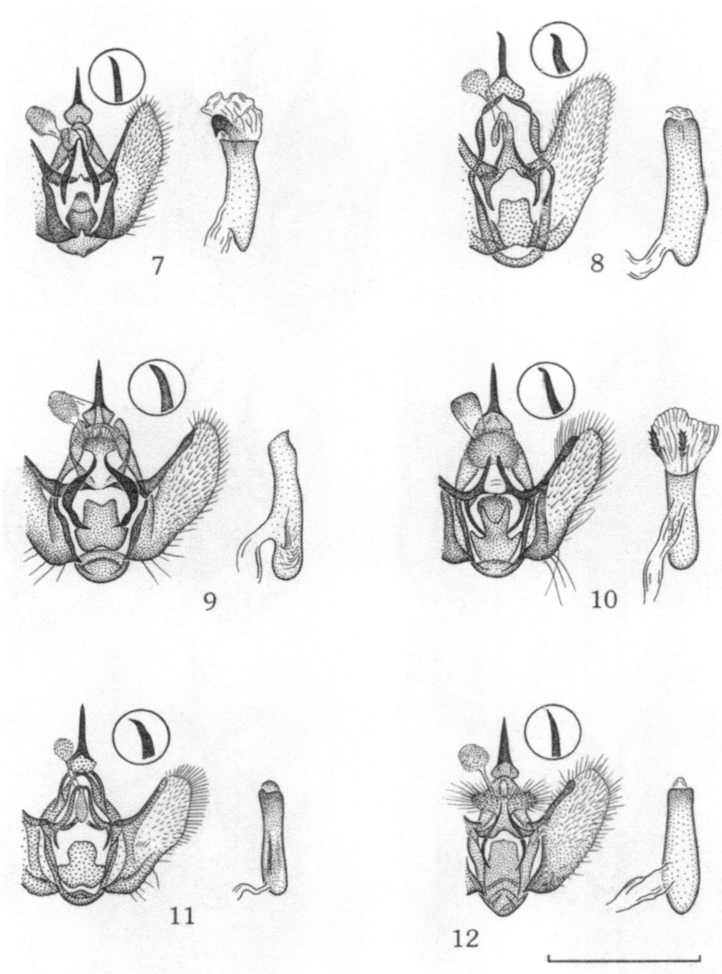

MALE GENITALIA (scale bar = 1 mm; lateral view of uncus enlarged)

7. *Perizoma hydrata* Tr., S. Poland [aedeagus lateral, with vesica semi-everted]
8. *Perizoma lugdunaria* H.-S., Austria [aedeagus lateral, with vesica uneverted]
9. *Perizoma bifaciata* Hw., NW. European Russia [aedeagus lateral, with vesica uneverted]
10. *Perizoma flavosparsata* Wagn., C. Spain [aedeagus dorsal, with vesica everted]
11. *Perizoma minorata* Tr., Austria [aedeagus lateral, with vesica uneverted]
12. *Perizoma blandiata* Den. & Schiff., NW. European Russia [aedeagus lateral, with vesica uneverted]

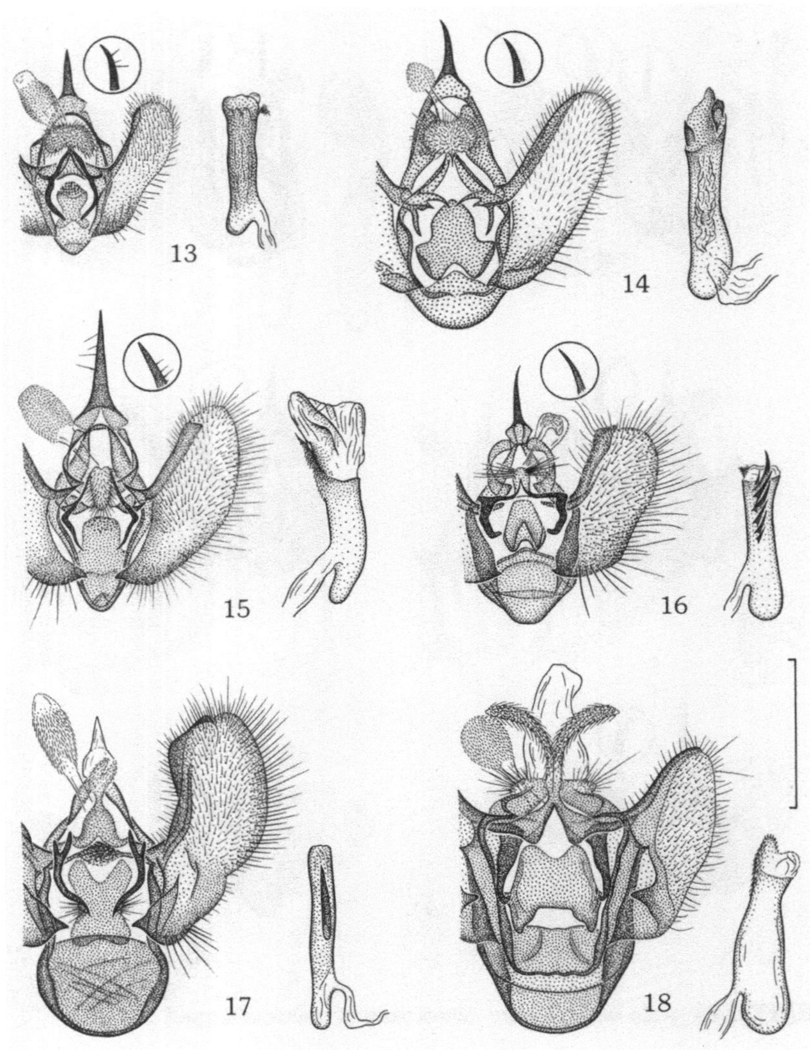

MALE GENITALIA (scale bar = 1 mm; lateral view of uncus enlarged)

13. *Perizoma albulata* Den. & Schiff., NW. European Russia [aedeagus lateral, with vesica uneverted]
14. *Perizoma flavofasciata* (Thnbg.), NW. European Russia [aedeagus dorsal, with vesica semi-everted]
15. *Perizoma obsoletata* H.-S., Austria [aedeagus lateral, with vesica everted]
16. *Perizoma incultaria* H.-S., Slovenia [aedeagus lateral, with vesica uneverted]
17. *Martania taeniata* Steph., NW. European Russia [aedeagus lateral, with vesica uneverted]
18. *Gagitodes sagittata* F., NW. European Russia [aedeagus lateral, with vesica uneverted]

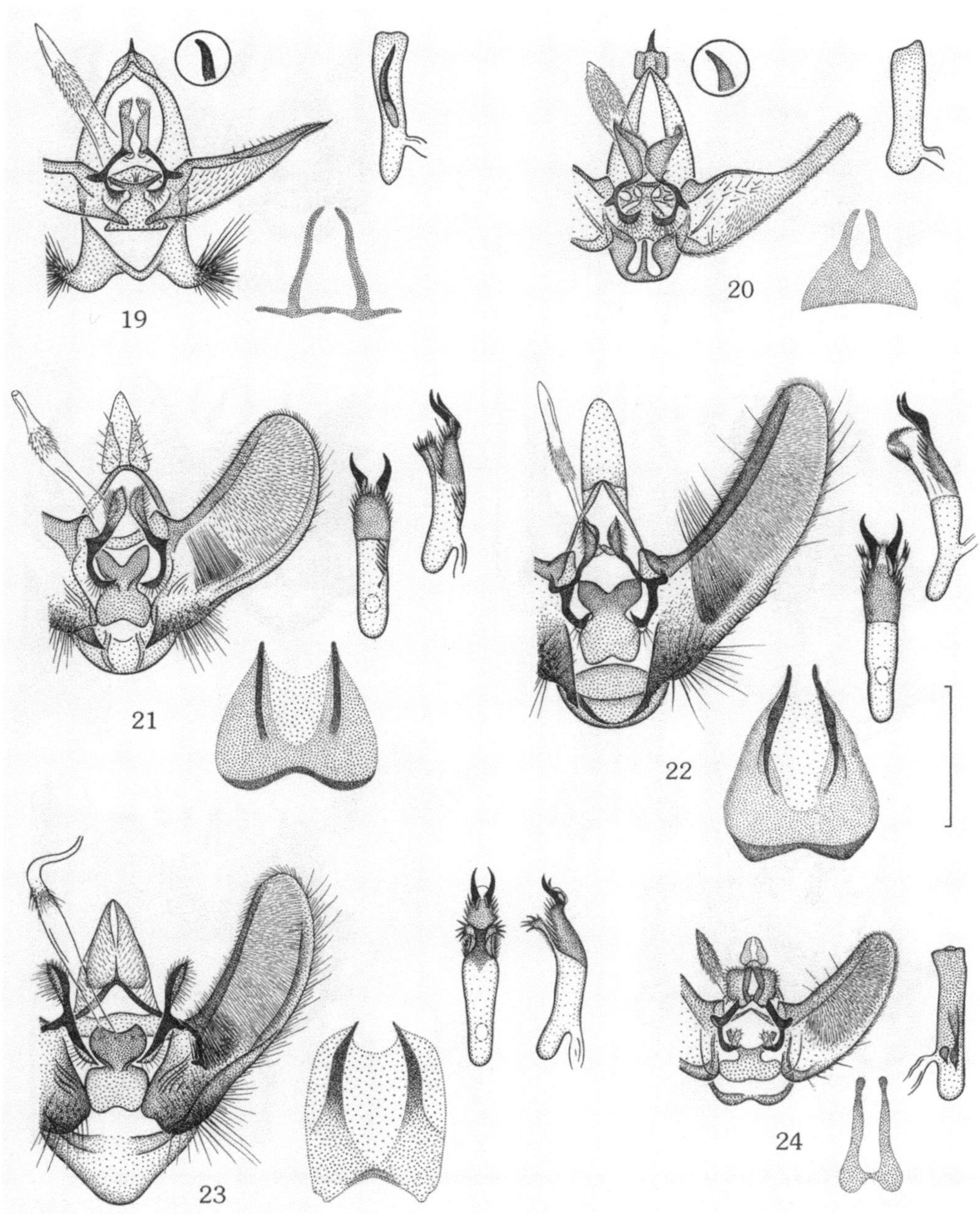

MALE GENITALIA (scale bar = 1 mm; with sternum A8; lateral view of uncus enlarged)

19. *Gymnoscelis rufifasciata* Hw., W. Azerbaijan [aedeagus lateral, with vesica uneverted]
20. *Chloroclystis v-ata* Hw., NE. Georgia [aedeagus lateral, with vesica uneverted]
21. *Pasiphila chloerata* Mab., NW. European Russia [aedeagus ventral and lateral, with vesica uneverted]
22. *Pasiphila rectangulata* L., NW. European Russia [aedeagus ventral and lateral, with vesica uneverted]
23. *Pasiphila debiliata* Hbn., NW. European Russia [aedeagus ventral and lateral, with vesica uneverted]
24. *Eupithecia haworthiata* Dbld., E. Germany [aedeagus lateral, with vesica uneverted]

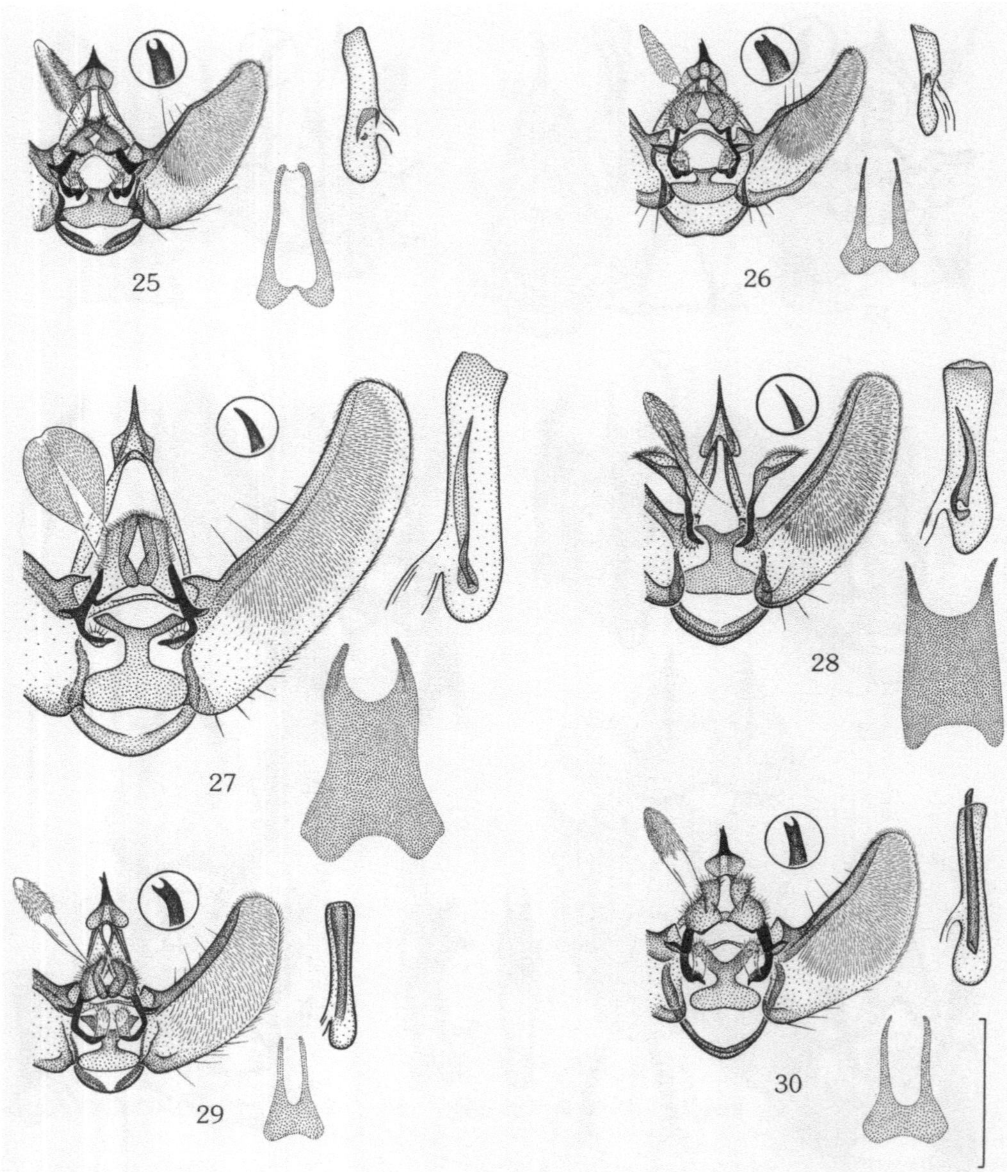

MALE GENITALIA (scale bar = 1 mm; with sternum A8; lateral view of uncus enlarged)

25. *Eupithecia tenuiata* Hbn., NW. European Russia [aedeagus lateral, with vesica uneverted]
26. *Eupithecia inturbata* Hbn., NW. European Russia [aedeagus lateral, with vesica uneverted]
27. *Eupithecia abietaria* Goeze, NW. European Russia [aedeagus lateral, with vesica uneverted]
28. *Eupithecia analoga* Djak., NW. European Russia [aedeagus lateral, with vesica uneverted]
29. *Eupithecia linariata* Den. & Schiff., NW. European Russia [aedeagus lateral, with vesica uneverted]
30. *Eupithecia pulchellata* Steph., United Kingdom (England) [aedeagus lateral, with vesica uneverted]

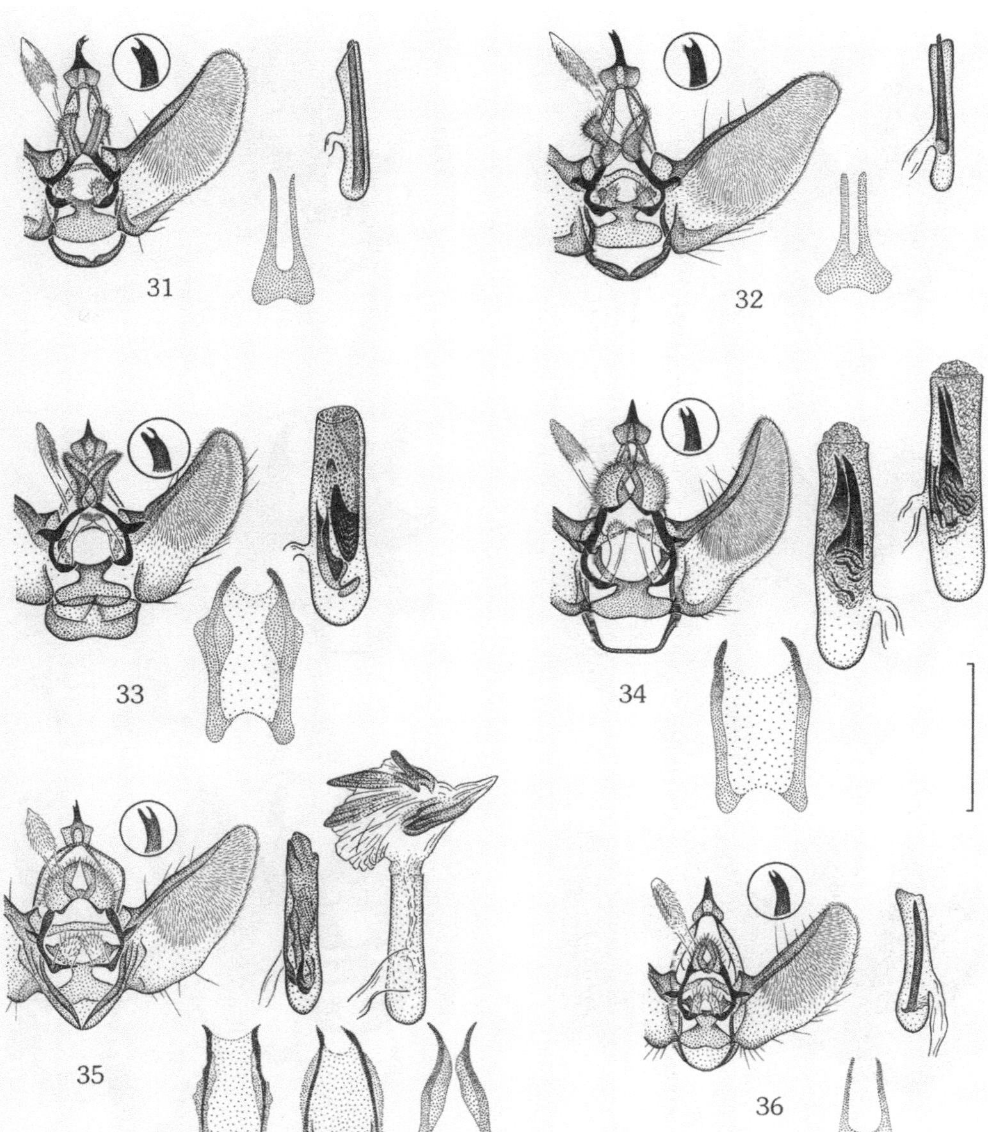

MALE GENITALIA (scale bar = 1 mm; with sternum A8; lateral view of uncus enlarged)

31. *Eupithecia pyreneata* Mab., Austria [aedeagus lateral, with vesica uneverted]
32. *Eupithecia laquaearia* H.-S., Sicily [aedeagus lateral, with vesica uneverted]
33. *Eupithecia ultimaria* Bsd., C. Italy [aedeagus lateral, with vesica uneverted]
34. *Eupithecia usbeca* Viid., 'Tura' (Central Asia) [aedeagus two lateral, with vesica uneverted]
35. *Eupithecia minusculata* Alph., S. European Russia [aedeagus lateral and dorsal, with vesica uneverted and everted]
36. *Eupithecia pantellata* Mill., Sicily [aedeagus lateral, with vesica uneverted]

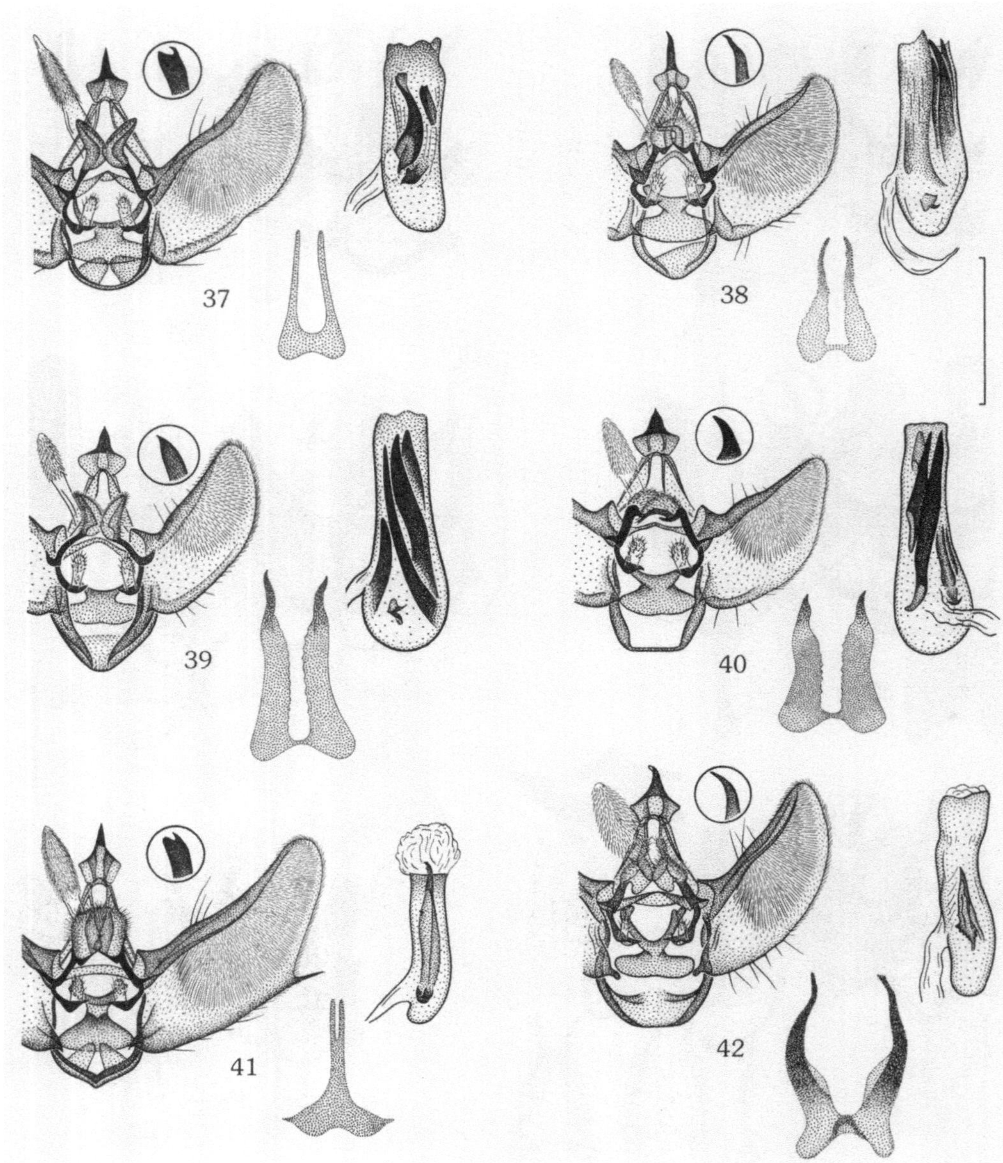

MALE GENITALIA (scale bar = 1 mm; with sternum A8; lateral view of uncus enlarged)

37. *Eupithecia plumbeolata* Hw., NW. European Russia [aedeagus lateral, with vesica uneverted]
38. *Eupithecia pygmaeata* Hbn., NW. European Russia [aedeagus lateral, with vesica uneverted]
39. *Eupithecia undata* Frr., Austria [aedeagus lateral, with vesica uneverted]
40. *Eupithecia fennoscandica* Knb., NE. European Russia (Polar Ural) [aedeagus dorsal, with vesica uneverted]
41. *Eupithecia silenata* Assm., S. Poland [aedeagus lateral, with vesica semi-everted]
42. *Eupithecia carpophagata* Stgr., C. Italy [aedeagus lateral, with vesica uneverted]

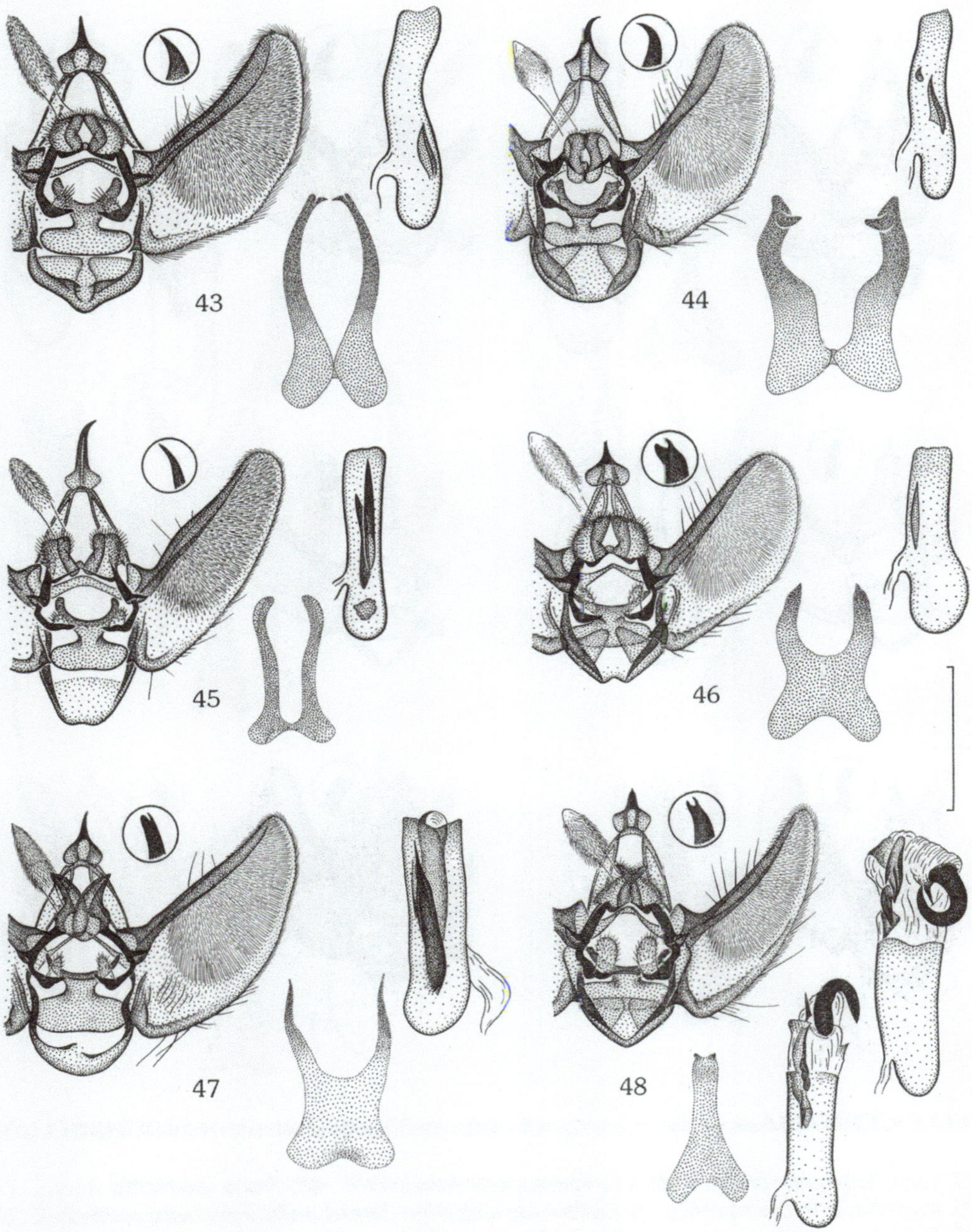

MALE GENITALIA (scale bar = 1 mm; with sternum A8; lateral view of uncus enlarged)

43. *Eupithecia venosata* F., NW. European Russia [aedeagus lateral, with vesica uneverted]
44. *Eupithecia schiefereri* Boh., S. Ukraine (Crimea) [aedeagus lateral, with vesica uneverted]
45. *Eupithecia silenicolata* Mab., SE. France [aedeagus lateral, with vesica uneverted]
46. *Eupithecia alliaria* Stgr., C. European Russia [aedeagus lateral, with vesica uneverted]
47. *Eupithecia dissertata* Püng., Slovakia [aedeagus lateral, with vesica uneverted]
48. *Eupithecia cocciferata* Mill., S. France [aedeagus two lateral, with vesica semi-everted and everted]

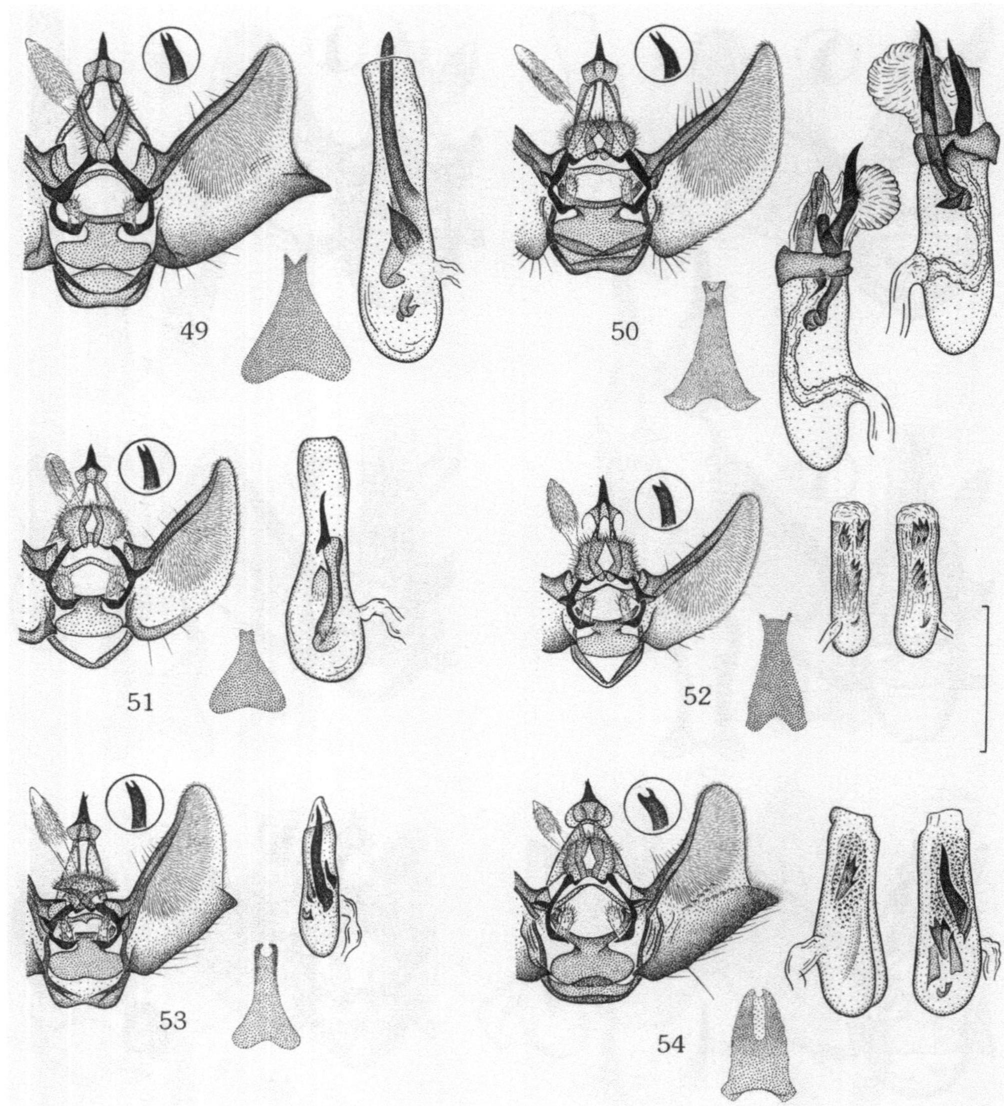

MALE GENITALIA (scale bar = 1 mm; with sternum A8; lateral view of uncus enlarged)

49. *Eupithecia abbreviata* Steph., C. France [aedeagus lateral, with vesica uneverted]
50. *Eupithecia lentiscata* Mab., S. Greece [aedeagus two lateral, with vesica semi-everted]
51. *Eupithecia dodoneata* Gn., S. France [aedeagus lateral, with vesica uneverted]
52. *Eupithecia reisserata* Pinker, N. Asian Turkey (paratype) [aedeagus two lateral, with vesica uneverted]
53. *Eupithecia massiliata* Mill., NE. Algeria [aedeagus ventral, with vesica uneverted]
54. *Eupithecia extremata* F., Corsica [aedeagus two lateral, with vesica uneverted]

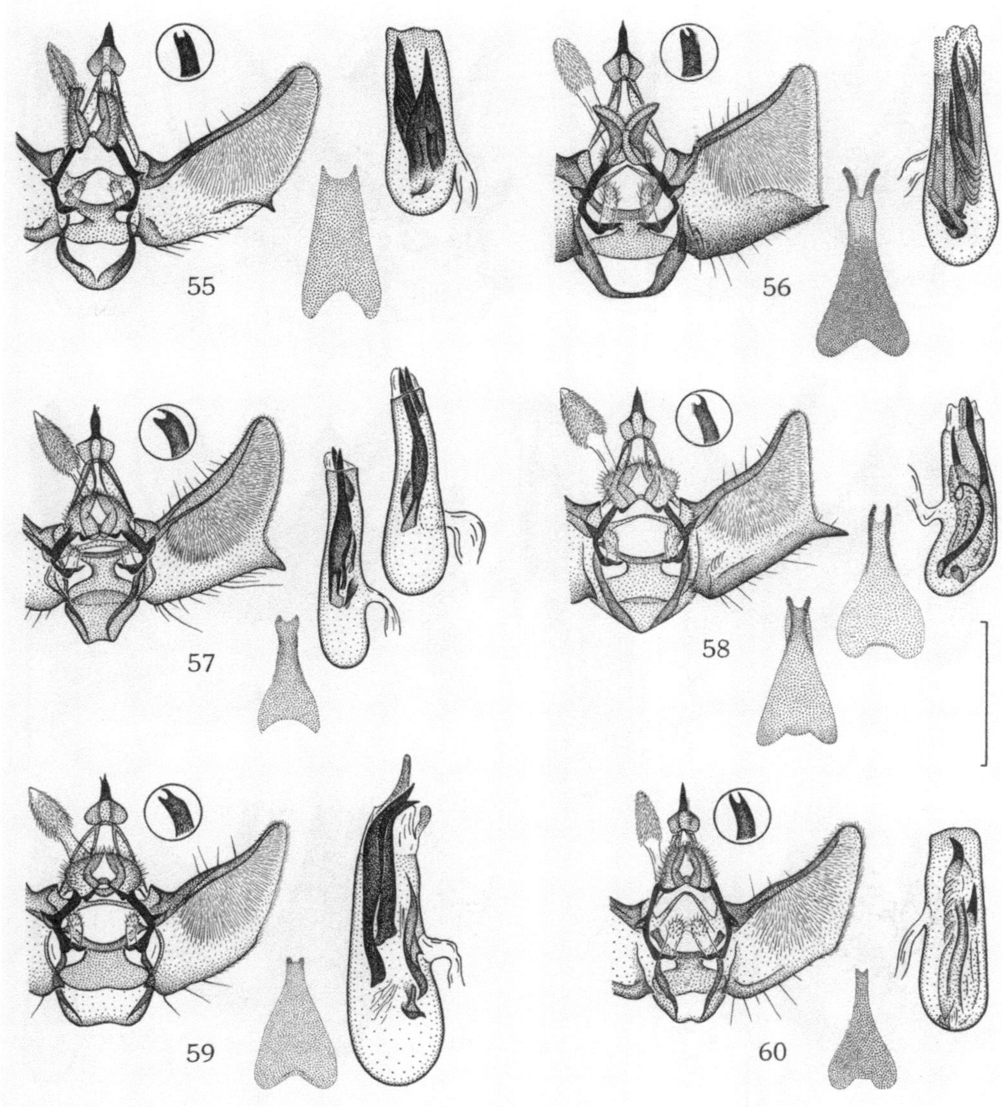

MALE GENITALIA (scale bar = 1 mm; with sternum A8; lateral view of uncus enlarged)

55. *Eupithecia scopariata* Rmbr., S. France [aedeagus lateral, with vesica uneverted]
56. *Eupithecia pusillata* Den. & Schiff., NW. European Russia [aedeagus ventral, with vesica uneverted]
57. *Eupithecia sardoa* Dtze., Corsica [aedeagus lateral and ventral, with vesica uneverted]
58. *Eupithecia ericeata* Rmbr., Austria (reared) [aedeagus lateral, with vesica uneverted; two diverse sternum A8]
59. *Eupithecia phoeniceata* Rmbr., N. Spain [aedeagus lateral, with vesica uneverted]
60. *Eupithecia oxycedrata* Rmbr., S. France [aedeagus lateral, with vesica uneverted]

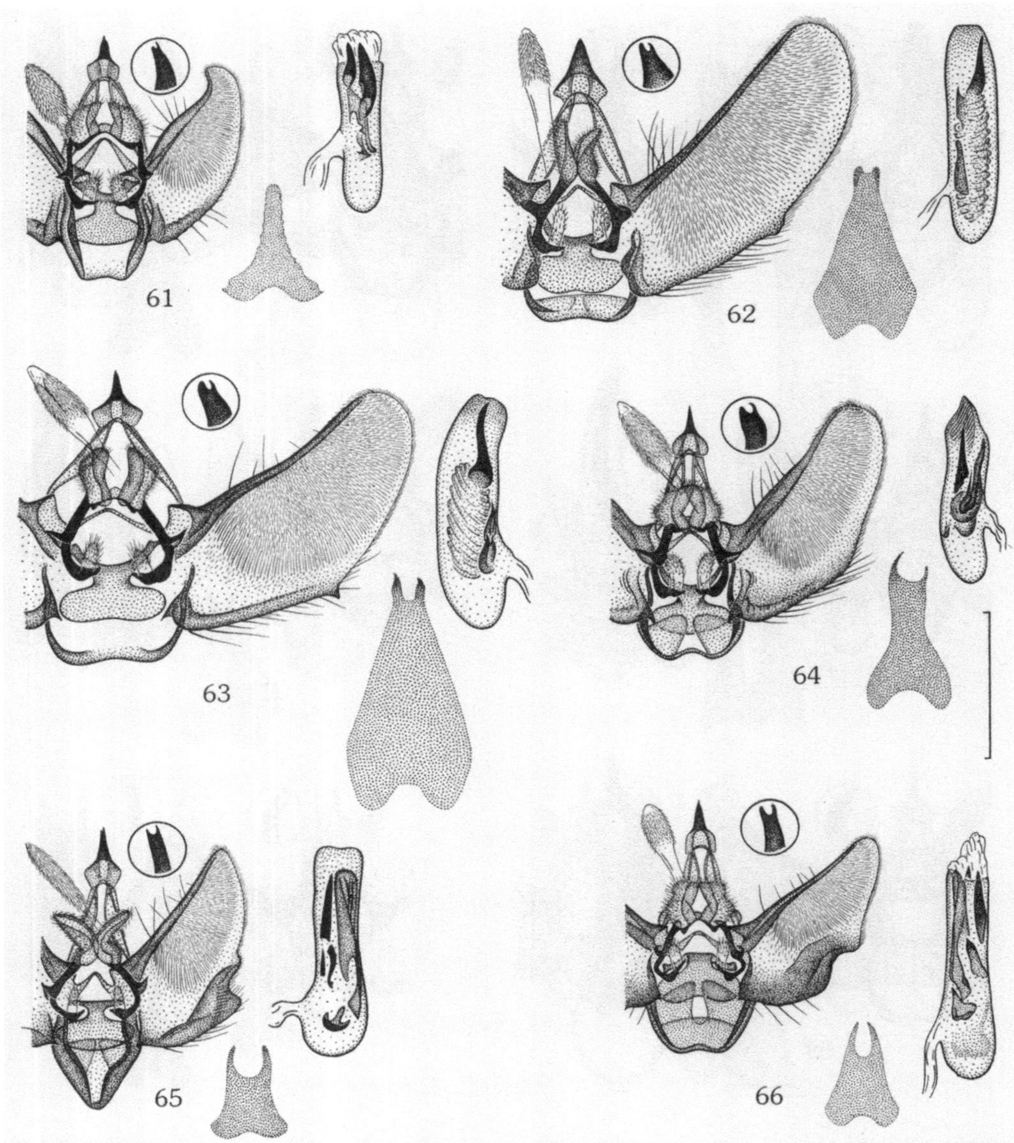

MALE GENITALIA (scale bar = 1 mm; with sternum A8; lateral view of uncus enlarged)

61. *Eupithecia rosmarinata* Mill., S. France [aedeagus lateral, with vesica uneverted]
62. *Eupithecia barteli* Dtze., NW. Kazakhstan (Uralsk) [aedeagus lateral, with vesica uneverted]
63. *Eupithecia tripunctaria* H.-S., N. Germany [aedeagus lateral, with vesica uneverted]
64. *Eupithecia virgaureata* Dbld., N. European Russia [aedeagus lateral, with vesica uneverted]
65. *Eupithecia tantillaria* Bsd., NW. European Russia [aedeagus lateral, with vesica uneverted]
66. *Eupithecia liguriata* Mill., SE. France [aedeagus lateral, with vesica uneverted]

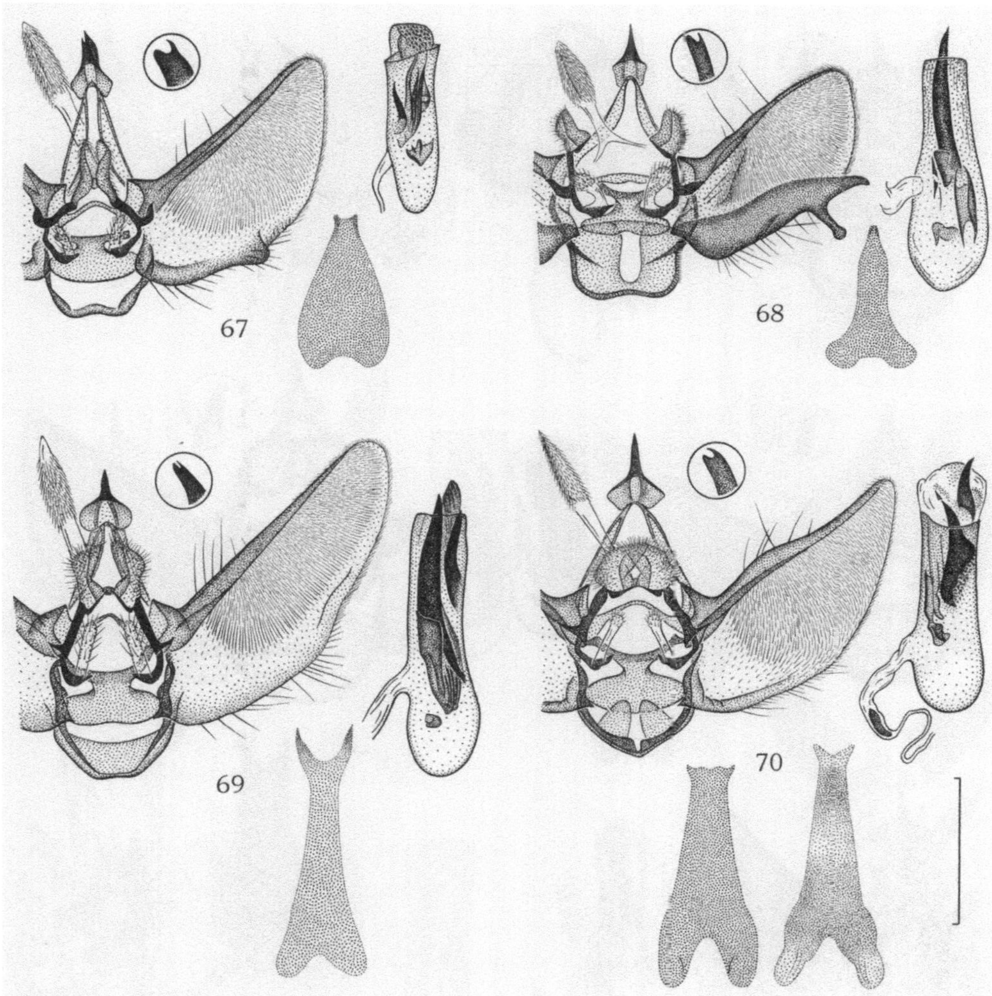

MALE GENITALIA (scale bar = 1 mm; with sternum A8; lateral view of uncus enlarged)

67. *Eupithecia lariciata* Frr., NW. European Russia [aedeagus lateral, with vesica uneverted]
68. *Eupithecia lanceata* Hbn., NW. European Russia [aedeagus lateral, with vesica uneverted]
69. *Eupithecia groenblomi* Urb., S. Finland [aedeagus lateral, with vesica uneverted]
70. *Eupithecia selinata* H.-S., S. European Russia [aedeagus lateral, with vesica semi-everted]

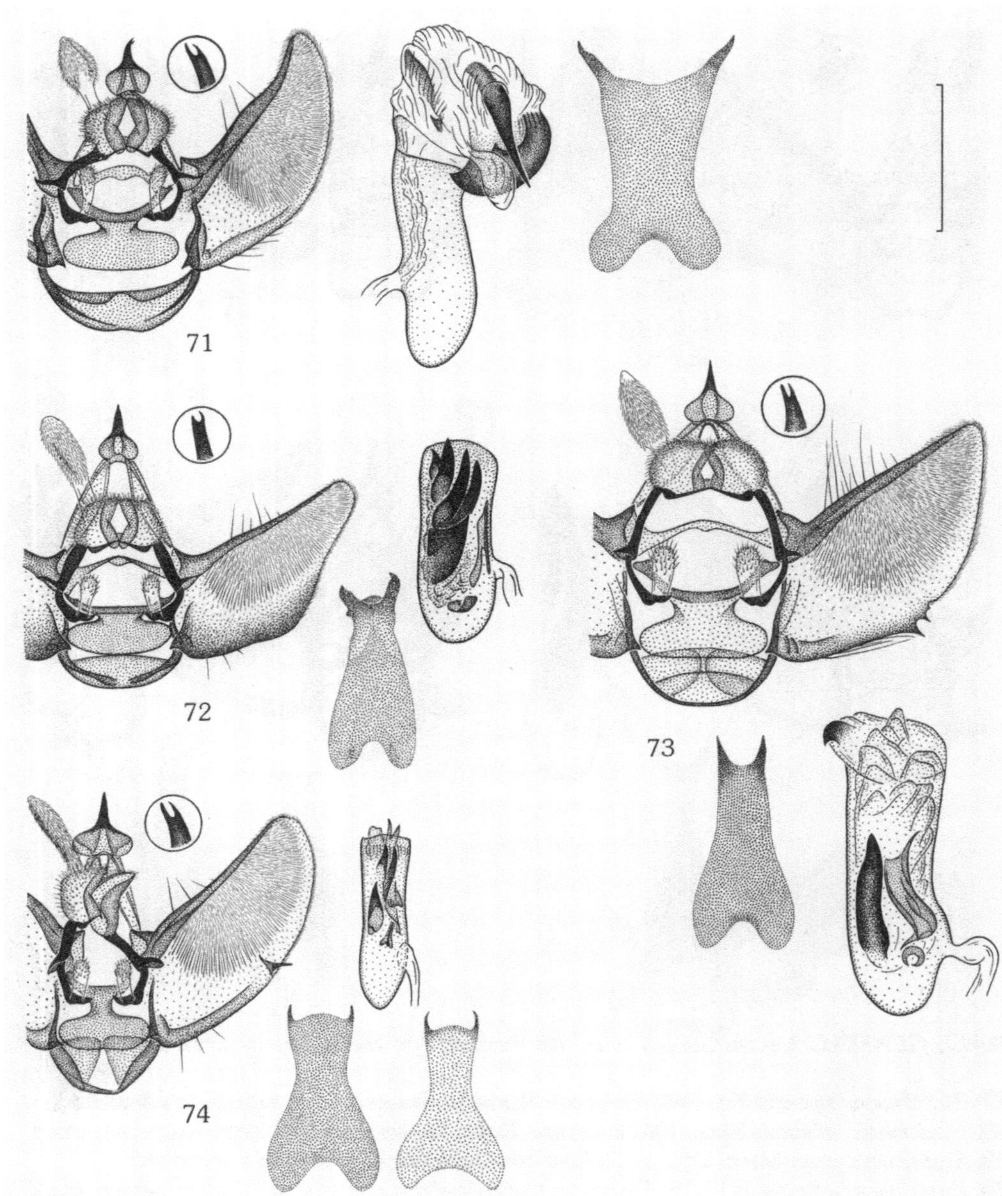

MALE GENITALIA (scale bar = 1 mm; with sternum A8; lateral view of uncus enlarged)

71. *Eupithecia actaeata* Wald., W. Germany [aedeagus lateral, with vesica everted]
72. *Eupithecia quercetica* Prt., N. Asian Turkey [aedeagus lateral, with vesica uneverted]
73. *Eupithecia egenaria* H.-S., S. Ukraine (Crimea) [aedeagus lateral, with vesica semi-everted]
74. *Eupithecia repentina* Vojnits & Laever, S. European Russia [aedeagus lateral, with vesica uneverted; two diverse sternum A8]

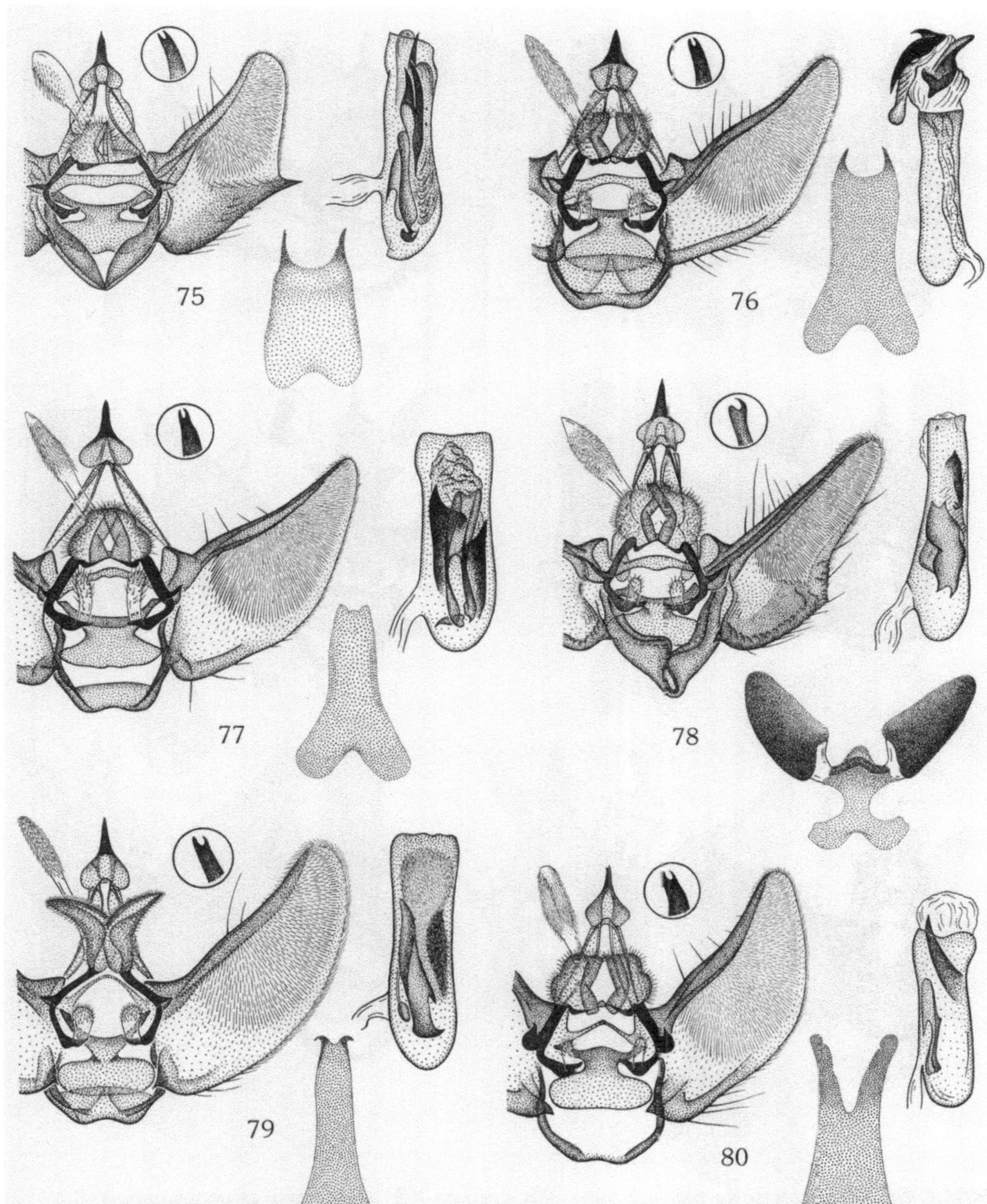

MALE GENITALIA (scale bar = 1 mm; with sternum A8; lateral view of uncus enlarged)

75. *Eupithecia orana* Dtze., S. Spain [aedeagus lateral, with vesica uneverted]
76. *Eupithecia pimpinellata* Hbn., S. European Russia [aedeagus lateral, with vesica everted]
77. *Eupithecia karadaghensis* Mir., S. Ukraine (Crimea) [aedeagus lateral, with vesica uneverted]
78. *Eupithecia suboxydata* Stgr., W. Asian Russia (Tjumen) [aedeagus lateral, with vesica uneverted]
79. *Eupithecia simpliciata* Hw., NW. European Russia [aedeagus ventral, with vesica uneverted]
80. *Eupithecia sinuosaria* Ev., NW. European Russia [aedeagus lateral, with vesica uneverted]

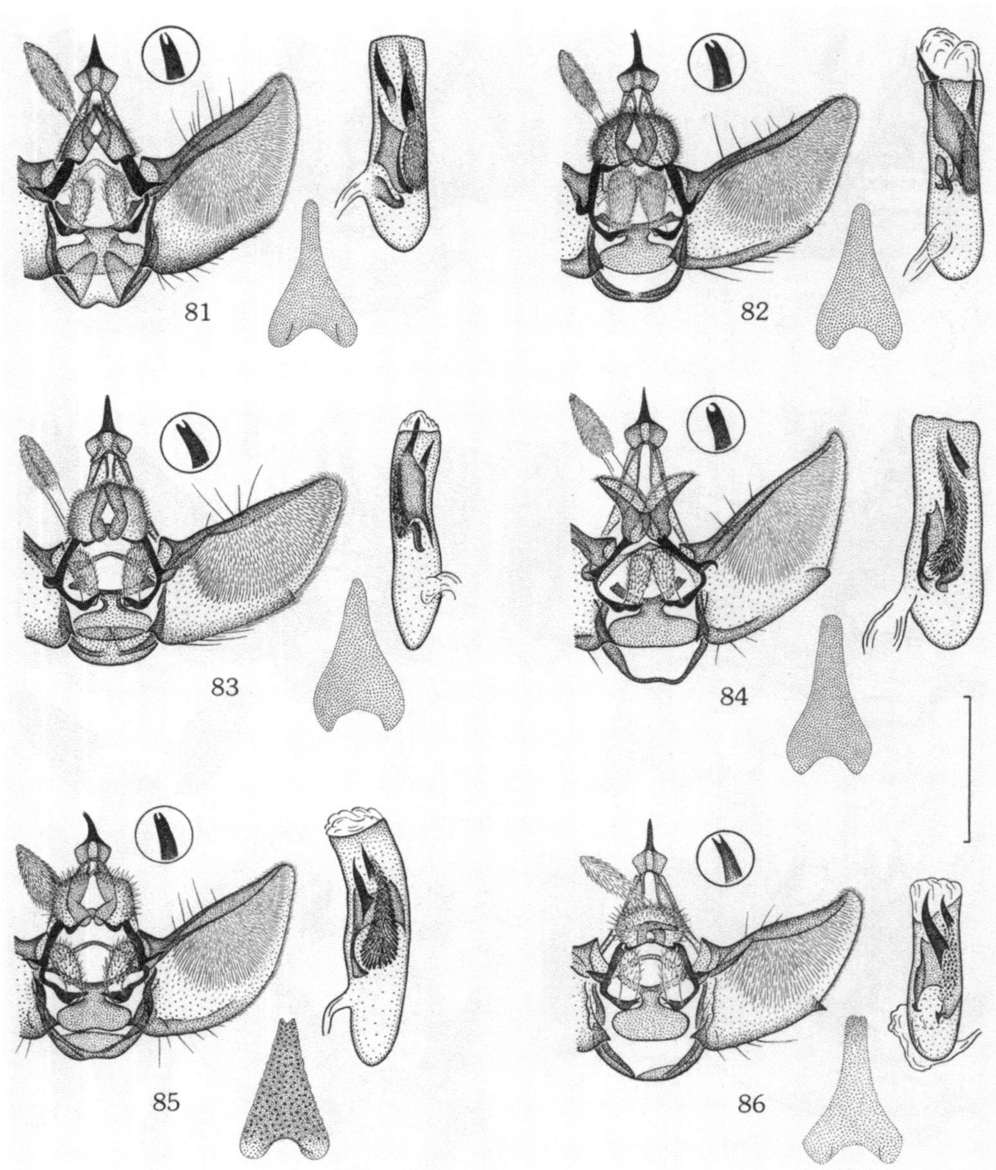

MALE GENITALIA (scale bar = 1 mm; with sternum A8; lateral view of uncus enlarged)

81. *Eupithecia gelidata hyperboreata* Stgr., N. European Russia [aedeagus lateral, with vesica uneverted]
82. *Eupithecia nanata* Hbn., NW. European Russia [aedeagus dorsal, with vesica semi-everted]
83. *Eupithecia innotata* Hufn., C. European Russia [aedeagus dorsal, with vesica uneverted]
84. *Eupithecia ochridata* Schütze & Pinker, Macedonia (paratype) [aedeagus lateral, with vesica uneverted]
85. *Eupithecia unedonata* Mab., C. Spain [aedeagus lateral, with vesica uneverted]
86. *Eupithecia chalikophila* Whli., S. Spain [aedeagus ventral, with vesica uneverted]

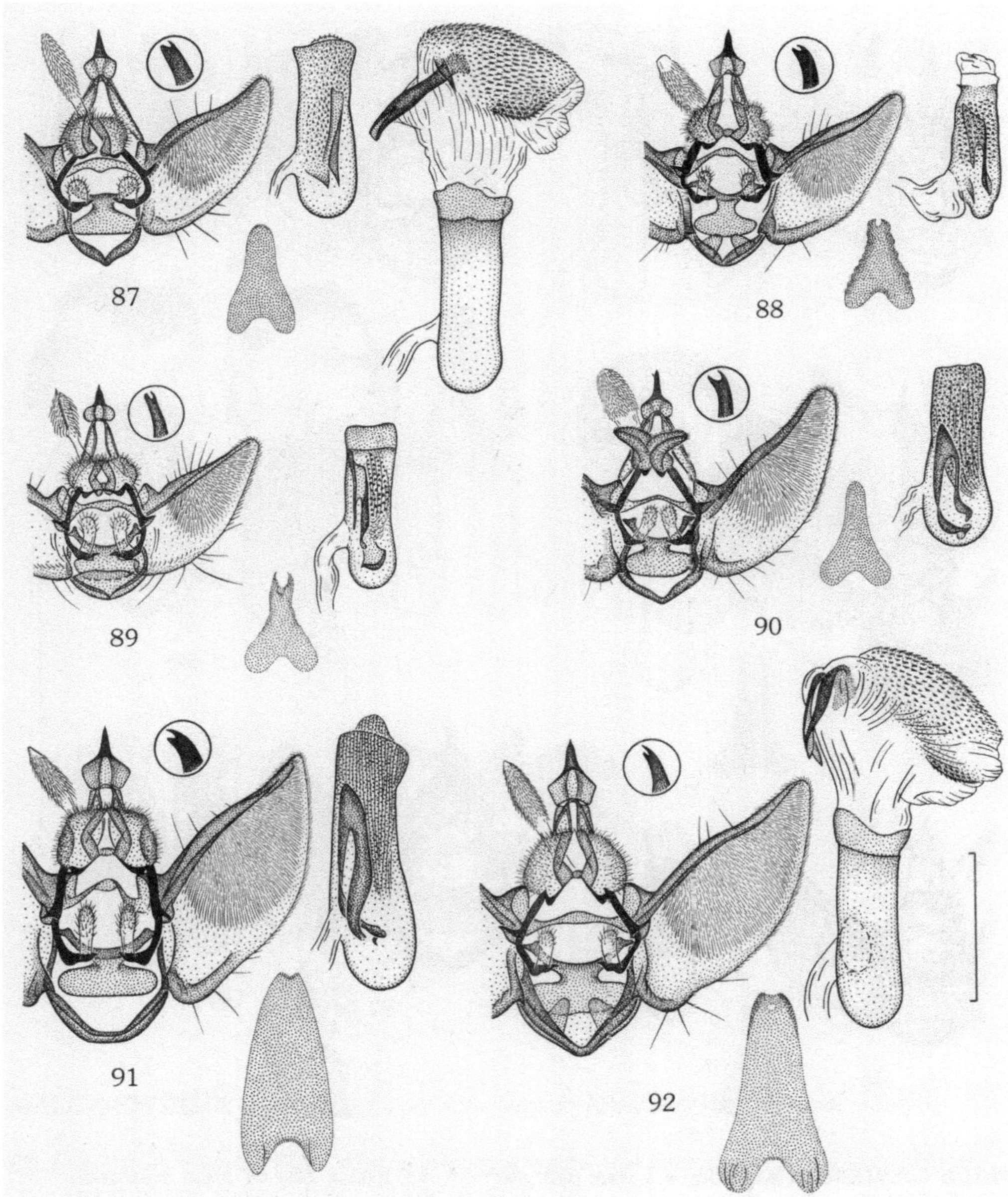

MALE GENITALIA (scale bar = 1 mm; with sternum A8; lateral view of uncus enlarged)

87. *Eupithecia graphata* Tr., Hungary [aedeagus two lateral, with vesica uneverted and everted]
88. *Eupithecia riparia* H.-S., S. Greece [aedeagus lateral, with vesica uneverted]
89. *Eupithecia thurnerata* Schütze, SW. Bulgaria [aedeagus lateral, with vesica uneverted]
90. *Eupithecia gemellata* H.-S., Sicily [aedeagus lateral, with vesica uneverted]
91. *Eupithecia mystica* Dtze., S. Ukraine (Crimea) [aedeagus lateral, with vesica uneverted]
92. *Eupithecia cooptata* Dtze., SE. France [aedeagus ventral, with vesica everted]

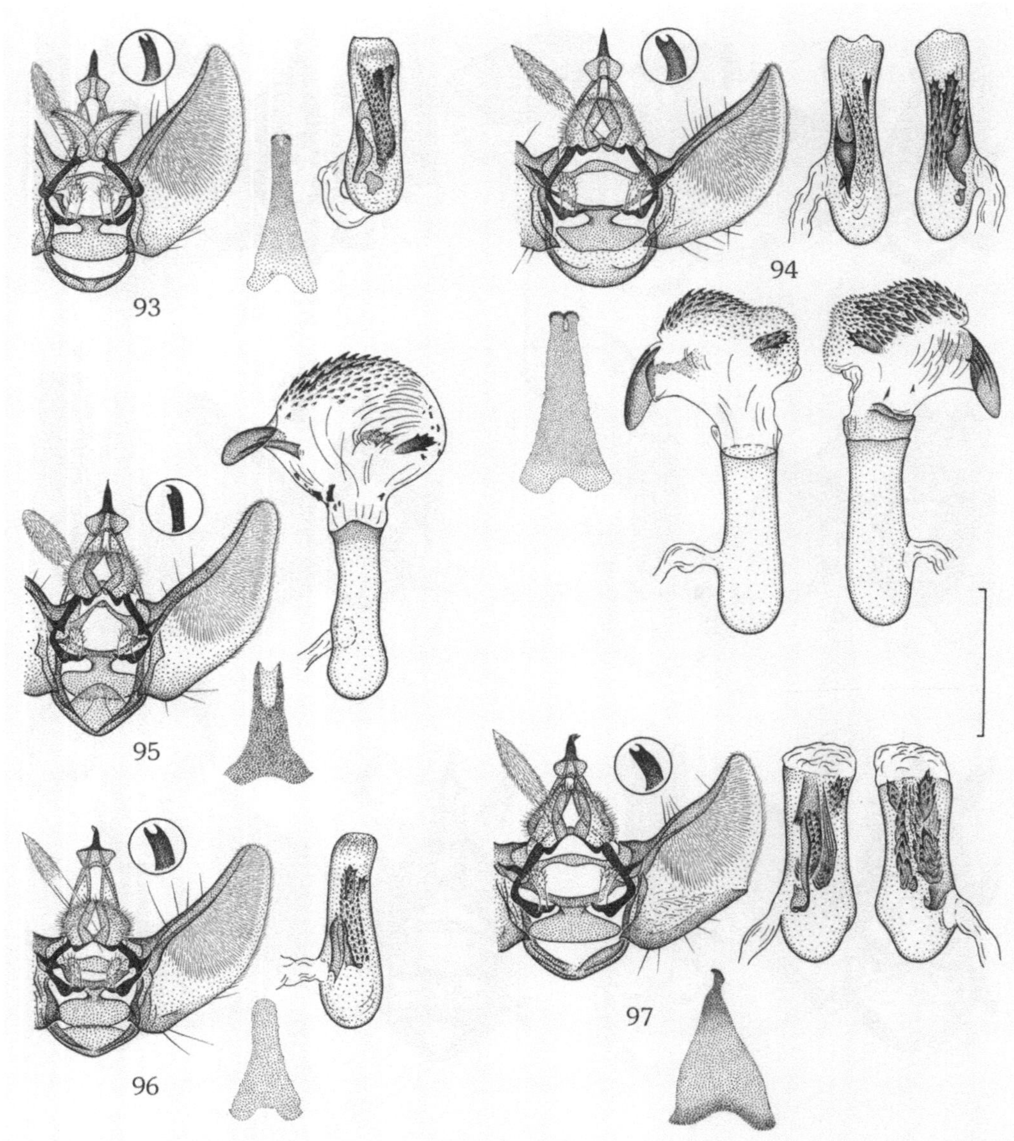

MALE GENITALIA (scale bar = 1 mm; with sternum A8; lateral view of uncus enlarged)

93. *Eupithecia spissilineata* Metzn., Azerbaijan [aedeagus ventral, with vesica uneverted]
94. *Eupithecia cuculliaria* Rbl., Macedonia [aedeagus four lateral, with vesica uneverted and everted]
95. *Eupithecia unitaria* H.-S., S. Spain [aedeagus ventral, with vesica everted]
96. *Eupithecia breviculata* Donz., Sicily [aedeagus lateral, with vesica uneverted]
97. *Eupithecia cerussaria* Led., N. Asian Turkey [aedeagus two lateral, with vesica uneverted]

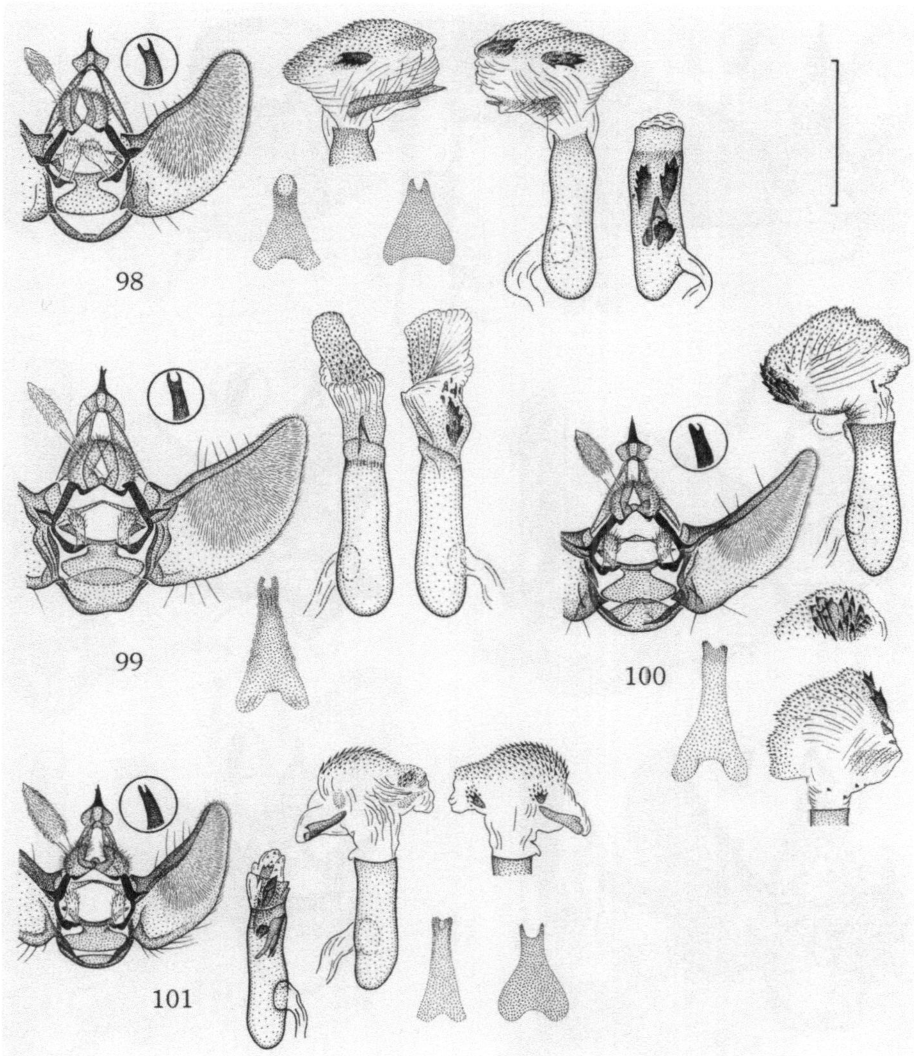

MALE GENITALIA (scale bar = 1 mm; with sternum A8; latera. view of uncus enlarged)

98. *Eupithecia irriguata* Hbn., Austria [aedeagus ventral and lateral, with vesica everted from two sides and uneverted; two diverse sternum A8]
99. *Eupithecia antalica* Mir., S. Asian Turkey [aedeagus two lateral, with vesica semi-everted]
100. *Eupithecia indigata* Hbn., NW. European Russia [aedeagus lateral, with vesica everted from three sides]
101. *Eupithecia conterminata* (Lienig & Zeller), NW. European Russia [aedeagus lateral and ventral, with vesica semi-everted and everted from two sides; two diverse sternum A8]

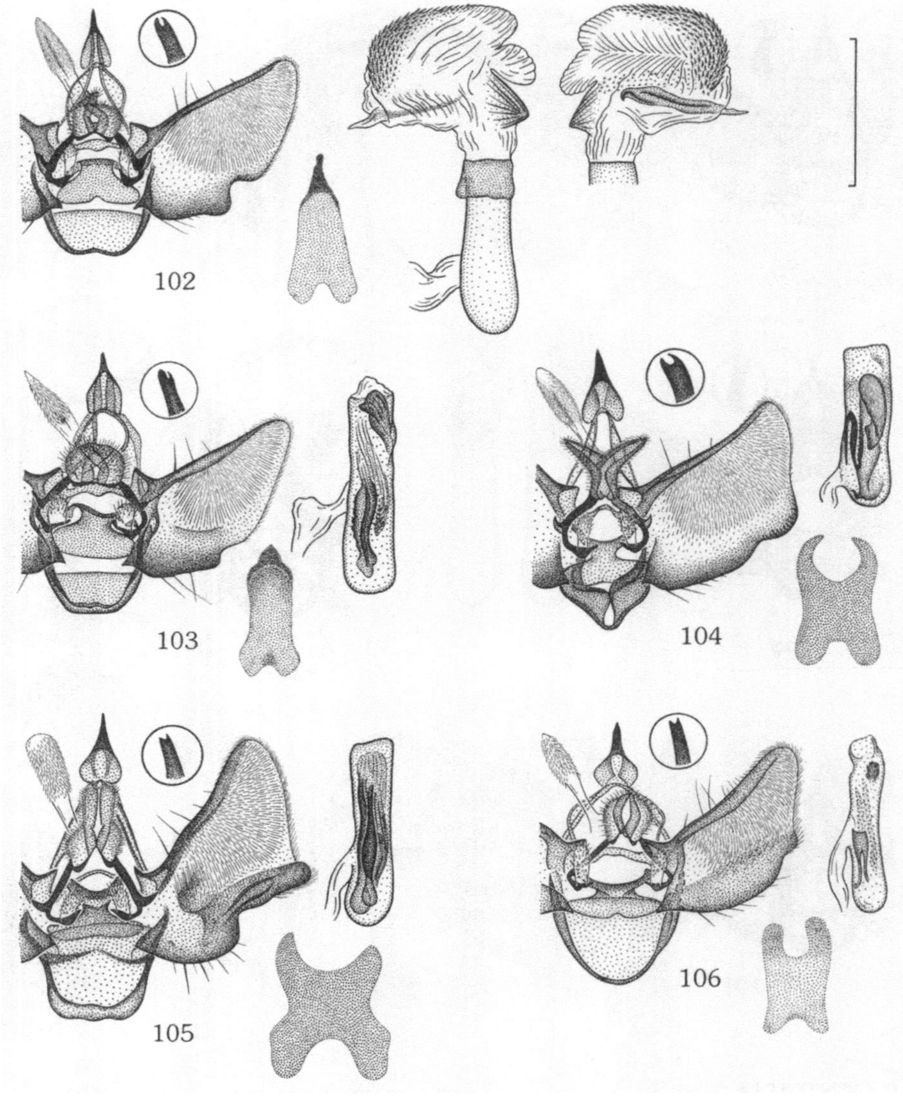

MALE GENITALIA (scale bar = 1 mm; with sternum A8; lateral view of uncus enlarged)

102. *Eupithecia distinctaria* H.-S., Austria [aedeagus ventral and dorsal, with vesica everted from two sides]
103. *Eupithecia senorita* Mir., SW. France [aedeagus ventral, with vesica uneverted]
104. *Eupithecia extraversaria* H.-S., S. European Russia (N. Caucasus) [aedeagus lateral, with vesica uneverted]
105. *Eupithecia centaureata* Den. & Schiff., NW. European Russia [aedeagus lateral, with vesica uneverted]
106. *Eupithecia limbata tomillata* Chrét., C. Spain [aedeagus lateral, with vesica uneverted]

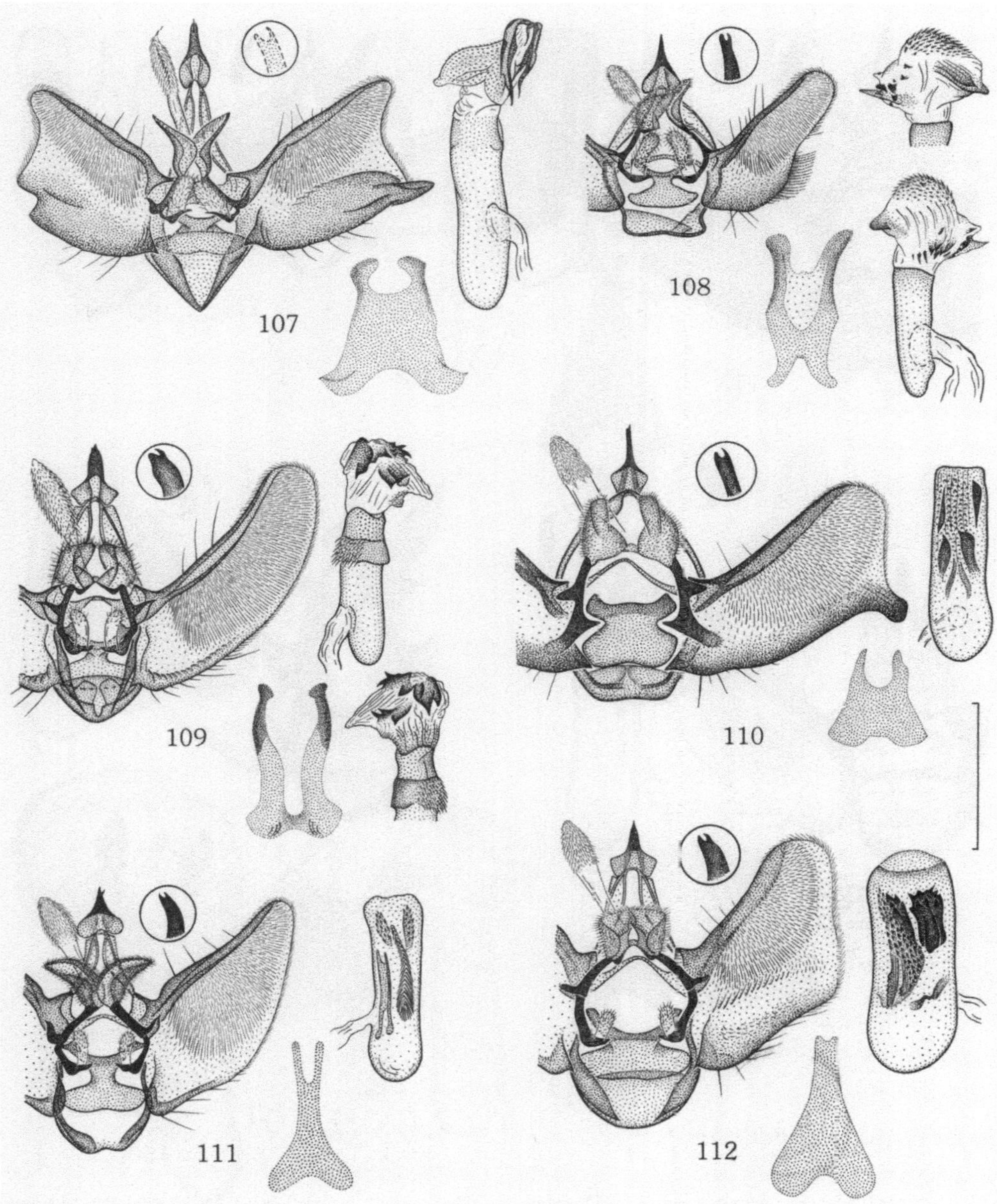

MALE GENITALIA (scale bar = 1 mm; with sternum A8; lateral view of uncus enlarged)

107. *Eupithecia nachadira* Brandt, N. Iran [aedeagus ventral, with vesica everted]
108. *Eupithecia mekrana* Brandt, SW. Asian Russia [aedeagus dorsal and ventral, with vesica everted from two sides]
109. *Eupithecia variostrigata* Alph., S. Ukraine (Crimea) [aedeagus two lateral, with vesica everted from two sides]
110. *Eupithecia insigniata* Hbn., Austria [aedeagus ventral, with vesica uneverted]
111. *Eupithecia trisignaria* H.-S., S. Ukraine (Crimea) [aedeagus lateral, with vesica uneverted]
112. *Eupithecia gueneata* Mill., Hungary [aedeagus ventral, with vesica uneverted]

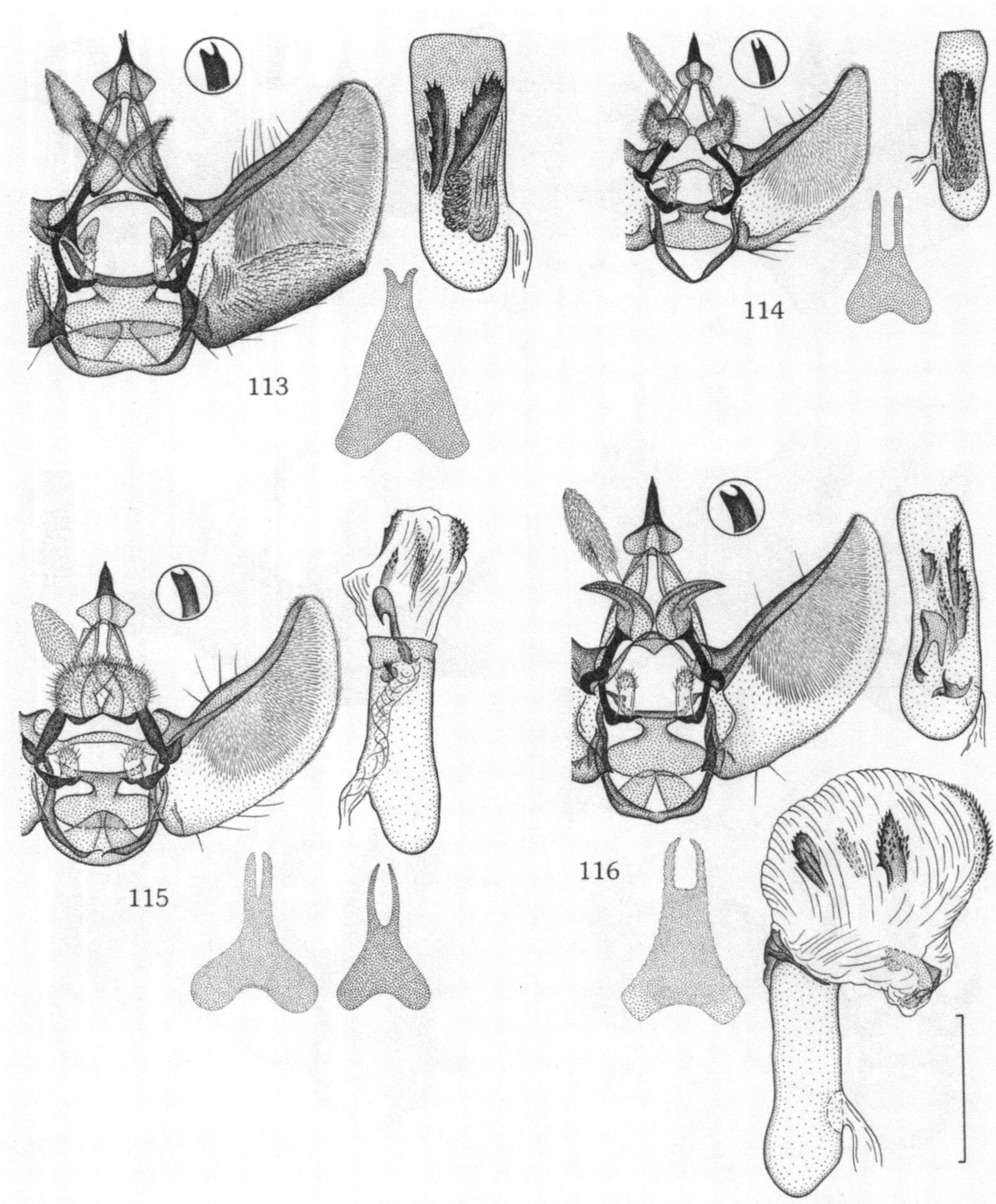

MALE GENITALIA (scale bar = 1 mm; with sternum A8; lateral view of uncus enlarged)

113. *Eupithecia gratiosata* H.-S., Turkmenistan [aedeagus lateral, with vesica uneverted]
114. *Eupithecia scalptata* Chr., Turkmenistan (lectotype) [aedeagus lateral, with vesica uneverted]
115. *Eupithecia veratraria* H.-S., Austria [aedeagus lateral, with vesica semi-everted; two diverse sternum A8]
116. *Eupithecia cretaceata fenestrata* Mill., SE. France [aedeagus ventral and lateral, with vesica uneverted and everted]

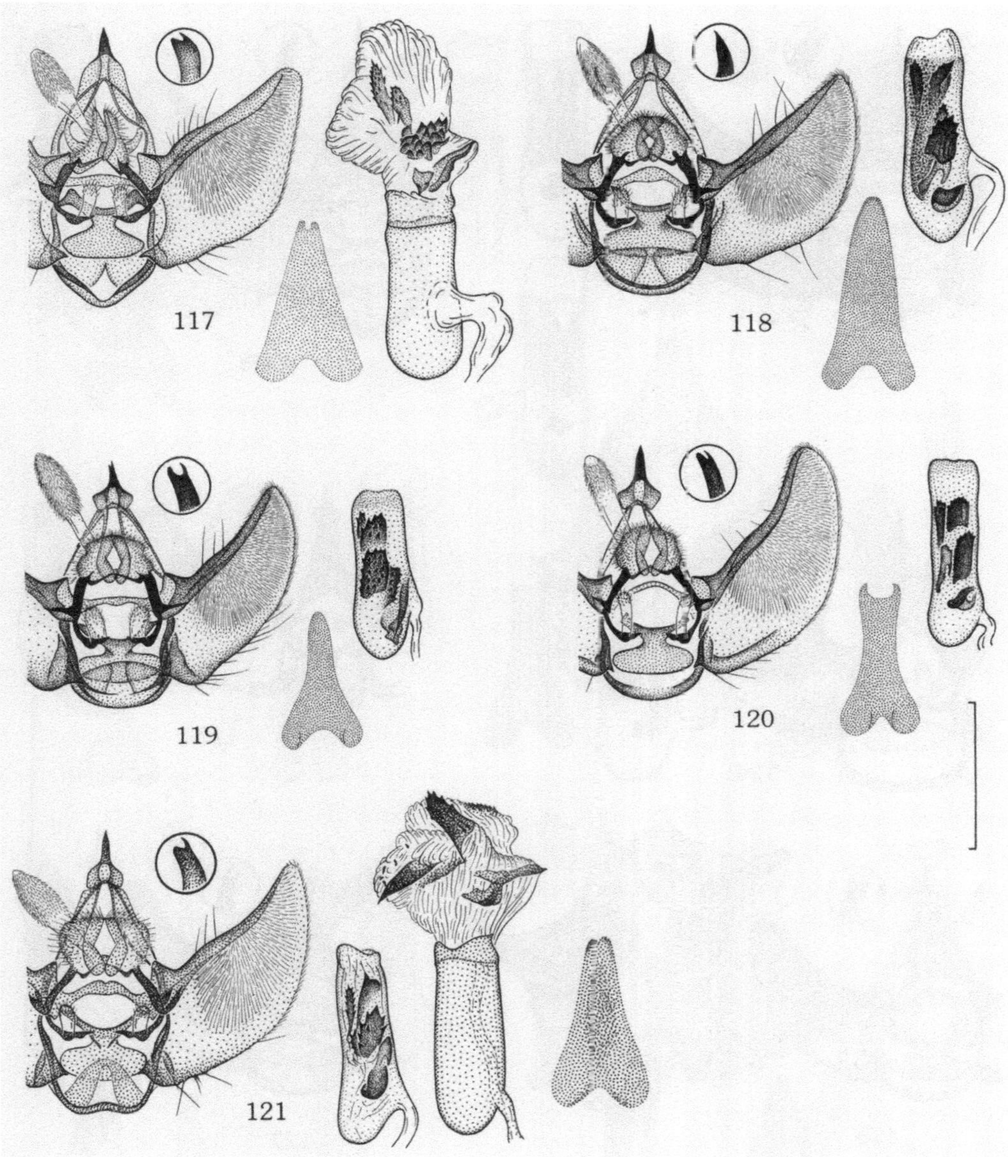

MALE GENITALIA (scale bar = 1 mm; with sternum A8; lateral view of uncus enlarged)

117. *Eupithecia praealta* Whli., S. Spain [aedeagus dorsal, with vesica everted]
118. *Eupithecia intricata* Zett., NW. Poland [aedeagus lateral, with vesica uneverted]
119. *Eupithecia satyrata* Hbn., S. Germany [aedeagus lateral, with vesica uneverted]
120. *Eupithecia cauchiata* Dup. W. Germany (reared) [aedeagus lateral, with vesica uneverted]
121. *Eupithecia pernotata* Gn., S. European Russia [aedeagus lateral, with vesica uneverted and everted]

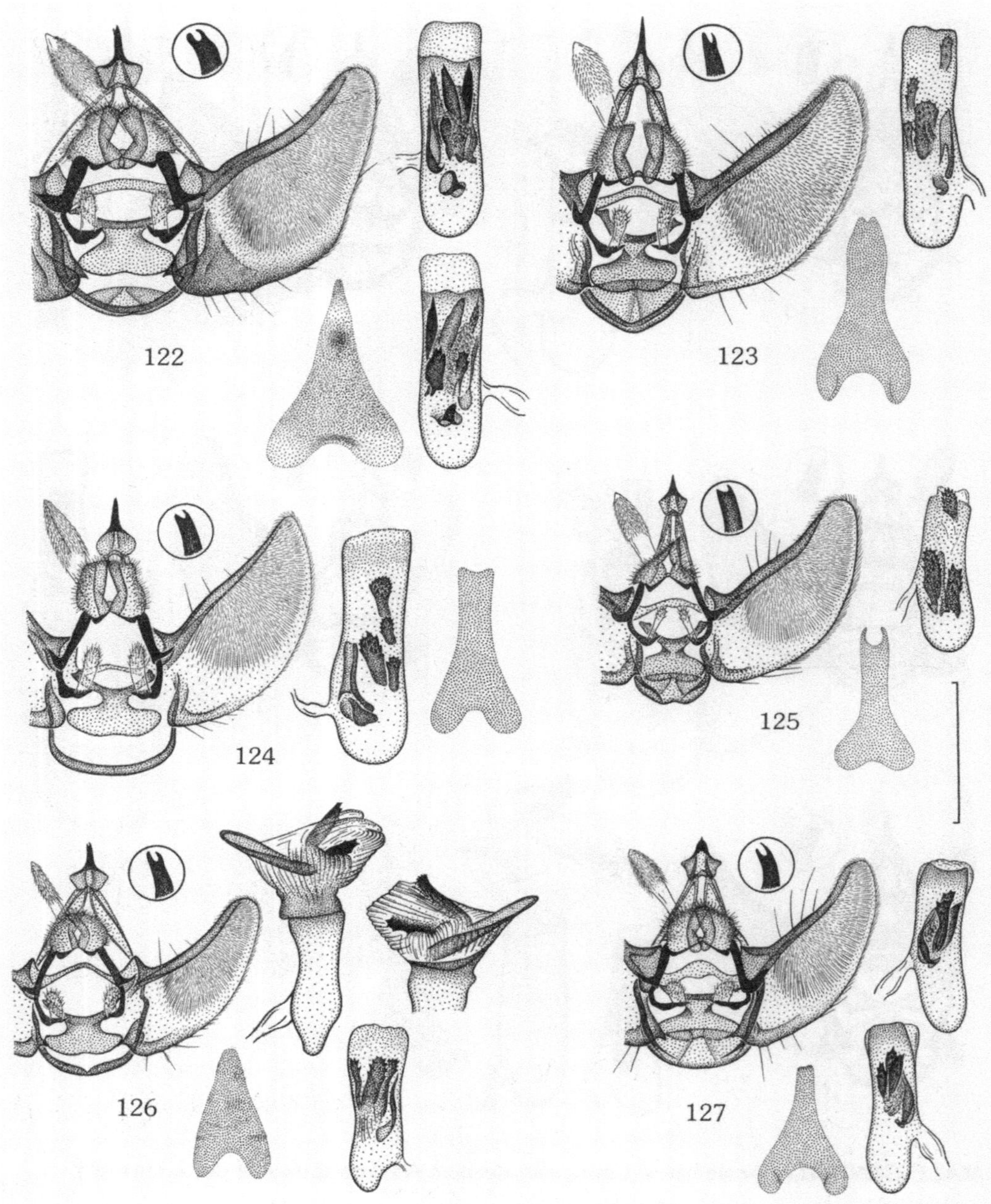

MALE GENITALIA (scale bar = 1 mm; with sternum A8; lateral view of uncus enlarged)

122. *Eupithecia extensaria* Frr., E. Ukraine [aedeagus two lateral, with vesica uneverted]
123. *Eupithecia absinthiata* Cl., NW. European Russia [aedeagus lateral, with vesica uneverted]
124. *Eupithecia expallidata* Dbld., Austria [aedeagus lateral, with vesica uneverted]
125. *Eupithecia valerianata* Hbn., Lithuania [aedeagus lateral, with vesica uneverted]
126. *Eupithecia assimilata* Dbld., NW. European Russia [aedeagus two lateral, with vesica everted and uneverted]
127. *Eupithecia vulgata* Hw., W. European Russia [aedeagus two lateral, with vesica uneverted]

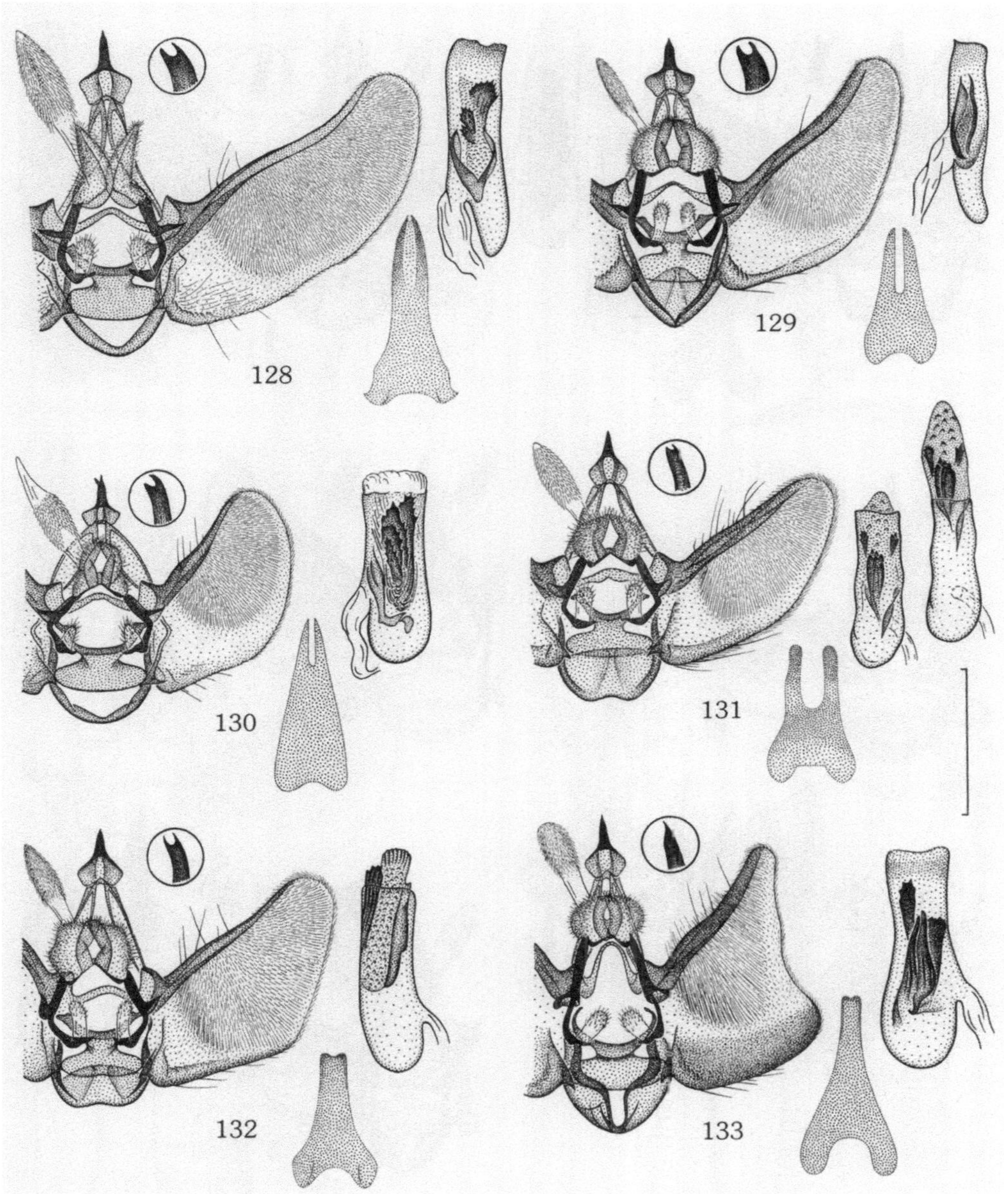

MALE GENITALIA (scale bar = 1 mm; with sternum A8; lateral view of uncus enlarged)

128. *Eupithecia biornata* Chr., S. European Russia [aedeagus lateral, with vesica uneverted]
129. *Eupithecia sutiliata* Chr., S. Kazakhstan [aedeagus lateral, with vesica uneverted]
130. *Eupithecia immundata* (Lienig & Zeller), NW. European Russia [aedeagus lateral, with vesica uneverted]
131. *Eupithecia thalictrata* Püng., Switzerland [aedeagus two ventral, with vesica uneverted and semi-everted]
132. *Eupithecia addictata* Dtze., S. Far East Russia [aedeagus lateral, with vesica semi-everted]
133. *Eupithecia exiguata* Hbn., NW. European Russia [aedeagus lateral, with vesica uneverted]

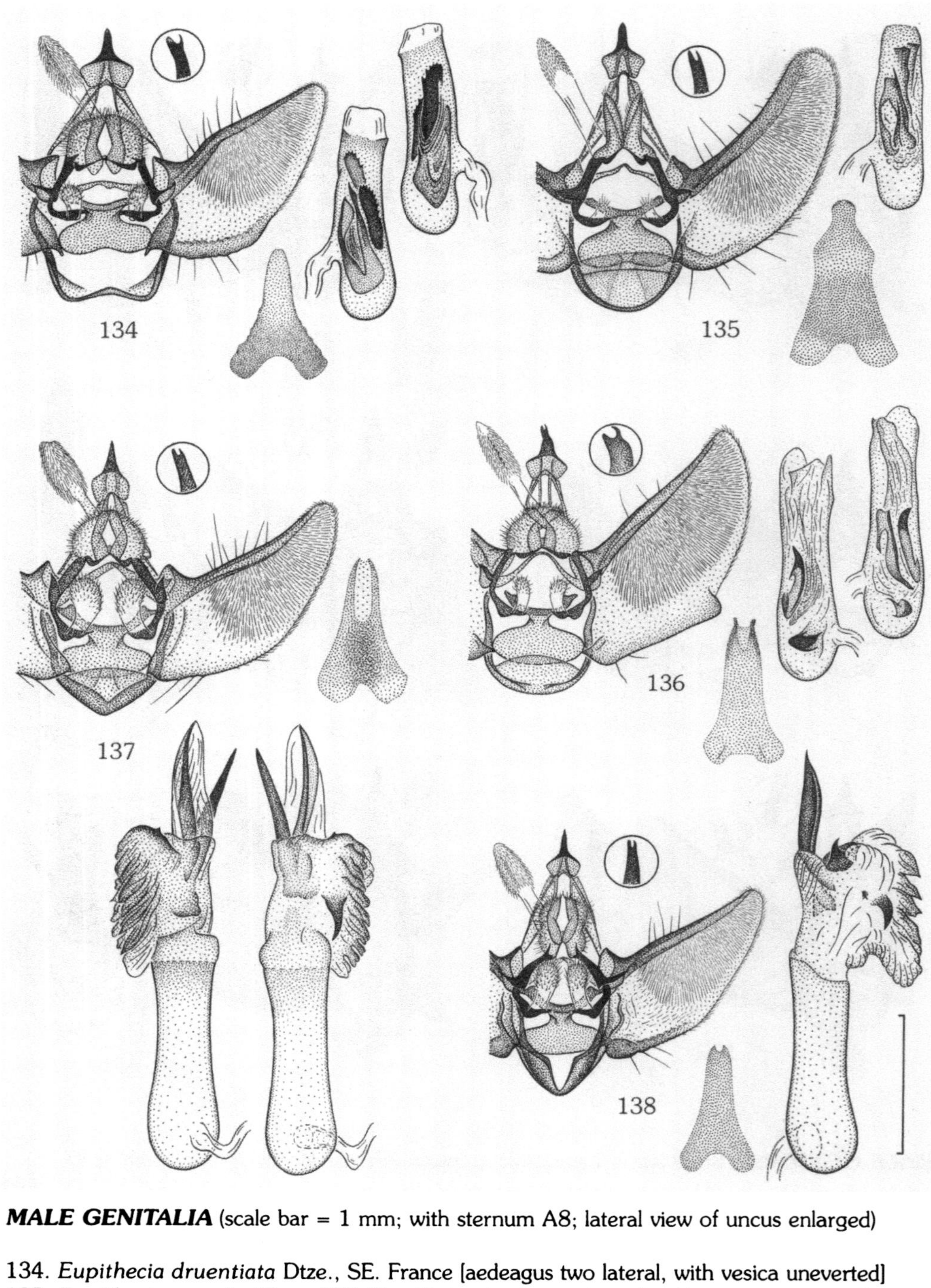

MALE GENITALIA (scale bar = 1 mm; with sternum A8; lateral view of uncus enlarged)

134. *Eupithecia druentiata* Dtze., SE. France [aedeagus two lateral, with vesica uneverted]
135. *Eupithecia moecha* Dtze., NW. Kazakhstan [aedeagus lateral, with vesica uneverted]
136. *Eupithecia denotata* Hbn., S. Ukraine (Crimea) [aedeagus two lateral, with vesica uneverted]
137. *Eupithecia weissi* Prt., SW. France [aedeagus dorsal and ventral, with vesica everted]
138. *Eupithecia pauxillaria* Bsd., Austria [aedeagus ventral, with vesica everted]

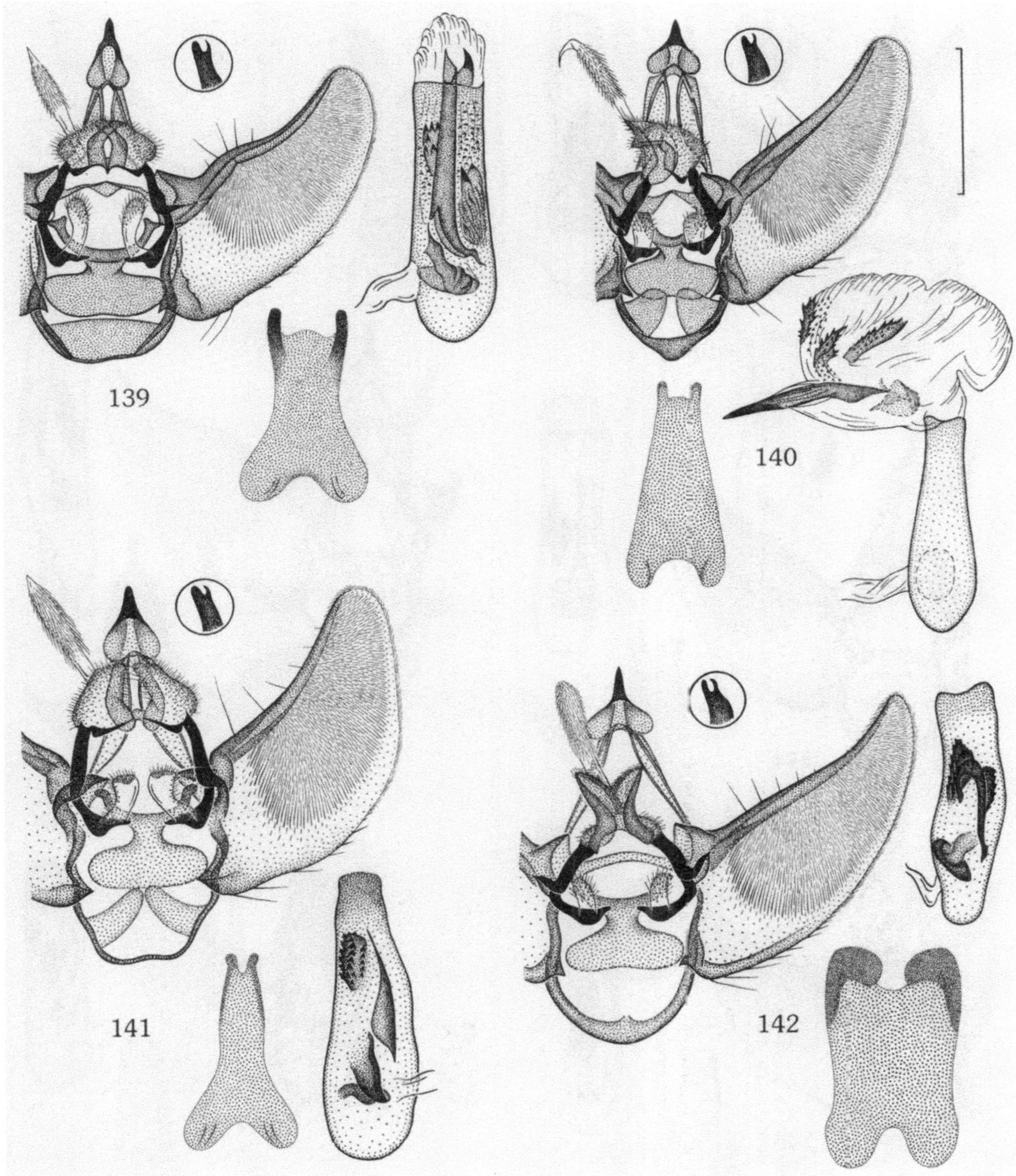

MALE GENITALIA (scale bar = 1 mm; with sternum A8; lateral view of uncus enlarged)

139. *Eupithecia santolinata* Mab., C. Spain [aedeagus lateral, with vesica uneverted]
140. *Eupithecia millefoliata* Rössl., Austria [aedeagus ventral, with vesica everted]
141. *Eupithecia spadiceata* Zerny, NW. Azerbaijan [aedeagus dorsal, with vesica uneverted]
142. *Eupithecia icterata* Vill., NW. European Russia [aedeagus ventral, with vesica uneverted]

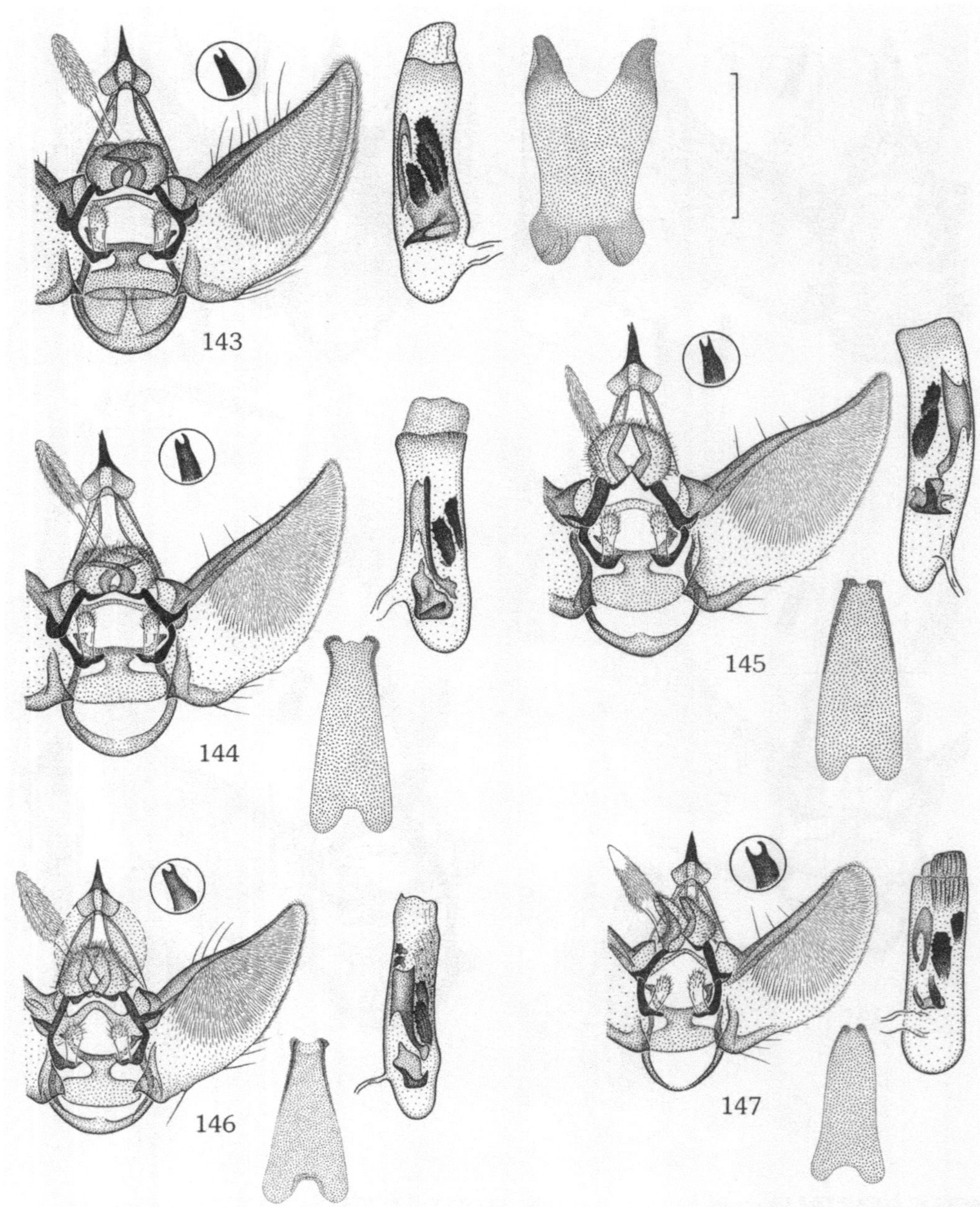

MALE GENITALIA (scale bar = 1 mm; with sternum A8; lateral view of uncus enlarged)

143. *Eupithecia succenturiata* L., NW. European Russia [aedeagus lateral, with vesica uneverted]
144. *Eupithecia semigraphata* Bruand, NE. Georgia [aedeagus lateral, with vesica uneverted]
145. *Eupithecia impurata* Hbn., Austria [aedeagus lateral, with vesica uneverted]
146. *Eupithecia poecilata* Püng., Sardinia [aedeagus lateral, with vesica uneverted]
147. *Eupithecia fuscicostata* Chr., W. Azerbaijan [aedeagus dorsal, with vesica uneverted]

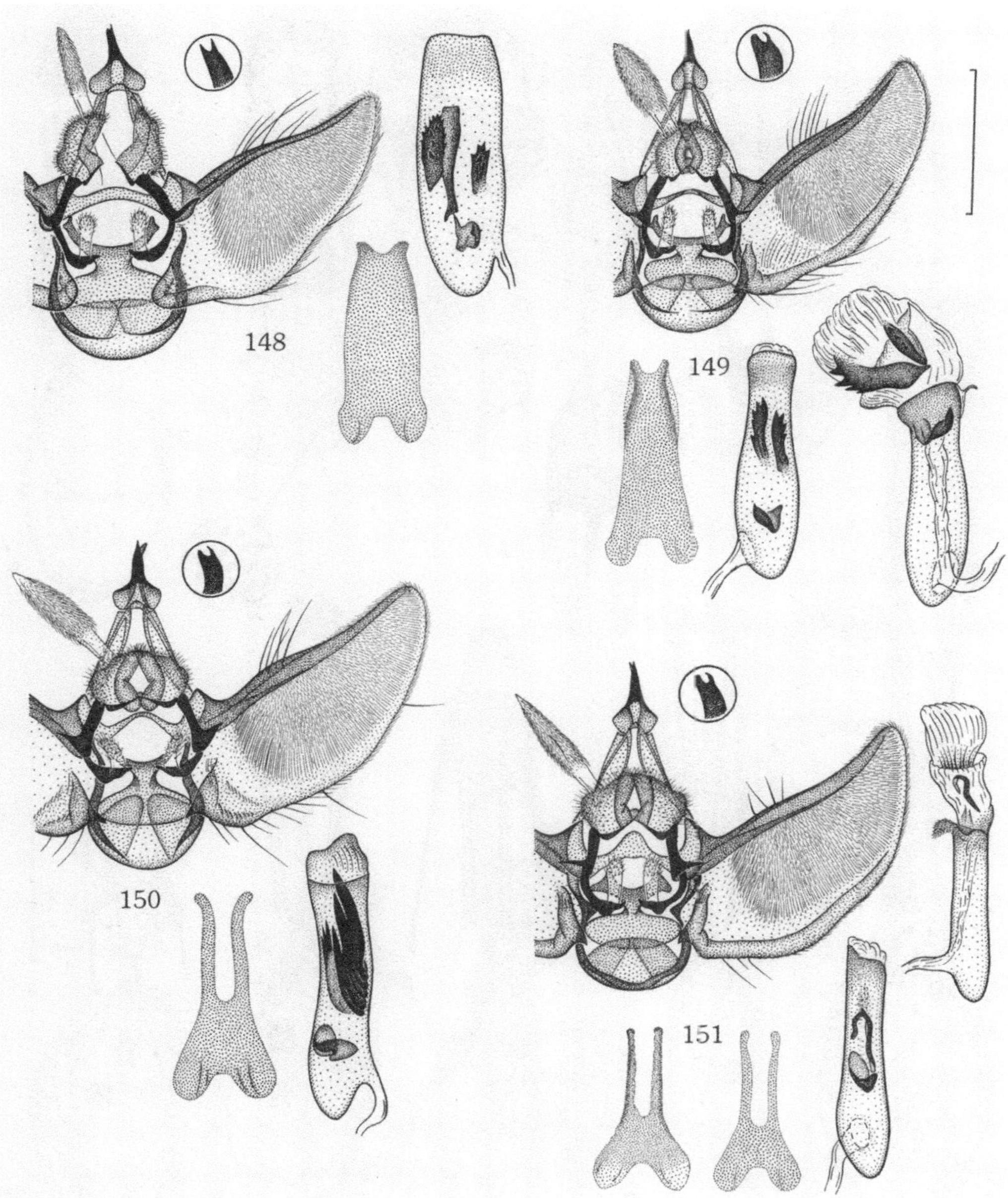

MALE GENITALIA (scale bar = 1 mm; with sternum A8; lateral view of uncus enlarged)

148. *Eupithecia denticulata* Tr., Austria [aedeagus lateral, with vesica uneverted]
149. *Eupithecia subumbrata* Den. & Schiff., NW. European Russia [aedeagus lateral and dorsal, with vesica uneverted and everted]
150. *Eupithecia orphnata* Pet., Austria [aedeagus lateral, with vesica uneverted]
151. *Eupithecia subfuscata* Hw., W. Bulgaria [aedeagus lateral and ventral, with vesica uneverted and semi-everted; two diverse sternum A8]

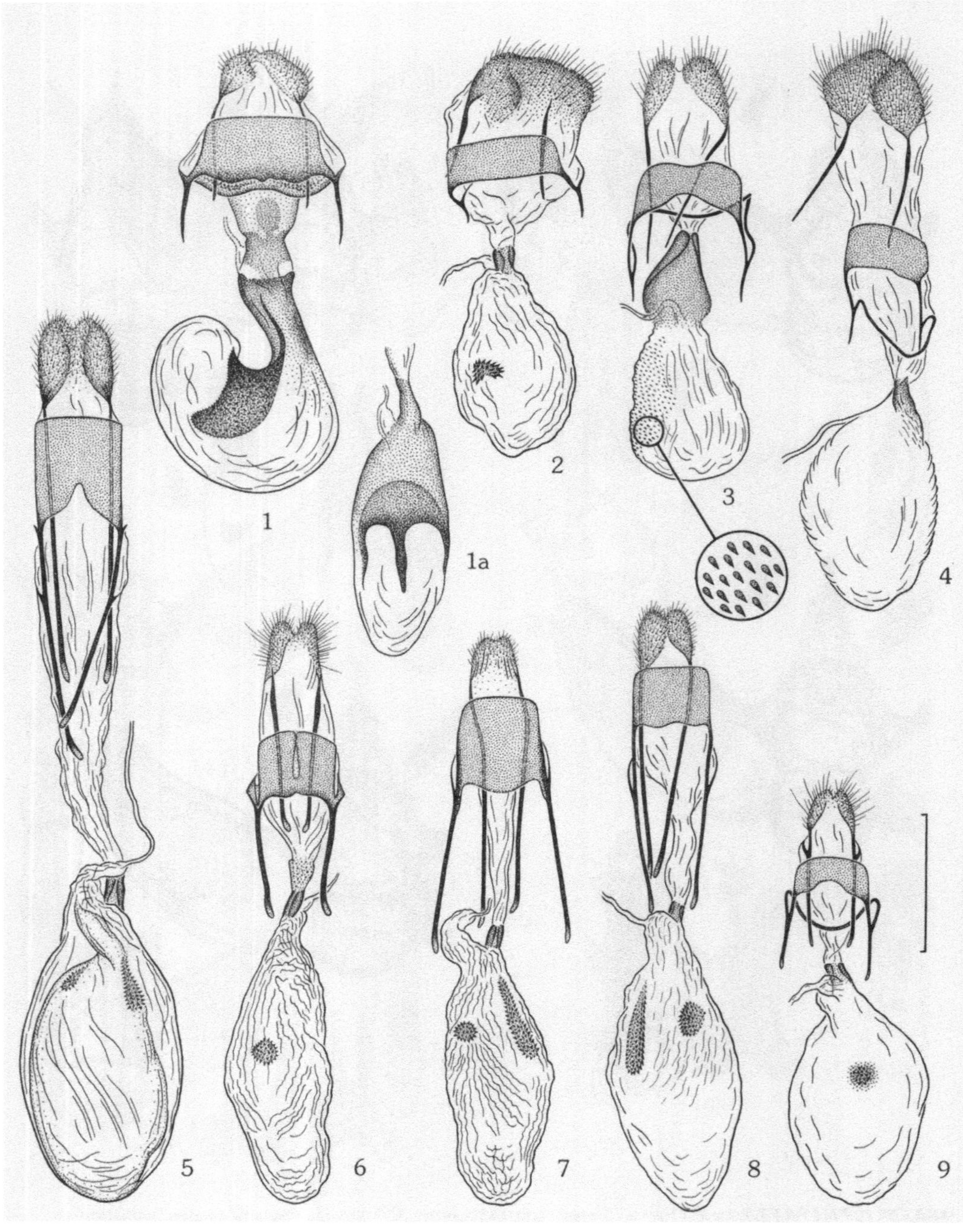

FEMALE GENITALIA (scale bar = 1 mm)

1. *Pseudobaptria bogumilaria* Rbl., Bosnia
1a. Ditto, bursa copulatrix from lateral side
2. *Mesotype didymata* L., NW. European Russia
3. *Mesotype parallelolineata* Retz., NW. European Russia
4. *Mesotype verberata* Scop., S. Germany
5. *Perizoma affinitata* Steph., NW. European Russia
6. *Perizoma alchemillata* L., NW. European Russia
7. *Perizoma hydrata* Tr., S. Poland
8. *Perizoma lugdunaria* H.-S., Austria
9. *Perizoma bifaciata* Hw., NW. European Russia

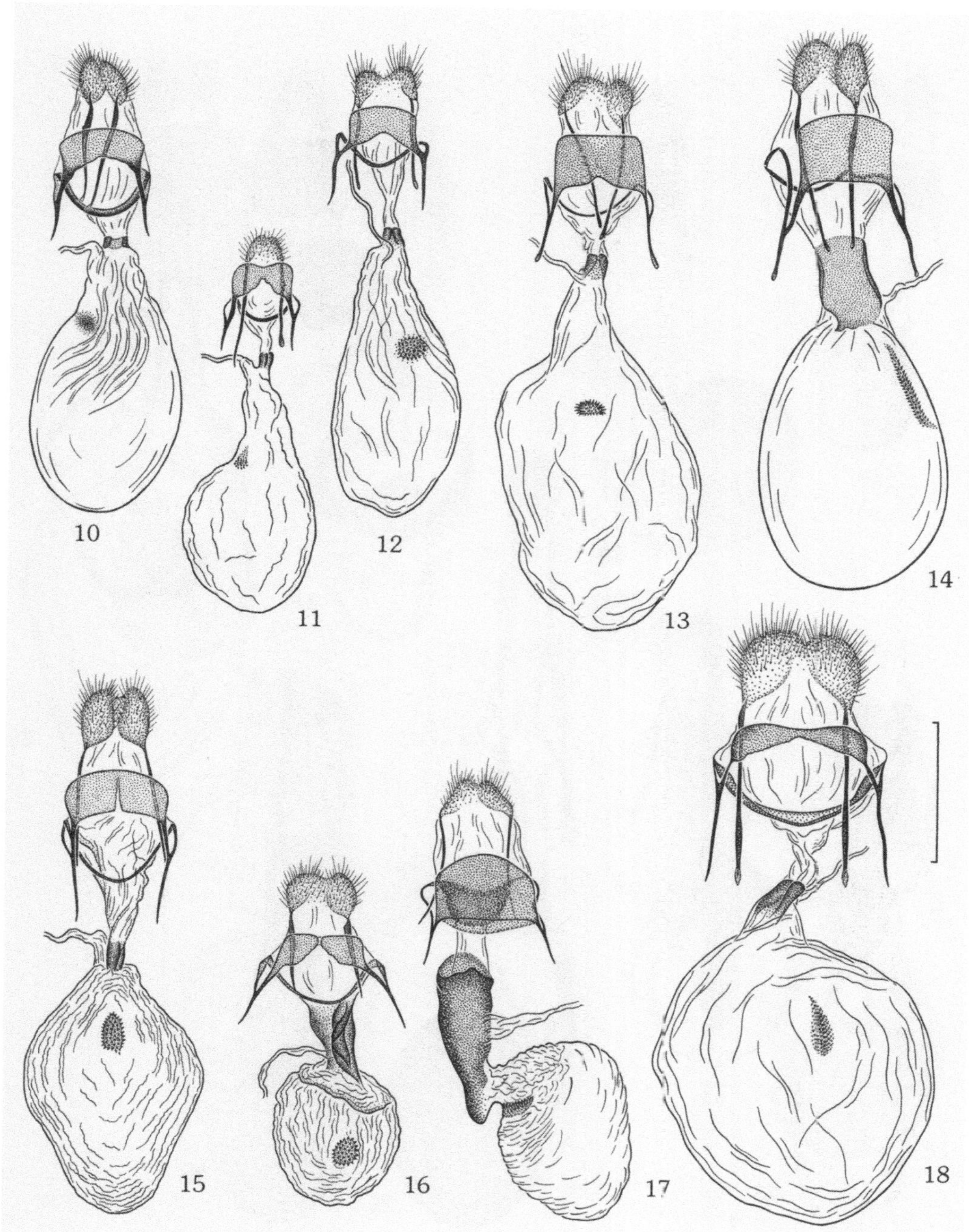

FEMALE GENITALIA (scale bar = 1 mm)

10. *Perizoma flavosparsata* Wagn., C. Spain
11. *Perizoma minorata* Tr., Austria
12. *Perizoma blandiata* Den. & Schiff., NW. European Russia
13. *Perizoma albulata* Den. & Schiff., NW. European Russia
14. *Perizoma flavofasciata* (Thnbg.), NW. European Russia
15. *Perizoma obsoletata* H.-S., Switzerland
16. *Perizoma incultaria* H.-S., Austria
17. *Martania taeniata* Steph., NW. European Russia
18. *Gagitodes sagittata* F., NW. European Russia

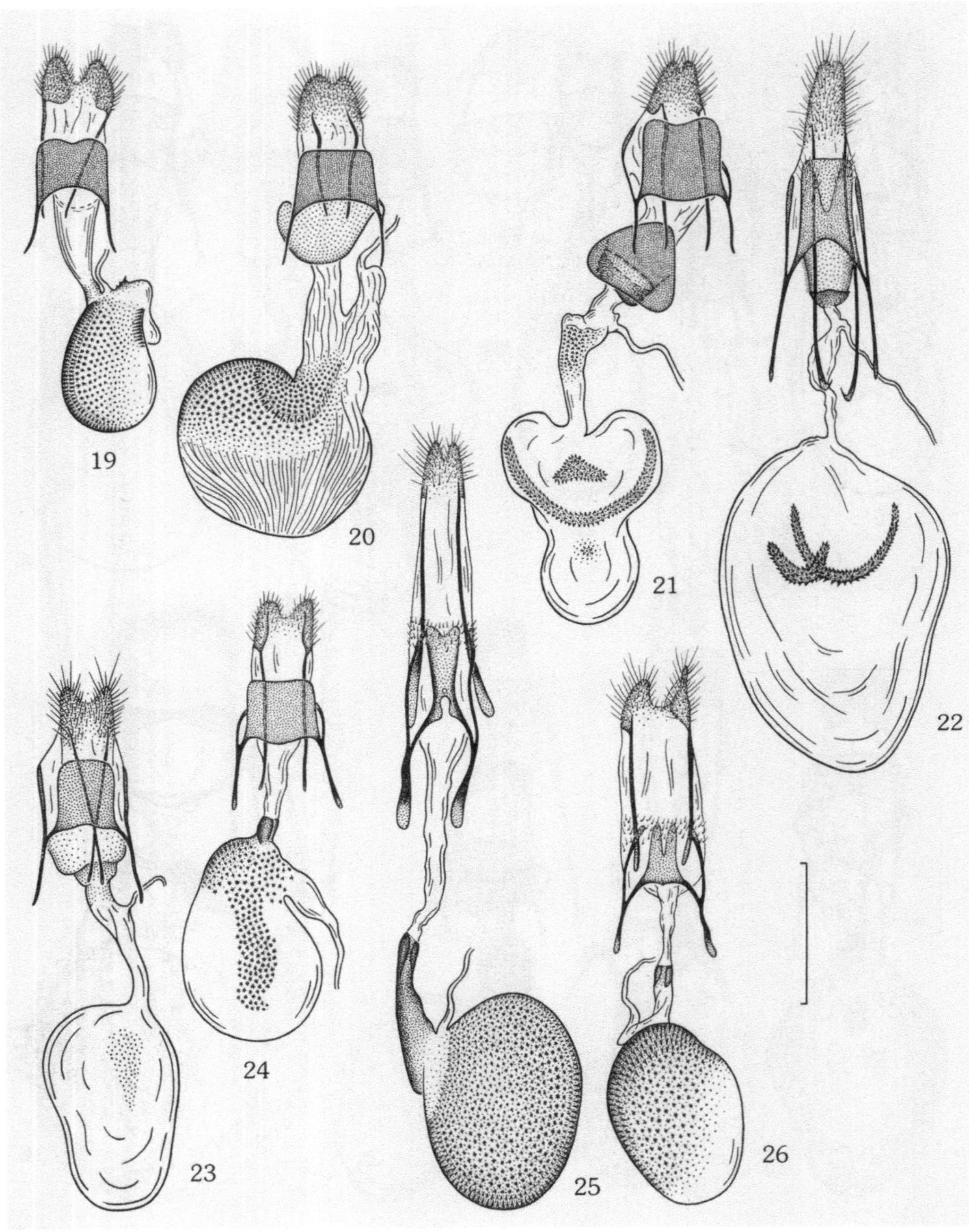

FEMALE GENITALIA (scale bar = 1 mm)

19. *Gymnoscelis rufifasciata* Hw., W. Azerbaijan
20. *Chloroclystis v-ata* Hw., W. Azerbaijan
21. *Pasiphila chloerata* Mab., NW. European Russia
22. *Pasiphila rectangulata* L., NW. European Russia
23. *Pasiphila debiliata* Hbn., NW. European Russia
24. *Eupithecia haworthiata* Dbld., Austria
25. *Eupithecia tenuiata* Hbn., NW. European Russia
26. *Eupithecia inturbata* Hbn., C. European Russia

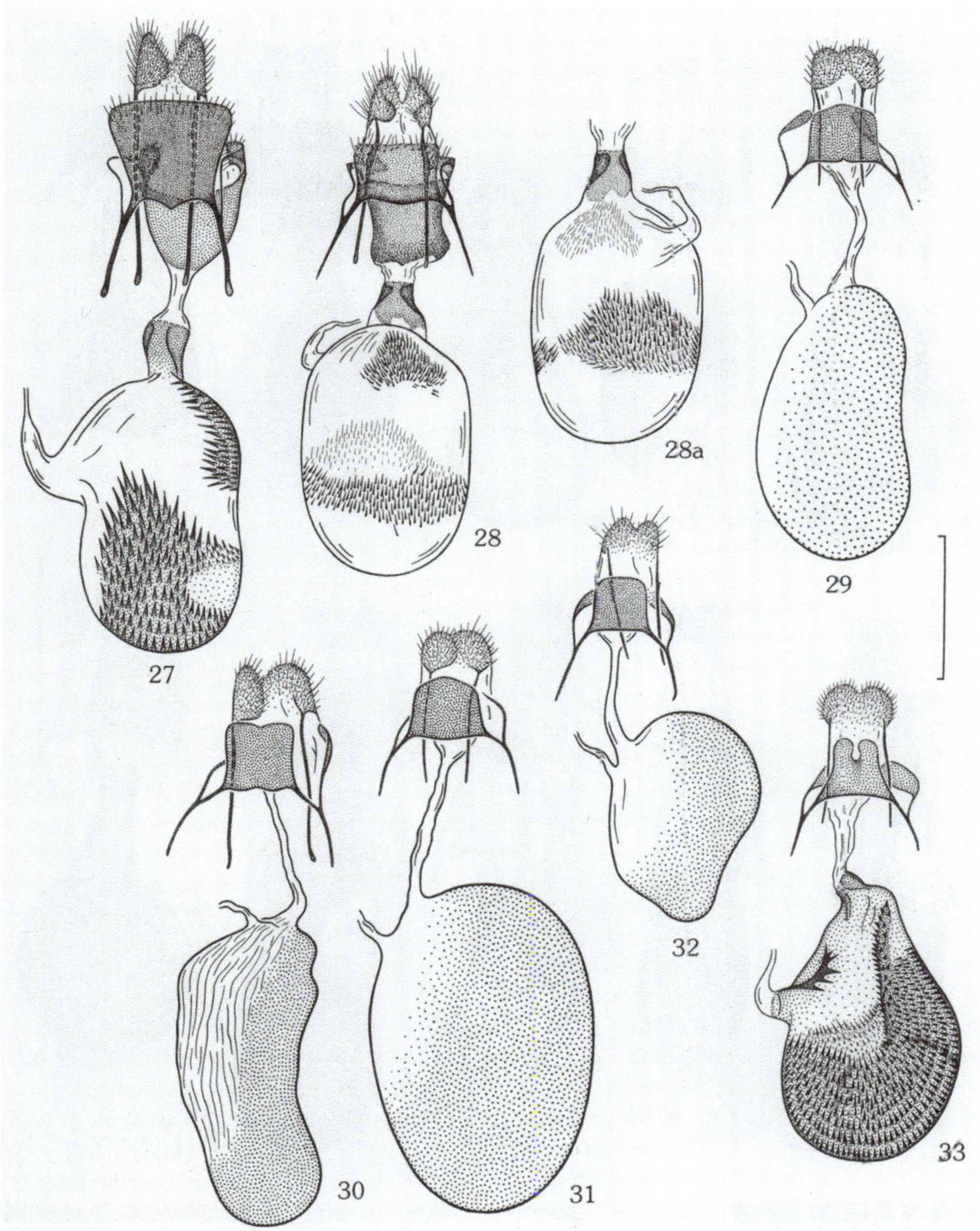

FEMALE GENITALIA (scale bar = 1 mm)

27. *Eupithecia abietaria* Goeze, NW. European Russia
28. *Eupithecia analoga* Djak., Asian Russia (S. Siberia) (lectotype)
28a. Ditto, bursa copulatrix from other side
29. *Eupithecia linariata* Den. & Schiff., NW. European Russia
30. *Eupithecia pulchellata* Steph., United Kingdom (England)
31. *Eupithecia pyreneata* Mab., Austria
32. *Eupithecia laquaearia* H.-S., S. Poland
33. *Eupithecia ultimaria* Bsd., Austria

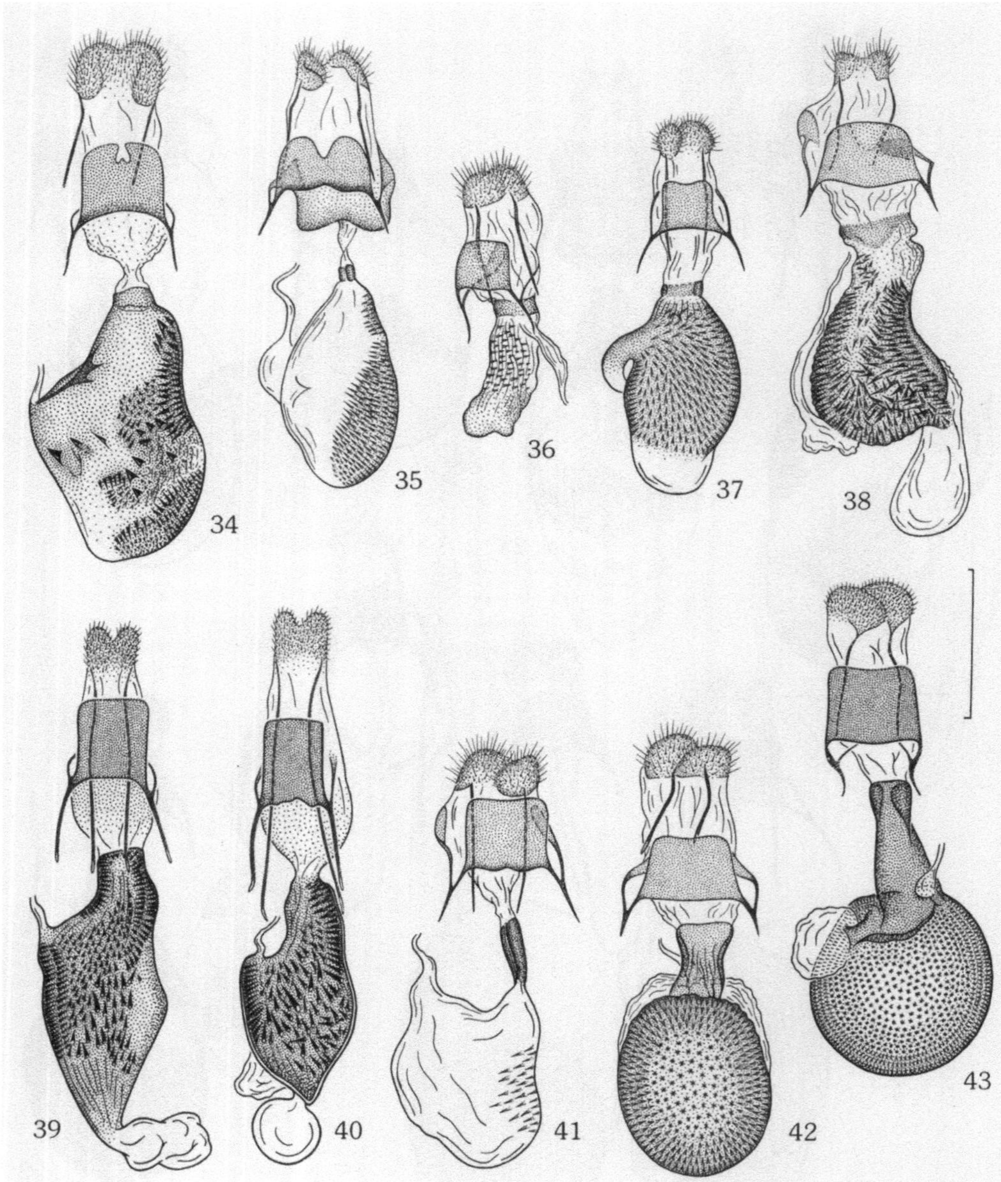

FEMALE GENITALIA (scale bar = 1 mm)

34. *Eupithecia usbeca* Viid., Uzbekistan (paratype)
35. *Eupithecia minusculata* Alph., NW. China (holotype)
36. *Eupithecia pantellata* Mill., Sicily
37. *Eupithecia plumbeolata* Hw., Germany
38. *Eupithecia pygmaeata* Hbn., NW. European Russia
39. *Eupithecia undata* Frr., Austria
40. *Eupithecia fennoscandica* Knb., N. European Russia
41. *Eupithecia silenata* Assm., S. European Russia (N. Caucasus)
42. *Eupithecia carpophagata* Stgr., C. Italy
43. *Eupithecia venosata* F., NW. European Russia

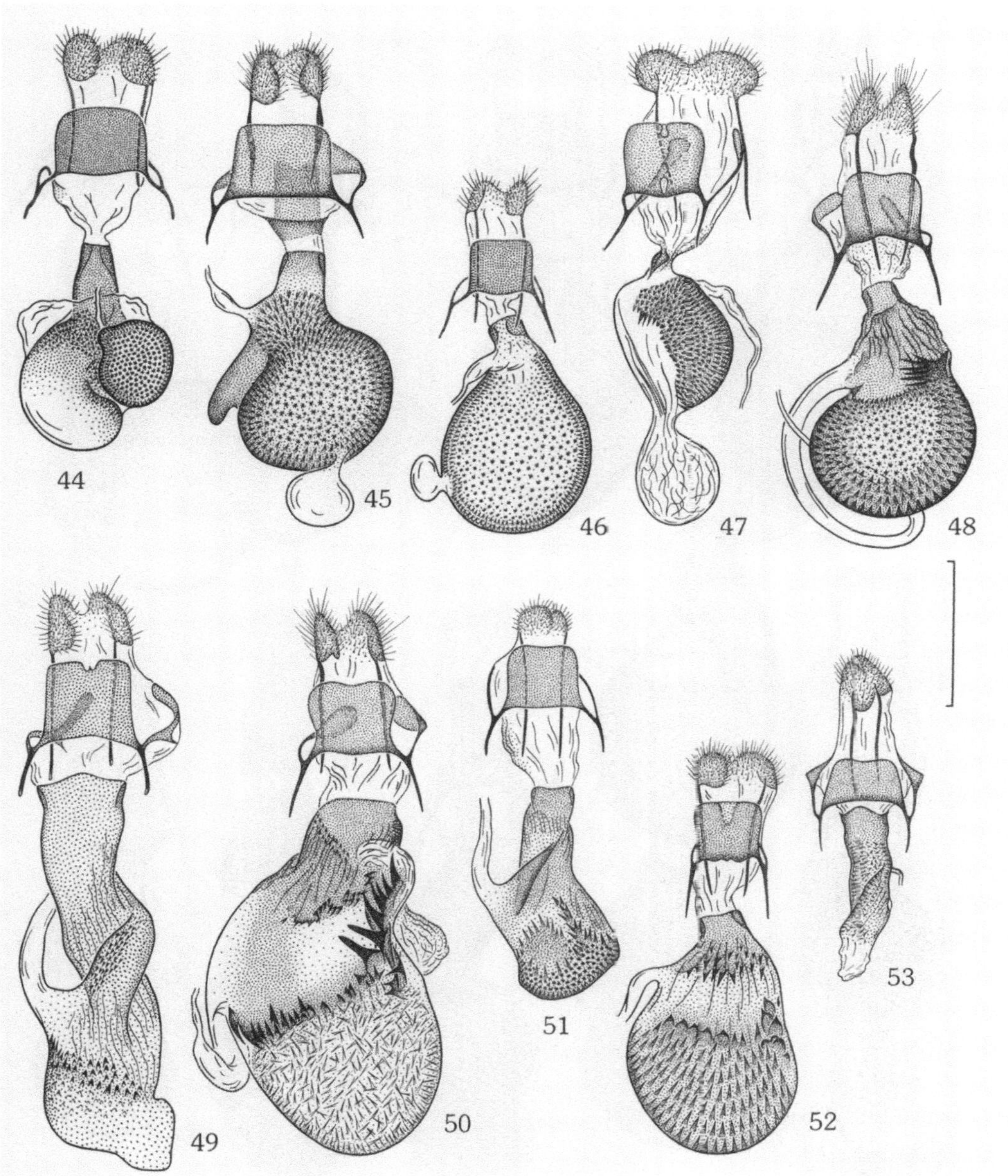

FEMALE GENITALIA (scale bar = 1 mm)

44. *Eupithecia schiefereri* Boh., S. Ukraine (Crimea)
45. *Eupithecia silenicolata* Mab., S. European Russia
46. *Eupithecia alliaria* Stgr., S. Ukraine (Crimea)
47. *Eupithecia dissertata* Püng., SE. France.
48. *Eupithecia cocciferata* Mill., S. France
49. *Eupithecia abbreviata* Steph., Germany
50. *Eupithecia lentiscata* Mab., S. Greece
51. *Eupithecia dodoneata* Gn., S. France
52. *Eupithecia reisserata* Pinker, N. Asian Turkey (paratypus)
53. *Eupithecia massiliata* Mill., N. Spain

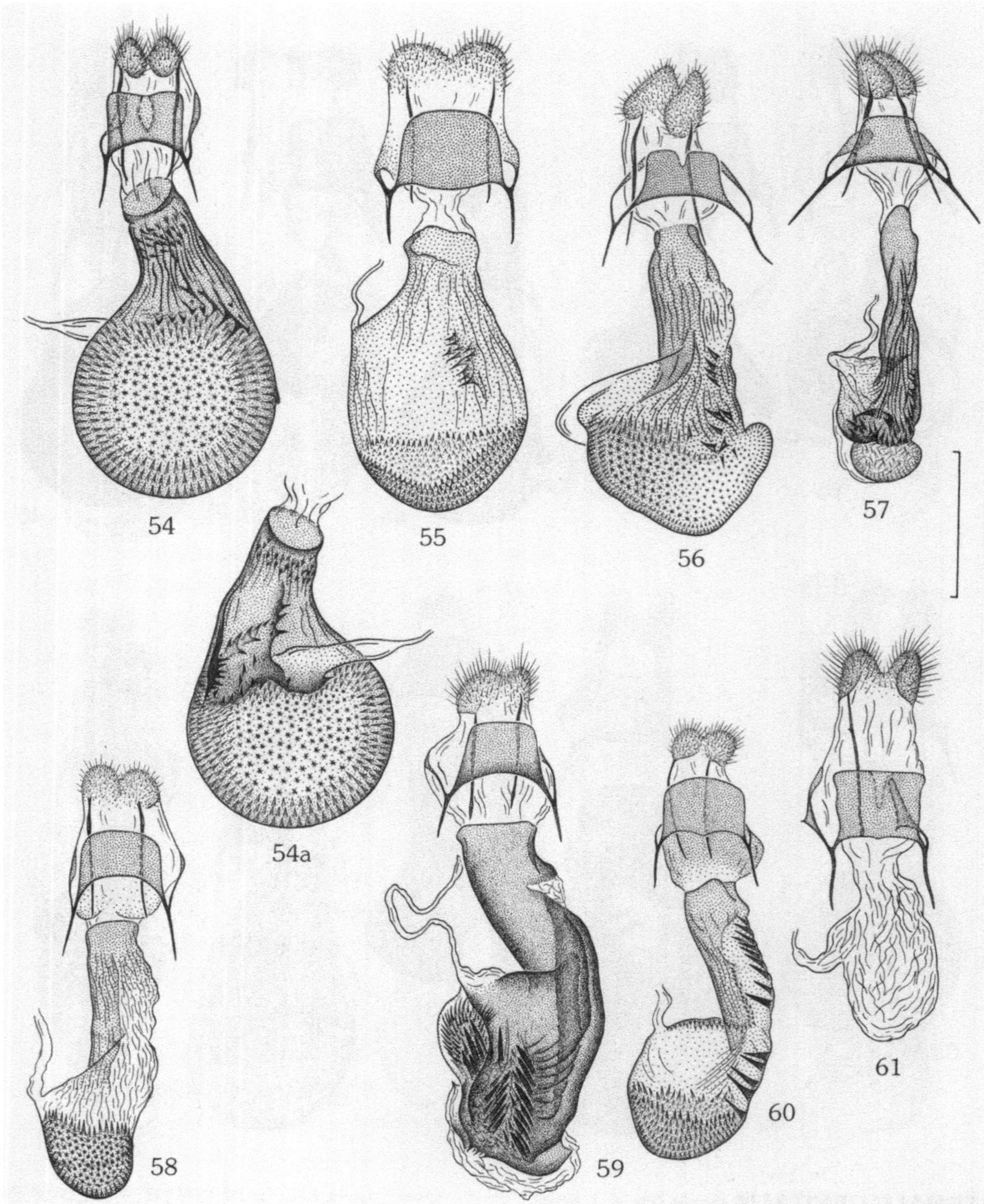

FEMALE GENITALIA (scale bar = 1 mm)

54. *Eupithecia extremata* F., Corsica
54a. Ditto, bursa copulatrix from other side
55. *Eupithecia scopariata* Rmbr., S. France
56. *Eupithecia pusillata* Den. & Schiff., NW. European Russia
57. *E. sardoa* Dtze., NW. Tunisia
58. *Eupithecia ericeata* Rmbr., France
59. *Eupithecia phoeniceata* Rmbr., N. Spain
60. *Eupithecia oxycedrata* Rmbr., France
61. *Eupithecia rosmarinata* Mill., SW. France

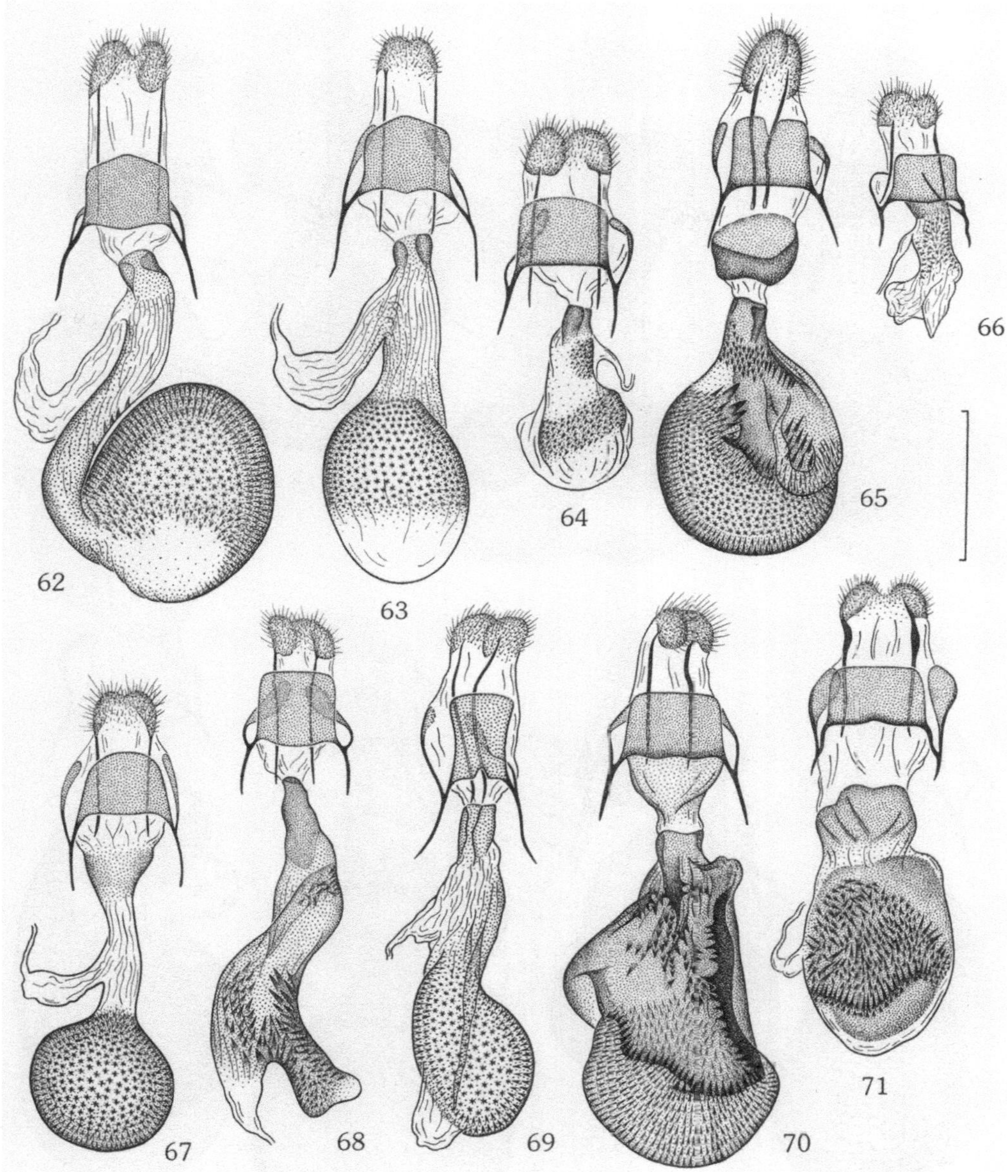

FEMALE GENITALIA (scale bar = 1 mm)

62. *Eupithecia barteli* Dtze., NW. Kazakhstan (Uralsk); (lectotype)
63. *Eupithecia tripunctaria* H.-S., S. Germany
64. *Eupithecia virgaureata* Dbld., S. European Russia
65. *Eupithecia tantillaria* Bsd., NW. European Russia
66. *Eupithecia liguriata* Mill., SE. France
67. *Eupithecia lariciata* Frr., S. Asian Russia (S. Siberia)
68. *Eupithecia lanceata* Hbn., NW. European Russia
69. *Eupithecia groenblomi* Urb., S. Finland
70. *Eupithecia selinata* H.-S., E. European Russia
71. *Eupithecia actaeata* Wald., Hungary

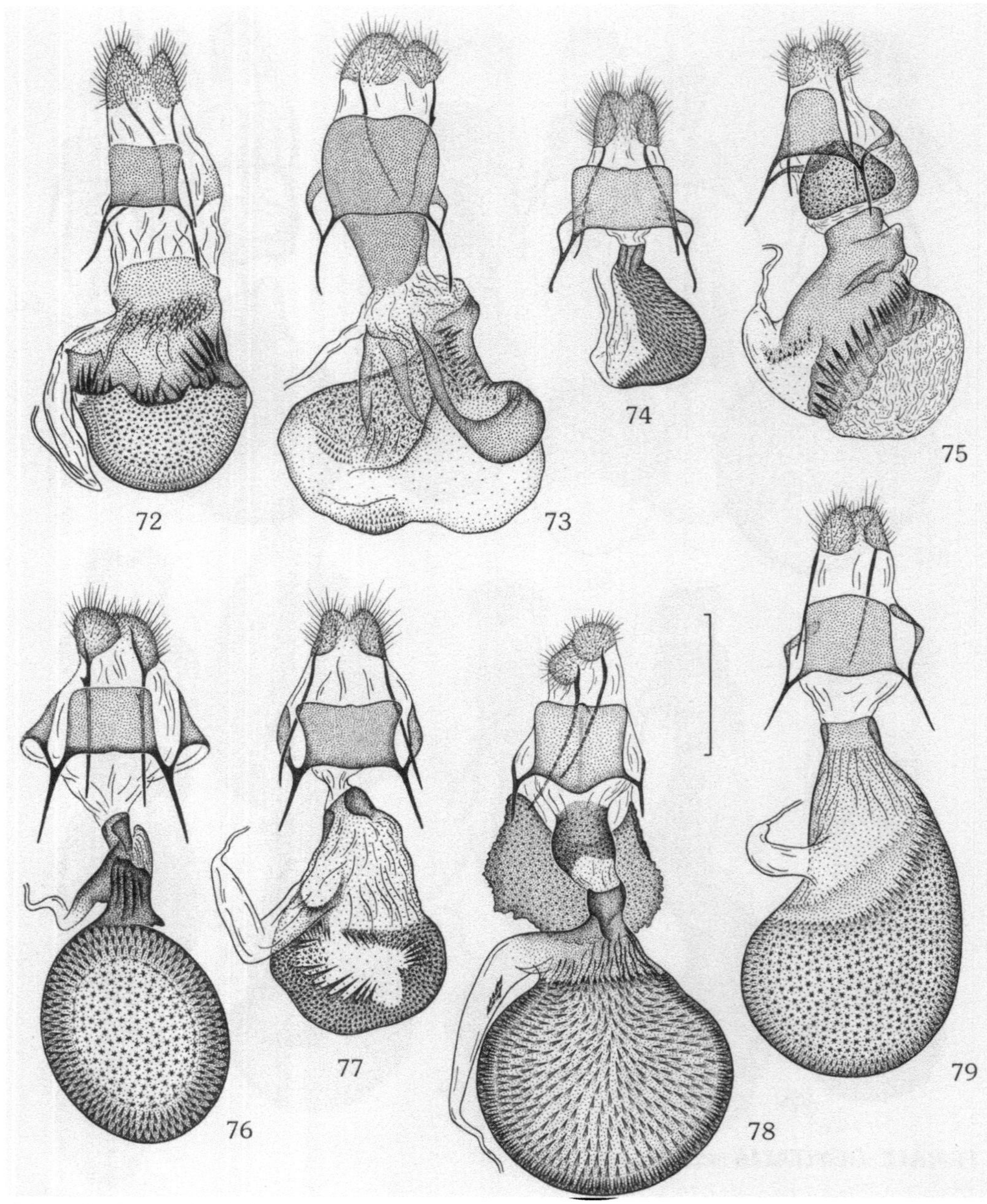

FEMALE GENITALIA (scale bar = 1 mm)

72. *Eupithecia quercetica* Prt., N. Asian Turkey
73. *Eupithecia egenaria* H.-S., SE. European Russia
74. *Eupithecia repentina* Vojnits & Laever, S. Far East Russia
75. *Eupithecia orana* Dtze., S. Spain
76. *Eupithecia pimpinellata* Hbn., E. European Russia
77. *Eupithecia karadaghensis* Mir., S. Ukraine (Crimea)
78. *Eupithecia suboxydata* Stgr., SE. European Russia
79. *Eupithecia simpliciata* Hw., NW. European Russia

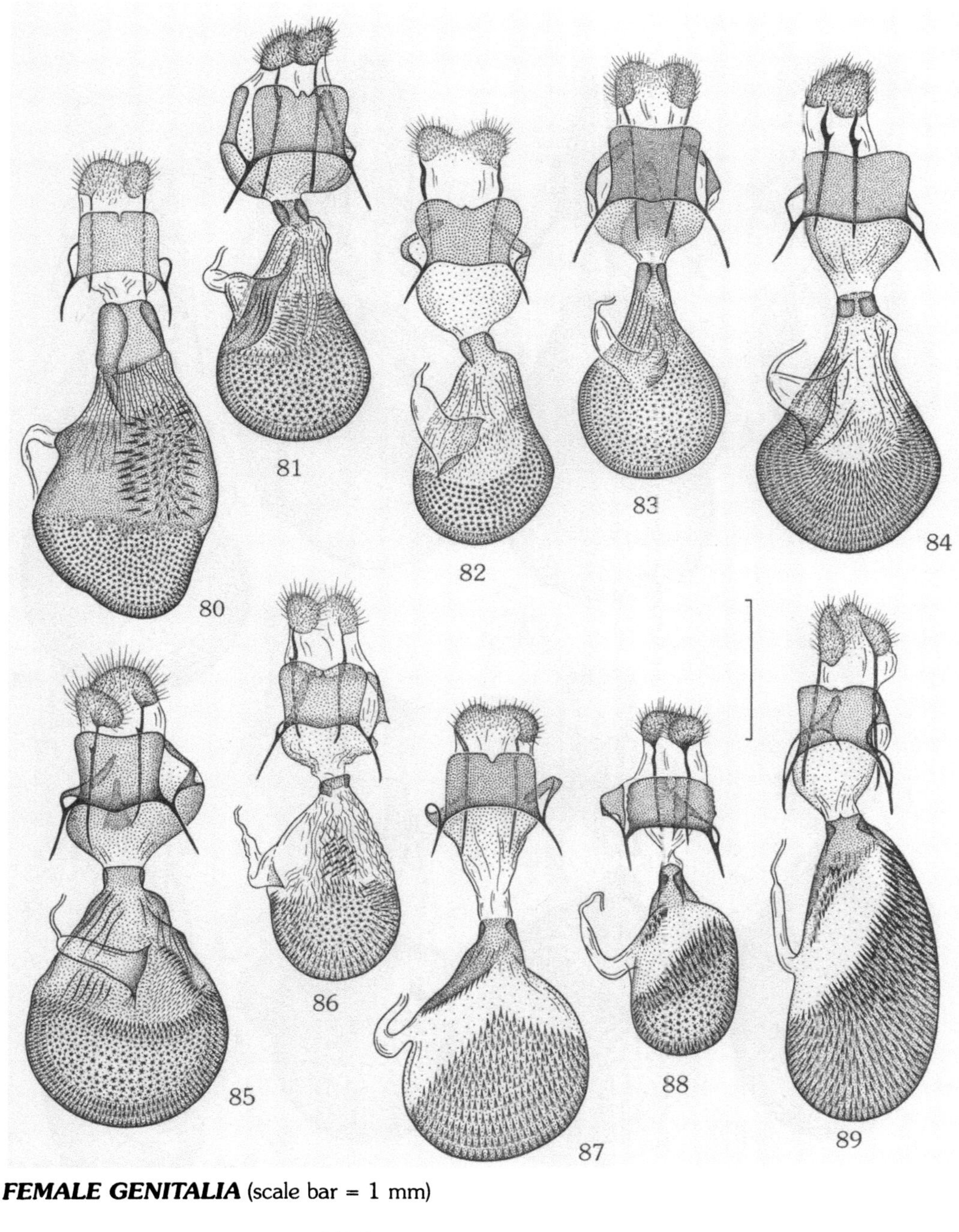

FEMALE GENITALIA (scale bar = 1 mm)

80. *Eupithecia sinuosaria* Ev., NW. European Russia
81. *Eupithecia gelidata hyperboreata* Stgr., N. European Russia
82. *Eupithecia nanata* Hbn., NW. European Russia
83. *Eupithecia innotata* Hufn., NW. European Russia
84. *Eupithecia ochridata* Schütze & Pinker, Macedonia
85. *Eupithecia unedonata* Mab., N. Spain
86. *Eupithecia chalikophila* Whli., S. Spain
87. *Eupithecia graphata* Tr., Hungary
88. *Eupithecia riparia* H.-S., S. Greece
89. *Eupithecia thurnerata* Schütze, Macedonia

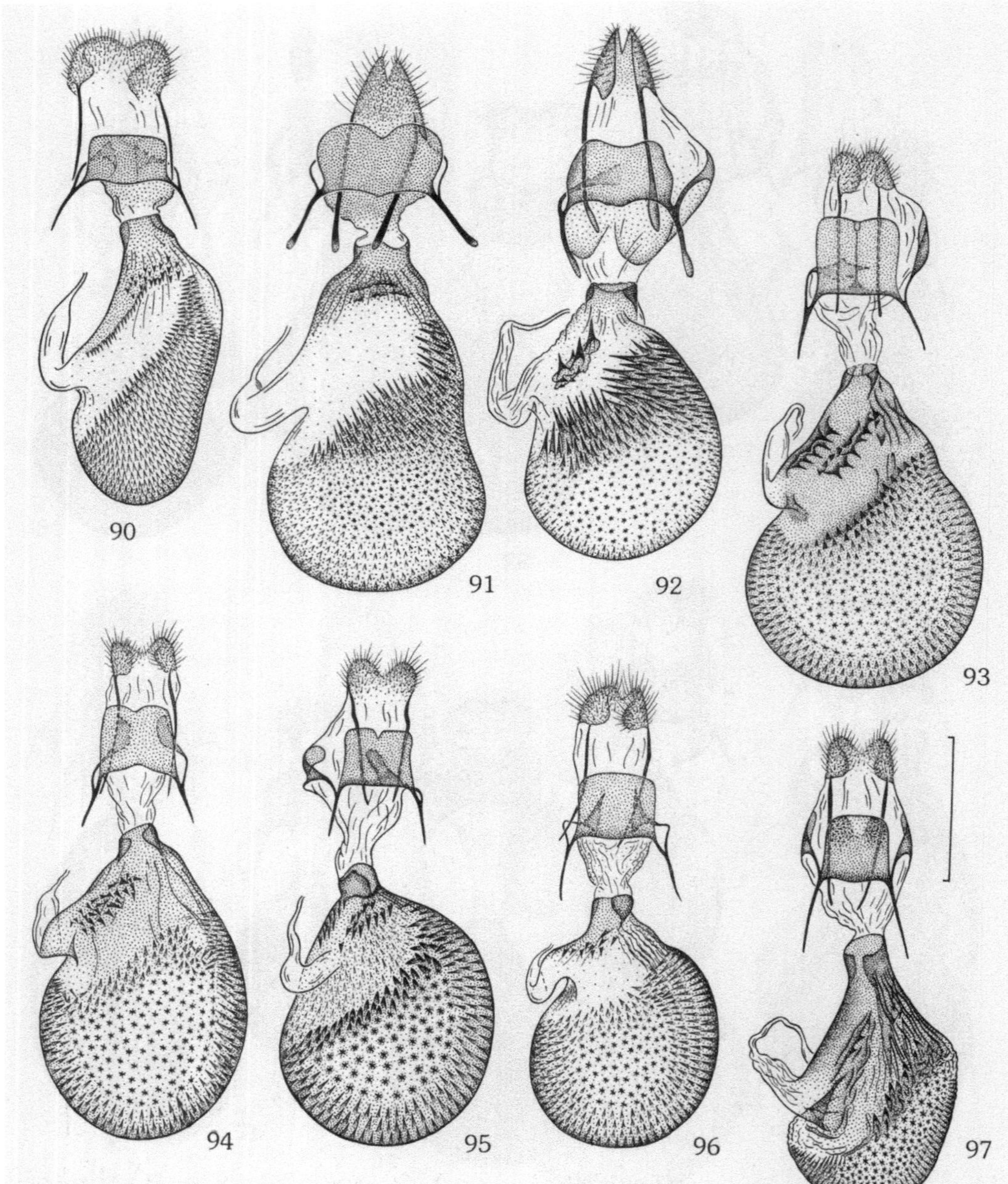

FEMALE GENITALIA (scale bar = 1 mm)

90. *Eupithecia gemellata* H.-S., Sicily
91. *Eupithecia mystica* Dtze., S. Ukraine (Crimea)
92. *Eupithecia cooptata* Dtze., C. Spain
93. *Eupithecia spissilineata* Metzn., Azerbaijan
94. *Eupithecia cuculliaria* Rbl., Macedonia
95. *Eupithecia unitaria* H.-S., S. Spain
96. *Eupithecia breviculata* Donz., Sicily
97. *Eupithecia cerussaria* Led., N. Asian Turkey

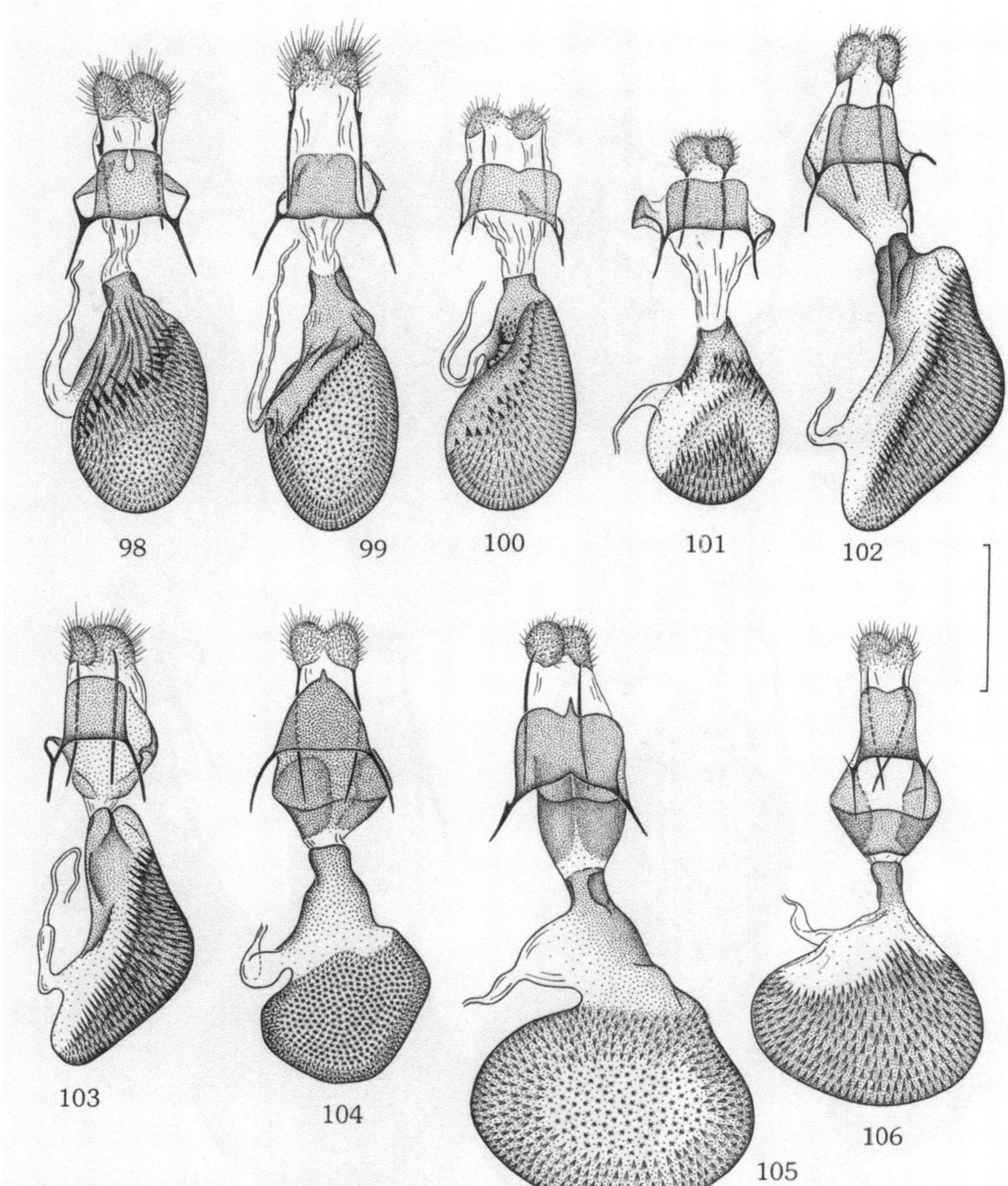

FEMALE GENITALIA (scale bar = 1 mm)

98. *Eupithecia irriguata* Hbn., Austria
99. *Eupithecia antalica* Mir., S. Asian Turkey
100. *Eupithecia indigata* Hbn., NW. European Russia
101. *Eupithecia conterminata* (Lienig & Zeller), NW. European Russia
102. *Eupithecia distinctaria* H.-S., Hungary
103. *Eupithecia senorita* Mir., S. Spain
104. *Eupithecia extraversaria* H.-S., Germany
105. *Eupithecia centaureata* Den. & Schiff., NW. European Russia
106. *Eupithecia limbata* Stgr., N. Asian Turkey

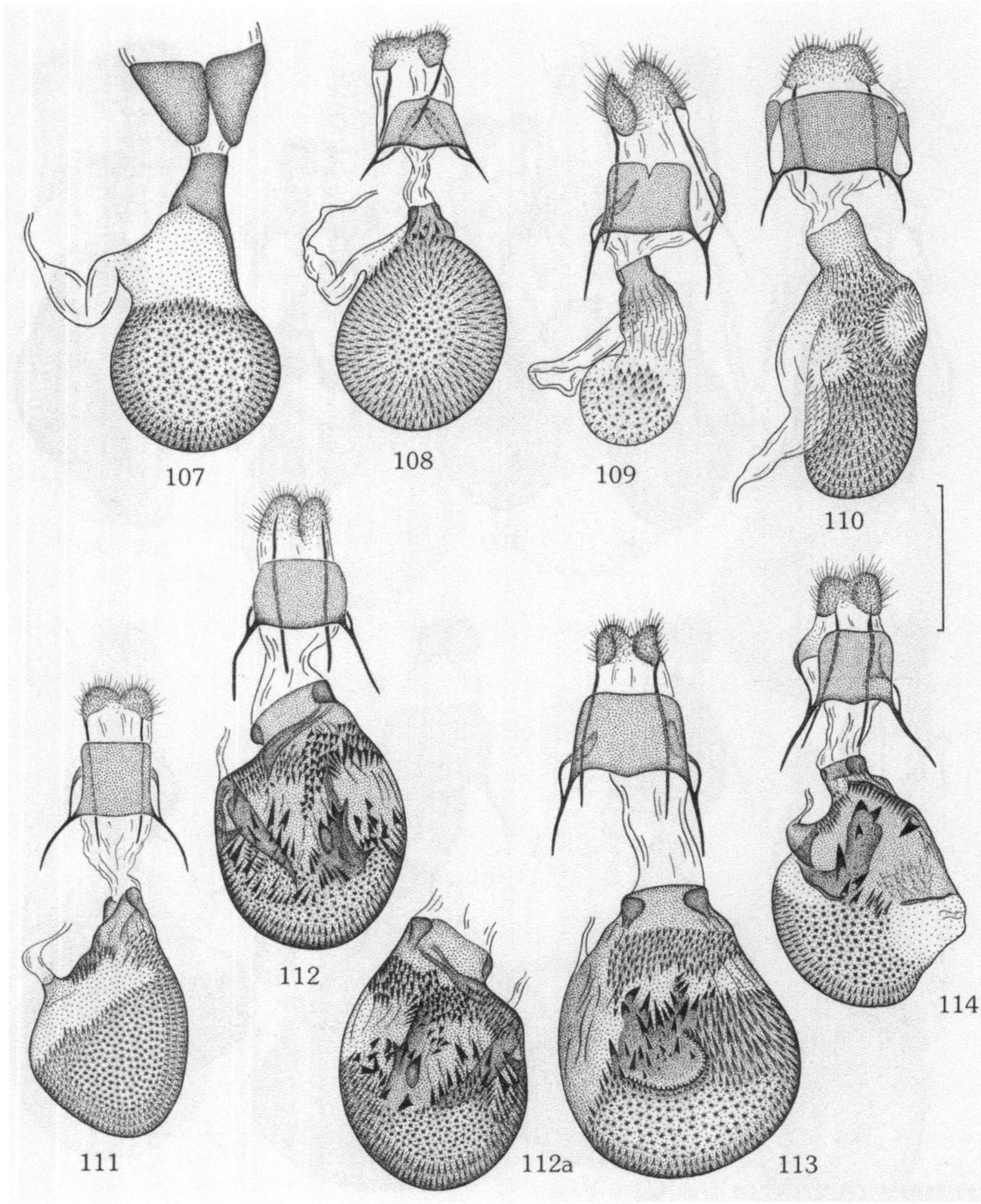

FEMALE GENITALIA (scale bar = 1 mm)

107. *Eupithecia nachadira* Brandt, S. Ukraine (Crimea)
108. *Eupithecia mekrana* Brandt, SW. Asian Russia
109. *Eupithecia variostrigata* Alph., S. Ukraine (Crimea)
110. *Eupithecia insigniata* Hbn., Hungary
111. *Eupithecia trisignaria* H.-S., C. European Russia
112. *Eupithecia gueneata* Mill., SE. France (reared)
112a. Ditto, bursa copulatrix from other side
113. *Eupithecia gratiosata* H.-S., W. Azerbaijan
114. *Eupithecia scalptata* Chr., S. Azerbaijan (paralectotype)

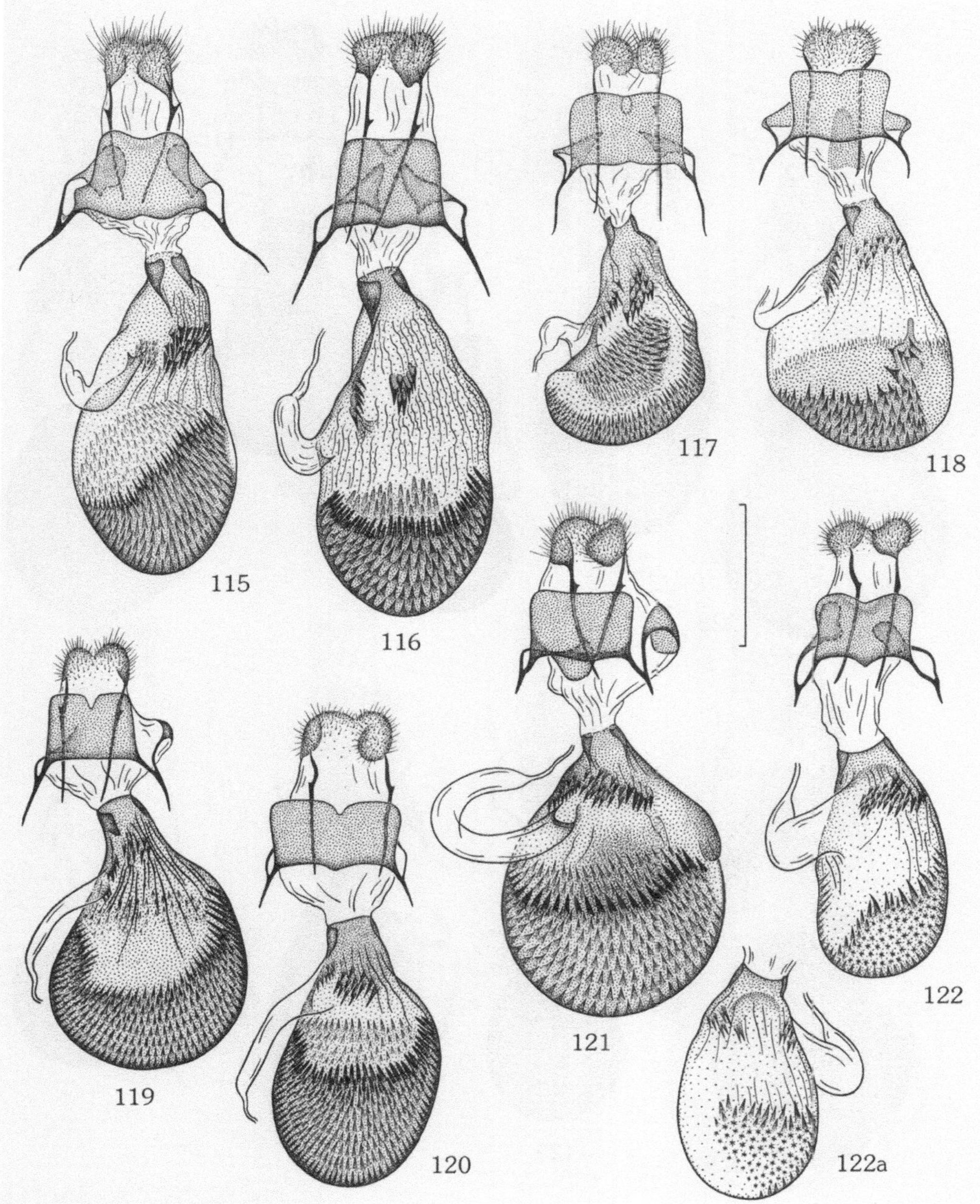

FEMALE GENITALIA (scale bar = 1 mm)

115. *Eupithecia veratraria* H.-S., Austria
116. *Eupithecia cretaceata fenestrata* Mill., Slovenia
117. *Eupithecia praealta* Whli., S. Spain
118. *Eupithecia intricata* Zett., NW. European Russia
119. *Eupithecia satyrata* Hbn., NW. European Russia
120. *Eupithecia cauchiata* Dup., W. Germany (reared)
121. *Eupithecia pernotata* Gn., S. European Russia
122. *Eupithecia extensaria* Frr., E. Asian Turkey
122a. Ditto, bursa copulatrix from other side

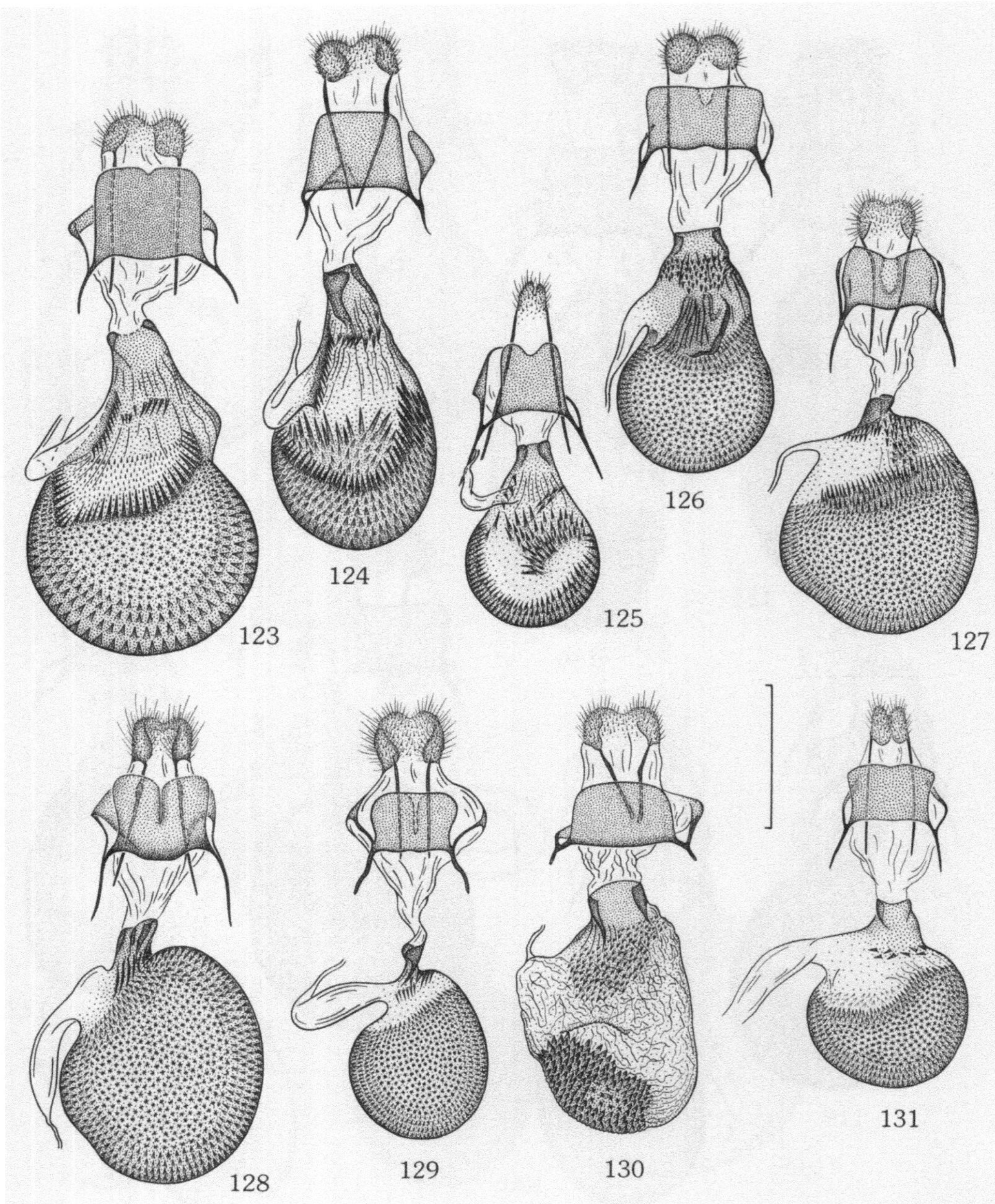

FEMALE GENITALIA (scale bar = 1 mm)

123. *Eupithecia absinthiata* Cl., NW. European Russia
124. *Eupithecia expallidata* Dbld., Austria
125. *Eupithecia valerianata* Hbn., W. Germany
126. *Eupithecia assimilata* Dbld., NW. European Russia
127. *Eupithecia vulgata* Hw., NW. European Russia
128. *Eupithecia biornata* Chr., S. European Russia
129. *Eupithecia sutiliata* Chr., S. Greece
130. *Eupithecia immundata* (Lienig & Zeller), NW. European Russia
131. *Eupithecia thalictrata* Püng., Switzerland

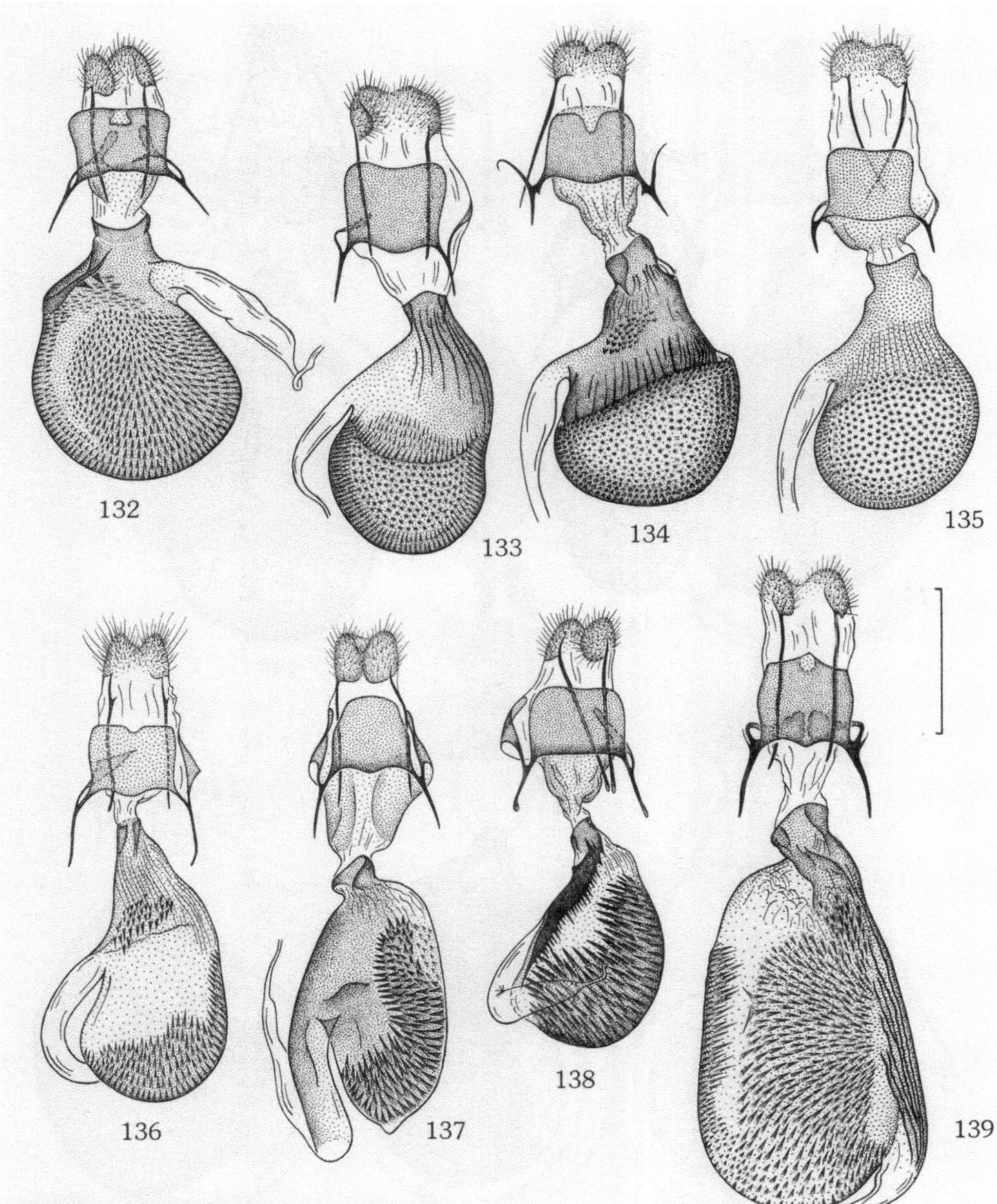

FEMALE GENITALIA (scale bar = 1 mm)

132. *Eupithecia addictata* Dtze., W. Asian Russia (Tjumen)
133. *Eupithecia exiguata* Hbn., NW. European Russia
134. *Eupithecia druentiata* Dtze., SE. France (reared)
135. *Eupithecia moecha* Dtze., NW. Kazakhstan
136. *Eupithecia denotata* Hbn., NW. European Russia
137. *Eupithecia weissi* Prt., N. Spain
138. *Eupithecia pauxillaria* Bsd., Austria
139. *Eupithecia santolinata* Mab., S. Spain

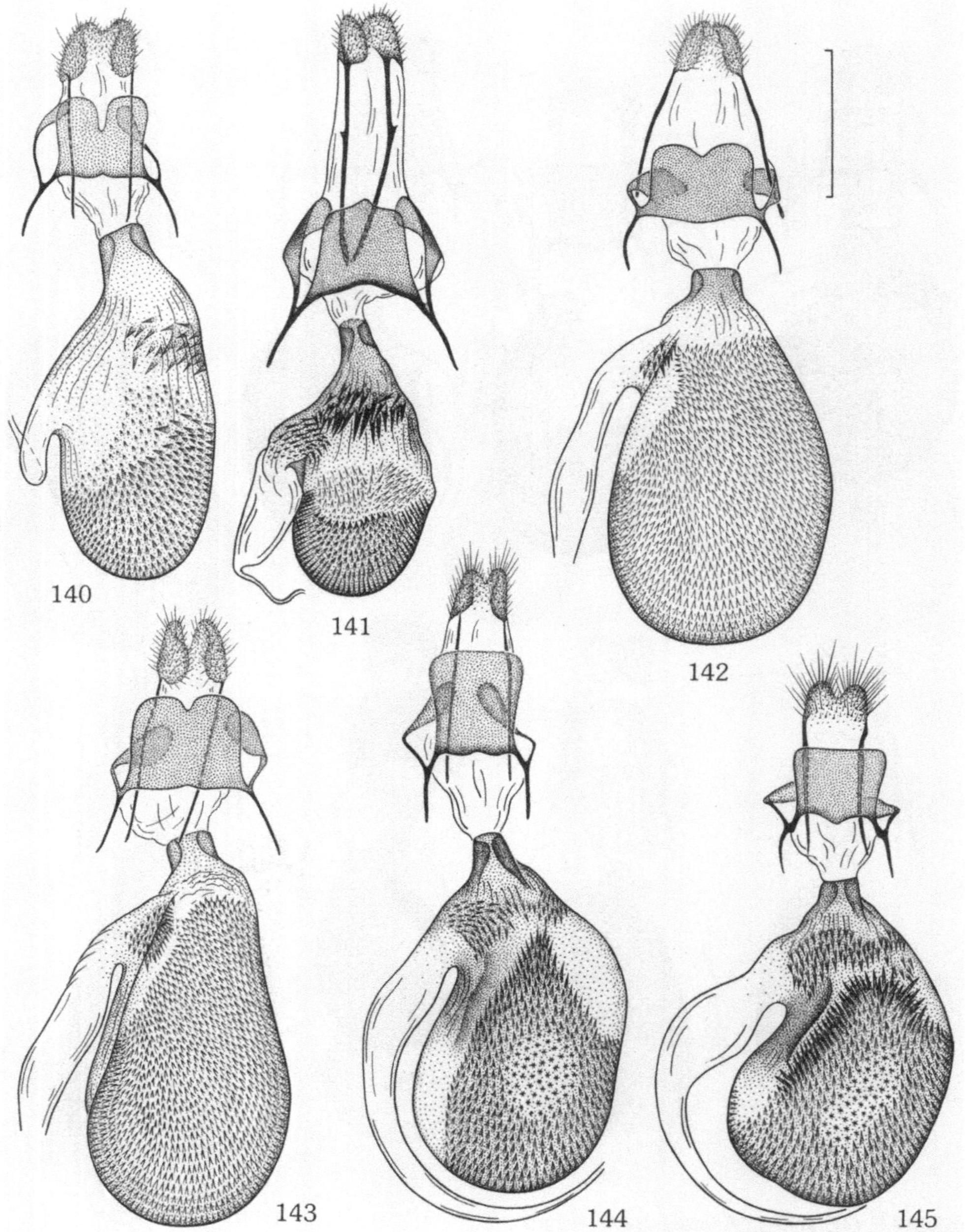

FEMALE GENITALIA (scale bar = 1 mm)

140. *Eupithecia millefoliata* Rössl., S. European Russia
141. *Eupithecia spadiceata* Zerny, S. Ukraine (Crimea)
142. *Eupithecia icterata* Vill., NW. European Russia
143. *Eupithecia succenturiata* L., NW. European Russia
144. *Eupithecia semigraphata* Bruand, E. Georgia
145. *Eupithecia impurata* Hbn., Austria

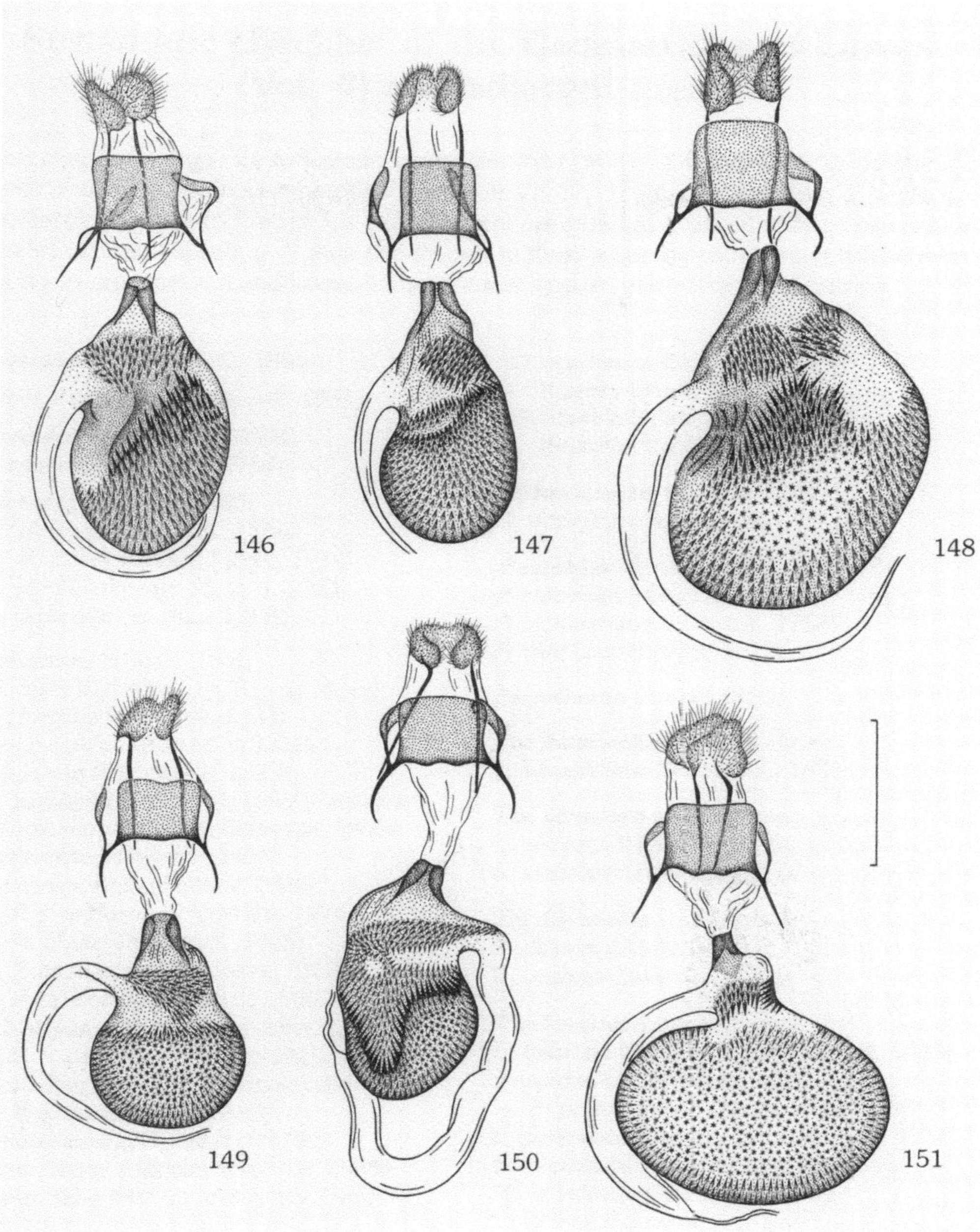

FEMALE GENITALIA (scale bar = 1 mm)

146. *Eupithecia poecilata* Püng., Sardinia
147. *Eupithecia fuscicostata* Chr., S. Russia (N. Caucasus)
148. *Eupithecia denticulata* Tr., Hungary
149. *Eupithecia subumbrata* Den. & Schiff., NW. European Russia
150. *Eupithecia orphnata* Pet., Austria
151. *Eupithecia subfuscata* Hw., SW. Bulgaria

Systematic checklist of the Geometridae of Europe (Vol. 4) and adjacent areas

The checklist shows, in systematic order, the taxa (from subfamily level to subspecific level) of the European Geometridae treated in Vol. 4. 53 species occurring in areas immediately adjacent to Europe (as defined here) are inserted (marked with an asterisk) in the list with an indication of their distribution in these adjacent countries. It will be interesting to see whether, and when, some of these species will be found in Europe.

Larentiinae DUPONCHEL, 1845

Tribus **Perizomini** HERBULOT, 1961

Pseudobaptria INOUE, 1982
P. bogumilaria (REBEL, 1904)

Mesotype HÜBNER, 1825
M. didymata (LINNAEUS, 1758)
  M. d. hethlandica (REBEL, 1910)
M. parallelolineata (RETZIUS, 1783)
M. verberata (SCOPOLI, 1763)

Perizoma HÜBNER, 1825
**P. gigas* WILTSHIRE, 1976 [Asia Minor]
P. affinitata (STEPHENS, 1831)
P. alchemillata (LINNAEUS, 1758)
P. hydrata (TREITSCHKE, 1829)
**P. parahydrata* ALBERTI, 1969 [Caucasus]
P. lugdunaria (HERRICH-SCHÄFFER, 1855)
P. bifaciata (HAWORTH, 1809)
P. flavosparsata (WAGNER, 1926)
  P. f. uniformata MENTZER, 1974
P. minorata (TREITSCHKE, 1828)
  P. m. ericetata (STEPHENS, 1831)
  P. m. norvegica PROUT, 1906
P. blandiata (DENIS & SCHIFFERMÜLLER, 1775)
  P. b. perfasciata PROUT, 1915
P. albulata (DENIS & SCHIFFERMÜLLER, 1775)
  P. a. thules (WEIR, 1880)
P. flavofasciata (THUNBERG, 1792)
P. obsoletata (HERRICH-SCHÄFFER, 1838)
P. incultaria (HERRICH-SCHÄFFER, 1848)

Martania MIRONOV, 2000
M. taeniata (STEPHENS, 1831)

Gagitodes WARREN, 1893
G. sagittata (FABRICIUS, 1787)

Tribus **Eupitheciini** PIERCE, 1914

Gymnoscelis MABILLE, 1868
G. rufifasciata (HAWORTH, 1809)
**G. insulariata* (STAINTON, 1859) [Canary Islands]
**G. lundbladi* PROUT, 1939 [Madeira, Canary Islands]

Chloroclystis HÜBNER, 1825
C. v-ata (HAWORTH, 1809)

Pasiphila MEYRICK, 1883
P. chloerata (MABILLE, 1870)
P. rectangulata (LINNAEUS, 1758)
P. debiliata (HÜBNER, 1817)

Eupithecia CURTIS, 1825

The ***haworthiata*** species-group
E. haworthiata DOUBLEDAY, 1856

The ***tenuiata*** species-group
E. tenuiata (HÜBNER, 1813)
E. inturbata (HÜBNER, 1817)

The ***abietaria*** species-group
E. abietaria (GOEZE, 1781)
E. analoga DJAKONOV, 1926

The ***linariata*** species-group
E. linariata (DENIS & SCHIFFERMÜLLER, 1775)
E. pulchellata STEPHENS, 1831
  E. p. intermedia DIETZE, 1913
E. pyreneata MABILLE, 1871
E. laquaearia HERRICH-SCHÄFFER, 1848
**E. albidulata* STAUDINGER, 1892 [Great Caucasus]

The ***ultimaria*** species-group
E. ultimaria BOISDUVAL, 1840
E. usbeca VIIDALEPP, 1992
E. minusculata ALPHÉRAKY, 1882
**E. inconspicuata* BOHATSCH, 1893 [Asia Minor, Transcaucasus]

The ***undata*** species-group
E. pantellata MILLIÈRE, 1875

E. plumbeolata (HAWORTH, 1809)
E. p. reikjavikaria (STAUDINGER, 1871)
E. p. lutosaria BOHATSCH, 1893
E. pygmaeata (HÜBNER, 1799)
**E. nigritaria* STAUDINGER, 1879 [Asia Minor]
**E. puengeleri* DIETZE, 1913 [Great Caucasus]
E. undata (FREYER, 1842)
E. fennoscandica KNABEN, 1949
**E. boryata* (REBEL, 1906) [Canary Islands]
**E. gomerensis* (REBEL, 1917) [Canary Islands]

The ***silenata*** species-group
E. silenata ASSMANN, 1848

The ***venosata*** species-group
E. carpophagata STAUDINGER, 1871
E. venosata (FABRICIUS, 1787)
E. v. ochracae GREGSON, 1886
E. v. fumosae GREGSON, 1887
E. v. hebridensis CURTIS, 1944
E. v. plumbea HUGGINS, 1962
E. schiefereri BOHATSCH, 1893
E. silenicolata MABILLE, 1866
**E. amasina* BOHATSCH, 1893 [Asia Minor]
**E. achyrdaghica* WEHRLI, 1929 [Asia Minor]
**E. rosai* PINKER, 1962 [Canary Islands]
**E. pinkeri* MIRONOV, 1991 [Asia Minor]
E. alliaria STAUDINGER, 1870
E. dissertata (PÜNGELER, 1905)

The ***interruptofasciata*** species-group
E. cocciferata MILLIÈRE, 1864
E. abbreviata STEPHENS, 1831
**E. dubiosa* DIETZE, 1910 [Levant]
E. lentiscata MABILLE, 1869
E. dodoneata GUENÉE, 1858
E. reisserata PINKER, 1976
E. massiliata MILLIÈRE, 1865
**E. maerkerata* SCHÜTZE, 1938 [Jordan, Saudi Arabia]
E. extremata (FABRICIUS, 1787)
E. scopariata (RAMBUR, 1833)
**E. rusicadaria* DIETZE, 1910 [Algeria, Tunisia]
E. pusillata (DENIS & SCHIFFERMÜLLER, 1775)
E. p. scoriata STAUDINGER, 1857
E. sardoa DIETZE, 1910
E. ericeata (RAMBUR, 1833)
E. phoeniceata (RAMBUR, 1834)
E. oxycedrata (RAMBUR, 1833)
E. rosmarinata DARDOIN & MILLIÈRE, 1865
**E. rhoisata* CHRÉTIEN, 1917 [North Africa]

The ***tripunctaria*** species-group
E. barteli DIETZE, 1908
E. tripunctaria HERRICH-SCHÄFFER, 1852

The ***fletcherata*** species-group
E. virgaureata DOUBLEDAY, 1861

The ***tantillaria*** species-group
E. tantillaria BOISDUVAL, 1840
E. liguriata MILLIÈRE, 1884

The ***lariciata*** species-group
E. lariciata (FREYER, 1842)

The ***lanceata*** species-group
E. lanceata (HÜBNER, 1825)

The ***egenaria*** species-group
E. groenblomi URBAHN, 1969
E. selinata HERRICH-SCHÄFFER, 1861
E. actaeata WALDERDORFF, 1869
**E. kunzi* PINKER, 1976 [Asia Minor]
E. quercetica PROUT, 1938
**E. tenerifensis* (REBEL, 1906) [Canary Islands]
**E. latipennata* PROUT, 1914 [Madeira]
E. egenaria HERRICH-SCHÄFFER, 1848
E. repentina VOJNITS & DE LAEVER, 1978
**E. schuetzeata* PINKER, 1961 [Canary Islands]
**E. stertzi* (REBEL, 1911) [Canary Islands]
E. orana DIETZE, 1910
**E. tricerata sperlichi* HAUSMANN, 1991 [Jordan]
**E. jizlensis muelleri* HAUSMANN, 1991 [Jordan]
E. pimpinellata (HÜBNER, 1813)
E. karadaghensis MIRONOV, 1988

The ***suboxydata*** species-group
E. suboxydata STAUDINGER, 1897

The ***simpliciata*** species-group
E. simpliciata (HAWORTH, 1809)

The ***sinuosaria*** species-group
E. sinuosaria (EVERSMANN, 1848)

The ***innotata*** species-group
E. gelidata hyperboreata STAUDINGER, 1861
E. nanata (HÜBNER, 1813)
E. n. zebrata WOLFF, 1929
E. n. gelidatoides WARNECKE, 1951
E. innotata (HUFNAGEL, 1767)
E. ochridata SCHÜTZE & PINKER, 1968
E. unedonata MABILLE, 1868
E. chalikophila WEHRLI, 1926

The ***graphata*** species-group
**E. brunneata* STAUDINGER, 1900 [Asia Minor]
E. graphata (TREITSCHKE, 1828)
E. g. setaceata DIETZE, 1903

E. g. sproengertsi DIETZE, 1910
E. g. olympica TULESCHKOV, 1951
E. g. hesperia WEHRLI, 1926
E. g. albofasciata STAUDINGER, 1879
E. riparia HERRICH-SCHÄFFER, 1851
**E. furcata* STAUDINGER, 1848 [Asia Minor]
**E. elissa* DIETZE, 1910 [Tunisia]
**E. farinosa* DIETZE, 1913 [Algeria]
E. thurnerata SCHÜTZE, 1958
**E. novata* DIETZE, 1904 [Asia Minor]
E. gemellata HERRICH-SCHÄFFER, 1861
E. mystica DIETZE, 1910
E. cooptata DIETZE, 1904
**E. subsequaria* HERRICH-SCHÄFFER, 1852 [Asia Minor]
**E. tesserata* BRANDT, 1938 [Jordan, Iran]

The ***breviculata*** species-group
E. spissilineata (METZNER, 1846)
E. cuculliaria (REBEL, 1901)
E. unitaria HERRICH-SCHÄFFER, 1851
E. breviculata (DONZEL, 1837)
E. cerussaria (LEDERER, 1855)

The ***irriguata*** species-group
E. irriguata (HÜBNER, 1813)
E. i. eriguata STAUDINGER, 1871
E. i. staudingeri BOHATSCH, 1893
E. antalica MIRONOV, 2001
**E. marasa* WEHRLI, 1932 [Asia Minor]
E. indigata (HÜBNER, 1813)
E. conterminata (LIENIG & ZELLER, 1846)

The ***distinctaria*** species-group
E. distinctaria HERRICH-SCHÄFFER, 1848
E. senorita MIRONOV, sp. n.

The ***centaureata*** species-group
E. extraversaria HERRICH-SCHÄFFER, 1852
E. centaureata (DENIS & SCHIFFERMÜLLER, 1775)
E. limbata STAUDINGER, 1879
E. l. tomillata (CHRÉTIEN, 1904)
**E. schwingenschussi* ZERNY, 1934 [Morocco]
**E. deverrata* DIETZE, 1910 [Tunisia]
**E. lecerfi* PROUT, 1928 [Morocco, Algeria]
E. nachadira BRANDT, 1941
**E. desertorum* DIETZE, 1910 [Algeria, Tunisia]

The ***despectaria*** species-group
E. mekrana BRANDT, 1941
**E. tenellata* DIETZE, 1906 [Morocco to Saudi Arabia]

The ***variostrigata*** species-group
E. variostrigata ALPHÉRAKY, 1876
**E. falkneri* VOJNITS, 1978 [Algeria]

The ***nevadata*** species-group
E. insigniata (HÜBNER, 1790)

The ***gueneata*** species-group
E. trisignaria HERRICH-SCHÄFFER, 1848
E. gueneata MILLIÈRE, 1862
**E. syriacata* STAUDINGER, 1878 [Asia Minor, Levant]
E. gratiosata HERRICH-SCHÄFFER, 1861
E. scalptata CHRISTOPH, 1885

The ***cretaceata*** species-group
E. veratraria HERRICH-SCHÄFFER, 1848
E. v. eynensata DE GRASLIN, 1863
E. v. arctica VIIDALEPP, 1974
E. cretaceata fenestrata MILLIÈRE, 1874
**E. subfenestrata* STAUDINGER, 1892 [Caucasus, Transcaucasus]

The ***satyrata*** species-group
E. praealta WEHRLI, 1926
**E. pfeifferi* WEHRLI, 1929 [Caucasus, Asia Minor]
E. intricata (ZETTERSTEDT, 1839)
E. i. hibernica MERE, 1964
**E. tribunaria* HERRICH-SCHÄFFER, 1852 [Great Caucasus]
E. satyrata (HÜBNER, 1813)
E. s. curzoni GREGSON, 1884
E. cauchiata (DUPONCHEL, 1830)
E. pernotata GUENÉE, 1858
E. p. enictata PELLMYR & MIKKOLA, 1984
E. extensaria (FREYER, 1845)
E. e. occidua PROUT, 1914
E. absinthiata (CLERCK, 1759)
E. expallidata DOUBLEDAY, 1856
E. valerianata (HÜBNER, 1813)
**E. arenbergeri* PINKER, 1976 [western Caucasus, Asia Minor]

The ***russeliata*** species-group
E. assimilata DOUBLEDAY, 1856
E. vulgata (HAWORTH, 1809)
E. v. cyrneata SCHAWERDA, 1933
E. biornata CHRISTOPH, 1867
E. sutiliata CHRISTOPH, 1876
E. immundata (LIENIG & ZELLER, 1846)

The ***addictata*** species-group
E. thalictrata (PÜNGELER, 1902)
E. addictata DIETZE 1908

The ***druentiata*** species-group

E. exiguata (HÜBNER, 1813)

E. druentiata DIETZE, 1902

E. moecha DIETZE, 1904

E. denotata (HÜBNER, 1813)

**E. pseudocastigata* PINKER, 1976 [Asia Minor]

The ***millefoliata*** species-group

E. weissi PROUT, 1938

E. pauxillaria BOISDUVAL, 1840

E. santolinata MABILLE, 1871

E. millefoliata RÖSSLER, 1866

E. spadiceata ZERNY, 1933

The ***semigraphata*** species-group

E. icterata (DE VILLERS, 1789)

E. succenturiata (LINNAEUS, 1758)

**E. tarfata* (LUCAS, 1907) [Algeria]

**E. marginata* STAUDINGER, 1892 [Cyprus, Asia Minor, Caucasus]

E. semigraphata BRUAND, 1851

E. impurata (HÜBNER, 1813)

E. poecilata PÜNGELER, 1888

E. fuscicostata CHRISTOPH, 1887

E. denticulata (TREITSCHKE, 1828)

**E. lacteolata* DIETZE, 1906 [western Caucasus]

E. subumbrata (DENIS & SCHIFFERMÜLLER, 1775)

E. orphnata PETERSEN, 1909

E. subfuscata (HAWORTH, 1809)

Incertae sedis

**E. arenitincta* PROUT, 1938 [Algeria]

**E. glaisi* LUCAS, 1937 [Algeria]

**E. insignifica* ROTHSCHILD, 1914 [Algeria]

**E. ogilviata* (WARREN, 1905) [Azores]

**E. yathomi* HAUSMANN, 1991 [Jordan]

References

AARTSEN, VAN B. (1964): Macrolepidoptera in 1962. - Ent. Ber. 24 (1): 1-2.

AARVIK, L. & A. BAKKE (1999): The moth (Lepidoptera) fauna of coastal spruce forest at Almdalen Forest Reserve, Namsos, and continental forest in Lierne, Central Norway. - Norw. J. Ent. 46 (1): 9-17.

AARVIK, L., BERGGREN, K.. & L.O. HANSEN (2000): Catalogus Lepidopterorum Norvegiae. Lepid. arbeidsgruppe Zool. Mus., Univ. Oslo, Norsk Inst. Skogforskn., Ås. Oslo, 2000, 192 pp.

ABOS CASTEL, F. (1980): Lepidópteros de la Provincia de Huesca (I). - Shilap Revta lepid. 8 (29): 41-43.

ABOS CASTEL, F. (1981): Lepidópteros de la Provincia de Huesca. Zona 3.- Cuencas de los rios Isabena y Noguera Ribagorzana. (II Parte). - Shilap Revta lepid. 9 (34): 141-144.

ABOS CASTEL, F.P. (1986): Lepidópteros del valle pirenaico de Roncal (Navarra). - Shilap Revta. lepid. 14 (56): 47-58.

ABOS CASTEL, F.P. (1988): Lepidópteros de la provincia de Huesca. Addenda a los capítulos publicados con anterioridad sobre el tema en SHILAP Revta. lepid. - Shilap Revta. lepid. 16 (64): 311-330.

ABOS CASTEL, F.P. (1990): Lepidópteros de la provincia de Huesca. Addenda segunda a los capítulos publicados con anterioridad sobre el tema en SHILAP Revta. lepid. - Shilap Revta. lepid. 18 (70): 159-168.

ADAMCZEWSKI, S. (1939): Przyczynek do znajomosci fauny motyli Podola Polskiego [Contribution a la connaissance des Lepidopteres de la Podolie Polonaise]. - Fragm. Faun. Mus. Zool. Polon., Warszawa 4 (9): 159-209.

ADAMCZEWSKI, S. (1951): Luskoskrzydle srodmiescia Warszawy (Lepidoptera) [The moths of Warsaw-City (*Lepidoptera*)]. - Fragm. Faun. Mus. Zool. Polon., Warszawa 6 (5): 111-128.

ADAMEK, A. (1940): Prispevek k soustavnemu vyzkumu fauny Lepidopter na Olomoucku. Nachtrag zur Falterfauna des Bezirkes Olmutz. - Ent. listy 3: 11-13.

AGENJO, R. (1950): Acerca de cuatro *Cidaria (sensu latus)* descritas de España, de las que dos pasan a sinonimia de las otras y una constituye genero nuevo. - EOS 26 (3-4): 321-339, lams. IX-X.

AGENJO, R. (1953): *Geometridae* espanoles de la coleccion D.Hospital, de Barcelona. - EOS 29: 289-328, lams. X-XI.

AGENJO, R. (1971): Ocho generos y veinte especies de *Geometridae* nuevos para Espana (Lep.). - Graellsia 27: 3-21, pls 1-4.

AHOLA, M., SILVONEN, K. & J. VILEN (1983): Kosken HI (EH) pitajan suurperhoset vuosina 1969-1982. - Notulae Ent. 63 (4): 145-175.

AHOLA, M. & L. KOHONEN (1985): A list of *Macrolepidoptera* collected in north-eastern Spain in June 1982. - Shilap Revta lepid. 13 (50): 113-123.

AHOLA, M., KAITILA, J.-P., NUPPONEN, K., JUNNILAINEN, J., OLSCHWANG, V.N. & YU.E. MIKHAILOV. (1997): Materialy k poznaniju babochek Urala. Nauchnye rezultaty russko-finskoj lepidopterologicheskoj ekspedicii na Juzhnom Urale v 1996 g. *Macrolepidoptera*. [Materials to knowledge of *Macrolepidoptera* of the Urals. The scientific results of Russian-Finish lepidopterological expedition to the Southern Urals in 1996]. Uspehi entomologii na Urale. Ekaterinburg (1997): 98-104.

ALBERTI, B. (1965): Über *Eupithecia subfenestrata* STGR. (Lep., Geometridae). – Dt. ent. Z. 12 (4-5): 375-377.

ALBERTI, B. & J. SOFFNER (1962): Zur Kenntnis der Lepidopteren-Fauna Sud- und Sudostrusslands. - Mitt. münchn. ent. Ges. 52: 146-198.

ALEXINSCHI, A. (1931): Fauna Macrolepidopterelor Besarabiei de Nord (judetul. Hotin). - Mem. Sect. Stiint. Acad. Rom., Bucuresti, ser. 3, 7 (6): 119-183.

ALEXINSCHI, A. (1942): Elemente noua pentru cunoaşterea faunei Lepidopterelor din Bucovina (Radauti). - Analele Moldovei 1 (1): 1-18.

ALEXINSCHI, A. (1952): Contributiuni la cunoaşterea Macrolepidopterelor noi sau rare pentru fauna R.P.R. - Buletin Ştiintific Sect. ştiinte biol., agron., geol. şi geografice 4 (1): 15-22.

ALEXINSCHI, A. & M. PEIU (1958a): Noi contributiuni la cunoaşterea faunei Lepidopterelor regiunii Iaşi. - Stud. cercetari ştiintifice (Biol. şi ştiinte Agricole) 9 (1): 57-67.

ALEXINSCHI, A. & M. PEIU (1958b): Contributii la cunoaşterea faunei Lepidopterelor din Moldova (Part. V-a). - Stud. cercetari ştiintifice (Biol. şi ştiinte Agricole) 9 (2): 245-251.

ALEXINSCHI, A. & M. PEIU (1960): Contributii la cunoaşterea Lepidopterelor din Moldova (Part. VI-a). - Stud. cercetari ştiintifice (Biol. şi ştiinte Agricole) 11 (2): 293-300.

ALEXINSCHI, A. & M. PEIU (1962): Contributii la cunoaşterea faunei Lepidopterelor din Moldova şi regiunea Iaşi (Part. VII-a). - Stud. cercetari ştiintifice (Biol. şi ştiinte Agricole) 13 (1): 69-78.

ALPHÉRAKY, S. (1876): Cheshuekrylye okrestnostej Taganroga. [Lépidoptères des environs de Taganrog]. - Horae Soc. ent. Ross. 10: 35-53.

ALPHÉRAKY, S. (1903): Cheshuekrylye okrestnostej Taganroga. III [Lepidopteren der Umgebung von Taganrog]. - Horae Soc. ent. Ross. 38: 558-618.
ANDERSEN, S. (1982): Revision of European species of *Siphona* Meigen (Diptera: Tachinidae). - Ent. scand. 13: 149-172.
ANDERSEN , S. (1996): The *Siphonini* (Diptera: Tachinidae) of Europe. - Fauna ent. scand. 33: 1-148.
ANDRES, A. (1913): Verzeichnis der bis jetzt in Aegypten beobachteten Schmetterlinge. – Bull. Soc. ent. d'Egypte, année 1912 (3 fasc.): 53-114.
ANDRES, A. & A. SEITZ (1924): Die Lepidopteren-Fauna Ägyptens. – Senckenbergiana, Frankfurt a.M. 6 (1/2): 13-83.
ANIKIN, V.V., SACHKOV, S.A., ZOLOTUHIN, V.V. & E.M. ANTONOVA. (2000): 'Fauna lepidopterologica Volgo-Uralensis' 150 years later: changes and additions. Part 3. *Geometridae* (Insecta, Lepidoptera). - Atalanta 31 (1/2): 293-326.
ANTONOVA, E.M. (1972a): K sostavu fauny babochek pjadenitz (Lepidoptera, Geometridae) Berezinskogo Gosudarstvennogo zapovednika i ih ekologicheskomu raspredeleniju. - Berezinskij zapovednik 2: 140-159 (in Russian).
ANTONOVA, E.M. (1972b): K poznaniju fauny i ekologii babochek pjadenitz (Geometridae, Lepidoptera) Mordovskogo zapovednika. - Trudy Mordovsk. Gos. Zapovedn. im. P.G. Smidovitcha 6: 121-133 (in Russian).
ANTONOVA, E.M. (1977): K poznaniju fauny pjadenitz (Geometridae, Lepidoptera) zapovednika 'Galich'ja Gora' (Lipetzkaja obl.). In: Novye i maloizvestnye vidy nasekomyh Evropejskoj chasti SSSR, Leningrad, pp. 41-46 (in Russian).
ANTONOVA, E.M. (1985): Pjadenitzy Bashkirii (Lepidoptera, Geometridae) i ih zoogeograficheskie osobennosti. - Sborn. trudov Zool. mus. MGU 23: 183-197 (in Russian).
ANTONOVA, E.M. (1989a): Pjadenitzy (Lepidoptera, Geometridae) Gorkovskoj oblasti, II. Podsemejstva *Larentiinae* i *Ennominae*. In: Nazemnye i vodnye ekosistemy, Corkij, 1989, pp. 51-60 (in Russian).
ANTONOVA, E.M. (1989b): Fauna i zoogeographia pjadenitz (Geometridae, Lepidoptera) Il'menskogo Gosudarstvennogo zapovednika. In: Nasekomye v ekosistemah Sibiri i Dal'nego Vostoka. Issledovanija po faune Sovetskogo Sojuza. Moskva (MGU press), pp. 180-191 (in Russian).
ANTONOVA, E.M. (1994): Pjadenitzy (Geometridae). In: CheshuekrylyePrioksko-Terrasnogo zapovednika (*Lepidoptera*). - Flora i fauna zapovednikov 55, Moscow, 1994, pp. 23-31 (in Russian).
ANTONOVA, E.M. & L.V. BOLSHAKOV (2001): Pjadenitzy (Lepidoptera, Geometridae) Tul'skoj oblasti. Dopolnenie (s vkljucheniem dannyh po sopredel'nym rajonam Kaluzhskoj, Moskovskoj i Orlovskoj oblastej). [The Geometrid moths (Lepidoptera) of Tula Area. Addition (with including the data on the neighbouring districts of Kaluga, Moscow and Orel Areas]. - In: Biologicheskoe raznoobrazie Tul'skogo kraja na rubezhe vekov, Tula, 1: 53-63.
ANTONOVA, E.M. & M.A. KLEPIKOV (2000): Pjadenitzy (Lepidoptera: Geometridae) Jaroslavskoj oblasti [The Geometrid Moths (*Lepidoptera*) of Yaroslavl Area]. - Russ. Ent. J. 9 (1): 79-90.
ANTONOVA, E.M., MIRONOV, V.G., SVIRIDOV, A.V. & I.V. SHMYTOVA (1999): Novye i redkie nahodki raznousyh cheshuekrylyh (Lepidoptera: Heterocera) dlja bassejna verhnej Oki [New and rare records of moths (*Lepidoptera: Heterocera*) for the upper Oka River Basin]. - Russ. Ent. J. 8 (1): 53-56.
ANTONOVA, E.M. & A.N. SMETANIN (1978): Vidovoj sostav, trophicheskie svjazi i zoogeographicheskie osobennosti pjadenitz zapovednika 'Trostjanetz' (Geometridae, Lepidoptera). [Species composition, throphic relationships and specific zoogeographical features of loopermoths (Geometridae, Lepidoptera) in the Nature Reserve 'Trostyanets']. - Bjul. Mosk. obsch. ispytatelej prirody (otd. biol.) 83 (1): 52-61.
ANTONOVA, E.M., SVIRIDOV, A.V. & V.I. ROSTSHINENKO (1989): Fauna pjadenitz i sovok (Geometridae, Noctuidae) v Udmurtii. In: Fauna i oekologia zhivotnyh UASSR i prilezhastshih rajonov. Izhevsk, pp. 108-121 (in Russian).
ANTONOVA, E.M. & A.M. TIKHOMIROV (1992): Regionalnyj kadastr pjadenitz Ivanovskoj oblasti (*Archiearinae, Hemitheinae, Sterrhinae* i *Larentiinae*). Voprosy inventarizatzii fauny, 1992, Ivanovo, pp. 17-26 (in Russian).
ARNSCHEID, W.R. (2000): Die Macrolepidopteren-Fauna Westliguriens (Riviera dei Fiori und Ligurische Alpen in Oberitalien (Insecta, Lepidoptera). - Neue ent. Nachr. 47: 1-310.
AUBERT, J.F. (1953): Revision des types et de la collection F. de Rougemont (1). - Revue franc. Lepid. 14 (9-10): 108-115.
BAKKE, A. (1955): Insects Reared from Spruce Cones in Northern Norway 1951. - Norsk entom. Tidsskr. 9 (3-4): 152-212.
BALOGH, I. (1967a): A Bukk-hegyseg lepkefaunajanak kritikai vizsgalata I. - Folia ent. hung. (series nova) 20 (9): 95-165.
BALOGH, I. (1967b): A Bukk-hegyseg lepkefaunajanak kritikai vizsgalata II. - Folia ent. hung. (series nova) 20

(24): 521-588.

BALSAC, H.H. de & M. CHOUI (1984): Les Lepidopteres de la Gaume franco-belge (Esquisse zoogeographique et liste des especes) Addenda. - Alexanor 13(6): 253-273.

BARANIAK, E. & J. SOSINSKI. (2000): Miernikowcowate (*Lepidoptera, Geometridae*) Wielkopolskiego Parku Narodowego [Geometrid Moths (Lepidoptera, Geometridae) of the Wielkopolski National Park]. - Badania Fizjograficzne nad Polska Zachodnia, Ser. C (Zoologia), 47: 29-44.

BARBOUR, D.A. (1976): *Macrolepidoptera* of Banffshire. – The Entomologist's Rec. and J. Var. 88 (1): 1-11.

BARRETT, C.G. (1877): Description of *Eupithecia albipunctata*, var. *angelicata*. – The Entomologist's Monthly Mag. (1876-1877) 8: 278.

BARRETT, C.G. (1902): The *Lepidoptera* of the British Islands. Heterocera (Geometridae). Vol. 8. London, 431 pp., pls. 329-376.

BARRETT, C.G. (1904): The *Lepidoptera* of the British Islands. Heterocera (Geometrina - Pyralidina). Vol. 9. London, 454 pp., pls. 377-424.

BARTEL, M. (1902): Lepidopteren des südlichen Urals. - Dt. ent. Z., Iris 15: 183-230.

BARTOL, B., BARTOL, V. & S. MICHIELI. (1964): Beitrag zur Kenntnis der Makrolepidopterenfauna der adriatischen Insel Krk (Veglia). - Nachrichtenbl. Bayer. ent. München 13 (6): 55-59.

BATKOWSKI, S., PALIK, E. & R. SZPOR (1972): Motyli wieksze Tatr Polskich. - Polskie Pismo Ent. 42 (3): 637-688.

BAYNES, E.S.A. (1957): The *Lepidoptera* of Tory Island, Co. Donegal. - The Entomologist 90 (1135): 310-313.

BAYNES, E.S.A. (1964): A revised catalogue of Irish *Macrolepidoptera* (Butterflies and Moths). E.W. Classey Ltd., Hampton, 110 pp.

BAYNES, E.S.A. (1970): Supplement to a revised catalogue of Irish *Macrolepidoptera* (Butterflies and Moths). E.W. Classey Ltd., Hampton, 28 pp.

BENDER, R. (1963): Beitrage zur Lepidopterenfauna der Insel Rhodos. - Z. Wien. ent. Ges. 48 (74, 1): 11-20.

BERARD, R. & Cl. DUFAY (1989): *Eupithecia conterminata* (LIENIG et ZELLER, 1846), espece nouvelle pour la faune francaise (Lep., Geometridae Larentiinae). - Alexanor 16 (1): 49-51.

BERGMANN, A. (1955): Die Grossschmetterlinge Mitteldeutschlands. Band 5 (1/2), Spanner. Urania-Verlag, Leipzig/Jena, 1267 pp.

BESHKOV, S. (2001): Noschnite i dnevnoaktivnite noschni peperudi (*Lepidoptera*: *Macrolepidoptera* bez *Hesperioidea* & *Papilionoidea*) na Krasnenskija prolom [Nocturnal and Day-flying Moths (*Macrolepidoptera*, excluding *Hesperioidea* & *Papilionoidea*) of Kresna Gorge, SW Bulgaria] - In: BERON, P. (ed.): Biodiversity of Kresna Gorge (SW Bulgaria): 255-287.

BESHKOV, S. & K. MISJA. (1995): A Contribution to the knowledge of the *Lepidoptera* fauna of Albania. 1. Some materials from the collection of K.Misja in the Natural History Museum Tirana and some results of the collecting trip of Beshkov during 1992 (Lepidoptera, Macrolepidoptera). - Atalanta 26 (1/2): 345-363.

BESHKOV, S., MISJA, K. & S. ABADJIEV. (1996): A Contribution to the knowledge of the *Lepidoptera* fauna of Albania. 3. Some results of the collecting in April 1994 and some materials from the collection of the Natural History Museum of Tirana (Lepidoptera, Macrolepidoptera, Heterocera). - Atalanta 27 (3/4): 623-648.

BESHKOV, S. & B. PETROV. (1996): A catalogue of the Bulgarian *Lepidoptera* species reported and collected from the caves and galleries in Bulgaria (Insecta, Lepidoptera). - Atalanta 27 (1/2): 433-448.

BESHKOV, S., NOWACKI, J. & K. PALKA. (1999): Contribution to the knowledge on the distribution of *Macrolepidoptera* in Bulgaria. - Wiad. ent., Poznan, 18 (3): 169-185.

BESHKOV, S. & S. ABADJIEV. (2000): The butterfly and moth fauna of the areas around the lakes of Durankulak, Ezerets, Shabla and Shablenska Tuzla in NE Bulgaria (Insecta: Lepidoptera). - Atalanta 31 (3/4): 543-573.

BESHKOV, S. & B. GOATER. (2000): Macrolepidoptera and Microlepidoptera (Alucitidae and Pyralidae) recorded in Bulgaria, 12-24 September 1995 (Lepidoptera). - Hist. nat. bulg. 12: 41-58.

BETTAG, E. (1966): *Eupithecia egenaria* in der Pfalz (Lep., Geometridae). - Ent. Z. 76 (16): 188-189.

BETTAG, E. (1970): Untersuchungen an drei in der Pfalz vorkommenden *Eupithecia*-Arten (Lep. Geom.). – Pfälzer Heimat 21 (4): 142-148.

BIELEWICZ, M. (1966): Motyle (Lepidoptera) Kamiennej Gory w Ligocie Dolnej pow. Strzelce Opolskie. - Roczn. Muz. Gornosl. Bytom, Przyroda, 3: 1-72.

BIELEWICZ, M. (1973): Motyle wieksze (Macrolepidoptera) Bieszczadów Zachodnich i Pogórza Przemyskiego. - Rocznik Muz. Górnoslaskiego w Bytomiu, Przyroda, 7: 1-170.

BIERMANN, H. (1990): Beitrag zur Schmetterlingsfauna der Insel Montecristo im Toskanischen Archipel (Lepidoptera). - Nachr. ent. Ver. Apollo, Frankfurt a/Main, N.F. 11 (3): 179-184.

BILOSOR, M. (1931): Beiträge zur Lepidopterenfauna Podoliens. - Mem. Classe Sci. Nat. et Techn. 5: 127-206.

BINDER, A. (1994): Beitrag zur Schmetterlingsfauna Böhmens (Insecta, Lepidoptera). - Neue ent. Nachr. 32: 1-150.

BJØRN, P., JELNES, J.E. & O. KARSHOLT (1976): *Eupithecia selinata* Herrich-Schäffer, 1861 (Lepidoptera, Geometridae) i Danmark. – Ent. Meddr 44: 19-22.

BLÁZQUEZ, A., HERNÁNDEZ-ROLDÁN, J., NIETO, M.A. & A. GARCÍA-SANTANO. (2000): Nuevos datos sobre la fauna de macroheteróceros de la provincia de Cáceres (España) III (Insecta: Lepidoptera). - Shilap Revta. lepid. 28 (110): 173-186.

BLESZYNSKI, S. (1965): Klucze do Oznaczania Owadow Polski, Czesc 27. Motyle - *Lepidoptera*, Zeszyt 46 b, Miernikowce-*Geometridae*, Podrodzina *Hydriomeninae*, Warszawa, 305 pp.

BLESZYNSKI, S., RAZOWSKI, J. & R. ZUKOWSKI. (1965): Fauna motyli Pienin. [The Lepidopterous fauna of the Pieniny Mountains]. - Acta zool. Cracov. 10 (5): 375-493.

BOHATSCH, O. (1883): Die Eupithecien Oesterreichs-Ungarns. - Wien. ent. Ztg. 2 (8): 185-188, 227-230.

BOHATSCH, O. (1891): Beitrage zur Lepidopteren-fauna Slawoniens. Jahresber. - Wien. ent. Ver. 2: 31-50.

BOHATSCH, O. (1893): Mitteilungen über Eupithecien. - Dt. ent. Z., Iris (1893) 6 (1): 1-35.

BOISDUVAL, J.A. (1840): Genera et Index methodicus europaeorum lepidopterorum. Parisiis, 238 pp.

BOLDT, R. (1929): Geometriden-Raupenfang. – Intern. ent. Z. 23 (20): 243-248; (23): 273-277; (24): 285-293.

BOLLAND, F. (1980): Province de Huesca: Paradis entomologique (suite). - Shilap Revta lepid. 8 (31): 193-196.

BOLLAND, F. (1984): Essai de recensement des Lepidopteres de la Region de Montserrat et du Port d'el Bruc. - Shilap Revta lepid. 12 (48): 291-294.

BOLSHAKOV, L.V. (1999): Cheshuekrylye muzeja-zapovednika 'Jasnaja Poljana' i ego blizhajshih okrestnostej. Tula. 57 pp.

BOLTE, K.B. (1990): Guide to the *Geometridae* of Canada (*Lepidoptera*). VI. Subfamily *Larentiinae*. 1. Revision of the genus *Eupithecia*. - Mem. ent. Soc. Can. 151: 1-253.

BONNES, M. & W. SCHMUTZ (1993): Erganzungen zur Schmetterlingsfauna der Balearen (*Lepidoptera*) nebst einigen Netzfluglerdaten (*Neuropteroidea*). - Ent. Z. 103 (8): 129-152.

BOORNE, M. (1970): *Eupithecia phoeniceata* (RAMBUR) (*Lep., Geometridae*) in Dorset. - The Entomologist's Gaz. 21 (1): 2.

BORKHAUSEN, M.B. (1794): Naturgeschichte der Europäischen Schmetterlinge nach systematischer Ordnung. 5. Theil: der Phalaenen dritte Horde: Spanner. – Frankfurt, 1- 572.

BOURGOGNE, J. (1986): Lepidopteres observes en Lorraine, principalement dans la region de Nancy (M.-et-M.). - Alexanor 14 (8): 363-374.

BRADLEY, J.D. & D.S. FLETCHER. (1958): Lepidoptera records from the Isle of Rhum (Inner Hebrides). - The Entomologist 91 (1141): 126-131.

BRANDT, W. (1938): Beitrag zur Lepidopteren-Fauna von Iran. Neue Gattungen, Arten und Formen (Macrolepidoptera). – Ent. Rund. 55 (50): 584-588; 597-602; (60): 701-702.

BRANDT, W. (1941): Beitrag zur Lepidopteren-Fauna von Iran (4). Einige neue Geometriden. - Mitt. münchn. ent. Ges. 31 (3): 864-886, Taf. xxviii-xxx.

BROMBACHER, E. (1923): Beiträge zur elsässischen Lepidopterenfauna unter hauptsächlicher Berücksichtigung der näheren Umgebung von Strassburg. - Intern. ent. Z. 6: 45-46.

BRUSSEAUX, G. (1995): Septieme contribution a la connaissance des Lepidopteres de Corse. Six nouvelles especes inedites pour l'ile (*Lepidoptera, Geometridae, Sterrhinae* et *Larentiinae*). - Alexanor 19 (3): 165-166.

BRUSSEAUX, G. (1997): Inventaire des Lépidoptères de la foret de Notre-Dame (Val-de-Marne), II (*Lepidoptera, Rhopalocera* et *Heterocera*). - Alexanor 20 (3): 131-140.

BRUSSEAUX, G. (1998): Inventaire des Lépidoptères de l'ancienne carrière de Belle-Assise, en for?t de Ferri?res (Seine-et-Marne) (*Lepidoptera, Rhopalocera* et *Heterocera*). - Alexanor 20 (7-8): 393-411.

BRUSSEAUX, G. & N. LAVENU. (1991): Contribution à la connaissance de la faune de l'Île-de-France: Inventaire des Lépidoptères de la forêt de Notre-Dame (Val-de-Marne) (*Lepidoptera, Rhopalocera* et *Heterocera*). - Alexanor 17 (4): 217-228, 2 fig.

BRUSSEAUX, G. & M. JACQUIN (1996): Contribution a la connaissance de la faune de l'Ile-de-France: Inventaire des Lepidopteres des coteaux d'Avron (Seine-Saint-Denis) (*Lepidoptera, Rhopalocera* et *Heterocera*). - Alexanor 19 (5): 299-315.

BRUUN, H.H. (1992): Changes in Species Composition of the Moth and Butterfly Fauna on Houtskar in the Archipelago of SW Finland during the years of 1954-1989 (*Lepidoptera: Hesperiidae-Noctuidae*). - Acta Acad. Aboensis, Ser. B, 52 (1): 1-49.

BUBAČEK, O. (1926): Sammelergebnisse in den Pyrenäen. – Verh. zool.-bot. Ges. Wien. 192 (4/5) 84-85: (6)-(10).

BUCHSBAUM, U. & H. LÖBEL (2000): Checkliste der *Lepidoptera* Thüringens (CLT). Check-Listen Thüringer Insekten und Spinnentiere, Teil 8. Jena, 176 pp.

BUCKLER, W. (1899): The Larvae of the British Butterflies and Moths, 8: i-xii, 1-120, pls. 128-147. London.
BUDASHKIN, Yu.I. & I.Yu. KOSTJUK (1987): Pjadenitzy (*Lepidoptera, Geometridae*). In: Cheshuekrylye Karadagskogo zapovednika. - Flora i Fauna zapovednikov SSSR. Moscow, pp. 22-31.
BURESCH, I. & D. ILTSCHEW (1915): Vtori prinos k'e peperudnata fauna na Trakia i Makedonia i s'sednite im zemi. [Zweiter Beitrag zur Erforschung der Lepidopterenfauna von Trazien-Mazedonien und der Nachbarlander]. - Trav. Soc. Bulg. Sci. Nat. 8: 151-197.
BURESCH, I. & K. TULESCHKOW. (1932-1943): Schmetterlingsfauna Bulgariens. Die horizontale Verbreitung der Schmetterlinge (Lepidoptera) in Bulgarien. Macrolepidoptera. Teil IV, V. Geometriformes. Sofia, pp. 349-486.
BURGERMEISTER, F. (1964): Makrolepidopteren aus dem Raume Dubrovnik (Suddalmatien, FVR Jugoslavien). - Z. Wien. ent. Ges. 49 (75, 9/10): 137-152.
BURMANN, K. (1957): Beiträge zur Kenntnis der Lepidopterenfauna Tirols. 1. Neufunde von Macrolepidopteren aus Nordtirol. - Z. Wien. ent. Ges. 42 (68, 12): 181-192.
BUSTILLO, G.M.R. (1979): Origen y Evolucion de los Lepidopteros de la P. Iberica (II Parte). - Shilap Revta. lepid. 7 (25): 7-22.
BUSTILLO, G.M.R. & M.A. VARELA (1981): Catalogo sistematico de los lepidopteros Ibericos, I, *Macrolepidoptera*. Madrid. 499 pp.
BUSZKO, J. (2000): Atlas motyli Polski. Cześć III. Falice, wycinki, miernikowce (*Thyatiridae, Drepanidae, Geometridae*). Warszawa. 518 pp.
BUSZKO, J. & A. MALKIEWICZ. (1994): *Eupithecia ochridata* PINKER, 1968 - nowy dla fauny Polski gatunek miernikowca (*Lepidoptera, Geometridae*). - Wiad. Ent. 13 (4): 249-253.
BUSZKO, J. & J. NOWACKI (2000): The Lepidoptera of Poland. A Distributional Checklist. – Polish Ent. Monographs, Poznan & Torun, 1: 1-178.
CALBERLA, H. (1890): Die Macrolepidopterenfauna der römischen Campagna und der angrenzenden Provinzen Mittel-Italiens. - Dt. ent. Z., Iris 3: 47-94.
CAMA, A. (1994): Les *Eupithecia* Curtis d'Indre-et-Loire (*Lepidoptera, Geometridae*). - Bull. ent. Tourangelle 15 (1-4): 6-10.
CAMA, A. & J. PELLETIER. (1984): Etude faunistique des Lepidopteres de l'Indre-et-Loire (suite) (*Heterocera*). - Alexanor 13 (8): 355-364.
CAMPBELL, J.L. (1971): Macro-Lepidoptera Cannae. The Butterflies and Moths of Canna. - The Entomologist's Rec. and J. Var. 83 (1): 6-12.
CANDIDO MENDES d'AZEVEDO. (1903): Lepidopteros de Portugal, I. Lepidopteros da regiao de S.Fiel (Beira Baixa). - Broteria 2: 41-80.
CANOVAI, R. & G. LOI (1993): Catalogo dei Geometridi (Lepidoptera Geometridae) presenti nella collezione della sez. Entomologia Agraria del 'Dipartimento di Coltivazione e Difesa delle specie Legnose' dell'Università degli studi di Pisa. - Fristula ent., N. S. 16 (29): 41-64.
CAPUȘE, I. & A. KOVÁCS. (1987): Catalogul colectiei de Lepidoptere 'László Diószeghy' de la Muzeul Judetean covasna, Sfîntu Gheorghe. [Catalogue de la collection de Lépidoptčres 'László Dioszeghy' du Musée départamental covasna Sfîntu Gheorghe]. - Inst. de Spéologie 'Émile Racovitza', Bucarest, 397 pp.
CAPUTA, A. (1970): *Lepidoptera* prirodnej rezervacie Devinska Kobyla. - Ent. Probl. 8: 55-153.
CAPUTA, A. (1974): Sezonna dynamika a druhove zlozenie imag piadivkovitych (*Lep., Geometridae*) v prirodnej rezervacii Bab. - Ent. problemy 12: 19-149.
CAPUTA, A. (1976): Vyznamnejsie nalezy motylov (*Lepidoptera*) v prirodnej rezervacii Rozsutec. - Biologia 31 (11): 875-882.
CARADJA, A. (1893): Beitrag zur Kenntniss der Grossschmetterlinge des 'Département de la Haute-Garonne'. - Dt. ent. Z., Iris 6: 161-240.
CARADJA, A. (1895): *Eupithecia gratiosata* HS. und ihre Raupe. – Soc. ent. 10 (12): 89-90.
CARADJA, A. (1896): Die Grossschmetterlinge des Konigreiches Rumanien. - Dt. ent. Z., Iris 9 (1): 1-112.
CARNELUTTI, J. (1989): VI. Prispevek k favni Lepidopterov Slovenije. - Biol. Vestn. 37 (1): 1-9.
CARNELUTTI, J. (1992): Rdeči seznam ogroženih metuljev (*Macrolepidoptera*) v Sloveniji. [The Red List of Endangered *Macrolepidoptera* in Slovenia]. - Varstvo Narave 17: 61-104.
CARNELUTTI, J. & S. MICHIELI (1958): Beitrag zur Lepidopterenfauna der Crna Gora. – Fragm. Balcan., Mus. Maced. Sci. Nat. 2 (10): 67-81.
CARNELUTTI, J. & S. MICHIELI (1960): II. Prispevek k favni Lepidopterov Slovenije. - Biol. Vestn. 7: 113-124.
CARRARA, G. (1928): I Macrolepidotteri del territorio di Trieste. - Atti del Mus. civico di storia Natur. di Trieste 11 (1): 63-116.
CHAPELON, P. (1992): Contribution a l'inventaire des Geometrides de Saone-et-Loire (*Lepidoptera, Geometridae*). - Alexanor 17 (8): 451-466.

CHAVIGNERIE, B. DE LA (1857): Observations sur quelques Lépidoptères d'Islande. - Annls Soc. ent. France, Ser. 3, 5: 5-12.
CHRÉTIEN, P. (1902): Histoire naturelle de l'*Eupithecia liguriata* Mill. - Le Naturaliste, (2 série) 24 (375): 239-241.
CHRÉTIEN, P. (1905): Les chenilles des Santolines. - Le Naturaliste, (2 série) 27 (435): 89-91; (437): 113-114; (438): 129-131; (439): 143-144.
CHRÉTIEN, P. (1906): L'Histoire naturelle de *Larentia verberata* Sc. - Le Naturaliste, (2 série) 28 (462): 125-126.
CHRÉTIEN, P. (1910): Note sur les premiers états de deux espèces d'*Eupithecia* (*Lep., Geometridae*) de Tunisie. - Bull. Soc. ent. France 19: 351-353.
CHRÉTIEN, P. (1917): Contribution à la connaissance des Lépidoptères du Nord de l'Afrique. Notes biologiques et critiques. - Annls Soc. ent. France 85 (3): 369-502.
CHRISTOPH, H. (1867): Beschreibung einiger neuer Schmetterlinge aus der Umgegend von Sarepta. - Stett. ent. Ztg. 28 (4-6): 233-240.
CIFUENTES, J. (1989): Algunos macroheteroceros interesantes para el Norte de Espana. - Shilap Revta. lepid. 17 (67): 277-280.
CIFUENTES, J. (1990): Nueva contribucion al conocimiento de la distribucion de los heteroceros de Navarra. - Shilap Revta. lepid. 18 (70): 181-185.
CLAUSEN, R. & LA TOUR-DE-PEILZ (1945): Observations sur la phalène anguleuse *Chloroclystis rectangulata* L. - Mitt. schweiz. ent. Ges. 19 (11): 611- 626.
CLEVE, K. (1970): Das Vordringen von *Eupithecia sinuosaria* EV. in Mitteleuropa (*Lep. Geometridae*). - Mitt. ent. Ges. BRD 29 (1): 6-9.
COCKAYNE, E.A. (1951): Two new subspecies of British *Geometridae* (*Lep.*). - The Entomologist 84 (1058): 154-155.
CORLEY, M.F.V. (1984): One generation or two? - Entomologist's Gazette 35 (2): 76-77.
CORLEY, M.F.V., GARDINER, A.J., CLEERE, N. & P.D. WALLIS (2000): Further additions to the *Lepidoptera* of Algarve, Portugal (*Insecta: Lepidoptera*). - Shilap Revta. lepid. 28 (111): 245-319.
CORNELSEN, H. (1920): Eine neue Form von *Eupithecia palustraria* Dbl. (*pygmaeata* Hb.). - Intern. ent. Z. 14 (11): 87-88.
COURTOIS, J.-M. (1987): Les Lépidoptères *Eupitheciini* du pays Messin (*Lep., Geometridae, Larentiinae*). - Alexanor 15 (2): 81-86.
CREWE, H.H. (1861a): Description of the Larva of *Eupithecia tripunctaria*. – The Zoologist 19 (230): 7567-7568.
CREWE, H.H. (1861b): Description of the Larva of *Eupithecia trisignaria*. – The Zoologist 19 (230): 7568-7569.
CREWE, H.H. (1861c): Notes on *Eupithecia* Larvae. - The Entomologist's Annual (1861): 126-146.
CREWE, H.H. (1862): Notes on *Eupithecia* Larvae. - The Entomologist's Annual (1862): 38-49.
CREWE, H.H. (1863): Notes on some of the Genus *Eupithecia*. - The Entomologist's Annual (1863): 116-128.
CREWE, H.H. (1864a): Notes on the larva, pupa, and food-plant of *Eupithecia pulchellata*. – The Entomologist's Monthly Mag. 1: 95.
CREWE, H.H. (1864b): Description of the larva and pupa of *Eup. lariciata*. – The Entomologist's Monthly Mag. 1: 141-142.
CREWE, H.H. (1864c): Occurence of *Eupithecia campanulata,* Herrich-Schäffer, in Bucks. Description of the larva and pupa. – The Entomologist's Monthly Mag. 1: 142-143.
CREWE, H.H. (1865): Notes on Eupitheciae. - The Entomologist's Annual (1865): 117-127.
CREWE, H.H. (1872a): Description of the larva of *Eupithecia subciliata*. – The Entomologist's Monthly Mag. 9: 16-17.
CREWE, H.H. (1872b): Description of the larva of *Eupithecia pygmaeata*. – The Entomologist's Monthly Mag. 9: 65.
CREWE, H.H. (1872c): Description of the larva of *Eupithecia togata*. – The Entomologist's Monthly Mag. 9: 114-115.
CUE, P. (1961): The Discovery of the Larva of *Eupithecia innotata* Hufn. in Britain on Sea Buckthorn, with Notes on its Habits, etc. – The Entomologist's Rec. J. Var. 73 (10): 210-211.
CUELLO i SUBIRANA, J. (1981): Els Lepidòpters de les Illes Balears. Cens provisional. - Treb. Soc. Cat. Lep. 4: 33-53.
CULOT, J. (1919-1920): Noctuelles et Géomètres d'Europe. Vol. 4, Géomètres 2: 1-167. (Reprint edition, 1987) Apollo Books, Svendborg.
CUNI y MARTORELL, M. (1874): Catálogo metódico y razonado de los Lepidópteros que se encuentran en los

alrededores de Barcelona, en los pueblos cercanos y otros lugares de Cataluna. Barcelona, 232 pp.
DADD, E.M. (1906a): A puzzling group of Eupitheciids. - The Entomologist's Rec. J. Var. 18: 259-260.
DADD, E.M. (1906b): Another puzzling group of Eupitheciids. - The Entomologist's Rec. J. Var. 18: 261.
DAMPF, A. (1908): Beitrag zur Lepidopterenfauna des Wilnaschen Gouvernements (Russisch Littauen). - Horae Soc. ent. Ross. 38: 523-557.
DANIEL, F. (1968): Die Makrolepidopteren-Fauna des Sausalgebirges in der Südsteiermark. - Mitt. Abt. Zool. und Bot. Landesmus. 'Joanneum' in Graz, 30: 1-176, IV Tafn.
DANIEL, F., FORSTER, W. & L. OSTHELDER. (1951): Beiträge zur Lepidopterenfauna Mazedoniens. - Veröff. Zool. Staatssaml. Münch. 2: 1-78.
DANIEL, F. & J. WOLFSBERGER. (1957): Die Fohrenheidegebiete des Alpenraumes als Refugien warmeliebender Insekten. II. Der Sonnenberghang bei Naturns im Vintschgau (Südtirol). - Mitt. münchn. ent. Ges. 47: 21-121.
DANNEHL, F. (1927): Beiträge zur Lepidopteren-Fauna Südtirols. - Ent. Z., Frankfurt a. M., 41 (1): 17-31; (13): 272-284; (15): 309-319.
DANNEHL, F. (1933): Neues aus meiner Sammlung (*Macrolepidoptera*). - Ent. Z., Frankfurt a. M., 47 (15): 123-124, 139-140.
DANTART, J. (1988): Datos para el conocimiento de los Geométridos Ibéricos (II): Siete especies nuevas para la fauna Ibérica (*Lep. Geometridae*). - Treb. Soc. Cat. Lep. 8: 59-67.
DANTART, J. (1990): Los Macroheteróceros estivales de Sant Jordi Desvalls (Gerona) (*Lepidoptera: Macroheterocera*). - Treb. Soc. Cat. Lep. 10 (1989): 65-92.
DANTART, J. (1991): Datos para el conocimiento de los Geométridos Ibéricos (III): Sobre la distribución de Algunas especies en el ne Ibérico (*Lepidoptera: Geometridae*). - Treb. Soc. Cat. Lep. 11 (1990-1991): 57-82.
DANTART, J. (1997): Dades per al coneixement dels Geomètrids Ibèrics (IV): *Eupithecia undata* (FREYER, 1840), espècie nova per a la fauna Ibèrica i altres citacions interessants (*Lepidoptera: Geometridae*). - Treb. Soc. Cat. Lep. 14 (1995-1996): 115-135.
DANTART, J. (2000): Llista sistemàtica dels geomètrids de Catalunya (*Lepidoptera: Geometridae*). - Treb. Soc. Cat. Lep. 15: 121-182.
DANTART, J. & P. ROCHE. (1992): Aproximación a un catálogo de los macroheteróceros de Andorra (III) (*Lepidoptera: Geometridae*). - Shilap Revta. lepid. 20 (78): 125-139.
DANTART, J. & E. REQUENA (1993): Aproximació a la Fauna de Geomètrids de la Comarca d'Anoia (*Lepidoptera: Geometridae*). - Treb. Soc. Cat. Lep. 12 (1992): 61-94
DAPPORTO, L. & F. FABIANO (2000): Notes on some interesting Geometridae collected in Tuscany (Italy). – Nota lepid. 23 (2): 185-190.
DARGE, P. (2001): Contribution à l'étude de la faune entomologique de Bourgogne (deuxième note): quelques données sur les Eupithecia (*Insecta, Lepidoptera, Geometridae, Larentiinae*). – Bull. Soc. ent. Mulhouse 57 (3): 51-55.
DAYONG, X. & Z. HONGFU (1999): Fauna Sinica. *Insecta*. Vol. 15. *Lepidoptera* (*Geometridae, Larentiinae*). Science press, Beijing, China, 1083 pp. & 25 pls.
DEGTJAREVA, V.I. (1973): Dendrophilnye cheshuekrylye Hissarskogo khrebta i Hissarskoj doliny. Dushanbe, Donish, 194 pp.
DERFF, Ch.L. (1994): Note sur deux Geometres observees a Valence (Drome): *Perizoma lugdunaria* H.-S. et *Gypsochroa renitidata* Hb. (*Lep., Geometridae, Larentiinae*). - Alexanor 18 (7): 400.
DERZHAVETS, Yu.A., IVANOV, A.I., MIRONOV, V.G., MISTSHENKO, O.A., PRASOLOV, V.N. & S.Yu. SINEV (1986): Spisok cheshuekryluh (*Macrolepidoptera*) Leningradskoj oblasti [The list of *Macrolepidoptera* of the Leningrad region]. - Trudy VEO 67: 186-270.
DIETZE, C. (1870): Beschreibung der Raupe von *Eupithecia irriguata* Hüb. - Stett. ent. Ztg. 30 (7-9): 336-337.
DIETZE, C. (1871a): *Eupithecia silenata* und *trisignaria*. - Stett. ent. Ztg. 32 (4-6): 139-140.
DIETZE, C. (1871b): Beschreibung der Raupe von *Eupithecia laquearia* HS. fig. 181 u. 82, *perfidata* Mn., *merinata* Gn. - Stett. ent. Ztg. 32 (4-6): 207-211.
DIETZE, C. (1871c): Beschreibung der Raupe von *Eupithecia cauchyata* Dup., *austerata* Freyer. - Stett. ent. Ztg. 32 (7-9): 278-279.
DIETZE, C. (1872): Beiträge zur Kenntniss der Arten der Gattung *Eupithecia* Curt. - Stett. ent. Ztg. 33 (4-6): 184-204.
DIETZE, C. (1874a): Beiträge zur Kenntniss der Arten der Gattung *Eupithecia* Curt. - Stett. ent. Ztg. 35 (4-6): 209-221.
DIETZE, C. (1874b): Beiträge zur Kenntniss der Arten der Gattung *Eupithecia* Curt. - Stett. ent. Ztg. 35 (7-9): 270-277.

DIETZE, C. (1875a): Beiträge zur Kenntniss der Arten der Gattung *Eupithecia* Curt. - Stett. ent. Ztg. 36 (1-3): 69-76.

DIETZE, C. (1875b): Beiträge zur Kenntniss der Arten der Gattung *Eupithecia* Curt. - Stett. ent. Ztg. 36 (4-6): 236-256.

DIETZE, C. (1877): Beiträge zur Kenntniss der Arten der Gattung *Eupithecia* Curt. - Stett. ent. Ztg. 38 (1-3): 98-100.

DIETZE, K. (1900): Beiträge zur Kenntniss der Eupithecien. - Dt. ent. Z., Iris (1900) 13 (1): 95-102.

DIETZE, K. (1901): Beiträge zur Kenntniss der Eupithecien. - Dt. ent. Z., Iris (1900) 13 (2): 306-327.

DIETZE, K. (1902): Beiträge zur Kenntniss der Eupithecien. - Dt. ent. Z., Iris (1901) 14 (2): 251-264.

DIETZE, K. (1903): Von der spanischen zur italienischen Mittelmeergrenze. Sammaltage in Südfrankreich; nebst Bericht über die vorgefundenen Eupithecien. - Dt. ent. Z., Iris (1902) 15 (2): 231-268.

DIETZE, K. (1904): Beiträge zur Kenntnis der Eupithecien. - Dt. ent. Z., Iris (1903) 16 (2): 331-387.

DIETZE, K. (1906): Beiträge zur Kenntnis der Eupithecien. - Dt. ent. Z., Iris (1906) 19 (2): 55-67.

DIETZE, K. (1908): Beiträge zur Kenntnis der Eupithecien. - Dt. ent. Z., Iris (1908) 21 (2/3): 153-201, Taf. II, III.

DIETZE, K. (1910): Biologie der Eupithecien. I. Berlin, Taf. 1-82.

DIETZE, K. (1913): Biologie der Eupithecien. II. Berlin, 173 S., 86 Taf.

DJAKONOV, A.M. (1968): Cheshuekrylye (*Macrolepidoptera*) Leningradskoj oblasti [The *Macrolepidoptera* of the Leningrad oblast']. - Trudy Leningr. obshch. estestvoisp. 74 (4): 1-116.

DOMINGUEZ, M. & J. BAIXERAS. (1989): Aportació al Millor coneixement dels *Drepanidae* Comstock, 1893 i *Geometridae* Leach, 1815, Valencians (*Lepidoptera*). – Treb.Soc. Cat. Lep. 9 (1988): 59-67.

DOMINGUEZ, M. & J. BAIXERAS. (1990): Estudio de la familia *Geometridae* en el sistema Ibérico turolense (Insecta: Lepidoptera). – Teruel 80-81 (1): 217-242.

DOUBLEDAY, H. (1850): Descriptions of the genera *Hypenodes*, *Eupithecia* and *Spilonota* recently discovered in Britain. - The Zoologist 8 (App.., Art. 8): 55-56.

DOUBLEDAY, H. (1858a): Occurence of *Eupithecia pernotata* in England. - The Zoologist 17: 5963.

DOUBLEDAY, H. (1858b): Description of a British *Eupithecia* new to Science. - The Zoologist 17: 6103.

DOUBLEDAY, H. (1861): Notes on new or little known British Eupitheciae. - The Zoologist 19 (230): 7566-7567.

DOUBLEDAY, H. (1866): A Synonymic List of British Butterflies and Moths. 2 ed. with suppl., London, 42 pp.

DOUWES, P. (1974): Intressantare fynd av *Macrolepidoptera* i Sverige 1973. - Ent. Tidskr. 95 (3-4): 190-191.

DRAUDT, M. (1900): Die Raupe von *Eupithecia conterminata* Z. - Dt. ent. Z., Iris 13: 328-331.

DRAUDT, M. (1905): Zur Kenntnis der Eupithecien-Eier. - Dt. ent. Z., Iris 18: 280-320.

DROZDA, A. (1962): Fauna Motyli okolic Raciborza. - Rocznik Muz. Gornoslaskie w Bytomiu (Przyroda) 1: 79-131.

DRUET, J. & M. FONTAINE (1938): Faune lepidopterologique de la region de Charleroi. - Lambillionea 11: 216-220.

DUFAY, Cl. (1954): Notes sur deux *Eupithecia*. - Bull. Mens. Soc. Linn. Lyon 23 (9): 228.

DUFAY, Cl. (1955): Les Lépidoptères du Perigord noir: I.- 'Macrolepidopteres' de la Region des Eyzies (Dordogne). - Revue franc. Lepidopt. 15 (4): 89-102.

DUFAY, Cl. (1961): Faune terreste et d'eau douce des Pyrénées-Orientales. Fasc. 6. Lépidoptères. I. Macrolépidoptères. - Vie et Milieu, Suppl. 12 (1): 1-154.

DUFAY, Cl. (1965-1966): Contribution a la connaissance du peuplement en Lépidoptères de la Haute-Provence. - Bull. Mens. Soc. Linn. Lyon 34: 197-212; 35 (8): 377-392.

DUFAY, Cl. (1972): Sur la geonemie de divers Lépidoptères rares ou nouveaux pour certaines regions (*Pyralidae, Geometridae, Notodontidae*). - Alexanor 7 (5): 219-223.

DUFAY, Cl. (1973): Lépidoptères de la Macedoine Yougoslave. Six éspèces nouvelles pour la Macedonie. - Entomops, Nice 30: 187-192.

DUFAY, Cl. (1976): *Eupithecia dissertata* Püngeler dans le Vercors (Drome). - Alexanor 9 (5): 237-239.

DUFAY, Cl. & R. MAZEL (1981): Les Lépidoptères des Pyrénées-Orientales. Suppl. a la faune de 1961. - Vie et Milieu 31 (2): 183-191.

DUFAY, Cl. & R. MAZEL (1981a): Les Lépidoptères des Pyrénées-Orientales. Suppl. a la faune de 1961 (2e partie). - Vie et Milieu 31 (3-4): 329-337.

DURCK, H. & H. REISSER (1934): Beitrag zur Lepidopterenfauna des Rifgebirges von Spanisch-Marokko. - EOS 9 (3-4): 211-300.

ELMQUIST, H. (1978): Det svenska fyndet av *Chloroclystis v-ata* Hw. (*Lep., Geometridae*). - Ent. Tidskr. 99 (3-4): 109-110.

ELMQUIST, H., HELLBERG, H., IMBY, L. & G. PALMQVIST (1994): Fjärilsfaunan i Sveriges nordligaste fjällmassiv, Pältsan och Duoibal - unik och hotad? - Ent. Tidskr. 115 (1-2): 1-10.

EMBACHER, G. (1998a): Prodromus der Grossschmetterlingsfauna des Landes Salzburg - Ergänzungen und Korrekturen 2 (*Insecta: Lepidoptera*). - Z. Arb.Gem. Öst. Ent. 50 (3-4): 85-90.

EMBACHER, G. (1998b): Ein Fund von *Gymnoscelis rufifasciata* (Haworth, 1809) im Nationalpark Hohe Tauern, Land Salzburg (*Lepidoptera: Geometridae*). - Z. Arb.Gem. Öst. Ent. 50 (3-4): 91-94.

EMBACHER, G. (2000): Kleiner Beitrag zur Lepidopterenfauna Griechenlands (*Insecta: Lepidoptera*). - Z. Arb.Gem. Öst. Ent. 52 (3-4): 65-70.

ENGLUND, M., JOKELA, H., ELO, O. & O. BLOMSTER (1999): Uton perhoset (The *Lepidoptera* of Uto Island). - Baptria 24 (1): 26-38.

ERSCHOV, N.G. (1881): Spisok cheshuekrylyh, najdennyh v S.-Peterburgskoj gubernii, proverennyj i ispravlennyj po 1-e Dekabrja 1880 g. (Catalogus Lepidopterorum agri Petropolitani). - Trudy Russ. ent. obshch. 12 (1880-1881): 199-221.

ERLACHER, S.-I. (1993): Vorkommen und Lebensweise von *Perizoma lugdunarium* (Herrich-Schäffer) (*Insecta, Lepidoptera, Geometridae*). - Rudolstädter nat. hist. Schr. 5 (1993): 17-21.

ERLACHER, S.-I. (1998a): Weitere Vorkommen von *Perizoma lugdunarium* in Thüringen (*Insecta, Lepidoptera, Geometridae*). - Rudolstädter nat. hist. Schr. 9 (1998): 35-37.

ERLACHER, S.-I. (1998b): Wiederfunde verschollen geglaubter Spannerarten für die Thüringer Fauna (*Lep., Geometridae*). - Ent. Nachr. Ber. 42 (1/2): 45-49.

ERLACHER, S.-I. & J. GELBRECHT (1994): Zum gegenwartigen Kenntnisstand des Vorkommens von *Eupithecia innotata* (HUFNAGEL, 1767) und *Eupithecia ochridata* PINKER, 1968 in Ostdeutschland (*Lep., Geometridae*). - Ent. Nachr. Ber. 39 (2): 115-120.

EVERSMANN, E. (1844): Fauna Lepidopterologica Volgo-Uralensis. Casani, Typis universitatis, 633 pp.

EXPÓSITO HERMOSA, A. (1978): Catálogo Provisional de la Familia *Geometridae* Leach, 1815 (Fauna Ibérica). I. - Shilap Revta. lepid. 6 (21): 37-44.

EXPÓSITO HERMOSA, A. (1985): Apuntes sobre la familia *Geometridae* (sensu stricto). - Shilap Revta lepid. 13 (49): 27-31, 16 figs.

EXPÓSITO HERMOSA, A. (1987): Apuntes sobre la familia *Geometridae* Leach, 1815. - Shilap Revta. lepid. 15 (57): 55-58.

FAILLIE, L. & R. PASSIN. (1983): Les Lepidopteres de la Sarthe (suite). - Alexanor 13 (3): 98-117.

FAJČIK, J. & F. SLAMKA. (1996): Motýle strednej Európy. I. zväzok. Určovanie-Rozširenie a Stanovište Motýl'a-Bionómia. *Drepanidae, Geometridae, Lasiocampidae, Endromidae, Lemoniidae, Saturniidae, Sphingidae, Notodontidae, Lymantriidae, Arctiidae*. (Die Schmetterlinge Mitteleuropas, I. Band. Bestimmung-Verbreitung-Flugstandort-Bionomie). Bratislava, 1996. pp. 1-113, tab. I-XX.

FALCK, P. & LYNGGÅRD, B. (1996): *Eupithecia pulchellata* Stph. og *Condica conducta* Wlk. fundet i Danmark. - Lepidoptera, København (N.S.) 7 (1): 25-28.

FASSNIDGE, W. (1926a): List of the Macrolepidoptera of Hampshire & the Isle of Wight. - The Entomologist's Rec. J. Var. 38 (3): (17)-(42).

FASSNIDGE, W. (1926b): A Month's Collecting in the Pyrenees. - The Entomologist's Rec. J. Var. 38 (4): 49-52; (5): 77-78; (7-8): 107-109.

FAVRE, E. (1899): Faune des Macro-Lépidoptères du Valais et des régions Limitrophes. Schaffhouse, Typ. Bolli & Böcherer, 318 pp.

FAVRE, E. (1905): Supplément à la Faune des Macrolépidoptères du Valais. – Mitt. schweiz. ent. Ges. 11 (Suppl.): 1-39.

FAZEKAS, I. (1977): Adatok dél-Dunántúl *Eupitheciini*-Faunájának elterjedéséhez és fenológiájához (Lep.: Geometridae). - Janus Pannonius Múz. Évkönyve, Pécs (1975-1976) 20-21: 49-56.

FAZEKAS, I. (1979a): A Matra-hegyseg nagylepke-faunaja I. *Geometridae: Eupithecia* Curt. - Folia Hist.-Nat. Mus. Matr. 5: 63-75.

FAZEKAS, I. (1979b): *Eupithecia silenicolata zengoensis* ssp. nova (Lepidoptera: Geometridae). – Linneana Belgica 7 (11): 406-410.

FAZEKAS, I. (1982): Daten zur Verbreitung der *Eupithecia*-Fauna der Balkanhalbinsel (Geometridae). - Nota lepid. 5 (4): 143-153.

FEICHTENBERGER, E.W. (1962): Die Macrolepidopteren-Fauna des Stubachtales (Salzburg, Hohe Tauern). - Z. Wien. ent. Ges. 47 (9): 142-149.

FEICHTENBERGER, E. (1965): Die norwegische Lepidopterenfauna am Polarkreis, mit Bezug auf die der Ostalpen. - Z. Wien. ent. Ges. 50 (76, 6-8): 80-118.

FERGUSON, D.C. (1955): The *Lepidoptera* of Nova Scotia. Part 1 (*Macrolepidoptera*). - Bull. Nova Scotia Mus. Sci. 2: 159-375.

FERGUSON, D.C. (1972): The occurrence of *Chloroclystis rectangulata* (L.) in North America (Geometridae). - Journ. Lepid. Soc. 26 (4): 220-221.

FIBIGER, M. (1976): Den nordiske form *Eupithecia virgaureata* Dbld. ab. *altenaria* Stdgr. fundet i Danmark - samt lidt om *Orthosia gothica* L. ab. *gothicina* H.S. - Lepidoptera, København 3 (2): 41-44.
FIBIGER, M. (1979a): *Eupithecia absinthiata* Cl./*goossensiata* Mab., én eller to arter? – Lepidoptera, København 3 (7): 203-206.
FIBIGER, M. (1979b): *Eupithecia innotata* Hfn. og *Eupithecia innotata* ssp. *fraxinata* Crewe. - Lepidoptera, København 3 (8): 256-257.
FIBIGER, M. & P. SVENDSEN (1979): Larven af *Eupithecia egenaria* H.S. 1848 fundet i Danmark - Lepidoptera, København 3 (8): 227-229.
FIBIGER, M. & P. SVENDSEN (1981): Danske natsommerfugle. Ændringer i den danske natsommerfuglefauna i perioden 1966-1980. Dansk faunistisk bibliotek. Bind 1. Scandinavian Science Press Ltd. Klampenborg, 272 pp., 6 pls.
FILIPJEV, I. (1910): Dopolnitelnyj spisok *Macrolepidoptera* Novgorodskoj gubernii. - Russ. ent. obozr. 10 (3): 154-160.
FIORI, A. & R. GALASSI (1955): Specie di Lepidotteri raccolte darante i mesi invernali degli anni 1951-1956 sul Colle della Guardia (Bologna). - Boll. Ist. ent. Univ. Bologna 21: 297-311.
FIORI, A. & R. GALASSI (1957): Specie di Lepidotteri raccolte darante i mesi invernali sul 'Colle della Guardia' (Bologna). (I Addenda). - Boll. Ist. ent. Univ. Bologna 22: 399-405.
FIUMI, G. (1988): *Eupithecia ochridata* Pinker, una specie nuova per la fauna italiana (*Lepidoptera, Geometridae*). – Giorn. ital. ent. 4 (18): 1-3.
FLAMIGNI, C. (1990): Note su alcune specie di Geometridi nuove o poco conosciute per l'Appennino settentrionale (*Lepidoptera, Geometridae*). - Giorn. ital. ent. 5 (25): 103-107.
FLAMIGNI, C., BASTIA, G. & L.DAPPORTO (2002): Nuove segnalazioni e note critiche sui Geometridi di Emilia, Romagna e Toscana. II parte (Insecta Lepidoptera Geometridae: Larentiinae). – Quad. Studi Nat. Romagna 16: 37-76.
FLORES, H. (1945): Contribution al conocimiento de los Lepidopteros Madrilenos. Faunula de la villa de Alcobendas. - Graellsia 3 (4): 133-153.
FLORES, H. (1981): Contributión al conocimiento de los heteróceros de Catalunya, Parte III. Familia *Geometridae*. - Treb. Soc. Cat. Lep. 4: 13-16.
FOLTIN, H. (1954): Die Macrolepidopterenfauna der Hochmoore Oberösterreichs. - Z. Wien. ent. Ges. 39 (65, 3): 98-115.
FOLTIN, H. (1963a): Beitrag zur Fauna von Oberosterreich. Erganzung zur Lepidopteren-Fauna des Muhlviertels. - Z. Wien. ent. Ges. 48 (74, 5): 93-96.
FOLTIN, H. (1963b): Beitrag zur Fauna von Oberosterreich. Erganzung zur Lepidopteren-Fauna des Muhlviertels. - Z. Wien. ent. Ges. 48 (74, 8): 133-140.
FOHST, P. (1991): Ein Beitrag zur Makrolepidopterenfauna des sudlichen Teils des Lago d'Iseo in der Lombardei (prov. Bergamo). - Atalanta 22 (2/4): 271-351.
FORBES, W.T.M. (1948): Lepidoptera of New York and neighboring states. II. – Mem. Cornell Univ. agric. Exp. Sta. 274: 1-263, 255 figs.
FORSTER, W. & T.A. WOHLFAHRT. (1975-1981): Die Schmetterlinge Mitteleuropas. Spanner (*Geometridae*). Franckh'sche Verl., Stuttgart. Bd. 5 (26): 97-128, Taf. 9-12; (27): 129-192, Taf. 13-16; (28): 193-240, Taf. 17-20.
FREER, R. (1892): Seasonal Variation of Larvae. - The Entomologist's Rec. J. Var. 3 (12): 279.
FRENCH, R.A. (1951): *Lepidoptera* at light in a Hertfordshire wood in July, 1949. - The Entomologist 84 (1054): 49-55.
FREYER, C. F. (1842): Neuere Beitrage zur Schmetterlinge, Augsburg, 4: 1-170, pls. 289-384.
FRIEDRICH, E. (1986): Breeding Butterflies and Moths - a practical Handbook for British and European Species. B.H. & A. Harley Ltd., Colchester, Essex, England, 176 pp.
FUCHS, A. (1873): Zur Naturgeschichte einiger Eupithecien. - Stett. ent. Ztg. 34 (7-9): 342-344.
FUCHS, A. (1875): Lepidopterologische Beobachtungen. - Stett. ent. Ztg. 36 (4-6): 225-236.
GAEDIKE, R. & W. HEINICKE (1999): Verzeichnis der Schmetterlinge Deutschlands. Entomofauna Germanica, Bd 3, Dresden, Ent. Nachr. Ber. 5: 1-216.
GALVAGNI, E. (1954): Schmetterlinge aus dem westlichen Alpengebiet Niederosterreichs. - Z. Wien. ent. Ges. 39 (65, 6): 227-237.
GANEV, J. (1983a): Die Schmetterlingsfauna der Hochgebirge Bulgariens. I Ossogovo Gebirge (*Lepidoptera, Noctuidae* und *Geometridae*). - Atalanta 14 (1): 60-79.
GANEV, J. (1983b): Systematic and synonymic list of Bulgarian *Geometridae* (*Lepidoptera*). - Phegea 11 (2): 31-42.
GANEV, J. (1984): Beitrag zur Kenntnis der Tribus *Eupitheciini* in Bulgarien (*Lepidoptera, Geometridae*). -

Nachrichtenbl. Bayer. Ent. 33 (3): 84-85.

GANEV, J. (1985): Catalogue of the *Eupitheciini* (*Lep. Geometridae*) from Bulgaria. - Shilap Revta. lepid. 13 (49): 63-67; (50): 137-142; (51): 217-220.

GARTHE, E. (1976): Geometriden (Schmetterlingsfamilie der Spanner im Bamberger Umland. - Ent. Arbgem. naturforsch. Ges. Bamberg, pp. 173-218.

GELBRECHT, J. (1988): Zur Schmetterlingsfauna von Hochmooren in der DDR. - Ent. Nachr. Ber. 32 (2): 49-56.

GELBRECHT, J. (1999): Die Geometriden Deutschlands - eine Übersicht über die Bundesländer (*Lep.*). - Ent. Nachr. Ber. 43 (1): 9-26.

GELBRECHT, J. (2001): Aktuelle Verbreitung und ökologische Ansprüche der an Gelbe Wiesenraute (*Thalictrum flavum*) gebundenen *Lamprotes c-aureum* (KNOCH, 1781) und *Perizoma sagittata* (FABRICIUS, 1787) in Brandenburg und angrenzenden Gebieten (Lepidoptera, Noctuidae et Geometridae). - Märkische Ent. Nachr. 3 (1): 1-10, 2 Tafeln.

GELBRECHT, J. & B. MÜLLER (1987): Kommentiertes Verzeichnis der Spanner der DDR nach dem Stande von 1986 (*Lep., Geometridae*). - Ent. Nachr. Ber. 31 (3): 97-106.

GERSTBERGER, M. (1979): Beitrag zur Kenntnis einiger südeuropäischer Eupithecien-Arten (*Lepidoptera: Geometridae*). - Z. ArbGem. Österr. Ent. 30 (3/4): 129-132.

GERSTBERGER, M. (1985): Beitrag zur Kenntnis südeuropäischer und kanarischer Eupithecien (*Lep.: Geometridae*). - Z. ArbGem. Öst. Ent. 36 (3/4): 101-104.

GILJAZOVA, E.V. (1997): Cheshuekrylye (*Lepidoptera*) Laplandskogo zapovednika. [The *Lepidoptera* of the 'Laplandskij' Nature Reserve]. - Flora i Fauna zapovednikov, Moskva, 63: 44-58.

GOATER, B. (1973): Some further observations on Shetland Lepidoptera, 1972. – The Entomologist's Gaz. 24 (1): 7-12.

GOATER, B. (1974): The Butterflies and Moths of Hampshire and the Isle of Wight (being an account of the whole of the Lepidoptera). E.W. Classey, Ltd. Faringdon, Berks, 439 pp.

GOATER, B. (1992): The Butterflies and Moths of Hampshire and the Isle of Wight: additions and corrections. - UK Nature Conserv. 7: i-vi, 1-266.

GOATER, B., SKINNER, B. & al. (1981): An identification Guide to the British Pugs (Lepidoptera: Geometridae). British Entomological and Natural History Society, 56 pp., 4 pls.

GOGALA, M., SIVEC, J. & J. CARNELUTTI (1982): Kartierung der slowenischen Entomofauna. - Acta ent. Jugosl. 18 (1-2): 27-34.

GOMBOC, S. (1999): Preliminary results of *Lepidoptera* fauna investigations in prekmurje (NE Slovenia). - Ent. Croat. 4 (1-2): 29-55.

GÓMEZ DE AIZPURÚA, C. (1974): Catálogo de los Lepidópteros que integran la Colección Científica del Norte de España. Caja de Ahorros Provincial de Guipúzcoa. San Sebastián, 448 pp.

GÓMEZ DE AIZPURÚA, C. (1988): Atlas provisional de los *Lepidoptera* (*Heterocera*) de Alava, Bizkaia y Guipuzcoa. T. IV. Vitoria-Gasteiz, 366 pp.

GÓMEZ DE AIZPURÚA, C. (1989): Biologia y morfologia de las orugas. *Lepidoptera, Geometridae*. - Boll. de Sanidad Vegetal 7 (15): 1-224.

GÓMEZ DE AIZPURÚA, C. (1994): Nota sobre la morfología de cuatro orugas (*Insecta: lepidoptera*), sin describir anteriormente. - Bol. San. Veg. Plagas 20: 955-962.

GOOSSENS, M.TH. (1869): Description de Chenilles d'Eupithecia. – Annls Soc. ent. France 9 (4 serie): 515-520.

GOOSSENS, M.TH. (1871a): Des variations chez les Chenilles. – Annls Soc. ent. France 1 (5 serie): 111-118.

GOOSSENS, M.TH. (1871b): Description de Chenilles. – Annls Soc. ent. France 1 (5 serie): 287-294.

GRASLIN, M.A. DE. (1863): Note des Lépidoptères observés dans les Pyrénées-Orientales. - Annls Soc. ent. France (4e série) 3: 330-364.

GRAVES, P.P. (1926): Heterocera from Macedonia, Gallipoli and Central Greece. - The Entomologist's Rec. J. Var. 38 (11): 152-158; (12): 165-170.

GRELIER, Y. (1989): Dix ans de chasses de nuit a Marsaneix (Dordogne) (*Lepidoptera*). - Alexanor 16 (4): 201-211.

GRILLO, N. & P. PARENZAN (1995): Contributi alla conoscenza della Lepidotterofauna siciliana IV. *Geometridae*. - Phytophaga 6 (1995): 111-129.

GRILLO, N., PARENZAN, P. & F. PORCELLI (1996): Segnalazioni faunistiche di Geometridi e Nottuidi per l'Italia settentrionale e centrale (*Lepidoptera*). - Entomologica, Bari, 30: 167-184.

GRÖNBLOM, Th. (1936): Verzeichnis der Gross-Schmetterlinge Finnlands mit Rucksicht auf ihre Verbreitung in den verschiedenen Provinzen. - Acta Soc. Fauna et Flora Fennica 58 (5): I-IV, 1-44, 1 map.

GRÖNBLOM, Th., JALAS, I., KAISILA, J., KROGERUS, H. & E. SUOMALAINEN (1962): Catalogus Lepidopterorum Fenniae et regionum adiacentium. I. *Macrolepidoptera*. Helsingfors, 28 pp.

GROSS, C. (1925): Beitrag zur Kenntnis der Lepidopteren-Fauna des mittleren rechtsseitigen Wolga-Gebietes. - Intern. ent. Ver., Frankfurt a/M, 38: 53-95.

GUENÉE, M.A. (1858): In BOISDUVAL, M.M. & M.A. GUENÉE. Histoire Naturelle des Insectes. Species général des Lépidoiptères. Uranides et Phalénites, T. 10 (2). Paris, 584 pp.

GUMPPENBERG, C.F.v. (1888-1890): Systema Geometrarum zonae temperatioris septentrionalis. Systematische Bearbeitung der Spanner der nördlichen gemässigten Zone, II, III. - Nova Acta Ksl. Leop.-Carol. Dt. Akad. Naturforsch. 54 (4): 125-255, 269-431.

GYÖRFI, J. (1956): Nadelholzzapfen- und Nadelholzsamenschädlinge und ihre parasiten. - Acta Agronom. Acad. Sci. Hung. 6 (3-4): 321-373.

HABELER, H. (1964): Beitrag zur Nachtschmetterlingsfauna der Kanzel bei Graz. - Mitt. naturwiss. Ver. Steiermark 94: 31-59.

HABELER, H. (1969): Die Grossschmetterlinge von Graz und seiner Umgebung V. - Mitt. naturwiss. Ver. Steiermark 99: 143-180.

HABELER, H. (1971): Die Grossschmetterlinge von Graz und seiner Umgebung VI. - Mitt. naturwiss. Ver. Steiermark 100: 301-379.

HABELER, H. (1976): Beitrag zur Lepidopterenfauna Dalmatiens. - Acta ent. Jugosl. 12 (1-2): 67-87.

HABELER, H. (1998): Neue und bemerkenswerte arten fur die Schmetterlingsfauna Kroatiens von der Insel Krk (*Lepidoptera*). - Ent. Croat. 3 (1-2): 33-44.

HABICH, O. & H. REBEL (1883): Ein Beitrag zur Lepidopteren-Fauna Nieder –Oesterreichs. – Wien. ent. Ztg. 2: 31-34.

HACKER, H. (1981): Beitrag zur Lepidopterenfauna des nordlichen Frankischen Jura. Teil 2: *Geometridae*. - Atalanta 12 (4): 260-284.

HACKER, H. & W. WOLF (1982): Contribution to the *Lepidoptera*-fauna of Spain. II. *Heterocera* of a three-week visit in autumn 1981. - Shilap Revta. lepid. 10 (40): 267-273.

HACKER, H. & J. LUKASCH (1984): Geometriden und Noctuiden aus der Sammlung J. Lukasch - ein Beitrag zur Fauna von Griechenland (*Lepidoptera*). - Spixiana 7 (1): 51-61.

HACKER, H. & G. DERRA (1985): Zweiter Beitrag zur Heterocerenfauna (*Lepidoptera*) Griechenlands. - Atalanta 16 (1/2): 114-157.

HAGGETT, G. (1951): *Eupithecia intricata* Zett. f. *arceuthata* Frr. (*Lep., Geometridae*). - The Entomologist 84 (1054): 58-60.

HAGGETT, G. (1963): *Eupithecia innotata* HUFNAGEL and *Eupithecia fraxinata* CREWE (*Lep., Geometridae*). – The Entomologist's Gaz. 14 (1): 13-23.

HAGGETT, G.M. (1980): The occurrence and foodplant in Britain of *Eupithecia insigniata* (Huebner) (*Lepidoptera: Geometridae*). – The Entomologist's Gaz. 31 (4): 221-226.

HAGGETT, G. (1968): Larvae of the British *Lepidoptera* not figured by BUCKLER, Part VIII. - Proc. Trans. Br. ent. nat. Hist. Soc. 1 (2): 57-109, 10 pls.

HAGGETT, G.M. (1981): *Eupithecia egenaria* H.-S. - A Recent Arrival? - The Entomologist's Rec. J. Var. 93 (11-12): 236.

HAGGETT, G.M. & J.M. CHALMERS-HUNT (1961): *Eupithecia innotata* Hufnagel on Sea Buckthorn (Hippophae rhamnoides). – The Entomologist's Rec. J. Var. 73 (10): 211-212.

HANSEN, H. (1969-1979): *Eupitheciinae*. - Lepidoptera, København 1 (8): 163-169; 1 (9): 183-190; 2 (8): 231-233; 3 (8): 254-255.

HARDY, D. & A. KENNARD (1960): *Lepidoptera* in Iceland, 1958. - The Entomologist 93: 2-8.

HARKORT, W. & H.-J. WEIGT (1969): Beobachtungen zur Schmetterlingsfauna im Raum Dortmund-Hagen-Iserlohn. Teil 3: Spanner. - Dortm. Beitr. Landeskd. 3: 19-67.

HARPE, DE LA J.C. (1853): Faune Suisse. Lépidoptères. - IV Partie. Phalénides. - Neue Denkschrift. allgem. Schweiz. Ges. ges. Naturwiss. (Nouveaux Mém. Soc. Helv. Sci. Natur.) 13: 1-160, 1 pl.

HARTIG, F. (1938): I Macrolepidotteri di Madonna di Campiglio. - Mem. Soc. ent. ital. 16: 232-270.

HARTIG, F. (1951): In HARTIG, F. & H.G. AMSEL. Lepidoptera Sardinica, Parte I. *Macrolepidoptera*. - Fragm. Ent. 1 (1): 1-159.

HAUSMANN, A. (1989): *Eupithecia dodoneata* GUENÉE, 1857, neu für die Fauna Südbayerns (*Lepidoptera, Geometridae*). - NachrBl. bayer. Ent. 38 (4): 114-116.

HAUSMANN, A. (1990): Die Bedeutung des genauen Lichtfallen-Standortes fur die Aussagekraft des Fangergebnisses (*Lepidoptera, Macroheterocera*). - Atalanta 21 (3/4): 301-312.

HAUSMANN, A. (1991): Beitrag zur Geometridenfauna Palästinas: Die Spanner der KLAPPERICH-Ausbeute aus Jordanien (*Lepidoptera, Geometridae*). - Mitt. münch. ent. Ges. 81: 111-163.

HAUSMANN, A. (1993): Contributo alla conoscenza della fauna siciliana: i rappresentanti della famiglia *Geometridae* presso la Collezione dello Zoologische Staatssammlung di Monaco (*Insecta: Lepidoptera*). -

Naturalista sicil., (4) 17 (1-2): 83-101.
HAUSMANN, A. (1994): Beitrag zur Geometridenfauna Zyperns. - Z. Arb.Gem. Österr. Ent. 46 (3/4): 81-98.
HAUSMANN, A. (1995): Neue Geometriden-Funde aus Zypern und Gesamtübersicht über die Fauna (*Lepidoptera, Geometridae*). - Mitt. münch. ent. Ges. 85: 79-111.
HAUSMANN, A. (1997): The Geometrid Moths of various entomological collections in Israel (*Lepidoptera, Geometridae*). - Entomofauna, Z. f. Ent. 18 (1): 1-20.
HAUSMANN, A. (1999): Falsification of an entomological rule: Polymorphic genitalia in Geometrid moths. - Spixiana 22 (1): 83-90.
HAUSMANN, A. (2001): The Geometrid Moths of Europe, vol. 1 (Introduction; *Archiearinae, Orthostixinae, Desmobathrinae, Alsophilinae, Geometrinae*). Apollo Books, Stenstrup. 282 pp. & 8 colour pls.
HAUSMANN, A. & U. AISTLEITNER (1998): Beitrag zur Kenntnis der Spanner-Fauna Spaniens (*Lepidoptera, Geometridae*). - NachrBl. bayer. Ent. 47 (3/4): 101-105.
HAUSMANN, A. & P. PARENZAN (1990): Neue und interessante Geometridenarten für die Süditalien-Fauna (*Lepidoptera, Geometridae*). - Entomofauna, Z. für Entomologie 11 (29) 497-503.
HAWKINS, C.N. (1942): Insectivorous habit of a larva of *Eupithecia oblongata*. - The Entomologist 75: 27.
HEINANEN, V.L. (1947): Havaintoja Lahden seudun suurperhosfaunasta. - Acta ent. Fenn. 2: 1-72.
HELLINS, J. (1867): Cannibalism of the larva of *Eupithecia minutata*. - The Entomologist's Monthly Mag. 3: 191.
HELLINS, J. (1872): Larvae of *Eupithecia pygmaeata*. - The Entomologist's Monthly Mag. 9: 42.
HELLMANN, F., BROCKMANN, E. & P.M. KRISTAL (1999): I Macrolepidotteri della Valle d'Aosta. Monografie-2. Saint-Pierre-Valle d'Aosta, Aosta, 284 pp.
HENNINGS, D. & W. VORBRUGGEN (1987): Zur Macrolepidopterenfauna des Aostatals (N-Italien). - Ent. Z. 97 (15): 209-224.
HERBULOT, C. (1961): Captures de *Geometridae* faites par P.Dardenne dans le département de l'Yonne. - Alexanor 2 (1): 19-21.
HERBULOT, C. (1961-1962): Mise à jour de la liste des *Geometridae* de France. - Alexanor 2 (4): 117-124; (5): 147-154.
HERBULOT, C. (1962): Sur quelques *Geometridae* des Llanos de Urgel (Aragon). - Alexanor 2 (6): 229-230.
HERBULOT, C. (1963): *Eupithecia thalictrata* Püngeler espèce nouvelle pour la France (*Geometridae*). - Alexanor 3 (1): 39.
HERBULOT, C. (1965a): Lépidoptères *Geometridae* du Tibesti. - Lambillionea 63 (5-8): 25-40.
HERBULOT, C. (1965b): Deuxieme note sur les *Eupithecia* de Buré (Meurthe & Moselle). - Lambillionea 64 (9-10): 33-35.
HERBULOT, C. (1966): Note sur quelques *Geometridae* des Monts de Vaucluse. - Lambillionea 65 (5-6): 35-41.
HERBULOT, C. (1968): Sur Cinq *Geometridae* de Corse. - Alexanor 5: 244-246.
HERBULOT, C. (1978): Liste des *Geometridae* de Corse qui ne se trouvent pas en France continentale. - Alexanor 10 (6): 245-248.
HERBULOT, C. (1982a): Contribution a l'inventaire des Lepidopteres du departement de l'Aisne. *Geometridae* pris a Puisieux-et-Clanlieu. - Lambillionea 82 (5-6): 34-40.
HERBULOT, C. (1982b): Les types de *Geometridae* décrits par PIERRE CHRÉTIEN (*Lepidoptera*). - Alexanor 12 (5): 207-208.
HERBULOT, C. (1987a): Les types de Lepidopteres Geometrides decrits ou figures par PIERRE RAMBUR. - Muscellanea Ent. 51 (2): 33-40.
HERBULOT, C. (1987b): *Eupithecia pauxillaria* Boisduval, espèce réhabilitée (*Lepidoptera, Geometridae*). - Alexanor 15 (2): 87-88.
HERBULOT, C. (1987c): Sur quelques *Eupithecia* rarement signalés de France (*Lep. Geometridae*). - Lambillionea 87 (11-12): 132-134.
HERING, E.M. (1935-1937): Die Blattminen Mittel- und Nord-Europas einschliesslich Englands. Verl. Gustav Feller. Neubrandenburg. 631 S.+ 500 Taf.
HERING, E.M. (1957): Bestimmungstabellen der Blattminen von Europa einschliesslich des Mittelmeerbeckens und der Kanarischen Inseln. Bd. I-III. Berlin. 1185 S.+ 86 Taf.
HERNANDEZ-ROLDAN, J., BLAZQUEZ, A., MARTIN-DIAZ, J. & M.A. NIETO (1999): Nuevos datos sobre la fauna de macroheteroceros de la provincia de Caceres (Espana) II (*Insecta: Lepidoptera*). - Shilap Revta. lepid. 27 (107): 327-348.
HERRENSCHMIDT, F. (1986): Contribution a la connaissance de la faune des Macrolepidopteres de la Drome meridionale (*Lepidoptera*). - Alexanor 14 (8): 339-362.
HERRICH-SCHÄFFER, G.A.W. (1847): Systematische Bearbeitung der Schmetterlinge von Europa, zugleich als Text, Revision und Supplement zu Jakob Hübner's Sammlung europäischer Schmetterlinge. Bd 3. Die

Spanner. Regensburg, 184 pp & 91 pls.
HERRICH-SCHÄFFER, G.A.W. (1861a): Die Arten der Spannergattung Eupithecia. – Correspondenzblatt für Sammler von Insecten, insbesondere von Schmetterlingen 2 (16): 121-125; (17): 129-133.
HERRICH-SCHÄFFER, G.A.W. (1861b): Neue Schmetterlinge aus Europa und den angrenzenden Laendern. Bd. 3. Regensburg., 32 pp.
HERTING, B. (1960): Biologie der westpaläarktischen Raupenfliegen Dipt., Tachinidae. - Monogr. Angew. Ent., Hamburg/Berlin, 16: 1-188.
HERTING, B. (1976): A Catalogue of Parasites and Predators of terrestrial arthropods. Sect. A, Host or prey/enemy. Vol. 7 *Lepidoptera*, Part 2 (*Macrolepidoptera*). Commonwealth agricultural Bureaux. (Commonwealth institute of Biol. Control), G. Britain, 221 pp.
HEYDEMANN, F. (1926): Beitrag zur Macrolepidopteren-Fauna Pommerns. - Intern. ent. Z. 19 (47): 370-373; (48): 377-379.
HEYDENREICH, (Dr.). (1851): Lepidopterorum Europaeorum Catalogus Methodicus. Systematisches Verzeichniss der Europaeischen Schmetterlinge. 3 Ausgabe. Leipzig, 131 pp.
HICKS, M.E. & J.W. HALLE (1998): Lepidoptera of St. Agnes Isles of Scilly. A Systematic list and analysis of the species recorded on St. Agnes 1992-1997. A brochure, 74 pp.
HODGES, R.W. (ed.). (1983): Check List of the Lepidoptera of America North of Mexico, including Greenland. London, XXIV + 284 pp.
HODGKINSON, J.B. (1895): *Emmelesia taeniata*. - The Entomologist 28 (384): 141-142.
HOEGH-GULDBERG, O. (1986): Macrolepidopteros espanoles en el Naturhistorisk Museum de Aarhus (Dinamarca) (III) Macroheteroceros. - Shilap Revta. lepid. 14 (54): 77-79.
HOFMANN, E. (1871): Naturgeschichte der *Cidaria incultaria* HS. – Stett. ent. Ztg. 32 (1-3): 43-44.
HOLMBERG, H. (1976): *Perizoma minorata norvegica* (Prout) Utsjoelta. - Baptria 1 (3-4): 57.
HOLMBERG, H. (1988-1994): Lapin suurperhoskesä 1987-1993. - Baptria 13 (2): 29-31, 14 (3): 67-71; 15 (2): 47-51, 17 (4): 89-95, 18 (2): 67-72, 19 (3): 77-83.
HORAK, A. (1979): Pidalky Sumperska (*Lepidoptera, Geometridae*). - Cas. slez. Muz. Opava [A] 28 (1): 59-67.
HORMUZAKI, C.F. (1899): Die Schmetterlinge (*Lepidoptera*) der Bukowina. - Verh. zool.-bot. Ges. Wien 49 (1): 32-72; (2-3): 73-86.
HORMUZAKI, C.F. (1904): Nachtrage zur Lepidopterenfauna der Bukowina. - Verh. zool.-bot. Ges. Wien 54 (6/7): 422-447.
HORN, W., KAHLE, I., FRIESE, G. & R. GAEDIKE (1990): Collectiones entomologicae. - Akad. Landw. DDR, Berlin, 2 vols., 573 pp.
HORVÁTH, G.J. (1993): Adatok a Szigetköz lepkefaunájának ismeretéhez (*Lepidoptera*). - Folia ent. hung. 54: 170-185.
HORVÁTH, G.J. (1997): Újabb adatok a Szigetköz lepkefaunájának ismeretéhez (*Lepidoptera*). - Folia ent. hung. 58: 238-247.
HRUBY, K. (1964): Prodromus Lepidopter Slovenska. SAV, Bratislava, 962 pp.
HUEMER, P. & G. TARMANN (1993): Die Schmetterlinge Österreichs (*Lepidoptera*). Systematisches Verzeichnis mit Verbreitungsangaben für die einzelnen Bundesländer. Tiroler Landesmuseum Ferdinandeum, Innsbruck, 224 pp.
HUGGINS, H.S. (1953): The *Lepidoptera* of Glengarriff, Co. Cork (Part II). - The Entomologist 86 (10): 242-254.
HUGGINS, H.S. (1962): A New Subspecies of *Eupithecia vulgata* Haworth (*Lepidoptera, Hydriomenidae*). - The Entomologist 95 (1185): 45-46.
HUISMAN, K.J. & R. DE VOS (2001): *Eupithecia inturbata* nieuw voor Nederland (Lepidoptera: Geometridae). - Ent. Ber., Amsterdam 61 (9): 130-131.
HYDEN, N. (1990): *Baptria tibiale* och *Eupithecia actaeata* (*Lep., Geometridae*) i Nordvästeuropautbredning, biologi, äggläggning och habitatval. - Ent. Tidskr. 111 (1-2): 1-15.
HYDEN, N. (1998): *Eupithecia millefoliata* (*Geometridae*) i NV Europa 1940-1998. - Ent. Tidskr. 119 (2): 65-76.
INOUE, H. (1977): Catalogue of the *Geometridae* of Japan (*Lepidoptera*). - Bull. Fac. domestic Sci., Otsuma Wom. Univ. 13: 227-346.
INOUE, H. (1979): Revision of the genus *Eupithecia* of Japan, Part 1 (*Lepidoptera, Geometridae*). - Bull. Fac. domestic Sci., Otsuma Wom. Univ. 15: 157-224.
INOUE, H. (1980): Revision of the genus *Eupithecia* of Japan, Part 2 (*Lepidoptera, Geometridae*). - Bull. Fac. domestic Sci., Otsuma Wom. Univ. 16: 153-213.
INOUE, H. (1998): Four unrecorded species of *Eupithecia* (*Geometridae, Larentiinae*) from Japan. - Jap. Heterocerists' J. 200: 433-437.

INOUE, H. (2000): Descriptions of males of *Eupithecia caliginea* Butler and *E. fujisana* Inoue with a list of species-group taxa of *Eupithecia* changed and newly joined after 1982 to Japan (*Geometridae*). - Yugato 160: 49-53.

International Code of Zoological Nomenclature (1999): Fourth edition, adopted by the International Union of Biological Sciences. The International Trust for Zoological Nomenclature, c/o The Natural History Museum, London, XXX + 306 pp.

IPPOLITO, F. & P. PARENZAN (1998): I Macrolepidotteri del Demanio Forestale di Santo Stefano Quisquina (Agrigento) Contributi alla conoscenza della Lepidotterofauna della Sicilia. VI. - Phytophaga 8: 57-84.

ISSEKUTZ, L. (1962): Faunistische Beitrachtungen eines Lepidopterologen. - Z. Wien. ent. Ges. 47 (73, 1): 11-20.

ITÄMIES, J., PYÖRNILÄ, M., KUUSELA, K. & A. PYÖRNILÄ (1980): Flight periods of geometrid moths in central and northern Finland (*Lepidoptera*). - Notulae Ent. 60: 77-87.

IVANOV, A.I., MATOV, A.Yu., KHRAMOV, B.A., GRIGOREV, G.A., MIRONOV, V.G. & O.A. MISHCHENKO (1999): Artenverzeichnis der *Macrolepidoptera* von Sankt-Petersburg und des Sankt-Petersburger Gebietes nach Aufsammlungen in der Jahren 1960-1998 (*Insecta, Lepidoptera*). - Atalanta 30 (1/4): 293-356.

IVINSKIS, P. (1993): Check-list of Lithuanian *Lepidoptera*. [Lietuvos drugiu sarašas]. Ekol. Inst., Vilnius, 210 pp.

IVINSKIS, P., KARALIUS, S., MANIKAS, A., PACEVIČIUS, V. & G. ŠVITRA (1988): 3 novyh i 62 redkih dlja fauny Litovskoj SSR vidov pjadenitz, obnaruzhennyh v 1968-1985 gg. [3 new and 62 rare to the Lithuanian SSR Geometridae species, found in 1968-1985]. Novye i redkie dlja Litovskoj SSR vidy nasekomyh. Soobstsh. i opisanija 1988 goda, Vilnjus, pp. 33-44.

JABLONKAI, J. (1972): A Matra-hegyseg lepkefaunaja. - Folia hist.-nat. Mus. Matr. 1: 9-41.

JOANNIS, DE J. (1909): Contribution à l'étude de Lépidoptères du Morbihan. - Annls Soc. ent. France 77: 689-838.

JONES, J.R.J.L. (1951): An annotated check list of the Macro-lepidoptera of British Columbia. - Occas. Pap. No.1, Ent. Soc. B.C.: 1-148.

JOSEPH, Ch. & P. JOSEPH (1991): Contribution a l'etablissement de la liste des Lepidopteres du departement du Jura, II (*Lep., Zygaenoidea* et *Geometroidea*). - Alexanor 17 (2): 79-87.

JUGAN, D. & Ch. JOSEPH (1988): Contribution a la connaissance des Macrolepidopteres de Haute-Saone (*Lepidoptera*). - Alexanor 15 (6): 323-381.

JÜRIVETE, U., KAITILA, J., KESKÜLA, T., NUPPONEN, K., VIIDALEPP, J. & E. ÕUNAP (2000): Eesti Liblikad, Katalog (Estonian Lepidoptera, Catalogue). Tallinn, 151 pp.

JUUL, K. (1948): Nordens Eupithecier. Gravers Andersens Forlag, Aarhus. 147 pp., 13 pls.

JUUL, K. (1958): Ny dansk sommerfugl. *Eupithecia millefoliata* Rössler (*Geom., Lep.*). - Flora og Fauna 64: 31-33.

KAABER, S. (1980): Om artsretten for *Eupithecia goossensiata* Mab. og *Eup. fraxinata* Crewe. - *Lepidoptera*, København, 3 (10): 307-311.

KAABER, S. (1981): II. *Eupithecia innotata* (Hfn.) og dens danske former. - Lepidoptera, København, 4 (2): 41-49.

KAILA, L. (1985): Lapin suurperhosista kesällä 1985. - Baptria 10 (4): 115-121.

KAILA, L. (1989): *Eupithecia ochridata* Pinker new to northern Europe: morphological and biological studies on the *E. innotata* complex (*Lepidoptera, Geometridae*). - Notulae Ent. 69 39-45.

KAILA, L. (1990): *Epirrhoe rivata* (Hb.) ja *Eupithecia cauchiata* (Dup.) Suomesta (*Lepidoptera, Geometridae*). - Baptria 15 (1): 1-7.

KAISILA, J. (1947): Die Makrolepidopterenfauna des Aunus-Gebietes. - Acta ent. Fenn. 1: 1-112.

KARSHOLT, O., KRISTENSEN, N.P., KAABER, S., LARSEN, K., NIELSEN, E.S., PALM, E., SCHNACK, K. (ed.), SKOU, P. & B. SKULE (1985): Katalog over de danske Sommerfugle (Catalogue of the Lepidoptera of Denmark). - Ent. Meddr, København, 52 (2-3): 1-163.

KARSHOLT, O. & E.S. NIELSEN (1976): Systematisk fortegnelse over Danmarks sommerfugle. (Catalogue of the Lepidoptera of Denmark). Scand. Sci. Press Ltd., Klampenborg, Denmark, 128 pp.

KARSHOLT, O. & E.S. NIELSEN (1986): The *Lepidoptera* described by C.P. Thunberg. - Ent. scand. 16 (4): 433-463.

KARSHOLT, O. & P.S. NIELSEN (1998): Revideret katalog over de danske Sommerfugle: 1-144. Entomologisk Forening & Lepidopterologisk Forening, København.

KARSHOLT, O. & P. SKOU (1987): Sommerfugle (*Lepidoptera*) fra Skallingen. - Ent. meddr 54 (1): 67-92.

KASY, F. (1978): Die Schmetterlingsfauna des Naturschutzgebietes Hackelsberg, Nordburgenland. - Z. Arb.Gem. Österr. Ent. 30 (Suppl.): 1-44.

KASY, F. (1987): Die Schmetterlingsfauna des naturschutzgebietes 'Glaslauterriegel-Heferlberg' südlich von Wien. - Z. Arb.Gem. Österr. Ent. 38 (Suppl.): 1-35.

KAWRIGIN, W.N. (1894): Verzeichniss der im St. Petersburger Gouvernement gefundenen Schmetterlinge (Catalogus Lepidopterorum gubernii Petropolitani). St. Petersburg, 57 pp.

KAZLAUSKAS, R. (1984): Lietuvos drugiai. Vilnius, 'Mokslas', 190 pp.

KITT, M. (1912): Über die Lepidopterenfauna des Ötztales. - Verh. zool.-bot. Ges. Wien 62 (8-10): 320-416.

KITT, M. (1932): Über die Lepidopterenfauna des Ötztales. - Verh. zool.-bot. Ges. Wien 82 (1-4): 82-111.

KNABEN, N. (1949): *Eupithecia fennoscandica* n. sp. (*Lepid., Geometridae*). - Ent. Tidskr. 70 (1-2): 77-81, pl. I.

KNABEN, N. (1977): The *Eupithecia* group (*Lep., Geometridae*) in Norway. - Norw. J. Ent. 24 (1): 43-82.

KNAGGS, H.G. (1864): Notes on New and Rare species of Lepidoptera (excepting Tineina), for 1863. – The Entomologist's Annual (1864): 118-139.

KNUDSEN, K. (1981): *Eupithecia virgaureata* Dbld. klækket fra æg. - Lepidoptera, København 4 (2): 51-53.

KNILL-JONES, S. A. (1998): The sloe pug *Chloroclystis chloerata* (Mabille) (*Lep.: Geometridae*) new to the Isle of Wight. – The Entomologist's Rec. J. Var. (Surrey), 110: 232.

KOCH, M. (1984): Wir bestimmen Schmetterlinge. Teil IV Spanner (*Geometridae*). Neumann-Neudamm Verlag, Leipzig, pp. 531-760.

KOKOT, A. (1993): Nowe dla fauny Polski i rzadko spotykane gatunki miernikowców (*Lepidoptera: Geometridae*). - Acta ent. Siles. 1: 5-7.

KOKOT, A. & D. WASILUK (1999): Nowe stanowisko *Eupithecia extremata* (FABRICIUS, 1787) (*Lepidoptera: Geometridae*) w Polsce. [New record of *Eupithecia extremata* (FABRICIUS, 1787) (*Lepidoptera: Geometridae*) from Poland]. - Wiad. ent., Poznan, 18 (2): 126.

KOLB, L. (1935): Geometridae. In: DANIEL, F., KOLB, L. & OSTHELDER. Weitere Beiträge zur Lepidopteren-Fauna der Ungarischen Tiefebene. - Dt. ent. Z., Iris 49: 26-52.

KONTIOKARI, S. (1999): Kahden ojitetun suonperhoslajiston kehitys Etela-Pohjanmaalla. - Baptria 24 (2): 73-94.

KONTUNIEMI, T. (1941): Beiträge zur Kenntnis der finnischen *Eupithecia*-Arten (*Lep., Geometridae*). - Ann. ent. Fenn. 7 (2): 81-85.

KOSCHWITZ, U., KRAUS, W. & E. BLUM (1985): Contribution to the Butterfly-fauna of Albarracin (Prov. Teruel) and adjacent areas (I). - Shilap Revta. lepid. 13 (50): 125-130.

KOSTJUK, I.Yu. (1985): Novye dannye k faune pjadenitz (*Lepidoptera, Geometridae*) Kanevskogo zapovednika i ego okrestnostej. [New data to the fauna of Geometrid Moths (*Lepidoptera, Geometridae*] of Kanevskij Nature Reserve and adjacent territories]. - In: 'Problemy obschej i molekuljarnoj biologii', Kiev 4: 78-80.

KOSTJUK, I.Yu. (1989): Novye nahodki pjadenitz roda *Eupithecia* (*Lepidoptera, Geometridae*) v Krymu. - Vestn. zool., 3: 75.

KOSTJUK, I.Yu., SHESHURAK, P.N., PLYUSHCH, I.G. & E.M. GALKINA (1998): Raznousye cheshuekrylye (*Lepidoptera: Heterocera*) Chernigovskoj oblasti Ukrainy. Chast 3. Pjadenitzy (*Geometridae*). Nezhin, 'Nauka-servis', 158 pp.

KOSTROWICKI, A.S. (1953): Studia nad fauna motyli wzgórz kserotermicznych nad dolna Nida. [Studies on *Lepidoptera* of xerothermic hills in the valley of lower Nida]. - Fragm. Faun. Mus. Zool. Polon., Warszawa, 6 (16): 263-447, tabl. 2, 3.

KOUTSAFTIKIS, A. (1973): Vergleichend-zoogeographische Untersuchung einiger Heterocerenfamilien der nordagaischen Inseln Thasos, Samotraki und Limnos. - Ann. Mus. Goulandris 1: 185-238.

KOVACS, L. (1953): A magyarorszagi nagylepkek es elterjedesuk (Die Gross-Schmetterlinge Ungarns und ihre Verbreitung). - Folia Ent. Hung. (series nova) 6 (2): 76-164, 1 map.

KOVACS, L. (1956): A magyarorszagi nagylepkek es elterjedesuk. II. - Folia Ent. Hung. (series nova) 9 (4): 89-140.

KOVACS, L. (1958): Valtozasok a magyarorszagi nagylepkek adataiban a Fauna Regni Hungariae, illetoled Abafi-Aigner lepkekonyvenek megjelenese ota. - Folia ent. hung. (series nova) 11 (19): 309-364.

KOZLOV, M.V. & J. JALAVA (1994): Lepidoptera of the Kola Peninsula, northwestern Russia. - Ent. Fenn. 5 (2): 65-85.

KRAHL, M. (1998): Erster Fund von *Perizoma sagittata* F. (Lep., Geometridae) in Sachsen. - Ent. Nachr. Ber. 42 (1/2): 97.

KRAMPL, F. (1973): Píd'alky južních Čech. (Die Spanner Südböhmens (Geometridae, Lepidoptera). - Prírodoved. čas. jihočes., Česke Budejovice, 13 (Suppl. 1): 1-74.

KRAMPL, F. (1989): K rozšírení nekolika druhu rodu *Eupithecia* CURTIS v Československu (Lepidoptera, Geometridae). - Čas. Slez. Muz. Opava [A] 38: 231-243.

KRAMPL, F. (1991): Nové nálezy peti druhu rodu *Eupithecia* CURTIS v Československu s poznamkami k jejich biologii a rozšírení (Lepidoptera, Geometridae). [New records of five species of the genus *Eupithecia* CURTIS in Czechoslovakia, with notes to their biology and distribution (Lepidoptera, Geometridae)]. - Sbor. Jihočes. Muz. v Čes. Budejovicích, Prír. Vedy 31: 5-19.

KRAMPL, F. (1992): Boreal macro-moths in central Europe (Czechoslovakia) and their eco-geographical characteristics (Lepidoptera: Geometridae, Noctuidae, Notodontidae). - Acta ent. bohemoslov. 89: 237-262.

KRAMPL, F. (1994): *Eupithecia addictata*, an Asian pug moth confused in Europe with *E. thalictrata* (Lepidoptera: Geometridae). - Eur. J. Ent. 91: 335-348.

KRAMPL, F. & J. MAREK (1977): Neue und bedeutende Funde der *Eupithecia*-Arten in der Tschechoslowakei (Lepidoptera, Geometridae). - Acta ent. bohemoslov. 74: 103-114.

KRAMPL, F. & J. MAREK (1986): Three Lepidopteran species from the Belanské Tatry Mts. new for Czechoslovakia (Lepidoptera: Incurvariidae, Coleophoridae, Geometridae). - Biologia [Bratislava] 41 (6): 619-625.

KRAMPL, F. & J. MAREK (1988): K poznani fauny pidalek Belianskych Tater (Lepidoptera, Geometridae). - Zbornik TANAP 28: 159-189.

KRAUS, W. (1999): Beobachtungen zur Macrolepidopterenfauna der Iberischen Halbinsel. Teil 2: Artenliste *Drepanidae* bis *Notodontidae*. - Nachr. ent. Ver. Apollo, N.F. 20 (2): 231-263.

KROGERUS, H. (1962): Överraskande beteende hos larven av *Dyserga coronata*. - Notulae Ent. 42 (3): 97.

KROGERUS, H. (1996): *Chloroclystis v-ata relicta* ssp. n. (Lepidoptera, Geometridae), a new taxon from SW Finland, doomed to go extinct? - Ent. Fenn. 7 (2): 63-66.

KRULIKOWSKY, L. (1908): Neues Verzeichnis der Lepidopteren des Gouvernements Kazan (östl. Russland). - Dt. ent. Z., Iris 21: 202-272.

KRULIKOWSKY, L. (1909): K svedenijam o faune cheshuekrylyh Wologodskoj gubernii. - Russ. ent. obozr. 1-2: 65-79.

KRULIKOWSKY, L. (1926): Zur Lepidopterenfauna der Ukraine. - Trav. Mus. Zool. Kiev 1: 83-94.

KUDERNATSCH, F. (1965): Lepidopteren-Fauna von Gloggnitz und Umgebung. Ergänzung des Prodromus, Zone 6. - Z. Wien. ent. Ges. 50 (76, 12): 209-222.

KUDRNA, O. & M. WIEMERS (1990): Lepidopterology in Europe. In: KUDRNA, O. (ed.): Butterflies of Europe, vol. 2: Introduction to Lepidopterology. Aula-Verlag Weisbaden, 560 pp.

KUDYN, N. & L. CERVINKA (1995): *Lepidoptera*. In: MOCEK, B. et al.: Prispevek k entomofaunistickemu vyzkumu CHKO Zelezne hory (vysledky 23. entomologickych dnu CSE v roce 1992). [Contribution to the knowledge of insects fauna in Zelezne hory Mts.]. - Acta Mus. Reginaehradecensis, Ser. A, 24: 135-153.

KULAK, A.V. (2001): Novye dannye o pjadenitzah triby *Eupitheciini* Pierce, 1914 (*Lepidoptera, Geometridae*) Belarusi. [New data on the Geometrid moths of the tribe *Eupitheciini* Pierce, 1914 (Lepidoptera, Geometridae) of Belorussia]. - Vesci NAN Belarusi, Ser. bijal. navuk 3: 104-107.

KULAK, A.V. & I.A. SOLODOVNIKOV (2001): Novye i redkie dlja Belarusi vidy pjadenitz triby *Perizomini* (Lepidoptera, Geometridae). [New and scarce species of Geometrid moths of the tribe *Perizomini* (Lepidoptera, Geometridae) for Belorussia]. - Vesnik VDU 2 (20): 123-125.

KULLBERG, J. (2000): Timjamipikkumittari *Eupithecia distinctaria* HERRICH-SCHÄFFER, 1848 Suomelle uusi laji. [The Thyme pug *Eupithecia distinctaria* Herrich-Schäffer, 1848 new to Finland]. - Baptria 25 (1): 1-3.

KUMAKOV, A.P. & Yu.P. KORSHUNOV (1979): Cheshuekrylye Saratovskoj oblasti. [The *Lepidoptera* of Saratov Province]. Izdat. Saratov. Univ., Saratov, 240 pp.

KUSKOW, D. (1933): Falterfauna der estländischen Hochmoore. - Beitr. Kunde Est. 18 (3): 1-51.

KUSNEZOW, N. (1898): K faune *Macrolepidoptera* Pskovskoj gubernii. [Beitrag zur Kenntniss der Grossschmetterlinge des Gouvernements Pleskau]. - Trudy Russ. ent. obsch. 33: 85-131.

KUTENKOVA, N.N. (1989): Cheshuekrylye zapovednika 'Kivach'. [The *Lepidioptera* of 'Kivach' Nature Reserve]. - Flora i Fauna zapovednikov SSSR. Moskva, 61 pp.

LAASONEN, E.M. & M. FIBIGER (1982): *Eupithecia orphnata* W.PETERSEN, 1909. Nye oplysninger om nomenklatur, morfologi, bionomi og udbredelse (*Lepidoptera: Geometridae*). - Ent. Meddr. 49: 85-92.

LAASONEN, E.M., LAASONEN, L., ALBRECHT, A. & L. HULDEN (1996): Urho Kekkosen kansallispuiston perhoset [*Lepidoptera* of Urho Kekkonen National Park, northeastern part of Eastern Kemi Lapland (Lkor), Finland]. - Baptria 21 (2): 53-73.

LAEVER, E. DE. (1956a): Note sur *Eupithecia cooptata* Dietze. - Revue franc. Lepid. 15 (6): 158-160.

LAEVER, E. DE. (1956b): Études sur le Genre *Eupithecia* Curtis. - Bonn. zool. Beitr. 7 (1-3): 237-247.

LAFONTAINE, J.D. & K. MIKKOLA (1987): Lås-och-nyckel systemen i de inre genitalierna av *Noctuidae* (Lepidoptera) som taxonomiska kännetecken. [Lock-and-key systems in the inner genitalia of *Noctuidae* (Lepidoptera) as a taxonomic character]. - Ent. Meddr 55 (2/3): 161-167.

LAINÉ, M. (1977): Macrolépidoptères de Normandie. II - Hétérocères (*Hepialoidea, Cossoidea, Zygaenoidea, Geometroidea*). - Ann. Mus. Du Havre 9: 1-58.

LAINÉ, M. (1982): *Perizoma didymata* L. en Normandie et bioclimatologie (*Lepidoptera, Geometridae*). - Alexanor 12 (7): 299-302.

LAJONQUIERE, E. DE. (1965-1967): Le tour d'Espagne entomologique. - Alexanor 4 (1): 33-37, 4 (3): 121-128,

5 (2): 59-66.
LANGE, E. (1924): Die Eupithecien der Freiberger Gegend. - Dt. ent. Z., Iris 38: 40-50; 159-180.
LASAN, M. (1997): Nekaj novih vrst v favni metuljev (*Lepidoptera*) Slovenije. - Acta ent. Slovenica 5 (1): 51-58.
LASAN, M. (2000): Nove vrste v favni metuljev (*Lepidoptera*) Slovenije. - Acta ent. Slovenica 8 (2): 137-146.
LASTUKHIN, A.A. (1990): K faune babochek roda *Eupithecia* Curt. v Chuvashii. Problemy rekreatzionnyh nasazhdenij, Cheboksary, 2: 133-135.
LASTUKHIN, A.A. (1998): K faune i fenologii pjadenitz (*Lepidoptera, Geometridae*) Tschuvaschskoj Respubliki. In: Ent. Issled. v Tschuvaschii, Tscheboksary, 1: 60-68.
LASTUKHIN, A. (2000): Eine kommentierte Artenliste die Blütenspanner der Tschuvaschia (Osteuropa, Russland) (*Lepidoptera, Geometridae, Eupitheciini* et *Perizomini*). - Atalanta 31 (1/2): 251-263.
LAŠTUVKA, Z. (1994): Motýli rozšireného území CHKO Pálava. [Lepidoptera of the Protected Landscape Area Pálava]. Univ. Agr. Brno, 118 pp.
LAŠTUVKA, Z. (1998): Seznam motýlu České a Slovenské republiky. [Checklist of Lepidoptera of the Czech and Slovak Republics (Insecta, Lepidoptera)]. Konvoj, Brno, 117 pp.
LAŠTUVKA, Z., LAŠTUVKA, A., LISKA, J., MAREK, J., SKYVA, J. & J. VAVRA (1992): Lepidopteran faunistic records from Czechoslovakia. - Acta ent. bohemoslov. 89 (6): 466-472.
LÁSZLÓ, G.M. & B. HERCZIG (1995): A general survey of the *Geometridae* (*Lepidoptera*) fauna of the Checheno-Ingoush Republic (Russia, NE Caucasus). - Folia ent. hung. 56: 75-84.
LATTIN, DE G. (1951): Türkiye Kelebekleri Hakkinda II. [Türkische Lepidopteren. II]. - Rev. Fac. Sci. l'Univ. d'Istanbul, Ser. B, 16: 45-73.
LATTIN, DE G. (1967): Grundriss der Zoogeographie. - Hochschullehrbücher für Biologie 12: 1-602.
LEHMANN, L. (1990): Beitrag zur Makrolepidopterenfauna Südwestbulgariens mit Bemerkungen zur subspezifischen Gliederung (*Lepidoptera*). - Atalanta 21 (1/2): 121-138.
LEMPKE, B.J. (1958): *Eupithecia valerianata* Hb. (*Lepidoptera, Geometridae*). - Ent. Ber. 18 (8): 143-145.
LEMPKE, B.J. (1969): Catalogus der Nederlandse *Macrolepidoptera* (Vijftiende Supplement). - Tijdschr. Ent. 112 (2): 15-79, 8 pls.
LERAUT, P.J.A. (1980): Liste systématique et synonymique des Lépidopteres de France, Belgique et Corse. - Alexanor (Suppl.), Paris, 334 pp.
LERAUT, P.J.A. (1997): Liste systématique et synonymique des Lépidoptères de France, Belgique et Corse (deuxieme edition). - Alexanor (Suppl.), Paris, 526 pp.
LEVERTON, R. (1998): *Eupithecia indigata* Hb. (*Lep.*: *Geometridae*) larvae eating aphids. – The Entomologist's Rec. J. Var. (Surrey), 110: 80-81.
LHOMME, L. (1923-1935): Catalogue des Lépidoptères de France et de Belgique; fam. xviii *Geometridae*. Tome I: Macrolépidoptères (2ème partie): 378-622. Le Carriol, Douelle (Lot).
LIEBIG, W.-H. & J. GELBRECHT. (1989): Zur Biologie und aktuellen Verbreitung von *Eupithecia egenaria* HERRICH-SCHÄFFER, 1848 in der DDR. - Ent. Nachr. Ber. 33 (4): 171-175.
LINDROTH, C.H. (1928): Zur Land-Evertebratenfauna Islands I. - Göteborgs K. Vetenskabs- o. Vitt. Samh. Handl., 5 följden, Ser. B, 1 (4): 1-52.
LINDROTH, C.H. (1931): Die Insektenfauna Islands und ihre Probleme. - Zool. Bidr. Uppsala 13: 1-599.
LOCKE, H. (1891): Entomologische Excursionen in dem Ortlergebiete. - Jahr. Wien. ent. Ver. 1: 15-19.
LORIMER, R.I. (1970): Orkney *Lepidoptera* 1868-1968. - The Entomologist's Gaz. 21: 73-101.
LORIMER, R.I. (1983): The Lepidoptera of the Orkney Islands. E.W. Classey Ltd., Faringdon, 103 pp.
LUNAK, R. (1936): Die Biologie von *Eupithecia egenaria* H.-S. - Z. Österrh. ent.-ver. Wien 21 (2/3): 15-17.
LUQUET, G.Chr. (1982): Macrolepidopteres recoltes dans les environs de La Ferte-Mace (Orne): Quelques complements aux 'Macrolepidopteres de Normandie' du Dr Marcel Laine. I. Rhopalocera, Bombycoidea et Geometroidea (2-e partie). - Linneana Belgica 8 (9): 399-416.
LUQUET, G.Chr. (1991): Nouvelles captures de *Perizoma lugdunaria* H.-S. en Île-de-France (*Lep., Geometridae, Larentiinae*). - Alexanor 17 (1): 51-56.
LUQUET, G.Chr. (1992): Nouvelles données sur la géonémie de *Perizoma lugdunaria* H.-S. en France (*Lep., Geometridae, Larentiinae*). - Alexanor 17 (8): 471-478.
LUTRAN, G., MAZEL, R., PESLIER, S. & Ch. TAVOILLOT (1998): Mise à jour au 31 décembre 1997 de la liste des Lépidoptères des Pyrénées-Orientales (*Insecta, Lepidoptera*). - Alexanor 20 (6): 325-332.
LVOVSKY, A.L. (1971): Materialy po faune cheshuekrylyh (*Macrolepidoptera*) Astrakhanskoj oblasti. - Ent. obozr. 50 (4): 800-810.
LVOVSKY, A.L. (1994): Cheshuekrylye nasekomye (*Insecta, Lepidoptera*) v predelah Sankt-Peterburga. - Izv. Kharkov ent. Soc. 2 (1): 5-48.
MABILLE, P. (1867): Notices sur les Lépidoptères de la Corse. - Annls Soc. ent. France 6 (4 sér.): 545-564.
MABILLE, P. (1868): Énumération monographique des Eupithécies de l'île de Corse (1e partie). - Annls Soc. ent.

France 7 (4 sér.): 642-658.
MABILLE, P. (1869): Énumération monographique des Eupithécies de l'île de Corse (2e partie). - Annls Soc. ent. France 9 (4 sér.): 64-80.
MABILLE, P. (1872): Recherches et observations lépidoptèrologiques. - Annls Soc. ent. France 2 (5 sér.): 489-502.
MACKAY, M.R. (1951): Species of *Eupithecia* Reared in the Forest Insect Survey in British Columbia (*Lepidoptera: Geometridae*). - The Can. Entomologist 83 (4): 77-91.
MALICKY, H. (1957): Lepidopterenfunde aus Niederosterreich. - Z Wien. ent. Ges. 42 (68, 7): 97-99.
MALICKY, H. (1965a): Eine Lepidopterenliste aus Vorarlberg. - Z. Arb.Gem Österr. Ent. 17 (1, 2): 9-23.
MALICKY, H. (1965b): Die lepidopterologischen Verhaltnisse des Hugellandes zwischen Bodensee und Alpstein (Nordostschweiz). - Z. Arb.Gem. Österr. Ent. 17 (1, 2): 24-62.
MALKIEWICZ, A. (1999): Dwa nowe dla Polski gatunki miernikowców (*Lepidoptera, Geometridae*) z Tatr. - Wiad. Ent. 18 (1): 23-28.
MALKIEWICZ, A. & W. KUBASIK (1998): Interesujace gatunki miernikowców (*Lepidoptera: Geometridae*) z poludniowej Polski. - Wiad. Ent. 17: 126.
MALKIEWICZ, A. & J. SOSINSKI (1999): Systematic annotated check-list of Polish *Geometridae* (*Lepidoptera*). - Polskie Pismo Ent. 68 (2): 197-208.
MANN, J. (1855): Die Lepidopteren, gesammelt auf einer entomologischen Reise in Corsika im Jahre 1855. - Verh. zool.-bot. Ver. Wien 5: 528-572.
MAREK, J. (1974): Tri druhy Lepidopter nove pro CSSR (*Noctuidae, Geometridae*). - Biologia 29 (5): 431-435.
MARIANI, M. (1939): Fauna Lepidopterorum Siciliae (Catalogo ragionato). - Mem. Soc. ent. Ital. 17 (2): 129-187.
MARIANI, M. (1941): Fauna Lepidopterorum Italiae. Parte I. Catalogo ragionato dei Lepidotteri d'Italia. - Giorn. Sci. Nat. Econ. Palermo 42 (3): 1-236.
MARINI, M. & M. TRENTINI (1985): Osservazioni sui Lepidotteri dell'Appennino Lucchese. IV. *Geometridae*. - Boll. Ist. ent. Univ. Bologna 39: 17-36.
MATVEEV, V.A., SÖDERMAN, G., LUNDSTEN, K.-E., MIKKOLA, K., POPOV, A.I., SAFIN, M.G., MATVEEV, I.V. & N.I. GOL'TSOVA (1999): The Moth Fauna (*Macrolepidoptera*) of the specially protected areas of the Republic of Mari El. Joshkar-Ola: The 'Bolshaya Kokshaga' State Nature Reserve, 27 pp.
MAY, H. (1891): Macro-Lepidopteren gesammelt im Gebiete des Schneeberges in Niederosterreich. Jahresber. - Wien. ent. Ver. 1: 21-24.
MAYR, E. (1963): Animal Species and Evolution. The Belknap Press of Harvard Univ. Press, Cambridge, Massachusetts, 797 pp.
MAYR, E. (1969): Principles of Systematic Zoology. McGraw-Hill Book Company. 428 pp.
MAYR, E. (1977): Evolution and the Diversity of Life. The Belknap Press of Harvard Univ. Press, Cambridge, Massachusetts and London, England. Second printing, 721 pp.
MAZEL, R. & S. PESLIER (1997): Cartographie des Lépidoptères des Pyrénées-Orientales (*Geometridae*). - Rev. l'Ass. Roussillonnaise d'Ent. (R.A.R.E.), 6 (Suppl.): 1-115.
MCDUNNOUGH, J.H. (1949): Revision of the North American species of the genus *Eupithecia* (*Lepidoptera, Geometridae*). - Bull. Amer. Mus. Nat. Hist. 93 (8): 533-734, pl. 26-32.
MCGUFFIN, W.C. (1958): Larvae of the Nearctic *Larentiinae* (*Lepidoptera: Geometridae*). – The Can. Entomologist 90 (Suppl. 8): 1-104.
MEIER, H. (1955): Neue und interessante Macrolepidopteren-funde aus dem Murtal in Obersteiermark. - Z. Wien. ent. Ges. 40 (66, 9): 248-264.
MÉNDEZ GARNICA, J.M. (1983): Contribución al conocimiento de los Lepidópteros del norte de Burgos (Cuadricula UTM VN-65). Heteroceros - I. - Shilap Revta. lepid. 11 (42): 133-140.
MÉNDEZ GARNICA, J.M. (2001): Resultado de 20 años de muestreo y seguimiento de lepidópteros en la Cerca (Burgos). - Est. Mus. Cienc. Nat. de Álava 16: 179-201.
MENTZER, E. (1974): Taxonomische Bemerkungen über *Perizoma flavosparsata* (Wagner) nebst Beschreibung einer neuen Unterart, *uniformata* ssp. n., aus Jugoslawien (*Lep., Geometridae*). - Ent. Tidskr. 95 (3-4): 192-197.
MENTZER, E. (1981): Neue und bemerkenswerte Funde von Lepidoperen in Jugoslawien. - Acta ent. Jugosl. 17 (1-2): 137-141.
MENTZER, E. & A. MOBERG (1992): *Eupithecia herrenschmidti* sp. n. from Spain (*Lepidoptera: Geometridae*). - Nota lepid. 14 (4): 312-319.
MERE, R.M. (1963): *Eupithecia egenaria* H.-S. (Lep.) in the British Isles. – The Entomologist's Gaz. 14 (1): 23.
MERE, R.M., BRADLEY, J.D. & E.C. PELHAM-CLINTON (1964): *Lepidoptera* in Ireland. May-June, 1962. – The Entomologist's Gaz. 15 (2): 66-92.

MÉSZÁROS, Z. & A. VOJNITS (1966): Adatok az Északnyugati Kárpátok *Lepidoptera*-faunájához. Rovartani Közlemények. - Folia ent. hung., Ser. Nova 19 (25): 455-478.

MÉSZÁROS, Z., RONKAY, L., HERCZIG, B., SZEOKE, K. & C. SZABOKY (1985): Datos para el conocimiento de la fauna de los Lepidopteros de Bulgaria (3, aParte). - Shilap Revta lepid. 12: 67-70, 199-201; 13: 297-299; 14: 67-72.

METZGER, A. (1891): Beitrag zur Lepidopteren-Fauna von Weyr in Oberösterreich. - Jahr. Wien. ent. Ver. 2: 13-17.

METZGER, A. (1892): Beitrag zur Lepidopteren-Fauna von Friesach in Karnthen. Jahresber. - Wien. ent. Ver. 3: 27-35.

MEYER, M. & M. HELLERS (1990): Les Lepidopteres de Madere. Note preliminare: Les Macrolepidopteres observes en mai 1989 (*Lepidoptera, Geometridae, Sphingidae, Noctuidae, Papilionidae*). - Nota lepid. 13 (2-3): 147-159.

MEYRICK, E. (1892): On the classification of the Geometrina of the European Fauna. Part 1. - Trans. Ent. Soc. London (1892): 53-140.

MEYRICK, E. (1928): A Revised Handbook of British Lepidoptera. London, 914 pp.

MICHIELI, S. (1963): Beitrag zur Kenntnis der Makrolepidopterenfauna Mazedoniens (NR Makedonija). - Acta Mus. Maced. Scient. nat. 9 (2): 15-32.

MIKAT, M. (1998): Zajimave nalezy pid'alkovitych (*Lepidoptera, Geometridae*) z vychodnich Cech. [Some interesting records of *Geometridae* (*Lepidoptera*) in Eastern Bohemia]. - Acta Mus. Regionaehrad., Ser. A, 26: 55-58.

MIKAT, M. & L. MARSIK. (1997): Prispevek k poznani fauny motylu navrhovaneho chraneneho uzemi 'Na Plachte' v Hradci Kralove (*Lepidoptera*). Contribution to the knowledge of fauna of butterflies (*Lepidoptera*) of protected area 'Na Plachte' (Hradec Kralove, Czech Republic). - Acta Mus. Reginaehrad., Ser. A, 25: 163-192.

MIKKOLA, K. (1982): *Eupithecia expallidata* (Lepidoptera, Geometridae) new to Fennoscandia, with notes on the taxonomy of the *E. absinthiata* complex. - Notulae Ent. 62 (1): 13-23.

MIKKOLA, K. (1985): The Geometroidea and Noctuoidea described by Carl Clerck (*Lepidoptera*). - Ent. scand. 16: 121-129.

MIKKOLA, K. (1994): Inferences about the function of genitalia in the genus *Eupithecia*, with description of a new organ (*Lepidoptera, Geometridae*). - Nota lepid. (Suppl. 5): 73-78.

MIKKOLA, K., JALAS, I. & O. PELTONEN (1985): Suomen Perhoset. Mittarit 1. Tampere, 260 pp.

MIKKOLA, K., JALAS, I. & O. PELTONEN (1989): Suomen Perhoset. Mittarit 2. Recallmed, Tampere, 280 pp.

MILLER, E., ZUBOWSKI, N. & A. RUSCINSKI. (1930): Materiale pentru fauna entomologica din Bessarabia. *Macrolepidoptera*. Suppl. 3. - Bul. Muz. Nat. Ist. Nat. Chişinau, 2 (3): 97-130.

MILLER, E., ZUBOWSKI, N. & A. RUSCINSKI. (1932): Materialen zur Kenntnis der entomologischen Fauna Bessarabiens. *Macrolepidoptera*. Nachtrag IV. - Bul. Muz. Nat. Ist. Nat. Chişinau, 4: 25-38.

MILLIÈRE, P. (1869-1874): Iconographie et description de Chenilles et Lépidoptères inédits. - Annls Soc. Linn. Lyon 3 (23-35): 1-488, 101-154 pls.

MIRONOV, V.G. (1988): Materialy k faune pjadenitz roda *Eupithecia* (*Lepidoptera, Geometridae*) Kryma, s opisaniem novogo vida. - Vestn. zool. 4: 15-21.

MIRONOV, V.G. (1989): Obzor pjadenitz gruppy *gueneata* roda *Eupithecia* Curt. (*Lepidoptera, Geometridae*). - Trudy Zool. Inst. AN SSSR 200: 110-135.

MIRONOV, V.G. (1990): Sistematicheskij katalog pjadenitz triby *Eupitheciini* (*Lepidoptera, Geometridae*) fauny SSSR, I. - Ent. obozr. 69 (3); 656-670.

MIRONOV, V.G. (1991): Sistematicheskij katalog pjadenitz triby *Eupitheciini* (*Lepidoptera, Geometridae*) fauny SSSR, II. - Ent. obozr. 70 (1): 157-167.

MIRONOV, V.G. (2000): Sistematika pjadenitz triby *Perizomini* (*Lepidoptera, Geometridae, Larentiinae*). [Systematics of the Geometrid Moth tribe *Perizomini* (*Lepidoptera, Geometridae, Larentiinae*)]. - Ent. obozr. 79 (1): 112 -122.

MIRONOV, V. (2001a): New and scarce European *Eupithecia* species (*Geometridae*). - Nota lepid. 24 (1/2): 21-28.

MIRONOV, V.G. (2001b): Novye vid i podvidy pjadenitz roda *Eupithecia* Curt. (Lepidoptera, Geometridae). s Kavkaza i iz Zakavkaz'ja. - [New species and subspecies of geometrid moths of the genus *Eupithecia* Curt. (Lepidoptera, Geometridae) from Caucasus and Transcaucasia]. - Trudy Zool. Inst. RAN 291: 101-110.

MIRONOV, V.G. (2001c): Tzvetochnye pjadenitzy gruppy vidov '*irriguata*' s opisaniem *Eupithecia antalica* sp. n. iz Gretzii i Turtzii (*Lepidoptera, Geometridae*). [Pugs of the species group '*irriguata*' with a description of *Eupithecia antalica* sp. n. from Greece and Turkey (Lepidoptera, Geometridae)]. - Trudy Zool. Inst. RAN 291: 111-118.

MIRONOV, V.G. (2001d): Problems in the study of the tribe *Eupitheciini* (Lepidoptera: Geometridae, Larentiinae). In: HAUSMANN, A. & R. TRUSCH (ed.). Proceedings of the FORUM HERBULOT 2001 Neotropical Geometridae: Approaches to a Modern System on Genus and Tribe Level (8.3.-9.3.2001). - Spixiana 24 (3): 199-200.

MITTERNDORFER, W. (1960): Ein kleiner Beitrag zur Lepidopterenfauna Niederosterreichs. (Zone 17 des Prodromus). - Z. Wien. Ent. Ges. 45 (71, 2): 17-18.

MLADINOV, L. (1980): Lepidoptera iz dolinegornjeg toka rijeke Kupe IV. *Geometridae*. - Acta ent. Jugosl. 16 (1-2): 69-91.

MOBERG, A. (1983): A new subspecies of *Eupithecia atlanticata* Pinker (*Lep. Geometridae*) from the Iberian Peninsula. - Shilap Revta. lepid. 11 (42): 141-142.

MONTEIRO, T. & J. PASSOS DE CARVALHO (1984): Lepidópteros do Algarve. - Anais Fac. Ciênc. Porto 64 (1-4): 95-219.

MORRIS, F.O. (1871): A Natural History of British Moths, Vol. 1, London, 253 pp. + 30 pls.

MOUCHA, J. & I. NOVAK. (1962): Ergebnisse der Erforschung der Lepidopteren-Fauna der Slowakei (1946-1960). II. Teil. - Acta faun. ent. Mus. Nat. Pragae 8: 95-137.

MÜLLER, B. (1996): *Geometridae*. In: KARSHOLT, O. & J. RAZOWSKI. The *Lepidoptera* of Europe. A Distributional Checklist. Apollo Books, Stenstrup, 380 pp.

MÜLLER, B. & J. GELBRECHT. (1992): Veränderungen in der Spannerfauna der DDR seit 1945 (*Lep., Geometridae*). - Nota lepid. (Suppl. 3): 70-81.

MURASE, M. (2000): Larvae of Geometridae found on the flowers in Wakayama Prefecture II. - Yugato 161: 97-98.

NEMTZEV, V.V., ANTONOVA, E.M. & A.V. SVIRIDOV. (1991): Cheshuekrylye Darvinskogo zapovednika. - Flora i Fauna zapovednikov SSSR. Moscow, 1991, 49 pp.

NESTOROVA, E. (1998): Catalogus faunae Bulgaricae, 2. Lepidoptera: Geometridae. – Acad. Sci. Bulg. Ist. Zool. Pensoft Publ. Sofia – Moscow, 194 pp.

NIELSEN, F.J. (1998): *Eupithecia millefoliata* Rössl. fundet talrigt i Danmark som larve. Arten overses let som imago og findes bedst som larve om efteråret. - Lepidoptera, København 7 (6): 196-201.

NIELSEN, K. (1969): *Perizoma taeniata* Stph. - ny måler for Danmark. - Lepidoptera, København 1 (8): 157-158.

NIESIOLOWSKI, W. & R.J. WOJTUSIAK (1950): Przyczynek do znajomosci motyli okolic Lubasza pod Szczucinem [Contribution to the knowledge of the Lepidopterous Fauna of Lubasz near Szczucin (South Poland)]. - Polsk. Pismo Ent. 19 (3-4): 256-276.

NIPPEL, F. (1993): Ein unerwarteter Nachweis von *Eupithecia tenuiata* Hübner 1813 (*Lepidoptera: Geometridae*). - Ent. Z. 103 (8): 142-144.

NORDSTRÖM, F. (1943): Forteckning over Sveriges storfjarilar [Catalogus Insectorum Sueciae. III. *Macrolepidoptera*]. - Opusc. Ent. 8 (3-4): 59-120.

NORDSTRÖM, F. (1949): Två för landet nya *Eupithecia*-arter. - Ent. Tidskr. 70 (1): 113.

NORDSTRÖM, F. (1953): Catalogus Insectorum Sueciae. III. *Macrolepidoptera*. - Opusc. Ent. 18 (1): 75-87.

NORDSTRÖM, F., WAHLGREN, E. & A. TULLGREN (1941): Svenska Fjarilar (Macrolepidoptera). Stockholm, IV & 353 pp. & 49 Tafn.

NORGAARD, I. (1964): Nye storsommerfugle i Danmark siden 1949. - Naturens Verden 1 (1964): 21-33.

NOVÁK, I. (1970): *Eupithecia alliaria* Stgr. - nove zjištený druh píd'alky v Československu (*Lep., Geometridae*). - Zprávy čs. spol. entomologické pri ČSAV 6: 49-50.

NOWACKI, J., PALKA, K. & J. SOSINSKI (1993): Nowe dla fauny Bieszczacow Zachodnich gatunki motyli (*Lepidoptera*) [*Lepidoptera* new to the fauna of the West Bieszczady Mts.]. - Wiad. Ent. 12 (1): 45-49.

NOWICKI, M.S. (1860): Enumerato Lepidopterorum Haliciae Orientalis. Leopoli, 269 pp. + XVI pp.

NYST, R.H. (1987): Contribution à l'établissement de la liste des Macrolépidoptères de la Côte-d'Or et du Var. - Alexanor 15 (1): 3-19.

OBRAZTSOV, N.S. (1935): Untersuchungsergebnisse der nächtlichen Lepidopterenfauna in der Stadt Nikolaew. - Trav. Mus. Zool. Kiev 15: 141-161.

OBRAZTSOV, N.S. (1936): Materialien zur Lepidopterenfauna des Parkes von Vessjolaja Bokovenjka (Ukraine). - Folia Zool. et Hydrobiol., Riga, 9 (1): 29-57.

OKYAR, Z. & N. AKTAÇ (1997): Türkiye *Lepidoptera* faunasi için iki yeni tür: *Calostigia corydalaria* (Graeser, 1888) (*Lepidoptera: Geometridae*); *Catocala sponsa* (Linnaeus, 1767) (*Lepidoptera: Noctuidae*). - Türk. ent. derg. 21 (4): 275-282.

OKYAR, Z. & N. AKTAÇ (1999): Trakya Bölgesi *Geometridae* Türlerinin Taksonomik ve Faunistik Yönden Araştirilmasi (Faunistic and Taxonomic Studies on the *Geometridae* Species of Turkish Thrace). - Tr. J. Zool. 23 (1): 99-132.

ÓLAFSSON, E. (1991): Íslenskt skordýratal. - Fjölrit náttúrufræðistofnunar 17: 1-69.
OLIVIER, A. (2000): CHRISTIAN FRIEDRICH FREYER's 'Neuere Beitrage zur Schmetterlingskunde mit Abbildungen nach der Natur': an analysis, with new data on its publication dates (*Insecta, Lepidoptera*). - Beitr. Ent. 50 (2): 407-486.
OPHEIM, M. (1979): *Eupithecia cauchiata* Dup. rediscovered in Norway. - Atalanta Norvegica, Oslo 3 (5): 129-130.
OSTHELDER, L. (1929): Die Schmetterlinge Südbayerns und der angrenzenden nördlichen Kalkalpen. I. Teil. Die Grossschmetterlinge. 3. Heft (Spanner). - Münchn. ent. Ges. 19: 377-537.
PACLT, J. (1976): The nomenclature of Lepidoptera in the second edition of 'Kloet and Hincks'. Part 1: Family-group names. - Arch. Sci. 28 (2): 177-181.
PAGENSTECHER, A. (1902): In: RÖMER, F. & F. SCHAUDINN: Die arktische Lepidopterenfauna. Fauna Arctica, II (2). Jena, pp. 195-400.
PALMQVIST, G. (1978): Intressantare fynd av *Macrolepidoptera* i Sverige 1977. - Ent. Tidskr. 99 (2): 65-67.
PALMQVIST, G. (1981): Intressanta fynd av *Macrolepidoptera* i Sverige 1980. - Ent. Tidskr. 102 (2-3): 99-104.
PALMQVIST, G. (1982): Intressanta fynd av *Macrolepidoptera* i Sverige 1981. - Ent. Tidskr. 103 (2): 89-95.
PALMQVIST, G. (1983): Intressanta fynd av *Macrolepidoptera* i Sverige 1982. - Ent. Tidskr. 104 (2): 55-58.
PALMQVIST, G. (1984): Intressanta fynd av *Macrolepidoptera* i Sverige 1983. - Ent. Tidskr. 105 (3): 81-88.
PALMQVIST, G. (1985): Intressanta fynd av *Macrolepidoptera* i Sverige 1984. - Ent. Tidskr. 106 (2-3): 65-70.
PALMQVIST, G. (1988): Intressanta fynd av *Macrolepidoptera* i Sverige 1987. - Ent. Tidskr. 109 (2): 59-64.
PALMQVIST, G. (1989): Intressanta fynd av *Macrolepidoptera* i Sverige 1988. - Ent. Tidskr. 110 (3): 96-102.
PALMQVIST, G. (1990): Intressanta fynd av *Macrolepidoptera* i Sverige 1989. - Ent. Tidskr. 111 (3): 61-68.
PALMQVIST, G. (1991): Intressanta fynd av *Macrolepidoptera* i Sverige 1990. - Ent. Tidskr. 112 (3): 73-78.
PALMQVIST, G. (1996): Intressanta fynd av storfjärilar (*Macrolepidoptera*) i Sverige 1995. - Ent. Tidskr. 117 (1-2): 35-47.
PALMQVIST, G. (1998): Intressanta fynd av storfjärilar (*Macrolepidoptera*) i Sverige 1997. - Ent. Tidskr. 119 (1): 13-27.
PALMQVIST, G. (1999): Intressanta fynd av storfjärilar (*Macrolepidoptera*) i Sverige 1998. - Ent. Tidskr. 120 (1-2): 59-74.
PALMQVIST, G. (2000): Intressanta fynd av storfjärilar (*Macrolepidoptera*) i Sverige 1999. - Ent. Tidskr. 121 (1-2): 31-45.
PARENZAN, P. (1988): Nuove segnalazioni di *Geometridae* (*Lepidoptera*) per l'Italia Meridionale. - Entomologica, Bari, 23: 139-160.
PARENZAN, P. (1994): Contributi alla conoscenza della Lepidotterofauna dell'Italia meridionale: XVII. *Geometridae*. - Entomologica, Bari, 28: 99-246.
PARENZAN, P. & A. HAUSMANN (1992): Nuovi interessanti reperti di Geometridi (*Lepidoptera*) in Italia Meridionale. - Entomofauna, Z. für Entomologie 13 (8): 157-172.
PASSOS DE CARVALHO, J. (1984): Contribuição para o conhecimento da Lepidopterofauna do Parque Nacional da Peneda-Gerês. Família *Geometridae*. Parque Nacional da Peneda-Gerês, 1-40 pp.
PASSOS DE CARVALHO, J. & M.F.V. CORLEY (1995): Additions to the *Lepidoptera* of Algarve, Portugal (*Insecta: Lepidoptera*). - Shilap Revta. lepid. 23 (91): 191-230.
PATOČKA, J. (1960): Die Tannenschmetterlinge der Slowakei. Mit berücksichtigung der fauna Mitteleuropas. - Verlag Slow. Akad. Wiss., Bratislava, 214 S.
PATOČKA, J. (1981): Ein weiterer beitrag zur puppentaxonomie der Mitteleuropäischen *Larentiinae* (*Lepidoptera, Geometridae*). - Biológia (Bratislava) 36 (8): 583-593.
PATOČKA, J. (1995): Die Puppen der Spanner (*Lepidoptera, Geometridae*) Mitteleuropas. Unterfamilie *Larentiinae*, Tribus *Hydriomenini*, *Rheumapterini*, *Euphyini*, *Operophterini* und *Perizomini*. - Bonn. zool. Beitr. 45 (3-4): 231-258.
PATOČKSA, J. (1996): Die Puppen der mitteleuropäischen Blütenspanner. - Dortm. Beitr. Landeskde. naturwiss. Mitt., 30: 185-283.
PAYNE, F.W.G. (1895): A catalogue of the *Macro-Lepidoptera* of Derbyshire. - The Entomologist 28 (385): 170-173.
PELLMYR, O. (1983): *Eupithecia orphnata* (*Lepidoptera, Geometridae*): information on the bionomy of the imago. - Ent. Meddr. 50: 47-48.
PELLMYR, O. & K. MIKKOLA (1984): *Eupithecia pernotata enictata* ssp. n. from southern Finland (*Lepidoptera, Geometridae: Larentiinae*). - Notulae Ent. 64: 54-56.
PELTONEN, E., KANGAS, J. & J. KAISILA (1970): Zur Bionomie von *Eupithecia groenblomi* Urb. (Lep., Geometridae). - Ann. ent. Fenn. 36 (2): 70-75.
PEREGOVITS, L. (1995): Data to the *Macrolepidoptera* fauna of Transylvania (Romania). I. Faunistical data. -

Folia ent. hung. 56: 163-178.

PETERSEN, W. (1910): Ein Beitrag zur Kenntnis der Gattung *Eupithecia* Curt. - Dt. ent. Z., Iris (1909) 22 (4): 203-314, 28 pls.

PETRY, A. (1933): *Eupithecia breviculata* in Süd-Baden. - Int. ent. Z. (Guben) 27: 351-353.

PIERCE, F.N. (1914): The Genitalia of The Group *Geometridae* of the *Lepidoptera* of the British Islands. Liverpool, XXIX & 88 pp.+ 48 Pl.

PIESZCZEK, G.H.A. (1911): Nachtrag zur Fauna von Judenburg in Steiermark. - Jahres-Bericht Wien. ent. Ver. 21: 67-75.

PINKER, R. (1957): Beitrag zur Lepidopterenfauna des südlichen Burgenlandes. - Z. Wien. ent. Ges. 43 (Bd. 69, 8): 147-151.

PINKER, R. (1958): Beobachtungen in der Treska Schlucht in Macedonien; Beschreibung einer neuen *Eupithecia* Curt. - Fragm. Balcan., Mus. Maced. Sci. Nat. 2 (12): 91-96.

PINKER, R. (1961): Interessante und neue Funde und Erkenntnisse für die Lepidopterenfauna der Kanaren. - Z. Wien. ent. Ges. 46 (Bd 72, 5): 69-72.

PINKER, R. (1962): Interessante und neue Funde und Erkenntnisse für die Lepidopterenfauna der Kanaren I. - Z. Wien. ent. Ges. 47 (Bd 73, 11): 169-179.

PINKER, R. (1969): Interessante und neue Funde und Erkenntnisse für die Lepidopterenfauna der Kanaren IV. - Z. Wien. ent. Ges. 53 (Bd 79, 7-12): 65-93.

PINKER, R. (1968): Die Lepidopterenfauna Mazedoniens. III. *Geometridae*. - Posebno Izd. Prir. Muz. Skopje 4: 1-72.

PINKER, R. (1969): Neue und interessante Lepidopteren aus Madeira und den Azoren mit faunistischen Hinweisen auf die Kanaren. - Z. Wien. ent. Ges. 54 (Bd 80, 9-12): 101-131, taf. 12-14.

PINKER, R. (1976): Über kleinasiatische Eupithecien und deren zusammenhänge mit jenen aus Macedonien (*Lep., Geometridae*). - Z. ArbGem. Österr. Ent. 28 (1-3): 1-6.

PIZZETTI, L. & V. RAINERI (1998): Segnalazioni di alcuni Geometridi della collezione Barajon interessanti per la fauna italiana (*Lepidoptera Geometridae*). - Boll. Soc. ent. Ital. 130 (2): 159-166.

POHIER, F. & Ph. LEFORT (1997): Lepidoperes recueillis sur l'Hautil (Yvelines) par Gustave Trachier entre 1941 et 1974 (*Insecta, Lepidoptera*). - Alexanor 20 (4): 195-213.

POPESCU-GORJ, A. (1964): Catalogue de la Collection de Lépidoptères 'Prof. A.Ostrogovich'. - Du Mus. d'Hist. nat. 'Grigore Antipa', Bucarest, 293 pp.

POPESCU-GORJ, A. (1987): La liste systematique revisee des especes de Macrolepidopteres mentionnees dans la faune de Roumanie. Mise a jour de leur classification et nomenclature. - Trav. Mus. hist. nat. 'Grigore Antipa' 29: 69-123.

POPESCU-GORJ, A. (1995): Lepidopterans from the surroundings of the town Sinaia and from Bucegi Mountains (Romania). - Trav. Mus. hist. nat. 'Grigore Antipa' 35: 161-220.

POPESCU-GORJ, A., OLARU, V. & I. DRAGHIA (1972): L'entomofaune du 'grind' de Caraorman (Delta du Danube). Ord. *Lepidoptera*. - Trav. Mus. hist. nat. 'Grigore Antipa' 12: 181-206.

POPESCU-GORJ, A. & F. KÖNIG (1976): Contributions a la connaissance de la faune du department Vrancea. Ord. *Lepidoptera*. - Trav. Mus. hist. nat. 'Grigore Antipa' 17: 303-307.

PREISSECKER, F. (1932): *Tephroclystia thalictrata* Püng. Versammlung am 8. April 1932. - Verh. zool.-bot. Ges. Wien 82 (1-4): (10)-(11).

PRENTICE, R.M. (1963): Forest Lepidoptera of Canada recorded by the Forest Insect Survey. 3. Lasiocampidae, Thyatiridae, Drepanidae, Geometridae. – Can. Dep. Forest. Publ. 1013: 1-543.

PRINS, DE W. O. (1983): Systematische naamlijst van de Belgische Lepidoptera. Entomobrochure (Antwerpen), 4: 1-57.

PRINS, DE W. O. (1984): Some faunistic remarks on the Spanish Lepidoptera-fauna (Part III). - Shilap Revta lepid. 12 (46): 131-134.

PRINS, DE W. O. (1998): Catalogue of the Lepidoptera of Belgium. – Doc. Trav. l'Inst. Royal Sci. Nat. Belg. 92: 1-236.

PRIOR, G. (1977): Observations on the larva of *Eupithecia insigniata* Hubn. (The Pinion Spotted Pug). - Proc. Trans. Br. ent. nat. Hist. Soc. 10: 103-104.

PROLA, C. & T. RACHELI (1980): I Geometridi dell'Italia centrale. Parte II. Larentiinae, Ennominae. - Boll. Ist. ent. Univ. Bologna 35: 29-108.

PROUT, L.B. (1904): On some northern Spanish Geometrides. - The Entomologist's Rec. J. Var.. 16: 284-289.

PROUT, L.B. (1907): Notes on the Genus *Eupithecia*. - The Entomologist 40 (531): 169-175; (532): 206-211; (533): 220-222.

PROUT, L.B. (1912-1916): Die spannerartige Nachtfalter. In A. SEITZ: Die Gross-Schmetterlinge des Palaarktischen Faunengebietes. Bd. 4. Stuttgart, 479 S.+ 25 Taf.

PROUT, L.B. (1930): On the japanese Geometridae of the Aigner collection. – Novit. zool. 35 (4): 289-337.

PROUT, L.B. (1934-1954): *Brephinae, Oenochrominae, Geometrinae, Sterrhinae, Larentiinae*. In A. SEITZ: Die Gross-Schmetterlinge des Palaarktischen Faunengebietes. Bd. 4, Suppl. Stuttgart, 766 S.+ 18 Taf.

PRUFFER, J. (1947): Studia nad motylami Wilenszczyzny. Torun, 490 pp.

PRUFFER, J. & E. SOLTYS (1974): Motyle ziemi Chelminskiej i terenow sasiednich. - Studia Soc. Sci. Torunensis, Sect. E (Zool.) 10 (1): 1-88.

PUECHREDON, E.B. DE. (1954): Lepidopteres de la Tremezzina (Lac de Come). - Revue franc. Lepid. 14 (15-17): 211-217.

PÜNGELER, R. (1889): Lepidopterologische Mittheilungen aus der Schweiz. – Stett. ent. Ztg. 50 (4-6): 143-151.

PÜNGELER, R. (1902): *Tephroclystis thalictrata* n. sp. - Dt. ent. Z., Iris (1902) 15: 137-140, Taf. 6: 20, 21.

QUINETTE, J.-P. & N. LEPERTEL (1993): Inventaire preliminaire des Macrolepidopteres du departement de la Monche (*Lepidoptera, Rhopalocera* et *Heterocera*). - Alexanor 18 (3): 157-177.

RACHELI, T., PROLA, C. & P. PROVERA (1978): Note su alcuni Lepidotteri dell'Italia centrale. - Boll. Assoc. rom. ent. 33: 19-39.

RADTKE, A. (1998): Zwei Feuchtgebiete an der Wetter bei Butzbach (BRD: Hessen) als Lebensraum auentypischer Grossschmetterlinge (*Insecta: Lepidoptera*). - Mitt. Inter. ent. Ver. 23 (1/2): 41-52.

RAEBEL, P.H. & S. TOLL (1962): Fauna Motyli Slaska. Miernikowce (*Lepidoptera, Geometridae*). - Rocznik Muz. Gornoslaskie w Bytomiu (Przyroda) 1: 5-78.

RAGUSA, E. (1889): Note Lepidotterologiche. - Il Naturalista Sicil. 8 (10-11): 228-229 & Tav. III.

RAINERI, V. (1985): Ricerche sui Geometridi delle Alpi Liguri (*Lepidoptera*). - Boll. Soc. ent. ital., Genova, 117 (4-7): 102-112.

RAINERI, V. (1986): Ricerche sui Geometridi delle Alpi Liguri. Parte seconda (*Lepidoptera*). - Boll. Soc. ent. ital., Genova, 118 (1-3): 25-30.

RAINERI, V. (1992a): I Geometridi raccolti in Spagna durante le campagne di ricerca dell'Istituto di Zoologia dell'Universita di Genova (*Lepidoptera*). - Boll. Soc. ent. ital., Genova, 124 (1): 3-11.

RAINERI, V. (1992b): Aggiornamento sui Geometridi raccolti nella Penisola Iberica durante una campagna di ricerca dell'Istituto di Zoologia dell'Universita di Genova (*Lepidoptera*). - Mem. Soc. ent. ital. 71 (2): 653-658.

RAINERI, V. (1994): Contributo alla conoscenza dei Geometridi italiani (*Lepidoptera, Geometridae*). - Boll. Soc. ent. ital., Genova, 125 (3): 259-262.

RAINERI, V. (1995): Breve nota su alcuni interessanti Geometridi raccolti in Grecia (*Lepidoptera, Geometridae*). - Boll. Soc. ent. ital., Genova, 126 (3): 277-284.

RAINERI, V. (1996): La collezione di Geometridi del conte Turati conservata al Museo Regionale di Scienze Naturali di Torino (Lepidoptera Geometridae). - Boll. Mus. reg. Sci. nat. Torino 14 (1): 179-247.

RAINERI, V. & S. ZANGHERI (1995): Lepidoptera Drepanoidea, Axioidea, Geometroidea. In: MINELLI, A., RUFFO, S. & S. LA POSTA (eds.), Checklist delle specie della fauna Italiana, 90. - Ed. Calderini, Bologna, 23 pp.

RAKOSY, L. & W. WEBER (1986): Die Gross-Schmetterlinge von Sighisoara (Schassburg) und Umgebung, Siebenburgen, Rumanien (*Lepidoptera*). - Atalanta 16 (3/4): 315-392.

RAMBUR, M. (1833): Suite du Catalogue des Lépidoptères de l'Ile de Corse. – Annls Soc. ent. France (4 ser.) 2 (1): 5-59.

RATZEL, U. (1995): Beitrag zur Kenntnis der Biologie und Ökologie von *Eupithecia tenuiata* (Hübner 1813) (*Lepidoptera: Geometridae: Larentiinae*). - Ent. Z. 105 (9): 177-179.

RAZOWSKI, J. & E. PALIK. (1969): Fauna motyli okolic Krakowa. - Acta zool. Cracov. 14 (11): 217-310.

RÉAL, P. (1972): Table synonymique et biogéografique des lépidoptères de France. *Geometridae*. - Publs Sect. Biol. Ecol. anim. Stn Bonnevaux 8: 64-107.

RÉAL, P. (1987): Trois espèces peu ou pas connues en Basse-Provence: *Satyrium pruni* L., *Eupithecia gueneata* Millière, *Cucullia formosa* Rogenhofer (*Lep. Lycaenidae, Geometridae* et *Noctuidae*). - Alexanor 15 (3): 131-140.

RÉAL, P. (1991): *Eupithecia conterminata* (LIENIG & ZELLER) dans le Doubs. - Ent. gallica 2 (2): 71-72.

REBEL, H. (1892): Zwei Geometriden-Classificationen. – Stett. ent. Ztg. 53 (7-9): 247-268.

REBEL, H. (1901): Neue Lepidopteren aus Bosnien und der Hercegovina. - Verh. zool.-bot. Ges. Wien 51 (10): 798-804.

REBEL, H. (1902): Lepidopteren aus Morea gesammelt von Herrn Martin Holtz im Jahre 1901. - Berl. ent. Z. 47 (1-2): 83-110.

REBEL, H. (1903a): Lepidopteren aus Morea. - Berl. ent. Z. 48: 243-249.

REBEL, H. (1903b): Studien über die Lepidopterenfauna der Balkanländer. I. Teil. Bulgarien und Ostrumelien. - Ann. k. k. Naturhist. Hofmuseums, Wien, 18: 123-347 + 1 Taf.

REBEL, H. (1904a): Studien uber die Lepidopterenfauna der Balkanlander. II. Teil. Bosnien und Hercegovina. -

Ann. k. k. Naturhist. Hofmuseums, Wien, 19 (2-3): 97-377.
REBEL, H. (1904b): Sistematski spisak lepidoptera (lepirova) Bosne i Hercegovine. - Glasnik zemaljskog Muz. u Bosni i Hercegovini 16 (3): 375-416.
REBEL, H. (1905a): Lepidopteren aus Morea. II. Nachtrag. - Berl. ent. Z. 50: 291-314.
REBEL, H. (1905b): In: Penther, A. & E. Zederbauer. Ergebnisse einer naturwissenschaftlichen Reise zum Erdschias-Dagh (Kleinasien). Lepidopteren. - Ann. k. k. Naturhist. Hofmuseums, Wien, 20: 189-219.
REBEL, H. (1910): Beitrag zur Lepidopterenfauna der Ionischen Inseln. - Verh. zool.-bot. Ges. Wien 60: 418-431.
REBEL, H. (1911a): Versammlung am 3. Februar 1911. - Vehr. zool.-bot. Ges. 61: (79)-(90).
REBEL, H. (1911b): Versammlung am 3. November 1911. - Verh. zool.-bot. Ges. 61: (174)-(181).
REBEL, H. (1909-1912): Lepidopteren aus dem Gebiete des Monte Maggiore in Istrien. - Jahres Bericht Wien. ent. Ver. 21: 97-110, 22: 227-240, 23: 177-205.
REBEL, H. (1911d): Lepidopteren aus dem Gebiete des Triglav und der Crna Prst in Krain. III. - Nachtrag. Jahres-Bericht Wien. ent. Ver. 21: 111-144.
REBEL, H. (1913): Studien uber die Lepidopterenfauna der Balkanlander. III. Teil. Sammelergebnisse aus Montenegro, Albanien, Mazedonien und Thrazien. - Ann. k. k. Naturhist. Hofmuseums, Wien, 27: 281-334.
REBEL, H. (1916): Die Lepidopterenfauna Kretas. - Ann. k. k. Naturhist. Hofmuseums, Wien, 30: 66-172.
REBEL, H. (1923a): Ersten Stände von *Larentia lugdunaria* HS. - Verh. zool.-bot. Ges. Wien 72 (1): (9)-(10).
REBEL, H. (1923b): Versammlung am 2. Dezember 1921. - Verh. zool.-bot. Ges. Wien (1922) 72 (1): (14)-(18).
REBEL, H. (1924): Versammlung am 3. November 1922. - Verh. zool.-bot. Ges. Wien (1923) 73 (1): (1)-(7).
REBEL, H. (1926): Lepidopteren von den Balearen. - Dt. ent. Z., Iris 40 (2/3): 135-146.
REBEL, H. (1930): Uber eine Lepidopterenausbeute von den Pityusen. - Verh. zool.-bot. Ges. Wien 80 (1-2): (1)-(5).
REBEL, H. (1939): Zur Lepidopterenfauna Cyperns. - Mitt. münchn. ent. Ges. 29: 487-564.
REBEL, H. & H. ZERNY (1934): Die Lepidopterenfauna Albaniens. - Denkschrift. Akad. Wissenschaften Wien, Math. Naturwiss. Kl. 103: 37-161.
REDONDO, V.M. (1986): Introduccion al conocimiento de los Macrolepidopteros del Valle Medio del Ebro en Aragon (2a Parte). - Shilap Revta. lepid. 14 (54): 17-31.
REDONDO, V.M. & F.J. GASTÓN (1999): Los *Geometridae* (*Lepidoptera*) de Aragón (España). - Soc. ent. Aragon. (Monografias S.E.A.), Zaragoza, 3: 1-130.
REIPRICH, A. (1972): Verzeichniss der Falterarten aus dem Naturschutzgebiet Slovensky raj und der Umgebung von Spisska Nova Ves (Ergebnisse der Erforschung aus den Jahren 1945-1969). - Ent. problemy, Bratislava, 10: 5-104.
REIPRICH, A. (1977): Doplnky k Prodromu lepidopter Slovenska. - Ent. problemy, Bratislava, 14: 13-69.
REIPRICH, A. & I. OKÁLI (1989): Dodatky k Prodromu Lepidopter Slovenska. 3. zväzok. - Biol. Práce, Bratislava, 141 pp.
REISSER, H. (1927): Sammelergebnisse aus Andalusien mit spezieller Berucksichtigung der Sierra Nevada. - Verh. zool.-bot. Ges. Wien. 77 (2): (64)-(68).
REISSER, H. (1928): Lichtfang in der Sierra Nevada. - Verh. zool.-bot. Ges. Wien. 78: (8)-(13).
REISSER, H. (1930): Herbstlicher Lichtfang von Lepidopteren in Corsica. - Verh. zool.-bot. Ges. Wien. 80 (1-2): 32-44.
REISSER, H. (1937): *Eupithecia ericeata* Rbr. in Niederosterreich. - Z. Öst. Ent.-Ver. 22: 48-50.
REISSER, H. (1958): Neue Heteroceren aus Kreta. - Z. Wien. ent. Ges. 43 (69, 7): 105-128.
REISSER, H. (1962): Weitere neue Heteroceren aus Kreta. - Z. Wien. ent. Ges. 47 (73, 12): 193-216.
REISSER, H. (1974): Zur Lepidopterenfauna Kretas: neue Zugange zum Artenbestand. - Ann. Mus. Goulandris 2: 137-140.
REISSER, H. & H. KAUTZ. (1926): Bericht uber eine Sammelreise nach Corsica und Beitrag zur dortigen Fauna. - Verh. zool.-bot. Ges. Wien. 76: 1-25.
REMM, H. & J. VIIDALEPP (1986): Catalogus Macrolepidopterorum Estoniae. [Eesti Suurliblikate Nimestik]. Tartu. 34 pp.
REPO, S. (1995): Makrotiedonannot 1994. - Baptria 20 (1): 23-29.
REPO, S. & J. KULLBERG (1996): Makrotiedonannot 1995. - Baptria 21 (1): 1-12.
REZBANYAI, L. (1981): Grundriss der Eupithecien - Fauna des Nord-Bakong-Gebirges in Ungarn (Lep.: Geometridae). - Veszprem meg. Muz. Kozl. (Termesz.) 16: 167-177.
REZBANYAI-RESER, L. (1989a): *Eupithecia abietaria* Gze. (= *pini* Retz.) und *analoga* Djak. (= *bilunulata* Zett.): eine Berichtigung zu Forster & Wohlfahrt 1977 (1981) (Lepidoptera: Geometridae). - Ent. Z. (Frankfurt/Essen) 99 (4): 47-48.
REZBANYAI-RESER, L. (1989b): Ein Musterbeispiel der rezenten Arealerweiterung: *Eupithecia sinuosaria*

EVERSMANN, 1848, bis zur Sudostschweiz vorgedrungen (Lepidoptera, Geometridae). - Atalanta 19 (1/4): 39-50.

REZBANYAI-RESER, L. (1991): Zwolf Tage Lichtfallenfang in Calas de Mallorca (Malorca: Balearen, Spanien), Ende September 1980 (Macrolepidoptera). - Ent. Z. 101 (10): 173-192.

REZBANYAI-RESER, L. (1993): Zur Insektenfauna der Umgebung von Lauerz, Kanton Schwyz. 1. Sägel (455 m) und Schuttwald (480 m). II. *Lepidoptera* 1: *'Macrolepidoptera'* ('Grossschmetterlinge') (Fortsetzung). - Ent. Ber. Luzern 29: 1-28.

REZBANYAI-RESER, L. (1997): Herbstliche Lichtfallen- und Tagfange von Gross- Schmetterlingen an drei Orten in Mittelitalien (Marche, Toscana, Lazio) (Lepidoptera: Macrolepidoptera). - Ent. Z. 107 (9): 362-390.

REZBANYAI-RESER, L. (1998): Bemerkungen und Berichtigungen zur Macrolepidopterenfauna des schweizerischen Nationalparks und seiner Umgebung von PICTET, 1942 (Lepidoptera). - Mitt. schweiz. ent. Ges. 71 (1-2): 177-186.

REZBANYAI-RESER, L. & S. WHITEBREAD (1987): *Eupithecia sinuosaria* EVERSMANN, 1848, neu für die Schweiz (Lep., Geometridae). - Mitt. ent. Ges. Basel 37 (3): 120-122.

REZBANYAI-RESER, L., BLOCHLINGER, H., HOPPE, H., SCHAFFER, E. & J. SCHMID (1998): Zur Weiterverbreitung von *Eupithecia sinuosaria* (Eversmann, 1848) in der Schweiz (*Lepidoptera, Geometridae*). - Atalanta 28 (3/4): 309-314.

REZBANYAI-RESER, L. & E. MAIER (2000): Les Eupitheciini (*Lepidoptera: Geometridae*) de la collection du Musée cantonal de zoologie, Lausanne. - Bull. romand d'Entomol. 18 (2): 147-167.

RIBBE, C. (1909-1912): Beitrage zu einer Lepidopteren-Fauna von Andalusien. - Dt. ent. Z., Iris 23: 1-395.

RIEMIS, A. (1994): *Geometridae* of Turkey 3. A provisional list of the *Geometridae* of Turkey (*Lepidoptera*). - Phegea 22 (1): 15-22.

RILEY, A.M. (1985): *Eupithecia ultimaria* BOISDUVAL (*Lepidoptera: Geometridae*): A Pug new to the British list. – The Entomologist's Gaz. 36: 259-261.

RILEY, A.M. (1986): A rewiew of the status of *Eupithecia goossensiata* Mab. (The Ling Pug) and *E. absinthiata* Cl. (The Wormwood Pug) (Lepidoptera: Geometridae). – The Entomologist Rec. J. Var. 98: 85-89.

RINDGE, F.H. (1952): Taxonomic and life history notes on North American *Eupithecia* (Lepidoptera: Geometridae). - Amer. Mus. Novit. 1569: 1-27.

RÖBER, J. (1897): Die Schmetterlinge des Taurus. - Ent. Nachr. 23 (17-18): 257-288.

ROESLER, U. (1966): Auf Lepidopterenfang in Damuls (Vorarlberg). - Ent. Z. 76 (5): 49-51.

ROMANOFF, N.M. (1887): Les Lépidoptères de la Transcaucasie. III partie. - Mém. Lep. Romanoff 3: 1-49.

RONDOU, J.-P. (1934): Catalogue des Lépidoptères des Pyrénées. *Geometridae*. - Annls Soc. ent. France 103 (3-4): 257-315.

RONKAY, L. & C. SZABOKY (1981): Investigations on the *Lepidoptera* fauna of the Zemplen Mts. (NE Hungary). I. The valley of Kemence stream. - Folia ent. hung. 42 (2): 167-184.

ROSS, D.A. & D. EVANS (1956): Annotated list of forest insects of British Columbia. Part III, *Eupithecia* spp. (*Geometridae*). - Proc. Ent. Soc. of B.C. 52: 36-38.

ROTHSCHILD, W. (1914): A preliminary account of the Lepidopterous fauna of Guelt-es-Stel, Central Algeria. - Novit. Zool. 21: 299-357.

ROUGEMONT, DE F. (1903): Catalogue des Lépidoptères du Jura neuchâtelois. - Bull. Soc. neuchat. sci. nat. 29-31: 1-366.

RUNGS, Ch.E.E. (1981): Catalogue raisonné des Lepidoptères du Maroc. Inventaire faunistique et observations écologiques. Tome II. - Trav. Inst. Sci. Série Zool. 40: 1-365.

RUNGS, Ch.E.E. (1988): Liste-inventaire systématique et synonymique des Lépidoptères de Corse. Contribution à la connaissance de la faune entomologique de la Corse: Lepidoptera. Liste-inventaire des espèces de l'île. - Alexanor 15 (Suppl.): [1]-[86].

RYRHOLM, N. (1994): Intressanta fynd av storfjärilar (*Macrolepidoptera*) i Sverige 1993. - Ent. Tidskr. 115 (1-2): 37-44.

RYRHOLM, N. (1995): Intressanta fynd av storfjärilar (*Macrolepidoptera*) i Sverige 1994. - Ent. Tidskr. 116 (1-2): 31-45.

RYRHOLM, N. & A. OHLSSON (1999): Intressanta fynd av fjällfjärilar i Sverige 1998. - Ent. Tidskr. 120 (1-2): 43-53.

SACHKOV, A.S., ANTONOVA, E.M. & A.V. SVIRIDOV (1996): Cheshuekrylye (*Lepidoptera*). Bespozvonochnye Zhiguljevskogo zapovednika. - Flora i fauna zapovednikov, 61, Moscow, 1996, pp. 48-132.

SAGNES, P. (1978): Capture de *Perizoma sagittata* (FABRICIUS) dans les Hautes-Alpes (*Lep. Geometridae, Larentiinae*). - Alexanor 10 (6): 276.

SAGNES, P. (1987): Quelques compléments à propos de *Satyrium pruni* Linné, *Eupithecia gueneata* Millière, et sur la localité de Celles-les-Bains (Ardèche) (*Lep., Lycaenidae* et *Geometridae*). - Alexanor 15 (4): 219-

220.
SÄLZL, M. (1935): Spanner (XIX. Geometridae). In: METSCHL, C. & M. SÄLZL. Die Schmetterlinge der Regensburger Umgebung. I. Teil Grossschmetterlinge. - Dt. ent. Z. Iris 49: 58-132.
SAMMUT, P.M. (1984): A Systematic and Synonymic List of the Lepidoptera of the Maltese Islands. - Neue ent. Nachr. 13: 1-124.
SANCHEZ EGUIALDE, D. & J. CIFUENTES (1990): Contribution al conocimiento de los Geometroidea de Tudela y la Ribera Sur de Navarra. Otras citas de Geometroidea en la Peninsula Iberica. - Shilap Revta. lepid. 18 (69): 73-82.
SATO, R. (2002): Taxonomic notes on *Pseudobaptria corydalaria* (Graeser) (Geometridae, Larentiinae) and its allies from Russia, Japan, China and Europe, with description of one new species from China. – Tinea 17 (1): 33-41.
SATTLER, K. & W.G. TREMEWAN (1973): The entomological Publications of PIERRE MILLIÈRE (1811-1887). - Bull. Brit. Mus. (Nat. Hist.), Hist. ser. 4 (4): 221-280; 3 pls.
SAVENKOVS, N.V. (1984): Nekotorye dannye po faune i ekologii cheshuekrylych Latvii. [Some data on fauna and ecology of *Lepidoptera* of the Latvia]. - Latv. Ent. 27: 46-63.
SAVENKOVS, N.V. (1991): Novye i redkie vidy cheshuekrylyh fauny Latvii, sobrannye preimustshestvenno v 1989 godu. [New and rare *Lepidoptera* species in the fauna of Latvia collected mainly in 1989]. - Latv. Ent. 34: 33-39.
SAVENKOV, N., ŠULCS, I., KERPPOLA, S. & L. HULDÉN (1996): Checklist of Latvian Lepidoptera - Latvijas Taurizu Katalogs. - Baptria 21 (3a): 1-71.
SCALERCIO, S. (1999): Macrolepidotteri notturni catturati nel Vincese (Toscana-Italia) (*Lepidoptera*). - Mem. Soc. ent. ital. 77: 311-316.
SCHÄFFER, W. (1977): Zur Verbreitung von *Eupithecia breviculata* DONZ. in Südbaden (*Lep. Geom.*). - Veröff. f. Naturschutz und Landschaftspflege in Baden-Württemberg 46: 75-76.
SCHAWERDA, K. (1913): Siebenter Nachtrag zur Lepidopterenfauna Bosniens und der Herzegowina. - Verh. zool.-bot. Ges. Wien. 63 (3-4): 141-178.
SCHAWERDA, K. (1914): Achter Nachtrag zur Lepidopterenfauna Bosniens und der Herzegowina. - Verh. zool.-bot. Ges. Wien. 64: 349-378.
SCHAWERDA, K. (1932): Eine neue Geometride. - Inter. ent. Z. 26 (26): 283-285.
SCHAWERDA, K. (1938): Kufstein und das Kaisergebirge. - Dt. ent. Z. Iris 52 (2): 51-72.
SCHMIDLIN, A. (1964): Übersicht über die europäischen Arten der Familie *Geometridae* (Lep.). - Mitt. ent. Ges. Basel, N. F. 14 (4/5): 77-137.
SCHMIDT-KOEHL, W. (1966): Neue Beitrage zur Macrolepidopteren-Fauna Oberandalusiens aus den Provinzen Granada, Almeria und Malaga (Südspanien). - Ent. Z. 76 (1/2): 1-15.
SCHNACK, K. (1978): *Eupithecia cauchiata* Dup. - fundet i Danmark. - Lepidoptera, København 3 (6): 169-176
SCHNACK, K. (1985): Katalog over de danske Sommerfugle. - Ent. Medd. 52 (2-3): 1-163.
SCHNEIDER, C. (1933): *Cidaria affinitata* Steph. ab *Jenischi* ab. nov. (*Lep., Geom.*). - Ent. Z., Frankfurt-M., 47 (2): 18.
SCHROTH, M. (1984): Die Makrolepidopteren aus dder Umgebung von Hanau am Mai (Hessen). - Nachr. ent. Ver. Apollo, Frankfurt a/Main, Suppl. 3: 1-83.
SCHROTH, M. (1985): 1.Nachtrag zu den 'Makrolepidopteren aus der Umgebung von Hanau am Main (Hessen). - Nachr. ent. Ver. Apollo, Frankfurt a/Main, N.F., 6 (1): 21-33.
SCHÜTZE, E. (1952): Eupithecien-Studien II. Zur Frage der Artberechtigung von *Eupithecia pyreneata* Mab. - Z. Lepid. 2 (3): 177-187.
SCHÜTZE, E. (1954): Eupithecien-Studien IV (*Lep., Geometr.*). Die *absinthiata*-Gruppe. - Ent. Z., Stuttgart, 64 (15): 169-175.
SCHÜTZE, E. (1956a): Zur Genitalmorphologie der Gattung *Eupithecia* Curt. Eupithecien-Studien VI. - Ent. Z. 66 (2): 17-21.
SCHÜTZE, E. (1956b): *Eupithecia*-Studien VII und VIII. Die *semigraphata*-Gruppe. - Z. Wien. ent. Ges. 41 (Bd 67, 11): 306-320; (12): 328-339 & Taf. 33-35.
SCHÜTZE, E. (1957): Über *Eupithecia breviculata* Donz. und *extremata* F. insbesondere in Deutschland. Eupithecien-Studien IX. (*Lep. Geom.*). - Nachrichtenbl. Bayer. Ent. 6 (7): 65-69.
SCHÜTZE, E. (1958): *Eupithecia*-Studien X. Die *graphata*-Gruppe (Lepidoptera, Geometridae). - Ent. Z., Frankfurt a/Main, 68 (5): 49-64; (6): 71-72, (7): 82-88, Taf. I.
SCHÜTZE, E. (1959): Neue Eupithecien aus Griechenland. Eupithecien-Studien XIII. - Mitt. münchn. ent. Ges. 49: 35-38.
SCHÜTZE, E. (1960): Alte und neue Eupithecien aus Iran (Lepidoptera, Geometridae). Eupithecien-Studien XV. - Mitt. münchn. ent. Ges. 50: 1-23, Taf. I-XIII.

SCHÜTZE, E. (1961a): Über einige nordafrikanische Eupithecien (Lep., Geom.). Eupithecien-Studien XIV. - Dt. ent. Z., N. F. 8 (1/2): 147-151.
SCHÜTZE, E. (1961b): Neue Eupithecien aus Arabien. Eupithecien-Studien XVI. - Ent. Z. 71 (6): 62-68.
SCHÜTZE, E. (1961c): Weitere Eupithecien aus Iran und Arabien (Lep., Geom.). - Mitt. münchn. ent. Ges. 51: 58-71.
SCHWINGENSCHUSS, L. (1930): Lepidopterologische Ergebnisse einer Herbstreise nach Andalusien. - Verh. zool.-bot. Ges. Wien. 80 (1-2): 1-31.
SCHWINGENSCHUSS, L. (1952): Kleiner Beitrag zur Lepidopterenfauna von Herzogenburg und Umgebung. - Z. Wien. ent. Ges. 37 (63, 9-11): 169-173.
SCHWINGENSCHUSS, L. (1953): Beitrag zur Lepidopterenfauna von Niederosterreich: St. Peteri. d. Au, Seitenstetten und Umgebung. - Z. Wien. ent. Ges. 38 (64, 9): 251-255.
SCHWINGENSCHUSS, L. (1955): Nachträge und Berichtungen zu den Lepidopterenfaunen des Libanon, Grossen Atlas, Iran und Neubeschreibungen. - Z. Wien. ent. Ges. 40 (Bd 66, 8): 225-231.
SCHWINGENSCHUSS, L. (1962): Nachträge, Ergänzungen und Berichtungen zur 'Lepidopterenfauna von Albarracin in Aragonien' von H. Zerny. - Z. Wien. ent. Ges. 47 (Bd 73, 1): 4-11.
SCHWINGENSCHUSS, L. & F. WAGNER. (1927): Beitrag zur Macro-Lepidopteren-Fauna Süddalmatiens insbesondere der Umgebung Gravosa's. - Z. Österr. Ent.-Ver. 12 (5): 45-50; (6): 62-64; (7): 68-72; (8): 73-75.
SCIARETTA, A. & V. RAINERI (2001): Nuove conoscenze faunistiche sui Geometridi del Molise (Italia centrale) (*Lepidoptera, Geometridae*). – Boll. zool. agr. Bachic. Ser. II 33 (1): 19-41.
SCOBLE, M.J. (ed.), PITKIN, L.M., PARSONS, M., HONEY, M.R. & B.R. PITKIN (1999): Geometrid Moths of the World: a catalogue (*Lepidoptera, Geometridae*). Vol. 1 and 2. CSIRO Publishing and Apollo Books, Stenstrup. 1016 pp. (+ 129 pp. of Index).
SEPPÄNEN, E.J. (1970): Suomen suurperhostoukkien ravintokasvit. - Animalia Fennica 14: 1-179.
SETTELE, L. (1973): Die Grossschmetterlinge vom Kaiserstuhl und der naeheren Umgebung. - Mitt. ent. Ges. Basel, N.F. 23: 29-74.
SEVEN, S. (1991): Trakya *Lepidoptera* faunasi üzerine bibliyografik araştirmalar (Bibliographical Studies on the *Lepidoptera* Fauna of Turkey-in-Europe). - Priamus 6 (1/2): 1-95.
SHENEFELT, R.D. (1969): Hymenopterorum Catalogus. Pars 4. *Braconidae* 1 (*Hybrizoninae, Euphorinae, Cosmophorinae, Neoneurinae, Macrocentrinae*). Uitgeverij Dr. W.Junk N.V. 's-Gravenhage, pp. 1-176.
SHENEFELT, R.D. (1970): Hymenopterorum Catalogus. Pars 6. *Braconidae* 3 (*Agathidinae*). Uitgeverij Dr. W.Junk N.V. 's-Gravenhage, pp. 307-428.
SHENEFELT, R.D. (1972): Hymenopterorum Catalogus. Pars 7. *Braconidae* 4 (*Microgasterinae, Apanteles*). Uitgeverij Dr. W.Junk N.V. 's-Gravenhage, pp. 429-668.
SHENEFELT, R.D. (1973): Hymenopterorum Catalogus. Pars 9. *Braconidae* 5 (*Microgasterinae & Ichneutinae*). Uitgeverij Dr. W.Junk N.V. 's-Gravenhage, pp. 669-812.
SHENEFELT, R.D. (1973a): Hymenopterorum Catalogus. Pars 10. *Braconidae* 6 (*Cheloninae*). Uitgeverij Dr. W.Junk N.V. 's-Gravenhage, pp. 813-936.
SHENEFELT, R.D. (1975): Hymenopterorum Catalogus. Pars 12. *Braconidae* 8 (*Exothecinae, Rogadinae*). Uitgeverij Dr. W.Junk B.V. 's-Gravenhage, pp. 1115-1262.
SHENEFELT, R.D. (1978): Hymenopterorum Catalogus. Pars 15. *Braconidae* 10 (*Braconinae, Gnathobraconinae, Mesostoinae, Pseudodicrogeniinae, Telengainae, Ypsistocerinae* plus *Braconidae* in general, major groups, unplaced genera and species). Dr. W.Junk B.V.-Publishers. The Hague - Holland, pp. 1425-1872.
SHLYKOV, O.V. (1988): Spisok cheshuekrylyh (*Macrolepidoptera*) Penzenskoj oblasti. [A List of *Macrolepidoptera* of the Penza Province]. - Ent. obozr. 67 (1): 48-61.
SICK, H. (1939): Beitrag zur Grossschmetterlingsfauna Ostholsteins. - Ent. Rund. 56 (33): 370-374.
SILVA CRUZ DA, M.A. (1967): Lepidópteros da Região de Abrantes. - Inst. zool. 'Dr. Augusto Nobre', Facul. Ciênc. do Porto 99: 9-35
SILVA CRUZ DA, M.A. & T. GONÇALVES (1966): Notas Lepidopterologicas. V. Contribuicao sistematica para o inventario dos Heteroceros portugueses. - Mem. Est. Mus. Univ. de Coimbra 298: 1-8.
SILVA CRUZ DA, M.A. & T. GONÇALVES (1974): Macrolepidópteros da Região do Alto Douro. - Inst. zool. 'Dr. Augusto Nobre', Facul. Ciênc. do Porto 127: 9-50, 1 map.
SILVA CRUZ DA, M.A. & T. GONÇALVES (1977): Catálogo sistemático dos Macrolepidópteros de Portugal. - Inst. zool. 'Dr. Augusto Nobre', Fac. Cienc. do Porto 133: 1-56.
SIMSON, E.C.L. (1980-1981): British Pugs. - The Entomologist's Rec. J. Var. 92 (11-12): 261-266; 93 (1): 7-10; 93 (2): 30-35.
SINGH, B. (1951): Immature stages of Indian Lepidoptera, No. 8. *Geometridae*. - Ind. For. Rec. (n.s.) 8 (7): 61-158.

SKALA, H. (1929a): Beitrag zur Grossschmetterlings-fauna Mährens und Schlesiens. - Ent. Z., Frankfurt a. M. 42 (24): 317-320.

SKALA, H. (1929b): *Tephroclystia nageli* SKALA und *Lythria purpuraria* L. in Mähren. - Z. Österr. Ent.-Ver., Wien 14 (11): 115.

SKALA, H. (1932): Die wichtigsten Neuentdeckungen an Grossschmetterlingen aus dem Gebiete der ehemal. österr.-ungar. Monarchie. - Ent. Z. 45 (8): 114-119.

SKINNER, B. (1984): Colour Identification Guide to Moths of the British Isles. (Macrolepidoptera). 267 pp., 42 pls. Viking, London.

SKOU, P. (1984): Nordens målere. Fauna bøger & Apollo bøger. København & Svenborg. 332 pp.

SKOU, P. (1986): The Geometroid Moths of North Europe (Lepidoptera: Drepanidae and Geometridae). - Entomonograph 6: 1-348, pls. 1-24.

SKULE, B. (1998): Oplevelser med larvefund og klækning af *Eupithecia pulchellata* STPH. - en dværgmåler med en særegen levevis. - Lepidoptera, København 7 (5): 158-164.

SLIWINSKI, Z. (1995): Miernikowce (*Geometridae - Lepidoptera*) Wyżyny Lódzkiej. - Biul. Ent. 6 (10): 3-6.

SLIWINSKI, Z., WIACKOWSKI, S. & B. MARCINIAK. (1991): Motyle (*Lepidoptera*) Swietokrzyskiego Parku Narodowego. - Fragm. Faun., Warszawa 35 (8): 123-145.

SÖDERMAN, G., LUNDSTEN, K.-E. & R. LEINONEN (1998) Luoteis-Venäjän yöperhosseurannan tulokset 1995-1997. [Results from the Moth Monitoring Scheme in Northwestern Russia 1995-97]. - Baptria 23 (4): 219-230.

SOFFNER, J. (1928): O 'Tephroclystiich' okoli Susickeho. - Cas. Cs. spol. ent. 25 (3-4): 33-40a.

SOFFNER, J. (1960): Schmetterlinge aus dem Riesengebirge. - Z. Wien. ent. Ges. 45 (Bd 71, 5/6): 70-91.

SOHN-RETHEL, O. (1929): Beiträge zur Heteroceren-Fauna Italiens. - Dt. ent. Z., Iris 43: 1-23.

SOPOTZKO, A. (1913): K faune cheshuekrylyh Tul'skoj gubernii. I. *Macrolepidoptera*. Tula, 11 pp.

SOTAVALTA, O. (1987): Provincial distribution of Finnish *Macrolepidoptera*. - Notulae Ent. 67: 187-205.

SOTAVALTA, O. (1995): Itä-Fennoskandian mittareiden (*Geometridae*) levinneisyys. [Distribution atlas of *Macrolepidoptera: Geometridae* of eastern Fennoscandia]. - Baptria 20 (1a): 1-64.

SPEIDEL, W. & M. HASSLER. (1989): Die Schmetterlingsfauna der südlichen algerischen Sahara und ihrer Hochgebierge Hoggar und Tassili n'Ajjer (Lepidoptera). - Nachr. ent. Ver. Apollo, Frankfurt, Suppl. 8: 1-156.

SPEIDEL, W. & M. HASSLER (1991): Die Schmetterlingsfauna der südlichen algerischen Sahara 1. Nachtrag. - Nachr. ent. Ver. Apollo, Frankfurt/Main, N.F. 12 (2): 12 -135.

SPESSIVTSEFF, P. (1924): Grankottmätarna (*Eupithecia abietaria* och *strobilata*) och deras skadegörelse. - Medd. från statens skogsförs. 21 (7): 295-310.

SPEYER, A. (1868): *Eupithecia actaeata* n. sp. - Stett. ent. Ztg. 30 (10-12): 395-400.

SPEYER, A. (1872): Lepidopterologische Notizen. - Stett. ent. Ztg. 33 (4-6): 169-175.

SPEYER, A. (1873): Lepidopterologische Notizen. - Stett. ent. Ztg. 34 (7-9): 357-365.

SPITZER, K., JAROŠ, J. & I. SVENSSON (1991): Geographical variation in food plant selection of *Eupithecia gelidata* Möschler, 1860 (Lepidoptera, Geometridae). - Ent. Fenn. 2 (1): 33-36.

SPITZER, K. & J. JAROŠ (1993): *Lepidoptera* associated with the Cervene Blato bog (Central Europe): Conservation implications. - Eur. J. Ent. 90 (3): 323-336.

SPULER, A. (1910): Die Schmetterlinge Europas. 2. Stuttgart 523 pp.

STADNITZKY, G.V. (1971): Vrediteli semjan eli. [Spruce seed pests]. 'Lesn. prom.', Moscow, 48 pp.

STADNITZKY, G.V., YURCHENKO, G.I., SMETANIN, A.N., GREBENSHCHIKOVA, V.P. & M.V. PRIBYLOVA (1978): Vrediteli shishek i semjan khvojnykh porod. [Conifer cone and seed pests]. 'Lesn. prom.', Moscow, 168 pp.

STAINTON, H.T. (1855): Lepidoptera. New British species since 1835. – The Entomologist's Annual (1855): 26-61.

STANESCU, M. (1995): The catalogue of 'Ludovic Beregszaszy Lepidopteran collection (*Insecta: Lepidoptera*). - Trav. Mus. hist. nat. 'Grigore Antipa' 35: 221-346.

STANESCU, M. (1997a): *Eupithecia innotata* Hufnagel, 1767 (*Lepidoptera: Geometridae*) - first mention in Iceland. - Trav. Mus. natl. hist. nat. 'Grigore Antipa' 39: 87-89.

STANESCU, M. (1997b): The updated systematic and synonymic list of the *Geometridae* (*Insecta: Lepidoptera*) of Romania. - Trav. Mus. natl. hist. nat. 'Grigore Antipa' 39: 109-160.

STANESCU, M. & D. RUSTI (1997): New data concerning the Lepidoptera (*Insecta*) of Maramures (Romania). - Trav. Mus. natl. hist. nat. 'Grigore Antipa' 37: 89-111.

STAUDER, H. (1929a): *Eupithecia stabiensis* Stauder, nova spec. – Ent. Anzg. 9: 105-107.

STAUDER, H. (1929b): *Eupithecia pliniata* Stauder, nova species – Ent. Anzg. 9: 191-193.

STAUDINGER, O. (1857): Reise nach Island zu entomologischen Zwecken unternommen. - Stett. ent. Ztg. 18 (7-9): 209-289.

STAUDINGER, O. (1870): Beschreibung neuer Lepidopteren des europäischen Faunengebiets. – Berl. ent. Z. 14:

97-132.
STAUDINGER, O. (1871): Beitrag zur Lepidopterenfauna Griechenlands. - Horae Soc. ent. Ross. 7: 3-304.
STAUDINGER, O. (1885): Einige neue Lepidopteren des europäischen Faunengebietes. - Stett. ent. Ztg. 46 (7-9): 349-352.
STAUDINGER, O. (1892a): Neue Arten und Varietaten von palaarktischen Geometriden. - Dt. ent. Z., Iris 5: 141-260.
STAUDINGER, O. (1892b): Lepidopteren aus Tunis. - Dt. ent. Z., Iris 5: 277-300, Taf. 3.
STAUDINGER,O. & M. WOCKE (1871): Catalog der Lepidopteren des europ. Faunengebietes. - Dresden, 426 pp.
STAUDINGER, O. & H. REBEL (1901): Catalog der Lepidopteren des Palaearktischen Faunengebietes. Teil I. Berlin, XXXII & 411 pp.
STEKOL'NIKOV, A.A. (1993): Pjadenitzy (*Lepidoptera, Geometridae*) zapovednoj lesostepnoj dubravy 'Les na Vorskle'. [The Geometroid Moths (*Lepidoptera*) of the reserve oak-grove 'Forest on Worskla']. - Vestn. St.-Petersburg Gos. Univ., Serie 3 (Biologia) 1 (3): 22-29.
STEKOL'NIKOV, A.A. & V.G. MIRONOV. (2001): Tzvetochnye pjadenitzy (*Lepidoptera, Geometridae:* triba *Eupitheciini*) zapovednoj lesostepnoj dubravy 'Les na Vorskle' (Belgorodskaja obl., Russia). [The Pugs (*Lepidoptera, Geometridae: Eupitheciini*) of the forest-steppe reserve 'Forest on the Vorskla river' (Belgorod region, Russia)]. - Vestn. St.-Petersburg Univ., Ser. 3, 3 (19): 3-10.
STEPHAN, J. (1925): Die spannerartigen Nachtschmetterlinge und die Kleinschmetterlinge der Grafschaft Glatz. - Dt. ent. Z., Iris 39: 65-133.
STERLING, D.H. (1981): The Cloaked Pug: *Eupithecia abietaria* Goeze. - The Entomologist's Rec. J. Var. 93 (2): 29.
STERNECK, J. (1929): Prodromus der Schmetterlingsfauna Böhmens. Karlsbad, 297 pp.
STERZL, O. (1965): Funfzig Jahre lepidopterologische Forschung in Niederösterreich. - Z. Wien. ent. Ges. 50 (76, 12): 185-208.
STERZL, O. (1967): Prodromus der Schmetterlingsfauna von Niederösterreich (I. Teil der 2. Auflage). - Verh. zool.-bot. Ges. Wien 107: 75-193.
STEVENS, S. (1882): *Eupithecia ultimaria* Dup. - The Entomologist 15 (224): 18.
STRAKA, V. (1977): K poznaniu motylov Turcianskej kotliny a casti pril'ahlych Pohori (*Lepidoptera*). - Ent. problemy, Bratislava 14: 83-122.
STRAND, E. (1901): Beitrag zur Schmetterlingsfauna Norwegens. - Nyt Mag. Naturvid. 39: 25-72.
SUKHAREVA, I.L. & M.I. FALKOVITSH (1984): Cheshuekrylye (*Lepidoptera*) Bol'shogo Berjezovogo ostrova. [*Lepidoptera* of the Island Bolshoy Berezovy]. - Trudy Zool. Inst., Leningrad, 123: 73-115.
ŠULCS, A. (1968): Neue und wenig bekannte Arten der Lepidopteren-Fauna Lettlands. 4. Mitteilung. - Ann. Ent. Fenn. 34 (1): 14-30.
ŠULCS, A. (1973): Neue und wenig bekannte Arten der Lepidopteren-Fauna Lettlands. 5. Mitteilung. - Ann. Ent. Fenn. 39 (1): 1-16.
ŠULCS, A. & I. ŠULCS (1978): Neue und wenig bekannte Arten der Lepidopteren-Fauna Lettlands. 7. Mitteilung. - Notulae Ent. 58 (4): 141-150.
ŠULCS, A. & I. ŠULCS (1981a): Neue und wenig bekannte Arten der Lepidopteren-Fauna Lettlands. 8. Mitteilung. - Notulae Ent. 61 (2): 91-102.
ŠULCS, A. & I. ŠULCS (1981b): *Perizoma lugdunarium* in Lettland gefunden (*Lepidoptera, Geometridae*). - Notulae Ent. 61 (4): 179-181.
ŠULCS, A. & I. ŠULCS (1983): Neue und wenig bekannte Arten der Lepidopteren-Fauna Lettlands. 9. Mitteilung. - Notulae Ent. 63 (1): 37-48.
ŠULCS, A. & I. Šulcs (1987): Neue und wenig bekannte Arten der Lepidopteren-Fauna Lettlands. 11. Mitteilung. - Notulae Ent. 67 (3): 141-145.
ŠULCS, A. & J. VIIDALEPP (1972): Verbreitung der Gross-Schmetterlinge im Balticum. IV. Spanner (*Geometridae*). - Dt. ent. Z. 19: 151-209.
ŠULCS, A., VIIDALEPP, J. & P. IVINSKIS (1981): 1. Nachtrag zur Verbreitung der Gross-Schmetterlinge im Balticum (*Lepidoptera*). - Dt. ent. Z., N.F. 28 (1-3): 123-146.
SUMPICH, J. (1995): Synekologicke hodnoceni synuzie motylu podmacenych luk u Hlinska na Ceskomoravske vrchovine (*Lepidoptera*). - Acta Mus. Reginaehrad., Ser. A 24: 71-134.
SUMPICH, J. (1998): Faunisticke doklady z Podyji ve sbirce motylu (*Lepidoptera*) K.Jordana ulozene v Muzeu vychodnich Cech v Hradci Kralove. - Acta mus. Reginaehrad., Ser. A 26: 39-54.
SUOMALAINEN, E., KAISILA, J. & K. MIKKOLA (1980): Noteworthy records of Finnish Lepidoptera 1955-1974. I. Hesperioidea, Papilionoidea, Bombycoidea and Geometroidea. - Notulae Ent. 60 (1): 49-61.
SUTTON, S.L. (1963): South Caspian insect fauna 1961. I. Systematic List of *Lepidoptera* with notes. - Ann. & Mag. Nat. Hist., London, Ser. 13a 6 (66): 353-374.

SVENSSON, I. (1957): De senaste tio arens nytillskott av svenska storfjarilar (*Lep.*). - Opuscula Ent. 22 (2-3): 143-160.
SVENSSON, I. (1977): Förändringar i Sveriges storfjarilfauna en tredje tioårsperiod (*Lepidoptera*). - Ent. Tidskr. 98 (4): 113-122.
SVENSSON, I., ELMQUIST, H., GUSTAFSSON, B., HELLBERG, H. IMBY, L. & G. PALMQVIST (1994): Catalogus Lepidopterorum Sueciae. Code list L1. Stockholm (Naturhistoriska Riksmuseet & Entomologiska Föreningen), 302 pp.
SVIRIDOV, A.V., ANTONOVA, E.M., BLINUSHOV, A.E. & O.M. BUTENKO. (1998): Vysshie cheshuekrylye Okskogo zapovednika (annotirovannyj spisok vidov). - Flora i Fauna zapovednikov 70, Moscow, 1998, pp. 1-40.
ŠVITRA, G. (1988): 3 novyh i 16 redkih dlja Litovskoj SSR vidov pjadenitz, obnaruzhennyh v 1979-1986 gg. [3 new and 16 rare to the Lithuanian SSR Geometridae species, found in 1979-1986]. Novye i redkie dlja Litovskoj SSR vidy nasekomyh. Soobstsh. i opisanija 1987 goda. Vilnjus, 61-63 pp.
SZÉKELY, L. (1989): Adatok Marosvasarhely (Kozep-Erdely) videke nagylepkefaunajanak ismeretehez (*Lepidoptera: Macrolepidoptera*). - Folia ent. hung. 50: 137-145.
SZÉKELY, L., HANESCHLÄGER, P. & R. GÖRBE (1998): Contributii la studiul faunei de lepidoptere din Câmpia Munteniei I. [Contributions to the study of the Fauna of Lepidopterans in the Plain of Muntenia (Romania). I]. - Bul. inf. Soc. lepid. rom. 9 (3-4): 177-186.
TAMAS, V. (1981): Adatok a nagylepkek elterjedesehez Erdelyben II. (*Lepidoptera*). - Folia ent. hung. 42 (1): 227-251.
TAMMARU, T., VIRTANEN, T., RUOHOMÄKI, K. & L. KAILA (1996): *Eupithecia veratraria* Herrich-Schäffer, 1850 new to Scandinavia, with the redescription of *E. veratraria* ssp. *arctica* VIIDALEPP, 1974 (*Lepidoptera, Geometridae*). - Nota lepid. 18 (3/4): 297-303.
TAMMARU, T., VIRTANEN, T. & K. RUOHOMÄKI. (1996): *Eupithecia veratraria* (*Geometridae*) Pohjois-Norjassa. - Baptria 21 (3): 123-126.
TANNERT, R. (1983): Feststellungen zur Macrolepidopterenfauna Nurnberg Sud-Ost -Langwasser, Fischbach, Brunn- (*Lepidoptera*). - Atalanta 14 (2): 146-157.
TAUTEL, Cl. (1996): L'Alsace, patrie de la premiere capture francaise d'*Eupithecia conterminata* (LIENIG & ZELLER, 1846) (*Lepidoptera, Geometridae*). - Alexanor 19 (6): 365-366.
TAUTEL, Cl. (2000): *Eupithecia santolinata* MABILLE. Un géomètre méconnu découvert dans les Pyrénées-Orientales. - Rev. l'Assoc. Roussillonnaise d'Entl. (R.A.R.E.) 9 (2): 37-43.
TAYLOR, G.W. (1907): The *Eupithecia* of Eastern North America. – The Can. Entomologist 39 (8): 276-280.
TEOBALDELLI, A. (1976): I Macrolepidotteri del Maceratese e dei Monti Sibillini (Appennino Umbrio-Marchigiano). - Note ed appunti sper. di Ent. Agr., Perugia, 16: 1-346.
TEOBALDELLI, A. (1979): Lepidopteres captures en Val d'Aoste. - Alexanor 11 (4): 145-152.
TEOBALDELLI, A. (1994): Nuovi rinvenimenti di Macrolepidotteri nel territorio Marchigiano (*Lepidoptera*). - Boll. Soc. ent. ital., Genova 126 (1): 69-74.
THIELE, V. & I. CÖSTER (1999): Zur Kenntnis der Schmetterlingsfauna verschiedenr Flusstaltypen in Mecklenburg-Vorpommern (*Lep.*). - Ent. Nachr. Ber. 43 (2): 87-99.
THOMPSON, W. R. (1944): A Catalogue of the Parasites and Predators of Insect Pests. Sect. 1 (Parasite Host Catalogue), Part 5 (Parasites of the *Lepidoptera*). The Imperial Parasite Service. Belleville, Ontario, Canada, 1-130 pp.
THOMPSON, W. R. (1945): A Catalogue of the Parasites and Predators of Insect Pests. Sect. 1 (Parasite Host Catalogue), Part 6 (Parasites of the *Lepidoptera*). The Imperial Parasite Service. Belleville, Ontario, Canada, 131-258 pp.
THOMPSON, W. R. (1946): A Catalogue of the Parasites and Predators of Insect Pests. Sect. 1 (Parasite Host Catalogue), Part 7 (Parasites of the *Lepidoptera*). The Imperial Parasite Service. Belleville, Ontario, Canada, 259-385 pp.
THOMPSON, W. R. (1946a): A Catalogue of the Parasites and Predators of Insect Pests. Sect. 1 (Parasite Host Catalogue), Part 8 (Parasites of the *Lepidoptera*). The Imperial Parasite Service. Belleville, Ontario, Canada, 386-523 pp.
THURNER, H.J. (1926): List of Geometers of the British Islands with The Named Varieties and the Synonyms in general use. - The Entomologist's Rec. J. Var. 38 (3): (1)-(20); (5): (21)-(24); (7-8): (25)-(28).
THURNER, J. (1938): Die Schmetterlinge der Ochrid-Gegend in Macedonien. - Izvest. tzarsk. prirodonauchni inst., Sofia [Mitt. Konigl. naturwiss. Inst. Sofia] 11: 121-179.
THURNER, J. (1967): Lepidopteren aus Morea. Einweiterer Beitrag zur Fauna des Peloponnes (Griechenland). - Z. Wien. ent. Ges. 52 (Bd 78, 6-8): 50-58.
THURNER, J. (1972): Die Schmetterlinge Karntens und Osttirols, Macros. II Nachtrag. - Carinthia II 161/81: 91-106.

TOMIČ, D., VASIČ, K., CARNELUTTI, J., ZEČEVIČ, M. & R. KRANJČEV (1990): Fauna Durmitora, Part 3. *Heterocera*, II. *Geometridae (Insecta, Lepidoptera)*. Crnogorska Akad. Nauka i Umjetnosti, Titograd, pp. 159-202.

TOMIČ, D., MIHAJLOVIČ, Lj. & M. RISTIČ (1994): Fauna arborikolnih sovica (*Lepidoptera, Noctuidae*) i zemljomerki (*Lepidoptera, Geometridae*) Deliblatskog Peska. - Bull. Nat. Hist. Mus., Belgrade B48: 147-164.

TOULGOET, H. de. (1966): Quinze jours de Chasse au Maroc. - Alexanor 4 (5): 197-207.

TRUSCH, R. & B. MÜLLER (2001): Zur Geometridenfauna des Dahar-Berglandes in Süd-Tunesien (*Lepidoptera: Geometridae*). - Ent. Z. 111 (5): 133-140.

TULESHKOV, K. (1951): Faunistichni prouchvanija na pl. Olimp v Grcija (*Insecta-Lepidoptera*). [Études faunistiques d'Olympe de la Grèce (*Insecta, Lepidoptera*)]. - Izv. Zool. Inst. 1: 305-328.

TURATI, E. (1913): Un Record Entomologico. Materiali per una faunula dei Lepidotteri della Sardegna. - Atti Soc. Ital. Sci. Nat., Milano, 51 (3-4): 265-365.

TURATI, E. & R. VERITY (1911): Faunula Valderiensis nell'Alta Valle del Gesso (Alpi Marittime). *Heterocera*. - Bull. Soc. ent. Ital., Genova, 43: 168-233.

TURATI, C.E. & A. FIORI (1930): Lepidotteri di Rodi. - Mem. Soc. ent. ital., Genova 9: 196-214.

TURNER, J. (1926): List of Geometres of the British Islands with the Named Varieties and the Synonyms in general use. - The Entomologist's Rec. and J. Var. 38 (3): (1)-(28).

TUTT, J.W. (1906a): A puzzling group of Eupitheciids. - The Entomologist's Rec. J. Var. 18: 157-158, 260.

TUTT, J.W. (1906b): Practical Hints relating to the Eupitheciids. - The Entomologist's Rec. J. Var. 18: 179-182, 201-204, 218-222.

TUTT, J.W. (1906c): Another puzzling group of Eupitheciids. - The Entomologist's Rec. J. Var. 18: 261-263.

UHERKOVICH, A. (1976): Adatok a Del-Dunantul nagylepkefaunajahoz (*Makrolepidoptera*) [Beitrage zur Kenntnis der Gross-Schmetterling-Fauna (*Macrolepidoptera*) von Sud-Transdanubien]. - Folia ent. hung. 29 (1): 119-137.

UHERKOVICH, A. (1981): Data to the *Macrolepidoptera* Fauna of South Transdanubia (*Lepidoptera*) II. - Folia ent. hung. 42 (2): 239-252.

URBAHN, E. (1966): Ergebnisse der Albanien-Expedition 1961 des Deutschen Entomologischen Institutes. 51. Beitrag. Lepidoptera: Geometridae. - Beitr. Ent. 16 (3/4): 407-446.

URBAHN, E. (1967): Das neue Geometridensystem von C. HERBULOT und Anmerkungen dazu (Lep. Geometridae). - Ent. Ber., Berlin 1 (1967): 33-52.

URBAHN, E. (1969): *Eupithecia groenblomi* - eine neue *Geometridae* (Lep.). - Ann. Ent. Fenn. 35 (4): 186-190.

URBAHN, E. (1975): *Perizoma lugdunaria* HS. - neu für die DDR (Lep., Geom.). - Ent. Nachr. 19 (11): 179-181.

URBAHN, E. & H. URBAHN (1939): Die Schmetterlinge Pommerns mit einem vergeichenden Überblick über den Ostseeraum. - Stett. ent. Ztg. 100: 185-826.

URBAHN, E. & H. URBAHN (1981): Schwierigkeiten um *Eupithecia innotata* HUFN. 1767 (Lep. Geom.). - Nota lepid. 4 (1-2): 47-51.

USKOV, M.V., SVIRIDOV, A.V. & E.M. ANTONOVA (2000): Lepidopterofauna Vladimirskoj oblasti. Vyp. 1. Raznousye cheshuekrylye severnoj chasti Meshchery Vladimirskoj oblasti (*Lepidoptera: Macroheterocera*). [The lepidopterous fauna of Vladimir Area. Issue 1. Moths of the Northern part of Meshchera of Vladimir Area (*Lepidoptera: Macroheterocera*). Annotated species list]. Vladimir (2000), 40 pp.

UTKIN, N.A. (1999): Prostejshie i bespozvonochnye Kurganskoj oblasti. Spisok izvestnyh vidov. [The Protozoa and Invertebrates of animals of the Kurgan region. The list of known species]. Kurgan: Kurgan State University publishing house, 1999. 363 pp.

VALLE, K.J. (1946): Suurperhoset (Macrolepidoptera). IV Mittarit (Geometrae). – Animalia Fennica 5: 1-370, 22 pls.

VALLETTA, A. (1973): The Moths of the Maltese Islands. Progress Press, Malta, 118 pp.

VALLHONRAT, F. (1989): Aproximació a la Fauna de Geomètrids de Susqueda i Les Guilleries (*Lepidoptera: Geometridae*). - Treb. Soc. Cat. Lep. 9 (1988): 23-32.

VARIS, V., JALAVA, J. & J. KYRKI (1987): Check-list of Finnish *Lepidoptera*. Suomen perhosten luettelo. - Notulae Ent. 67 (2): 49-118.

VARIS, V. (ed.), AHOLA, M., ALBRECHT, A., JALAVA, J., KAILA, L., KERPPOLA, S., & J. KULLBERG (1995): Checklist of the Finnish Lepidoptera - Suomen perhosten luettelo. - Sahlbergia 2 (1): 1-80.

VASILENKO, S.V. (1998): Pyadenitzy (*Lepidoptera, Geometridae*) Kurganskoj oblasti. In: Bespozvonochnye zhivotnye Juzhnogo Zaural'ja i sopredel'nyh territorij. Mater. Vseross. konf., 14-16 April 1998. Kurgan: Kurgan State Univ. publ. house: 82-86.

VIEILLEDENT, P. (1905): Lepidopteros da Regiao de Setubal. - Broteria 4 (4): 185-206.
VIEIRA, V. (1997): Lepidoptera of the Azores Islands. - Bol. Mus. Mun. Funchal, 49 (273): 5-76.
VIERTL, A. (1898): Két Geometra-hernyó. – Rovartani Lapók (Budapest) 5: 79-80.
VIGNAL, P. & M. VINTÉJOUX (1991): Additions à la liste des Macrolépidoptères de Corrèze (*Lepidoptera*). - Alexanor 17 (4): 229-231.
VIIDALEPP, J. (1967): Materialy k faune cheshuekrylyh Litovskoj SSR. - Tartu riikliku ulikooli Toimetised, Zoologia-alaseid Toid 4 (194): 48-59.
VIIDALEPP, J. (1974): Die *Eupithecia*-Arten aus Sibirien in der Sammlung der Staats-Universitat zu Tartu (*Lepidoptera, Geometridae*). - TRU Toimetised 8 (327): 86-93.
VIIDALEPP, J. (1977): Spisok pyadenitz (*Lepidoptera, Geometridae*) fauny SSSR, II. - Ent. obozr. 56 (3): 564-576 (in Russian).
VIIDALEPP, J. (1978): Spisok pyadenitz (*Lepidoptera, Geometridae*) fauny SSSR, III. - Ent. obozr. 57 (4): 752-761 (in Russian).
VIIDALEPP, J. (1995): Catalogus Macrolepidopterorum Estoniae. [Eesti suurliblikate nimestik levikutabelina]. Tallinn-Tartu, 64 pp.
VIIDALEPP, J. (1996): Checklist of the *Geometridae* (Lepidoptera) of the former U.S.S.R. Apollo Books Aps., Stenstrup, Denmark. 111 pp.
VIIDALEPP, J. & V. MIRONOV (1988a): Pyadenitzy roda *Eupithecia* CURT. (*Lepidoptera, Geometridae*) Dal'nego Vostoka SSSR. I. [Pugs of the Soviet Far East (*Lepidoptera, Geometridae: Eupithecia*). I]. - Izv. AN Est. SSR (Biologia) 37 (3): 200-214.
VIIDALEPP, J. & V. MIRONOV (1988b): Pyadenitzy roda *Eupithecia* CURT. (*Lepidoptera, Geometridae*) Dal'nego Vostoka SSSR. II. [Pugs of the Soviet Far East (*Lepidoptera, Geometridae: Eupithecia*). II]. - Izv. AN Est. SSR (Biologia) 37 (4): 281-293.
VIIDALEPP, J. & V. MIRONOV (1990): Pyadenitzy roda *Eupithecia* CURT. (*Lepidoptera, Geometridae*) Dalnego Vostoka SSSR. III. [Pugs of the Soviet Far East (*Lepidoptera, Geometridae: Eupithecia*). III]. - Izv. AN Est. SSR (Biologia) 39 (2): 112-120.
VIIDALEPP, J. & H. REMM (1996): Eesti Liblikate määraja. Tallinn 'Valgus'. 444 pp. & 40 pls.
VIIDALEPP, J., JÜRIVETE, U., KRUUS, M., LINDT, A., LINDT, R. & T. RUBEN (2001): Uusi vaksikuliike Eesti fauna nimestikus. [New species of Geometridae in the checklist of Estonian fauna]. - Lepinfo 12: 6-11.
VILHEMSEN, F. (2001): Ny dobbeltart i Danmark, *Eupithecia innotata* Hfn. og *Eupithecia ochridata* PINKER. - Lepidoptera, København 8 (2): 65-70.
VINTÉJOUX, M. (1976): Contribution a l'etude des Lépidoptères de Corrèze (suite et fin). - Alexanor 9 (6): 279-287.
VIVES MORENO, A. (1994): Catalogo sistematico y sinonimico de los Lepidopteros de la peninsula Iberica y Baleares (*Insecta: Lepidoptera*). Madrid, 775 pp.
VOJNITS, A. (1957): Adatok a Csovanyos (Borzsony-hegyseg) lepkefaunajahoz. - Folia ent. hung. (series nova) 10 (19): 395-406.
VOJNITS, A. (1967): New Data Concerning the Eupithecid Fauna of Hungary (*Lepidoptera, Geometridae*). - Folia ent. hung. (series nova) 20 (26): 605-613.
VOJNITS, A.M. (1969a): A new Geometrid species: *Eupithecia catharinae* sp. nov. (*Lepidoptera: Geometridae*). - Acta Zool. Acad. Sci. Hung. 15 (1-2): 231-236.
VOJNITS, A.M. (1969b): *Eupithecia szelenyii* sp. nov. (*Lepidoptera: Geometridae*). - Acta Zool. Acad. Sci. Hung. 15 (3-4): 463-466.
VOJNITS, A.M. (1970): Adatok Magyarorszag *Eupithecia* faunajahoz (*Lep.*) I. - Folia ent. hung., Ser. nova 23 (4): 125-132.
VOJNITS, A.M. (1972): New Taxa in the *Eupithecia haworthiata* - Group (*Lepidoptera: Geometridae*). (Studies on Palaearctic *Eupithecia* Species I). - Ann. Hist.-Nat. Mus. Natn. Hung. 64: 299-302.
VOJNITS, A.M. (1973): Az *Eupithecia unedonata* MAB. törpearaszoló magyarországi előfordulása (Lep.: Geometridae) (Nagylepkefaunánk újdonságai I.) [The Occurence of *Eupithecia unedonata* MAB. in Hungary (Novelties in Macrolepidoptera Fauna of Hungary I.)]. - Folia ent. hung. 26 (1): 225-226.
VOJNITS, A.M. (1974a): Adatok a Központi-vagy Magas-Börzöny nagylepke faunájához. 1. - Folia Hist.-nat. Mus. Matr. 2: 31-43.
VOJNITS, A.M. (1974b): A New Geometrid Species: *Eupithecia wettsteini* sp. nov. from Hungary (*Lepidoptera*). - Folia ent. hung. 27 (2): 235-237.
VOJNITS, A.M. (1975a): *Geometridae: Eupithecini* II. (*Lepidoptera*). - Acta Zool. Acad. Sci. Hung. 21 (3-4): 447-453.
VOJNITS, A.M. (1975b): Die Eupithecien-Typen in der Zoologische Sammlung des Bayerischen Staates. - Folia ent. hung. 28 (1): 217-222.

VOJNITS, A.M. (1976a): New species of the *Eupithecia bohatschi* group from China (*Lepidoptera: Geometridae*). - Acta Zool. Acad. Sci. Hung. 22 (1-2): 197-211.
VOJNITS, A.M. (1976b): A revision of the species group *Eupithecia sinuosaria* EV. (*Lepidoptera: Geometridae*). - Acta Zool. Acad. Sci. Hung. 22 (3-4): 415-423.
VOJNITS, A.M. (1977a): New *Eupithecia* species and subspecies from Asia and North Africa (*Lepidoptera: Geometridae*). - Acta Zool. Acad. Sci. Hung. 23 (1-2): 227-236.
VOJNITS, A.M. (1977b): *Geometridae Eupithecini* III. (*Lepidoptera*). - Acta Zool. Acad. Sci. Hung. 23 (3-4): 461-485.
VOJNITS, A.M. (1982): A revision of the '*Eupithecia innotata* group', I (*Lepidoptera, Geometridae*). - Ann. Hist.-Nat. Mus. Natn. Hung. 74: 217-239.
VOJNITS, A.M. (1977c): Az *Eupithecia szelenyii* VOJNITS, 1969 = *E. ochridata* PINKER, 1968. – Folia ent. hung. (series nova) 30 (1): 186-187.
VOJNITS, A.M. (1984a): A revision of the '*Eupithecia innotata* group', II. (*Lepidoptera, Geometridae*). - Boll. Mus. Reg. Sci. Nat. Torino 2 (1): 199-220.
VOJNITS, A.M. (1984b): Berichtigungen und Ergänzungen zur Geometridae-Fauna Ungarns. – Articulata 2 (5): 110-118.
VOJNITS, A.M. (1985): Revision der Paläarktischen *Eupithecini*-Arten I. - Acta Zool. Acad. Sci. Hung. 31 (1-3): 267-273.
VOJNITS, A.M. (1987a): Falces and clavulus, hitherto disregarded parts of the male genitalia in *Eupithecia* species (*Lepidoptera, Geometridae*). - Ann. Hist.-Nat. Mus. Nat. Hung. 79: 179-184.
VOJNITS, A.M. (1987b): The main evolutional lines of the tribe *Eupitheciini* (*Lepidoptera: Geometridae*). - Folia ent. hung. 48: 241-242.
VOJNITS, A.M. (1987c): Notes on *Eupithecia* (Lepidoptera: Geometridae). – Articulata 3 (1): 33-39.
VOJNITS, A. & E. DE LAEVER (1973): Revision of the '*Eupithecia suboxydata-subbrunneata*' group (*Lepidoptera: Geometridae*). - Acta Zool. Acad. Sci. Hung. 19 (3-4): 427-444.
VOJNITS, A. & L. RONKAY (1981): The *Geometridae* of the Hortobagy National Park (*Lepidoptera*). - Natur. Hist. Nat. Parks of Hungary, Budapest 1: 197-201.
VOJNITS, A.M. & E. SZABÓ (1988): *Eupithecia domogledana* sp. nov. from Mount Domogled, Rumania, and a redescription of *Eupithecia salami* BRANDT, 1938 (*Lepidoptera: Geometridae*). - Nota lepid. 11 (2): 158-163.
VOLDRICH, M. (1963): La faune des papillons des montagnes de Sumava Centrale. - Acta faun. ent. Mus. Nat. Pragae 9 (75): 5-55.
VORBRODT, K. (1921): Die Schmetterlinge der Schweiz (4. Nachtrag). - Mitt. schweiz. ent. Ges. 13 (3-4): 175-206.
VORBRODT, K. (1925): Die Schmetterlinge der Schweiz (5. Nachtrag). - Mitt. schweiz. ent. Ges. 13 (8): 434-466.
VORBRODT, C. (1928): Die Schmetterlinge von Zermatt. - Dt. ent. Z. Iris 42 (1/2): 7-80; 81-130.
VORBRODT, C. (1931a): Tessiner und Misoxer Schmetterlinge. - Mitt. schweiz. ent. Ges. 14 (6): 199-396.
VORBRODT, C. (1931b): Lichtfang am Luganersee. - Ent. Z., Frankfurt a.M. 44 (22): 336-338.
VORBRODT, K. & J. MÜLLER-RUTZ. (1914): Die Schmetterlinge der Schweiz. Bd 2, 726 pp., Bern.
WAGNER, F. (1914): Eine neue Tephroclystia aus Tunis. – Wien. ent. Ver. 24: 175-179, 1 Taf.
WAGNER, F. (1926): Ueber einige neue und wenig bekannte Lepidopteren aus Spanien. - Z. Öst. ent. Ver. 11 (12): 111-115.
WAGNER, F. (1929): Weiterer Beitrag zur Lepidopteren-Fauna Inner-Anatoliens. - Mitt. münchn. ent. Ges. 19 (10-12): 175-206 & Taf. II.
WAGNER, F. (1930): Zweiter (III). Beitrag zur Lepidopteren-Fauna Inner-Anatoliens. - Int. ent. Z. 23: 545-559.
WAGNER, F. (1931): Neue heteroceren aus Kleinasien. - Int. ent. Z. 25: 367-371.
WAGNER, F. (1932): Vierter (V). Beitrag zur Lepidopteren-Fauna Inner-Anatoliens. - Int. ent. Z. 26: 178-183.
WAGNER, F. (1935): Ueber Sammeln und Zucht der Eupithecien (*Tephroclystia*)-Raupen. - Z. Österr. Ent.-Ver. 20 (5): 37-43.
WALDERDORFF, H.G. (1869): *Eupithecia actaeata* HS., eine neue Spannerart. - Corr-Bl. Zool. Mus. Ver. Regensburg 23: 82-90.
WALKER, F. (1862): List of the specimens of Lepidopterous Insects in collection of the British Museum. Part XXIV (Geometrites). London, pp. 1021-1796.
WALKER, J.J. (1880): Notes on the *Lepidoptera* in the West of Ireland. - The Entomologist's Montly Mag. 17: 79-82.
WARDIKJAN, S.A. (1985): Atlas genital'nogo apparata pyadenitz (*Lepidoptera, Geometridae*) Armjanskoj SSR, Erevan, AN Arm. SSR: 5-135 (in Russian).

WARNECKE, G. (1937a): Die Grossschmetterlinge der nordfriesischen Insel Sylt. - Ent. Rund. 54 (17): 202-206.
WARNECKE, G. (1937b): Weitere Mitteilungen über die Verbreitung einiger Makrolepidopteren in Schleswig-Holstein. - Ent. Rund. 54 (26): 324-326.
WARNECKE, G. (1939): Namensverzeichnis der Grossschmetterlinge der Nordmark (Schleswig-Holsteins, Lübecks und Lauenburg, einschliesslich der nordelbischen Gebietsteile Hamburgs). - Ent. Rund. 56 (25): 277-280.
WARNECKE, G. (1951): Sommerfugle fra Island. Med Beskrivelse af to nye Lokalformer. - Flora og Fauna 57 (1): 41-44.
WARNECKE, G. & K. HOLDHAUS (1954): *Lepidoptera*. In: HOLDHAUS, K. Die Spuren der Eiszeit in der Tierwelt Europas. - Abh. zool.-bot. Ges. Wien. 18: 1-493.
WARREN, W. (1893): On new Genera and Species of Moths of the Family Geometridae from India, in the Collection of H.J. Elves. - Proc. Zool. Soc. London, 1893: 341-434.
WEBB, S. (1881): *Eupithecia ultimaria* Dup. - The Entomologist 14 (223): 300.
WEGNER, H. (1998): Die Geometriden-Arten Nordwestdeutschlands. - Bombus 3 (35-38): 137-149.
WEHRLI, E. (1924): Nice - St.Martin Vésubie - Digne. Ein Beitrag zur Lepidopterenfauna der Alpes Maritimes und der Basses Alpes. - Dt. ent. Z., Iris 38: 50-57; 59-98.
WEHRLI, E. (1926a): Neue Eupithecien aus dem andalusischen Hochgebirge. - Schweiz. ent. Anzeiger 5 (11): 1-4.
WEHRLI, E. (1926b): Ein Streifzug in die andalusischen Gebirge. - Dt. ent. Z., Iris 40 (2/3): 113-129.
WEHRLI, E. (1927a): Einige interessante Acidalien und Eupithecien aus den Sierra Nevada (Esp.). - Int. ent. Z. (Guben) 37: 340-346.
WEHRLI, E. (1927b): Ein weiterer Streifzug in die andalusischen Gebirge. 22.VI. bis 13.VII.1926. Beitrag zur Geometridenfauna Andalusiens. (*Lepid*.). - Dt. ent. Z., Iris 41 (1): 49-80.
WEHRLI, E. (1929): Zwei neue Eupithecien aus der Ausbeute des Herrn E. Pfeiffer, München, aus dem Taurus (Lepidoptera Geometridae). - Mitt. münchn. ent. Ges. 19 (10-12): 324-328.
WEHRLI, E. (1932): Neue Geometriden-Arten und –Rassen (Lep. Het) von der Maras-Expedition L. Osthelder u. E. Pfeiffer, München. - Mitt. münchn. ent. Ges. 22 (1): 3-11.
WEHRLI, E. (1934): In: OSTHELDER, L. & E. PFEIFFER. Lepidopterenfauna von Marasch in türkisch Nordsyrien. - Mitt. münchn. ent. Ges. 24 (2): 25-55.
WEIDNER, H. (1974): Die Entomologischen Sammlungen des Zoologischen Instituts und Zoologischen Museums der Universität Hamburg. Teil. XI. Insecta VIII. *Lepidoptera*. - Mitt. Hamb. Zool. Mus. und Inst. 70: 181-266.
WEIGT, H.-J. (1976): Die Blütenspanner Westfalens (Lepidoptera Geometridae). Teil 1. Die Imagines und ihre Verbreitung. - Dortm. Beitr. Landeskde 10: 61-152.
WEIGT, H.-J. (1977): Die Blütenspanner Westfalens (Lepidoptera Geometridae). Teil 2. Die Raupen und ihre Futterpflanzen. - Dortm. Beitr. Landeskde 11: 41-98.
WEIGT, H.-J. (1979): Blütenspanner-Beobachtungen 1: *Eupithecia actaeata* (Lep.: Geometridae). - Ent. Z. 89 (3): 17-23.
WEIGT, H.-J. (1980a): Blütenspanner-Beobachtungen 2. Bemerkungen zur *Eupithecia - absinthiata* Gruppe (Lepidoptera Geometr.). - Abh. Landesmus. Naturkde Münster in Westfalen 42 (1): 31-50.
WEIGT, H.-J. (1980b): Blütenspanner-Beobachtungen 4 (Lepidoptera, Geometridae). - Dortm. Beitr. Landeskde 14: 3-83.
WEIGT, H.-J. (1981): Zum Status von *Chloroclystis v-ata* HAWORTH und Ihre Stellung zu den Gattungen *Gymnoscelis* MABILLE und *Calliclystis* DIETZE. Blütenspanner-Beobachtungen 5 (Lepidoptera, Geometridae). - Dortm. Beitr. Landeskde 15: 59-61.
WEIGT, H.-J. (1982): Blütenspanner-Beobachtungen 7 (Lepidoptera, Geometridae). - Dortm. Beitr. Landeskde 16: 3-10.
WEIGT, H.-J. (1985): Blütenspanner-Beobachtungen 8 (Lepidoptera Geometridae). Vorkommen und Lebensweise von *Eupithecia pygmaeata* HÜBNER, 1799 (*palustraria* DOUBLEDAY, 1850). - Dortm. Beitr. Landeskde 19: 9-18.
WEIGT, H.-J. (1986): Blütenspanner-Beobachtungen 9 (*Lepidoptera Geometridae*). Über die Lebensweisen von *Eupithecia venosata* FABRICIUS, 1787 und *Eupithecia schiefereri* BOHATSCH, 1893. - Dortm. Beitr. Landeskde 20: 5-18.
WEIGT, H.-J. (1987): Die Blütenspanner Mitteleuropas (*Lepidoptera, Geometridae: Eupitheciini*). Teil 1: Biologie der Blütenspanner. - Dortm. Beitr. Landeskde 21: 5-57.
WEIGT, H.-J. (1988): Die Blütenspanner Mitteleuropas (*Lepidoptera, Geometridae: Eupitheciini*). Teil 2: *Gymnoscelis rufifasciata* bis *Eupithecia insigniata*. - Dortm. Beitr. Landeskde 22: 5-81.
WEIGT, H.-J. (1989): Blütenspanner-Beobachtungen 10 (*Lepidoptera, Geometridae*). Eine neue Unterart von *Eupithecia impurata* HÜBNER, 1813. - Dortm. Beitr. Landeskde 23: 39-46.

WEIGT, H.-J. (1990): Die Blütenspanner Mitteleuropas (*Lepidoptera, Geometridae: Eupitheciini*). Teil 3: *Eupithecia sinuosaria* bis *pernotata*. - Dortm. Beitr. Landeskde 24: 5-100.
WEIGT, H.-J. (1991): Die Blütenspanner Mitteleuropas (*Lepidoptera, Geometridae: Eupitheciini*). Teil 4: *Eupithecia satyrata* bis *indigata*. - Dortm. Beitr. Landeskde 25: 5-106.
WEIGT, H.-J. (1993): Die Blütenspanner Mitteleuropas (*Lepidoptera, Geometridae: Eupitheciini*). Teil 5: *Eupithecia pimpinellata* bis *lanceata*. - Dortm. Beitr. Landeskde 27: 5-108.
WEISERT, F. (1998): Österreichische Turkmenistan-Expedition 1993 und 1996. - Z. Arb.Gem. Öst. Ent. 50 (3-4): 119-123.
WICHRA, J. (1968): *Eupithecia orphnata* Bohatsch a *Eupithecia egenaria* /HS/ (Lep. Geometridae) v Československu. - Zprávy Čs. spol. entomologické pri ČSAV 4: 65-66.
WIESER, C. & P. HUEMER (1999): Rote Listen der Schmetterlinge Kärntens (Insecta: Lepidoptera). In: ROTTENBURG, T., WIESER, C., MILDNER, P. & W.E. HOLZINGER. Rote Listen gefährdeter Tiere Kärntens. - Naturschutz in Kärnten 15: 133-200.
WILTSHIRE, E.P. (1936): More notes on the Early Stages of Syrian Lepidoptera. – The Entomologist's Rec. and J. Var. 48 (10, suppl.): (9)-(11).
WILTSHIRE, E.P. (1939): In: ELLISON, R.E. & E.P. WILTSHIRE. The Lepidoptera of the Lebanon: with notes on their season and distribution. - Trans. Royal ent. Soc. London 88 (1): 1-56.
WILTSHIRE, E.P. (1948): Middle East Lepidoptera: IX. Part I: New records for Cyprus. - Ent. Rec. 60: 80-84.
WILTSHIRE, E.P. (1949): The Lepidoptera of the Kingdom of Egypt. Part II. - Bull. Soc. Fouad ler Ent. 33: 381-460, pls. 8-9.
WILTSHIRE, E.P. (1957): The Lepidoptera of Iraq. Nicholas Kaye Limited. Government of Iraq (Ministry of Agriculture), 162 pp. & 17 pls.
WILTSHIRE, E.P. (1976): La repartition en Bretagne et en Normandie des Lepidopteres de la biocenose du Genevrier et du Cypres (*Geometridae, Noctuidae*). - Alexanor 9 (8): 349-360.
WILTSHIRE, E.P. (1985a): Notes on *Cucullia scrophulariphaga* RAMBUR (Noctuidae) and on a few other interesting Sardinian Lepidoptera seen on the Costa Smeralda in mid-May 1984. - Nota lepid. 8 (2): 161-171.
WILTSHIRE, E.P. (1985b): Middle East Lepidoptera XLIV. *Eupithecia penultimaria* sp. n. (*Geometridae*) from central Iraq. – The Entomologist's Gaz. 36: 263-267.
WILTSHIRE, E.P. (1986): In: BÜTTIKER, W. & F. KRUPP. *Lepidoptera* of Saudi Arabia: Fam. *Cossidae*, *Sesiidae*, *Metarbelidae*, *Lasiocampidae*, *Sphingidae*, *Geometridae*, *Lymantriidae*, *Arctiidae*, *Nolidae*, *Noctuidae* (*Heterocera*); Fam. *Satyridae* (*Rhopalocera*) (Part 5). - Fauna of Saudi Arabia 8: 262-323.
WNUKOWSKY, W. (1929): 3. Einige Nomenklatur - Notizen über die paläarktischen Lepidopteren. - Zool. Anz., Leipzig 83 (9/10): 221-224.
WOJTUSIAK, J. (1966): Fauna makrolepidopterologiczna Wzniesienia Gubalowskiego i jej charakterystyka zoogeograficzna [Macrolepidopterous Fauna of the Gubalowka Elevation and its Zoogeographical Charakteristic]. - Acta zool. Cracov. 11 (4): 75-109 + 1 tab.
WOLF, W. (1981): Die sogenannten Grossschmetterlinge des nordöstlichen Bayreuther Umlandes (*Insecta, Lepidoptera*). Ein Beitrag zur Kenntnis des Obermain-Hügellandes. - Ber. Naturwiss. Ges. Bayreuth 17: 145-254.
WOLF, W. (1988): Systematische und synonymische Liste der Spanner Deutschlands unter besonderer Berücksichtigung der DENIS & SCHIFFERMÜLLER'schen Taxa (*Lepidoptera: Geometridae*). - Neue ent. Nachr. 22: 1-78.
WOLFF, N.L. (1929a): *Lepidoptera*. - The Zoology of the Faroes 39, 2 (1): 1-38.
WOLFF, N.L. (1929b): Synopsis of the *Lepidoptera* of Iceland. - Ent. Meddr. 16: 339-365.
WOLFF, N.L. (1935): Sommerfugle fra Frøslev i Sonderjylland 1934. - Ent. Meddr. 19: 172-177.
WOLFF, N.L. (1936): Sommerfugle fra Frøslev i Sonderjylland 1935. - Ent. Meddr. 19: 443-451.
WOLFF, N.L. (1964): The *Lepidoptera* of Greenland. - Meddr. Grønland (København) 159 (11): 1-74, 21 pls.
WOLFF, N.L. (1970): Revideret fortegnelse over Færøernes sommerfugle (A revised list of the *Lepidoptera* known from the Faroe Islands). - Ent. Meddr. 38 (1): 3-14.
WOLFF, N.L. (1971): *Lepidoptera*. - The Zoology of Iceland (Copenhagen and Reykjavik), 3 (45): 193 pp.+ 15 pls.
WOLFSBERGER, J. (1945-1949): Neue und interessante Macrolepidopterenfunde aus Sudbayern und den angrenzenden nordlichen Kalkalpen. - Mitt. münchn. ent. Ges. 35-39: 308-329.
WOLFSBERGER, J. (1950): Neue und interessante Macrolepidopterenfunde aus Südbayern und den angrenzenden nordlichen Kalkalpen. - Mitt. münchn. ent. Ges. 40: 207-236.
WOLFSBERGER, J. (1954-1955): Neue und interessante Macrolepidopterenfunde aus Sudbayern und den angrenzenden nordlichen Kalkalpen. - Mitt. münchn. ent. Ges. 44-45: 300-347.
WOLFSBERGER, J. (1965): Die Macrolepidopteren-Fauna des Gardaseegebietes. - Mem. Mus. Civ. Storia nat.

Verona 13: 1-385.
WOLFSBERGER, J. (1971): Die Macrolepidopteren-Fauna des Monte Baldo in Oberitalien. - Mem. Mus. Civ. Storia nat. Verona, (Fuori Serie), 4: 1-336 + 20 tavv.
WOLFSBERGER, J. (1974): Die Macrolepidopteren-Fauna des Gardaseegebietes (I Nachtrag). - Mem. Mus. Civ. Storia nat. Verona 1: 167-193.
WORM-HANSEN, J.G. (1935): Additional notes on the Lepidopterous Fauna of San Cataldo (Southern Italy). - Ent. Meddr 19 (6): 277-285.
WORMS, DE C.G.M. (1962): The Macrolepidoptera of WILTSHIRE. Publ. Wiltshire Arch. and Nat. Hist. Soc. and T. Buncle & Co. Ltd., Arbroath, Scotland, 177 pp. + 9 pls.
WOSKRESSENSKY, N.M. (1927): Zur Lepidopterenfauna des Gouv. Poltawa. - Trav. Mus. Zool. Kiev 3: 119-145.
YAKOVLEV, B.P. (1961): Vrediteli shishek i semjan eli [The pests of the cones and seeds of spruce]. Gos. izdat. Karelsk. ASSR, Petrozavodsk, 48 pp.
ZANGHERI, P. (1969): Repertorio della Flora e Fauna della Romagna, 3 (Regno Animale). – Mem. Mus. Civ. Storia nat. Verona, Fuori ser. 1: 855-1414.
ZANGHERI, S. (1975): La Lepidotterofauna delle Prealpi Venete sulla base della Collezione A.Ancilotto. - Boll. Ist. ent. Univ., Bologna 31: 215-250.
ZERNY, H. (1927a): Die Lepidopterenfauna von Albarracin in Aragonien. - EOS 3 (3-4): 299-488.
ZERNY, H. (1927b): Die Lepidopteren-Fauna von Algeciras und Gibraltar in Sud-Andalusien. - Dt. ent. Z., Iris 41 (2): 83-146.
ZERNY, H. (1933): Lepidopteren aus dem nordlichen Libanon. - Dt. ent. Z., Iris 47: 60-109.
ZERNY, H. (1934): Neue Heteroceren aus dem Grossen Atlas in Marokko. - Z. Österr. Ent.-Ver. Wien 19 (5): 33-39; (6): 43-48; (7/8): 52-53, Taf. 5-7.
ZERNY, H. (1935): Die Lepidopterenfauna des Grossen Atlas in Marokko und seiner Randgebiete. - Mém. Soc. Sci. nat. Maroc 42: 1-163 & 2 pls.
ZIMSEN, E. (1964): The type material of I.C. FABRICIUS. Munksgaard. Copenhagen, 656 pp.
ZOCCHI, R. (1971): Contributo alla conoscenza dell'entomofauna delle Tamerici in Italia. Redia, Firenze 52: 31-129.
ZUKOWSKY, B. (1937): Reise ans Schwarze Meer und Herculesbad (Lep.). - Ent. Rund. 54 (48): 573-576.
ZUKOWSKY, B. (1941): Sivas und Akschehir 1937 (Lepidoptera). - Ent. Z. Frankfurt a/Main 54 (34): 266-272.

Index to scientific names
of the Systematic Account, Vol. 4

The index gives page reference to the main treatment. Numbers in semibold refer to species numbers and figures.